Lecture Notes in Computer Science 4976

Commenced Publication in 1973
Founding and Former Series Editors:
Gerhard Goos, Juris Hartmanis, and Jan van Leeuwen

Yanchun Zhang   Ge Yu   Elisa Bertino
Guandong Xu (Eds.)

Progress in WWW Research and Development

10th Asia-Pacific Web Conference, APWeb 2008
Shenyang, China, April 26-28, 2008
Proceedings

 Springer

Volume Editors

Yanchun Zhang
Guandong Xu
Victoria University, School of Computer Science and Mathematics
Melbourne, VIC 8001, Australia
E-mail: yanchun.zhang@vu.edu.au, xu@csm.vu.edu.au

Ge Yu
Northeastern University, Department of Computer Science and Engineering
Shenyang 110004, China
E-mail: yuge@mail.neu.edu.cn

Elisa Bertino
Purdue University, Department of Computer Science
West Lafayette, IN 47907, USA
E-mail: bertino@cs.purdue.edu

Library of Congress Control Number: 2008924178

CR Subject Classification (1998): H.2-5, C.2, D.2, I.2, K.4, J.1

LNCS Sublibrary: SL 3 – Information Systems and Application, incl. Internet/Web
and HCI

ISSN   0302-9743
ISBN-10   3-540-78848-4 Springer Berlin Heidelberg New York
ISBN-13   978-3-540-78848-5 Springer Berlin Heidelberg New York

Springer is a part of Springer Science+Business Media

springer.com

© Springer-Verlag Berlin Heidelberg 2008
Printed in Germany

Typesetting: Camera-ready by author, data conversion by Scientific Publishing Services, Chennai, India
Printed on acid-free paper   SPIN: 12247257   06/3180   5 4 3 2 1 0

Message from the Conference Co-chairs

This volume is the published record of the 10th Asia Pacific Conference on Web Technology (APWeb 2008), held in Shenyang, China, April 26-28, 2008.

APWeb has been a premier forum in the Asia-Pacific region on theoretical and practical aspects of Web technologies, database systems, information management and software engineering. Previous APWeb conferences were held in Beijing (1998), Hong Kong (1999), Xi'an (2000), Changsha (2001), Xi'an (2003), Hangzhou (2004), Shanghai (2005), Harbin (2006), and Huang Shan (Yellow Mountain) (2007). It was our great pleasure to have hosted this year's APWeb conference in Shenyang, which is the center for communication, commerce, science and culture in the northeastern part of China.

APWeb 2008 attracted more than 160 papers from 19 countries and regions. This made the work of the Program Committee rather challenging. We are grateful to the Program Co-chairs, namely, Yanchun Zhang, Ge Yu, and Elisa Bertino, who worked very hard to ensure the quality of the paper review process. Special thanks are due to Yanchun Zhang for taking care of many organizational issues. Without him the conference would not have been so successful. We would like to express our gratitude to Ling Feng and Keun Ho Ryu, Tutorial/Panel Co-chairs, and Haixun Wang, Industrial Chair, for their devotion in designing an attractive overall conference program. Moreover, we are thankful to Masaru Kitsuregawa, Wei-Ying Ma, Xuemin Lin, and Yong Shi for their keynote/invited lectures, which were the highlights of this year's conference.

Alongside the main conference, three high-quality workshops were arranged by Yoshiharu Ishikawa and Jing He, the Workshop Co-chairs. The Workshop on Information-explosion and Next Generation Search was organized by Katsumi Tanaka; the Workshop on Business Intelligence and Data Mining was run by Yong Shi, Guangyan Huang, and Jing He; and the Workshop on Health Data Management was offered by Chaoyi Pang and Qing Zhang. Moreover, a Doctoral Consortium on Data Engineering and Web Technology Research was organized to promote the research of doctoral students. All of them attracted many participants.

In addition to the afore-mentioned program officers, the success of APWeb 2008 is attributed to many other people. Especially, we would like to thank Guoren Wang and Bin Zhang, Local Arrangements Co-chairs; Guandong Xu, Publication Chair, and Toshiyuki Amagasa and Hua Wang, Publicity Co-chairs. We also thank the APWeb Steering Committee and WISE Society for their continuous support.

We hope that you will find the papers in this volume intellectually stimulating. We believe APWeb 2008 will lead to further developments in advanced Web engineering technologies and research activities, not only in the Asia-Pacific region but also in the international research arena.

April 2008   Hiroyuki Kitagawa
                                                              Kam Fai Wong

Preface

The rapid development of Web applications and the flux of Web information require new technologies for the design, implementation and management of information infrastructure on the Web. This volume contains papers selected for presentation at the 10th Asia Pacific Conference on Web Technology (APWeb 2008), which was held in Shenyang, China during April 26-28, 2008. APWeb is an international conference series on WWW technologies and is the primary forum for researchers, practitioners, developers and users from both academia and industry to exchange cutting-edge ideas, results, experience, techniques and tools on WWW-related technologies and new advanced applications.

APWeb 2008 received 169 submissions from 19 countries and regions world-wide, including USA, Australia, Japan, Korea, China, Hong Kong, Taiwan, UK, Germany, India, France, Turkey, Pakistan, Switzerland, New Zealand, Iran, Macao, Malaysia, and Tunisia. After a thorough review process in which each paper was reviewed and recommended by at least three Program Committee (PC) members or external reviewers, the APWeb 2008 PC selected 48 regular research papers (acceptance ratio 28%) and 15 short papers (acceptance ratio 9%). This volume also includes four invited/keynote papers. The keynote lectures were given by Masaru Kitsuregawa (University of Tokyo), Wei-Ying Ma (Microsoft Research Asia), Xuemin Lin (University of New South Wales) and Yong Shi (China Academy of Sciences). The abstracts of two tutorials presented by Xiaofang Zhou (University of Queensland) and Souvav Bhowmick (Nanyang Technological University, Singapore) are also included in these proceedings.

The conference was co-organized by Northeastern University, China, and Victoria University, Australia, and it was also financially sponsored by the National Natural Science Foundation of China and the Science and Technology Innovation Platform of Information Infrastructure Key Techniques of the 985 Program.

We wish to present our gratitude to the APWeb Steering Committee, the APWeb 2008 Organizing Committee, and the PC members and many external reviewers for their dedication in promoting the conference and for their expertise in selecting papers. We also wish to thank all authors for submitting high-quality work to the conference.

We would like to thank the General Chairs Hiroyuki Kitagawa and Kam-Fai Wong for the leadership; the Workshop Chairs Yoshiharu Ishikawa and Jing He for coordinating and organizing a high-quality workshop program together with the workshop organizers; the Tutorial Chairs Ling Feng and Keun Ho Ryu for the soliciting and selecting two excellent tutorials; and the Publicity Chairs Toshiyuki Amagasa and Hua Wang for actively promoting the event.

Last but not least, we would like to thank the local Arrangements Committee, led by Guoren Wang, and many colleagues and volunteers significantly contributed to the preparation of the conference for their enormous help. In particular, we thank Zhibin Zhao, Xiangguo Zhao and Donghong Han for their support and help in registration, accommodation and local arrangements, and Yanan Hao for his great efforts in maintaining the paper review system and communication with authors.

February 2008

Yanchun Zhang
Ge Yu
Elisa Bertino
Guandong Xu

Conference Organization

Conference Co-chairs

Hiroyuki Kitagawa, University of Tsukuba, Japan
Kam-Fai Wong, Chinese University of Hong Kong

Program Committee Co-chairs

Yanchun Zhang, Victoria University, Australia
Ge Yu, Northeastern University, China
Elisa Bertino, Purdue University, USA

Workshop Co-chairs

Yoshiharu Ishikawa, Nagoya University, Japan
Jing He, Chinese Academy of Sciences, China

Tutorial/Panel Co-chairs

Ling Feng, Tsinghua University, China
Keun Ho Ryu, Chungbuk National University, Korea

Industrial Chair

Haixun Wang, IBM T.J. Watson Research Center, USA

Publication Chair

Guandong Xu, Victoria University, Australia

Publicity Co-chairs

Toshiyuki Amagasa, University of Tsukuba, Japan
Hua Wang, University of Southern Queensland, Australia

Local Arrangements Co-chairs

Guoren Wang, Northeastern University, China
Bin Zhang, Northeastern University, China

APWeb Steering Committee

Xiaoming Li (Peking University)
Xuemin Lin (University of New South Wales)
Maria Orlowska (Ministry of Sciences and Higher Education, Poland)
Kyu-Young Whang (KAIST)
Jeffrey Yu (Chinese University of Hong Kong)
Yanchun Zhang (Victoria University)
Xiaofang Zhou (University of Queensland) Chair

APWeb Steering Committee Liaison

Xiaofang Zhou, University of Queensland, Australia

WISE Society Liaison

Qing Li, City University of Hong Kong

Program Committee

James Bailey, Australia
Rafae Bhatti, USA
Sourav Bhowmick, Singapore
Haiyun Bian, USA
Klemens Boehm, Germany
Athman Bouguettaya, USA
Stephane Bressan, Singapore
Ji-Won Byun, USA
Akmal Chaudhri, USA
Lei Chen, Hong Kong
Reynold Cheng, Hong Kong
Byron Choi, Singapore
Gao Cong, UK
Bin Cui, China
Alfredo Cuzzocrea, Italy
Xiaoyong Du, China
Yaokai Feng, Japan
Ling Feng, China
Jianhua Feng, China
Eduardo Fernandez, USA
Gabriel Fung, Hong Kong
Hong Gao, China
Zhiguo Gong, Macao

Guido Governatori, Australia
Stephane Grumbach, France
Giovanna Guerrini, Italy
Jingfeng Guo, China
Mohand-Said Hacid, France
Michael Houle, Japan
Ela Hunt, Switzerland
Yoshiharu Ishikawa, Japan
Renato Iannella, Australia
Yan Jia, China
Panagiotis Kalnis, Singapore
Murat Kantarcioglu, USA
Markus Kirchberg, New Zealand
Flip Korn, USA
Manolis Koubarakis, Greece
Chiang Lee, Taiwan
Chen Li, USA
Qing Li, Hong Kong
Zhanhuai Li, China
Xuemin Lin, Australia
Tieyan Liu, China
Qing Liu, Australia
Mengchi Liu, Canada

Hongyan Liu, China
Weiyi Liu, China
Jiaheng Lu, USA
Emil Lupu, UK
Liping Ma, Australia
Lorenzo Martino, USA
Weiyi Meng, USA
Xiaofeng Meng, China
Mukesh Mohania, India
Miyuki Nakano, Japan
Federica Paci, Italy
Vasile Palade, UK
Chaoyi Pang, Australia
Jian Pei, Canada
Zhiyong Peng, China
Evaggelia Pitoura, Greece
Weining Qian, China
Wenyu Qu, Japan
Cartic Ramakrishnan, USA
KeunHo Ryu, Korea
Monica Scannapieco, Italy
Albrecht Schmidt, Denmark
Markus Schneider, USA
HengTao Shen, Australia
Jialie Shen, Singapore
Derong Shen, China
Timothy Shih, Taiwan
Anna Squicciarini, USA
Peter Stanchev, USA
Kian-Lee Tan, Singapore

Nan Tang, Hong Kong
Changjie Tang, China
David Taniar, Australia
Jianyong Wang, China
Tengjiao Wang, China
Guoren Wang, China
Junhu Wang, Australia
Lizhen Wang, China
Wei Wang, Australia
Daling Wang, China
Haixun Wang, USA
Hua Wang, Australia
Min Wang, USA
Wei Wang, China
Jitian Xiao, Australia
Jianliang Xu, Hong Kong
Jun Yan, Australia
Dongqing Yang, China
Jian Yang, Australia
Xiaochun Yang, China
Jian Yin, China
Cui Yu, USA
Jeffrey Yu, Hong Kong
Lihua Yue, China
Rui Zhang, Australia
Xiuzhen Zhang, Australia
Qing Zhang, Australia
Peixiang Zhao, Hong Kong
Aoying Zhou, China
Qiang Zhu, USA

Table of Contents

Wireless, Sensor Networks and Grid

XML and Query Processing and Optimization

Privacy, and Security

Information Extraction, Presentation and Retrieval

P2P, Agent Systems

Ontology, Semantic Web and Web Applications

Data Streams, Time Series Analysis and Data Mining

Web Mining and Web Search

Workflow and Middleware

Socio-Sense: A System for Analysing the Societal Behavior from Long Term Web Archive*

Masaru Kitsuregawa[1], Takayuki Tamura[1,2],
Masashi Toyoda[1], and Nobuhiro Kaji[1]

[1] Institute of Industrial Science, The University of Tokyo,
4-6-1 Komaba, Meguro-ku, Tokyo, 153-8505 Japan
{kitsure,tamura,toyoda,kaji}@tkl.iis.u-tokyo.ac.jp
[2] Information Technology R&D Center, Mitsubishi Electric Corporation,
5-1-1 Ofuna, Kamakura-shi, Kanagawa, 247-8501 Japan

Abstract. We introduce Socio-Sense Web analysis system. The system applies structural and temporal analysis methods to long term Web archive to obtain insight into the real society. We present an overview of the system and core methods followed by excerpts from case studies on consumer behavior analyses.

1 Introduction

Socio-Sense is a system for analysing the societal behavior based on exhaustive Web information, regarding the Web as a projection of the real world. The Web is inundated with information issued from companies, governments, groups, and individuals, and various events in the real world tend to be reflected on the Web very quickly. Understanding the structure of the cyber space and keeping track of its changes will bring us deep insight into the background and the omen of real phenomena. Such insight cannot be achieved with current search engines, which mainly focus on providing plain facts.

The system has been developed from the ground up in the following directions:

- Web archive consisting of 9 years' worth of Japanese-centric Web contents, which enable long term historical analyses.
- Web structural analysis methods based on graph mining algorithms and natural language processing techniques. By grouping topically related Web pages, one can browse and navigate the cyber space at a macroscopic level. On the other hand, microscopic information such as product reputations can also be identified.
- Web temporal analysis methods to capture events in the cyber space such as emergence, growth, decay, and disappearance of some topic, or split and merger among topics.

* Part of this research was supported by the Comprehensive Development of e-Society Foundation Software program of the Ministry of Education, Culture, Sports, Science and Technology of Japan.

Y. Zhang et al. (Eds.): APWeb 2008, LNCS 4976, pp. 1–8, 2008.

Fig. 1. Display wall at the frontend of the system

The above elements have been integrated into the system to conduct case studies assuming corporate users' needs, such as tracking of reputations of brands or companies, grasping of consumer preferences, and analysis of consumers' lifestyles.

2   System Overview

At the base of the system is the Web archive, which consists of Japanese-centric Web contents[1] and their derivatives accumulated in a bunch of storage devices. The archived contents span 9 years now.

The Web archive started as a mere collection of yearly snapshots obtained from each run of a batch-mode crawler, and has evolved towards a general temporal database, where new versions of each Web page are independently appended. The associated crawler, which keeps running on a bunch of servers, now operates in the continuous mode, estimating update intervals of Web pages to visit them adaptively. As a result, the minimum time resolution between versions has been reduced to a day.

The URL-time index of the Web archive supports tracking of history of a URL, and crosscutting of whole URLs at arbitrary times. Contents of different periods can be uniformly searched with full text queries. Thus, history of ocurrence frequency of specific words can be easily obtained.

Though the Web archive supports exporting of its subset in one of general archive formats such as tar, the system tightly couples the Web archive with an analysis cluster to avoid overhead of moving around huge amount of data. With this parallel scanning mechanism, contents are extracted from the Web

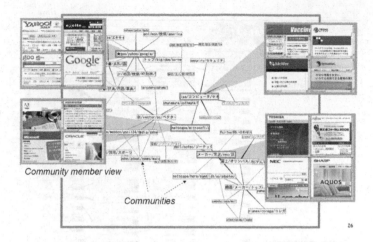

Fig. 2. Community chart on "computers"

archive and dispatched on the fly to one of the cluster nodes, where an instance of application-specific content processing loop is running. The system also takes care of load balancing among the cluster nodes.

The results of the analyses are significantly reduced in size compared with their input, but they tend to be still too complicated to present on space-limited desktop screens. Thus, we built a display wall with 5k x 3k pixels to visualize complex results nicely. Figure 1 shows the display wall showing the results from structural and temporal analyses, which are described next.

3   Web Structural Analysis

Topically related Web pages tend to be connected with relatively large number of hyper-links and reside topologically near in the Web graph. Leveraging this property, we obtained sets of related pages by extracting dense subgraphs from the whole Web space. We call each set of related pages a Web community. Subgraphs dense enough to comprise Web communities are commonly observed in various areas, from home pages of companies in the same category of industry, to personal pages mentioning the same hobbies.

After having extracted the Web communities exhaustively, communities were linked to each other according to the sparse part of the Web graph. This resulted in an associative graph with communities as nodes and degrees of relationship among communities as edges. This high level graph is called a community chart and serves as a map of the cyber space in terms of communities. Figure 2 is a subset of the community chart which relates to a term "computer." Each rectangle represents a community and related communities are connected with links. Member pages are also shown for four communities, namely computer hardware

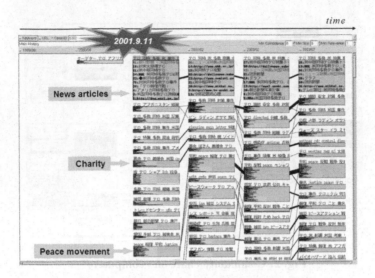

Fig. 3. Evolution of communities on "terror"

vendor community, software vendor community, security information/vendor community, and portal/search engine community (from right-bottom to left-top). This can be regarded as inter-industry relationship exposed on the Web. A graphical frontend is provided to explore the community chart, modifying the visible communities and their layout interactively.

In addition to the relationships among Web pages, it is also important to analyze textual data in the Web pages themselves. One challenge is reputation extraction. The Web text contains consumer-originated reputations of products and companies, which are useful for marketing purpose. However, extracting reputations is not trivial. It requires huge lexicon that exhaustively lists up affective words and phrases, and it is costly or even impractical to build such lexicon by hand. To tackle this problem, we employed linguistic patterns and a statistical measure in order to automatically build the lexicon from the Web archive[3]. Using this lexicon we developed a reputation extraction tool. In this tool, reputations of a query are extracted from the archive and displayed to users. Besides original texts, the number of positive/negative reputations and facets on topic are also presented. These functions provide users brief overview of the result.

4 Web Temporal Analysis

By arranging community charts derived for different times side-by-side, we can track the evolution process of the same communities. Linking communities of different times can be accomplished by regarding the member URL set as the

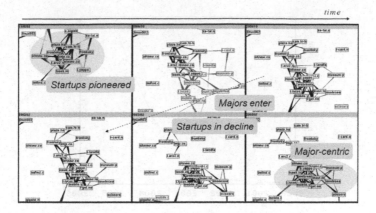

Fig. 4. Structural evolution inside of a community

identity of a community. Some URLs join and leave a community over time, and sometimes communities split and merge. The lack of counterpart of a community implies emergence or disappearance of the community.

Figure 3 shows a screenshot of a tool for visualizing the evolution process of communities. Each column corresponds to different times and each rectangle represents a community with its member URLs shown inside. Inter-community links between adjacent time slots are depicted instead of links within each time slice. This example reveals that right after the September 11 attacks, terror-related communities emerged abruptly. The method can be applied to investigate emeregence of new information, transitions of topics, and sociological phenomena.

The above methods for Web structural and temporal analyses can be combined to visualize the evolution of graph structures themselves[2]. This is most useful for Web graphs at page granularity in that subgraphs not dense enough to form a community can be captured. We can observe the characteristics of graph structures at embryonic stage of community formation and at stage of community growth.

Figure 4 shows evolution of the graph structure inside Japanese mobile search engine communities. Each of 6 panes displays the graph structure at the corresponding time. Each pane is layed out in a "synchronized" manner, where corresponding pages (nodes) are located at similar positions in each pane. We can easily identify what has happened at each stage by interactively manipulating the graphs. At the early stage, search services for mobile phones in Japan were mainly provided by startups or individuals. We can observe that, however, after major companies entered the industry, the center of the community has gradually moved to such companies.

A lot of new words are born and die every day on the Web. It is interesting to observe and analyze dynamics of new words from linguistic perspective. To analyze the dynamics of words, we estimate the frequency of new words in each

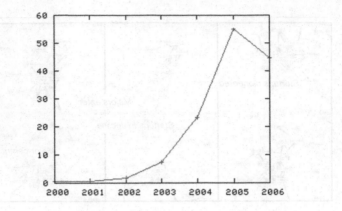

Fig. 5. Evolution of new verb *gugu-ru* (google in Japanese)

year. Because Japanese does not have word separator and it is often difficult for
conventional technique to accurately estimate the freqnency of new words, we
employed Support Vector Machine (SVM) to extract new verbs and adjectives
from the Web. Since verbs and adjectives usually inflect regularly, character
n-gram was used as features of SVM.

Figure 5 shows evolution of new verb *gugu-ru* (google in Japanese). The y-axis
represents the normalized frequency in the Web archive. We can see that *gugu-ru*
has become popular in recent years although it was not frequently used in 1999.

5   Consumer Behavior Analysis

Prevalence of blogs drastically reduced the burden for individuals to express
their opinions or impressions, and blogs have been recognized as an influential
source for decision making of individuals because blogs have agility and reality in
contrast to information originated in companies and mass media. Companies also
start recognizing significance of blogs as a tool for grasping consumers' behavior
and for communicating with consumers more intimately.

Because of this situation, we applied the methods for Web structural and temporal analyses to analysis of consumer behavior based on blog information[1]. As
a consequence, we obtained a visualization tool for inter-blog links. This tool can
visualize link structure at arbitrary time, which can be intuitively adjusted with
a slide bar. We can easily replay the temporal evolution of link relationships[4].

Considering inter-blog links as an indication of topic diffusion, we can observe how word-of-mouth information has spread out via blogs. For example, we
succeeded in identifying a source blog for a book which got drastic popularity
gain through WOM. Figure 6 shows temporal changes in links to the source blog
(circled). We can observe that, over time, the site gets more and more links.

---

[1] This part of work was done through a joint effort with Dentsu, Inc. and Senshu
University.

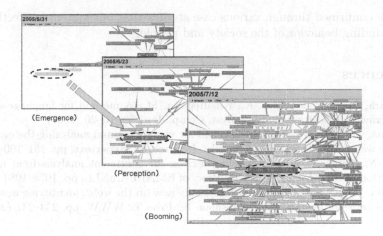

Fig. 6. Popularity evolution of a WOM source

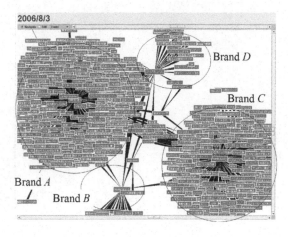

Fig. 7. Comparison of commodity brands

For companies, it's more important to figure out how they and their products are perceived. Figure 7 shows a link structure among blog entries and Web sites of commodity brands in the same industry. Blogs mentioning (linking to) a brand gather around the brand's site. Thus, attractive brands can be easily identified (in this case, brands A and C).

6　Summary

Socio-Sense system was overviewed. The combination of long term Web archive, graph mining algorithms, and natural language processing techniques enabled the system to figure out structure and evolution process of the cyber space.

We have confirmed through various case studies that our system is effective for understanding behavior of the society and people.

References

1. Tamura, T., Somboonviwat, K., Kitsuregawa, M.: A method for language-specific web crawling and its evaluation. Syst. Comp. Jpn. 38, 10–20 (2007)
2. Toyoda, M., Kitsuregawa, M.: A system for visualizing and analyzing the evolution of the web with a time series of graphs. In: Proc. of Hypertext, pp. 151–160 (2005)
3. Kaji, N., Kitsuregawa, M.: Building lexicon for sentiment analysis from massive collection of HTML documents. In: Proc. of EMNLP-CoNLL, pp. 1075–1083 (2007)
4. Toyoda, M., Kitsuregawa, M.: What's really new on the web?: identifying new pages from a series of unstable web snapshots. In: Proc. of WWW, pp. 233–241 (2006)

Building Web-Scale Data Mining Infrastructure
for Search

Wei-Ying Ma

Microsoft Research Asia
wyma@microsoft.com

Abstract. The competition in search has been driving great innovation and investment on next generation Internet services, with the goal of providing the computing platform for Internet economies on a global scale. Different from traditional Internet services, search involves myriad offline computations to analyze data at a very large scale, and an infrastructure for "scale" experiments is often required to evaluate the effectiveness of newly invented algorithms in a simulated "real" environment. In this talk, I will first review a variety of new trends in computational economies on the Internet in which search and online advertising have become the driving forces in building underlying computing infrastructure. Then, I will introduce current efforts at Microsoft Research Asia on building this new infrastructure. I will also discuss how these efforts are influencing the design of next-generation search engines from an architecture stand-point. Some advanced search technologies based on the use of this infrastructure and deeper data mining on the Web will also be demonstrated.

Keywords: Web search, data mining, and information retrieval.

Y. Zhang et al. (Eds.): APWeb 2008, LNCS 4976, p. 9, 2008.
© Springer-Verlag Berlin Heidelberg 2008

Aggregate Computation over Data Streams

Xuemin Lin and Ying Zhang

The University of News South Wales, NSW 2052, Australia
{lxue, yingz}@cse.unsw.edu.au

Abstract. Nowadays, we have witnessed the widely recognized phenomenon of high speed data streams. Various statistics computation over data streams is often required by many applications, including processing of relational type queries, data mining and high speed network management. In this paper, we provide survey for three important kinds of aggregate computations over data streams: frequency moment, frequency count and order statistic.

1 Introduction

In recent years, we have witnessed the widely recognized phenomenon of high speed data streams. A data stream is a massive real-time continuous sequence of data elements. The typical applications include sensor network, stock tickers, network traffic measurement, click streams and telecom call records. The main challenge of these applications is that the data element arrives continuously and the volume of the data is so large that they can hardly be stored in the main memory (even on the local disk) for online processing, and sometimes the system has to drop some data elements due to the high arriving speed. The data in the traditional database applications are organized on the hard disk by the Database Management System(DBMS) so the queries from the users can be answered by scanning the indices or the whole data set. Considering of the characteristics of the stream applications, it is not feasible to simply load the arriving data elements onto the DBMS and operate on them because the traditional DBMS's are not designed for rapid and continuous loading of individual data element and they do not directly support continuous queries that are typical of data stream applications [6]. Therefore, in order to support the emerging data stream applications, many works on data stream systems and related algorithms have been done by researchers in various communities and it still remains an active research area nowadays.

As mentioned in [6], following characteristics make data streams different from the conventional relational models :

- The data elements in the stream arrive online and the system has no control over the order in which data elements arrive to be processed, either within a data stream or across data streams. Moreover, the system can not predicate the arriving rate of the data elements.
- Data streams are potentionally unbounded in size. The stream elements are usually relational tuples, but sometimes might be semi-structured data like XML and HTML documents or more complex objects.

Y. Zhang et al. (Eds.): APWeb 2008, LNCS 4976, pp. 10–25, 2008.

– Once an element from data streams has been processed, it is discarded or archived. Then it can not be retrieved easily unless it is explicitly stored in the main memory, which typically is small relative to the size of the data stream.

there has been tremendous progress in building Data Stream Manage Systems(DSMSs) as evidenced by many emerging DSMSs like NiagaraCQ [17], STREAM [86,6], statStream [98], Gougar [94], Aurora [14], Telegraph [59], Borealis [2], Gigascope [32] etc.

Besides the works of building Data Stream Manage Systems (DSMS's), various data stream algorithms have been proposed from various communities like database, network, computation theory and multimedia. Among various computations over data streams, the aggregate computation plays an important role in many realistic applications such as network traffic management system and sensor network. Following are some important aggregate queries studied in the network traffic management system:

– How much HTTP traffic went on a link today from a given range of IP address? This is an example of a slice and dice query on the multidimensional time series of the flow traffic log [81].
– Find out the number of flows (with or without aggregation) which exceed a threshold T [39]. In many applications, the knowledge of these large flows is suffice.
– What are the $1 - \phi, 1 - \phi^2, 1 - \phi^3 \ldots 1 - \phi^k$ ($0 < \phi < 1$) quantiles of the TCP round trip times to each destination? This is important information to gauge the performance of the network in details [26].
– Identifying the *superspreaders* in the network. A superspreader is defined to be a host that contacts at least a given number of distinct destinations within a short time period. Superspreaders could be responsible for fast worm propagation, so it is important to detect them at early stage [91].

The rest of the paper is organized as follows. Section 2 presents the computational model of the data stream algorithms. Then three important aggregate computations over data stream are introduced in sections 3,4 and 5. Particularly, section 3 and section 4 give survey on the frequency moment and frequency count computation over data streams respectively. In section 5, we first introduce the rank computation over data stream with uniform and relative error metrics. Then the top-k computation follows. Some future work on the aggregate computation over data streams are proposed in section 6.

2 Computation Model of Data Stream Algorithms

Because of the unique characteristics of data stream applications, following issues are critical for data stream algorithms:

2.1 Processing Time

In many data stream applications, data elements arrive at a high speed so it is essential for the system to reduce the per record processing time. Otherwise the system might get congested and many elements will be dropped without being processed since usually there is no enough space to keep all of the elements. The arriving rate might burst in some applications, so the consideration of buffering and load-shedding is also required. Likewise, the query response time is another critical issue as a short response time is one of key requirements in many real time data stream applications like network monitoring and stock data analysis.

2.2 Space Usage

Since the size of the data stream is potentially unbounded, it is infeasible to keep all of the stream data elements. Moreover, although the processing time of secondary storage device has been significantly improved in recent years, it might be unacceptable even the system simply keeps every incoming stream element. So many steam algorithms are confined to the main memory without accessing the disk. Consequently, only a synopsis of the data stream can be kept to support user queries. Usually the space used is at most poly-logarithmic in the data size. Sampling, histogram, wavelet and sketch are widely used techniques to *summarize* the stream data.

2.3 Accuracy

It has been shown that in order to get exact answers for some important complex statistics like median and the number of distinct value, a linear space is required. As the synopsis maintained by the system must be very small in size, usually poly-logarithmic in the size of the data stream, the approximation is a key ingredient for stream algorithms. In many applications the exact answer is not crucial, so an approximate answer is sufficient. The system needs to make a trade-off between accuracy and storage space. Hopefully, the algorithm's performance in terms of accuracy will decrease gracefully when there is less memory available.

3 Frequency Moment Computation

Suppose a data stream consists of elements $\{a_1, a_2, \ldots, a_m\}$ which arrive sequentially and a_j is a member of $U = \{1, 2, \ldots, n\}$. Let f_i denote the number of occurrences of i in the data stream. The k-th frequency moment of the data set, denoted by F_k, is defined by $\sum_{i=1}^{n} f_i^k$. Frequency moments play an important role in many applications as they can capture the demographic information of the data set. Particularly, F_0 is the number of distinct elements appearing in the data sequence and F_1 is the length of the sequence. While F_2 is the self-join size (also called surprise index) of the data set and F_∞ is the maximal f_i. In [4], Alon *et al.* present the seminal work to study the problem of frequency moment computation against data streams. It is shown that the exact computation of F_k ($k \neq 1$) needs space linear to the data set size. A general frame work is proposed to estimate the F_k in [4] and many works [21,62,47] have been done in the literature to improve the space(time) efficiency and theoretical bounds. The range efficient computation of frequency moment is studied in [11] and their result is improved by Buriol *et al.* in [13]. The problem of frequency moment computation over sliding window is studied in [35].

Compared with other F_ks, much more attention has been given to the computation of F_0 because of its wide applications. Flajolet and Martin [46] develop the well known *FM* algorithm to estimate the F_0 of the dataset with one scan. A new algorithm is proposed by Durand and Flajolet [37] to improve the space complexity. As algorithms in [46, 37] assume the existence of hash functions with some ideal properties which are hard to construct in practise, Alon *et al.* build on similar technique but only require random pairwise independent hash functions. An adaptive distinct sampling technique is developed by Gibbons *et al.* in [49, 48]. In [10], three new algorithms are proposed to improve the space and time efficiency of previous work. In the context of graphic theoretic applications, Cohen [18] develops a size-estimation framework to explore the closure and reachability of a graph. The *linear counting* algorithm is proposed by Whang *et al.* in [92] based on the *bitmap counting* technique to estimate the cardinality in the database applications. The result is further improved in [41] based on an *adaptive bitmap* technique. Moreover, range efficient algorithm for estimating F_0 on stream data is proposed by Yossef *et al.* in [11]. Recently, [1] improves the time efficiency of the algorithm. And the same problem is investigated under sliding window model by [35] and [50].

4 Frequency Counting

Frequency counting is one of the most basic statistics of a data set because it can mark the most influential part of elements, especially in the skewed data distribution. It has many applications including network monitoring, traffic management, click stream monitoring, telephone call recording and sensor readings. Ideally, we would like to keep track of the top k elements with the highest frequency for desired value of k (top-k elements) or the elements with frequency exceeding a pre-given threshold (*heavy hitters*). For simplicity, we call them *frequent* elements. Exact computation of *frequent* elements against data stream in a small space (sub-linear to N) is infeasible. Rather, various approximate algorithms are proposed to address this problem in the context of data streams.

Misra *et al.* [78] present the first deterministic algorithm for finding ϵ-approximate frequent elements, which uses $O(\frac{1}{\epsilon})$ space and $O(1)$ amortised processing time. Recently [36] and [65] improve the algorithm by reducing the processing time to $O(1)$ in the worst case. Their algorithms guarantee to find all of the frequent candidates in the first pass, but the second pass is required to identify real frequent ones from the candidates. In many data stream applications, it is infeasible to rescan the data. By combining the hashing and sampling techniques, Estan and Verghese [40] present a novel algorithm to identify flows which exceed a pre-defined threshold. In [74], a deterministic algorithm called *lossy counting* is presented for ϵ-approximate frequency elements. Only one pass is required in their algorithm and the worst working space used is $O(\frac{1}{\epsilon} \log(\epsilon N))$. An adaptive sampling algorithm called *sticky sampling* is also proposed in [74]. Although *stick sampling* only uses $O(\frac{1}{\epsilon})$ space in the worst case, it is shown in [74] that *lossy counting* is more space efficient in practise. Metwally *et al.* [77] present *space saving* algorithm which can support approximation of top-k elements and *heavy hitters* with a unified data structure called *Stream Summary*.

Recently, [53] investigates how to efficiently compute the frequent elements in the data stream with the graphics processors. And Bandi *et al.* [9] study the problem under a special networking architectures called Network Processing Units(NPUs). Base on the *lossy counting* [74] and *space saving* [77] techniques, two TCAM-conscious algorithms are proposed to provide efficient solutions.

The algorithms above can not work under *Turnstile Model*. Consequently some *sketch* based algorithms are proposed to address this problem. As a variance of *AMS* sketch in [4], the *count sketch* technique is introduced by Charikar *et al.* [16] to find k elements whose frequencies are at least $(1-\epsilon)$ times the frequency of the k-th most frequent elements, with probability at $1 - \delta$ and space $O(\frac{1}{\epsilon^2} \log \frac{n}{\delta})$. In [28], a novel algorithm called *group test* is proposed by Cormode *et al.* to further reduce the space requirement by re-examining the previous algorithm in [16]. Based on the main idea of the *bloom filter*, which is a classical data structure to support membership queries with certain probabilistic guarantees, [39, 19] extend the *bloom filter* to find frequent elements in the data stream. The *Minimal Increase* and *Recurring Miminum* techniques are introduced to improve the accuracy of their algorithm. Recently, following the basic idea of previous algorithms, a new *count min* sketch is presented by Cormode *et al.* in [29]. Their algorithm significantly improve theoretical space efficiency of the previous results by reducing a factor of $O(\frac{1}{\epsilon})$. Their essential idea is to estimate the frequency of each element in the domain by maintaining hash based multiple counters. A similar sketch called *hCount* is independently developed by Jin *et al.* in [63].

The problem has been studied in various applications and many new algorithms are introduced to support different scenarios.

- As the hierarchy structure is widely employed in various online applications such as network management, text mining and XML data summarisation, Cormode *et al.* [24] develop novel algorithm to find out the *hierarchical heavy hitters (HHHs)* based on a layered structure. In [25], the algorithm is extended by the same authors to support identifying the *HHHs* for the hierarchical multidimensional data. The lower bound of the space complexity for this problem is studied in [60].
- [52] proposes an efficient algorithm to find the frequent elements against the sliding windows over online packet streams. Based on a layered structure, new algorithm with bounded space is introduced by Arasu and Manku in [5]. Recently, improvement work is done by Lee and Ting in [68]. Both [5] and [68] study the problem over the variable sliding window as well.
- With the development of the sensor network, various efficient algorithms have been proposed to collect or monitor the frequent elements with tree model [85, 73, 72] as well as multipath model [20, 72, 58]. One of the major concern of these work is the communication cost(total cost and maximal cost between any two nodes) for the computation. In addition to finding frequent elements in a *snapshot* fashion, the problem of continuous monitoring

frequent elements in the network also attracts much attention in the literature [7, 23, 66, 31].
- In some applications, it is desirable to find the *distinct frequent elements*. For
  instance, in order to identify the potential attacks in the network system such
  as the Distributed Denial of Service(DDoS) and warm, one of the important
  approaches is to find out the sources that contact many *distinct* destinations,
  which is called *superspread*. This problem has been studied in a number of
  papers [41, 91, 30, 8].

5   Order Statistic Computation

Among various statistics, order statistics computation is one of the most challenging, and is employed in many real applications, such as web ranking aggregation and log mining [3, 38], sensor data analysis [55], trends and fleeting
opportunities detection in stock markets [6, 71], and load balanced data partitioning for distributed computation [76, 84].

In this section, we will introduce existing works on two kinds of order statistic
oriented queries: rank queries and top-k ranked queries. Although the top-k
ranked query can be regarded as a special case of rank query where the rank is
limited between 1 and k, they have different focuses. For the rank query, we can
only provide approximate solution in the context of data stream while the exact
solution is required for the top-k ranked query. Moreover, the rank function is
pre-given for the rank query problem while usually we need to support *ad-hoc*
rank function for the later problem.

5.1   Rank Query

A rank query is essentially to find a data element with a given rank against
a monotonic order specified on data elements. And it has several equivalent
variations [57, 30]. Rank queries over data streams have been investigated in the
form of *quantile* computation. A ϕ-*quantile* ($\phi \in (0, 1]$) of an ordered set of N
data elements is the element with rank $\lceil \phi N \rceil$.

Rank and quantile queries have many applications including query optimization, finding association rule, monitoring high speed networks, trends and fleeting
opportunities detection in the stock markets, sensor data analysis, webranking
aggregation, log mining and query visualisation etc. Simple statistics such as the
mean and variance are both insufficiently descriptive and highly sensitive to data
anomalies in real world data distributions, while quantiles can summarize the
distribution of massive data more robustly. Several applications employ quantile
algorithms as a foundation, like counting inversions [57] and maintaining reverse
nearest neighbour aggregates [67] in the context of data streams.

It has been shown in [80] that the space required for any algorithm to compute
the exact rank query with p passes is $\Omega(N^{\frac{1}{p}})$, where N is number of elements.
Clearly, it is infeasible to do the exact rank computation in data stream applications where data is massive in size and fast in arriving speed. Consequently,
approximate computation of rank queries over data stream has receive a great
attentions in the recent years [80].

In this subsection, we will introduce the space and time efficient techniques of continuously maintaining data summaries to support the rank(quantile) queries in various data stream models.

Suppose an element x may be augmented to (x, v) where $v = f(x)$ (called "value") is to rank elements according to a monotonic order v and f is a pre-defined function. Without loss of generality, we assume $v > 0$ and the monotonic order is always an increasing order. We study the following rank queries over a data stream S.

Rank Query: given a rank r, find the rank r element in S.

Suppose that r is the given rank in a rank query and r' is the rank of an approximate solution. We could use the constant-based absolute error metric, say $|r - r'| \leq \epsilon$ for any given ϵ. It is immediate that such an absolute error precision guarantee will lead to the space requirement $\Omega(N)$ even for the an offline computation where $N = |S|$. So two kinds of error metrics have been used in the recent works.

Uniform Error. $\frac{r' - r}{N} \leq \epsilon$.
Relative Error. $\frac{r' - r}{r} \leq \epsilon$.

An answer to a rank query regarding r is *uniform ϵ-approximate* if its rank r' has the precision $|r - r'| \leq \epsilon N$. And it is *relative ϵ-approximate* if its rank r' has the precision $|r - r'| \leq \epsilon r$. In the following part, we will introduce the techniques of continuously maintaining a synopsis over data stream S such that at any time, the synopsis can provide a (relative or uniform) *ϵ-approximate* answer for a given rank query. The focus of the techniques is to minimize the maximal memory space required in such a continuous synopsis maintenance procedure. The processing time per element and query response time are also important issues.

Uniform Error Techniques. In [80], Munro and Paterson present a one pass algorithm to provide the uniform ϵN approximate answer for quantile query. A binary tree structure is employed in their paper and the work space required is $O(\frac{1}{\epsilon} \log^2(\epsilon N))$. Manku *et al.* [75] improve the previous algorithm in terms of the working space. They reduce the constant factor significantly by applying a more sophisticated merge strategy. Then they propose a space efficient randomized algorithm in [76] to further reduce the space bound to $O(\frac{1}{\epsilon} \log^2 \frac{1}{\epsilon\delta})$ by applying an adaptive sampling approach. Then with probability at least $1 - \delta$, their algorithm can achieve ϵN approximation. Moreover, their algorithm can compute the quantiles without the advanced knowledge of the length of the data stream. They also show that further space deduction can be achieved by feeding the sample set to any deterministic quantile algorithm.

Greenwald and Khanna [54] propose the best known deterministic quantile algorithm, called GK algorithm, for the *Cash Register Model*, with $O(\frac{1}{\epsilon} \log(\epsilon N))$ working space in worst case. Following the space reduction framework of [76], a randomized algorithm with space $O(\frac{1}{\epsilon} \log(\frac{1}{\epsilon\delta}))$ can be immediately developed. The GK algorithm has been widely employed as a building block in many quantile related works [67, 69, 5, 55].

As to the *Turnstile Model*, Gilbert *et al.* [51] propose the first algorithm to ϵ-approximate the rank query with probability at least $1 - \delta$. Their algorithm is based on estimating *range-sums* of the data stream over *dyadic* intervals with $O(\frac{1}{\epsilon^2} \log^2 |U| \log \frac{\log |U|}{\delta})$ working space, where U is the size of the element domain. The data structure they used for estimating *range-sums* of the data stream is called *Random subset sums sketch*, which can be directly replaced by the *Count-Min sketch* proposed in [29]. Then an immediate improvement over the space complexity follows with $O(\frac{1}{\epsilon} \log^2 |U| \log \frac{\log |U|}{\delta})$, which is the currently best known space bound in the *Turnstile model*. Applications of their algorithms include the telecommunication transaction monitoring and query optimization in the DBMS.

Lin *et al.* [69] propose the first space and time efficient algorithm to continuously maintain order statistics against the *count-based* sliding window model. Their techniques are based on a combination of GK-algorithm [54] and *exponential histogram* technique in [35]; They considered the rank queries over *fixed sliding windows* as well as *variable sliding windows*. And their space bound is $O(\frac{\log \epsilon^2 N}{\epsilon} + \frac{1}{\epsilon^2})$ and $(\frac{1}{\epsilon^2} \log^2(\epsilon N))$ for *fixed sliding windows* and *variable sliding windows* respectively. Based on a more sophisticated interval-tree like structure, Arasu and Manku [5] improve the space bound in [69].

With the development of the sensor network, various statistic computation algorithms on the sensor network have been developed by various communities. Greenwald and Khanna [55] study the problem of power-conserving computation of order statistics in sensor networks. They show that the tree model of the sensor network model is at least as hard as stream model. Their algorithm enforces that the largest load difference between any two nodes will not exceed $O(\log(\epsilon N))$ in order to achieve ϵ-approximation. The maximal load for each node is bounded by $O(\frac{\log^2 n}{\epsilon})$. Shrivastava *et al.* [85] improve the maximal load to $O(\frac{\log n}{\epsilon})$ based on a novel *Q-digest* data structure.

Instead of answering rank queries against a *snapshot* of the data set like [55, 85], Cormode *et al.* [22] investigate the problem of continuous tracking of complex aggregates (e.g quantile) and data-distribution summaries over collections of distributed streams. In order to achieve highly communication- and space-efficient solutions, they combine a local tracking at remote sites and simple prediction models for local site behaviour.

In [70], novel techniques are proposed to efficiently process a massive set of continuous rank queries where the *Continuous Queries* are issued and run continuously to update the query results along with the updates of the underlying datasets.

In [56], Guha *et al.* investigate the importance of the ordering of a data stream, without any assumptions about the actual distribution of the data. The quantile computation is used as a sample application. They prove some theoretical space bounds for the quantile algorithm over the data streams with *adversary* and *completely random* order. And their space efficient technique enforces a finer rank error guarantee $|r - r'| = O(r^{0.5+\epsilon})$. [53] shows how to efficiently compute the quatiles over the data stream with the graphics processors.

Relative Error Techniques. Using the relative error metric to measure approximation is not only of theoretical interest but also very useful in many applications. For instance, as shown in [26], finer error guarantees at higher ranks are often desired in network management. This is because IP traffic data often exhibits skew towards the tail and it is exactly in the most skewed region where user wants relative rank error guarantees, to get more precise information about changes in values. Relative error is also motivated by the problem of approximately *counting inversions* of a data stream [57].

The problem of finding approximate quantiles with relative error guarantee is first studied by Gupta and Zane [57], who develop a one-scan randomized technique with $O(\frac{1}{\epsilon^3} \log^2 N)$ space requirement for approximately counting inversions, by maintaining an order sketch with the relative rank error guarantee ϵ. However, their technique requires advanced knowledge of (an upper bound on) N to do one-scan sampling. This potentially limits its applications. Cormode *et al.* [26] study the related problem of computing *biased quantiles*, that is, the set of quantiles $\Phi = \{\phi_i = \phi_0^i : 1 \leq i \leq k\}$, for a fixed k and some ϕ_0, which are estimated with precision $\epsilon \phi_i N$. [26] gives an algorithm to approximate such biased quantiles with deterministic error guarantees which performs very well against many real data sets. While the problem of computing biased quantiles focuses on the relative rank error guarantee bounded by a minimum quantile $\phi_0^k N$, the rank query addresses relative error guarantees at *all* ranks, no matter how small ϕ is. As shown in [26], the application of their technique to the arbitrary rank queries leads to a linear space requirement $\Omega(N)$ in the worst case; this can render the deterministic technique impracticable in applications where small space usage is imperative.

In [96], We developed a novel, one-scan randomized algorithm ("MR") which guarantees the precision ϵ of relative rank errors with confidence $1 - \delta$ and requires $O(\frac{1}{\epsilon^2} \log \frac{2}{\delta} \log \epsilon^2 N)$ space. We also develop an effective one-scan space compression technique. Combined with the above one-scan randomized technique, it leads to a more space-efficient one-scan randomized algorithm ("MRC") which guarantees the average space requirement $O(\frac{1}{\epsilon} \log(\frac{1}{\epsilon} \log \frac{2}{\delta}) \frac{\log^{2+\alpha} \epsilon N}{1-1/2^\alpha})$ (for $\alpha > 0$), while the worst case space requirement remains $O(\frac{1}{\epsilon^2} \log \frac{2}{\delta} \log \epsilon^2 N)$. Recently, Cormode*et al.* [27] develop a novel deterministic algorithm to approximate the rank queries with relative error. The *Q-digest* structure in [85] is extended in their work, and the space required by their algorithm is $O(\frac{\log |U|}{\epsilon} \log \epsilon N)$. As shown in [27], their algorithm outperforms the randomized algorithms. However, their solution is restricted to a fixed value domain U. The space efficient deterministic algorithm with relative error guarantee remains open for the applications where the domain size of the data elements is unlimited.

Duplicate-insensitive. In many data stream applications, duplicates may often occur due to the projection on a subspace if elements have multiple attributes. For example, in the stock market a deal with respect to a particular stock is recorded by the transaction ID (TID), volume (*vol*), and average price (*av*) per share. To study purchase trends, it is important to estimate the number of different types of deals (i.e. deals with the same vol and the same av are regarded as the same type of deal) with their total prices (i.e. *vol*av*) higher (or lower) than a given value. It is also interesting to know the total price (of a deal) ranked as a median, or 25th percentile, or 10th, or 5th percentile, etc. among all different types of deals. These two types of rank queries are equivalent [27,57]; To accommodate such queries, we need to project each deal transaction (TID, vol, av) on (vol, av) and then summarize the distribution of **distinct**(vol, av)s according to a decreasing (or increasing) order of vol^*av. In this application, the data elements to be summarized are mapped from (TID, vol, av) to (vol, av) by the projection. Consequently, any generated duplicates (vol, av) must be removed to process such rank queries. Moreover, relative (or biased) rank error metrics need to be used to provide more accurate results towards heads (or tails depending on which monotonic order is adopted). Note that the generality of rank queries (quantiles) remains unchanged in this application since two different types of deals (i.e., (vol, av)s) may also have the same values of vol^*av. The unique challenge is to detect and remove the effect of duplicates without keeping every element.

Duplicates may also occur when data elements are observed and recorded multiple times at different data sites. For instance, as shown in [26,30] the same packet may be seen at many tap points within an IP network depending on how the packet is routed; thus it is important to discount those duplicates while summarising data distributions by rank queries (quantiles). Moreover, to deal with possible communication loss TCP retransmits lost packets and leads to the same packet being seen even at a given monitor more than once. Similarly, in order to achieve high fault-tolerance against communication errors in a sensor network a popular mechanism is to send data items by multi-paths [20, 72, 82] which will create duplicates.

In such distributed applications, continuously maintaining order sketches for processing rank queries may be conducted either centrally at one site or at a set of coordinating sites depending on the computing environment and the availability of software and hardware devices. Nevertheless, in those situations a crucial issue is to efficiently and continuously maintain a small space sketch with a precision guarantee at a single site, by discounting duplicates.

The *FM* technique [46] has been first applied in [12, 20, 82] to develop duplicate-insensitive techniques for approximate computing *sum*, *count* (number of sensor nodes), *average* to achieve high communication fault-tolerance.

In [82], Nath *et al.* present a duplicate-insensitive in-network quantile computation algorithm to cope with multi-path communication protocol. For each element, a random number between [0, 1] is drawn, which determines if the element will remain in the quantile sample; this combines with the element ID to remove duplicates generated by the multipass communication. As the uniform sampling technique does not guarantee to draw the same random number for the duplicated element, the technique in [82] can only handle the duplicates generated in communication rather than duplicates in data streams.

In [72], Manjhi, Nath and Gibbons propose an effective adaption paradigm for in-network aggregates computation over stream data with the aim to minimize communication costs and to achieve high fault-tolerance. A duplicate-insensitive technique for approximately computing quantiles can be immediately obtained by a combination of their tree-based approximation technique and the existing *distinct counting* technique in [10]. It can be immediately applied to a single site, where a data stream has duplicated elements, with the uniform precision guarantee $|r' - r| \leq \epsilon n$ by confidence $1 - \delta$ and space $O(1/\epsilon^3 \log 1/\delta \log m)$, where m is the maximal possible number of distinct elements.

In [30], Cormode and Muthukrishnan present a *Distinct range sums* technique by applying the FM [46] technique on the top of their *count-min* sketch [29]. The technique can be immediately used to approximately process rank query with the uniform precision guarantee $|r' - r| \leq \epsilon n$, confidence $1 - \delta$, and space $O(\frac{1}{\epsilon^3} \log \frac{1}{\delta} \log^2 m)$. Independently, Hadjieleftheriou, Byers and Kollios [58] develop two novel duplicate-insensitive techniques based on [85] and [29] to approximately compute quantiles in a distributed environment. Applying their techniques to a single site immediately leads the uniform precision guarantee $|r' - r| \leq \epsilon n$ by confidence $1 - \delta$ and space $O(\frac{1}{\epsilon^3} \log \frac{1}{\delta} \log m)$.

Very recently, we develop the first space- and time- efficient, duplicate-insensitive algorithms [97] to continuously maintain a sketch of order statistics over data stream to enforce relative ϵ-approximation. They not only improve the existing precision guarantee (from uniform ϵ-approximation to relative ϵ-approximation) but also reduce the space from $O(\frac{1}{\epsilon^3} \log \frac{1}{\delta} \log m)$ to $O(\frac{1}{\epsilon^3} \log \frac{1}{\delta} \log m)$ where m is the element domain size.

5.2   Top-k Ranked Query

Instead of finding records with arbitrary rank, in many applications users are typically interested in the k records with highest ranks, where $k \ll N$ and N is the number of records. Moreover, the ranking(preference) function might be proposed by users at query time. Providing efficient answers to such top-k ranked queries has been a quite active topic and has many important applications involving multi-criteria decision making.

In many applications the volume of the dataset is extremely large while users are usually only interested in a limited number of answers regarding to their preference functions, so it becomes necessary to pre-process the data to speed up the performance. Many related works have been done in the literature, and they can be classified into three categories: distributed index [42,43,44,45], view based index [61,95,34] and minimal space index [15,90,87,93,64,79,89]. However, only a few work [79,88,33] investigate the problem in the context of data streams.

In [79], Mouratidis *el al.* study the problem of continuous monitoring of top-k queries over sliding windows. Based on the concept of K-skyband introduced in [83], it is shown that only the tuples in the K-skyband can be answers for any top-k ranked query with monotonic preference function where $k < K$. In [79], elements are indexed by a regular grid in main memory. Two algorithms, *TMA* and *SMA*, are proposed to continuously monitor the top-k answers for those queries. In [79], a tuple can be regarded as a two dimensional data point. One dimension is the score of the tuple (the rank function is pre-given) and another is its timestamp. The *SMA* algorithm is proposed to continuously maintain the K-skyband against the stream data. The K-skyband of a dataset is the points which can be dominated by at most K-1 other points, Clearly skyline is a special instance of skyband with $K=1$. The basic idea of *SMA* is to maintain a *dominance count(DC)* for each tuple t where the DC is the number of tuples which dominate t, One tuple can be immediately discarded once its DC exceeds K since it will not be touched by any top-k query with $k \leq K$.

In [88], Tao *et al.* show how to continuously maintain the K-skyband against multidimensional data indexed by the R Tree, and the concept of *dominance count* is also employed in [89] as well. With a branch-and-bound search strategy proposed in [88], [89] can efficiently retrieve answers for the top-k ranked queries.

Recently, based on the novel concept of the geometric arrangement, Das *et al.* [33] further improve the efficiency of the top-k algorithms over data streams. Instead of continuously maintaining the K-skyband of the stream data, new tuple pruning algorithm is proposed in the paper such that the cost of minimal candidate set maintenance is significantly reduced.

6 Future Work

Although there are still many problems remaining open for the data stream algorithms, recently a great attention is given to the aggregate computation over probabilistic data streams. In many important applications such as environmental surveillance, market analysis and quantitative economics research, uncertainty is inherent because of various factors including data randomness and incompleteness, limitations of measuring equipments, delayed data updates, etc. Meanwhile, those data are created rapidly so it is worthwhile to investigate various computations over the probabilistic data streams. The main challenge is to design space and time efficient algorithms to handle the uncertain data which might arrive rapidly.

References

1. Aduri, P., Tirthapura, S.: Range efficient computation of f_0 over massive data streams. In: ICDE, pp. 32–43 (2005)
2. Ahmad, Y., Berg, B., Çetintemel, U., Humphrey, M., Hwang, J.-H., Jhingran, A., Maskey, A., Papaemmanouil, O., Rasin, A., Tatbul, N., Xing, W., Xing, Y., Zdonik, S.B.: Distributed operation in the borealis stream processing engine. In: SIGMOD, pp. 882–884 (2005)
3. Ajtai, M., Jayram, T.S., Kumar, R., Sivakumar, D.: Approximate counting of inversions in a data stream. In: STOC, pp. 370–379 (2002)
4. Alon, N., Matias, Y., Szegedy, M.: The space complexity of approximating the frequency moments. In: STOCK, pp. 20–29 (1996)
5. Arasu, A., Manku, G.S.: Approximate counts and quantiles over sliding windows. In: PODS, pp. 286–296 (2004)
6. Babcock, B., Babu, S., Datar, M., Motwani, R., Widom, J.: Models and issues in data stream systems. In: PODS (2002)
7. Babcock, B., Olston, C.: Distributed top-k monitoring. In: SIGMOD, pp. 28–39 (2003)
8. Bandi, N., Agrawal, D., Abbadi, A.E.: Fast algorithms for heavy distinct hitters using associative memories. In: IEEE International Conference on Distributed Computing Systems(ICDCS), p. 6 (2007)
9. Bandi, N., Metwally, A., Agrawal, D., Abbadi, A.E.: Fast data stream algorithms using associative memories. In: SIGMOD, pp. 247–256 (2007)

10. Bar-Yossef, Z., Jayram, T.S., Kumar, R., Sivakumar, D., Trevisan, L.: Counting distinct elements in a data stream. In: Randomization and Approximation Techniques, 6th International Workshop, RANDOM, pp. 1–10 (2002)
11. Bar-Yossef, Z., Kumar, R., Sivakumar, D.: Reductions in streaming algorithms, with an application to counting triangles in graphs. In: SODA, pp. 623–632 (2002)
12. Bawa, M., Molina, H.G., Gionis, A., Motwani, R.: Estimating aggregates on a peer-to-peer network. Technical report, Stanford University (2003)
13. Buriol, L.S., Frahling, G., Leonardi, S., Marchetti-Spaccamela, A., Sohler, C.: Counting triangles in data streams. In: PODS, pp. 253–262 (2006)
14. Carney, D., Çetintemel, U., Cherniack, M., Convey, C., Lee, S., Seidman, G., Stonebraker, M., Tatbul, N., Zdonik, S.B.: Monitoring streams - a new class of data management applications. In: Bressan, S., Chaudhri, A.B., Li Lee, M., Yu, J.X., Lacroix, Z. (eds.) CAiSE 2002 and VLDB 2002. LNCS, vol. 2590, pp. 215–226. Springer, Heidelberg (2003)
15. Chang, Y.-C., Bergman, L.D., Castelli, V., Li, C.-S., Lo, M.-L., Smith, J.R.: The onion technique: Indexing for linear optimization queries. In: SIGMOD, pp. 391–402 (2000)
16. Charikar, M., Chen, K., Farach-Colton, M.: Finding frequent items in data streams. In: Widmayer, P., Triguero, F., Morales, R., Hennessy, M., Eidenbenz, S., Conejo, R. (eds.) ICALP 2002. LNCS, vol. 2380, pp. 693–703. Springer, Heidelberg (2002)
17. Chen, J., DeWitt, D.J., Tian, F., Wang, Y.: Niagaracq: A scalable continuous query system for internet databases. In: SIGMOD, pp. 379–390 (2000)
18. Cohen, E.: Size-estimation framework with applications to transitive closure and reachability. J. Comput. Syst. Sci. 55(3), 441–453 (1997)
19. Cohen, S., Matias, Y.: Spectral bloom filters. In: Proceedings of the ACM SIGMOD International Conference on Management of Data, pp. 241–252 (2003)
20. Considine, J., Li, F., Kollios, G., Byers, J.W.: Approximate aggregation techniques for sensor databases. In: ICDE, pp. 449–460 (2004)
21. Coppersmith, D., Kumar, R.: An improved data stream algorithm for frequency moments. In: SODA, pp. 151–156 (2004)
22. Cormode, G., Garofalakis, M.N.: Sketching streams through the net: Distributed approximate query tracking. In: VLDB, pp. 13–24 (2005)
23. Cormode, G., Garofalakis, M.N., Muthukrishnan, S., Rastogi, R.: Holistic aggregates in a networked world: Distributed tracking of approximate quantiles. In: SIGMOD, pp. 25–36 (2005)
24. Cormode, G., Korn, F., Muthukrishnan, S., Srivastava, D.: Finding hierarchical heavy hitters in data streams. In: VLDB, pp. 464–475 (2003)
25. Cormode, G., Korn, F., Muthukrishnan, S., Srivastava, D.: Diamond in the rough: Finding hierarchical heavy hitters in multi-dimensional data. In: SIGMOD, pp. 155–166 (2004)
26. Cormode, G., Korn, F., Muthukrishnan, S., Srivastava, D.: Effective computation of biased quantiles over data streams. In: ICDE, pp. 20–31 (2005)
27. Cormode, G., Korn, F., Muthukrishnan, S., Srivastava, D.: Space- and time-efficient deterministic algorithms for biased quantiles over data streams. In: PODS, pp. 263–272 (2006)
28. Cormode, G., Muthukrishnan, S.: What's hot and what's not: tracking most frequent items dynamically. In: PODS, pp. 296–306 (2003)
29. Cormode, G., Muthukrishnan, S.: An improved data stream summary: The count-min sketch and its applications. In: Farach-Colton, M. (ed.) LATIN 2004. LNCS, vol. 2976, pp. 29–38. Springer, Heidelberg (2004)
30. Cormode, G., Muthukrishnan, S.: Space efficient mining of multigraph streams. In: PODS, pp. 271–282 (2005)

31. Cormode, G., Muthukrishnan, S., Zhuang, W.: What's different: Distributed, continuous monitoring of duplicate-resilient aggregates on data streams. In: ICDE, p. 57 (2006)
32. Cranor, C.D., Johnson, T., Spatscheck, O., Shkapenyuk, V.: Gigascope: A stream database for network applications. In: SIGMOD, pp. 647–651 (2003)
33. Das, G., Gunoplulos, D., Koudas, N., Sarkas, N.: Ad-hoc top-k query answering for data streams. In: VLDB (2007)
34. Das, G., Gunopulos, D., Koudas, N., Tsirogiannis, D.: Answering top-k queries using views. In: VLDB, pp. 451–462 (2006)
35. Datar, M., Gionis, A., Indyk, P., Motwani, R.: Maintaining stream statistics over sliding windows (extended abstract). In: SODA, pp. 635–644 (2002)
36. Demaine, E.D., López-Ortiz, A., Munro, J.I.: Frequency estimation of internet packet streams with limited space. In: Möhring, R.H., Raman, R. (eds.) ESA 2002. LNCS, vol. 2461, pp. 348–360. Springer, Heidelberg (2002)
37. Durand, M., Flajolet, P.: Loglog counting of large cardinalities (extended abstract). In: Di Battista, G., Zwick, U. (eds.) ESA 2003. LNCS, vol. 2832, pp. 605–617. Springer, Heidelberg (2003)
38. Dwork, C., Kumar, R., Naor, M., Sivakumar, D.: Rank aggregation methods for the web. In: WWW, pp. 613–622 (2001)
39. Estan, C., Varghese, G.: New directions in traffic measurement and accounting. In: Proceedings of the conference on Applications, technologies, architectures, and protocols for computer communications(SIGCOMM) (2002)
40. Estan, C., Varghese, G.: New directions in traffic measurement and accounting: Focusing on the elephants, ignoring the mice. ACM Trans. Comput. Syst. 21(3), 270–313 (2003)
41. Estan, C., Varghese, G., Fisk, M.: Bitmap algorithms for counting active flows on high speed links. In: ACM SIGCOMM Conference on Internet Measurement, pp. 153–166 (2003)
42. Fagin, R.: Combining fuzzy information from multiple systems. In: PODS, pp. 216–226 (1996)
43. Fagin, R.: Fuzzy queries in multimedia database systems. In: PODS, pp. 1–10 (1998)
44. Fagin, R.: Combining fuzzy information from multiple systems. J. Comput. Syst. Sci. 58(1), 83–99 (1999)
45. Fagin, R., Lotem, A., Naor, M.: Optimal aggregation algorithms for middleware. In: PODS (2001)
46. Flajolet, P., Martin, G.N.: Probabilistic counting algorithms for data base applications. J. Comput. Syst. Sci. 31(2), 182–209 (1985)
47. Ganguly, S., Cormode, G.: On Estimating Frequency Moments of Data Streams. In: Charikar, M., Jansen, K., Reingold, O., Rolim, J.D.P. (eds.) RANDOM 2007 and APPROX 2007. LNCS, vol. 4627, pp. 479–493. Springer, Heidelberg (2007)
48. Gibbons, P.B.: Distinct sampling for highly-accurate answers to distinct values queries and event reports. In: VLDB, pp. 541–550 (2001)
49. Gibbons, P.B., Tirthapura, S.: Estimating simple functions on the union of data streams. In: SPAA, pp. 281–291 (2001)
50. Gibbons, P.B., Tirthapura, S.: Distributed streams algorithms for sliding windows. In: SPAA, pp. 63–72 (2002)
51. Gilbert, A.C., Kotidis, Y., Muthukrishnan, S., Strauss, M.: How to summarize the universe: Dynamic maintenance of quantiles. In: VLDB, pp. 454–465 (2002)
52. Golab, L., DeHaan, D., Demaine, E.D., López-Ortiz, A., Munro, J.I.: Identifying frequent items in sliding windows over on-line packet streams. In: ACM SIGCOMM Conference on Internet Measurement, pp. 173–178 (2003)

53. Govindaraju, N.K., Raghuvanshi, N., Manocha, D.: Fast and approximate stream mining of quantiles and frequencies using graphics processors. In: SIGMOD, pp. 611–622 (2005)
54. Greenwald, M., Khanna, S.: Space-efficient online computation of quantile summaries. In: SIGMOD, pp. 58–66 (2001)
55. Greenwald, M., Khanna, S.: Power-conserving computation of order-statistics over sensor networks. In: PODS, pp. 275–285 (2004)
56. Guha, S., McGregor, A.: Approximate quantiles and the order of the stream. In: PODS, pp. 273–279 (2006)
57. Gupta, A., Zane, F.: Counting inversions in lists. In: SODA, pp. 253–254 (2003)
58. Hadjieleftheriou, M., Byers, J.W., Kollios, G.: Robust sketching and aggregation of distributed data streams. Technical report. Boston University (2005)
59. Hellerstein, J.M., Franklin, M.J., Chandrasekaran, S., Deshpande, A., Hildrum, K., Madden, S., Raman, V., Shah, M.A.: Adaptive query processing: Technology in evolution. IEEE Data Eng. Bull. 23(2), 7–18 (2000)
60. Hershberger, J., Shrivastava, N., Suri, S., Tóth, C.D.: Space complexity of hierarchical heavy hitters in multi-dimensional data streams. In: PODS, pp. 338–347 (2005)
61. Hristidis, V., Koudas, N., Papakonstantinou, Y.: Prefer: A system for the efficient execution of multi-parametric ranked queries. In: SIGMOD, pp. 259–270 (2001)
62. Indyk, P., Woodruff, D.P.: Optimal approximations of the frequency moments of data streams. In: STOCK, pp. 202–208 (2005)
63. Jin, C., Qian, W., Sha, C., Yu, J.X., Zhou, A.: Dynamically maintaining frequent items over a data stream. In: CIKM, pp. 287–294 (2003)
64. Jin, W., Ester, M., Han, J.: Efficient processing of ranked queries with sweeping selection. In: Jorge, A.M., Torgo, L., Brazdil, P.B., Camacho, R., Gama, J. (eds.) PKDD 2005. LNCS (LNAI), vol. 3721, pp. 527–535. Springer, Heidelberg (2005)
65. Karp, R.M., Shenker, S., Papadimitriou, C.H.: A simple algorithm for finding frequent elements in streams and bags. ACM Trans. Database Syst. 28, 51–55 (2003)
66. Keralapura, R., Cormode, G., Ramamirtham, J.: Communication-efficient distributed monitoring of thresholded counts. In: SIGMOD, pp. 289–300 (2006)
67. Korn, F., Muthukrishnan, S., Srivastava, D.: Reverse nearest neighbor aggregates over data streams. In: Bressan, S., Chaudhri, A.B., Li Lee, M., Yu, J.X., Lacroix, Z. (eds.) CAiSE 2002 and VLDB 2002. LNCS, vol. 2590, pp. 814–825. Springer, Heidelberg (2003)
68. Lee, L.K., Ting, H.F.: A simpler and more efficient deterministic scheme for finding frequent items over sliding windows. In: PODS, pp. 290–297 (2006)
69. Lin, X., Lu, H., Xu, J., Yu, J.X.: Continuously maintaining quantile summaries of the most recent n elements over a data stream. In: ICDE, pp. 362–374 (2004)
70. Lin, X., Xu, J., Zhang, Q., Lu, H., Yu, J.X., Zhou, X., Yuan, Y.: Approximate processing of massive continuous quantile queries over high-speed data streams. IEEE Trans. Knowl. Data Eng. 18(5), 683–698 (2006)
71. Manganelli, S., Engle, R.: Value at risk models in finance. In: European Central Bank Working Paper Series No. 75 (2001)
72. Manjhi, A., Nath, S., Gibbons, P.B.: Tributaries and deltas: Efficient and robust aggregation in sensor network streams. In: SIGMOD, pp. 287–298 (2005)
73. Manjhi, A., Shkapenyuk, V., Dhamdhere, K., Olston, C.: Finding (recently) frequent items in distributed data streams. In: ICDE, pp. 767–778 (2005)
74. Manku, G.S., Motwani, R.: Approximate frequency counts over data streams. In: Bressan, S., Chaudhri, A.B., Li Lee, M., Yu, J.X., Lacroix, Z. (eds.) CAiSE 2002 and VLDB 2002. LNCS, vol. 2590, pp. 346–357. Springer, Heidelberg (2003)

75. Manku, G.S., Rajagopalan, S., Lindsay, B.G.: Approximate medians and other quantiles in one pass and with limited memory. In: SIGMOD, pp. 426–435 (1998)
76. Manku, G.S., Rajagopalan, S., Lindsay, B.G.: Random sampling techniques for space efficient online computation of order statistics of large datasets. In: SIGMOD, pp. 251–262 (1999)
77. Metwally, A., Agrawal, D., Abbadi, A.E.: Efficient computation of frequent and top-k elements in data streams. In: Eiter, T., Libkin, L. (eds.) ICDT 2005. LNCS, vol. 3363, pp. 398–412. Springer, Heidelberg (2004)
78. Misra, J., Gries, D.: Finding repeated elements. Sci. Comput. Program. 2(2), 143–152 (1982)
79. Mouratidis, K., Bakiras, S., Papadias, D.: Continuous monitoring of top-k queries over sliding windows. In: SIGMOD, pp. 635–646 (2006)
80. Munro, J.I., Paterson, M.: Selection and sorting with limited storage. Theor. Comput. Sci. 12, 315–323 (1980)
81. Muthukrishnan, S.: Data streams: algorithms and applications. In: SODA, pp. 413–413 (2003)
82. Nath, S., Gibbons, P.B., Seshan, S., Anderson, Z.R.: Synopsis diffusion for robust aggregation in sensor networks. In: SenSys, pp. 250–262 (2004)
83. Papadias, D., Tao, Y., Fu, G., Seeger, B.: Progressive skyline computation in database systems. ACM Trans. Database Syst. 30(1), 41–82 (2005)
84. Poosala, V., Ioannidis, Y.E.: Estimation of query-result distribution and its application in parallel-join load balancing. In: VLDB, pp. 448–459 (1996)
85. Shrivastava, N., Buragohain, C., Agrawal, D., Suri, S.: Medians and beyond: new aggregation techniques for sensor networks. In: SenSys, pp. 239–249 (2004)
86. STREAM stream data manager, http://www-db.stanford.edu/stream/sqr
87. Tao, Y., Hadjieleftheriou, M.: Processing ranked queries with the minimum space. In: Dix, J., Hegner, S.J. (eds.) FoIKS 2006. LNCS, vol. 3861, pp. 294–312. Springer, Heidelberg (2006)
88. Tao, Y., Hristidis, V., Papadias, D., Papakonstantinou, Y.: Branch-and-bound processing of ranked queries. Inf. Syst. 32(3), 424–445 (2007)
89. Tao, Y., Xiao, X., Pei, J.: Efficient skyline and top-k retrieval in subspaces. IEEE Trans. Knowl. Data Eng (to appear, 2007)
90. Tsaparas, P., Palpanas, T., Kotidis, Y., Koudas, N., Srivastava, D.: Ranked join indices. In: ICDE, pp. 277–288 (2003)
91. Venkataraman, S., Song, D.X., Gibbons, P.B., Blum, A.: New streaming algorithms for fast detection of superspreaders. In: NDSS (2005)
92. Whang, K.-Y., Zanden, B.T.V., Taylor, H.M.: A linear-time probabilistic counting algorithm for database applications. ACM Trans. Database Syst. 15(2), 208–229 (1990)
93. Xin, D., Chen, C., Han, J.: Towards robust indexing for ranked queries. In: VLDB, pp. 235–246 (2006)
94. Yao, Y., Gehrke, J.: The cougar approach to in-network query processing in sensor networks. SIGMOD Record 31(3), 9–18 (2002)
95. Yi, K., Yu, H., Yang, J., Xia, G., Chen, Y.: Efficient maintenance of materialized top-k views. In: ICDE, pp. 189–200 (2003)
96. Zhang, Y., Lin, X., Xu, J., Korn, F., Wang, W.: Space-efficient relative error order sketch over data streams. In ICDE, page 51 (2006)
97. Zhang, Y., Lin, X., Yuan, Y., Kitsuregawa, M., Zhou, X., Yu, J.X.: Summarizing order statistics over data streams with duplicates. In: ICDE, pp. 1329–1333 (2007)
98. Zhu, Y., Shasha, D.: Statstream: Statistical monitoring of thousands of data streams in real time. In: Bressan, S., Chaudhri, A.B., Li Lee, M., Yu, J.X., Lacroix, Z. (eds.) CAiSE 2002 and VLDB 2002. LNCS, vol. 2590, pp. 358–369. Springer, Heidelberg (2003)

A Family of Optimization Based Data Mining Methods

Yong Shi[1,3,*], Rong Liu[1,2], Nian Yan[3], and Zhenxing Chen[3]

[1] Research Center on Fictitious Economy and Data Sciences, Chinese Academy of Sciences
100080 Beijing, China
[2] School of Mathematical Science, Graduate University of Chinese Academy of Sciences
100049 Beijing, China
[3] College of Information Science and Technology, University of Nebraska at Omaha
Omaha NE 68132, USA
liu.rong@163.com, yshi@unomaha.edu yshi@gucas.ac.cn,
nyan@mail.unomaha.edu, zchen@mail.unomaha.edu

Abstract. An extensive review for the family of multi-criteria programming data mining models is provided in this paper. These models are introduced in a systematic way according to the evolution of the multi-criteria programming. Successful applications of these methods to real world problems are also included in detail. This survey paper can serve as an introduction and reference repertory of multi-criteria programming methods helping researchers in data mining.

Keywords: multi-criteria programming; data mining; classification; regression; fuzzy programming; credit scoring; network intrusion detection.

1 Introduction

In the past five decades, linear programming, which is formulated with a single criterion, has been widely used in many real world problems. However, this great advancement has some inherent limitations. In most practices, there are generally multiple objectives which need to optimize. Further more, these objectives may even be conflicting with each other. The Bias-Variance dilemma (Wassaman 2005) in statistical inference and Fitness-Generality trade-off (Han and Kamber, 2000) in data mining are two famous examples which can not be well addressed within the scope of single criterion programming. In statistical inference, the best estimator or learner is the one with lowest bias and lowest variance. Unfortunately, these two criteria can not be minimized in the same time, because the decrease of the bias will increases the variance. In data mining, the classifier with low training error (fitness) and good generalizability (performance on unseen data) is highly desired. However, the over-fitness will almost sure weaken the generalizability of the classifier. Moreover, unlike the Bias and Variance criteria which can be integrated together as Risk (= $Bias^2$+Variance) under the lost function of squared error, the fitness and generalizability can not be simply combined together since they have different measurements. Another example which will be investigated in detail throughout this paper is the classification or the so-called discriminate analysis in statistical inference.

---

[*] Corresponding author.

Y. Zhang et al. (Eds.): APWeb 2008, LNCS 4976, pp. 26–38, 2008.

The purpose of classification is to separate data according to some criteria. There are two commonly used criteria among them. The first one is the overlapping degree with respect to the discriminate boundary. The lower of this degree the better the classification is. Another one is the distance from a point to the discriminate boundary. The larger the sum of these distances the better the classification is. Accordingly, the objective of a classification is to minimize the sum of the overlapping degree and maximize the sum of the distances (Freed and Glover, 1981; Freed and Glover, 1986). Unfortunately, these two criteria can not be optimized simultaneously because they are contradictory to each other. Thus, the multi-criteria programming can be used to overcome this kind of problems in a systematical way.

It has been thirty years since the first appearance of the multi-criteria linear programming. During these years, the multi-criteria programming has been not only improved in theoretical foundations but also applied successfully in real world problems. The data mining is such an area where the multi-criteria program got great achievement in. Initialed by Shi et al. (2000), the model and ideal of multi-criteria programming have been widely adopted by the researches for classification, regression, etc. On this basis, Jing et al. (2004) introduced the fuzzy approach in the multi-criteria programming to address the uncertainty in data. Using a different norm to measure the overlapping degree and distance, Kou (2006) presented the Multiple Criteria Quadratic Programming for data mining. Kou et al (2002) proposed Multi-Group Multiple Criteria Mathematical Programming aimed to handle the multi-group classification. To extend the application of multi-criteria programming, Zhang et al. (2008) developed a regressing method based on this technique. The development of this family of multi-criteria data mining technique is described in Figure 1.

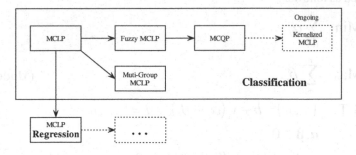

Fig. 1. Evolution of multi-criteria models

These multi-criteria data mining techniques also have yielded fruitful results in diverse real world applications. Kou (2006) applied the Multiple Criteria Quadratic Programming to Credit Card Risk Analysis and get comparable results with some sophisticated methods. Classification of HIV-1 Mediated Neuronal Dendritic and Synaptic Damage is another successful example of the multi-criteria data mining techniques (Zheng et al., 2003). Kou et al. (2004) introduced this technique to Network Surveillance and Intrusion Detection System. The multi-criteria data mining has also been applied to prediction bankruptcy by Kwak et al. (2005). Zhang et al. (2007a, 2007b) employed the Multiple-Criteria Linear and Quadratic Programming for VIP E-Mail Behavior Analysis.

In respect of the abundance of the variations of this method and the diversity of their applications, a survey paper will be helpful to understand and use this family of data mining techniques more easily. The particular objective of this paper is to closely survey the multi-criteria data mining techniques and their applications. The arrangement is as follows: Models with different variations are described in section 2; Section 3 will introduce the applications of these techniques; At the end of this paper, conclusions and future works are presented.

2 Multi-criteria Data Mining Methods

The variations of multi-criteria linear programming methods in data mining are briefly introduced and analyzed in this section.

2.1 Multiple Criteria Linear Programming

In linear discriminate analysis, the data separation can be achieved by two opposite objectives. The first one is to maximize the minimum distances of observations from the critical value. The second objective separates the observations by minimizing the sum of the deviations (the overlapping) among the observations (Freed and Glover, 1981; Freed and Glover, 1986). However, it is theoretically impossible to optimize MMD and MSD simultaneously, the best tradeoff of two measurements is difficult to find. This shortcoming has been coped with by the technique of multiple criteria linear programming (MCLP) (Shi et al, 2000; Shi, 2001). The first MCLP model could be described as follows:

$$\text{Min} \quad \sum_{i=1}^{n} \alpha_i$$

$$\text{Max} \quad \sum_{i=1}^{n} \beta_i \qquad \qquad \textbf{(Model 1)}$$

$$\text{S.T.} \quad (x_i, w) = b + y_i(\alpha_i - \beta_i), \quad i = 1, \ldots, n$$

$$\alpha, \beta \geq 0$$

Here, α_i is the overlapping and β_i the distance from the training sample x_i to the discriminator $(w, x) = b$ (classification boundary). $y_i \in \{1, -1\}$ denotes the label of x_i and n is the number of samples. The weights vector w and the bias b are the unknown variables need to be tuned to optimize the two objectives. A visual description of this model is shown in Figure 2.

The **Model 1** is formulized as the Multiple Criteria Linear Programming which is difficult to optimize. In order to facilitate the computation, the compromise solution approach (Shi, 2000; Shi and Yu, 1989) can be employed to reform the above model so that we can systematically identify the best trade-off between $-\Sigma\alpha_i$ and $\Sigma\beta_i$ for an

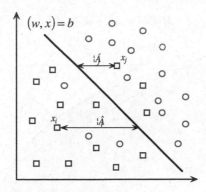

Fig. 2. The two criteria of classification

optimal solution. The "ideal value" of $-\Sigma\alpha_i$ and $\Sigma\beta_i$ are assumed to be $\alpha^* > 0$ and $\beta^* > 0$ respectively. Then, if $-\Sigma\alpha_i > \alpha^*$, we define the regret measure as $-d_\alpha^+ = \Sigma\alpha_i + \alpha^*$; otherwise, it is 0. If $-\Sigma_i\alpha_i < \alpha^*$, the regret measure is defined as $d_\alpha^- = \alpha^* + \Sigma\alpha_i$; otherwise, it is 0. Thus, we have (i) $\alpha^* + \Sigma\alpha_i = d_\alpha^- - d_\alpha^+$, (ii) $|\alpha^* + \Sigma\alpha_i| = d_\alpha^- + d_\alpha^+$, and (iii) $d_\alpha^-, d_\alpha^+ \geq 0$. Similarly, we derive $\beta^* - \Sigma\beta_i = d_\beta^- - d_\beta^+$, $|\beta^* - \Sigma\beta_i| = d_\beta^- + d_\beta^+$, and $d_\beta^-, d_\beta^+ \geq 0$. The two-class MCLP model has been gradually evolved as:

$$\text{Min} \quad d_\alpha^+ + d_\alpha^- + d_\beta^+ + d_\beta^-$$

$$\text{S.T.} \quad \alpha^* + \sum_{i=1}^{n}\alpha_i = d_\alpha^- - d_\alpha^+$$

$$\beta^* + \sum_{i=1}^{n}\beta_i = d_\beta^- - d_\beta^+ \qquad \text{(Model 2)}$$

$$(x_i, w) = b + y_i(\alpha_i - \beta_i), \quad i = 1, \ldots, n$$

$$\boldsymbol{\alpha}, \boldsymbol{\beta} \geq 0, \ d_\alpha^+, d_\alpha^-, d_\beta^+, d_\beta^- \geq 0$$

Here α^* and β^* are given, w and b are unrestricted. The geometric meaning of the model is shown as in Figure 3.

In order to calculate a huge data set, the Linux-based MCLP classification algorithm was developed to implement the above **Model 2** (Kou and Shi, 2002).

2.2 Multiple Criteria Quadratic Programming

Based on MCLP, the Multiple Criteria Quadratic Programming is later developed to achieve better classification performance and stability. The overlapping and distance are respectively represented by the nonlinear functions $f(\alpha)$ and $g(\beta)$. Given

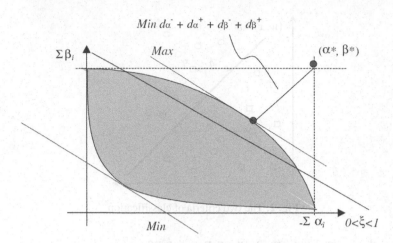

Fig. 3. Compromised and Fuzzy Formulations

weights w_α and w_β, let $f(\alpha) = \| \alpha \|^p$ and $g(\beta) = \| \beta \|^p$, the two criteria basic **Model 1** can be converted into a single criterion general non-linear classification model:

$$\text{Min} \quad w_\alpha \| \boldsymbol{\alpha} \|^p - w_\beta \| \boldsymbol{\beta} \|^p$$
$$\text{S.T.} \quad (x_i, w) = b + y_i(\alpha_i - \beta_i), \quad i = 1, \dots, n \qquad \textbf{(Model 3)}$$
$$\boldsymbol{\alpha}, \boldsymbol{\beta} \geq 0$$

On the basis of **Model 3**, non-linear classification models with any norm can be defined theoretically. Let

$$f(\alpha) = \alpha^T H \alpha = \sum_{i=1}^m \alpha_i^2 \text{ and } f(\beta) = \beta^T Q \beta = \sum_{i=1}^m \beta_i^2$$

where H and Q are predefined as identity matrices. We formulate a simple quadratic programming with 2-norm as in **Model 4**:

$$\text{Min} \quad w_\alpha \sum_{i=1}^n \alpha_i^2 - w_\beta \sum_{i=1}^n \beta_i^2$$
$$\text{S.T.} \quad (x_i, w) = b + y_i(\alpha_i - \beta_i), \quad i = 1, \dots, n \qquad \textbf{(Model 4)}$$
$$\boldsymbol{\alpha}, \boldsymbol{\beta} \geq 0$$

In order to reduce the number of variables involved in our model and thus simplify the computing. Let $\eta_i = \alpha_i - \beta_i$. According to our definition, $\eta_i = \alpha_i$ for all mis-classified records and $\eta_i = -\beta_i$ for all correctly separated records. To obtain strong convexity to the objection function, we add $\dfrac{W_b}{2} b^2$ to **Model 4**'s objective function.

The weight W_b is an arbitrary positive number and $W_b \ll W_\beta$. Model 5 becomes (Kou, 2006):

$$\text{Min} \quad \frac{1}{2}\|X\|_2^2 + \frac{W_\alpha}{2}\sum_{i=1}^m \eta_i^2 + W_\beta \sum_{i=1}^m \eta_i + \frac{W_b}{2}b^2$$

$$\text{S.T.} \quad (x_i, w) = b + y_i\eta_i, \quad i = 1,\ldots,n \qquad \textbf{(Model 5)}$$

$$\eta \geq 0$$

2.3 Multiple Criteria Fuzzy Linear Programming

Instead of identifying a compromise solution for the separation of data in MCLP, the fuzzy approach classifies the data by seeking a fuzzy (satisfying) solution obtained from a fuzzy linear program (FLP) (He et al., 2004). Let y_{1L} be MSD and y_{2U} be MMD, then one can assume that the value of Maximize$\Sigma\alpha_i$ to be y_{1U} and that of Minimize $\Sigma\alpha_i$ to be y_{2L}. The classification problem is equivalent to the following fuzzy linear program:

$$\text{Min} \quad \xi$$

$$\text{S.T.} \quad \xi \leq \frac{\sum \alpha_i - y_{1L}}{y_{1U} - y_{1L}}$$

$$\xi \leq \frac{\sum \beta_i - y_{2L}}{y_{2U} - y_{2L}} \qquad \textbf{(Model 6)}$$

$$(x_i, w) = b + y_i(\alpha_i - \beta_i), \quad i = 1,\ldots,n$$

$$\alpha, \beta \geq 0$$

Note that **Model 6** will produce a value of ξ with $0 \leq \xi < 1$. To avoid the trivial solution, one can set up $0 \leq \varepsilon < \xi$, for a given ε. Therefore, seeking Maximum ξ in the FLP approach becomes the standard of determining the classifications between Good and Bad records in the database. A graphical illustration of this approach can be seen from Figure 3, any point of hyper plane $0 < \xi < 1$ over the shadow area represents the possible determination of classifications by the FLP method.

2.4 Multi-group Multiple Criteria Mathematical Programming

The above models are concerned with two groups' case. Now suppose we have k groups, G_1, $G_2,\ldots,$ G_k, are predefined. $G_i \cap G_j = \Phi, i \neq j, 1 \leq i, j \leq k$ and $x_i \in \{G_1 \cup G_2 \cup \ldots \cup G_k\}$. A series of boundary scalars $b_1 < b_2 < \ldots < b_{k-1}$ can be set to separate these k groups. The boundary b_j is used to separate G_j and G_{j+1}. Let

$\mathbf{w} = (w_1, \ldots, w_m)^T \in R^m$ be a vector of real number to be determined. Thus, we can establish the following linear inequations (Kou et al., 2002):

$$(\mathbf{w}, \mathbf{x}_i) < b_1, \qquad \forall \mathbf{x}_i \in G_1; \tag{1}$$

$$b_{j-1} \leq (\mathbf{w}, \mathbf{x}_i) < b_j, \qquad \forall \mathbf{x}_i \in G_j; \tag{2}$$

$$(\mathbf{w}, \mathbf{x}_i) \geq b_{k-1}, \qquad \forall \mathbf{x}_i \in G_k; \tag{3}$$

$2 \leq j \leq k-1, 1 \leq i \leq n.$

A mathematical function f can be used to describe the summation of total overlapping while another mathematical function g represents the aggregation of all distances. The final classification accuracies of this multi-group classification problem depend on simultaneously minimize f and maximize g. Thus, a generalized bi-criteria programming method for classification can be formulated as:

$$\text{Min} \quad f$$

$$\text{Max} \quad g \qquad \qquad \textbf{(Generalized Model)}$$

$$\text{S.T.} \quad (1), (2), \text{and } (3)$$

Furthermore, to transform the bi-criteria problems of the generalized model into a single-criterion problem, weights $w_\alpha > 0$ and $w_\zeta > 0$ are introduced for $f(\alpha)$ and $g(\zeta)$, respectively. The values of w_α and w_ζ can be pre-defined in the process of identifying the optimal solution. As a result, the generalized model can be converted into a single-criterion mathematical programming model as: (**Model 7**)

$$\text{Min} \quad w_\alpha \sum_{j=1}^{k} \sum_{i=1}^{n} |\alpha_{i,j}|_p - w_\zeta \left(\sum_{j=1 \text{ or } j=k}^{n} \sum_{i=1}^{n} |\zeta_{i,j}|_p - \sum_{j=2}^{k-1} \sum_{i=1}^{n} \left| \frac{b_j - b_{j-1}}{2} - \zeta_{i,j} \right|_p \right)$$

$$\text{S.T.} \quad (x_i, w) = b_j + \alpha_{i,j} - \zeta_{i,j}, \quad 1 \leq j \leq k-1 \tag{4}$$

$$(x_i, w) = b_{j-1} + \alpha_{i,j-1} - \zeta_{i,j-1}, \quad 2 \leq j \leq k \tag{5}$$

$$\zeta_{i,j-1} \leq b_j - b_{j-1} \quad 2 \leq j \leq k \tag{a}$$

$$\zeta_{i,j} \leq b_{j+1} - b_j \quad 1 \leq j \leq k-1 \tag{b}$$

Here $\mathbf{x}_i$ is given, $\mathbf{w}$ and b_j are unrestricted, and $\alpha_{i,j}, \zeta_{i,j} \geq 0, 1 \leq i \leq n.$

(a) and (b) are defined as such due to the fact that the distances from any correctly classified data $(A_i \in G_j, 2 \leq j \leq k-1)$ to two adjunct boundaries b_{j-1} and b_j must be less than $b_j - b_{j-1}$. A better separation of two adjunct groups may be achieved by the following constraints instead of (a) and (b) because (c) and (d) set up stronger limitation on ζ_i^j:

$$\zeta_{i,j} \leq (b_j - b_{j-1})/2 + \varepsilon, 2 \leq j \leq k \tag{c}$$

$$\zeta_{i,j} \le (b_{j+1} - b_j)/2 + \varepsilon, 1 \le j \le k-1 \tag{d}$$

$\varepsilon \in \Re^+$ *is a small positive real number.*

Let $p = 2$, then objective function in Model 1 can now be a quadratic objective and we have: (**Model 8**)

$$\text{Min} \quad w_\alpha \sum_{j=1}^{k} \sum_{i=1}^{n} \left| \alpha_{i,j} \right|_p - w_\zeta \left(\sum_{j=1 \, or \, j=k} \sum_{i=1}^{n} \left| \zeta_{i,j} \right|_p - \sum_{j=2}^{k-1} \sum_{i=1}^{n} \left| \frac{b_j - b_{j-1}}{2} - \zeta_{i,j} \right|_p \right)$$

S.T.   (4), (5), (c), and (d)

Note that the constant $(\dfrac{b_j - b_{j-1}}{2})^2$ is omitted from the **Model 5** without any effect to the solution.

2.5 Regression Method by Multiple Criteria Linear Programming

In order to apply MCLP to regression problem, it is necessary to construct the data set which is fit for MCLP problem at first. The data set of the regression problem is $T = \{(x_1^T, y_1), (x_2^T, y_2), \ldots, (x_n^T, y_n)\}$, where $x_i \in R^m$ is the input variable, $y_i \in R$ is the output variable, which can be any real number. Similar to the D^+ and D^- data sets in section 3, the D_{MCLP}^+ and D_{MCLP}^- data sets for MCLP classification model is constructed. And with these data sets, MCLP regression model can be written as follows (Zhang et al. 2008) : (**Model 9**)

$$\text{Min} \quad \sum_{i=1}^{n} (\alpha_i - \alpha_i') - \text{Max} \quad \sum_{i=1}^{n} (\beta_i - \beta_i')$$

$$\text{S.T.} \quad x_{i1} w_1 + \cdots + x_{im} w_m + (y_1 + \varepsilon) w_{m+1} = b - \alpha_1 + \beta_1$$

$$\cdots \qquad\qquad\qquad\qquad\qquad for\ all \in G$$

$$x_{n1} w_1 + \cdots + x_{nm} w_m + (y_n + \varepsilon) w_{m+1} = b - \alpha_n + \beta_n$$

$$x_{i1} w_1 + \cdots + x_{im} w_m + (y_1 - \varepsilon) w_{m+1} = b + \alpha_1' - \beta_1'$$

$$\cdots \qquad\qquad\qquad\qquad\qquad for\ all \in B$$

$$x_{n1} w_1 + \cdots + x_{nm} w_m + (y_n - \varepsilon) w_{m+1} = b + \alpha_n' - \beta_n'$$

$$\alpha, \alpha', \beta, \beta' \ge 0$$

Aggregation of Good samples:

$$D_{MCLP}^+ = \{((x_i^T, y_i + \varepsilon)^T, +1), \ i = 1, \cdots, l\}$$

Aggregation of Bad samples:

$$D_{MCLP}^- = \{((x_i^T, y_i - \varepsilon)^T, -1), \ i = 1, \cdots, l\}$$

3 Applications

The Multiple Criteria Mathematical Programming approaches have been successfully used for many real world data mining applications, such as Credit Card Risk Analysis, Classification of HIV-1 Mediated Neuronal Dendritic and Synaptic Damage, Network Surveillance and Intrusion Detection System, Bankruptcy Prediction and etc. In this section, we introduce the examples of the utilization of the proposed models mentioned above.

3.1 Credit Card Risk Analysis

The credit card dataset is obtained from a major US bank. The data set consists of 65 attributes and 6000 records. There are two groups: current customer and bankrupt customer. The task is to analyze and predict the customers who have high or low potential risk of bankruptcy. Each record in these data has a class label to indicate its' financial status: either Normal or Bad. Bad indicates a bankrupt credit or firm account and Normal indicates a current status account. The result of MCQP (Model 4, 5, 6) (Kou, 2006) is compared with four well-known classification methods: SPSS linear discriminant analysis (LDA) (SPSS 2004), Decision Tree (See5.0), SVM light (Joachims, 2004), and LibSVM (Chang and Lin, 2001). A standard 10-fold cross-validation is performed in this application. As a result, the overall classification accuracies of LDA, See5, SVMlight, LibSVM and MCQP are 78.79%, 75.00%, 74.00%, 77.57% and 78.50% respectively. The experimental study indicates that MCQP can classify credit risk data and achieves comparable results with well-known classification techniques.

3.2 Classification of HIV-1 Mediated Neuronal Dendritic and Synaptic Damage

The ability to identify neuronal damage in the dendritic arbor during HIV-1-associated dementia (HAD) is crucial for designing specific therapies for the treatment of HAD. Zheng et al. (2003) used two-class model of multiple criteria linear programming to classify the HIV data (each with nine attributes) produced by laboratory experimentation and image analysis. The dataset includes four classes: (1) treatments with brain derived neurotrophic factor (BDNF), (2) glutamate, (3) gp120 or (4) non-treatment controls from the experimental systems. The HIV database contained data from 2112 neurons. Among them, 101 are from G1, 1001 from G2, 229 from G3, and 781 from G4. We applied the 5-fold cross-validation with 80% for training and 20% for testing. The classifications were conducted as {G1 vs. G2}, {G2 vs. G3}, {G2 vs. G4} and {G3 vs. G4}. The conclusion on this comparison is that the proposed MCLP method can be competitive in neuronal injury classification since it showed stronger performance than the back propagation neural networks classifier in three of four class pairs.

3.3 Network Surveillance and Intrusion Detection System

Network intrusion refers to inappropriate, incorrect, or anomalous activities aimed at compromise computer networks (Kou et al. 2004; Kou, 2006). The early and reliable detection of network attacks is a pressing issue of today's network security. Classification methods are one the major tools in network intrusion detection. The dataset used in this application is KDDCUP-99 data set which was provided by DARPA in 1998 for the evaluation of intrusion detection approaches (Stolfo et al. 2000). There are four main categories of attacks: denial-of-service (DOS); unauthorized access from a remote machine (R2L); unauthorized access to local root privileges (U2R); surveillance and other probing. Because U2R has only 52 distinct records, we test KDDCUP-99 as a four-group classification problem. The four groups are DOS (247267 distinct records), R2L (999 distinct records), and Probe (13851 distinct records), and normal activity (812813 distinct records). The overall classification accuracy, which is defined as the average of each class's accuracy, of See5, MCMP, and MCMP with kernel is 93.08%, 95.71%, and 97.2%, respectively. MCMP with kernel outperforms See5 and MCMP in every class, especially for Normal class. The high classification accuracy of MCMP with kernel leads to low false alarm rates.

3.4 Bankruptcy Prediction

Bankruptcy prediction is an interesting topic because many stakeholders such as bankers, investors, auditors, management, employees, and the general public are interested in assessing firms' risk of bankruptcy. The multiple criteria linear programming (MCLP) approach is used as a data mining tool for bankruptcy prediction. In (Kwak et al., 2006), the empirical results show that the previous Ohlson's (1980) predictor variables perform better than Altman's (1968) predictor variables using 1990s Japanese financial data based on the overall and Type I prediction rates of 77.70% and 70.27%, respectively, using Ohlson's (1980) predictor variables.

Kwak et al. (2005) also tested the use of multiple criteria linear programming (MCLP) model to data mining for bankruptcy prediction using U.S. data from a sample period of 1992 to 1998 and concluded that the MCLP approach performed better than either the MDA approach of Altman (1968) or the logit approach of Ohlson (1980). Kwak et al. (2005) report overall prediction rates of 88.56% and 86.76% for the Altman (1968) and Ohlson (1980) models, respectively, and Type I prediction rates of 46% and 43%, respectively.

3.5 VIP E-Mail Behavior Analysis

According to the statistic, as Chinese network advanced in the past few years, the total market size of Chinese VIP E-mail service has reached 6.4 hundred million RMB by 2005. This huge market dramatically enforced market competition among all E-mail service companies. The analysis for the pattern of lost customer account is hereby a significant research topic. Hopefully, the result can help the decision to reduce the customer loss rate.

To this end, Zhang et al. (2006) applied the MCLP (**model 1**) to the charged VIP E-mail service analysis. Through cross-validation process, they demonstrated that MCLP is highly accurate and stable on VIP E-mail dataset. These results from majority voting also showed that the accuracy of numeral voters does not exceed that of single excellent voter by noticeable margin. On these basses, They concluded that the application of MCLP to VIP E-Mail behavior classification is success and MCLP is accurate and dependable for categorizing VIP E-Mail dataset.

MQLC, a variation of **model 3**, has also been proposed by Zhang et al. (2007) to categorize the VIP E-Mail dataset. The results of cross-validation showed that the model is extremely stable for multiple groups of randomly generated training set and testing set. The comparison of MQLC and Decision Tree in C5.0 indicated that MQLC performs better than Decision Tree on small samples.

4   Conclusion and Future Work

This paper offers an abundance of models and applications of the multi-criteria programming in data mining. Different variations of multi-criteria programming models are proposed to solve the many problems, such as classification/prediction and regressions. The history of these optimization based data mining methods has been revisited and illustrated with mathematical formulations and real world applications.

The researchers and engineers in data mining can benefit a lot from this survey. In addition, it can also serve as a reference repertory of multi-criteria programming methods.

To extend the applications of multi-criteria data mining techniques, there are several problem are still in need of further investigation. The optimum of the parameter b is the one that demands extensive research since it is essential to improve the classification result. One possible solution to this problem is to augment the multi-criteria model to multi-criteria & multi-constraint model and hereby to obtain different b. A convex combination will lead to a better classifier. Many efforts have been devoted to generalize the multi-criteria data mining model to make it suitable for non-linear classifications. And there are some preliminary results for this problem, but they are over-complicated. It is a more reasonable method to directly kernelize the model by representing the weights with the linear combination of the input points. The great computational loads in this method need address. Another direction of great importance is to enhance the theoretical foundation of the multi-criteria data mining. A potential method can be used to this end is the decision risk estimation based on the statistical learning theory.

Acknowledgments. This work was partially supported by National Natural Science Foundation of China (Grant No.70621001, 70531040, 70501030, 10601064, 70472074), National Natural Science Foundation of Beijing (Grant No.9073020), 973 Project of Chinese Ministry of Science and Technology (Grant No.2004CB720103), and BHP Billiton Cooperation of Australia.

References

1. Altman, E.: Financial Ratios, Discriminant Analysis and the Prediction of Corporate Bankruptcy. The Journal of Finance 23(3), 589–609 (1968)
2. Chang, C.C., Lin, C.J.: LIBSVM: A library for support vector machines (2001), http://www.csie.ntu.edu.tw/~cjlin/libsvm

3. Freed, N., Glover, F.: Simple but powerful goal programming models for discriminant problems. European Journal of Operational Research 7, 44–60 (1981)
4. Freed, N., Glover, F.: Evaluating Alternative Linear Programming Models to Solve the Two-Group Discriminant Problem. Decision Sciences 17, 151–162 (1986)
5. Shi, Y., Wise, M., Luo, M., Lin, Y.: Data Mining in Credit Card Portfolio management: a multiple criteria decision making approach. In: Proceedings of international conference on multiple criteria decision making, Ankara, Turkey (2000)
6. Shi, Y.: Multiple Criteria and Multiple Constraint Level Linear Programming: Concepts. World Scientific Publishing Co, Singapore (2001)
7. Shi, Y., Peng, Y., Xu, W., Tang, X.: Data mining via multiple criteria linear programming: applications in credit card portfolio management. International Journal of Information Technology and Decision Making 1, 131–151 (2002)
8. Shi, Y., Yu, P.L.: Goal setting and compromise solutions. In: Karpak, B., Zionts, S. (eds.) Multiple Criteria Decision Making and Risk Analysis Using Microcomputers, pp. 165–204. Springer, Heidelberg (1989)
9. He, J., Liu, X., Shi, Y., Xu, W., Yan, N.: Classifications of Credit Cardholder Behavior by using Fuzzy Linear Programming. International Journal Of Information Technology And Decision Making 3(4), 633–650 (2004)
10. Joachims, T.: SVM-light: Support Vector Machine, (2004), http://svmlight.joachims.org/
11. Kou, G., Peng, Y., Shi, Y., Wise, M., Xu, W.: Discovering credit cardholders. Behavior by multiple criteria linear programming, Working Paper, College of Information Science and Technology, University of Nebraska, Omaha (2002)
12. Kou, G., Peng, Y., Yan, N., Shi, Y., Chen, Z., Zhu, Q., Huff, J., McCartney, S.: Network Intrusion Detection by Using Multiple-Criteria Linear Programming. In: Chen, J. (ed.) Proceedings of 2004 International Conference on Service Systems and Service Management, Beijing, China, July 19-21, 2004, pp. 806–809 (2004)
13. Kou, G., Shi, Y.: Linux based Multiple Linear Programming Classification Program (Omaha, NE, U.S.A., College of Information Science and Technology, University of Nebraska-Omaha) (2002), http://dm.ist.unomaha.edu/tools.htm
14. Kou, G.: Multi-Class Multi-Criteria Mathematical Programming and its Applications in Large Scale Data Mining Problems, PhD Dissertation, University of Nebraska Omaha (2006)
15. Kwak, W., Shi, Y., Cheh, J.J.: Firm Bankruptcy Prediction Using Multiple Criteria Linear Programming Data Mining Approach. In: Advances in Investment Analysis and Portfolio Management (2005)
16. Kwak, W., Shi, Y., Eldridge, S., Kou, G.: Bankruptcy prediction for Japanese firms: using Multiple Criteria Linear Programming data mining approach. International Journal of Business Intelligence and Data Mining 1(4), 401–416 (2006)
17. Ohlson, J.: Financial Ratios and the Probabilistic Prediction of Bankruptcy. Journal of Accounting Research 18(1), 109–131 (1980)
18. Wasserman, L.: All of Statistics: A Concise Course in Statistical Inference. Springer, New York (2004)
19. Han, J., Kamber, M.: Data Mining: Concepts and Techniques. Morgan Kaufmann, San Francisco (2000)
20. Zhang, D.L., Tian, Y.J., Shi, Y.: A Regression Method by Multiple Criteria Linear Programming. In: 19th International Conference on Multiple Criteria Decision Making (2008)

21. Zheng, J., Zhuang, W., Yan, N., Kou, G., Peng, H., McNally, C., Erichsen, D., Cheloha, A., Herek, S., Shi, C., Shi, Y.: Classification of HIV-1 Mediated Neuronal Dendritic and Synaptic Damage Using Multiple Criteria Linear Programming, Neuroinformatics (2003)
22. Zhang, P., Dai, J.R.: Multiple-Criteria Linear Programming for VIP E-Mail Behavior Analysis. In: ICDMW 2007.
23. Zhang, P., Zhang, J.L., Shi, Y.: A New Multi-Criteria Quadratic-Programming Linear Classification Model for VIP E-Mail Analysis. In: Shi, Y., van Albada, G.D., Dongarra, J., Sloot, P.M.A. (eds.) ICCS 2007. LNCS, vol. 4488, pp. 499–502. Springer, Heidelberg (2007)

Web Evolution Management: Detection, Monitoring, and Mining

Sourav S. Bhowmick[1] and Sanjay Madria[2]

[1] School of Computer Engineering, Nanyang Technological University, Singapore
assourav@ntu.edu.sg
[2] Department of Computer Science, University of Missouri-Rolla, USA
madrias@umr.edu

Abstract. We present an overview of a tutorial on Web evolution management - an approach to solve the management of dynamic nature of the Web. Web evolution management defines a set of techniques for detecting changes to Web data, monitoring these changes, and mining interesting patterns by analyzing evolutionary features of the data. This tutorial offers an introduction to issues related Web evolution management and a synopsis of the state of the art.

1   Tutorial Overview

The Web offers access to large amounts of heterogeneous information and allows this information to evolve any time and in any way. These evolutions take two general forms. The first is existence. Web pages (static as well as dynamic) and Web sites exhibit varied longevity pattern. The second is structure and content modification. Web pages replace its antecedent, usually leaving no trace of the previous document. These rapid and often unpredictable changes to the information create a problem of detecting, monitoring, and analyzing these evolutions. This is a challenging problem because information sources in the Web are autonomous and typical database approaches to detect these changes based on triggering mechanisms are not usable. Moreover, these information sources are semistructured or unstructured in nature and hence traditional evolution management techniques for structured data (relational) cannot be used effectively.

This tutorial offers an introduction to issues involving evolution management on the Web. We motivate the necessity for managing evolution and give an overview of the evolutionary features of the Web. Next, we identify various research issues involved in evolution management. Specifically, our discussion can be categorized into three main components: (a) detection of changes/evolutions (b) querying over evolutions and (c) analyzing or mining evolutions of data. The tutorial session also reveal various application domains of evolution management systems (such as social networks, blogs, and Web event detection). We conclude by identifying potential research directions in this area.

This tutorial is intended for both engineers and researchers. The former will learn about solutions to use and hard problems to avoid. The latter will get a snapshot of the research field and problems that are worth tackling next.

2   Full Description of the Tutorial

The tutorial consists of the following topics.

Y. Zhang et al. (Eds.): APWeb 2008, LNCS 4976, pp. 39–40, 2008.
© Springer-Verlag Berlin Heidelberg 2008

1. **Introduction and motivation:** This section includes a brief overview on the dynamic nature of the Web data and how it is affecting our information need. Then we motivate the need for evolution management in the context of the Web.
2. **Evolution management problem and characteristics:** In this section we formally define the meaning of evolution management in the context of the Web. We then identify the technical challenges associated with this problem. Finally, we present various characteristics of Web evolution.
3. **Change detection and representation:** The first step for evolution management is to have effective tools for detecting changes to Web pages. We discuss state-of-the art research efforts in detecting and represent changes to content and structure of Web page into two parts: main memory-based and external memory-based approaches.
   - *Main memory-based approaches:* As a Web page can be represented using a tree structure, we begin our discussion by introducing early research activities in detecting changes to tree-structured data (*MH-Diff, La-Diff, MM-Diff* algorithms). Next, we present recent activities in XML change detection (*X-Diff, XyDiff*). Then, different page-based approaches (such as *AIDE, HTML-Diff, WebCQ*) are presented. Finally, we discuss metrics to measure the quality of changes detected by these systems.
   - *External memory-based approaches:* It has been shown recently that main-memory based approaches are not suitable for detecting changes to large tree-structured document as it requires a lot of memory to keep the two versions of such documents in the memory. Consequently, we discuss recent efforts in designing scalable solution using external memory and relational databases. We highlight two major approaches in this arena: *XMDiff* and XANADUE. We also compare the performance of the above two change detection approaches.
4. **Querying evolutions:** Next, we discuss research activities in querying evolutions. Surprisingly, there has been very few work in this area. The most notable being the *Chorel* language for querying changes to semistructured data. In this section, we highlight the key features of this language.
5. **Mining evolution of Web data:** Finally, we explore works that propose novel data mining techniques to mine the changes to Web data in order to discover interesting evolutionary patterns. Specifically, we present the followings: (a) Mining evolutionary features of social networks; (b) Mining evolutions of blogs in predicting blogging behaviors; (c) Mining evolution of structure of Web sites, XML query and data; (d) Mining evolution of Web usage data.
6. **The road ahead:** We expose potential research issues in managing evolutions in the Web. We also throw some open questions for the audience to ponder about.

3   Speakers

Sourav S Bhowmick and Sanjay Madria have published several papers in the area of Web evolution management. One of Sourav's Web evolution management paper received *Best Interdisciplinary Paper Award* at the ACM CIKM 2004. Biographies of Sourav and Sanjay can be found at www.ntu.edu.sg/home/assourav and web.mst.edu/~madrias/, respectively.

Research and Practice in Data Quality

Shazia Sadiq, Xiaofang Zhou, and Ke Deng

ITEE School, The University of Queensland, QLD 4072 Australia
{shazia,zxf,dengke}@itee.uq.edu.au

1 Data Quality and Its Importance

According to Gartner, human data-entry errors, and lack of proper corporate data standards result in more than 25 percent of critical data used in large corporations to be flawed. While the issue of data quality is as old as data itself, it is now exposed at a much more strategic level, e.g. through business intelligence (BI) systems, increasing manifold the stakes involved. Corporations routinely operate and make strategic decisions based on remarkably inaccurate or incomplete data. This proves a leading reason for failure of high-profile and high-cost IT projects such as Enterprise Resource Planning (ERP), Customer Relationship Management (CRM), Supply Chain Management (SCM) and others. According to an industry survey [1], the presence of data quality (DQ) problems costs U.S. business more than 600 billion dollars per annum.

As the response to the urgent call to prevent semantic/syntactic data contamination, we can see a rising interest from research community and increasing presence of data quality related papers in prominent conferences and journals. At the same time the industry response has resulted in dedicated tools and technologies to provide reliable data quality control methods. In this tutorial, we will deliver a comprehensive and cohesive overview of the key issues and solutions regarding the Data Quality (DQ) problem. It will cover general aspects of the problem space as well as the most significant research achievements in key topics. Since the most recent tutorial [2], there have been several new developments.

In summary, the aims of the tutorial are to: (1) Presenting the full picture of data quality control in front of wider database research community in a concise and holistic way. (2) Discussing the unique set of challenges that the practices of data quality control are faced with, being aware of gaps in industry needs and existing solutions, and improving a better understanding of how to ideally cope with data quality problem. (3) Passing the insight knowledge and the latest advances in this domain and inspiring thinking with an interactive discussion on the major concerns and future trends.

2 Structure of Tutorial

- *Introduction:* We will briefly introduce the history of the DQ problem and its importance in various IT systems.

Y. Zhang et al. (Eds.): APWeb 2008, LNCS 4976, pp. 41–42, 2008.

- *Running Example:* Prior to formal discussion, examples based on the real world applications will be presented to help give an intuitive impression of various DQ problems and would assist to develop an understanding of their characteristics.
- *Problems Classification:* This section will present and discuss various flavours of DQ definitions and classifications. We will discuss the total data quality control which covers three disparate themes: organizational, architectural and computational.
- *Solutions Classification:* We will discuss the key techniques, research achievements and tools developed in each problem category discussed above. The most recent work will be covered.
- *Discussions:* The current DQ control systems have provided significant advances at the various levels thereby eliminating many standard errors. These solutions definitely improve the accuracy and integrity of data but only in limited scope. This section will identify the limits of current methods and present an outlook on the potential directions.

3   Intended Audience

This tutorial is tailored to suit researchers with interest in the data quality control from both academia and industry. They are business analyst, solution architects and database experts, statisticians, data managers, and practitioners.

4   Presenters Background

Shazia Sadiq is currently Senior Lecturer in ITEE School of UQ. Her main research interests are innovative solutions for enterprise information integration, which include in particular Business Process Management Systems, Service Oriented Architectures, Workflow Systems, advanced messaging technologies, and deployment of large scale distributed devices. Shazia is leading a large Australian Research Council funded project on Data Quality.

Xiaofang Zhou is a Professor of Computer Science at the University of Queensland. He is the Head of the Data and Knowledge Engineering Research Group and the Convenor of ARC Research Network in Enterprise Information Infrastructure (EII), and a Chief Investigator of ARC Centre of Excellence in Bioinformatics.

Ke Deng is a research fellow in ITEE School of UQ. His current research areas focus on the data quality control, including record linkage, similarity text join, etc.

References

1. Data Warehousing institute. Data Quality and the bottom line: archiving business success through a commitment to high quality data, http://www.dw-institute.com
2. Johnson, T., Dasu, P.: Data quality and data cleaning: an overview. SIGMOD tutorial (2003)

Detecting Overlapping Community Structures in Networks with Global Partition and Local Expansion

Fang Wei[1], Chen Wang[2], Li Ma[2], and Aoying Zhou[1]

[1] Department of Computer Science and Engineering, Fudan University,
Shanghai, China
[2] IBM Research Laboratory, Beijing, China
{weifang,ayzhou}@fudan.edu.cn,{chwang,malli}@cn.ibm.com

Abstract. The problem of discovering community structures in a network has received a lot of attention in many fields like social network, weblog, and protein-protein interaction network. Most of the efforts, however, were made to measure, qualify, detect, and refine "uncrossed" communities from a network, where each member in a network was implicitly assumed to play an unique role corresponding to its resided community. In practical, this hypothesis is not always reasonable. In social network, for example, one people can perform different interests and thus become members of multiple real communities. In this context, we propose a novel algorithm for finding overlapping community structures from a network. This algorithm can be divided into two phases: 1) globally collect proper seeds from which the communities are derived in next step; 2) randomly walk over the network from the seeds by a well designed local optimization process. We conduct the experiments by real-world networks. The experimental results demonstrate high quality of our algorithm and validate the usefulness of discovering overlapping community structures in a networks.

1 Introduction

Many network systems demonstrate the characteristics of community structure, e.g., social network and biological network. In general, a community (cluster or group) is a subset of tightly-linked vertices where the connections among them are relatively denser than those cross to the rest of the network.

Recently, people in many domains, such as physicists, applied mathematicians and computer scientists, pay lots of attention to the community structure discovery. Many algorithms [1,2,3,4,5] have been developed. Most of them were related to graph partitioning. They divided a network into disjoint groups by minimizing the cost of cut-edges.

A majority of those community detecting methods rule that a vertex in a network attributes to at most a community. Therefore, they would never find out overlapping community structures, even though it might be valuable for some cases. In real world, a person could serve as the member of multiple clubs

Y. Zhang et al. (Eds.): APWeb 2008, LNCS 4976, pp. 43–55, 2008.

in a social network; a paper could involve several topics in a citation network; a kind of gene could be concerned with diverse biological processes in a gene network, and so on.

Considering the example in Figure 1, we can partition the network (left one) into the communities (middle one) using the common community discovery algorithm. However, it seems to be more reasonable for the network to be partitioned into the communities (right one) because some vertices might serve as multiple roles.

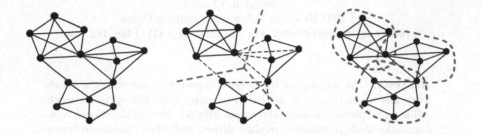

Fig. 1. Overlapping example

This phenomena arises a new issue in community discovery. Here, we hope to solve the problem with a new direction, i.e., finding out the overlapping communities from seed expansion. It is a reasonable idea because it naturally extends seeds by adding vertices without considering whether the vertex has been a member of another community. But the expansion methods depend on good seeds [6,7]. Inspired by this, we try to use some graph partitioning methods to find the seed groups from global view and extend the seed candidates with random walks techniques from locally-optimal view. Meanwhile, the modularity function Q [1] which is popular in networks is chosen as the measure of overlapping community structures. It quantifies the difference between a community and its random community that has same structure but random connections between the vertices. Following this, the problem about whether the community structures have better Q value after random expansion is a key issue. In the fourth section, the analysis is presented.

In this paper, we present a novel algorithm DOCS (Detecting Overlapping Community Structures) based on global partition and local expansion. Our contributions are highlighted as follows:

1. We apply the existing spectral partitioning method to generate seed groups for overlapping community structures. The classical algorithms help us find good seeds from overall network structure, which grasps the main bodies of community structures.
2. A locally-optimal expansion process based on modularity Q is introduced. For every scanned vertex, our algorithm computes their contribution to modularity Q. With the contribution property, this paper gives a theorem. Some useless vertices can be safely pruned based on the theorem.

3. In the real-world networks, we evaluate our DOCS algorithm from four aspects, and experimental results give enough examples to prove that overlapping is important to find the complete community structures.

The rest of the paper is organized into five sections. Section 2 introduces the preliminary knowledge. Our algorithm framework is described in Section 3. Section 4 gives the theoretic analysis to the expansion process. We evaluate our algorithm with six real-world datasets in Section 5. Finally, the conclusion and future work will be given in Section 6.

2 Preliminaries

2.1 Network Model

The network can be modeled as a graph G = (V, E) , where the V is composed of N vertices and the E is the set of links (edges) between vertices. The results in this paper are discussed in unweighted and undirected graphs.

Definition 1. *If* $A = (A_{ij})_{n \times n}$ *is the adjacent matrix of a network, then*

$$A_{ij} = \begin{cases} 1 & \text{if vertex } i \text{ and } j \text{ are connected,} \\ 0 & \text{otherwise.} \end{cases}$$

We often use $D = (D_{ij})_{n \times n}$ to represent the diagonal matrix. D_{ij} equals to $\sum_k A_{ik}$ when i is same as j , otherwise it is 0. The two matrixes are the basis for other matrixes, such as, the laplacian matrix $L = D - A$ and transition matrix $P = D^{-1}A$.

Definition 2. *The edge border of a community S is*

$$B(S) = \{\{u, v\} | \{u, v\} \in E, u \in S, v \notin S\}$$

and $|B(S)|$ *is the cutsize of the community.*

2.2 Community Weight Functions

There are many weight functions or metrics to evaluate the strength of community structures. For example, intensity ratio and edge ratio are defined in [6].

The most popular one is the modularity Q proposed by Newman and Girvan in [1], which becomes a kind of main criterion for evaluating the quality of community structures. It measures the difference between a community and its random community which has the same structure but random connections between the vertices. The modularity functions mentioned by [3,4,5] are its varieties.

For our algorithm chooses the spectral partitioning method to produce the seeds, we cite the Q formulation in [5]. If the network has a division of P_k (k is the number of communities), the modularity Q is:

$$Q(P_k) = \sum_{c=1}^{k} \left[\frac{A(V_c, V_c)}{A(V, V)} - \left(\frac{A(V_c, V)}{A(V, V)} \right)^2 \right]$$

where $A(V', V'') = \sum_{u \in V', v \in V''} w(u, v)$. The $w(u, v)$ is the weight of the edge between vertex u and v. For DOCS is discussed in the unweighted graph, we let $w(u, v)$ be one.

In formulation, $\frac{A(V_c, V_c)}{A(V, V)}$ measures the probability of the edges whose ends both lie in the same community c, and $\frac{A(V_c, V)}{A(V, V)}$ measures the probability of the edges that at least one of ends belong to the community c. The last one reflects the math expected value of the random community having same structure.

A good community structure should have a high value on modularity Q. But maximizing the modularity Q is NP-hard [16]. The optimal solutions need exponential-time. Many algorithms seek to the suboptimal partition solutions.

3   The Framework of DOCS Algorithm

Based on the spectral graph partitioning methods and random walks techniques, we propose a new DOCS algorithm. For the optimal Q is NP-hard, our algorithm tries to seek the locally-optimal Q. The main steps are outlined below.

1. Create the seed groups by choosing the partition results of the spectral bi-section method with the multi-level recursion.
2. Extend the seed groups with the lazy random walks techniques. At every time step, the algorithm scans in turn all the vertices linking to the community candidate. The expansion process stops if it satisfies the conditions
   a). The degree-normalized probabilities of scanned vertex is below a threshold.
   b). The overlapping rates among communities surpass the tolerance of users.
3. Choose a trade-off point between good modularity Q and overlapping rate, then add the relevant extended vertices to the seeds and get the overlapping community structures.

DOCS algorithm chooses the graph partitioning results as special seeds, which reduce the probability of deflecting to bad communities in expansion. However, if the total number of edges in a community is below three, we abandon it as a seed for the reason that the special community candidate only is a line or a vertex and cannot reflect the main structure of target communities. The first phase will be introduced in the following subsection.

After the t-th step expansion in the second phase, the contributing vertices are identified and the Q of community candidates are computed. The community candidates are updated by adding the new vertices.

The overlapping rate referred in the framework will be defined later.

3.1   Identifying Community Candidates

We have mentioned that the goal of identifying overlapping groups in networks is to find the relatively complete community structure or seek better value in community weight functions. So, it is wise to use some main bodies of communities as the seeds of target communities.

The main bodies of communities (here, named with community candidates) can be captured by some existing community-detecting algorithms, such as, Kernighan-Lin algorithm [12], spectral partitioning [5,13], hierarchical clustering [15] and divisive algorithm [1]. Spectral partitioning method is based on the spectral analysis of a suitably-defined matrix which exhibits the global structure of network. It is indicated in [4,5,13,14] that this method has good performance on finding communities in graphs. We adopt a spectral bisection method with multi-level recursion to identify the community candidates.

Given an unweighted and undirected graph, DOCS coarsens the original graph into a series of higher coarsening level graphs. At each coarsening graph, the Fiedler vector of the laplacian matrix is mapped into next coarser graph, which further approximates the Fiedler vector for the coarsest graph. In the highest coarsening level, DOCS uses the spectral bisection algorithm to choose a partition point based on the Fiedler vector which corresponds to the second eigenvector of the laplacian matrix. According to the positive or negative value in the eigenvector, the vertices are assigned to one part or another. The coarsest graph is split into two groups. The recursive partition is executed at each group until there is no partitioning clue. Finally, the partition is projected back to the original graph by going through a series of lower level graphs. The final partition results are chosen as the seed sets.

3.2   Expansion Process

Given a set of seed groups, we can use a fixed-depth or random walks techniques to extend. The fixed-depth expansion method corresponds to a Breadth First Search tree. Their vertices have same expansion weight. The random walks techniques extend the neighbors of a vertex at random. The sequence of walks forms a Markov chain. Lovász in [17] gives a complete explanation to the technique and indicates that its basic properties are determined by the spectrum of the graph. It is more efficient than fixed-depth expansion. Here, we present a novel expansion process based on the lazy random walks techniques with weak conductance guarantees [11].

The expansion process is scaled by time t. At each time step, the scanned vertices are sorted in descending order with the degree-normalized probabilities [11]. If a new vertex brings better Q change to the community candidates, it may be absorbed as a new member of the community structures. For instance, figure 2 shows the process that a new vertex is added to the elliptical seed group.

The broken lines connected with the seed group members are named as inseedlinks, and the total number of broken lines is denoted by $|IL|$. Similarly, the real line pointing to outside are named as outseedlinks and the total number of real lines are expressed by $|OL|$. Those lines effect the Q change of seed groups when the new vertex is added. Next, we use the modularity formulation to measure the Q change of seed groups.

If the initial seed group is expressed as S, the seed size is $|S|$ and cutsize is $|B(S)|$. Then, its initial Q value is:

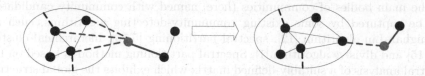

Fig. 2. Adding a neighborhood vertex to the elliptical seed group

$$Q_0 = |S|/|E| - ((|S| + |B(S)|)/|E|)^2$$

where $|E|$ is the total number of edges in a network.
After adding a new neighborhood vertex, the modularity Q becomes:

$$Q' = (|S| + |IL|)/|E| - ((|S| + |B(S)| + |OL|)/|E|)^2$$

The change value in Q is:

$$Q' - Q_0 = |IL|/|E| - (2(|S| + |B(S)|) \times |OL| + |OL|^2)/|E|^2$$

If the value is above zero, the new vertex is named as contributing vertex, otherwise it is called uncontributing one. The locally-optimal expansion process seeks the contributing vertices which improve the seed community structures.

After many steps, the mixing time of random walks is coming. If the graph is connected, the probabilities about vertices degree converges to a uniform stationary distribution ψ_v. The difference between them can be computed. Depending on the difference, we can stop the whole expansion process and some low-probability vertices can be safely pruned.

3.3 Overlapping Rate

In this paper, overlapping only is one approach to discover the complete community structures. We not only seek good Q value but also care the overlapping rate in real world. For example, the overlapping cost is considered in storage.

The overlapping rate can be defined by different user requirements. The usual one is:

Definition 3. *Given a set of communities $\{C_m, ..., C_j\}$ in a network, they overlap with community C_i. The overlapping rate of community C_i in the network is defined as the ratio:*

$$COR(C_i) = \frac{|C_i \cap C_m| \cup ... \cup |C_i \cap C_j|}{|C_i|}$$

where COR is the abbreviation of Community Overlapping Rate. $|C_i|$ represents the total number of edges in the community i, $|C_i \cap C_j|$ is the number of common edges in the communities i and j. In our experiments, we have a average on all the overlapping rates of the community candidates in a network and let Overrate denote the value.

4   Locally-Optimal Expansion Process Analysis

In this section, we will give a general analysis to the properties of DOCS expansion process. After the t-th step expansion, the Q value of the community candidates are computed. That is how to add the scanned vertices in the t-th time step to the seed groups. The seed groups have been updated in the $(t\text{-}1)$-th time step.

In scan process, the contributing vertices and uncontributing ones are distinguishingly flagged. The flags may improve the computing process. The following theorem presents this property.

Theorem 1. *If a vertex is not a contributing vertex to a community candidate in the t-th time step, it cannot bring better Q value to the computing of community candidate on modularity function.*

Proof. Given a seed group S at the t-th time step, $|S|$ and $|B(S)|$ are respectively the seed size and cutsize. The vertex u is its uncontributing vertex. We discuss the theorem in two kinds of situations.

Case 1: the current computing state is not added any vertex.

From the definition of uncontributing vertex, the value of $Q_u - Q_0 = |IL_u|/|E| - (2(|S| + |B(S)|) \times |OL_u| + |OL_u|^2)/\|E\|^2$ is smaller than zero. At the state, the uncontributing vertex cannot bring better Q value in the computing process.

Case 2: the current computing state has added some vertices.

The current modularity Q of the community candidate is: $Q' = (|S| + |IL'|)/|E| - ((|S| + |B(S)| + |OL'|)/|E|)^2$, where $|IL'|$ and $|OL'|$ are respectively the number of inseedlinks and outseedlinks. They are the new links in seed group by absorbing the vertices. If we add the vertex u, the candidate gets the new state Q'_u and the change is:

$$Q'_u - Q' = |IL_u|/|E| - (2(|S| + |B(S)| + |OL'|) \times |OL_u| + |OL_u|^2)/|E|^2$$
$$= Q_u - Q_0 - 2|OL'| \times |OL_u|/|E|^2$$

For $Q_u - Q_0 < 0$, we get $Q'_u - Q' < 0$. Hence, it cannot bring better Q change to the seed group and the uncontributing vertex u is useless in the computing process.

Based on the theorem, some useless vertices can be safely pruned in computing process.

From the analysis to above case 2 in the theorem, we also find that the contributing vertices in scan process not always bring the better Q change in computing process. For a contributing vertex v, it is useful to the computing unless the value of $Q_v - Q_0$ is above $2|OL'| \times |OL_v|/|E|^2$. So, the decision about whether adding a new contributing vertex to the seed group is effected by the contributing vertices which have been added to the computing process.

If the $|OL'|$ is zero, the community candidate is recursively back to the initial seed group. For the first time to add a new contributing vertex to the seed group, we choose the most contributing vertex which brings the best Q value to the seed group to seek the locally-optimal goal. According to this rule, the rest contributing vertices are chosen for bringing best Q change to the current community state. At every computing state, the seed group absorbs the new vertex which is the best contributor to the current expansion step. So, we say that the seed expansion is locally-optimal at the t-th step.

In above section, we have presented that the expansion process is stopped when the expansion is close to the stationary state. At every time step, we add the new vertices to the community candidates and compute the modularity Q of the current expansion step. There will be a series of Q at the whole expansion process. From the initial step to the mixing one, we choose the expansion step which has the best Q value as the final seed expansion state which gets locally-optimal goal.

5   Experimental Results

We present the experimental results of DOCS model in real-world datasets which have diverse sizes from hundreds vertices to millions ones. Table 1 shows the features of datasets. The experiments were performed on a single processor of a 3.2 GHz Pentium Xeon with 2GB RAM, running Window 2K.

Table 1. The detailed information in the six datasets

Data Name	Vertices	Edges	Data Source
Zachary's karate	34	78	http://www-personal.umich.edu/~mejn/netdata/
Football games	115	613	http://www-personal.umich.edu/~mejn/netdata/
NIPS coauthorships	1,063	2,083	http://www.cs.toronto.edu/~roweis/data.html
Protein interactions	1,458	1,948	http://www.nd.edu/ networks/resources.htm
KDD citations	27,400	352,504	http://www.cs.cornell.edu/projects/kddcup/
WWW	325,729	1,090,107	http://www.nd.edu/networks/resources.htm

We evaluate DOCS from four aspects at each time step: the change on modularity Q of whole graph(denoted as ΔQ), the Betterrate, the Bestrate and the Overrate.

The Betterrate is the ratio that the number of better community candidates to the total number of candidates in whole network. The better community candidate has larger Q value than the initial one. The initial candidate refers to the state without any expansion. We show the rate using the blue columns in figures.

At each time step, the Q value may decrease or increase. After many steps, there is a best expansion step. That is, the community candidate has best Q value at that time step.

The Bestrate expresses the proportion of the community candidates owning best Q value in the given step, which is represented by the green columns in figures.

The last one is the Overrate. It is the average for all the overlapping rates of single community candidates. The grey columns in figures describe the information. In the following experiments, the first step refers to the initial state without any expansion.

5.1   An Example of Overlapping Community Structure

Before analyzing our algorithm on above metrics, we present the example of overlapping community structure detected by DOCS in the Zachary's karate club network.

The Zachary's karate club network is a dataset familiar to us in social network analysis. In figure 3, the vertices represent the 34 members of the karate club and edges indicate the ties between members of the club. Many communities-detecting methods[1,2] separate vertex 1 with vertex 33 respectively represented the karate club's administrator and its principal teacher, which correspond to the actual divisions in the club for the dispute between them. They often partition the network into two or more parts which is disjoint. Their methods cut many edges and the edges may lose some potential information. The green thin lines in figure 3 represent the edges that are possibly cut. If a vertex may belong to different groups, there is an interesting discovery. Performing our DOCS, we find that the vertices·5,6,7 and 11 are connected densely with vertex 1. If we admit vertex 1 to belong to two parts, they get a relatively complete community structure and look like more reasonable. Perhaps, they are the potential ingroup which is showed by the red ellipse in figure 3.

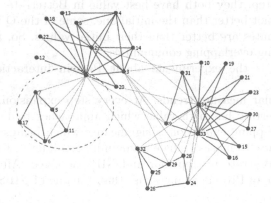

Fig. 3. An example of overlapping community structure detected by DOCS algorithm in Zachary's karate club. The green thin lines represent the edges that are possibly cut and the red ellipse gives a overlapping community structure.

5.2   The Experimental Analysis of Six Datasets

In order to clearly demonstrate the experimental results, we give a figure to every dataset. In the figures, we use two y-axes and one x-axis to represent bi-coordinate system. The Betterrate, Best-rate and Overrate are corresponding to the left y-axis which indicates the rate information satisfied the well-defined conditions. The ΔQ solely uses the right y-axis which reveals the change of Q value. The x-axis shows the time steps.

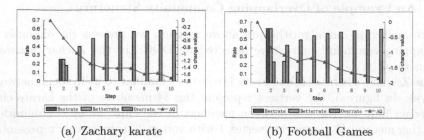

(a) Zachary karate   (b) Football Games

Fig. 4. The trend analyses of Karate and Football Games at every time step. The Bestrate, Betterrate and Overrate use the left y-axis. The ΔQ is below to zero corresponding to the right y-axis.

Figure 4 shows the four trends on karate and football datasets which are popular data in community-detecting problem. We can see that the ΔQ lines decreases sharply after several steps expansion. At the same time, the Overrate is above 0.5. The reason is that karate and football datasets have small graph size and every seed candidate has less diameter, which make every community candidate overlap its most space with others only in several steps.

In the second step, they both have best value in Betterrate. Though the Q of whole graph is not better than the initial one, some of the Q values in single community candidates are better than their initial values. So, the step two is important in finding overlapping communities.

Next, figure 5 is the experiments about Protein interactions and NIPS coauthorships.

They have similar measure trends. At every step expansion, ΔQ increases steadily and Betterrate has high value which approaches to 1. From this, we can see that overlapping expansion is significance which brings more complete community structure to most of single community candidates. Protein data has the top value of Bestrate in step four and NIPS at three. After several time steps, the Overrate of Protein changes fast than the one of NIPS.

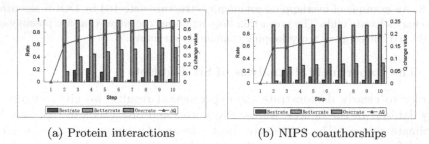

(a) Protein interactions   (b) NIPS coauthorships

Fig. 5. The trend analyses of Protein interactions and NIPS coauthorships at time steps. The Bestrate, Betterrate and Overrate are corresponding to the left y-axis. The ΔQ lonely uses the right y-axis.

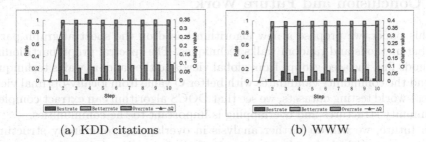

(a) KDD citations   (b) WWW

Fig. 6. The trend analyses of KDD citations and WWW at time steps. The Bestrate, Betterrate and Overrate are corresponding to the left y-axis. The Betterrate reaches to 1. The ΔQ lonely uses the right y-axis.

Last, figure 6 is related with KDD citations and WWW. The KDD citations has smooth change on ΔQ after finishing one step expansion. And, all the community candidates have better Q than the initial one. The Bestrate gets its top at the second step.

For WWW data, the blue Betterrate is similar to KDD citations, but the trend of ΔQ increases slowly not decreases as KDD data. The Overrate increases slightly and not is above 0.2 in the ten steps expansion. The Bestrate scatters at every step and the step three is its top in the figure.

From above experimental analysis, we can clearly see that the expansion with overlapping makes the single community candidate structures more complete. For the large-size data, it also brings the better Q value to the whole graph. Our DOCS algorithm validates the fact.

6 Related Work

There are some algorithms for finding overlapping community structures.

Baumes et al. present IS and IS^2 (improved edition) algorithms in [6,7]. The two algorithms iteratively scan vertices in graph until the weight function not increase. The algorithms both depend on having good seeds. Moreover, they cannot give theoretic analysis for the extending process and the process still can be improved. The RaRe and LA algorithms decompose the graph into disjoint components as initial clusters and then return each removed vertex to the clusters by the weight function. The replaced vertices can belong to several clusters.

Palla et al. [8] think a community can be interpreted as a union of smaller fully-connected subgraphs that share vertices. They present a clique percolation algorithm to find overlapping clusters. The algorithm focuses on densely connected-parts not sparse edges in network. It suits to detect highly overlapping cohesive clusters of vertices. CFinder[9] is its implemented system.

Zhang et al.[10] detect overlapping community structures from fuzzy clustering view which permits a vertex to have fuzzy membership by a probability. When the probability of vertex is larger than a defined community threshold, the vertex is added to the community.

7 Conclusion and Future Work

In this paper, we propose a new algorithm based on the spectral graph partitioning methods and random walks techniques. The spectral graph partitioning methods find the seed groups from global view and the random walks techniques extend the community candidates with better Q value from locally-optimal view. In real-world testing datasets, we see that DOCS algorithm can extract complete community structures and overlapping is important to the communities.

In future, we will do further analysis in overlapping community structures and compare with the overlapping results from fuzzy cluster view.

Acknowledgments. The work is done when the author visit the IBM China Research Laboratory. Thanks to Yue Pan for his encouragement and guidance.

References

1. Newman, M.E.J., Girvan, M.: Finding and evaluating community structure in networks. Physical Review 69, 26113 (2004)
2. Duch, J., Arenas, A.: Community detection in complex networks using extremal optimization. Phys. Rev. E 72, 27104 (2005)
3. Newman, M.E.J.: Finding community structure in networks using the eigenvectors of matrices. Physical Review E 74, 36104 (2006)
4. Newman, M.E.J.: Modularity and community structure in networks. PROC. NATL. ACAD. SCI. USA 103, 8577 (2006)
5. White, S., Smyth, P.: A spectral clustering approach to finding communities in graphs. In: SIAM International Conference on Data Mining (2005)
6. Baumes, J., Goldberg, M., Krishnamoorty, M., Magdon-Ismail, M., Preston, N.: Finding communities by clustering a graph into overlapping subgraphs. In: Proc. IADIS Applied Computing, pp. 97–104 (2005)
7. Baumes, J., Goldberg, M., Krishnamoorty, M., Magdon-Ismail, M.: Efficient identification of overlapping communities. In: Kantor, P., Muresan, G., Roberts, F., Zeng, D.D., Wang, F.-Y., Chen, H., Merkle, R.C. (eds.) ISI 2005. LNCS, vol. 3495, pp. 27–36. Springer, Heidelberg (2005)
8. Palla, G., Derényi, I., Farkas, I., Vicsek, T.: Uncovering the overlapping community structure of complex networks in nature and society. Nature 435, 814–818 (2005)
9. Adamcsek, B., Palla, G., Farkas, I., Derényi, I., Vicsek, T.: CFinder: locating cliques and overlapping modules in biological networks. Bioinformatics 22, 1021–1023 (2006)
10. Zhang, S.H., Wang, R.S., Zhang, X.S.: Identification of overlapping community structure in complex networks using fuzzy c-means clustering. Physica a-statistical mechanics and its application 374(1), 483–490 (2007)
11. Andersen, R., Lang, K.J.: Communities from seed sets. In: Proceedings of the 15th International World Wide Web Conference (2006)
12. Kernighan, B.W., Lin, S.: An efficient heuristic procedure for partitioning graphs. Bell System Technical Journal 49, 291–307 (1970)
13. Pothen, A., Simon, H., Liou, K.-P.: Partitioning sparse matrices with eigenvectors of graphs. SIAM J. Matrix Anal. Appl. 11, 430–452 (1990)

14. Simon, H.D.: Partitioning of unstructured problems for parallel processing. Computing Systmes in Engineering 2(2/3), 135–148 (1991)
15. Scott, J.: Social Network Analysis: A Handbook, 2nd edn. Sage, London (2000)
16. Brandes, U., Delling, D., Gaertler, M., Goerke, R., Hoefer, M., Nikoloski, Z.: Maximizing modularity is hard.Physics 0608255.
17. Lovász, L.: Random walks on graphs: A survey. In: Combinatorics, Paul Erdös is eighty, vol. 2, pp. 353–397, (Keszthely, 1993); Bolyai Soc. Math. Stud. 2, János Bolyai Math. Soc. Budapest (1996)

High Confidence Fragment-Based Classification Rule Mining for Imbalanced HIV Data*

Bing Lv, Jianyong Wang, and Lizhu Zhou

Database Group, Department of Computer Science and Technology,
Tsinghua University, Beijing, 100084, P.R. China
alvb05@gmail.com, {jianyong,dcszlz}@tsinghua.edu.cn

Abstract. In this paper, we study the problem of mining high confidence fragment-based classification rules from the imbalanced HIV data whose class distribution is extremely skewed. We propose an efficient approach to mining frequent fragments in different classes of compounds that can provide best hints of the characteristic of each class and can be used to build associative classification rules. We adopt the pattern-growth paradigm and define an efficient fragment enumeration scheme. Moreover, we introduce an improved instance-centric rule-generation strategy to mine the high-confidence fragment-based classification rules, which are very insightful and useful in differentiating one class from other classes. Experiments show that our algorithm can discover more interesting rules than the previous method and can facilitate the detection of new compounds with desired anti-HIV activity.

1 Introduction

Frequent pattern discovery in structured data has gained more and more attention in both scientific and industry areas, which can support a broad arrange of applications. In particular, researchers in bio-informatics and chemistry often need to analyze large collections of molecular compounds to find some desirable common features and regularities for different classes of compounds, which helps reveal some intrinsic behaviors of those molecular compounds. One typical example is the DTP AIDS antiviral screen program, which has checked tens of thousands of compounds for evidence of anti-HIV activity. When new compounds arrive, the traditional approach to discovering its properties is to conduct biological experiments and evaluate them. But with feature mining and classification techniques, the behaviors and activities of new compounds can be predicted according to the known features. Another example is the drug design progress, where biologists need to screen thousands of compounds to find evidence of activity against a specific disease. With the help of data mining tools, we can mine some useful features and build a classifier in order to guide the discovery or composition of desirable drug candidates.

* This work was supported in part by National Natural Science Foundation of China under Grant No. 60573061, Basic Research Foundation of Tsinghua National Laboratory for Information Science and Technology(TNList), Program for New Century Excellent Talents in University under Grant No. NCET-07-0491, State Education Ministry of China.

Y. Zhang et al. (Eds.): APWeb 2008, LNCS 4976, pp. 56–67, 2008.

The discovery of common features in a set of given instances makes sense in many cases. In contrast to some other data mining problems, given a database of compounds with known classes, we aim to analyze it to find some high confidence molecular fragment features, which can be used to build an accurate classifier. When a new molecular arrives, we can use the classifier to predict which class it belongs to.

In this paper, we present a fragment feature based approach to predicting the anti-HIV activity of new compounds by finding high confidence instance-centric classification rules for the imbalanced DTP AIDS database. We mine the frequent fragments in different classes of compounds that can provide best hints of the characteristics of the class they belong to, and then use these fragments to form association rules. Extensible performance study shows that our algorithm is effective in finding discriminating fragments for the imbalanced database and can facilitate the development of detecting new compounds with desired anti-HIV activity.

The rest of this paper is organized as follows. Section 2 reviews the related work. Section 3 introduces some notations, definitions, and the problem formulation. Section 4 describes the SMILES and SMARTS languages used for representing compounds and fragments. Section 5 presents our solution to efficiently and effectively mine the frequent fragments of molecular compounds. Section 6 discusses how to build the classification model in details. Then we show the experimental results in section 7. Finally we conclude the study in section 8.

2   Related Works

The discovery of molecular fragment features can be formulated as a frequent subgraph mining problem, which relies on graph and subgraph isomorphism theory. Various measures to mine substructures of graphs have been proposed. Early subgraph mining algorithms generate candidates and check whether they are frequent in a breadth-first search manner, and a typical example is FSG [5]. While algorithms adopting the depth-first search paradigm, such as MoFa [1] and gSpan [10], generate and test the candidates in a depth first way, and usually consume less memory, but are more efficient. Recently several algorithms [2,4] have been proposed to provide an automatic approach to computing and finding common features of compounds. They focus on efficient mining of frequent molecular substructures in chemical compounds.

This work also relates to associative classification algorithms, which extend the association rule mining to associative classification rule mining, and build classification model from the mined classification rules. Typical associative classifiers include CBA [7], CMAR [6], and CPAR [11]. Previous studies have shown that associative classification approaches can achieve better classification results than traditional rule-based approaches such as FOIL and C4.5. One drawback of the associative classifiers is that the number of initial classification rules is usually extremely large, which significantly increases the time of rule discovery and selection. Recently an alternate rule-based classifier, HARMONY [8], which adopts an *instance-centric* rule generation framework, is proposed and shown to be more effective and scalable than previous ones. It keeps for each training instance one of the highest confidence classification rules during the rule mining process, and later builds the classification model from the final set of high confidence rules. This work extends the *instance-centric* classification framework from itemset setting to a more complicated case, that is, fragment feature mining, and validates its utility in classifying imbalanced HIV data.

3   Notations and Definitions

In this section, we introduce some preliminary concepts, notations, and terms in order to simplify the discussion. We also give the problem formulation.

3.1   Preliminary Concepts

In this paper, we treat each chemical compound as a simple undirected, labeled graph without multi-edges and self-loops. An **undirected labeled graph** G can be represented by a 6-tuple, $G = (V, E, L_v, L_e, F_v, F_e)$, where $V = \{v_1, v_2, \ldots, v_k\}$ is the set of vertices, $E \subseteq V \times V$ is the set of edges in G, L_v is the set of vertex labels and L_e is the set of edge labels, respectively, $F_v : V \to L_v$ and $F_e : E \to L_e$ are mapping functions assigning the labels to the vertices and edges, respectively.

A graph G_1 is **isomorphic** to another graph G_2 iff there exists a bijection $f : V_1 \to V_2$ such that for any vertex $v \in V_1$, $f(v) \in V_2 \wedge L_v = L_{f(v)}$, and for any edge $(u, v) \in E_1$, $(f(u), f(v)) \in E_2 \wedge L_{(u,v)} = L_{(f(u),f(v))}$. G_1 is a **subgraph** of another graph G_2 iff $V_1 \subseteq V_2$ and $E_1 \subseteq E_2 \cap (V_1 \times V_1)$. Equivalently, G_2 is called a **supergraph** of G_1. If G_1 is graph isomorphic to a subgraph g in G_2, we say G_1 is **subgraph isomorphic** to G_2 and g is an **embedding** of G_1 in G_2.

A training graph instance can be denoted as $\langle tid, cid \rangle$, where tid represents the unique identifier and cid represents the class label. An input **training graph database** $D = \{t_1, t_2, \ldots, t_n\}$ consists of a set of n graphs, where t_i represents the ith instance. A class label set $C = \{c_1, c_2, \ldots, c_k\}$ contains all the k distinct class labels in D.

Let $I = \{t_1, t_2, \ldots, t_m\}$ be the set of all distinct vertices appearing in database D. A **subgraph pattern** p for D is a subgraph consisting of vertex set I_p and edge set E_p where $I_p \subseteq I \wedge E_p \subseteq (I_p \times I_p)$. A graph t_j is said to *cover* subgraph pattern p if t_j is a supergraph of p. The number of graphs in D which cover p is called the **support** of p, denoted by sup_p. Given a user-specified minimum support threshold min_sup, a subgraph pattern p is said to be *frequent* if $sup_p \geq min\_sup$.

An classification rule is an expression "$p \to c_i : sup_p^{c_i}, conf_p^{c_i}$", where p is the body, c_i is the head, $sup_p^{c_i}$ is the support, and $conf_p^{c_i} = \frac{sup_p^{c_i}}{sup_p}$ is the confidence of the rule, respectively. The corresponding classification rule of a frequent pattern p covered by graph instance t_j, "$p \to c_i : sup_p^{c_i}, conf_p^{c_i}$", is called a frequent covering rule of graph instance t_j. For a graph instance t_j, the frequent covering rules which have the highest $conf_p^{c_i}$ are called the **Highest Confidence Covering Rules** w.r.t t_j. The support and confidence of a Highest Confidence Covering Rule w.r.t graph instance t_j are denoted by $HCCR_{t_j}^{sup}$ and $HCCR_{t_j}^{conf}$ respectively.

3.2   Problem Definition

Given an input training chemical compound database D with a class label set C, and a minimum support threshold min_sup, our problem is to find a set of high quality graph-based classification rules for each class c_i, which can help classify new compounds whose class labels are not known.

4 Canonical Molecular Representation

A chemical compound is modelled as a labeled graph in order to reflect its inner structure in a way that each atom is treated as a vertex and each bond as an edge with labels to indicate the atom types and the bond types.

In mining frequent subgraph patterns, one key point is how to avoid redundant search, as the same subgraph patterns can be grown by gradually adding nodes and edges in different orders. To avoid such redundant search and make the algorithm efficient, an effective method is to design a canonical growing order and adopt a unique representation for each pattern [5,10]. In this section we introduce the SMILES and the SMARTS which can also help avoid redundant pattern generating and facilitate the mining process.

4.1 SMILES Representation of Compounds

First of all, in order to avoid the redundant search of subgraph patterns, we use the unique SMILES (Simplified Molecular Input Line Entry System) language as the canonical representation of a compound [9]. This is because the canonical SMILES representation has several advantages for mining compounds. Firstly, SMILES keeps compounds in a lineal format, which supports fast structure searching and is very compact for storage. It is said that a typical SMILES will take 50% to 70% less space than an equivalent connection table. Secondly, SMILES is a general representing language which is very widely used and can be well understood by chemists and biologists. Thirdly, SMILES just contains a simple vocabulary and a few grammar rules and is very vivid for description.

In SMILES representations, atoms and bonds are formed into a sequence, hydrogen atoms can be omitted, aromatic atoms are written in lowercase while others start with a capital, the single bonds and the aromatic bonds in some cases are not written, cycles need to be depicted by breaking one bond in the ring and adding a pair of numbers, and branch structures are represented within the brackets.

For example, the compound in Figure 1 (a) can be represented by a SMILES string "CC(C)(N)Cc1ccccc1", where "c1ccccc1" represents the benzene ring, four C's denote the non-aromatic carbon atoms, N denotes the nitrogen atom, and the brackets represent the branch-connected structure.

4.2 Fragment Pattern Encoding Using SMARTS

For most applications, connected substructures are useful enough to reveal some intrinsic characteristics of compounds. Moreover, the search space of mining disconnected substructures is too huge. Thus, we in this paper only consider fragments of connected substructures, which will be used to form high confidence classification rules. For fragment patterns we use the SMARTS (please see [3]) to describe them. SMARTS is a language derived from SMILES in order to specify substructures in compounds, which

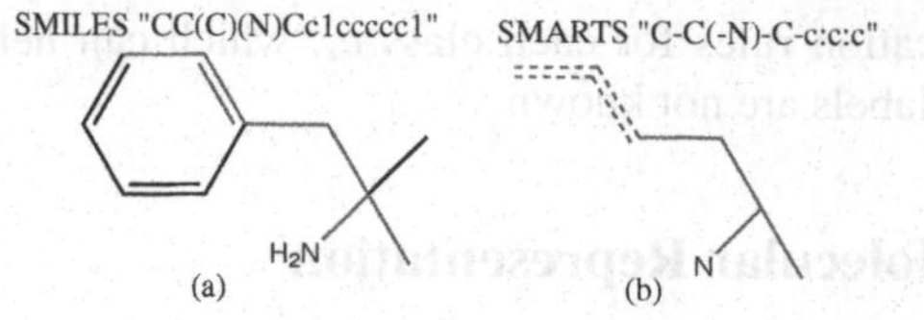

SMILES "CC(C)(N)Cc1ccccc1"   SMARTS "C-C(-N)-C-c:c:c"

H₂N
(a)
N
(b)

Fig. 1. An example compound and an example fragment

can be used to determine whether a fragment is covered by a compound. Then in the
frequent fragment enumeration process, we can figure out the frequency of a fragment
pattern, which is the number of compounds that cover the fragment, by checking the
SMARTS of this fragment through the SMILES's of the compounds.

For example, the SMARTS expression "C-C(-N)-C-c:c:c" represents the molecular
fragment shown in Figure 1 (b), which is covered by the compound in Figure 1 (a).

5   Fragment-Based Classification Rule Mining

In this section, we mainly discuss how to efficiently enumerate frequent fragments and
how to design some pruning techniques to prune some parts of search space which
are not promising in generating high quality fragment-based classification rules. In addition, we also discuss how to mine frequent fragments with class-specific supports,
which is especially useful for imbalanced databases.

5.1   Frequent Fragment Enumeration

We adopt the widely used projection-based mining framework as a basis to efficiently
enumerate the fragment features and get their corresponding classification rules. As
shown in Figure 2, each node in the traversal tree represents a fragment. Each fragment
node in the tree is a connected substructure of compounds with a support no less than
min_sup. Many previous algorithms have shown that depth-first search is more efficient
in mining long frequent patterns than breadth-first search. In this paper, our algorithm
traverses the tree of fragment nodes in a strict depth-first search manner.

We first compute the set of frequent bonds, B, in the input database. The root node
is assumed to be $\emptyset$ with no atom or bond. Each node in the fisrt level represents a

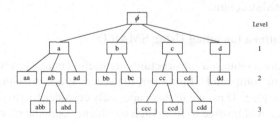

Fig. 2. An example search tree

fragment with only one bond in B. When traversing down one level we recursively extend the current fragment by adding a bond, which will either form a ring or add an atom that is currently not in the fragment substructure. During the mining process, a certain linear order of the bonds, $\prec_T$, is designed and used to avoid generating duplicate fragments. For each fragment node, we construct its pseudo projection, which can be efficiently done by recording a set of pointers that index the boundary positions where the fragment is embedded in the compounds. For the enumeration of the nodes in the next layer, we only need to scan these pointers of the fragment to compute its child frequent fragments. This will be explained more clearly by introducing the definition of *immediate extensible bonds*.

Definition 1. *(Immediate extensible bonds) Given a fragment f, a minimum support threshold min_sup, and an input compound database D, for each compound m in D that covers f, we check the boundary positions and record the bonds each of which appears in m but not currently in f and connects at least one atom in f. All these bonds are categorized according to the position of starting atom, the bond type and the ending atom type. Then the collection of bonds each of which occurs more than min_sup times are called immediate extensible bonds of f.*

Lemma 1. *(Immediate extension). Given a frequent fragment f and its projected database D_f, the complete set of frequent child fragments w.r.t. prefix f is equivalent to the set of fragments each of which extends f by a bond in the immediate extensible bonds of f.*

Proof. The proof is simple and due to the limited space we omit it here. ∎

Therefore, given a fragment f and its projected database D_f, the immediate extensible bonds of f, denoted by $IEB(f)$, can be used to extend f to generate bigger frequent child fragments.

Theorem 1. *(Needed extension). Given a frequent fragment f and one of its child fragment f_b extended from f by an immediate extensible bond b, let the immediate extensible bonds of f_b that exactly spread from b be $IEB(f_b, b)$, then the immediate extensible bonds of f and f_b hold the following relation:*

$$IEB(f_b) \subseteq (IEB(f) \bigcup IEB(f_b, b) - \{b\}).$$

Proof. First, with the definition of $IEB(f_b, b)$, we can divide the $IEB(f_b)$ into two subsets: $IEB(f_b, b)$ and $(IEB(f_b) - IEB(f_b, b))$. Then it follows that each bond in $(IEB(f_b) - IEB(f_b, b))$ should start from an atom in f, which is the same as the bonds in $IEB(f)$. According to the anti-monotonic property of frequent pattern mining and the definition of immediate extensible bonds, any bond that is not in $(IEB(f) - \{b\})$ (as b has already been extended into f_b) would not be an immediate extensible bond in $(IEB(f_b) - IEB(f_b, b))$, and any bond in $(IEB(f_b) - IEB(f_b, b))$ should also be an immediate extensible bond in $(IEB(f) - \{b\})$. Thus $(IEB(f_b) - IEB(f_b, b)) \subseteq (IEB(f) - \{b\})$ holds. So we have: $IEB(f_b) \subseteq (IEB(f) \bigcup IEB(f_b, b) - \{b\})$. ∎

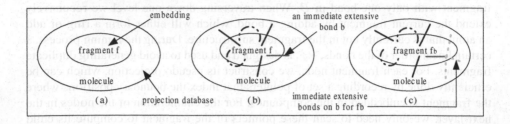

Fig. 3. An example of Theorem 1. Lines are immediate extensible bonds, and ellipses are compounds or embeddings of fragments.

An example is shown in Figure 3. Figure 3(a) shows a prefix fragment f and its simple projection database. Figure 3(b) shows the immediate extensible bonds of f. The bonds of dotted lines in Figure 3(c) constitute $IEB(f_b, b)$.

Based on Lemma 1 and Theorem 1, for a fragment f_b extended from a prefix fragment f, we only need to focus on the last extended bond b and figure out $IEB(f_b, b)$, which is the immediate extensible bonds on it. Then we can just check the bonds in $(IEB(f) - \{b\})$ and select those which are still frequent and immediate extensible as $(IEB(f_b) - IEB(f_b, b))$. Then we will easily get the $IEF(f_b)$ for fragment f_b and can use it to generate frequent child fragments for f_b. In a recursive way we can get the complete set of all the frequent fragments for the input compound database.

5.2 Pruning Strategies

Based on the preceding fragment enumeration framework, we can find the complete set of frequent fragments. However, as we introduced in Section 3, for each molecule we only concern the covering rules with the highest confidence. How to efficiently discover these highest confidence covering rules? Following we will discuss some useful pruning strategies to speed up the rule mining process.

Unpromising Fragment Pruning. According to [8], given a prefix fragment P, and any fragment f which can be used to extend P to get a bigger and also frequent fragment, $P \cup f$, for any class label c_i, the following equation holds:

$$conf_{P \cup f}^{c_i} = \frac{sup_{P \cup f}^{c_i}}{sup_{P \cup f}} \leq \frac{sup_{P \cup f}^{c_i}}{min\_sup} \leq \frac{sup_P^{c_i}}{min\_sup}.$$

With the upper bound on the right side of the equation, the unpromising conditional database pruning technique proposed in [8] can be adopted to prune the entire projected database of a fragment. That is, for any compound m_i with class label c_j in the projected database D_P of the current fragment P ($1 \leq i \leq |D_P|, 1 \leq j \leq k$), if the following equation always holds,

$$HCCR_{m_i}^{conf} \geq \min\{1, \frac{sup_P^{c_i}}{min\_sup}\}$$

it means that any fragment-based classification rule generated from a fragment with prefix P has no chance to become the highest confidence covering rule of any compound in D_P, then the prefix fragment P can be safely pruned.

Redundant SMARTS Pruning. Suppose there are two fragments f_A and f_B, and f_A is discovered earlier than f_B, if f_A and f_B have the same optimized SMARTS expression, which means they have the same embeddings to any compound sharing the same set of bonds and atoms with the same topology, then f_B is redundant with regard to f_A in the fragment enumeration process. Therefore, f_B and its child fragments can be pruned to facilitate the fragment enumeration. This can be efficiently implemented by maintaining a hash table for the optimized SMARTS expressions of all the fragments that have been mined. When we get a new fragment f, we first check if its optimized SMARTS expression has been inserted in the hash table, if so, we can safely stop growing f.

Support Equivalence Bond Pruning. The support equivalence item pruning strategy proposed in [8] can also be extended and applied in the fragment-based rule mining with some modifications. For any child fragment f_b which extends a bond b to f, if $sup_{f_b} = sup_f$ and f_b has no immediate extensible bonds starting from the ending atom of b, the child fragment f_b can be pruned to facilitate the enumeration for the following reason. From $sup_{f_b} = sup_f$ we know that for any class label $c_j (1 \leq j \leq k)$, $sup_{f_b}^{c_j} = sup_f^{c_j}$ holds. Given any other immediate extension e to f, we then have $sup_{f_{b,e}}^{c_j} = sup_{f_e}^{c_j}$ and $conf_{f_{b,e}}^{c_j} = conf_{f_e}^{c_j}$ for any class c_i, thus, $f_{b,e}$ must be a frequent child fragment of f_b. As f_b has no other immediate extensible bonds starting from the ending atom of b, so all the frequent child fragments extended from f_b will generate rules with the same confidence as those extended from f. Thus f_b can be safely pruned.

5.3 Class-Specific Support Threshold

Using a uniform minimum support may be suitable for frequent pattern mining without class label information. However, it may cause some problems for imbalanced databases with class information. In imbalanced databases, class distribution is skewed. Some classes may be quite large, while the others are small. On one hand, if we set a large value for minimum support, the algorithm will have difficulties in finding high quality rules for the small classes. On the other hand, if we set a small value for minimum support, it may lead to the overfitting problem for large classes. This suggests that class-specific support thresholds should be adopted for imbalanced databases.

To specify different minimum support values for different classes, a reasonable approach is to set them in proportion to their size. Suppose that min_sup corresponds to the minimum support of the smallest class, then we automatically compute the minimum supports for each class c_i from min_sup according to the class distribution, denoted as min_sup_i, by the following equation:

$$min\_sup_i = min\_sup \times (\frac{|c_i|}{min_{\forall j, 1 \leq j \leq k} |c_j|})^{\xi}.$$

Here, $|c_i|$ is the number of all the compounds of class c_i, and ξ ($\xi \geq 0$) is the *support differentia factor* which is set to 1 by default.

By assigning class-specific minimum support values for different sized classes, the pruning equation in Section 5.2 is still applicable, except that min_sup needs to be replaced with min_sup_i and the equation should be changed to the following form:

$$HCCR_{m_i}^{conf} \geq \min\{1, \frac{sup_P^{c_i}}{min\_sup_i}\}.$$

6 Fragment-Based Classification

By integrating the canonical representation of the compound structure and fragment pattern, the projection-based enumeration framework, and the search space pruning techniques, we can mine a set of fragment-based high confidence classification rules, which can be divided into k groups according to their class labels. When a new chemical compound arrives, we search the rules in each group to find the highest confidence matching rule in each group, and use the confidence of this rule in group i as the score for class c_i. We predict the class label of the new compound as the class with the highest score. This scoring function is simple and also effective in most cases, but may not be the best one for the imbalanced databases like DTP AIDS. As the class distribution in an imbalanced database is extremely skewed, it is usually easier to mine high confidence rules for a large class, while it is relatively difficult for a small class. Thus, the HARMONY-like model [8] biases the big classes. To mitigate this problem, we compute the final score, $FinalScore_{t_i}^{c_k}$, for a new compound ti and each class c_k using the following equation.

$$FinalScore_{t_i}^{c_k} = Score_{t_i}^{c_k} \times (\frac{conf_{c_{max}}^{avg}}{conf_{c_k}^{avg}})^{\delta}$$

Here, $Score_{t_i}^{c_k}$ is the value of the confidence of the first matching rule in group k, $conf_{c_k}^{avg}$ is the average confidence of the group of classification rules w.r.t. class label c_k, c_{max} is the class label with the highest average confidence, and δ ($\delta \geq 0$) is called the *score differentia factor*.

7 Experimental Results

We used the HIV data derived from the DTP AIDS antiviral Screen program to evaluate our approach. This database contains 41911 compounds which are divided into three classes, CA, CM and CI. There are a total of 417 compounds from class CA, 1072 compounds from class CM, and 40422 ones of class CI. All the experiments were conducted on a 2.8GHZ Pentium 4 machine with 1GB memory running Windows XP.

7.1 Effectiveness of High Confidence Rules Selecting

We first conducted experiments to evaluate the effectiveness of the pruning strategies. In Figure 4, we show the runtime comparison for mining high confidence fragments with and without pruning strategies. We can see that the algorithm using pruning methods takes much less time than the one without pruning, especially at lower support thresholds. This might be because that higher confidence fragments can be found at

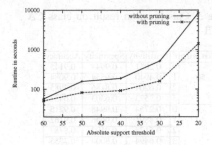

Fig. 4. Pruning effects (Runtime)

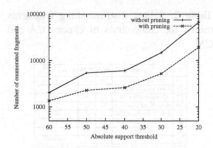

Fig. 5. Pruning effects (# enumerated fragments)

lower supports, which will subsequently be used to raise the currently maintained highest confidences and thus prune the search space. As we can notice from Figure 5, the number of enumerated fragments decreases significantly when the pruning methods are adopted, which illustrates that the pruning methods are very effective in pruning the unpromising parts of search space.

7.2 Interesting Distinguishing Fragments with High Quality

Distinguishing patterns, whose frequencies vary significantly from one class of data to another, often reveal ample difference among classes of data and are quite useful for prediction and building classifiers. Here we show the ability of our algorithm of finding distinguishing fragments among different classes. In these experiments we used all the compounds of class CA, CI and CM as the input database and applied the algorithm

Table 1. Salient distinguishing fragments of classes CA, CM and CI

No.	Fragment	#CA	#CM	#CI	$conf_{frag}^{CA}$
1	S=C-N-c:c:c:c(-Cl):c-C	32	0	0	1
2	S=C-N-c:c:c(:c-Cl)-C	32	0	0	1
3	S=C(-N-c:c:c:c-Cl)-c	32	0	0	1
4	S=C(-N-c:c:c:c-Cl):c:c	32	0	0	1
5	S=C(-N-c:c:c:c-Cl)-c:c:c	32	0	0	1
6	S=C(-N-c:c:c:c:c-C)-c:c:c	28	0	0	1
7	S=C(-N-c:c:c-C)-c:c:c	28	0	0	1
8	o:c(:c-C=S)-C	21	0	0	1
9	o:c:c:c-C(=S)-N-c	21	0	0	1
10	S=C(-N-c)-c:c:c	35	2	1	0.9211

No.	Fragment	#CA	#CM	#CI	$conf_{frag}^{CM}$
11	S(=O)-c:c:c:c:c(-N):c	75	82	160	0.2587
12	S-c:c:c:c:c:c-S	11	53	153	0.2442
13	O=C-c:c:c(:c:c:c-O)-C-c:c:c:c	20	59	164	0.2428
14	S-c:c:c:c:c(-N):c	76	86	218	0.2263
15	S-c:c:c(-N):c	57	76	209	0.2222

No.	Fragment	#CA	#CM	#CI	$conf_{frag}^{CI}$
16	c(:c(:c:c:c:c:c)-C-C)-C	1	7	680	0.9884
17	n(:c:c:c:c):c(:c)-C	3	14	1432	0.9883
18	N(-c1:c:c:c:c:c1)-C-C-C	1	26	2077	0.9872
19	O=C-C-C=O	4	26	2241	0.9868
20	O=C-C=C-c	2	30	2306	0.9868
21	O=C(-N-c:c:c:c:c)-C-C	4	31	2397	0.9856
22	O=C(-N-c:c:c:c)-C-C	5	31	2415	0.9853
23	n(:c:c:c:c):c(:c)-C	4	30	2266	0.9852
24	O=C-C-C-C=O	8	25	2172	0.9850

Fig. 6. The molecular structures of some typical fragments

Table 2. Salient distinguishing fragments mined for compounds of classes CA and CM

No.	Fragment	#CA	#CM	$conf_{frag}^{CA}$
25	S=C-N-c:c:c:c-Cl	39	0	1
26	Cl-c(:c:c:c-N-C):c-C	34	0	1
27	Cl-c:c(:c:c-N-C)-C	34	0	1
28	S=C(-N-c:c:c-Cl)-c	32	0	1
29	S=C-N-c:c:c(:c-Cl)-C	32	0	1
30	S=C(-N-c:c:c-C)-c	28	0	1
31	o:c(:c-C)-C	21	0	1
32	S=C(-N-c)-c	35	2	0.9459

No.	Fragment	#CA	#CM	$conf_{frag}^{CM}$
33	n(:n):c:n	2	58	0.9667
34	n:n:c:n:c	2	56	0.9655
35	S-S	3	61	0.9531
36	n:c(:n)-N	3	59	0.9516
37	N=C-C	4	64	0.9412
38	n:n	7	96	0.9320
39	n(:n:c):c	7	91	0.9286
40	S-N	9	84	0.9032

Table 3. Classification result on class CA and CI

min_sup	δ	Sensitivity	Specificity	Accuracy
50	0	0.2061	0.9948	0.9145
	1	0.4940	0.9554	0.9084
	2	0.6204	0.8186	0.7985
	3	0.7573	0.6303	0.6433
40	0	0.2754	0.9948	0.9215
	1	0.4891	0.9646	0.9162
	2	0.5178	0.9192	0.8783
	3	0.6494	0.8009	0.7855
30	0	0.4148	0.9946	0.9355
	1	0.5708	0.9660	0.9257
	2	0.5900	0.9225	0.8889
	3	0.7121	0.8099	0.7999
20	0	0.4703	0.9739	0.9226
	1	0.6045	0.9486	0.9135
	2	0.6213	0.9031	0.8744
	3	0.7481	0.7987	0.7936

to it by setting min_sup to 20 and δ to 1 by default. Our approach found 154 distinct fragments in total, among which 113 are for class CA, 41 are for class CM, and 102 are for class CI. To the best of our knowledge, most of these fragments have never been discovered by previous methods. Some salient fragments are listed in Table 1 (fragments are here represented as SMARTS strings), and a few of them are also depicted in Figure 6. These fragments are quite useful for prediction of class label. For example, the fragment 1 shown in Figure 6 appears in 32 compounds of the class CA, but never occur in the other two classes. Based on this observation we are able to infer that an unknown compound containing this fragment can be predicted as confirmed active and should be a member of class CA. We note that in Table 1, compared with those of classes CA and CI, the confidences of fragments in class CM are relative low. This might be caused by the fact that the anti-HIV activity of compounds of class CM is more than 50% and less than 100%, which act as compromise between class CA and class CI.

Compared with MOLFEA [4], our algorithm has several advantages. First, it can find distinguishing fragments with arbitrary structures, not just the linear structures MOLFEA can discover. Secondly, our algorithm can mine distinguishing fragments for all the classes simultaneously from multi-class data, while MOLFEA only deals with binary data. Moreover, our algorithm can detect some linear distinguishing fragments that MOLFEA cannot find(e.g the fragment 25 with SMART string "S=C-N-c:c:c:c-Cl"). Table 2 shows some salient distinguishing fragments our approach found but MOLFEA missed from the input compounds of classes CA and CM.

7.3   Classification Results

We used 10-fold cross validation method to evaluate the classification performance of our approach. We tested the sensitivity, specificity, and accuracy of the classification model for databases consisting of compounds of different classes, with min_sup varying from 20 to 50, δ varying from 0 to 3, and $\xi=1$ by default. Due to the limited space only the results of experiments on classes CA and CI are shown in Table 3.

From these results we can see that as we decrease min_sup, the sensitivity and accuracy tend to increase significantly, and the specificity tends to decrease marginally. This is because as the min_sup goes down, more distinguishing fragments with lower support for rare class CA are detected, which helps a lot in improving the sensitivity and overall accuracy of the classifier. However, if min_sup is set to a low value, it may lead to overfitting, especially for the large class, which impairs the specificity. We can also observe the effectiveness of the *score differentia factor* δ. With δ varying from 0 to 3, it does improve the sensitivity a lot, although in the meantime it hurts the specificity and the overall accuracy.

8 Conclusions

In this paper, we study the problem of mining high-quality fragment-based classification rules for imbalanced molecular database of the DTP AIDS antiviral screen program. We present an efficient approach to mining the fragments that can form rules with highest confidence for each compound. We also take into account the imbalanced class distribution of the database to improve the performance. Some pruning strategies are also introduced to speed up the mining process. Experiments show the efficiency of our method and its effectiveness for aiding classification, which surely will facilitate the detection of new compounds with desired anti-HIV activity.

References

1. Borgelt, C., Berthold, M.R.: Mining molecular fragments: Finding relevant substructures of molecules. In: ICDM 2002 (2002)
2. Dehaspe, L., Toivonen, H., King, R.D.: Finding frequent substructures in chemical compounds. In: KDD 1998 (1998)
3. James, C.A., Weininger, D., Delany, J.: Daylight theory manual - Daylight 4.71, Daylight Chemical Information Systems (2000), http://www.daylight.com
4. Kramer, S., Raedt, L.D., Helma, C.: Molecular feature mining in hiv data. In: KDD 2001 (2001)
5. Kuramochi, M., Karypis, G.: Frequent subgraph discovery. In: ICDM 2001 (2001)
6. Li, W., Han, J., Pei, J.: CMAR: Accurate and efficient classification based on multiple class-association rules. In: ICDM 2001 (2001)
7. Liu, B., Hsu, W., Ma, Y.: Integrating classification and association rule mining. In: Knowledge Discovery and Data Mining, pp. 80–86 (1998)
8. Wang, J., Karypis, G.: HARMONY: Efficiently mining the best rules for classification. In: Jonker, W., Petković, M. (eds.) SDM 2005. LNCS, vol. 3674, Springer, Heidelberg (2005)
9. Weininger, D.: SMILES 1. Introduction and encoding rules. Journal of Chemical Information and Computer Sciences 28, 31 (1988)
10. Yan, X., Han, J.: gSpan: Graph-based substructure pattern mining. In: ICDM 2002 (2002)
11. Yin, X., Han, J.: CPAR: Classification based on predictive association rules. In: SDM 2003 (2003)

Source Credibility Model for Neighbor Selection in Collaborative Web Content Recommendation

JinHyung Cho[1] and Kwiseok Kwon[2]

[1] School of Computer and Information Engineering, Dongyang Technical College
62-160 Gocheok-dong, Guro-gu, Seoul, 152-714, Korea
solver3@gmail.com
[2] Interdisciplinary Graduate Program of Technology and Management, Seoul National
University, San 56-1, Shillim-dong, Gwanak-gu, Seoul, 151-742, Korea
kskwon@gmail.com

Abstract. Since collaborative filtering (CF) based recommendation methods rely on neighbors as information sources, their performance depends on the quality of neighbor selection process. However, conventional CF has a few fundamental limitations that make them unsuitable for Web content services: recommender reliability problem and no consideration of customers' heterogeneous susceptibility on information sources. To overcome these problems, we propose a new CF method based on the source credibility model in consumer psychology. The proposed method extracts each target customer's part-worth on source credibility attributes using conjoint analysis. The results of the experiment using the real Web usage data verified that the proposed method outperforms the conventional methods in the personalized web content recommendation.

Keywords: collaborative filtering, source credibility, conjoint analysis, neighbor selection, recommender system, web content.

1 Introduction

With the explosive growth of digital content providing services on the Web, a suitable recommender system has become an important component for personalization of content delivery and its use is essential if the content providers are to remain competitive [1][2]. Collaborative filtering (CF) is one of the most successful recommendation techniques whose performance has been proved in various e-commerce applications [2][3]. However, conventional CF methods suffer from a few fundamental limitations. One of the limitations is the "recommender reliability problem." Another challenging issue is ability of recommendation mechanisms to be adaptable to environment where users have heterogeneous susceptibility on information sources [4].

In order to address these issues, it is necessary to reflect various consumer psychological factors of online users, which have been actively researched in offline consumer behavior studies. This study attempts to apply the source credibility model in consumer behavior theory to Web content recommendation method through the integration of technological and managerial perspectives. The proposed approach extracts

Y. Zhang et al. (Eds.): APWeb 2008, LNCS 4976, pp. 68–80, 2008.

each target customer's part-worth on source credibility attributes using conjoint analysis. The results of the experiment using the real Web usage data verified that the proposed method outperforms the conventional CF methods in the personalized web content recommendation.

2 Collaborative Filtering and Neighbor Group Selection

CF method identifies neighbors, a group of customers who had shown similar tastes to a target user for whom recommendations need to be generated, and recommends the items that those neighbors had liked in the past. CF method automates the "word of mouth (WOM)" that we usually receive from our family, friends, and coworkers [5]. Therefore, CF based recommender systems should form a group of recommenders (we term it a "neighbor group selection") that plays the role of an information sources for a target customer. The recommendation quality of CF depends on how credible and valuable neighbors (or information sources) are selected for the target customer, consequently.

Traditional CF method forms a recommender group that comprises the nearest neighbors whose preferences are similar to those of a target user. It selects the users that have most similar rating tendency to the target user as neighbors. Thus, these methods are referred to as similarity-based CF methods. However, similarity-based CF methods have a few fundamental problems. One of them is the recommender reliability problem, which is addressed in recent researches [6][7][8]. It means that a recommender might not be reliable for a given item or set of items, even though the recommender's and target user's preferences are similar [7]. In addition, an information source has various attributes and the level of a customer's value about the attributes of a source varies with reference to the characteristics of the user [9]. For example, when choosing a product or services, some customers rely on the opinions of users that have similar preferences, while others depend on the opinions of users that have expertise. That is, customers' value to the attributes of information sources is heterogeneous. However, conventional similarity-based CF cannot reflect this important heterogeneity since it only consider one attribute, similarity between users. Here, another fundamental problem of the similarity-based CF lies.

3 Source Credibility Model for Collaborative Filtering

Trust-based concepts (i.e., trustworthiness, expertise, reputation, reliability, credibility, and so on) have been identified as key components in the solution to secure transactions in e-commerce [10][11][12][13]. As a countermeasure to the abovementioned recommender reliability problem, they have also gained an increasing amount of attention in the field of CF research. Recently, a few trust-based CF methods [6][7][14] have been proposed. These techniques derive the neighbors' trust explicitly or implicitly and use it as an important criterion to select neighbors, which eases the recommender reliability problem.

Although a trust or expertise metric has been employed for trust-based CF in order to solve recommender reliability problems, two fundamental problems still exist.

First, there is no agreement on the definition of "trust" or "expertise." One study used "trust" to mean "trustworthiness," whereas another study used it as "expertise." Some papers have used the terms "trustworthy" and "credible" synonymously. Second, more fundamentally, there is not enough theoretical background to introduce a trust or expertise metric, in addition to the similarity metric, to CF recommendation systems. Is trust the only criterion for selecting credible neighbors? Are there any other metrics to consider? In other words, there is no managerial or psychological basis to prove which attributes of a human recommender as an information source are pertinent.

Therefore, we introduce source credibility model to recommender systems to explain the credibility of an information source consisting of multidimensional factors. Source credibility model has been proposed in the WOM communications studies of consumer psychology and marketing [15][16]. Credibility refers to a person's perception of the truth of a piece of information [15]. Previous studies have shown that source credibility has a positive, persuasive impact [17]. Credibility is seen as a multidimensional concept and a variety of studies have dealt with the discovery of its dimensions [15]. The identified key dimensions of the source credibility in the online environment are expertise (competency), trustworthiness, and similarity (co-orientation) [16]. We term these "attributes," and each attribute is defined as Table 1.

Table 1. Three key attributes of the source credibility model for neighbor selection

Attributes	Description
Expertise	The extent to which a source is perceived as being capable of providing correct information
Trustworthiness	The degree to which a source is perceived as providing information that reflects the source's actual feelings or opinions
Similarity	The degree to which a source is similar to the target audience members, or is depicted as having similar problems or other characteristics relating to use of a particular product or brand

We assert that if the source credibility model that consists of multidimensional attributes is applied to the neighbor selection in CF based recommender systems, the performance of the systems can be improved as compared to the existing ones that select neighbors by single dimension.

4   Neighbor Selection with Conjoint Analysis

As abovementioned, each consumer's relative importance about the attributes of information sources is varying with reference to his or her characteristics, differences in prior knowledge to the products, product involvements, and so on. Previous CF researches have not considered this variation. For example, existing trust-based CF selects neighbors by the weights that combine trust (expertise) and similarity as same relative importance.

This study takes heterogeneity in users' susceptibility to the credibility attributes into account. And, we regard that a neighbor's value to a target user as an information source is the sum of the target user's part-worth to each credibility attribute. If we identify these individual part-worths, now we can consider the heterogeneous user susceptibility to the credibility attributes. We use conjoint analysis in order to find the most valuable and suitable neighbors to the target user.

A conjoint analysis is an experimental procedure for assessing values that are called part-worths of attributes. It represents the most widely applied methodologies for measuring and analyzing consumer preferences [17][18][19]. It assumes that a unit is a bundle of attributes and the unit's value is the sum of the part-worths of the attributes [19]. The analysis was originated in marketing researches and has mainly applied to new product development. In the studies of recommender systems, there were lots of content-based filtering (CBF) researches that apply this concept to the product attributes [20]. However, there has been no research that employs the conjoint analysis to the neighbor selection problem in CF. We, firstly, adopt the conjoint approach to the neighbor selection problem with the source credibility model. The value of a human recommender as an information source u to target user a, Va,u, is related with the part-worths by a regression equation as follows:

$$V_{a,u} = \sum_{n=1}^{N} \sum_{m=1}^{Mn} p_{anm} \; x_{unm} + \varepsilon_{au} \tag{1}$$

Where,

$V_{a,u}$: the target user a's evaluation of user u (known)

p_{anm} : part-worth of attribute n level m to the target user a (unknown)

x_{unm} : 1, if the m-th level of the n-th attribute is present in information source u, and 0 otherwise (known)

The proposed method selects neighbors by reflecting this target user's part-worth. The highly evaluated users according to the part-worths of the target user will be selected as the most valuable neighbors to the target user. We assert that there exists heterogeneous susceptibility to credibility attributes between users, and to consider it can improve the performance of recommendation.

5 Collaborative Web Content Recommendation

The overall process of the proposed Web content recommendation method consists of the offline mining phase and the online recommendation phase (see Figure 1).

5.1 Offline Mining Phase

Step 1: Creating User Rating Profile
The user rating profile describes a user's preference with respect to each item by mining the web usage data collected in the Web content service site. In this study, each user rating profile is constructed according to the steps of the website's general content service usage (click-through, preview, and payment). The proposed method uses the relative usage frequency of a user as an implicit preference rating based on the assumption that if a user's usage frequency for a specific item is relatively high,

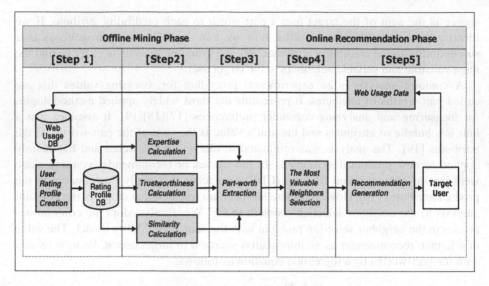

Fig. 1. Recommendation process

he or she has a high preference for it. We define the user rating profile $R_{u,i}$ by modifying the previous approach [21] related to web usage mining and making it suitable for Web content service as follows:

$$R_{u,i} = \frac{R_{u,i}^c - \min_{1 \le i \le m}(R_{u,i}^c)}{\max_{1 \le i \le m}(R_{u,i}^c) - \min_{1 \le i \le m}(R_{u,i}^c)} + \frac{R_{u,i}^v - \min_{1 \le i \le m}(R_{u,i}^v)}{\max_{1 \le i \le m}(R_{u,i}^v) - \min_{1 \le i \le m}(R_{u,i}^v)} + \frac{R_{u,i}^p - \min_{1 \le i \le m}(R_{u,i}^p)}{\max_{1 \le i \le m}(R_{u,i}^p) - \min_{1 \le i \le m}(R_{u,i}^p)} \tag{2}$$

Where, $R_{u,i}$: the rating profile of user u for item i

　　　　m : the number of items

$R_{u,i}^c$, $R_{u,i}^v$ and $R_{u,i}^p$ are the number of click-throughs, previews, and payments respectively, by a user for each item. The value of $R_{u,i}$ ranges from 0 to 3 and increases with increasing frequencies. The weights for each step are not the same although they look equal, as shown in Equation 2, since the customers who purchased a specific content not only clicked the web pages related to it but also previewed the content. Therefore, $R_{u,i}$, which is used in the subsequent phases, is the normalized and weighted sum of $R_{u,i}^c$, $R_{u,i}^v$ and $R_{u,i}^p$.

Step 2: Deriving Source Credibility Attributes
1) Expertise Measurement
We define expertise as the degree of a user's competency to provide an accurate prediction and exhibit high activity, on the basis of source credibility. Based on the definition, we devised a measure of expertise reflecting the users' activity and prediction competency. To measure the prediction competency, we employ and improve the equation used in previous research [7]. The equation assumes that user u's rating on an item i, $R_{u,i}$, for another user v is correct if the rating is within a range ε from v's

actual rating, $R_{v,i}$. While previous research [7] placed ε at 1.8, we set the value more strictly at 1.0 in this study. However, if the number of raters in an item is lower, the user's prediction competency for the item should be underestimated. Thus, we introduce the compensation value, λ, as n/10 where n is the number of raters, with the exception of user u, in an item when the number of raters in an item is less than 10; otherwise, it is 1. See Equation 3.

$$CORRECT_{u,i} = 0, \quad if \; |R_{u,i} - R_{v,i}| > \varepsilon$$
$$= \lambda, \quad if \; |R_{u,i} - R_{v,i}| \leq \varepsilon \tag{3}$$

Where,
λ: n/10 if n < 10, otherwise 1 (n: number of raters except user u)
ε: tolerance range

The expertise of user u ($EXPERTISE_u$) is calculated as the rate of the correct value with activity weighting, as shown in Equation 4. In order to obtain a higher value of expertise with more rating activities for more items, we add activity weighting to the measure in the previous research. If the number of ratings for items by the user considered is lower than that by the average of all the users, the user's activity weighting α_u is the number of ratings of the user considered/average number of ratings of all the users; otherwise, $\alpha_u = 1$.

$$EXPERTISE_u = \alpha_u \frac{\sum_{m_u} \sum_{n_i} CORRECT_{u,i}}{\sum_{m_u} \sum_{n_i} 1} \tag{4}$$

Where,
α_u: activity weighting
n_i: rating set with ratings in item i
m_u: item sets with the ratings of user u

2) Trustworthiness Measurement
Trustworthiness is defined as the degree to which a user is perceived as providing information that reflects his actual feelings or opinions, on the basis of source credibility. It is difficult to implicitly extract trustworthiness. We assume that if a user's ratings are similar to the average ratings of all the users in a community, he offers his true opinions. Let us consider, for example, the case when the averages of all the users in the community for five movies are 3-5-4-3-5, and the ratings of an individual user are 1-3-2-1-3. Even if his ratings are not exactly the same as the others, his rating tendency is very similar to the other users. In this case, the user is not believed to provide false representation. Thereby, a user's trustworthiness could be measured by the similarity between the user's ratings and the averages of the ratings given by the group that the user belongs to. The trustworthiness of user u ($TRUST_u$) is calculated by using Equation 5 and employing the absolute value of the Pearson correlation coefficient, which sets the range for trustworthiness from 0 to 1.

The significance weighting β is 1 if the number of ratings of a target user is over 50; otherwise, it is n/50, where n is the number of ratings of the user [22]. This provides a user who has many ratings for the items with a high trustworthiness value.

$$TRUST_u = \beta \times \left| \frac{\sum_{i=1}^{m}(R_{u,i} - \overline{R}_u)(R_{average,i} - \overline{R}_{average})}{\sqrt{\sum_{i=1}^{m}(R_{u,i} - \overline{R}_u)^2 \sum_{i=1}^{m}(R_{average,i} - \overline{R}_{average})^2}} \right|$$ (5)

Where,

$R_{u,i}$: rating of user u for item i

$R_{average,i}$: average rating of the users in the community for item i

β: significance weighting

m: number of user u's rated items

3) Similarity Measurement

Similarity is defined as the rating similarity between two users. To measure the rating similarity between the users, Pearson's correlation coefficient, which is the most widely used in conventional CF methods, is employed. The similarity between user u and another user v ($SIMILARITY_{u,v}$) is calculated using Equation 6. Significance weighting, γ, which is 1 if the number of co-rated items between the two users is over 50 and n/50 (where n is the number of co-ratings between the users) otherwise, is included in the equation [22]. This also assigns a high similarity to a user who has many co-ratings with the other user.

$$SIMILARITY_{u,v} = \gamma \times \frac{\sum_{i=1}^{m}(R_{u,i} - \overline{R}_u)(R_{v,i} - \overline{R}_v)}{\sqrt{\sum_{i=1}^{m}(R_{u,i} - \overline{R}_u)^2 \sum_{i=1}^{m}(R_{v,i} - \overline{R}_v)^2}}$$ (6)

Where,

$R_{u,i}$: rating of user u for item i

$R_{v,i}$: rating of user v for item i

γ: significance weighting

m: number of co-rated items

Step 3: Extracting Each User's Part-worths about the credibility attributes

In step 3, we extract the user-specific part-worths of the credibility attributes. We develop a model from the equation 1.

1) Value of an information source

The value of an information source can be measured in various aspects. Because CF methods, however, do not expose the provider to a target user, it is needed to estimate the value implicitly. We regard that the value of an information source to a target user depends on the distance of ratings of two users. If a user's rating is exactly same to a target user, the user's value to the target user is regarded to be high. Therefore, the value of an information source is estimated using mean-absolute-error (MAE) as shown in equation 7.

$$V_{a,u} = C - MAE_{a,u} = C - \frac{\sum_{i=1}^{m}|R_{a,i} - R_{u,i}|}{m}$$ (7)

Where,

$V_{a,u}$: the target user a's evaluation of user u

$R_{a,i}$: the rating of the user a for the item i

$R_{u,i}$: the rating of the user u for the item i
m : the number of co-rated items
C : a constant

C is a constant in order to make it bigger and positive the value of an information source whose MAE between the target user is small(Note: C is set to 10 at the experiment in this paper). To avoid give high value to a user who has small co-rated items with the target user, we include the users who have minimum 20 items co-rated with the target user.

2) Attributes
As abovementioned, we assume that an information source's credibility consists of expertise, trustworthiness, and similarity according to the source credibility model. Thus, the equation 1 can be rewritten as like below.

$$V_{a,u} = \sum_{m=1}^{Me} p_{aem}\, x_{uem} + \sum_{m=1}^{Mt} p_{atm}\, x_{utm} + \sum_{m=1}^{Ms} p_{asm}\, x_{usm} + \varepsilon_{au} \tag{8}$$

where,
$V_{a,u}$: the target user a's evaluation of user u (known)
$p_{aem},\, p_{atm},\, p_{asm}$: part-worth of expertise, trustworthiness, and similarity level m to the target user a (unknown)
$x_{uem},\, x_{utm},\, x_{usm}$: 1, if the m-th level of the expertise, trustworthiness or similarity is present in information source u, and 0 otherwise (known)
$M_e,\, M_t,\, M_s$: the number of levels of expertise, trustworthiness, and similarity
ε : random error with normal distribution

3) Levels
Because the variables of expertise, trustworthiness, and similarity calculated in offline mining phase are continuous ones, it is needed to separate them into some discrete levels that have some ranges. It is one of the important issues in conjoint analysis because users' preference may be differently captured or misunderstood with reference to the number and the range of levels [19].

One may instinctively notice that if the number of levels increases, users' preference can be captured better, but it also causes the complexity and the difficulties in finding unknown variables. This paper assumes that each attribute consists of 5 levels of "very high(1)", "high(2)", "medium(3)", "low(4)", and "very low(5)." It makes the equation 1, 7 and 8 finalized as equation 9.

$$V_{a,u} = \sum_{m=1}^{5} p_{aem}\, x_{uem} + \sum_{m=1}^{5} p_{atm}\, x_{utm} + \sum_{m=1}^{5} p_{asm}\, x_{usm} + \varepsilon_{au} \tag{9}$$

Expertise and trustworthiness ranges from 0 to 1, while similarity between users from -1 to 1 in this paper. The range of each level in each attribute may be issue that has very large rooms to be studied in this case. But the main concern of the paper does not lie in that point but in employing conjoint approach to neighbor selection of CF. In the experiment, we set each range of the level by considering the average, maximum, minimum and distribution of each attribute as depicted in Table 2.

Table 2. Conjoint model in this study

Attribute	Expertise		Trustworthiness		Similarity	
User a	Description	Part-worth	Description	Part-worth	Description	Part-worth
Level 1	very high (>0.7)	p_{ae1}	very high (>0.6)	p_{at1}	very high (>0.7)	p_{as1}
Level 2	High (0.7≥ >0.5)	p_{ae2}	High (0.6≥ >0.4)	p_{at2}	High (0.7≥ >0.5)	p_{as2}
Level 3	Medium (0.5≥ >0.3)	p_{ae3}	Medium (0.4≥ >0.2)	p_{at3}	Medium (0.5≥ >0.3)	p_{as3}
Level 4	Low (0.3≥ >0.1)	p_{ae4}	Low (0.2≥ >0.1)	p_{at4}	Low (0.3≥ >0.1)	p_{as4}
Level 5	very low (<0.1)	p_{ae5}	very low (<0.1)	p_{at5}	very low (<0.1)	p_{as5}

5.2 Online Recommendation Phase

In this phase, recommendation is generated in real-time by utilizing the values in source credibility and part-worths database. Here, we explain recommendation procedures.

Step 4: Selecting the Most Valuable Neighbors
The value of a user u to a target user a, $V_{a,u}$, is calculated by equation 8. For example, if a user u's expertise, trustworthiness, and similarity fall on very high, medium, and high, respectively, the user's value is the sum of p_{ae1}, p_{at3}, and p_{as2}. Then, top-N valuable users that have highest value from the viewpoint of the target user are selected as "the most valuable neighbors." We select top 20 valuable users in this study.

Step 5: Generating Recommendation
The rating of the target user for an item is estimated by the ratings of the neighbors for the item from the conventional CF equation below. We use similarity between two users as the sole weight for prediction because the CF prediction formula is optimized to the correlation coefficient for similarity.

$$R_{a,i}^{Estimated} = \overline{R_a} + \frac{\sum_{u=1}^{N} SIMILARITY_{a,u}(R_{u,i} - \overline{R_u})}{\sum_{u=1}^{N} SIMILARITY_{a,u}} \tag{10}$$

6 Performance Evaluation

6.1 Benchmark Systems

In order to prove the performance of the proposed credibility based-CF method, we compared this method to two existing CF methods. We used these three benchmark methods as follows:

(1) Similarity-based CF (SCF): Neighbor selection by similarity
(2) Trust-based CF (TCF): Neighbor selection by similarity and trust (expertise)

(3) Credibility-based CF (CCF): Neighbor selection based on source credibility model (The proposed method)

6.2 Data Set and Evaluation Metrics

We have evaluated the feasibility and advantages of our proposed method with a real Web usage data set and purchasing data from Paran.com (www.paran.com), which is a major Korean digital-content provider. Through data preparation and Web-usage mining, we obtained 295,910 ratings of 8,421 users for 245 items in the digital-comics content (digitalized comic books), logged from 1 July to 31 December 2005. We divided each data set into a modeling set and a validation set. The modeling set contained randomly selected items amounting to 80 percent of the total items; the validation set contained the remaining items. We calculated the source credibility and part-worths of the users in the calibration set, and verified the performance of the benchmark systems in the validation set.

To evaluate the performance of the systems, we employed two broad classes of rec-ommendation accuracy metrics. The first class is predictive accuracy metrics. Here, we use the MAE to compare each system's predictive accuracy. MAE is the absolute dif-ference between an actual and a predicted rating value. The second class is classifica-tion accuracy metrics. To evaluate how well the recommendation lists match the user's preferences, we employ the widely used precision, recall, and F1 measures.

6.3 Experimental Results

Table 3 shows the predictive and classification accuracy of each benchmark method. The MAE of CCF is approximately 5.2% lower than that of SCF, 9.9% lower than that of TCF at a significance level of 1%. Compared to SCF and TCF, CCF improves SCF by 16.4% and TCF by 7.27% with F1 values.

These results prove that our proposed method can significantly improve the rec-ommendation quality of the existing CF. We believe that the proposed method can improve the recommendation in the case of the items that need more expertise or trustworthiness, such as cars and digital cameras, which are known as "high-involvement products." Since a movie is a relatively low-priced item, people are dis-posed to the similarity of users in general. If an item under consideration is expensive and/or important, one tends to rely on the users with expertise or trustworthiness rather than similarity. In that case, the proposed method that considers part-worths on all the attributes will show its potential.

Table 3. Performance comparison

Method	MAE		Precision	Recall	F1
CCF[C]	0.3556	t-Value (p<0.01)	0.3571	0.3257	0.3407
SCF[S]	0.3750	-12.354	0.2785	0.3081	0.2926
IC-SI/S*	5.2%		28.2%	5.71%	16.4%
TCF[T]	0.3948	-24.075	0.3187	0.3166	0.3176
IC-TI/T*	9.9%		12.0%	2.87%	7.27%
* : The performance gain of CCF over each benchmark system					

Table 4. An example of a market-level part-worth table

Attribute	Expertise		Trustworthiness		Similarity	
Market Level	Description	Part-worth	Description	Part-worth	Description	Part-worth
Level 1	very high	34.0770	very high	16.4573	very high	30.1677
Level 2	high	34.0688	high	16.4564	High	30.1110
Level 3	medium	34.0795	medium	16.4546	medium	30.0954
Level 4	low	34.0896	low	16.4574	Low	30.0822
Level 5	very low	33.9777	very low	16.4456	very low	30.0586

If each level's part-worths of all the users in a group or a market are to be averaged, the group or the market-level part-worths can be found. Table 4 shows an example of the market-level part-worth. This part-worth table enables us to identify the most valuable users to the specific group or market. The users are regarded as the most valuable information sources in the group or market as a social network. Thereby, marketing staff in a company can find out the most effective information sources in a group or market segment for a specific product domain.

7 Conclusion

Currently, most CF recommendation methods do not consider the customers' heterogeneous susceptibility to recommendation sources' attributes. Our proposed method that employs source credibility model and customer's susceptibility to the credibility attributes. It was proven that the proposed method could improve the performance of the recommender system to a greater extent than could the existing CF method by experiments.

This study has several important contributions to the CF researches as follows: First, it has upgraded the existing CF's neighbor selection process by adopting source credibility model and conjoint analysis. This enabled CF to include multi-dimensional criteria and consider heterogeneous users' susceptibility to the attributes of the credibility and provide more adaptive and personalized recommendation. Second, this study has proposed an interdisciplinary research framework by incorporating the consumer psychological view to the existing CF based recommendation technique. Thereby, theoretical background about the credibility and trust related concept can be established. Thirdly, our proposed method makes a more adaptive and effective marketing strategy possible. With conjoint approach, e-commerce marketers can identify not only each customer's part-worth about the credibility attributes, but also the market level part-worth. It is possible to find the most valuable customers to the target market who may act as an important information source.

The directions for future research are as follows. First, since this study is the first attempt to apply conjoint analysis to select neighbor group for CF, there may be a few rooms to be improved. For instance, since the attributes of the credibility are continuous, the study about the optimized level separation is to be needed. Moreover, noncompensatory rule could be considered in the neighbor selection. Second, it is required to combine some data mining techniques which can automatically extract user credibility as well as user preference. Third, it will be interesting to expand it to other challenging e-commerce domain. Comparative research considering the characteristics of product such as product involvement is to be anticipated.

Acknowledgments. This work has been supported by the KESRI Research Project(R-2005-7-132), which is funded by MOCIE (Ministry of Commerce, Industry and Energy) in the Republic of Korea.

References

1. Garfinkel, R., Gopal, R., Tripathi, A., Yin, F.: Design of a Shopbot and Recommender System for Bundle Purchases. Decision Support Systems 42(3), 1974–1986 (2006)
2. Cho, J., Kwon, K., Park, Y.: Collaborative Filtering Using Dual Information Sources. IEEE Intelligent Systems 22(3), 30–38 (2007)
3. Herlocker, J.L., Konstan, J.A., Terveen, L.G., Riedl, J.T.: Evaluating Collaborative Filtering Recommender Systems. ACM Transactions on Information Systems 22(1), 5–53 (2004)
4. Bearden, W.O., Netemeyer, R.G., Teel, J.E.: Measurement of Consumer Susceptibility to Interpersonal Influence. J. Consumer Research. 15(4), 473–481 (1989)
5. Shardanand, U., Maes, P.: Social Information Filtering: Algorithms for Automating 'Word of Mouth'. In: Human Factors in Computing Systems Conference (CHI 1995), pp. 210–217. ACM Press, New York (1995)
6. Massa, P., Avesani, P.: Trust-aware Collaborative Filtering for Recommender Systems. In: 2nd Int'l Conf. Cooperative Information Systems (CoopIS 2004) (2004)
7. O'Donovan, J., Smyth, B.: Trust in recommender systems. In: 10th Int'l Conf. Intelligent User Interfaces (IUI 2005) (2005)
8. Riggs, T., Wilensky, R.: An Algorithm for Automated Rating of Reviewers. In: 1st ACM/IEEE-CS Joint Conf. Digital Libraries (JCDL 2001), pp. 381–387. ACM Press, New York (2001)
9. Duhan, D.F.: Influences on Consumer Use of Word-of-Mouth Recommendation Sources. J. the Academy of Marketing Science. 22, 283–295 (1997)
10. Jøsang, A., Ismail, R., Boyd, C.: A Survey of Trust and Reputation Systems for Online Service Provision. Decision Support Systems 43(2), 618–644 (2007)
11. Kim, D.J., Song, Y.I., Braynov, S.B., Rao, H.R.: A multidimensional trust formation model in B-to-C e-commerce: a conceptual framework and content analyses of academia/practitioner perspectives. Decision Support Systems 40(2), 143–165 (2005)
12. Manchala, D.W.: E-commerce trust metrics and models. IEEE Internet Computing 4(2), 36–44 (2000)
13. Resnick, P., Zeckhauser, R., Friedman, E., Kuwabara, K.: Reputation systems. Communications of the ACM 43(12), 45–48 (2000)
14. Weng, J., Miao, C., Goh, A.: Improving Collaborative filtering with Trust-based Metrics. In: Biham, E., Youssef, A.M. (eds.) SAC 2006. LNCS, vol. 4356, Springer, Heidelberg (2007)
15. Eisend, M.: Source Credibility Dimensions in Marketing Communication – A Generalized Solution, J. Empirical Generalizations in Marketing 10, 1–33 (2006)
16. Robertson, T.S., Zielinski, J., Ward, S.: Consumer Behavior. Scott, Foresman and Company (1984)
17. Carroll, J.D., Green, P.E.: Guest Editorial: Psychometric Methods in Marketing Research: Part I, Conjoint Analysis. J. Marketing Research. 32(4), 385–391 (1995)

18. Aggarwal, P., Vaidyanathan, R.: Eliciting Online Customers' Preferences: Conjoint vs Self-Explicated Attribute-Level Measurements. J. Marketing Management. 19, 157–177 (2003)
19. Green, P.E., Krieger, A.M., Wind, Y.: Thirty Years of Conjoint Analysis: Reflections and Prospects. Interfaces 31(3), 56–73 (2001)
20. Adomavicius, G., Tuzhilin, A.: Toward the Next Generation of Recommender Systems: A Survey of the State-of-the-Art and Possible Extensions. IEEE Transactions on Knowledge and Data Engineering 17(6), 734–749 (2005)
21. Cho, Y.H., Kim, J.K., Kim, S.H.: A Personalized Recommender System Based on Web Usage Mining and Decision Tree Induction. Expert Systems with Applications 23(3), 329–342 (2002)
22. Herlocker, J.L., Konstan, J.A., Borchers, A., Riedl, J.: An Algorithmic Framework for Performing Collaborative Filtering. In: 1999 ACM SIGIR (1999)

The Layered World of Scientific Conferences

Michael Kuhn and Roger Wattenhofer

Computer Engineering and Networks Laboratory, ETH Zurich, Switzerland
{kuhnmi,wattenhofer}@tik.ee.ethz.ch

Abstract. Recent models have introduced the notion of dimensions and hierarchies in social networks. These models motivate the mining of small world graphs under a new perspective. We exemplary base our work on a conference graph, which is constructed from the DBLP publication records. We show that this graph indeed exhibits a layered structure as the models suggest. We then introduce a subtraction approach that allows to segregate layers. Using this technique we separate the conference graph into a thematic and a quality layer. As concrete applications of the discussed methods we present a novel rating method as well as a conference search tool that bases on our graph and its layer separation.

1 Introduction

Imagine your research has drifted into a field unfamiliar to you, and you do not know where to publish. In such situations it is helpful to have a better understanding of the world of computer science conferences. Throughout this paper we will explore this world and present an application that bases on this exploration and seeks to assist people that are in situations as described before.

The starting point of our research are recent findings in the context of social networks. These findings emphasize the fact that nodes in natural graphs are interconnected for different reasons, such as common interests, close geographic distances, or family relations in case of friendship networks. Based on the researchers' social network we will introduce a similarity measure for conferences and setup a conference graph. Similar as in friendship networks or the worldwide-web, edges in this graph are caused by different reasons—we will refer to them as the layers of our graph. Such reasons surely are area of research, but maybe also the quality, geographic location, or the community behind the conference. Throughout this paper, we demonstrate that and how it is possible to isolate some of these layers in the case of the conference graph. In particular, we will show that:

- The social network behind conferences provides a good measure to relate conferences to each other.
- This measure consists of a thematic as well as a quality component—the major layers of our graph.
- The thematic layer can be identified by a mere analysis of publication titles.
- The quality layer can be partly isolated by subtracting the thematic component from the overall relationship.

Y. Zhang et al. (Eds.): APWeb 2008, LNCS 4976, pp. 81–92, 2008.

As a result of the layer separation, it becomes possible to explore the conference graph under different points of view. We introduce a novel idea for conference rating based on the quality layer of the graph. Afterwards, we present a collaborative conference search website that demonstrates the advantages of having independent notions of the thematic scope and the quality of a conference. It offers different ways to search for conferences and can be fine-tuned to match the quality and deadline restrictions of an author and thus greatly assists researchers finding themselves in situations as described in the very beginning of this paper.

2 Related Work

This section briefly reviews relevant literature in the context of our work, namely the mining of bibliometric data and social networks.

The analysis of publication records has been an active field of research for a long time. Clearly, one of the most attractive goals for publication database mining is automated conference and journal rating. Garfield's pioneering work in 1972 [1], which describes the use of citation analysis for this purpose, initiated a long—and still ongoing—controversy. On one hand, many authors point out the wide variety of problems of the citation indexing approach [2,3,4]. On the other hand, citation analysis is presumably still the best method to automatically rate scientific conferences and journals. Other measures that are used to indicate a venue's relevance are the acceptance rate as well as time delays, such as turn-around time, end-to-end time, or reference age [5]. It seems that the community behind a conference has so far not been taken into account for automated rating. We believe that this criterion should not be neglected and provide an idea to fill this gap.

The rating of venues is not the only motivation for research on bibliometric data. Other insights have been gained from publication databases. One closely related aspect is the characterization of authors (rather than venues). Various measures, such as closeness [6,7,8,9,10], betweenness [6,8,9], or AuthorRank [9] have been evaluated in this context. Also, many studies analyze the evolution of different properties [6,7,10,11,12]. For us, the publications of Lee et al. [12] and Smeaton et al. [10] are of particular interest, as they study the topical changes within a single conference over the years. Thereby they show that the analysis of publication titles, keywords, and abstracts is sufficient to extract the thematic scope of a venue—a fact that we will take advantage of.

Another perspective to looking at the thematic scope of venues is presented in [6]. By considering a common author of two venues an indicator for thematic similarity, a weighted graph is constructed that interrelates the most important conferences in the field of database research. We improve on this measure by incorporating some means of normalization and show that the thematic proximity is only one aspect contained in this weight.

Newer studies on social networks emphasize that many of these graphs exhibit some sort of social dimensions [13,14,15,16]. They state that there exist different catalysts for friendships, such as geography, family ties, or occupation. An observation that Killworth and Bernard [17] already made in 1978, when examining the different reasons by which a starter in a Milgram-like experiment would choose the next hop. Their findings show that most decisions are based on the geographic location and the occupation of the target. This result agrees with the findings of Dodds et al. [13] in a recent Internet-based small-world experiment. Based on this evidence, Watts et al. [16] developed a graph model based on different social dimensions. We will show that similar dimensions can also be found in our conference graph and refer to them as layers.

3 The Conference Graph

This section describes how the publication records of DBLP[1] can be used to generate a graph that interconnects scientific conferences. The graph construction bases on the social network behind these conferences. We basically assume, that the more common authors two conferences have, the more related they are. To avoid overestimating the similarity of massive events—they naturally have a large number of common authors—we improve on this idea by incorporating a normalization method: Consider two conferences, C_1 and C_2, that contain a total of s_1 and s_2 publications, respectively. Further, assume that there are k authors A_i $(i = 1, ..., k)$ that have published in both places and that author A_i has $p_{i,1}$ publications in conference C_1 and $p_{i,2}$ publications in conference C_2. We can now define the similarity $S(C_1, C_2)$ between C_1 and C_2 as follows:

$$S(C_1, C_2) = \sum_{i=1}^{k} \min\left(\frac{p_{i,1}}{s_1}, \frac{p_{i,2}}{s_2}\right)$$

Applying this similarity measure to all pairs of conferences results in the desired graph. The required information for this graph was extracted from the DBLP bibliographic repository. Any publications that appeared in a scientific conference between 1996 and 2006 have been taken into account. To reduce the amount of data, edges of extremely low weight that do not significantly contribute to the connectivity have been removed. To give a more concrete idea of the structure of this graph, Table 1 lists the 10 top edges for some sample conferences.

4 The Layers

This section introduces the idea of *layers* as the building blocks of our graph. These layers reflect, as we will see, different catalysts for edges. We will have a closer look at two of these layers, namely the thematic and the quality layer, throughout this section.

---

[1] http://dblp.uni-trier.de

Table 1. The 10 strongest links to the conferences KDD, AAAI and ECAI

KDD		AAAI		ECAI	
ICDM	0.69	IJCAI	0.76	IJCAI	0.53
SDM	0.58	ATAL	0.37	KR	0.29
PKDD	0.45	ICML	0.33	ATAL	0.27
PAKDD	0.40	AGENTS	0.32	AAAI	0.26
ICML	0.37	AIPS	0.31	AI*IA	0.24
DMKD	0.37	ECAI	0.26	JELIA	0.22
CIKM	0.36	KR	0.25	ECSQARU	0.21
SIGMOD	0.36	UAI	0.25	CP	0.19
ICDE	0.35	CP	0.23	IEA/AIE	0.19
VLDB	0.33	FLAIRS	0.20	KI	0.19

Proximity in the conference graph is not purely defined by the thematic similarity of venues as a careful look at Table 1 reveals. *ECAI* is, for example, typically said to be thematically closer to *AAAI* than *ATAL*, *ICML*, or *AGENTS*, which appear earlier in the *AAAI* top-10 list. We conclude that authors choose conferences not only because of the topic it covers. Other properties, such as quality, geographic location, or the community behind a venue also influence the author's decision. In fact, we believe that it is a weighted combination of all these factors that leads to a submission at a certain place. Exactly this combination is reflected by the conference graph presented in the previous section. The graph consists of different *layers*, where each layer represents one of these factors. This idea is illustrated in Figure 1.

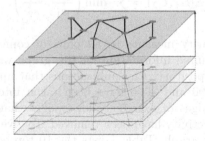

Fig. 1. The total graph can be seen as the sum of its layers

4.1 The Thematic Layer

Clearly, the thematic scope of a venue has a significant impact on its relationship to other venues. In the following, we present a technique that bases on publication title analysis and measures the thematic similarity of conferences. It thus allows to define the *thematic layer*, which is surely an ingredient of the social similarity measure, as a majority of authors mostly work in only one area and therefore submit papers to thematically similar venues.

Table 2. The 10 best matching keywords to KDD, AAAI and ECAI together with their TF-IDF score

KDD		AAAI		ECAI	
mining	0.051	learning	0.013	reasoning	0.011
data	0.013	planning	0.012	learning	0.010
discovery	0.013	robot	0.010	qualitative	0.009
clustering	0.013	reasoning	0.008	planning	0.008
association	0.010	knowledge	0.007	knowledge	0.008
sigkdd	0.010	search	0.007	logics	0.008
kdd	0.009	agent	0.006	logic	0.008
frequent	0.009	constraint	0.006	ecai	0.008
rules	0.009	ai	0.006	constraint	0.007
discovering	0.009	reinforcement	0.006	diagnosis	0.007

For each conference, we have extracted all the titles from DBLP and applied the well-known *term frequency - inverse document frequency (TF-IDF)* method (see [18] for some theoretic background) to identify the most relevant keywords. The TF-IDF score for a document increases proportionally to the number of occurrences of the keyword in the document (TF). However, words that have a high overall frequency are penalized (IDF). In our context, a document corresponds to a venue and the words stem from publication titles. Consequently, a document consists of all titles in a venue and the complete corpus consists of all venues in DBLP.

Once a score has been applied to all the keywords that appear in the conference's collection of titles, the scope of the conference can easily be estimated by looking at the most relevant terms. Table 2 shows some examples.

Using the keyword-lists seen before, we have implemented a simple algorithm that estimates the thematic relationship between two venues. It takes the top-50 keywords of each conference, and counts the number of keywords appearing in both lists, resulting in a score from 0 to 50 for each pair of conferences.[2]

Applying the thematic similarity function to each pair of venues results in a weighted undirected graph—the *thematic layer* of our graph. The corresponding neighborhood lists for our sample conferences are shown in Table 3.

4.2   The Quality Layer: Filtering by Subtraction

Section 2 briefly discussed the problem of conference rating and its difficulties. For computer sciences, the Citeseer Impact List tries to estimate the impact of venues based on citation analysis. Further, many researchers maintain hand-made lists that distinguish between tier-1, tier-2, and tier-3 conferences. Even though hand-made lists suffer from a subjective bias and citation analysis from other weaknesses (recall Section 2), tier-1 conferences typically have a high

---

[2] Surprisingly, this simple comparison function achieved slightly better results than the more commonly used cosine similarity approach.

Table 3. The 10 closest neighbors to KDD, AAAI and ECAI in the thematic layer, together with their thematic score

KDD		AAAI		ECAI	
ICDM	26	IJCAI	37	IJCAI	29
PKDD	23	ECAI	27	AAAI	27
PAKDD	21	FLAIRS	22	ICTAI	22
SDM	20	ICTAI	21	KI	21
Dis. Science	20	AIPS	17	FLAIRS	20
DMKD	18	Can-AI	16	Can-AI	19
ADMA	17	IEA/AIE	16	IEA/AIE	18
ISMIS	17	PRICAI	15	PRICAI	18
IDA	15	Aus-AI	15	KR	16
IDEAL	15	KI	14	Aus-AI	16

impact and, contrariwise, tier-3 conferences get low scores in the Citeseer list. We will refer to similarly classified conferences as conferences of similar quality.

Comparing the neighborhood tables for the total graph (Table 1) and the thematic layer (Table 3) shows, that the total graph is *not* purely defined by the thematic correlation of conferences. Looking at the total graph, an interesting observation is that conferences often considered to be of high quality (such as *KDD* and *AAAI*) tend to have other high quality conferences in their proximity. In contrast, the number of lower-tier conferences in the proximity of *ECAI*, which is mostly classified as tier-2, is significantly higher. This observation is illustrated in Table 4 that uses the impact value of the Citeseer Impact List[3] to classify the conferences.

We conclude that a single author tends to publish not only in venues of similar topic, but also in venues of similar quality, meaning that our graph contains a second major layer—the *quality layer*.

The observation that thematically weaker related nodes in a conference's proximity tend to be closer in quality suggests that the quality layer can be extracted using the information about the total graph and the thematic layer. In the following we will introduce a *layer subtraction approach* to demonstrate that such a layer separation can indeed be achieved. The approach bases on the assumption that the total graph is a linear combination of the single layers. As a result of the observations in the previous section we assume that the major layers of the conference graph are the thematic layer t and the quality layer q. This also matches our experience when selecting a conference: We make sure the publication matches the call for papers and we try to submit at a conference of reasonable quality. Other factors, such as geographic location, play a minor role in the decision. These factors (including noise) are thus subsumed into a remainder layer r. Consequently the total edge weight S becomes to $S = \alpha_1 \cdot t + \alpha_2 \cdot q + \alpha_3 \cdot r$, for some weights α_i, with $\alpha_1, \alpha_2 \gg \alpha_3$. Neglecting α_3 and setting $\alpha_2 = 1$ (α_2 can be chosen arbitrarily as it only results in a scaling of q) allows to extract the quality layer q as

$$q \approx S - \alpha_1 \cdot t,$$

---

[3] http://citeseer.ist.psu.edu/impact.html

Table 4. The 10 closest neighbors to AAAI (left) and ECAI (right) in the total graph and the thematic layer, together with the Citeseer impact value. Note that for AAAI, conferences in the total graph neighborhood that are not present in the thematic layer list (*italic*) all have relatively high impact value. The impact value of such conferences in the neighborhood of ECAI is considerably lower.

AAAI Total		AAAI Thematic		ECAI Total		ECAI Thematic	
IJCAI	1.10	IJCAI	1.10	IJCAI	1.10	IJCAI	1.10
ATAL	*1.51*	ECAI	0.69	KR	1.76	AAAI	1.49
ICML	*2.12*	FLAIRS	N/A	*ATAL*	*1.51*	ICTAI	0.25
AGENTS	*1.00*	ICTAI	0.25	AAAI	1.49	KI	0.41
AIPS	1.53	AIPS	1.53	*AI*IA*	*0.26*	FLAIRS	N/A
ECAI	0.69	Can-AI	0.26	*JELIA*	*0.72*	Can-AI	0.26
KR	*1.76*	IEA/AIE	0.09	*ECSQUARU*	*0.38*	IEA/AIE	0.09
UAI	*N/A*	PRICAI	0.19	*CP*	*1.04*	PRICAI	0.19
CP	*1.04*	Aus-AI	0.16	IEA/AIE	0.09	KR	1.76
FLAIRS	N/A	KI	0.41	KI	0.41	Aus-AI	0.16

· Note that the validity of the linear combination assumption greatly depends on the characteristics of the weight functions in the different layers. In [19] Fernandez et al. presented the idea of score distribution normalization for aggregation purposes. They suggest to shape the histograms of the independent score functions to match the "ideal" distribution prior to merging them by linear combination. For simplicity we assume a uniform weight distribution for both, the total as well as the thematic scores.

Observe that the subtraction approach generally allows to extract one out of L layers of a graph, if the remaining $L - 1$ layers are known. It seems that such a layered structure can often be observed—recall Section 2 and also think of recommendation systems that often build on similar co-occurrence structures as our graph. We thus believe that the layer-subtraction approach might be a valuable preprocessing step in various data-mining settings.

The next sections discuss how the quality of the filtering can be estimated by producing a conference rating and thereby show some evidence of the correctness of the proposed subtraction approach.

4.3 Interpolation Based Conference Rating

The proximity of a conference in the quality layer is supposed to contain mostly conferences of similar quality. This observation immediately leads to the idea of conference rating by interpolation: Provided some initial ratings are known, the tier of a conference can be estimated by looking at its proximity in the quality layer. Initial ratings can be retrieved from manually created lists (we use the one found at www.ntu.edu.sg/home/assourav/crank.htm and refer to it as *CS Rating List*) as well as from Citeseer's impact list. We have further introduced the *Citeseer Tier List*, which assigns a tier (1, 2, or 3) to each conference in the Citeseer Impact List. The borders between tiers have been chosen such that the number of incorrectly rated conferences with respect to the CS Rating List becomes minimal. The best that can be achieved is an error rate of 38.8%, which indicates how difficult the task of conference rating is.

We have then defined a heuristic to rate a conference C_0 as follows:

1. For all conferences in the CS Rating List or the Citeseer Tier List, set the initial rating to the value found in the lists. In case of conflicts, the CS Rating List is treated with priority. For any conference not in the lists, set the initial rating to *unrated*.
2. Overwrite the initial rating of C_0 with *unrated*. This step avoids that the rating function is biased towards the initial value.
3. Take the 30 shortest edges e_i adjacent to C_0 in the total graph, together with their values S_i and t_i. For all these edges, calculate $q_i = S_i - \alpha_1 \cdot t_i$ (for some value of α_1) and sort by q_i. We will call the resulting list the *filtered neighborhood list* of C_0: $N_f(C_0)$.
4. For the first 5 entries C_j ($j = 1..5$) in $N_f(C_0)$, calculate $N_f(C_j)$.
5. Return the median of all the rated conferences found within the first 5 entries in all the lists $N_f(C_j)$ ($j = 0..5$) as the rating of C_0.

Note that this conference rating method is in some sense natural. Many people would judge a venue based on people participating in it (or leading it). This information is implicitly contained in the total graph which forms the basis of the rating heuristic.

The quality of the heuristic can be estimated by comparing the calculated ratings to those found in the CS Rating List (which is presumably the most accurate list we dispose of). The optimal value of α_1 was scanned for by exhaustive search over some reasonable interval.[4] This is illustrated in Figure 2 which plots the error rate of the rating function with respect to the CS Rating List for different values of α_1. The figure clearly shows that the subtraction approach reduces the number of incorrect ratings and suggests that the optimal value of α_1 is somewhere between 0.5 and 1.

Arguing with error rates beyond 40% might at the first glance seem suspicious. However, the fact that approximately 75% of the input values (namely those that originate from the Citeseer Tier List) exhibit an error rate of approximately 40% themselves relativizes the high error rate produced by our algorithm.

Ignoring all the conferences rated as tier-2 either by the algorithm or the CS Rating List shows that the errors are not random. Dividing the number of conferences rated as tier-1 instead of tier-3 (and vice versa) by the number of conferences the algorithm rates as tier-1 or tier-3 results in an error rate of around 6.4% without thematic filtering and of 2.6% for the optimal value of α_1. (Note that dividing by the number of tier-1 and tier-3 conferences in the list would overestimate an algorithm that tends to rate conferences as tier-2.)

---

[4] Note that optimizing for α_1 using regression by comparing to a "quality relationship" between two conferences is likely to fail, as this quality relationship is very vague (i.e. can take only the values 0, 1, and 2).

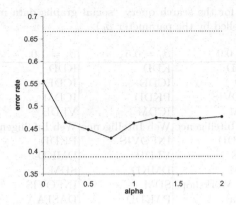

Fig. 2. The fraction of incorrectly rated conferences using our rating function versus the value of α_1. The dotted lines indicate the error rates for a random guess (0.667) and for the Citeseer Tier List (0.388), which is in about the best we can expect to reach as most of the initial rating values stem from this list.

These low values show three things:

1. The total edge weight is clearly influenced by the quality of conferences. This supports the assumption that the thematic and the quality layer are the two main layers of the graph.
2. The success of extracting the quality layer by subtraction of the thematic layer as shown in Figure 2 is confirmed.
3. Most of the around 43% of errors are minor errors. That is, they are wrong by only one tier. Severe errors are rare,they make up less than 3%.

Remark: The rating heuristic was developed for two reasons: To provide a complete rating list for the conference search application presented next, and to demonstrate the effect of subtraction filtering. It is thought as a proof-of-concept algorithm that neither has a strong mathematical foundation nor provides any guarantees on the results.

5   ConfSearch

In this section we will show that the previously discussed conference graph and its separation into different layers can directly be applied for conference search. For this purpose we have developed a website that is able to suggest conferences together with their most important attributes (try it at http://www.confsearch.org). The application offers four different search types:

- *Keyword Search*: Search by keywords provided.
- *Related Conference Search*: Explore the proximity of a given conference in the conference graph and return the closest neighbors.

Table 5. The results for the search query "social graphs data mining" for different quality weights (controlled by the parameter β_q)

$\beta_q = 0.0$	$\beta_q = 0.5$	$\beta_q = 1.0$
PKDD	KDD	KDD
KDD	ICDE	ICDE
INFOVIS	PKDD	ICDM
ICDM	ICDM	VLDB
Web Intelligence	Web Intelligence	Web Intelligence
PAKDD	INFOVIS	PKDD
ICDE	VLDB	DMKD
ICDM	DMKD	SDM
JSAI Workshops	SDM	INFOVIS
DaWaK	PAKDD	DASFAA

- *Author Search*: Search for the places a given author publishes most often.
- *General Search*: A weighted combination of the above search methods.

For all search types the application allows to sort the results by deadline, a criteria that has a considerable impact when deciding for one or the other venue. Motivated by the success of Wikipedia-like services, we follow a collaborative approach to gather conference deadlines as well as locations and website URLs. Our application can be seen as an improvement on the many lists with conference deadlines found in the Internet today: We basically cover the whole area of computer science and augment the typically static lists with sophisticated search options.

The *keyword search* bases on a score s_{ij} for each keyword-conference pair (where only keywords appearing in the query are considered), which is a slightly modified variant of the TF-IDF value presented in Section 4.1. Next, the scores s_{ij} of the conference-keyword pairs are combined to a single value S_i^* per conference C_i using the *p-norm* method introduced by Salton et al. [20]. The final score S_i results from the quality adjustment of S_i^* controlled by a user-settable parameter β_q: $S_i = S_i^* \cdot f(Q)^{\beta_q}$. The function f(Q) is defined on a per query basis to smoothly adapt to the different score distributions for different queries. The quality part Q is estimated using the heuristic presented in Section 4.3. Table 5 presents a keyword search example and the effect of quality filtering.

The *related conference search* operates directly on the conference graph. We simply return the closest nodes around a conference in terms of path-length. Again, a user settable parameter allows to control whether the thematic or the quality aspect should be emphasized. A visualization of the *AAAI* neighborhood in the thematic and the quality layer can be found in Figure 3. The increased amount of high quality nodes (dark) in the *AAAI*'s "qualitative proximity" indicates that *AAAI* itself is also likely to be of high quality. The search option on one hand allows to browse the conference graph and on the other hand might prove extremely helpful if you look for alternative places to submit, after a reject, for example, or because a deadline does not fit.

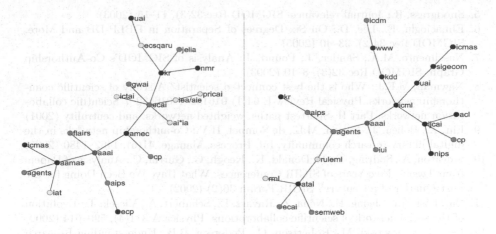

Fig. 3. The minimum spanning tree around *AAAI* in the thematic layer (left), and the quality layer (right). Darker nodes refer to higher tier venues.

6   Conclusion

Throughout this paper we have provided evidence for recent small-world models using the real-world data of a scientific conference graph. We have shown that this graph indeed consists of layers and demonstrated that these layers can be effectively combined. The combination assumption has led to the subtraction approach for layer segregation which provides an attractive preprocessing step when mining graphs. In our setting it was used to accentuate the different aspects of conference relations.

We have seen that the conference graph consists of two major layers—the thematic layer and the quality layer. We have then presented a novel rating method for scientific conferences that operates on the quality layer of the graph.

The separation of the two layers further builds the basis of the conference search application presented in the last section. Exploring the thematic layer allows to retrieve venues matching a user query. The sorting of the retrieved venues can then be adjusted using the information gained from the quality layer and thereby effectively fitted to the user's needs.

References

1. Garfield, E.: Citation Analysis as a Tool in Journal Evaluation. Science 178, 471–479 (1972)
2. Linde, A.: On the pitfalls of journal ranking by impact factor. Eur J Oral Sci 106(1), 525–526 (1998)
3. MacRoberts, M.H., MacRoberts, B.R.: Problems of citation analysis: A critical review. JASIS 40(5), 342–349 (1989)
4. Scharnhorst, A., Thelwall, M.: Citation and hyperlink networks. Current Science 89(9) (2005)

5. Snodgrass, R.: Journal relevance. SIGMOD Rec 32(3), 11–15 (2003)
6. Elmacioglu, E., Lee, D.: On Six Degrees of Separation in DBLP-DB and More. SIGMOD Rec 34(2), 33–40 (2005)
7. Nascimento, M.A., Sander, J., Pound, J.: Analysis of SIGMOD's Co-Authorship Graph. SIGMOD Rec 32(3), 8–10 (2003)
8. Newman, M.E.J.: Who is the best connected scientist? A study of scientific coauthorship networks. Physical Review E 64(1) 016132-1–016132-7 Scientific collaboration networks. Part II. Shortest paths, weighted networks, and centrality (2001)
9. Liu, X., Bollen, J., Nelson, M.L., de Sompel, H.V.: Co-authorship networks in the digital library research community. Inf. Process. Manage. 41(6), 1462–1480 (2005)
10. Smeaton, A., Sodring, T., McDonald, K., Keogh, G., Gurrin, C.: Analysis of Papers from Twenty-Five Years of SIGIR Conferences: What Have We Been Doing for the Last Quarter of a Century? SIGIR Forum 36(2) (2002)
11. Barabási, A.L., Jeong, H., Néda, Z., Ravasz, E., Schubert, A., Vicsek, T.: Evolution of the social network of scientific collaborations. Physica A 311(4), 590–614 (2002)
12. Lee, B., Czerwinski, M., Robertson, G., Bederson, B.B.: Understanding Research Trends in Conferences using PaperLens. In: CHI 2005: CHI 2005 extended abstracts on Human factors in computing systems, pp. 1969–1972. ACM Press, New York (2005)
13. Dodds, P., Muhamad, R., Watts, D.: An experimental study of search in global social networks. Science 301(5634), 827–829 (2003)
14. Goussevskaia, O., Kuhn, M., Wattenhofer, R.: Layers and Hierarchies in Real Virtual Networks. In: IEEE/WIC/ACM International Conference on Web Intelligence (WI), Silicon Valley, California, USA (2007)
15. Kleinberg, J.: Small-World Phenomena and the Dynamics of Information. In: Dietterich, T.G., Becker, S., Ghahramani, Z. (eds.) Advances in Neural Information Processing Systems, vol. 14, MIT Press, Cambridge (2002)
16. Watts, D.J., Dodds, P.S., Newman, M.E.J.: Identity and Search in Social Networks. Science 296, 1302–1305 (2002)
17. Killworth, P.D., Bernard, H.R.: The reversal small-world experiment. Social Networks 1(2), 159–192 (1978)
18. Robertson, S.: Understanding Inverse Document Frequency: On theoretical arguments for IDF. Journal of Documentation 5, 503–520 (2004)
19. Fernández, M., Vallet, D., Castells, P.: Probabilistic Score Normalization for Rank Aggregation. In: Lalmas, M., MacFarlane, A., Rüger, S.M., Tombros, A., Tsikrika, T., Yavlinsky, A. (eds.) ECIR 2006. LNCS, vol. 3936, pp. 553–556. Springer, Heidelberg (2006)
20. Salton, G., Fox, E.A., Wu, H.: Extended Boolean information retrieval. Commun. ACM 26(11), 1022–1036 (1983)

Mining Maximal Frequent Subtrees with Lists-Based Pattern-Growth Method*

Juryon Paik[1], Junghyun Nam[2], Jaegak Hwang[3], and Ung Mo Kim[1]

[1] Dept. of Computer Engineering, Sungkyunkwan University, Republic of Korea
quasa277@gmail.com, umkim@ece.skku.ac.kr
[2] Dept. of Computer Science, Konkuk University, Republic of Korea
jhnam@kku.ac.kr
[3] Electronics and Telecommunications Research Institute (ETRI), Republic of Korea
jghwang@etri.re.kr

Abstract. Mining maximal frequent subtrees remains at a preliminary state compared to the fruitful achievements in mining frequent subtrees. Thus, most of them are fundamentally complicated and this causes computational problems. In this paper, we present a conceptually simple, yet effective approach based on lists structures. The beneficial effect of the proposed approach is that it not only gets rid of the process for infrequent tree pruning, but also eliminates totally the problem of candidate subtrees generation. As far as we know, this is the first algorithm that discovers maximal frequent subtrees without any subtree generation.

1 Introduction

It is not trivial work to discover useful and common information from a collection of trees. With increasing demands, it is necessary to develop new or adjust existing extraction methods for tree data and various works have been proposed. The most common existing approaches to do are either *Apriori*-based [5,1,6,2,3] or frequent-pattern-growth(FP)-based [4,7]. However, Apriori-based algorithms suffer from two high computational costs: generating a huge number of candidate sets and scanning the database repeatedly for the frequency counting of candidate sets. To solve those problems, FP-growth method is developed, which adopts a divide-and-conquer strategy. The FP-growth based algorithms avoid candidate sets generation. However, this approach still has a severe problem. Instead of avoiding candidate subtrees generation, its goal is to discover all frequent subtrees which become infeasible for a large number of trees.

A more practical and scalable approach is to use maximal frequent subtrees, and handling the maximal frequent subtrees is an interesting challenge, and represents the core of this paper.

---

* This work was supported in part by the Ubiquitous Autonomic Computing and Network Project, 21st Century Frontier R&D Program funded by the Korean Ministry of Information and Communication, and by the Electronics and Telecommunications Research Institute (2007-0475-000).

Y. Zhang et al. (Eds.): APWeb 2008, LNCS 4976, pp. 93–98, 2008.

2   SEAMSON Algorithm

In this section, we introduce a new pattern-growth method SEAMSON (Scalable and Efficient Algorithm for Maximal frequent Subtrees extractiON) appropriate for use as the core of discovering valuable information from a database of rooted labeled trees where data not only exhibit heterogeneity but also are distributed.

2.1   Problem Definition

Embedded subtrees Given a tree $T = (r, N, E, \mathcal{L})$, we say that a labeled rooted tree $S = (r', N_S, E_S, \mathcal{L}')$ is included as an *embedded subtree* of T, denoted $S \precsim T$, iff (1) $N_S \subseteq N$, (2) for all edges $(u, v) \in E_S$ such that u is the parent of v, u is an ancestor of v in T, (3) the label of any node $v \in N_S$, $\mathcal{L}'(v) = \mathcal{L}(v)$. The tree T must preserve ancestor relation but not necessarily parent relation for nodes in S.

Support and frequent subtree Let $\mathcal{D} = \{T_1, T_2, \ldots, T_i\}$ be a set of trees and $|\mathcal{D}|$ be the number of trees in $\mathcal{D}$, where $0 < i \leq |\mathcal{D}|$. Given $\mathcal{D}$ and a tree S, the frequency of S with respect to $\mathcal{D}$, $freq_{\mathcal{D}}(S)$, is defined as $\Sigma_{T_i \in \mathcal{D}} freq_{T_i}(S)$ where $freq_{T_i}(S)$ is 1 if S is a subtree of T_i and 0 otherwise. The *support* of S with respect to $\mathcal{D}$, $sup_{\mathcal{D}}(S)$, is the fraction of the trees in $\mathcal{D}$ that have S as a subtree. That is, $sup_{\mathcal{D}}(S) = \frac{freq_{\mathcal{D}}(S)}{|\mathcal{D}|}$. A subtree is called *frequent* if its support is greater than or equal to a minimum value of support specified by users or applications. This user-specified minimum value is often called the *minimum support (minsup)*.

The problem of mining frequent subtrees is defined as to uncover all pattern trees S_j, such that $sup_{\mathcal{D}}(S_j) = \frac{\Sigma_{T_i \in \mathcal{D}} freq_{T_i}(S_j)}{|\mathcal{D}|} \geq minsup$, where $j > 0$. A *maximal frequent subtree* is one of frequent subtrees which none of its proper supertrees are frequent. Thus, there are fewer maximal frequent subtrees compared to the total number of frequent subtrees.

2.2   Transformation Phase

The process of transformation phase is initiated by scanning a trees database, $\mathcal{D}$, and generating a label-driven compact database. We refer to this database *label-dictionary*, and denote *L-dictionary*. In this label-driven db, each label plays an important role, which is usually performed by tree indexes ot transaction indexes. After the in-memory *L-dictionary* is constructed our approach does not require further database scanning.

For each distinct label in trees, a list is generated and stores label itself, pre-order traversal node indexes, parent positions, and tree indexes. The list is divided into two parts: head and body, and is named as *label-list*, abbreviated ℓ-*list*.

Definition 1 (head & body). *The part whose purpose is to clearly identify each ℓ-list is named as* head of ℓ-list *because its leads the whole list. The remained part concerns about how many times a label is occurred and hierarchy* *information of the label in original trees. Because this part directly follows its corresponding head part, it is called* **body** *of* **ℓ-list**.

Every head consists of a key field, satellite data of the key, and one link, which are a label, node indexes, and a link pointing to its body, respectively. Since ℓ-list is generated according to a label, labels place in key fields. The node indexes are those nodes whose labels are same as the label in the key field. The link field indicates a starting element of its corresponding body part.

A body of ℓ-list follows its head immediately. The main concerns of the body is to evaluate how many trees have the key and to find parents positions of the nodes in the head. The former is for dealing with frequencies of each label, while the latter is for handling hierarchical information of the label. To achieve such intentions the structure of body is a sequence of elements which is arranged in a linear order. Each element is an object with a tree index, one link field pointing to the next element, and one satellite data field.

Since only the trees which use the label in the corresponding head to their nodes should create elements, the total number of elements in a body indicates the number of trees that have the key of ℓ-list. This number of elements is used to get size of any ℓ-list, |ℓ-list|, and to determine if a given ℓ-list is frequent or not.

Fig. 1 shows an depicted example of how *L-dictionary* and its ℓ-lists are constructed and managed from the database $\mathcal{D}$. For easy distinction between nodes in different trees, we assign unique consecutive indexes in pre-order traversal. The bodies of ℓ-lists decide the frequencies of corresponding labels. For instance, the label A does not satisfy the given minimum value which is set $\frac{2}{3}$ because it has only 1 element in the body of ℓ_A-list.

2.3   Refinement Phase

The initially built *L-dictionary* contains at most same number of ℓ-lists as the number of labels, because the purpose of this *L-dictionary* is to store tree structures as list structures according to distinct node labels. Therefore, the unit ℓ-list is generated without considering the frequency of a label. Hence, even the label is used in only one tree, the ℓ-list for the label is generated and inserted into *L-dictionary*. Some ℓ-lists satisfy the given *minsup*, but some do not. As infrequent single-node trees are eliminated in conventional approaches, the ℓ-lists which do not confirm the condition also have to be removed from *L-dictionary*, because the initial *L-dictionary* is similar to the set of single-node trees.

A ℓ-list is said to have *frequent-head* iff its body contains more than or equal to the minimum number of elements. *Having frequent-head* technically means that the label of a current ℓ-list is frequently occurred among trees. The ℓ-lists do not have frequent-head cannot be further extended with other labels. Therefore, such ℓ-lists have to be filtered out from *L-dictionary*.

The label of parent node p has to be frequent in order that an extended subtree is qualified for being frequent. However, this is not guaranteed in the current *L-dictionary* because the filtration was performed on only the frequency of labels.

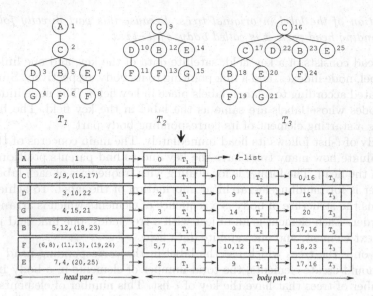

Fig. 1. Transformation phase to obtain *L-dictionary* and its units

Therefore, even if parent nodes are included in the ℓ-list which has frequent-head, it is not sure whether the labels of those parent nodes also correspond the ℓ-lists which have frequent-head. To resolve this problem we refine every parent node p in the filtered *L-dictionary* by following procedures: (1) a parent node in an element is verified by candidate_hash_table which was constructed with the ℓ-lists that had been filtered from *L-dictionary*, to detect if the node is assigned by unfrequent label or not. (2) If so, the node is marked 'replace' and its record is retrieved to search a node id assigned by frequent node labels. (3) Step (1) and (2) continues until the node id assigned by any frequent label is found. (4) The original parent node id is replaced by the found node which is actually an ancestor node of the original parent node. (5) Through step (1) to (3), if no any other nodes are frequent, the original parent node id is replaced by 0. After the replacement according to the procedure is done, the current *L-dictionary* is renamed as L^+-*dictionary* to distinguish with the previous one.

Parent nodes in an element which are assigned by the same label are removed from the element except only one node. The outcome of the end of the refinement phase is shown in Fig. 2. Note that the dummy list $\ell_/$-list is inserted to L^+-*dictionary* for later use in deriving a certain tree; this plays a root node of the tree.

2.4 Derivation Phase

Definition 2 (label-extension). *Extending an ℓ-list is to connect explicitly between a node having the label of ℓ-list.head and its parent node which is also labeled by one of labels of ℓ-lists, denoted ℓ-list + ℓ-list$_p$. The extension is*

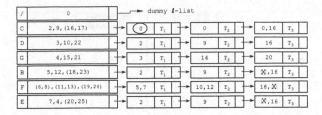

Fig. 2. L^+-*dictionary* after through the whole process of refinement

performed over the labels of ℓ-list not the nodes in L^+-dictionary. Thus, we call this extension as **label-extension**.

While the label-extension is committed to creating connection between ℓ-lists, one tree is constructed as a by product of the extension which has a root node corresponding to ℓ_I-list. This tree is called Potentially Maximal Pattern **tree** (**PMP-tree** in short), where edge has its own count to keep how much often it is occurred in the tree (the tree_header_table supports to build the tree; the detailed explanation is omitted due to the space). Based on those counts, the edge whose count is less than the given minimum value is deleted from the tree. After deleting such edges and rearranging the tree, the goal of this paper is produced.

3 Experimental Evaluation

The synthetic datasets are generated by the tree generation program whose underlying ideas are inspired by Termier [3] and Zaki [6]. The generator constructs a set of trees, $\mathcal{D}$, based on some parameters supplied by the user: $\mathcal{T}$: the number of trees in $\mathcal{D}$, L : the set of labels, f : the maximum branching factor of a node, d : the maximum depth of a tree, ρ : the random probability of one node in the tree to generate children or not, η : the average number of nodes in each tree in $\mathcal{D}$. We used the following default values for the parameters: the number of trees $\mathcal{T} = 10,000$, the number of labels $L = 100$, the maximal branch factor $f = 5$, and the maximum depth $d = 5$.

In our first experiment, we want to evaluate the scalability of our algorithm with varying minimum support and the number of trees $\mathcal{T}$, while other parameters are fixed as: $L = 100, f = 5, d = 5, \rho = 20\%, \eta = 13.8$ and 20.5 (when $\mathcal{T} = 10,000$ and $15,000$, respectively). Fig. 3 shows the result, where the minimum support is set from 10% to 0.0001%. We can find that the running time increases when the number of trees $\mathcal{T}$ increases, however, both running times are rarely affected by the decrease of the minimum support. With the threshold becoming smaller, there is no big difference in execution time for both datasets. This is because SEAMSON relies on the number of labels not the number of nodes. Thus it is very efficient for datasets with varying and growing tree sizes.

Fig. 3(b) shows the trends of memory usage at three phases. The transformation phase consumes the most amount of memory along with the growing

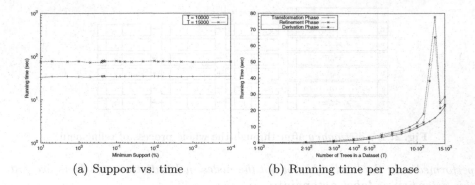

(a) Support vs. time   (b) Running time per phase

Fig. 3. Scalability and memory usage

number of trees. However, the derivation phase uses almost stabilized amount of memory during the experiments ranged from $\mathcal{T} = 10,000$ to $\mathcal{T} = 15,000$. Because the transformation phase is responsible for scanning an original trees dataset and converting them into list-based structures, the required memory size is getting larger along with the increasing of trees in a dataset $(\mathcal{T})$.

References

1. Asai, T., Abe, K., Kawasoe, S., Arimura, H., Satamoto, H., Arikawa, S.: Efficient Substructure Discovery from Large Semi-Strucutured Data. In: Proceedings of the 2nd SIAM International Conference on Data Mining, pp. 158–174 (2002)
2. Chi, Y., Yang, Y., Muntz, R.R.: Canonical Forms for Labeled Trees and Their Applications in Frequent Subtree Mining. Knowledge and Information Systems 8(2), 203–234 (2005)
3. Termier, A., Rousset, M.-C., Sebag, M.: TreeFinder: a First Step towards XML Data Mining. In: Proceedings of IEEE International Conference on Data Mining (ICDM 2002), pp. 450–457 (2002)
4. Wang, C., Hong, M., Pei, H., Zhou, H., Wang, W., Shi, B.: Efficient Pattern-Growth Methods for Frequent Tree Pattern Mining. In: Dai, H., Srikant, R., Zhang, C. (eds.) PAKDD 2004. LNCS (LNAI), vol. 3056, pp. 441–451. Springer, Heidelberg (2004)
5. Wang, K., Liu, H.: Schema Discovery for Semistructured Data. In: Proceedings of the 3rd International Conference on Knowledge Discovery and Data Mining (KDD 1997), pp. 271–274 (1997)
6. Zaki, M.J.: Efficiently Mining Frequent Trees in a Forest: Algorithms and Applications. IEEE Transactions on Knowledge and Data Engineering 17(8), 1021–1035 (2005)
7. Zou, L., Lu, Y., Zhang, H.: Mining Frequent Induced Subtrees by Prefix-Tree-Projected Pattern Growth. In: Yu, J.X., Kitsuregawa, M., Leong, H.-V. (eds.) WAIM 2006. LNCS, vol. 4016, pp. 18–25. Springer, Heidelberg (2006)

Mining the Web for Hyponymy Relations Based on Property Inheritance

Shun Hattori, Hiroaki Ohshima, Satoshi Oyama, and Katsumi Tanaka

Department of Social Informatics, Graduate School of Informatics, Kyoto University
Yoshida-Honmachi, Sakyo, Kyoto 606-8501, Japan
{hattori, ohshima, oyama, tanaka}@dl.kuis.kyoto-u.ac.jp

Abstract. Concept hierarchies, such as hyponymy and meronymy relations, are very important for various natural language processing systems. Many researchers have tackled how to mine very large corpora of documents such as the Web for them not manually but automatically. However, their methods are mostly based on lexico-syntactic patterns as not necessary but sufficient conditions of concept hierarchies, so they can achieve high precision but low recall when using stricter patterns or they can achieve high recall but low precision when using looser patterns. In this paper, property inheritance from a concept to its hyponyms is assumed to be necessary and sufficient conditions of hyponymy relations to achieve high recall and not low precision, and we propose a method to acquire hyponymy relations from the Web based on property inheritance.

1 Introduction

Conceptual hierarchies, such as hyponymy and meronymy relations, are very important for various natural language processing systems. For example, query expansion in information retrieval [1,2], WebQA [3], machine translation, object information extraction by text mining [4], and so forth.

While WordNet [5,6] and Wikipedia [7,8] are being manually constructed and maintained as lexical ontologies, many researchers have tackled how to extract concept hierarchies from very large corpora of text documents such as the Web not manually but automatically [9,10,11,12,13,14,15]. However, their methods are mostly based on lexico-syntactic patterns, e.g., "x such as y" and "y is a/an x", as sufficient but not necessary conditions of concept hierarchies. Therefore, they can achieve high precision but low recall when using stricter patterns or they can achieve high recall but low precision when using looser patterns.

To achieve high recall and not low precision, this paper proposes a method to extract hyponyms and hypernyms of a target concept from the Web based on property inheritance. Our method assumes property inheritance from a concept to its hyponyms to be necessary and sufficient conditions of hyponymy relations.

The remainder of the paper is organized as follows. First, Section 2 introduces one assumption on the relationship between hyponymy and property inheritance. Next, Section 3 and 4 describe our method of mining the Web for hyponyms and hypernyms based on it, respectively. Section 5 shows some experimental results to validate our method. Finally, we conclude the paper in Section 6.

Y. Zhang et al. (Eds.): APWeb 2008, LNCS 4976, pp. 99–110, 2008.
© Springer-Verlag Berlin Heidelberg 2008

2   Hyponymy and Property Inheritance

In this section, we introduce and discuss one assumption on the relationship between hyponymy and property inheritance before proposing our method of mining the Web for hyponymy relations (hyponyms and hypernyms of a target concept) based on property inheritance in the following sections.

Assumption 1. Let C be the universal set of concepts (words). We assume that if and only if a concept $x \in C$ is a superordinate (hypernym) of $y \in C$, in other words, a concept y is a subordinate (hyponym) of a concept x, then the set of properties that a concept y has, $P(y)$, completely includes the set of properties that a concept x has, $P(x)$:

$$\text{hyponym}(y, x) = \text{hypernym}(x, y) = \begin{cases} 1 & \text{if } P(x) \subseteq P(y) \text{ and } x \neq y, \\ 0 & \text{otherwise.} \end{cases}$$

$$P(c) = \{p_i \in P \mid \text{has}(p_i, c) = 1\},$$

where P stands for the universal set of N properties,

$$P = \{p_1, p_2, ..., p_N\},$$

and $\text{has}(p_i, c)$ indicates whether or not a concept $c \in C$ has a property $p_i \in P$,

$$\text{has}(p_i, c) = \begin{cases} 1 & \text{if a concept } c \text{ has a property } p_i, \\ 0 & \text{otherwise.} \end{cases}$$

In other words,

$$\text{hyponym}(y, x) = \begin{cases} 1 & \text{if } \sum_{p_i \in P} \text{has}(p_i, x) \cdot \text{has}(p_i, y) = \sum_{p_i \in P} \text{has}(p_i, x) \cdot \text{has}(p_i, x), \\ & \text{(i.e., } \boldsymbol{P}(x) \cdot \boldsymbol{P}(y) = \boldsymbol{P}(x) \cdot \boldsymbol{P}(x) \text{)} \\ 0 & \text{if } \sum_{p_i \in P} \text{has}(p_i, x) \cdot \text{has}(p_i, y) < \sum_{p_i \in P} \text{has}(p_i, x) \cdot \text{has}(p_i, x), \\ & \text{(i.e., } \boldsymbol{P}(x) \cdot \boldsymbol{P}(y) < \boldsymbol{P}(x) \cdot \boldsymbol{P}(x) \text{)} \end{cases}$$

$$\boldsymbol{P}(c) = (\text{has}(p_1, c), \text{has}(p_2, c), ..., \text{has}(p_N, c)).$$

It is very essential for hyponymy extraction based on the above assumption to calculate the binary value $\text{has}(p_i, c) \in \{0, 1\}$ of any concept and property accurately. However, it is not easy, and we can only use the continuous value $\text{has}(p_i, c) \in [0, 1]$ in this paper. Therefore, we suppose that the ratio of the number of properties that a concept y inherits from a concept x to the number of properties that a concept x has, $\frac{\sum_{p_i \in P} \text{has}(p_i, x) \cdot \text{has}(p_i, y)}{\sum_{p_i \in P} \text{has}(p_i, x) \cdot \text{has}(p_i, x)}$, can simulate how suitable a concept y is for a hyponym of a concept x, as an approximation of whether or not a concept y is a hyponym of a concept x, $\text{hyponym}(y, x)$. That is to say, we consider a concept y as a hyponym of a concept x when the ratio is enough near to one, while we consider that a concept y is not a hyponym of a concept x when the ratio is not near to one.

3   Hyponym Extraction Based on Property Inheritance

In this section, we propose a method to mine such a very large corpus of documents as the Web for hyponyms of a target concept, by using not only lexico-syntactic patterns as sufficient (but not necessary) conditions of hyponymy, but also property inheritance as its necessary and sufficient condition.

When a target concept $x \in C$ is given, our method executes the following three steps to extract its hyponyms from the Web. First, a set of candidates for its hyponyms, $C(x)$, is collected. Second, some weight of each pair of a property $p \in P$ and a concept c (the target concept x or its hyponym candidate $y \in C(x)$), has(p, c) is calculated by using Web search engine indices. That is, the property vector of each concept c, $P(c)$, is obtained. Last, some weight of each pair of the target concept x and its hyponym candidate $y \in C(x)$, hyponym(y, x), is calculated based on property inheritance, and then a set of its top k hyponym candidates ordered by their weight, $C_k(x)$, would be outputted to the users.

Step 1. Listing of candidates for hyponyms of a target concept

Our method has to collect a set of candidates for hyponyms of a target concept x, $C(x)$, as exhaustively as possible and enough precisely. If $C(x)$ is set to the universal set of concepts, C, its recall always equals to 1.0 (the highest value) but its precision nearly equals to 0.0 (too low value). Meanwhile, if $y \in C(x)$ is collected from some sort of corpus of documents by using too strict syntactic pattern such as "y is a kind of x", its precision is enough high but its recall is too low in most cases.

Therefore, our method collects a set of candidates for hyponyms of a target concept x, $C(x)$, from the Web by using not too strict but enough strict syntactic pattern such as "y is a/an x". In the after-mentioned experiments, any 2-gram noun phrase y where the syntactic pattern "y is a/an x" exists at least once in the title or summary text of the top 1000 retrieval results by submitting a phrase "is a/an x" as a query to Yahoo! Web search API [16] is inserted into $C(x)$ as a candidate for hyponyms of the target concept x.

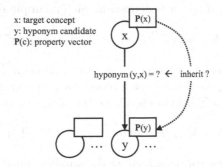

Fig. 1. Property Inheritance (PI) based hyponym extraction

Step 2. Extracting of typical properties of each concept

Our method has to obtain the property vector of each concept c (a target concept $x \in C$ or its hyponym candidate $y \in C(x)$), $\boldsymbol{P}(c)$. The weight of each pair of a property $p \in P$ and a concept c, $\mathrm{has}(p, c)$, is defined as follows:

$$\mathrm{has}(p, c) := \frac{\mathrm{df}(\text{``}c\text{'s } p\text{''})}{\mathrm{df}(\text{``}c\text{'s''}) + 1},$$

where $\mathrm{df}(q)$ stands for the number of documents that meet a query condition q in such a corpus as the Web. In the after-mentioned experiments, we calculate it by submitting each query to Yahoo! Web search API [16].

Note that $\mathrm{has}(p, c)$ is not a binary value $\{0, 1\}$ but a continuous value $[0, 1]$, so $\mathrm{has}(p, c)$ cannot indicate whether or not a concept c has a property p but it would indicate how typical a property p is of a concept c.

Step 3. Weighting of candidates for hyponyms of a target concept

In our hyponym extraction based on property inheritance, hyponym candidates of a target concept x, $C(x)$, that are collected in Step 1, have to be ordered based on how many properties a hyponym candidate $y \in C(x)$ inherits from the target concept x. The weight of a candidate y for hyponyms of a target concept x is defined as follows:

$$\mathrm{hyponym}_n^{PI}(y, x) := \frac{\sum\limits_{p_i \in P_n(x)} \mathrm{has}(p_i, x) \cdot \mathrm{has}(p_i, y)}{\sum\limits_{p_i \in P_n(x)} \mathrm{has}(p_i, x) \cdot \mathrm{has}(p_i, x)},$$

where $P_n(c)$ stands for a set of the top n typical properties of a concept c ordered by their weight $\mathrm{has}(p, c)$ that is calculated in Step 2. Note that if $n = N$, i.e., $P_n(x) = P_n(y) = P$, we cannot decide which x or y is subordinate to the other because of $\mathrm{hyponym}_n^{PI}(y, x) = \mathrm{hyponym}_n^{PI}(x, y)$.

In the after-mentioned experiments, we set n to 3 or 10 fixedly.

Meanwhile, in an existing hyponym extraction based on such a syntactic pattern as "y is a/an x", it is defined in such a simple manner as follows;

$$\mathrm{hyponym}^{SP}(y, x) := \mathrm{df}(\text{``}y \text{ is a/an } x\text{''}).$$

In the after-mentioned experiments, we use $\mathrm{hyponym}^{SP}(y, x)$ as a baseline to validate our hyponym extraction based on property inheritance.

Last, we define a hybrid of the above two kinds of weights as follows:

$$\mathrm{hyponym}_n^{HB}(y, x) := \log_2(\mathrm{hyponym}^{SP}(y, x) + 1) \cdot (\mathrm{hyponym}_n^{PI}(y, x) + \mu),$$

where μ stands for a constant holding the smallest positive nonzero value to prevent $\mathrm{hyponym}^{SP}(y, x)$ from being ineffective when $\mathrm{hyponym}_n^{PI}(y, x) = 0$. In the after-mentioned experiments, μ is set to $4.9 \cdot 10^{-324}$.

4   Hypernym Extraction Based on Property Inheritance

In this section, we propose a method to mine such a very large corpus of documents as the Web for hypernyms of a target concept, by using not only lexico-syntactic patterns as sufficient (but not necessary) conditions of hyponymy, but also property inheritance as its necessary and sufficient condition.

When a target concept $x \in C$ is given, our method executes the following three steps to extract its hypernyms from the Web. First, a set of candidates for its hypernyms, $C(x)$, is collected. Second, some weight of each pair of a property $p \in P$ and a concept c (the target concept x or its hypernym candidate $y \in C(x)$), has(p, c) is calculated by using Web search engine indices. That is, the property vector of each concept c, $P(c)$, is obtained. Last, some weight of each pair of the target concept x and its hypernym candidate $y \in C(x)$, hypernym(y, x), is calculated based on property inheritance, and then a set of its top k hypernym candidates ordered by their weight, $C_k(x)$, would be outputted to the users.

Step 1. Listing of candidates for hyponyms of a target concept

Our method has to collect a set of candidates for hypernyms of a target concept x, $C(x)$, as exhaustively as possible and enough precisely. On the same score as described in Step 1 of Section 3, we don't use the universal set of concepts, C, as $C(x)$, and we don't use too strict syntactic pattern such as "x is a kind of y" to collect it from the Web by text mining.

Therefore, our method collects a set of candidates for hypernyms of a target concept x, $C(x)$, from the Web by using not too strict but enough strict syntactic pattern such as "x is a/an y", to achieve enough high precision and not too low recall. In the after-mentioned experiments, any 1-gram noun phrase y where the syntactic pattern "x is a/an y" exists at least once in the title or summary text of at most 1000 retrieval results by submitting a phrase "x is a/an" as a query to Yahoo! Web search API [16] is inserted into $C(x)$ as a candidate for hypernyms of the target concept x.

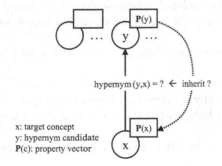

Fig. 2. Property Inheritance (PI) based hypernym extraction

Step 2. Extracting of typical properties of each concept

The same as described in Step 2 of Section 3. Our method has to obtain the property vector of each concept c (a target concept $x \in C$ or its hypernym candidate $y \in C(x)$), $P(c)$. The weight of each pair of a property $p \in P$ and a concept c, $has(p, c)$, is defined as follows:

$$has(p, c) := \frac{df(\text{``}c\text{'s } p\text{''})}{df(\text{``}c\text{'s''}) + 1},$$

where $df(q)$ stands for the number of documents that meet a query condition q in such a corpus as the Web. In the after-mentioned experiments, we calculate it by submitting each query to Yahoo! Web search API [16].

Note that $has(p, c)$ is not a binary value $\{0, 1\}$ but a continuous value $[0, 1]$, so $has(p, c)$ cannot indicate whether or not a concept c has a property p but it would indicate how typical a property p is of a concept c.

Step 3. Weighting of candidates for hyponyms of a target concept

In our hypernym extraction based on property inheritance, hypernym candidates of a target concept x, $C(x)$, that are collected in Step 1, have to be ordered based on how many properties the target concept x inherits from a hypernym candidate $y \in C(x)$. The weight of a candidate y for hypernyms of a target concept x is defined as follows:

$$hypernym_n^{PI}(y, x) := \frac{\sum_{p_i \in P_n(y)} has(p_i, y) \cdot has(p_i, x)}{\sum_{p_i \in P_n(y)} has(p_i, y) \cdot has(p_i, y)},$$

where $P_n(c)$ stands for a set of the top n typical properties of a target concept c ordered by their weight $has(p, c)$ that is calculated in Step 2. Note that $hypernym_n^{PI}(y, x)$ is equivalent to $hyponym_n^{PI}(x, y)$ defined in Step 3 of Section 3. In the after-mentioned experiments, we set n to 3 or 10.

Meanwhile, in a traditional hyponym extraction based on such a syntactic pattern as "x is a/an y", it is defined in such a simple manner as follows;

$$hypernym^{SP}(y, x) := df(\text{``}x \text{ is a/an } y\text{''}).$$

In the after-mentioned experiments, we use $hypernym^{SP}(y, x)$ as a baseline to validate our hypernym extraction based on property inheritance.

Last, we define a hybrid of the above two kinds of weights as follows:

$$hypernym_n^{HB}(y, x) := \log_2(hypernym^{SP}(y, x) + 1) \cdot (hypernym_n^{PI}(y, x) + \mu),$$

where μ stands for a constant holding the smallest positive nonzero value to prevent $hypernym^{SP}(y, x)$ from being invalid when $hypernym_n^{PI}(y, x) = 0$. In the after-mentioned experiments, μ is set to $4.9 \cdot 10^{-324}$.

5   Experiment

In this section, we show some experimental results to validate our method of mining the Web for hyponymy relations (hyponyms and hypernyms of a target concept) based on property inheritance, by comparing it with a traditional syntactic pattern based hyponymy extraction and the hybrid of them.

5.1   Experiment on Mining the Web for Hyponyms

We applied our method of mining the Web for hyponyms based on property inheritance to five kinds of concepts: "amphibian", "bird", "fish", "mammal", and "reptile". All of them are subordinate of "animal".

Table 1 shows the top 10 typical properties with their weight $\mathrm{has}(p,c)$ for each of five target concepts and "animal" (which is one of their common hypernyms). Note that their weights were calculated by using Yahoo! Web search API [16] in October of 2007. It shows that these subordinate concepts of "animal" inherit such a typical property of "animal" as "body", "head" and "life". The results seem to be also good for a meronymy extraction. For example, "beak" and "wings" seem to be characteristic components for "bird", "gills" is for "fish", and "milk" is for "mammal".

Table 1. Top 10 typical properties with their weight for each concept

k	amphibian		k	bird		k	fish	
1	skin	(0.01039)	1	eye	(0.44815)	1	mouth	(0.05737)
2	internal	(0.00825)	2	nest	(0.09907)	2	body	(0.02953)
3	chest	(0.00796)	3	foot	(0.01787)	3	tail	(0.02673)
4	life	(0.00782)	4	head	(0.01352)	4	head	(0.02240)
5	body	(0.00653)	5	**beak**	(0.01157)	5	belly	(0.01971)
6	habitat	(0.00604)	6	**wings**	(0.00821)	6	eye	(0.01626)
7	eye	(0.00467)	7	song	(0.00797)	7	ability	(0.01035)
8	assault	(0.00291)	8	body	(0.00701)	8	life	(0.01029)
9	descent	(0.00277)	9	nests	(0.00700)	9	**gills**	(0.01006)
10	heart	(0.00242)	10	life	(0.00647)	10	skin	(0.00819)

k	mammal		k	reptile		k	animal	
1	body	(0.02596)	1	habitat	(0.01304)	1	body	(0.04674)
2	notebook	(0.00846)	2	head	(0.01304)	2	life	(0.03774)
3	**milk**	(0.00688)	3	lair	(0.01170)	3	head	(0.02414)
4	blood	(0.00558)	4	body	(0.01141)	4	health	(0.02280)
5	heart	(0.00524)	5	skin	(0.00822)	5	behavior	(0.01665)
6	life	(0.00411)	6	eyes	(0.00785)	6	skin	(0.01467)
7	skeleton	(0.00408)	7	mouth	(0.00763)	7	people	(0.01345)
8	immune	(0.00363)	8	tail	(0.00563)	8	ability	(0.01253)
9	head	(0.00357)	9	hide	(0.00549)	9	owner	(0.01245)
10	brain	(0.00311)	10	cage	(0.00459)	10	neck	(0.01077)

Table 2. Top 5 hyponym candidates for each concept by Syntactic Pattern (SP) based vs. Property Inheritance (PI) based vs. Hybrid hyponym extraction. (Note that $|C(x)|$ stands for the number of candidates for hypernyms of a concept x, and $|S(x)|$ stands for the number of acceptable ones in $C(x)$).

\multicolumn{5}{c}{amphibian ($|C(x)| = 255$, $|S(x)| = 44$)}				
k	SP-based	PI-based (n=3)	PI-based (n=10)	Hybrid (n=3)	Hybrid (n=10)
1	frog	mink frog	bird	painted frog	bird
2	turtle	painted frog	chubby frog	man	frog
3	newt	sp100	mink frog	mink frog	animal
4	salamander	man	crested newt	frog	crested newt
5	toad	crested newt	vertebrate	animal	man

\multicolumn{5}{c}{bird ($|C(x)| = 434$, $|S(x)| = 140$)}				
k	SP-based	PI-based (n=3)	PI-based (n=10)	Hybrid (n=3)	Hybrid (n=10)
1	love	snake-bird	snake-bird	mind	mind
2	mind	mind	mind	ostrich	ostrich
3	chicken	ostrich	ostrich	snake-bird	snake-bird
4	eagle	crested crane	crested crane	cuckoo	cuckoo
5	ambition	cuckoo	cuckoo	eagle	eagle

\multicolumn{5}{c}{fish ($|C(x)| = 492$, $|S(x)| = 114$)}				
k	SP-based	PI-based (n=3)	PI-based (n=10)	Hybrid (n=3)	Hybrid (n=10)
1	fish tycoon	horse	horse	horse	horse
2	tycoon	intermediate host	natural fish	electric eel	whale
3	mother	electric eel	chilazon	lingcod	sea horse
4	school	chilazon	intermediate host	whale	lingcod
5	mccarthy school	lingcod	sea horse	intermediate host	electric eel

\multicolumn{5}{c}{mammal ($|C(x)| = 350$, $|S(x)| = 169$)}				
k	SP-based	PI-based (n=3)	PI-based (n=10)	Hybrid (n=3)	Hybrid (n=10)
1	cat	brazilian tapir	brazilian tapir	brazilian tapir	cow
2	whale	pilot whale	pilot whale	cow	brazilian tapir
3	dog	goat	goat	raccoon	vertebrate
4	leo	cow	cow	vertebrate	raccoon
5	panthera leo	animal species	vertebrate	goat	goat

\multicolumn{5}{c}{reptile ($|C(x)| = 283$, $|S(x)| = 65$)}				
k	SP-based	PI-based (n=3)	PI-based (n=10)	Hybrid (n=3)	Hybrid (n=10)
1	snake	dragon	horse	snake	snake
2	maher	serpent	dragon	dragon	dragon
3	bill maher	snake	snake	serpent	alligator
4	pet zoo	viper	alligator	alligator	serpent
5	zoo	alligator	serpent	turtle	animal

Table 2 compares the top 5 hyponym candidates for each target concept by syntactic pattern based, property inheritance based ($n = 3, 10$), or hybrid ($n = 3, 10$) hyponym extraction. It shows that our hyponym extraction based on property inheritance is superior to a traditional hyponym extraction based on such a syntactic pattern as "is a/an", and the hybrid of them seems to be somewhat superior to both of them.

Figure 3 shows precision-recall graphs for each target concept and the 21-point interpolated average for all of five target concepts by syntactic pattern based, property inheritance based ($n = 3, 10$), or hybrid ($n = 3, 10$) hyponym extraction. Our hyponym extraction based on property inheritance and the hybrid are completely superior to a traditional hyponym extraction based on such a syntactic pattern as "is a/an". It also shows that the bigger number of typical properties for the calculation of property inheritance from a target concept to its hyponym candidate, n, the more precise result, and that the hybrid ($n = 10$) is superior to the others in the middle recall range (about from 0.2 to 0.9).

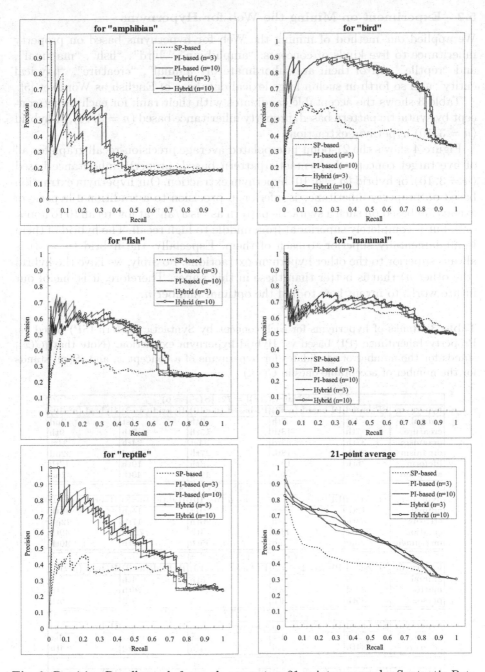

Fig. 3. Precision-Recall graph for each concept or 21-point average by Syntactic Pattern (SP) based vs. Property Inheritance (PI) based vs. Hybrid hyponym extraction

5.2   Experiment on Mining the Web for Hypernyms

We applied our method of mining the Web for hypernyms based on property inheritance to five kinds of concepts: "amphibian", "bird", "fish", "mammal", and "reptile". All of them are subordinate of "animal", "creature", "physical entity" and so forth in such a large lexical database of English as WordNet [5].

Table 3 shows the acceptable candidates with their rank for each target concept by syntactic pattern based, property inheritance based ($n = 3, 10$), or hybrid ($n = 3, 10$) hypernym extraction.

Figure 4 shows the 21-point interpolated average precision-recall graph for all of five target concepts by syntactic pattern based, property inheritance based ($n = 3, 10$), or hybrid ($n = 3, 10$) hypernym extraction. Our hypernym extraction based on property inheritance is inferior in precision to a traditional hypernym extraction based on such a syntactic pattern as "is a/an" at low recall (0 to about 0.2), but is completely superior at from middle to high recall. The hybrid of them is fundamentally superior to each of them. Especially, the hybrid ($n = 3$) is always superior to the other hypernym extractions. Possibly, we have the hybrid (the other n) that is better than these in this paper. Therefore, it is one of our future works to invent how to find the optimal number n.

Table 3. Ranks of hypernyms for each concept by Syntactic Pattern (SP) based vs. Property Inheritance (PI) based vs. Hybrid hypernym extraction. (Note that $|C(x)|$ stands for the number of candidates for hypernyms of a concept x, and $|S(x)|$ stands for the number of acceptable ones in $C(x)$.)

amphibian ($|C(x)| = 72$, $|S(x)| = 6$)					
hypernym	SP-based	PI-based (n=3)	PI-based (n=10)	Hybrid (n=3)	Hybrid (n=10)
animal	1st	7th	9th	1st	2nd
creature	7th	13th	15th	7th	9th
organism	37th	23rd	31st	24th	31st
organisms	70th	18th	27th	23rd	32nd
lifeform	36th	19th	13th	19th	17th
aeroplane	31st	42nd	46th	40th	42nd

bird ($|C(x)| = 144$, $|S(x)| = 4$)					
hypernym	SP-based	PI-based (n=3)	PI-based (n=10)	Hybrid (n=3)	Hybrid (n=10)
animal	12th	15th	21st	11th	16th
creature	38th	17th	20th	17th	22nd
creation	101st	71st	44th	72nd	49th
instrument	2nd	52nd	55th	37th	40th

fish ($|C(x)| = 123$, $|S(x)| = 3$)					
hypernym	SP-based	PI-based (n=3)	PI-based (n=10)	Hybrid (n=3)	Hybrid (n=10)
animal	7th	6th	5th	4th	2th
matter	33rd	79th	60th	76th	57th
person	35th	8th	12th	8th	9th

mammal ($|C(x)| = 142$, $|S(x)| = 2$)					
hypernym	SP-based	PI-based (n=3)	PI-based (n=10)	Hybrid (n=3)	Hybrid (n=10)
animal	2nd	1st	1st	1st	1st
creature	34th	3rd	4th	3rd	4th

reptile ($|C(x)| = 108$, $|S(x)| = 2$)					
hypernym	SP-based	PI-based (n=3)	PI-based (n=10)	Hybrid (n=3)	Hybrid (n=10)
animal	6th	7th	9th	2nd	3rd
organism	39th	34th	38th	36th	37th

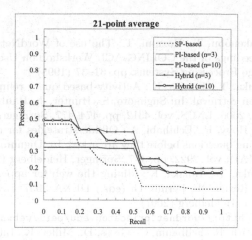

Fig. 4. Average precision-recall graph for five concepts by Syntactic Pattern (SP) based vs. Property Inheritance (PI) based vs. Hybrid hypernym extraction

6 Conclusion and Future Work

In this paper, to achieve high recall and not low precision in automatic acquisition of hyponymy relations, we assumed property inheritance from a concept to its hyponyms to be necessary and sufficient conditions of hyponymy relations, and then we proposed a method to mine the Web for hyponyms and hypernyms of a target concept based on property inheritance. The experimental results show that our hyponymy extraction based on property inheritance is superior to a commonly-used hyponymy extraction based on such a lexico-syntactic pattern as "is a/an", and the hybrid of them is more superior to each of them.

In the future, we plan to refine a method to extract typical properties for each concept, to invent how to find the optimal number of typical properties for the calculation of property inheritance from a target concept to its hyponym candidate, and moreover, to utilize coordinate terms [17] for more robust method.

Acknowledgments. This work was supported in part by (1) a Grant-in-Aid for JSPS (Japan Society for the Promotion of Science) Fellows (Grant#: 1955301); (2) a MEXT (Ministry of Education, Culture, Sports, Science and Technology) Grant-in-Aid for Scientific Research on Priority Areas: "Cyber Infrastructure for the Information-explosion Era," Planning Research on "Contents Fusion and Seamless Search for Information Explosion" (Project Leader: Katsumi Tanaka, A01-00-02, Grant#: 18049041); and (3) the Kyoto University Global COE (Center of Excellence) Program: Informatics Education and Research Center for Knowledge-Circulating Society (Program Leader: Katsumi Tanaka).

References

1. Mandala, R., Takenobu, T., Hozumi, T.: The use of WordNet in information retrieval. In: Proceedings of the COLING-ACL Workshop on Useage of WordNet in Natural Language Processing Systems, pp. 31–37 (1998)
2. Hattori, S., Tezuka, T., Tanaka, K.: Activity-based query refinement for context-aware information retrieval. In: Sugimoto, S., Hunter, J., Rauber, A., Morishima, A. (eds.) ICADL 2006. LNCS, vol. 4312, pp. 474–477. Springer, Heidelberg (2006)
3. Fleischman, M., Hovy, E., Echihabi, A.: Offline strategies for online question answering: Answering questions before they are asked. In: Dignum, F.P.M. (ed.) ACL 2003. LNCS (LNAI), vol. 2922, pp. 1–7. Springer, Heidelberg (2004)
4. Hattori, S., Tezuka, T., Tanaka, K.: Mining the web for appearance description. In: Wagner, R., Revell, N., Pernul, G. (eds.) DEXA 2007. LNCS, vol. 4653, pp. 790–800. Springer, Heidelberg (2007)
5. WordNet (2007), http://wordnet.princeton.edu/perl/webwn
6. Miller, G., Beckwith, R., Fellbaum, C., Gross, D., Miller, K.: Introduction to wordnet: An online lexical database. International Journal of Lexicography 3(4), 235–312 (1990)
7. Wikipedia (2007), http://www.wikipedia.org/
8. Völkel, M., Krötzsch, M., Vrandecic, D., Haller, H., Studer, R.: Semantic wikipedia. In: Proceedings of the 15th International Conference on World Wide Web (WWW 2006), pp. 585–594 (2006)
9. Hearst, M.A.: Automatic acquisition of hyponyms from large text corpora. In: Proceedings of the 14th International Conference on Computational Linguistics (COLING 1992), pp. 539–545 (1992)
10. Caraballo, S.A.: Automatic construction of a hypernym-labeled noun hierarchy from text. In: Proceedings of the 37th annual meeting of the Association for Computational Linguistics on Computational Linguistics (ACL 1999), pp. 120–126 (1999)
11. Sanderson, M., Croft, B.: Deriving concept hierarchies from text. In: Proceedings of the 22nd Annual International ACM SIGIR Conference on Research and Development in Information Retrieval, pp. 206–213 (1999)
12. Morin, E., Jacquemin, C.: Automatic acquisition and expansion of hypernym links. Computers and the Humanities 38(4), 363–396 (2004)
13. Shinzato, K., Torisawa, K.: Automatic acquisition of hyponymy relations from HTML documents. Japanese Journal of Natural Language Processing 12(1), 125–150 (2005)
14. Kim, H., Kim, H., Choi, I., Kim, M.: Finding relations from a large corpus using generalized patterns. International Journal of Information Technology 12(7), 22–29 (2006)
15. Ruiz-Casado, M., Alfonseca, E., Castells, P.: Automatising the learning of lexical patterns: An application to the enrichment of wordnet by extracting semantic relationships from wikipedia. Data & Knowledge Engineering 61(3), 484–499 (2007)
16. Yahoo! (2007), http://api.search.yahoo.co.jp/WebSearchService/V1/webSearch
17. Ohshima, H., Oyama, S., Tanaka, K.: Searching coordinate terms with their context from the web. In: Aberer, K., Peng, Z., Rundensteiner, E.A., Zhang, Y., Li, X. (eds.) WISE 2006. LNCS, vol. 4255, pp. 40–47. Springer, Heidelberg (2006)

Detecting Outliers in Categorical Record Databases Based on Attribute Associations

Kazuyo Narita and Hiroyuki Kitagawa

[1] Graduate School of Systems and Information Engineering
University of Tsukuba
1-1-1 Tennodai, Tsukuba, Ibaraki, Japan
narita@kde.cs.tsukuba.ac.jp
[2] Graduate School of Systems and Information Engineering
Center for Computational Sciences
University of Tsukuba
1-1-1 Tennodai, Tsukuba, Ibaraki, Japan
kitagawa@cs.tsukuba.ac.jp

Abstract. Outlier detection, a data mining technique to detect rare events, deviant objects, and exceptions from data, has been drawing increasing attention in recent years. Most existing outlier detection algorithms focus on numerical data sets. We target categorical record databases and detect records in which many attribute values are not observed even though they should occur in association with other attribute values in the records. To detect such records as outliers, we provide an outlier degree, which demonstrates sufficient detection performance in accuracy-evaluation experiments compared with the probabilistic approach used in a related work. We also propose an efficient algorithm for detecting such outlier records. Experiments using real data sets show that our method detects interesting records as outliers.

Keywords: outlier detection, association rules, categorical record data.

1 Introduction

With the growth of information technology, data accessible through the Internet has become increasingly voluminous and diverse. It is impossible for humans to check this much data by hand and identify important information. Thus, data mining, which systematically extracts useful information from data, has become essential. Outlier detection, a data mining technique to detect rare events, deviant objects, and exceptions from data, has been drawing increasing attention in recent years. Most existing outlier detection algorithms focus on numerical data sets [1, 2, 3, 4, 5, 6, 7, 8, 9, 10].

We target categorical record databases and detect records in which many attribute values are not observed even though they should occur in association with other attribute values in the records. To detect such records as outliers, we provide an outlier degree based on associations among attribute value sets.

Y. Zhang et al. (Eds.): APWeb 2008, LNCS 4976, pp. 111–123, 2008.

We derive a set of association rules with high confidence from an input record database as information about such attribute associations. The outlier degree we propose demonstrates sufficient detection performance in accuracy-evaluation experiments.

Beyond that, since a naive brute force algorithm for detecting outlier records based on attribute associations takes much time, we propose a more efficient algorithm, which includes two, speed increasing devices as well as a fast path for finding cases in which no outliers exist in the data set.

Experiments using real data sets have yielded three interesting results: (i) The outlier degree we derive presents sufficient accuracy compared with the probabilistic approach of a related work [14]. (ii) Our method detects practically interesting records as outliers. (iii) Our proposed algorithm detects outliers at least ten times faster than a brute force algorithm.

The rest of the paper is organized as follows: Section 2 mentions related work, where we introduce a study used in our experiments as the baseline for the accuracy evaluation. Section 3 explains preliminaries for this paper and defines terms and statements. Section 4 introduces a formula of the outlier degree based on the association among attribute value sets. Section 5 presents an algorithm for outlier detection, which includes a fast path and two speed increasing devices. Section 6 shows experimental results for real data sets. Section 7 summarizes the paper and introduces future work.

2   Related Work

Until now, much research on outlier detection has been proposed [1, 2, 3, 4, 5, 6, 7, 8, 9, 10]. However, most research has targeted numerical data and differs from our research, which targets categorical record databases. [11, 12, 13, 14] is a study of outlier or anomaly detection, which applies to categorical record databases.

Approaches proposed in [11,12] are based on a bayesian network. They build a probabilistic model and estimate the likelihood for each record. If the likelihood is unnaturally low, the corresponding record is regarded as an anomaly. Just as with our method, Chan and others propose an approach based on association rules. However, while our outlier degree is not based on statistics, they derive an anomaly score on the basis of the probabilistic concept [13]. The method [13] is used as the baseline for an accuracy comparison in [14]. According to results presented in [14], the method of [14] significantly outperforms [13]. Therefore we use the method of [14] as the baseline for our accuracy comparison.

As far as we know, the work [14] is the hottest in this study field. They detect anomalous records in a categorical database using a probabilistic method. Their idea is that a given test record is interesting as an anomaly if it has unusual combinations for an attribute values set. For a training database, their method counts the number of records containing each attribute value set and keeps the counts. With these counts, they calculate the rareness degree of each test record. For each combination of attribute value sets A and B contained in a test record, they figure the ratio of the marginal probabilities $\frac{P(A,B)}{P(A)P(B)}$. The minimal ratio is regarded as the rareness degree of the record. If this degree is less than or equal to a user-specified threshold α, the record is regarded as an anomaly. Marginal probabilities are estimated with the counts of records that contain A, B and $A \cup B$ respectively. The time and space costs for calculations are very large because the number of attribute value sets is huge, so they give parameter k as a maximum size of an attribute values set for the calculation.

3 Preliminaries

Here, we describe terms and statements used in this paper. A categorical record data DB is a set of records. $|DB|$ is the number of all records on DB. A record $t \in DB$ is a tuple. $t[A_i]$ indicates the value of the i-th attribute A_i on DB that t has. Assume n be the number of all attributes of DB.

Let A be an attribute and let a be a value of A. Then, a categorical record database can be regarded as a set of n-*itemsets*, where an *item* corresponds to a pair (A, a). For an itemset X, the *support* of X is $sup(X) = \frac{\{t | \forall (A,a) \in X, t[A]=a\}}{|DB|}$. For a *minimal support msup*, if $sup(X) \geq msup$, then X is called a *frequent itemset* (FI). A *Maximal frequent itemset* (MFI) is an FI having no super set in a set of all FIs. Let X and Y be itemsets such that $X \cap Y = \emptyset$, the form $X \to Y$ is called an *association rule*. The left hand itemset of a rule is called an *antecedent*, while the right hand is called the *consequent*. The rule's confidence is derived as $conf(X \to Y) = \frac{sup(X \cup Y)}{sup(X)}$. For a *minimal confidence mconf*, if $conf(X \to Y) \geq mconf$ then $X \to Y$ is called a *high-confidence rule*.

For an asymmetric attribute A, where a specific attribute value a occurs very frequently and the other values rarely occur, such a frequent value may hide really meaningful associations among attribute values. To tackle such sparce data, we can consider the following option when mapping a record into an itemset.

[Mapping Option] We ignore (A, a) in the mapping and consider only (A, b) where $b \neq a$.

4 Outlier Degree

We want to detect records in which many attribute values are not observed even though they should occur in association with other attribute values in the records. We use association rules with high confidence to determine the degree of associations among attribute values set included in a record. It is obvious that as $conf(X \to Y)$ is higher, the association between attribute values of X and those of Y is stronger. For a record t and a rule $X \to Y$ with high confidence, if $t[A]$ is a for all pairs $(A, a) \in X$ while $t[B]$ is not b for some pairs $(B, b) \in Y$, it can be thought to be rare. This is true because $t[B]$ should be b for all $(B, b) \in Y$, which tends to occur with X stochastically. We call such a rule an *unobserved association*.

Definition 1. *Unobserved Association*
Let R be a set of association rules and let Z be an itemset, a rule $X \rightarrow Y \in R$ is Z's unobserved association when $Y \not\subseteq Z$ although $X \subseteq Z$.

To determine the degree of outlier for a record t, we first search how many attribute values are not observed in t even though t should have them. For a record t, we focus on t's *associative cover*, which covers all attribute values in association with attribute values occurring in t.

Definition 2. *Associative Cover*
Let R be a set of association rules. An initial associative cover of t is derived as $C_t{}^0 = \{(A_i, t[A_i])\}$, which contains the complete information about attribute values in t. t's associative cover $C_t{}^+$ is derived as follows:

$$C_t{}^{i+1} = C_t{}^i \bigcup\nolimits_{X \rightarrow Y \in R, X \subseteq C_t{}^i} Y$$
$$C_t{}^+ = C_t{}^\infty \tag{1}$$

If this cover expands larger, it means that more attribute values in t have stronger association with other attribute values not in t. In other words, such a t can be regarded as rare.

Definition 3. *Outlier Degree*
t's outlier degree is defined as $od(t) = \frac{|C_t{}^+ - C_t{}^0|}{|C_t{}^0|}$. For a minimal outlier degree mod, if $od(t) \geq mod$ then the record t is detected as an outlier.

5   Detection Algorithms

Table 1 lists the naive algorithm for detecting outlier records. It calls function getOutliers listed in Table 2. With this brute force algorithm, for each record $t \in DB$, we have to check all association rules in a given R to make the associative cover according to Definition 2.

Table 1. Brute Force Algorithm

input: DB, $msup$, $mconf$, mod
output: a complete set of outliers O
1. mine a set of FIs F;
2. generate a set of high-confidence rules R;
3. $O = $ getOutliers(DB, R);

Table 2. getOutliers

Function getOutliers(a records set DB, a rules set R)
1. $O = \emptyset$;
2. foreach $t \in DB$
3.   make the initial associative cover $C_t{}^0$;
4. $i = 0$, $R' = R$;
5.   until $C_t{}^i$ stops growing
6. $TMP = \emptyset$;
7.   foreach $r = X \rightarrow Y \in R'$
8.   if $X \subseteq C_t{}^i$ then
9.   append Y to TMP and deleate r from R';
10. $C_t{}^{i+1} = C_t{}^i \cup TMP$, $i++$;
11. $od(t) = \frac{|C_t{}^i - C_t{}^0|}{|C_t{}^+|}$
12.   if $od(t) \geq mod$, append t to O;
13. return O

As Table 1 shows, our method requires three parameters: $msup$ for mining frequent itemsets, $mconf$ for generating high-confidence rules, and mod for outlier detection. Experimental results show that the distribution of outlier degrees changes depending on $msup$ and $mconf$. Given an inadequate mod for $msup$ and $mconf$, there will be no outliers. We want to find such cases without spending time and resources in vain. Consequently, we provide a fast path for finding the case of no outliers existing in DB.

In addition, the naive algorithm obviously takes $O(|DB| \times |R|^2)$, where both $|DB|$ and $|R|$ are generally vast. Thus our algorithm includes two, speed increasing devices, which decrease the number of records and rules, which are needed to calculate the outlier degree. Below, we first explain about the fast path, then present two devices and introduce the proposed algorithm.

5.1 Fast Path for Inadequate Parameters

We can find the case of no outliers existing in the data sets as soon as we obtain frequent itemsets, by focusing on the following property of outlier degree.

Property 1. For a given record t, the outlier degree $od(t)$ increases monotonically as $|C_t^+|$ increases.

Now, let us consider the case that $|C_t^0| = n$ for all records t. This is the case in which we do not use the Mapping Option.

Definition 4. *Maximum Increment*
Let F be a set of FIs for $msup$, the maximum increment is the itemset $I_{\geq 2}$ derived as $I_{\geq 2} = \{i | X \in F \wedge |X| \geq 2 \wedge i \in X\}$. The cardinality $max\_inc = |I_{\geq 2}|$ is called a maximum increment value.

Note that $C_t^+ - C_t^0 \subseteq I_{\geq 2}$ for each t.

Property 2. For a set F of FIs for $msup$, the outlier degree $od(t)$ of a record t is always less than mod when $max\_inc < \frac{mod}{1-mod}n$.

On the mining phase, we derive max_inc as well as a set of FIs. If $max\_inc < \frac{mod}{1-mod}n$, we can say that there is no outlier.

In the case where we use the Mapping Option, we can use the minimal cardinality $n_{min} = min_{t \in DB}|C_t^0|$ in place of n and check mix_inc in a similar way.

5.2 Pruning Candidate Records of Outliers

By deriving outlier degree upper bounds, we prune candidate records of outliers.

Definition 5. *Maximal Associative Cover*
Let M be a set of all MFIs for $msup$, and let $MC_t^0 = C_t^0$. For a record t, t's maximal associative cover MC_t^+ is derived as follows:

$$MC_t^{i+1} = MC_t^i \cup (\bigcup_{X \in M, X \cap MC_t^i \neq \emptyset} X)$$

$$MC_t^+ = MC_t^\infty \tag{2}$$

Property 3. Let M be a set of all MFIs for $msup$, and let MC_t^+ be the maximal associative cover of a record t for M. MC_t^+ is equal to t's associative cover C_t^+ with the same $msup$ and $mconf = 0$.

Because C_t^+ can grow most when $mconf = 0$ for $msup$, we derive an upper bound of an outlier degree as follows.

Definition 6. *Outlier Degree Upper Bound*
For $msup$ and $mconf$, let R be a set of high-confidence rules and let M be a complete set of MFIs. The outlier degree upper bound $od_{max}(t)$ of a record t is $od_{max}(t) = \frac{|MC_t^+ - C_t^0|}{|MC_t^0|}$.

Before outlier calculation, we check all records and prune such a record t that $od_{max}(t) < mod$, since $od(t) < mod$ holds for t by Property 1. Increased speed can be expected, because $|M|$ is generally much less than $|R|$.

Table 3 is an example record data set in which each record shows a habit of an individual animal. Table 4 represents its attributes. Given $msup = 30\%$, we derive MFIs as shown in Table 5. Assuming the Mapping Option, the initial associative cover of record t_3 is generated to be $C_{t_3}^0 = \{(\text{\# of Legs, 4}), (\text{Toothed, T}),$ $(\text{Eggs, T}), (\text{Aquatic, T})\}$. Thus, we know $MC_{t_3}^+$ and $C_{t_3}^+$ are as follows:

$MC_{t_3}^+ = \{(\text{\# of Legs, 4}), (\text{Toothed, T}), (\text{Eggs, T}), (\text{Aquatic, T}), (\text{Milk,T}), (\text{Airborne, T}),$

$(\text{\# of Legs, 6})\}$.

$C_{t_3}^+ = \{(\text{\# of Legs, 4}), (\text{Toothed, T}), (\text{Eggs, T}), (\text{Aquatic, T}), (\text{Milk,T}), (\text{Airborne, T})\}$.

Therefore, $od_{max}(t_3) = 0.43$, on another front, $od(t_3) = 0.33$

Table 3. Animal Data

ID	Tuples					
t_1	F	2	F	T	T	F
t_2	F	6	F	T	T	F
t_3	F	4	T	T	F	T
t_4	F	0	T	T	F	F
t_5	T	4	T	F	F	T
t_6	F	6	F	T	T	F
t_7	T	4	T	F	F	F
t_8	F	2	F	T	T	F
t_9	T	4	T	F	F	F
t_{10}	F	6	F	T	T	F

Table 4. Attributes of Animal Data

ID	Attributes	Values
A_0	Milk	boolean
A_1	# of Legs	multiple
A_2	Toothed	boolean
A_3	Eggs	boolean
A_4	Airborne	boolean
A_5	Aquatic	boolean

Table 5. MFIs

MFIs
{(Milk,T), (# of Legs, 4), (Toothed, T)}
{(# of Legs, 6), (Eggs, T), (Airborne, T)}

Table 6. Generated Rules

ID	Rules
$rule_0$	{(Toothed,T)}→{(# of Legs,4)}
$rule_1$	{(# of Legs,4)}→{(Milk,T)}
$rule_2$	{(# of Legs,4),(Toothed,T)}→{(Milk,T)}
$rule_3$	{(# of Legs,4)}→{(Milk,T),(Toothed,T)}
$rule_4$	{(Eggs,T)}→{(Airborne,T)}

5.3 Redundant Rules Removal

In R, which is a set of association rules for a given $msup$ and $mconf$, there are redundant rules to make associative covers. Again we take Table 3 as an example. Given $msup = 30\%$ and $mconf = 70\%$, we derive the set of high-confidence rules listed in Table 6, where rules with 100% confidence are ignored because they do not become unobserved associations nor do they contribute to the growth of associative covers.

In this case, $rule_1$ is redundant because the growth by $rule_1$ toward the associative cover is subsumed by $rule_3$. Also, $rule_2$ is a redundant rule in the light of $rule_3$.

Definition 7. *Redundant Rule*

Let R be a set of association rules, a rule $X \to Y \in R$ is a redundant rule on R when it has the other rule $V \to W \in R$, which satisfies at least one of following two conditions: (i) $V \subset X \wedge V \cup W = X \cup Y$, or, (ii) $V = X \wedge W \supset Y$.

We remove such redundant rules on the rule generation phase, and check only remaining rules for making associative covers. A set of such remaining rules is called a *minimal rule set*. In the case of Table 6, a minimal rule set is $\{rule_0, rule_3, rule_4\}$.

Let $C_t{}^+$ be t's associative cover generated by a set of association rules R and let $C_{t,min}{}^+$ be the one generated by a minimal rule set R_{min} for R. Note that $C_{t,min}{}^+ = C_t{}^+$.

Table 7. Proposed Algorithm

input: DB, $msup$, $mconf$, mod and n (or n_{min})
output: a complete set of outliers O

1. get a set of FIs F, that of MFIs M and max_inc;
2. /* fast path for the case of no outliers existing in DB */
3. if $max\_inc < \frac{mod}{1-mod}n$ then exit;
4. /* get a set of candidate records C */
5. foreach $t \in DB$
6.   make t's maximal associative cover $MC_t{}^+$;
7.   if $od_{max}(t) \geq mod$, append t to C;
8. if $C = \emptyset$ then exit;
9. /* get a minimal rule set R_{min} */
10. foreach $l_k \in F$ ($k \geq 2$)
11. $H_1 = \{\{h \in l_k\}|conf((l_k - \{h\}) \to \{h\}) \geq mconf\}$;
12. $R_{min} =$ genMinRule(l_k, H_1);
13. foreach no-marked $r_1 \in R_{min}$ and the other $r_2 \in R_{min}$
14.   if r_2 holds Condition (ii) in Definition 7 then mark r_1;
15. remove marked rules from R_{min};
16. $O =$ getOutliers(C, R_{min});

Table 8. genMinRule

Function genMinrule(FI l_k, set of m-itemsets H_m)

1. if $k > m + 1$
2. $TMP_1 = \emptyset$;
3. $H_{m+1} =$ apriori-gen(H_m);
4.   foreach $h_{m+1} \in H_{m+1}$
5. $r = h_{m+1} \to (l_k - h_{m+1})$;
6.   if $conf(r) \geq mconf$ then append r to TMP_1;
7.   else $H_{m+1} = H_m - h_{m+1}$;
8. $TMP_2 =$ genMinRule(l_k, H_{m+1});
9.   foreach $r_1 = X \to Y \in TMP_1$
10.   foreach $r_2 = V \to W \in TMP_2$
11.   if r_2 holds Condition (i) in Definition 7 then
12.   mark r_1;
13. return $TMP_1 \cup TMP_2$;

5.4   Outlier Records Detection Algorithm

Table 7 shows our proposed algorithm, which includes a fast path and the two, speed increasing devices explained above.

At line 1, the algorithm derives MFIs and max_inc as well as FIs. To derive MFIs concurrently while getting FIs, we improve the FP-growth algorithm [15] for implementation in which the MFI-tree and subset-checking method proposed in [17] are utilized. If $max\_inc < \frac{mod}{1-mod}n$, it exits (line 2). At line 5-7, candidate records of outliers are pruned. The manner to calculate outlier degree upper bounds is very similar to that of outlier degree calculation. The process for getting a minimal rule set corresponds to lines 10-15, which is based on the algorithm of [16].

While the function genMinRule listed in Table 8 is called recursively, we mark high-confidence rules, which are redundant for Condition (i) in Definition 7. The function apriori-gen called at line 3 in Table 8 is proposed in [16], which generates all possible $(m+1)$-itemsets from H_m. When the process exits the recursion, we again mark no-marked high-confidence rules, which are redundant for Condition (ii) in Definition 7 (line 13-14). Finally we remove all redundant rules marked so far and get a minimal rule set (line 15). At the last line, we call function getOutliers to calculate outlier degrees for only candidate records and derive a set of outlier records.

6 Experiments

We first introduce real data sets used in experiments, and then show results of the outlier records validation, accuracy comparison against [14] and runtime comparison between proposed algorithms and the brute force algorithm in that order.

6.1 Data Sets

We use three real data sets: Zoo, Mushroom and KDD Cup 99. All are derived from the UCI Machine Learning Repository [18].

Zoo is a simple database about the habits of animals; it contains 15 boolean-valued attributes and 2 non-binary attributes. It is almost binary and sparce. Each record corresponds to an individual animal classified into one of 7 classes: mammal, bird life, crawler, amphibia, fin, bug and arthropods other than bug. It has 101 records. We use 16 attributes excepting the animal name and classes. Moreover, to make runtime comparison easier, we prepare a new data set Zoo10000, which has 10,000 copies of each record in Zoo. Note that support values of each itemset on Zoo10000 equals those on Zoo.

Mushroom is a record data set about the botany of mushrooms. All attributes are nominal-valued. Each record belongs to class edible or class poisonous. Originally, the number of instances for the two classes is almost equal (4208 edible class records and 3916 poisonous class records). We call this original data set Original Mushroom. For the accuracy comparison, we regard poisonous class records as true outliers, delete the class attribute and decrease the number of poisonous records so that the ratio of poisonous class records is approximately 1%. We call this data set Mushroom.

KDD Cup 99 is a network connection data set, which includes a wide variety of intrusions simulated in a military network environment. Each record is classified into the normal class or one of the intrusion Classes, such as guess password, warezmaster and so on. The existing work [14], which we use as the baseline method for comparison of our detection performance, uses this data set for a detection accuracy experiment. Most attributes take continuous values, which are put into to 5 discrete levels. We selected records in the guess password class as the true outliers set and created a data set KDD Cup 99, which contains 97% normal records and 3% outlier records.

6.2 Distribution of Outlier Degrees

First, to observe distributions of outlier degrees, we calculate the outlier degrees for three data sets. They are presented in Figs. 1(a) to 1(c). We fix $msup$ as 10% for Zoo assuming the Mapping Option, 20% for Mushroom, and 95% for KDD

Cup 99, and calculate outlier degrees for some values of $mconf$. The horizontal axis of these figures corresponds to records sorted in the descending order of their outlier degrees. The vertical axis presents an outlier degree. From these figures, we know that as $mconf$ becomes larger, the variance becomes smaller, i.e., fewer records have a relatively large outlier degree value. In Fig. 1(b), outlier degrees for $mconf = 0\%$, i.e., the upper bounds, are very high and similar to each other. We think the reason is that cardinalities of associative covers are nearly equal to the number of attributes. In Fig. 1(c), the line for $mconf = 95\%$ perfectly corresponds with that for $mconf = 0\%$ because association among attribute values is extremely strong in KDD Cup 99.

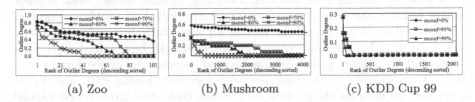

(a) Zoo   (b) Mushroom   (c) KDD Cup 99

Fig. 1. Outlier Degrees Distribution

6.3 Validation of Outliers

We apply our method (called the AA Method) to Zoo and validate whether detected records (animals) can really be regarded as outliers.

Table 9 shows the top 3 outliers in Zoo assuming the Mapping Option with $msup = 20\%$ and $mconf = 90\%$, where the first column denotes the corresponding animal name, the second column represents initial associative covers $C_t^{\,0}$ and the third column lists typical unobserved associations of $C_t^{\,0}$. Because the Mapping Option is assumed, false values are ignored here. Due to space limitations here, we denote an item for a boolean-valued attribute as just the attribute. Although the Crab fundamentally has 8 legs, it is thought that item (# of Legs, 4) is used to mean "4 pairs of legs." However, item (# of Legs, 4) normally means "4 legs." Usage to mean "4 pairs of legs" is obviously an exception. The Housefly and Moth have the same outlier degrees because their initial associative covers are the same. Because the unobserved association {Hair}→{Milk, Backbone, Breathes} is typical for mammals, other typical rules for mammals and bird life become unobserved associations. From the beginning, the bug class for the Housefly and Moth is a minor class having only 8 kinds of animals. Of them, bugs having item Hair are the Housefly, Moth and Wasp, which is the fourth outlier. Judging from the results, detected outliers are reasonable.

When we do not use the Mapping Option, we derive a record of Octopus with the highest outlier degree. A typical unobserved association is $\big\{$(Feathers, F), (Fins, F), (Catsize, T)$\big\} \longrightarrow \big\{$(Hair, T), (Eggs, F), (Milk, T), (Airborne, F), (Toothed, T), (Backbone, T), (Breathes, T), (Venomous, F)$\big\}$. The consequent typically indicates the mammal, which is the largest class in this data, while the antecedent can be observed in any class other than bird life and fin. That is, this unobserved association is not

Table 9. Top 3 Outliers in Zoo (AA Method)

Name	C_t^0	Typical Unobserved Associations
Crab	{Eggs, Aquatic, Predator, (# of Legs, 4)}	{(# of Legs, 4)}→{Toothed, Backbone, Breathes}
Housefly/Moth	{Hair, Eggs, Airborne, Breathes, (# of Legs, 6)}	{Hair}→{Milk, Backbone, Breathes}

quite meaningful and it is uncertain why the Octopus is detected as an outlier with the highest outlier degree. This suggests that the Mapping Option is useful for asymmetric attributes.

We also apply the AA Method to Original Mushroom with $msup = 6\%$ and $mconf = 90\%$. The top 112 records have the same outlier degree value (1.4% for all records). Among these outliers, 93 belong to class edible. None of these outliers includes the item (Odor, none). There are only 10% edible mushrooms that have odor. Moreover, about half of all high-confidence rules have item (Odor, none) in either the antecedent or consequent. In fact, initial associative covers of these outliers grow with unobserved associations of the form {(Class, edible),...}→{(Odor, none),...}, then unobserved associations having (Odor, none) in the antecedent are used for the growth. As a result, their associative covers expand more. The remaining 19 outliers ranked in number-one belong to the poisonous class. Their associative covers similarily expand because they do not have items such as (Stalk-shape, tapering) and (Stalk-root, bulbous), which are included in half of all high-confidence rules. We conclude, then, that our method can detect more rare records.

6.4   Accuracies Comparison

We evaluate our method (AA Method) against the probabilistic method of [14]. Similar to accuracy experiments in [14], we take two accuracy measures: the detection rate that is the fraction of detected true outliers to all true ones, and the detection precision that is the fraction of detected true outliers to all transactions detected as outliers. To see fair play, we provide a fixed parameters for each method, and get accuracy plot lines of top-l records ordered by the outlier degree which each method calculates, as changing the l value.

Fig. 2(a) presents the comparison for Mushroom, when $msup = 20\%$, $mconf = 95\%$, and $mod = 0.25$ are given for the AA Method and $k = 2$ and $\alpha = 0.003$ for the probabilistic method. Both the AA Method and Probabilistic Method show very good accuracy. In the best case, the AA Method derives about 98% of the detection rate and 100% of the detection precision, where, in fact, it fails to detect only one true outlier. Although the Probabilistic Method achieves both 100% accuracies, the performance gap between the AA Method and Probabilistic Method is very small.

Fig. 2(b) shows the experimental result for KDD Cup 99. We have given $msup = 95\%$, $mconf = 95\%$ and $mod = 0.15$ for the AA Method and have given $k = 2$ and $\alpha = 0.15$ for the Probabilistic Method. The reason that the plot line of the Probabilistic Method moves upwards partway is that a few records ranked top, which derive the highest outlier degree, are not true outliers, although true

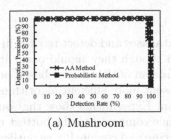

(a) Mushroom

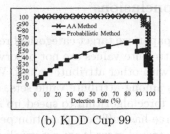

(b) KDD Cup 99

Fig. 2. Accuracy Comparison

outlier records also derive higher outlier degree. As shown in this figure, our method obviously outperforms the method of [14].

Although we give the other *msup* value and *mconf* for Mushroom and KDD Cup 99 respectively, the accuracy changes are not dramatic because attribute associations in both data are strong and the set of generated rules does not change significantly. However, it is obvious that our algorithm is generally sensitive to parameters, so finding proper parameters is important. We regard this problem as future work.

6.5   Runtime Comparison

Here, we verify that the proposed algorithm delivers faster detection. We compare 4 detecting methods. First is the brute force algorithm in Table 1 (BF). Second is an outlier detection that uses only a minimal rule set (MinR). Third is one that uses only a maximal associative cover (MaxC). Fourth is our complete algorithm utilizing both a minimal rule set and a maximal associative cover (MinR+MaxC). For Zoo, we use the Mapping Option. Here we omit the result for Mushroom due to space limitations.

Fig. 3 shows the processing time (logarithmic scale) as *msup* changes. For Zoo10000 (Fig. 3a), we provide $(mconf, mod) = (90\%, 0.6)$. For KDD Cup 99 (Fig. 3b), we give $(mconf, mod) = (95\%, 0.15)$.

BF takes much runtime and the availabilities of proposed algorithms increase as *msup* decreases. In both figures, MinR+MaxC runs fastset for some *msup*. Overall, MinR can decrease runtime much more than can MaxC.

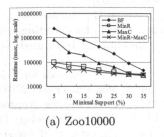

(a) Zoo10000

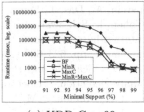

(a) KDD Cup 99

Fig. 3. Runtime Comparicon

7 Conclusions

In this paper, we target categorical record databases and detect records in which many attribute values are not observed even though they should occur in association with other attribute values. We provide an outlier degree based on associations among attribute value sets, and present more efficient algorithms than a brute force algorithm to speed up detection. Experiments show that our outlier degree has sufficient detection performance compared to the hottest related work, detected records are practically interesting and convincing as outliers, and the proposed algorithms are much faster than a brute force algorithm. A future task is to derive the most proper parameter values such as $msup$, $mconf$ and mod.

Acknowledgement. This research was supported in part by the grant-in-aid for scientific research from JSPS (#19024006).

References

1. Knorr, E.M., Ng, R.T.: Algorithms for Mining Distance-Based Outliers in Large Datasets. In: VLDB, pp. 392–403 (1998)
2. Ramaswamy, S., Rastogi, R., Shim, K.: Efficient Algorithms for Mining Outliers from Large Data Sets. In: SIGMOD Conference, pp. 427–438 (2000)
3. Aggarwal, C.C., Yu, P.S.: Outlier Detection for High Dimensional Data. In: SIGMOD Conference, pp. 37–46 (2001)
4. Arning, A., Agrawal, R., Raghavan, P.: A Linear Method for Deviation Detection in Large Databases. In: KDD, pp. 164–169 (1996)
5. Breunig, M.M., Kriegel, H.P., Ng, R.T., Sander, J.: LOF: Identifying Density-Based Local Outliers. In: SIGMOD Conference, pp. 93–104 (2000)
6. Jagadish, H.V., Koudas, N., Muthukrishnan, S.: Mining Deviants in a Time Series Database. In: VLDB, pp. 102–113 (1999)
7. Knorr, E.M., Ng, R.T.: Finding Intentional Knowledge of Distance-Based Outliers. In: VLDB, pp. 211–222 (1999)
8. Papadimitriou, S., Kitagawa, H., Gibbons, P.B., Faloutsos, C.: LOCI: Fast Outlier Detection Using the Local Correlation Integral. In: ICDE, p. 315 (2003)
9. Rousseeuw, P.J., Leroy, A.M.: Robust Regression and Outlier Detection. John Wiley and Sons, Chichester (1987)
10. Zhu, C., Kitagawa, H., Faloutsos, C.: Example-Based Robust Outlier Detection in High Dimensional Datasets. In: ICDM, pp. 829–832 (2005)
11. Bronstein, A., Das, J., Duro, M., Friedrich, R., Kleyner, G., Mueller, M., Singhal, S., Cohen, I.: Self-aware services: using Bayesian networks for detectinganomalies in Internet-based services. In: International Symposium on Integrated Network Management, pp. 623–638 (2001)
12. Pelleg, D.: "Scalable and Practical Probability Density Estimators for Scientific Anomaly Detection, " Doctoral Thesis of Carnegie Mellon University (2004)
13. Chan, P.K., Mahoney, M.V., Arshad, M.H.: "A Machine Learning Approach to Anomaly Detection," Technical Report of Florida Institute of Technology (2003)
14. Das, K., Schneider, J.G.: Detecting anomalous records in categorical datasets. In: KDD, pp. 220–229 (2007)

15. Han, J., Pei, J., Yin, Y.: Mining Frequent Patterns without Candidate Generation. In: Proceedings of the 2000 ACM SIGMOD International Conference on Management of Data, pp. 1–12 (2000)
16. Agrawal, R., Srikant, R.: Fast Algorithms for Mining Association Rules in Large Databases. In: VLDB, pp. 487–499 (1994)
17. Grahne, G., Zhu, J.: Efficiently Using Prefix-trees in Mining Frequent Itemsets. In: FIMI (2003)
18. UCI Machine Learning Repository mlearn/MLRepository.html, http://www.ics.uci.edu/~mlearn/MLRepository.html

An Energy-Efficient Multi-agent Based Architecture in Wireless Sensor Network

Yi-Ying Zhang, Wen-Cheng Yang, Kee-Bum Kim, Min-Yu Cui, Ming Xue,
and Myong-Soon Park*

Department of Computer Science and Engineering,
Korea University, Seoul, Korea
{zhangyiying, wencheng, givme, minwoo,xueming,
myongsp}@ilab.korea.ac.kr

Abstract. Wireless sensor network (WSN) containing thousands of tiny and low-power nodes can be used to monitor environment. An energy-efficient and reliable wireless communication architecture is usually required. In this paper, we propose a novel energy-efficient multi-agent based architecture (EEMA), which is based on a clustering algorithm and multi-agent system to reduce the redundant messages and filter most error messages. EEMA consists of three important agents: classification agent, error agent and filter agent which divide messages processing into three different phases separately and achieve different functions. The simulation results show that EEMA can outperform LEACH and ECDG. It can distinctly prolong the lifetime of the system and improve the system reliability.

Keywords: Agent, Wireless Sensor Network, energy efficient.

1 Introduction

Wireless sensor networks (WSN) are undergoing rapid progress and inspiring numerous applications more and more. However, challenges and difficulties still exist: (1) the sensor nodes own limited power, processing and sensing ability. (2) The sensor nodes are prone to failure because of lack of power, physical damage etc; (3) since the information generated by a single node is usually incomplete or inaccurate, and the applications need collaborative communication and computation among multiple sensors [1]. Therefore, the applications need an architecture which can efficiently sense and transact the data with less energy dissipation as well as fault tolerant [3].

In this paper, we present an intelligent architecture, an energy-efficient multi-agent based architecture (EEMA), which is based on clustering algorithm and multi-agent system. By reducing unnecessary or redundant messages as well as error messages, EEMA achieves good energy efficiency and network performance.

---

* The corresponding author.
This work was supported by the Second Brain Korea 21 Project.

Y. Zhang et al. (Eds.): APWeb 2008, LNCS 4976, pp. 124–129, 2008.
© Springer-Verlag Berlin Heidelberg 2008

The rest of the paper is organized as follows: In section 2, the related work is briefly reviewed. In section 3, we introduce the multi-agent architecture with three agents, classification agent, error agent and filter agent. In section 4, we give the simulation results for the multi-agent architecture. In section 5, we conclude the paper with a summary of the main ideas and directions for future work.

2 Related Work

Clustering algorithm has been fundamental and widely used technique in WSN. In [1], an energy efficient adaptive clustering protocol (LEACH) is presented, which employs a hierarchical clustering, and the nodes organize themselves into clusters and elect a node as cluster head (CH) for each cluster. The CHs collect and aggregate information from sensors and pass them to the base station. In this way it does realize a large reduction in energy consumption.

In [2], an efficient clustering of data gathering protocol (ECDG) is proposed, which groups nodes as clusters, builds a routing tree among distributed CHs and only lets the root node of the rooting tree communicate with the base station. ECDG uses a routing tree to route messages to base station by multi-hop. However, although these two architectures achieve data fusion to some degree, the data is not further processed, for example, by way of data classification, data filtering and error message detection. It therefore causes too much redundant information which is transmitted in the network, which means the energy efficiency and reliability of WSNs cannot be satisfied.

3 Multi-agent Architecture

This section introduces the main construction of multi-agent architecture. As shown in Figure 1, the multi-agent architecture is divided into three phase: classification phase, authentication phase and aggregation phase. Firstly, the classification agent interacts with the sensor nodes and disseminates the messages in the first phase. Then, the error agent performs error message judgment and disposal at authentication phase. Finally, at aggregation phase, the filter agent focuses on filtering the duplicate messages.

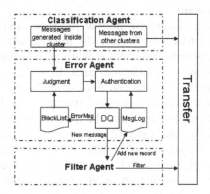

Fig. 1. The Architecture of Multi-Agent

3.1 Classification Agent

The classification agent is responsible for message acquisition and classification. The CH has to be powerful enough to receive each incoming message to determine whether the message should be processed or forwarded. We divide the messages into three categories: (1) Generated by the CH itself; (2) Generated by other non-CH nodes in the same cluster; (3) Transmitted from other CHs in different clusters.

For the third type messages, our solution is to forward them towards base station directly, because these messages from other clusters are usually important and already filtered by the CH. We need not to consume the precious energy to deal with them again. But for the first two types of messages, which are generated by the nodes in the cluster, we must authenticate validity of them. All of these messages are gathered at the cluster head, and are authenticated there. Disseminated by the classification agent, those packets flow to the error agent and filter agent. We show how to build nearly optimal aggregation structures that can further deal with network reliability and redundancy for the error and duplication of data by exploiting probabilistic techniques.

3.2 Error Agent

In error agent, we judge and deal with the exceptional messages called outliers. The outliers are different from other earlier messages. In order to distinguish them, error agent involves two components: judgment and authentication. Based on the facts that nodes in sensor networks often encounter spatiotemporal correlation [7], we can assume that the messages generated in the same cluster normally are almost the same, which means that in normal cases the probability that there are different messages is very small or rare. In other words, in WSNs, especially in the same cluster, multiple sensors which share the wireless medium have similar messages to send at almost the same time, because they all generate messages in response to external events.

On the other hand, in the error-prone wireless sensor network, its own fault or outside interference might cause the nodes to send error messages. We design two components to distinguish the outliers based on the above knowledge. We further introduce the error frequency to judge the type of messages and find the real reasons that cause the errors.

3.2.1 Judgment Component

This component is used to distinguish normal messages and exceptional messages. We design a table named *BlackList* to store the information of nodes which will be repealed or have been repealed. The nodes contain an information table structured as *<ID, Frequency, BoolStatus>*. All the messages from those nodes repealed in the *BlackList* are rejected. In the tri-tuple, the *BoolStatus* field is a Boolean flag for repealing status. If *BoolStatus* is true, it indicates that this node has failed. Its messages are not considered to be credible and never be used again and vice versa. Rather, its messages are influenced by other external interference. So we only filter these messages and increase the *Frequency* value. The *Frequency* field is a symbol of the amount of error messages that the node has sent. When it exceeds a certain value, we are sure that the node has failed. We set the *BoolStatus* value as true, and update the *BlackList* Simultaneously.

When the messages arrive, the error agent calls the judgment engine to evaluate whether the messages coming from those nodes are included in *BlackList* or not. If they belong to *BlackList* and *BoolStatus* is true, we think that node has been a failure. It means that its messages are not credible and all the messages from this node will be filtered out. At the same time this node will be informed by the CH to stop sending messages. Otherwise, the CH considers only the message is wrong by external reasons, and filters the messages without informing the nodes. At the same time, the *Frequency* field value is increased by one unit. Those messages not belonging to *BlackList* are sent to authentication component.

3.2.2 Authentication Component

This function focuses on the correctness of messages. We designed another table named *MsgLog*, to store the latest messages records, which reserve the information of messages sent correctly recently. The table has the structure of a tri-tuple as *<ID, Content, Frequency>*. Utilizing the *MsgLog*, we can get the historical information about the node's *ID*, content of message, and frequency. We can utilize *MsgLog* to judge whether those messages are authentic or not.

If the message is different from the historical messages, it will be categorized to erroneous tendency. This kind of messages is divided into two types: genuine error messages and new messages. In either case, the agent waits for a certain threshold of time to obtain sufficient information to distinguish the messages. In the wireless sensor network, according to temporal and spatial correlation, the possibility that the individual node simultaneously send different messages is very little. Based on the above facts, we design a buffer for Delay Queue (DQ), which gathers limited new coming messages. If the outlier does not belong to the DQ, the message is authenticated to be erroneous. The authentication component would drop these messages and record the sensor node ID to the *BlackList*.

The algorithm of authentication:

```
1. Get new Messagex;
2. If Messagex∈ MsgLog Then retransmit Messagex to Fil-
   ter Agent;
3. Get messages to DQ during a time period;
4. xCount ← 0      //Initialize the counter of Messages
      for i=1 to DQ.Length{
            Get BufferMessage[i] ∈ DQ;
            If ( Messagex.content simulates to
            BufferMessage[i].content and
            Messagex.ID ≠ BufferMessage[i].ID ) Then
         xCount ++;
      }
   If xCount >0 then
      Messagex is new message;
      Add to MsgLog and transmit it.
   Else
      Messagex is doubted as error message;
      Add to BlackList and update frequency;
5. End
```

3.3 Filter Agent

Because data transmission over wireless costs hundreds to thousands of times more energy than performing local computation on the same data [8], The filter agent focuses on filtering proportional redundancy messages to minimize communications overhead and reduce collision as well as on dealing with the amount of message and data aggregation. Actually, neither all sensing nodes need to report, nor all messages are needed to be sent. Therefore, we just need to send sufficient amount of messages for the robust sensor network. Below are some notations used for calculating messages amount.

P_e is the error probability of the WSNs transmission; C_i is the ith cluster in WSNs; H_i is the hops that from CH to base station; P_t is the correct probability from CH to base station.

We can get the number of hops for every C_i to base station in the early stages of cluster formation [6], we call it as H_i. And then we can get that the correct probability P_t is about $(1 - P_e)^{H_i}$ by multiple hops. Under the Taylor series theory, the correct probability is about $1 - P_e H_i$ for easily computing, namely $P_t \approx 1 - P_e H_i$ [4]. In accordance with the probability and statistics, the C_i at least sends $\left\lceil \dfrac{1}{1 - P_e H_i} \right\rceil$ messages for guaranteeing at least one message arrives at the base station successfully. After sending in a sufficient number of messages, the CH informs the cluster nodes within the cluster that they can retard to send message until a new cycle starts. Through this mechanism, we can prune many excessive messages and reduce the communication overhead. The status tables in the nodes record the trails of probing and recalling messages, and also provide a way to suppress redundant retransmissions.

4 Experiments and Results

To demonstrate the proposed multi-agent architecture, we have designed a sensor network simulation incorporating ECDG. We choose a 100m*100m network, where 100 nodes are deployed. We use the network's lifetime as the evaluation criteria. Network's lifetime means the network lose the ability to fulfill its mandate.

Figure 2 shows our EEMA architecture performs better than ECDG and LEACH. The EEMA's lifetime is about 1.2 times longer than ECDG's lifetime and about 1.4 times than LEACH's lifetime. This indicates that EEMA can achieve much higher efficiency in energy consumption. From the figure, we can also notice that the EEMA's line is smoother than other lines. It means that energy consumption is evenly distributed to all the nodes rather than being concentrated on one or few nodes, thereby reducing the risk that some nodes may die too early. Figure.3 shows that our EEMA can reduce more error messages than LEACH and ECDG, which is an indication of higher information reliability and further reduction of unnecessary messages.

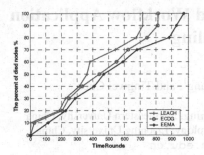
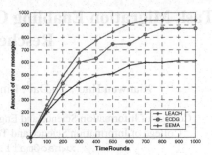

Fig. 2. Network's lifetime   **Fig. 3.** Amount of Error messages

5 Conclusion

In this paper, we present a novel, scalable and intelligent architecture, an Energy-Efficient and Multi-Agent based architecture (EEMA), which not only can reduce the amount of messages, but also improve the reliability of WSN. The simulation shows, with EEMA architecture, WSN can perform much better than LEACH and ECDG in terms of energy-efficiency and reliability.

References

1. Bandyopadhyay, S., Coyle, E.J.: An Energy Efficient Hierarchical Clustering Algorithm for Wireless Sensor Networks. In: Proceeding of IEEE INFOCOM 2003, San Francisco (April 2003)
2. Fu, Z., Yang, Y., Park, M.-S.: Efficient Clustering of Data Gathering in Wireless Sensor Networks. In: The 8th International Conference on Electronics, Information, and Communication (ICEIC 2006), Ulaanbaatar, Mongolia, June 2006, pp. 351–354 (2006)
3. Tynan, R., Marsh, D., O'Kane, D., O'Hare, G.M.P.: Agents for wireless sensor network power management. In: Parallel Processing. ICPP 2005 Workshops. International Conference, June 2005, pp. 413–418 (2005)
4. http://en.wikipedia.org/wiki/Taylor_series
5. Heo, N., Varshney, P.K.: An intelligent deployment and clustering algorithm for a distributed mobile sensor network. In: IEEE International Conference on Systems, Man and Cybernetics (October 2003)
6. Sharma, N.K., Kumar, M.: An Energy Constrained Multi-hop Clustering Algorithm for Wireless Sensor Networks. In: Lorenz, P., Dini, P. (eds.) ICN 2005. LNCS, vol. 3420, pp. 706–713. Springer, Heidelberg (2005)
7. Jamieson, K., Balakrishnan, H., Tay, Y.C.: Sift: a MAC Protocol for Event-Driven Wireless Sensor Networks'. In: Römer, K., Karl, H., Mattern, F. (eds.) EWSN 2006. LNCS, vol. 3868, pp. 260–275. Springer, Heidelberg (2006)

Task Migration Enabling Grid Workflow Application Rescheduling

Xianwen Hao[1], Yu Dai[1], Bin Zhang[1], and Tingwei Chen[2]

[1] College of Information Science and Engineering, Northeastern University, China
[2] College of Information Science and Technology, Liaoning University, China
xwhao@mail.neu.edu.cn

Abstract. This paper focuses on the task migration enabling grid workflow application rescheduling problem, presents a reduced task graph model, and implements a performance oriented rescheduling algorithm based on immune genetic algorithm. The experiment shows that, compared with Adaptive Heterogeneous Earliest Finish Time static rescheduling algorithm and the classical dynamic Max-Min scheduling algorithm, the performance advantage of the proposed rescheduling algorithm is obvious, on the one hand because of the performance contribution of global optimization and task migration, and on the other hand because of the efficiency contribution of task graph reduction and immune genetic algorithm's convergent speed. It also shows that task migration improves grid application's adaptability of dynamics further.

1 Introduction

Rescheduling is one of the key parts in the grid static scheduling strategy. This paper focuses on the task migration enabling grid workflow application rescheduling problem, presents a reduced task graph model, and implements a performance oriented rescheduling algorithm based on immune genetic algorithm (IGA), due to the main problems in the current research works [1-3]. The experiments show that the rescheduling performance advantage is obvious compared with related works, such as the Adaptive Heterogeneous Earliest Finish Time (AHEFT) rescheduling algorithm and the classical dynamic Max-Min scheduling algorithm, because of the optimal performance advantage of IGA and the efficiency contribution of task graph reduction and the faster convergent speed of IGA compared with general genetic algorithm (GA). The experiments also show that by task migration support, grid application gets further performance improvement and greater adaptability of dynamics, which extends the rescheduling process to the dynamics scope that other algorithms can't touch.

This paper is organized as follows: section 2 describes task migration enabling grid workflow application rescheduling problem; section 3 presents the rescheduling algorithm based on IGA; section 4 shows the experimentation and analyzes the performance, and section 5 concludes the whole paper.

Y. Zhang et al. (Eds.): APWeb 2008, LNCS 4976, pp. 130–135, 2008.

2 Task Migration Enabling Grid Workflow Application Rescheduling Problem

In this paper, we assume the heterogeneous processors connected in a full connected topology in which all inter-processor communications are assumed to perform without contention. It is also assumed that computation can be overlapped with communication and task executions can be non-preemptive.

The details of processor and network resource model see paper [2].

With each task as a node, communication and precedence constraints between tasks as a directed edge, the grid workflow application can be expressed as a Directed Acyclic Graph (DAG). Given that popular label methods are all unsuitable for the grid resource heterogeneity and task migration, a new method is proposed below.

If a task could be migrated, it is named a Migratory Task in this paper. In general, the task migration cost includes the checkpoint written cost, task staged cost and checkpoint reading cost, which are all calculated by their implementation policy. In this paper, we assume the checkpointing is triggered periodically itself, and the state is saved locally. So the Migration Cost is defined as $cm_i = \sum_{s_j \in pred(s_i)} c_{ji}$.

For task migration, migratory task nodes in task graph in EXECUTING state (defined as follows) should be split into two nodes each time when rescheduling is going on, according to the task finished rate: a TRANSFFRING task s_f and a WAITING task s_r. Fig.1 shows a sample split progress.

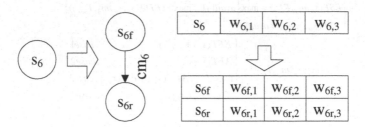

Fig. 1. Migratory task graph split process

To improve the efficiency of the scheduling algorithm based on graph, a reduction rule is introduced.

Rule 1 Reduction Rule: Delete any finished task node which children tasks are all finished. If there is more than one node with none zero in-degree, select any of them as the final entry node, and add directed edge from it to all the others. Finally, set all the finished task nodes' weights and edges' weights among them to zero.

Fig.2 shows the task graph reduction process.

It keeps the assumption that there exists only one entry node in the task graph through adding directed arcs, and also keeps the reductive equivalence of the task graph through making the weight be 0 which will also keeps the consistency of the scheduling and rescheduling algorithms.

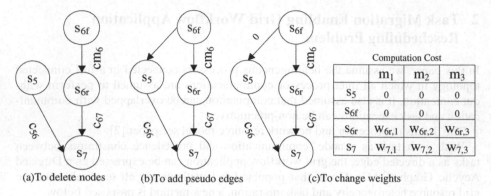

Computation Cost	m_1	m_2	m_3
s_5	0	0	0
s_{6f}	0	0	0
s_{6r}	$W_{6r,1}$	$W_{6r,2}$	$W_{6r,3}$
s_7	$W_{7,1}$	$W_{7,2}$	$W_{7,3}$

(a)To delete nodes   (b)To add pseudo edges   (c)To change weights

Fig. 2. Task graph reduction process

The Earliest Finish Time (EFT) is a usual goal of application-oriented scheduling. Aim at the reduced task graph, the application's EFT is equivalent to the exit task's EFT. But the weights of node are varying, so the EFT can only be gotten when the task-resource mapping and the order of task execution on the same resource is settled, which is expressed by T-RAG(details see paper [2]).

The earliest finish time of a T-RAG g equals to the earliest finish time of the exit task node s_{exit}, which is calculated by the following equation:

$$EFT(s_u, m_j, T_{now}) = w_{u,j} + EST(s_u, m_j, T_{now})$$

$$\begin{cases} EST(s_u, m_j, T_{now}) = \max \left\{ avail[j], \max_{s_v \in pred(s_u)} (EDA(s_v, s_u, m_j, T_{now})) \right\} \\ EST(s_{entry}, m_j) = avail[j] \end{cases}$$

$$EDA(s_v, s_u, m_j, T_{now}) = \begin{cases} EFT(s_v) + c_{vu}/cr_{vu} & , case1 \\ AFT(s_v) & , case2 \\ \max(T_{now}, AFT(s_v) + c_{vu}/cr_{vu}) & , case3 \\ T_{now} + c_{vu}/cr_{vu} & , case4 \end{cases} \quad (1)$$

Where,

Case1: s_v isn't a FINFISH or TRANFERRING state task;

Case2: s_v is TRANFERRING, and s_v was scheduled to resource m_j;

Case3: s_v is TRANFERRING, and s_v wasn't scheduled to resource m_j, but its output data was scheduled to resource m_j;

Case3: s_v is TRANFERRING, s_v wasn't scheduled to resource m_j, and its output data wasn't scheduled to resource m_j too;

Different task-resource mappings produce different RT-RAGs, and the objective of scheduling is to select the best RT-RAG.

3 Problem Solving Based on Immune Generic Algorithm

This paper takes zero as the antigen and uses the text gene coding method which is based on the task execution order. Each gene position (e.g. K) contains 2-vection: one is for the number of task s_n and another is for the number of resource mi, which means that the task s_n is assigned to resource mi and the execution order is K. If the executed order of the sub-tasks assigned to the same resource is oppo-site to the order in such table, the antibody is an illegal solution which will be re-moved from the set of the potential solutions.

1. To compute the affinity
The affinity between antibodies is expressed as the Hamming Distance. Equation (2) can be used to compute the adaptation degree between antibody and antigen.

$$ax_v=1/(1+EFT(v)) \tag{2}$$

Where, EFT(v) can be calculated by equation (1).

2. To compute the antibody concentration
The concentration [7] of antibody v means the ratio of antibodies similar or same as v in the whole population. The concentration of antibody v is:

$$C_v = \frac{1}{N}\sum_{w=1}^{N} ac_{v,w} \tag{3}$$

Where, $ac_{v,w} = \begin{cases} 1 & \text{when } ay_{v,w} \ge Tac \\ 0 & \text{other} \end{cases}$, Tac is the predefined threshold.

3. To promote and restrain the antibody generation
The expected propagation rate of antibody v is computed as follows.

$$e_v=ax_v/c_v \tag{4}$$

4. To use crossover and mutation operations
For the limitation of the coding way of the antibody in this paper, crossover and muta-tion need some special methods to take during the crossover and mutation process, which are shown in Fig.3.

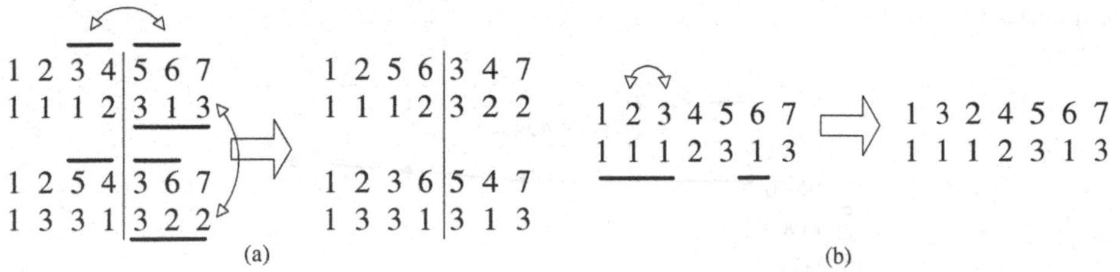

```
1 2 3 4|5 6 7        1 2 5 6|3 4 7
1 1 1 2|3 1 3        1 1 1 2|3 2 2      1 2 3 4 5 6 7  ⟹  1 3 2 4 5 6 7
                                        1 1 1 2 3 1 3      1 1 1 2 3 1 3
1 2 5 4|3 6 7        1 2 3 6|5 4 7      ─────────────      ─────────────
1 3 3 1|3 2 2        1 3 3 1|3 1 3
        (a)                                        (b)
```

Fig. 3. Illustration of crossover and mutation operation

4 Experimentation and Analysis

In order to evaluate the performance and stability of the scheduling algorithm pro-posed in this paper, we use parametric randomly generated DAGs in the experiment. For the purpose of fair comparison, we directly follow the heterogeneous computation modeling approach defined in [3] to generate representative DAG test cases. To model the dynamic change and heterogeneity of resources, we introduce several additional parameters.

The new parameters as listed below:

β: The computing resource heterogeneous factor. A higher value of β suggests the bigger difference of resource capability.

δ: The network heterogeneous factor.

γ: Interval of resource change, expressed as multiples of average computation cost of DAG average computation cost..

The value set for each parameter is listed in Table 1.

Table 1. Parameters of random generated DAGs and resources

parameter	value
V	20,50,80
α	0.5,1,2
out-degree	0.1, 0.5,1
CCR	0.1,0.5,1,5,10
β	0.2,0.6,1,1.4,1.8
δ	2,3,5,8,10
γ	100,50,10,5,1,0.8,0.5,0.2,0.1
N	10,20,30,40,50

The anticipated contribution of our rescheduling algorithm compared with related works includes performance improvement due to resource available cycles saving, and the broader dynamic adaptability directly benefiting from task migration. So the experimentations compare the performance among the proposed IGA algorithm, IGA without Task Migration Support (IGA without TMS), the AHEFT algorithm [1] and the classical dynamic scheduling algorithm Max-Min [13] with different dynamic degree.

The results of the experimentations show that the average makespan of Max-Min, AHEFT, IGA without task migration and IGA are 16860, 7791, 6095 and 5186 respectively.

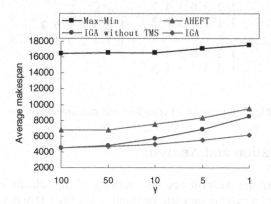

Fig. 4. Performance comparison under weak dynamics

Fig. 4 shows the average makespan of these four algorithms under different situations that the resources have different dynamic degrees.

When the resource dynamics getting even stronger, without task migration support, Max-Min, AHEFT and IGA without TMS cannot get the available schedule any more. But the proposed IGA still works. When $\gamma > 1$, the algorithms have no results except IGA still working to $\gamma = 0.1$. If $\gamma < 0.8$, the performance decreases rapidly that's because under strong resource dynamics, the possibility of a task finish on a single resource is very low, the global optimized performance of IGA has totally lost, but the running time of IGA will be increased. When γ equals to 0.1, the makespan is 3.5 times of that when $\gamma = 1$.

5 Conclusions

This paper presents a reduced task graph rescheduling model, and implements a performance oriented immune GA algorithm, which gains obvious performance advantage and greater adaptability of dynamics.

For that the appropriate initial parameters will improve the performance of the algorithm, how to initialize the value of parameters is a work in the future, or to find a better heuristic alternative.

References

1. Yu, Z., Shi, W.: An Adaptive Rescheduling Strategy for Grid Workflow Applications. In: The 21st International Proceedings of Parallel and Distributed Processing Symposium (IPDPS 2007), pp. 1–8. IEEE Press, New York, USA (2007)
2. Chen, T., Zhang, B., Hao, X., Dai, Y.: Task Scheduling in Grid Based on Particle Swarm Optimization. In: Proceedings of The 5th International Symposium on Parallel and Distributed Computing (ISPDC 2006), pp. 238–245. IEEE Press, New York, USA (2006)
3. Topcuoglu, H., Harir, S., Wu, M.-Y.: Performance-effective and low-complexity task scheduling for heterogeneous computing. IEEE Trans. on Parallel and Distributed Systems 13(03), 260–274 (2002)
4. Jianning, L., Haizhong, W.: Scheduling in Grid Computing Environment Based on Genetic Algorithm. Journal of computer research and development 41(12), 2195–2199 (2004)
5. Sakellariou, R., Zhao, H.: A low-cost rescheduling policy for efficient mapping of workflows on grid systems. Scientific Programming 12(4), 253–262 (2004)
6. Imamagic, E., Radic, B., Dobrenic, D.: An approach to grid scheduling by using condor-G matchmaking mechanism. In: The 28th International Conference on Information Technology Interfaces (ITI 2006), pp. 625–632. IEEE Press, New York, USA (2006)
7. Meshref, H., VanLandingham, H.: Artificial immune systems: application to autonomous agents. In: The 2000 IEEE International Conference on Systems, Man, and Cybernetics, vol. 1, pp. 61–66. IEEE Press, New York, USA (2000)

A Dependent Tasks Scheduling Model in Grid*

Tingwei Chen[1], Bin Zhang[2], and Xianwen Hao[2]

[1] College of Information Science and Technology, Liaoning University, P.R. China 110036
twchen@lnu.edu.cn
[2] College of Information Science and Engineering, Northeastern University, P.R. China 110004
{zhangbin,xwhao}@mail.neu.edu.cn

Abstract. In Grid computing, an application will be decomposed into a set of dependent tasks. In the Grid environment where resources have different capability and resources are interconnected over the world, the dependence among tasks affects the scheduling strategy greatly. This paper uses a Task-Resource Assignment Graph (T-RAG) to represent a potential resource assignment plan. And a dependent tasks scheduling model based on Best Task-Resource Assignment Graph (BT-RAG) construction is proposed which maps the dependent tasks scheduling problem into a graph construction problem. The BT-RAG is obtained and such graph is the optimal scheduling plan which determines the resource assignment plan and the execution order of tasks. Finally, the task scheduling algorithm based on the proposed scheduling model is implemented. Compared with HEFT algorithm, the proposed algorithm shows better performance in the situation of a large body of data transported among tasks.

1 Introduction

In the Grid computing [1], an application is always decomposed into a set of tasks among which there exist dependent relations. And such relations become a new challenge for Grid task scheduling. Currently, the research area of Grid task scheduling is focused on meta-task and batch task [2, 3]. Although such research efforts can solve the problem of heterogeneity and availability of Grid resources to some extent, they are not adaptable to task scheduling in which the dependent relations between tasks need to be considered. The meta-task and batch task are the special forms of Grid tasks which can not embody the relations of data and prior constraints. And the traditional DAG task graph only shows the computing quantity, communicating quantity and prior relations of tasks which can only be used for scheduling model in homogeneous environment [4]. In the Grid environment, for the heterogeneity and distribute property of resources, not only is the computing ability of resources different, but also is the bandwidths of internet connection different. Thus, the traditional task scheduling model based on DAG task graph is not suitable to Grid environment.

For the dependent tasks scheduling problem in the case of Grid environment, this paper uses a Task-Resource Assignment Graph, T-RAG in short, to relate dependent

---

* This research is supported by the National Key Technologies Research and Development programming in the 10th Five-year (2004BA721A05) of P.R. China.

Y. Zhang et al. (Eds.): APWeb 2008, LNCS 4976, pp. 136–147, 2008.

tasks with Grid resources. The nodes of this graph represent the pair of task and resource which represent that the resource is assigned to the task. The arcs of this graph represent the data transportation quantity between two dependent tasks and the internet bandwidth between resources assigned to such two tasks. For a T-RAG, the scheduling length of application can be obtained according to critical path of the graph. The minimum scheduling length of T-RAG is defined as Best T-RAG (BT-RAG). Aiming at shortest scheduling length, this paper proposes a dependent tasks scheduling model which is centered on generating BT-RAG. This scheduling model transforms the dependent task scheduling problem to the problem of generating BT-RAG. Finally, the heuristic algorithm for generating BT-RAG is implemented. Compared with HEFT algorithm [5], the dependent task scheduling algorithm based on proposed heuristic algorithm shows better performance in the situation of a large body of data transported among tasks.

2 Related Works

The aim of task scheduling is to assign resource for each task of the graph in order to satisfy the prior constraints of tasks and make the scheduling length shortest. The classic list scheduling algorithm is introduced and analyzed by R.L.Graham [6]. The basic idea of list algorithm is computing weight of task node according to computing quantity of tasks and the constraint relations among tasks. Then, the weights are sorted to form a list. Finally, the following steps will be executed until all the tasks are assigned suitable resources. (1) Select the nodes with highest prior level from the list ;(2) select suitable resource for such task.

Currently, many heuristic algorithms are all based on this idea such as HLF (Highest Level First), LP (Longest Path), LPT (Longest Processing Time) and CP (Critical Path) [7-9]. The differences of these algorithms are how to compute the weight of task nodes and select strategies of these tasks.

The list scheduling algorithm is proposed in the hypothesis of there not being communication costs among tasks in the homogeneous system [10]. In the traditional parallel computing system, after a short time, each processing machine returns some results, the hypothesis is true. But in the Grid, the differences of resources capability and the cost of data transportation among tasks change the base of traditional list scheduling algorithm. Thus, such algorithm can not be applied to Grid environment directly.

In order to make the traditional list scheduling algorithm suitable to Grid environment, some research efforts renew the traditional algorithm such as ILHA (Iso-Level Heterogeneous Allocation) algorithm proposed by Radulescu [11]. For the difference of capability of resources, this algorithm assigns tasks proportional according to computing capability of resources in order to achieve the load balance. For the time of transporting data among tasks, this algorithm assumes that when parent task node is executed successfully, the data will be transported to the resource assigned to child task. Obviously, for it has a hypothesis of data transportation time, the algorithm ignores the different data transportation of different resources. Thus, this algorithm has its disadvantage.

Additionally, Haluk proposes a HEFT (Heterogeneous Earliest Finish Time) algorithm [5] which aims at extending scheduling of homogeneous environment to heterogeneity system. When computing weights in this algorithm, it considers both the execution time of tasks and communication time of tasks. The weight of nodes (rank (n_i)) can be computed as (1).

$$rank \ (n_i) = \overline{w}_i + \max_{n_j \in \ pred \ (n_i)} \ (\overline{c_{i,j}} + rank \ (n_j)) \tag{1}$$

Where, $pred(n_i)$ is the set of direct successor of n_i, $\overline{c_{i,j}}$ represents the average communication cost between n_i and n_j, $\overline{w}_i$ is the average computing cost of task n_i. According to (1), although this algorithm considers the communication cost, this algorithm is not suitable to Grid computing for during computing, average computing time and average communication time is used which hides the affection brought by difference of resources.

In short, for dependent task scheduling problem in the homogeneous system, for that there does not exist difference of computing capability and communication cost, the task scheduling algorithm is centered on task graph and mainly considers the prior constraints among tasks. For the difference of computing capability and communication cost, the scheduling algorithm based on list scheduling and its renewed algorithms modeled difference of computing capability (e.g. difference of communication cost) as an average value which can not show the difference of computing capability in essence. The dependent task scheduling should also consider the dependent relations among tasks and differences of resources.

3   Grid Dependent Tasks Scheduling Model Based on BT-RAG

Grid dependant tasks scheduling model includes three parts: (1) The application's expression model which describes the application's structure and resource requirements; (2) Grid resource model which describes the available resources of the application;(3) The goal function which describes the target of the scheduling process.

3.1   The Description of an Application

In a Grid application T which is composed of K tasks, $w(s_u)$ expresses the computation amount of task s_u , $a(s_u)$ expresses the lowest resource requirements of task s_u, e_{uv} is the directed edge from s_u to s_v, which express the dependant relationship of them, and s_v only can be executed after s_u. c_{uv} expresses the data transfer amount between task s_u and s_v. Nodes expressing task and directed edges between nodes expressing the dependent relationship, and the weight of edge expressing data transfer amount, a Grid application can expressed as a Task Graph (Fig.1).

3.2   Grid Application Oriented Resource Description

All the registered resources in Grid system can be called Grid Resource, which is expressed by $M_G=(m,w_G,t_G)$,in which m is the identifier of resource, w_G is compute

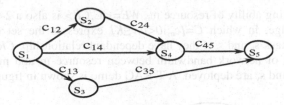

Fig. 1. Grid task graph

ability of resource m, t_G is the ordered set of available time slots of resource m. $t_G=\{<t_{i0},t_{i1}>|i0<i1,i<j->i1<j0\}$. Using absolute time to describe Grid resources is not convenient for comparing and computing. Thus, we use a formalization approach to represent the time of resources.

If the absolute time interval of executing Grid application T is represented as $[t_s,t_c]$, and the absolute time interval of resources m_i being available is $[t_{i0},t_{i1}]$, the application T oriented standard time interval of m_i can be represented as $t_T(m_i)$. We call this standard time interval is the application relative time.

$$t_T(m_i) = \begin{cases} [0, t_c - t_s], t_{i0} \leq t_s \, and \, t_{i1} \geq t_c \\ [t_{i0} - t_s, t_{i1} - t_s], t_{i0} > t_s \, and \, t_{i1} < t_c \\ [0, t_{i1} - t_s], t_{i0} < t_s \, and \, t_{i1} < t_c \\ [t_{i0} - t_s, t_c - t_s], t_{i0} > t_s \, and \, t_{i1} > t_c \end{cases} \qquad (2)$$

If the standard time interval of application T is $t(T)$, we have $t(T)= [0, t_c\text{-}t_s]$. In this paper, time is the application relative time.If the minimum capability of task required in Grid application T is w_{Tnin}, the execution time of T is $t(T)$, and for Grid resources $M_G=(m,w_G,t_G)$, there exists $(w_G>w_{Tnin}) \cap (\forall t_T(m)\text{->} t_T(m)\cap t(T)\neq\varnothing, t_T(m)\in t_G)$, we call m be the application available resources of application T.

Definition 1 Grid Application Oriented Resource Description. $GR=(T,M_T,CR_T)$ is a triples, in which T is a Grid application, M_T is the Application Available Resource of T, $M_T=\{m_T \mid m_T\S T\}$; $CR=(cr_{ij})_{n\times n}$ is the network resources matrix of n resources in M_T, cr_{ij} is the network bandwidth between resource i and resource j.

3.3 Goal Function

In this paper, based on the expression model of tasks and resources, with task-resource mapping regarding as node, the constraint relationship between tasks as edge, the bandwidth and the data transfer amount as the edge's weight, a task-resource assignment graph is composed.

Definition 2 Task-Resource Assignment Graph(T-RAG). $T\text{-}RAG=<V,E,WV,WE>$ is a 4-tuple,in which V is the set of node ,expressed by $V=\{(s_u,m_j)|u<=K;j<=N\}$,in which (s_u,m_i) expresses the mapping from task s_u to resource m_i; $;E=\{e_{ij}|0<i,j\leq K;i\neq j\}$expresses the communication relationship and privilege constraint; $WV=<W,A>$ is a 2-tuple,expressing the weight of node, in which $W=\{w(s_u)|0<u\leq K\}$ express task s_u's computing amount, $A=\{a(m_i)|0<i<N\}$ expresses

the set of computing ability of resource m_i, $WE=<C,CR>$ is also a 2-tuple,expressing the weight of edge, in which $C=\{c_{uv}|0<u,v\leq K\}$ expresses the set of data transfer amount between task s_u and s_v if they have dependant relationship. $CR=\{cr_{ij}|0<i,j<N\}$ expresses the set of network bandwidth between resource m_i and m_j ,on which dependant tasks s_u and s_v are deployed. A T-RAG demo is shown in figure 2.

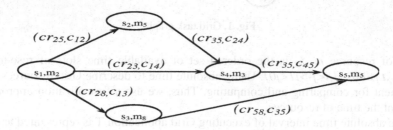

Fig. 2. Task-Resource assignment graph

G expresses the set of all the T-RAGs, when TG_T mapping to GR_T, and g is a single T-RAG in G. In this paper, the node which has zero in-degree is called entry node, and the node which has zero out-degree is called exit node. Obviously an effective Grid T-RAG has the characteristic of AOE network.

P is the path $s_e{\rightarrow}s_0{\rightarrow}s_1{\rightarrow}...{\rightarrow}s_c$ which starts from the entry node s_e and ends to the exit node s_c in T-RAG. Since there are multiple Task Reachable Paths from same s_e to s_c, p expresses a single one. The set of edges in p is defined as $p_e=\{e|e \in p\}$;The set of nodes in p is defined as $p_v=\{v|v \in p\}$. The path length $L(p)$ is defined as the the sum of tasks execution time and communication time.

$$L(p) = \sum_{V \in p_v} L(V) + \sum_{E \in p_e} L(E) \tag{3}$$

in which $L(V)$ is the execution time of a node. If task s_u is assigned to resource m_i, $L(V) = w(s_u)/a(m_i)$; $L(E)$ is the data transfer time between two nodes. If s_u,s_v are assigned to resources m_i and m_j separately, $L(E) = c_{uv}/cr_{ij}$,in which c_{uv} is the data transfer amount between task s_u and s_v, and cr_{ij} is the network bandwidth between resource m_i and m_j.

Definition 3 Longest Task Reachable Path. The execution time $T(g)$ is the max value of any Task Reachable Path Length of g in T-RAG G, $T(g) = \max[L(p)]$.In this paper, the Longest Task Reachable Path of g is defined as the max Task Reachable Path Length of g.

Definition 4 Best Task-Resource Assignment Graph. The Best Task-Resource Assignment Graph (BT-RAG) of a Grid application $T(BT\text{-}RAG_T)$is defined as the T-RAG which's Longest Task Reachable Path is the shortest of all the T-RAGs.

The scheduling target of application T is obtain a BT-RAG$_T$, so, the target function of dependent task scheduling can described as:

If the Task Graph of T is $TG_T=<S_T,E_T>$, and resource description is $GR_T=(T,M_T,CR_T)$, the Grid Application T 's Scheduling Goal Function is defined as:

$$f(T) = \min_{g \in G} \left\{ \max_{p \in g} \left[\sum_{\substack{s_u \in p \\ m_i \in p}} \frac{w(s_u)}{a(m_i)} + \sum_{\substack{c_{uv} \in p \\ cr_{ij} \in p}} \frac{c_{uv}}{cr_{ij}} \right] \right\} \qquad (4)$$

3.4 The Scheduling Process of Dependant Grid Tasks Based on BT-RAG

The Scheduling process of dependant Grid tasks based on BT-RAG is showed in figure 3. The process is divided into 4 phases:

Phase (1) is like the traditional developing process of application system. The developer describes an application into a Task Graph, in which the dependant relationship between tasks and resource requirement of single task is described.

Phases (2) (3) (4) are three phases of the application runtime, and executed one by one when the user's request arrives every time. Phase (2) is the phase to describe resources, which is dynamic unlike Task Graph. The available time slot of Grid resource and expected finished time of Grid application can both make the description of resources change. Phase (3) is the phase to get the BT-RAG, based on which the assignment and the execution order are both determined. Phase (4) is a phase to analyze the BT-RAG and generate a schedule for executing engine.

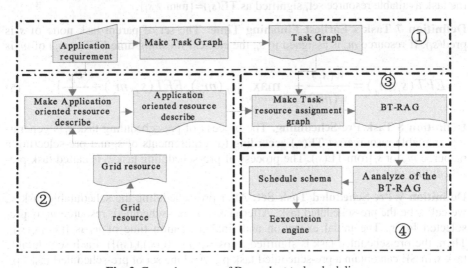

Fig. 3. General process of Dependent task scheduling

From the process of dependent task scheduling, it is known that the core of scheduling model is the Best Task-Resource Graph. So how to get the BT-RAG is a key sub-process, and in next section the BT-RAG generate algorithm is introduced.

4 The BT-RAG Generating Algorithm

For Grid application T, the direct approach of obtaining the Best Task-Resources Assignment Graph BT-RAG$_T$ is through enumerating. From definition of T-RAG, it is known that for the application composed by k tasks, the number of T-RAG is $\prod_{i=1}^{k} Sum\ (s_u)$, where $Sum(s_u)$ is the number of resources satisfying the require-ment of task s_u. With the extension of the scalability of problem, the number of T-RAG will be increased exponentially. Thus, this paper proposes an effective heuristic algorithm to solve this problem. First, the definition of parameters related with the proposed heuristic algorithm is introduced.

Definition 5 Schedulable Task and Schedulable Task Set. The parent node of task node s_i in TG$_T$ is represented as pred(s_i) and the task s_j which is scheduled already is represented as Δs_j. If (pred(s_i)={Φ})∨(Δs_j, εs_j∈pred(s_i), pred(s_i)≠{Φ}), s_i is the schedulable task. And the set of schedulable tasks is the schedulable task set, signified as SE={s_i| (pred(s_i)={Φ})∨(Δs_j, εs_j∈pred(s_i), pred(s_i)≠{Φ}),s_i∈s_T}.

Definition 6 Task Available Resource and Task Available Resource Set. If task s_i requires that the minimum capability of resources be w$_{min}$(s_i). And the available re-sources of application T is M$_T$=(m,w_T,t_T). If w$_T$>w$_{min}$(s_i), resource m is the task avail-able resource of task s_i, signified as m $⨏$ s_i. The set of task available resources of s_i is the task available resource set, signified as TE(s_i)={m|m $⨏$ s_i}.

Definition 7 Task's Earliest Finishing Time. The set of parent task node of s_i is pred(s_i). If resource m_u is assigned to s_i, the earliest finishing time EFT(s_i,m_u) of m_u is

$$EFT(s_i,m_u) = \frac{w(s_i)}{w(m_u)} + \max_{s_j\in pred(s_i)} \{t_T(m_u), EFT(s_j,m_v)+\frac{c_{ij}}{cr_{uv}}\}. \quad (5)$$

Definition 8 Task Pre-Scheduling. The process of pre-scheduling task s_i is generat-ing available resource set TE(s_i) according to requirements of s_i and pre-selecting a resource m_u for s_i from TE(s_i). The process of pre-scheduling task s_i is called task pre-scheduling of s_i.

Definition 9 Pre-Scheduled Task Set. After pre-scheduling the schedulable task s_i, we call s_i be the pre-scheduled task. After task s_i is pre-scheduled, resource m_u is pre-selected for s_i. The initial execution and final execution time of m_u is [t$_s$(s_i),t$_c$(s_i)]. Then, the pre-scheduled task is signified as ps$_i$=(s_i,m_u, [t$_s$(s_i),t$_c$(s_i)]). Each schedulable task s_i in SE can obtain a pre-scheduled task ps$_i$. And the set of pre-scheduled tasks ps$_i$ is pre-scheduled task set, signified as PS.

Definition 10 Weight of Pre-Scheduled Task Node. The weight of pre-scheduled task node is defined as a sum of computing time and data transportation time of pre-selected resource m_u of pre-scheduled task ps$_i$.

$$v(ps_i) = \sum_{s_j\in pred\ (s_i)} \frac{c_{ij}}{cr_{uv}} + \frac{w(s_i)}{w(m_u)} \quad (6)$$

Based on the parameters as above defined, the process of generating BT-RAG$_T$ in the proposed algorithm can be described as:

(1) Obtain current schedulable task set SE of TG$_T$;
(2) Pre-scheduling each schedulable task in SE to obtain pre-scheduled task set PS;
(3) Select a pre-scheduled task ps$_i$ from PS to generate a node of BT-RAG$_T$;
(4) Delete this task node from TG$_T$, and if the number of nodes of TG$_T$ is above 0, goto (1), else, finish.

In the above algorithm, the pre-scheduling process of (2) and the selection of pre-scheduled task from task set all adopt the heuristic idea. In the following, the heuristic rule and heuristic algorithm used in the proposed algorithm will be introduced.

4.1 Task Pre-scheduling Rule and Algorithm

The process of pre-scheduling task is selecting a resource for schedulable task s_i in order to obtain a pre-scheduled task ps$_i$. For there existing many resource mapping plan from s_i to its available resource set TE (s_i), in order to improve the effectiveness of algorithm and make the resource plan be close to the best one, this paper defines the task scheduling rules and based on these rules implements the task scheduling algorithm.

Rule 1 Task Pre-Scheduling Rule. The resource which can make the task finished earliest is the pre-selected resource of this task. That is to say, from the available task set TE in which resources are all satisfied requirements of s_i, select a resource m_u which enables $\min_{m_u \in TE} (EFT(s_i, m_u))$.

Algorithm 1. Task Pre-Scheduling Algorithm—PreTaskSchedule
Input: Schedulabe Task s$_i$,GR$_T$
Output: Pre-scheduled task ps$_i$
1 Initialize: TE= {Φ}, Temp= {Φ};
2 According to requirements of s_i, generate task available resource set TE(s_i);
3 Compute the earliest finishing time EFT(s_i,m_u) of each resource m_u in TE;
4 If the pre-occupation time of m_u is in the available time interval, then put $[s_i, m_u, EFT(s_i, m_u)]$ into Temp;
5 Sort the element in Temp order by EFT(s_i,m_u), let ps$_i$ equal to value of first element in Temp.

4.2 Pre-scheduled Task Selecting Rule and Algorithm

The process of selection of pre-scheduled task is selecting a pre-scheduled task from PS and assigning a resource for it in order to form a node of BT-RAG$_T$. This paper uses the computing capability and cost of data transportation as the standard for selection of pre-scheduled task, defines the rule of selection the pre-scheduled task and based on this rule implements the selection algorithm of pre-scheduled task.

Rule 2 Pre-Scheduled Task Selecting Rule. In the pre-scheduled task set PS, select the pre-scheduled task node which has the maximum weight as the node of BT-RAG$_T$.

Algorithm 2. Pre-Scheduled Task Selecting Algorithm--PreTaskSelect
Input: PS, TG_T, GR_T
Output: the node of $BT\text{-}RAG_T$
1 Compute weight of all the pre-scheduled task ps_i in PS;
2 Compare $v(ps_i)$, select the pre-scheduled task $ps_i=(s_i,m_u)$ which has maximum of $v(ps_i)$ as the node of BT-RAG;
3 Delete ps_i from PS;
4 Update GR_T, modify the available time interval of resource m_u;
5 Delete s_i from SE;
6 Delete pre-scheduled task from PS whose pre-selected resource is m_u.

5 Experiments and Analysis

To test the performances of the algorithm, we compare our algorithm to the HEFT algorithm [5]. We implement a simulation platform which is composed of simulating grid environment, scheduling algorithm library and performance evaluation module. On the platform, first we simulate the tasks and resources, then we use algorithms in the scheduling algorithm library to generate schedule schema, in the last we use performance evaluation module to evaluate the performances of the algorithms and schedule plans.

5.1 Simulation Environment

Task graph generator is able to generate different task graphs according to different parameters:

(1) K: the number of the tasks in the task graph. K is the initial value.
(2) α: parameter of the shape of the task graph. The height of the task graph is generated by the integer part of random normal distribution variable with the average value of $\sqrt{k}\big/\alpha$, the number of the tasks on each level of the graph is generated by integer part of random normal distribution variable with the average value of $\alpha\times\sqrt{k}$.

(3) β: the ratio of data transportation to computation in the task graph. Small value of β means there are more computations and big value of β means there are more transportations.

Resource generator mainly simulates the attributes of the resource of the Grid. There setup three parameters:

(1) n: resources in the Grid.
(2) γ: deviation of the capability of computing of the resources. The deviation stands for the range of the capability of computing of the resources, for a random value a, the value of the capability of computing is a random value in $[\dfrac{a}{\gamma}, a\gamma]$.

(3) δ: deviation of the band width. Deviation of the band width stands for the range of the band width of the resources, for a random value b, the value of the band width

of the resources is a random value in $[\dfrac{b}{\delta}, b\delta]$.

By changing these parameters we can simulate different grid environments and tasks.

5.2 Results and Analysis

Experiment 1. Impact of data transportation among tasks on schedule length under the circumstances of the same capability of computing and same bandwidth.

This experiment compares the impact of data transportation between tasks on schedule length of the two algorithms under the same capability of computing and same bandwidths. The result shows up in figure 4.

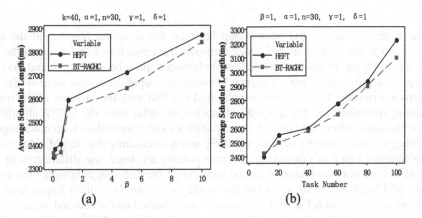

Fig. 4. Scheduling length in the situation of same resource capability and network bandwidth

In figure (a) shows the relationship between the schedule length and the task number. In figure (b) shows the relationship between the schedule length and the task number when the quantity of data transportations are very large. From the figures (Figure 4) we can find out that under different circumstances, there is not much difference between the schedule lengths of the two algorithms.

Experiment 1 shows that in the homogeneous environment, whether there are data transportations or not, by comparing with HEFT algorithm, BT-RAGHC algorithm is able to replace the HEFT algorithm, so the algorithm purposed by us is feasible.

Experiment 2. Impact of data transportation between tasks on schedule length of algorithms under the circumstances of the different capability of computing and same ban widths.

This experiment compares the impact of data transportation between tasks on schedule length of the two algorithms under the circumstances of the different capability of computing and different band widths. The result shows up in figure 5.

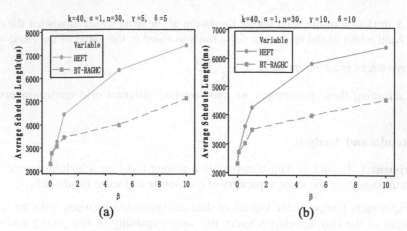

Fig. 5. Scheduling length in the situation of different resource capability and network bandwidth

The figure (a) shows the relationship between the schedule length and the task number when the maximum difference between the capabilities and bandwidths are 25 times. In figure (b) shows the relationship between the schedule length and the task number when the maximum difference between the capabilities and bandwidths are 100 times. From the figures above we can find out that with the increase of the transportation, obviously the BT-RAGHC algorithm is better than the HEFT algorithm. And with larger difference between the capabilities and bandwidths such advantage is even bigger. This is mainly because in HEFT when calculating the weight value of the nodes average value of computing and transporting are used, the differences of the capabilities are not considered. With the increase of the differences between the capabilities and bandwidths, the errors of the weight value are becoming larger, thus the result are not good. In BT-RAGHC resources are selected and scheduled according to there actual capabilities, thus the differences between the capabilities and bandwidths don't have much impact on the results.

Experiment 2 shows that in the environment where the differences between the capabilities and bandwidths are large, with much data transportations, BT-RAGHC algorithm purposed by us is able to avoid the impact of the differences. Comparing to the HEFT algorithm, BT-RAGHC algorithm is able to remarkably shorten the schedule length, so it is very effective.

Experiments above show that the BT-RAGHC algorithm purposed by this paper is more adaptable. BT-RAGHC solves the problem better when the capabilities and bandwidths are different, and there are much data transportations. Thereby the Grid task schedule module based on task-resource assignment graph construction is sound and effective.

6 Conclusions and Future Work

This paper is researched on dependent task scheduling problem. A dependent task scheduling model is proposed which is centered on task-resource assignment graph and maps the task scheduling problem into a BT-RAG generating problem. Finally, a task scheduling algorithm based on this model is implemented. Compared with HEFT algorithm, the proposed algorithm shows better performance in the situation of a large body of data transported among tasks and the scheduling working in the heterogeneity environment.

This paper only considers the bandwidth factor. The future work will consider other two factors--Network delay and the packet loss rate—which affect the scheduling. Additionally, in the case of invalidation or exception of resources, how to provide effective quality assurance for users is a area needed to study. In this paper, we hypothesize that in the process of execution of tasks, there are no data transported. This hypothesis limits the range of task scheduling model. In the future, we will study the dependent task scheduling model with large body of transported data during the process of execution of tasks.

References

1. Foster, I., Kesselman, C.: The Grid: Blueprint for a Future Computing Infrastructure. Morgan Kaufmann, San Francisco (1998)
2. Vadhiyar, J.D.: A Metascheduler For The Grid. In: Proceedings of the 11th IEEE International Symposium on High PerformanceDistributed Computing, Edinburgh, Scotland, pp. 343–351. IEEE Computer Society, Los Alamitos (2002)
3. Faycal, B., Jean-Patrick, G., Laurent, L., et al.: Designing and Evaluating an Active Grid Architecture. Future Generation Computer Systems, Advanced Grid Technologies 21(2), 315–330 (2005)
4. Freund, R.F., Siegel, H.J.: Heterogeneous processing. Computer 26(6), 18–27 (1993)
5. Topcuoglu, H., Harir, S., Wu, M.-Y.: Performance-effective and low-complexity task scheduling for heterogeneous computing. IEEE Trans on Parallel and Distributed Systems 13(1), 260–274 (2002)
6. Graham, R.L.: Bounds for certain multiprocessing anomalies. Bell System Technical Journal 45, 1563–1581 (1966)
7. Adam, T.L., Chandy, K.M., Dickson, J.R.: A Comparison of List Scheduling for Parallel Processing Systems. Comm. ACM 12(17), 685–690 (1974)
8. Wu, M.Y., Gajski, D.D.: Hypertool: A Programming Aid for Message-Passing Systems. IEEE Trans. Parallel and Distributed Systems 3(1), 330–343 (1990)
9. Kruatrachue, B., Lewis, T.: Grain Size Determination for Parallel Processing. IEEE Software, 23–32 (1988)
10. Macey, B.S., Zomaya, A.Y.: A Performance Evaluation of CP List Scheduling Heuristics for Communication Intensive Task Graphs. In: International Parallel Processing Symposium (IPPS 1998), Orlando, Florida, March 30-April 3 (1998)
11. Beaumont, O., Boudet, V., Robert, Y.: The Iso-Level Scheduling Heuristic for Heterogeneous Processors. In: PDP 2002, pp. 335–342 (2002)

A Hierarchical Replica Location Approach Based on Cache Mechanism and Load Balancing in Data Grid*

Baoyan Song[1], Yanying Mao[1], Yan Wang[1], and Derong Shen[2]

[1] School of Information, Liaoning University, Shenyang, China 110036
bysong@lnu.edu.cn
[2] School of Information Science and Engineering, Northeastern University,
Shenyang,China 110004

Abstract. According to data grid characteristics and data-sharing characteristics in distributed environments, we present *RepliLoc*, a hierarchical replica location approach based on Cache mechanism and load balancing. In *RepliLoc,* considering storage load and computational load balancing, replica location is classified into community layer and community alliance layer location. On community layer, a circular information dissemination mechanism based on WS model is proposed to locally locate replica. On the other layer, we propose a replica location approach based on prefix matching and DHT. In addition, Cache mechanism is used by both the two layers to shorten query latency. A series of experiments proved that our approach is efficiency and effectiveness.

1 Introduction

A data grid connects a collection of geographically distributed computers and storage resources that may be located around the world, which enables users to share data and other resources. In data grid, Replication of data can avoid access bottleneck and help to reduce access latency and bandwidth consumption. A Replica Location Service is a core component of one grid data management architecture, which provides a mechanism for registering the replicas and discovering them according to given parameters.

Recently, replica location mechanism has attracted much attention of scholars and specialists around the world, and they have proposed some better strategies. Giggle [1] proposed by Globus and Europe Data Grid is a framework for constructing scalable replica location service, but membership services have not been implemented fully. PRLS [2] proposed by Min CAI etc is a p2p replica location service based on a Distributed Hash Table (DHT), but local clustering property in data grid environment is not considered. Ridrop [3] proposed by Xiaodong You etc is a replica location Mechanism based on DHT and small-world theory, but using of Gossip is apt to generate bandwidth and computational redundancies.

---

* This work is supported by the National Natural Science Foundation of China(60673139) and the Educational Department of Liaoning Province Scientific Research Projects of College & University(20060349) and Shenyang Scientific and Technology Program (061833).

Y. Zhang et al. (Eds.): APWeb 2008, LNCS 4976, pp. 148–153, 2008.

Objects location technique for p2p application has been a topic of extensive research in recent years. The structured p2p overlay network[4] builds a search-efficient indexing structure that provides good scalability and search performance. Considering the nature of data grid, we can not apply it to replica location directly. But they can lead us to solve some issues of replica location.

In data grid environments, data is heterogeneous, distributes widely around internet, Data-sharing characteristics exploit small-world characteristics [5-7], which are temporal locality, geographical locality and spatial locality. Considering these characteristics and p2p objection location technology, we present a novel replica location approach, a hierarchical replica location approach based on Cache mechanism and load balancing, named *RepliLoc*.

In *RepliLoc*, we focus on two issues: storage load balancing and computational load balancing. Storage load balancing is achieved via rough evenly distributing replica index maps among nodes by using DHT and prefix matching. Computational load balancing can be accomplished via localizing replica location by data community. Replica location is classified into two layers: replica location on community layer and community alliance layer. On community layer, we present a circular information dissemination mechanism based on small world model (WS model) to accomplish replica location. On the other layer, we propose a replica location approach based on prefix matching and DHT. In addition, Cache mechanism is used by both the two layers to shorten query latency. A series of experiments proved that our approach is efficiency and effectiveness.

2 *RepliLoc* Architecture

In data grid environment, *Resource Logical Name* (RLN) is a unique logical identifier for desired data content. Each RLN has one or more physical copies, which identified by *Replica Physical Address* (RPA) to specify its location on a storage site. The replica location is to determine one or more RPAs of a given RLN.

According to data grid and data-sharing characteristics in distributed environments, we design a two-layer

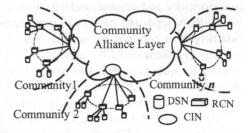

Fig. 1. Two-layer Architecture of *RepliLoc*

architect-ture. As shown in Fig.1. The low-layer is the community layer, and the high-layer is the community alliance layer. The hemicycle in dotted lines identifies a data community. The cloud formation identifies a community alliance layer. Each community includes some *Data Storage Nodes* (DSN), some *Replica Catalog Nodes* (RCN), and two *Catalog Index Nodes* (CIN). When every RCN registers into a community, it is identified by a monotony increasing decimal number given by *RepliLoc*. Then all RCNs are organized into the circle by its identifier in ascending order. All communities connected by CINs in a certain way to form Community alliance layer.

A DSN is any node joined in data grid. Physical data is stored at these DSNs. RCNs provide an interface for client to register replica information when creating one or more new copies on DSNs and discover them. A CIN aggregates replica information stored at all RCNs in one community, and then generates an index by distributing their replica index maps (*RLN*, *CIN*) among CINs.

3   Replica Distribution and Location on Community Layer

According to two-layer architecture of *RepliLoc*, replica distribution and location need two steps: replica distribution and location on community layer and on community alliance layer. When replicas are created on DSN, users need register their information in RCN which DSN belongs to. After recorded them, RCN sends them to CIN in the same community. The process of distributing replica on community layer is simple. So we only describe replica location process in details in this section.

- Cache Mechanism on Community Layer

In data grid, data-sharing environments exploit the small world topology. In *RepliLoc*, a data community can be seen as a small world. Small world phenomenon can be described by small world model (WS model)[5-7]. According to the small world properties, we employ Cache mechanism in replica location. That is, each RCN caches some other RCNs that were accessed recently. When disseminating replica queries, local RCN disseminates queries to these RCNs to improve probability of hitting the target. After Cache table is filled, the oldest RCNs can be updated appropriately by the new one.

- Replica Location on Community layer

On community layer, our circular information dissemination mechanism based on WS model is described as follows: When querying replicas, RCN sends replica queries clockwise and counterclockwise to two nearest neighbors on either side. Meanwhile, RCN sends the queries to some remote RCNs. After received these queries, two neighbors continue to sending them to their two neighbors in their directions respectively. Remote RCNs generate a new dissemination headstream and continue sending the queries to their two nearest neighbors in different directions of circle and their new remote RCNs in the same way. Finally, queries are disseminated along double circles till find the replicas or visit all RCNs in current community.

The remote RCNs must meet following condition: one is that remote RCNs are selected from cache RCNs, the other is that these selected Cache RCNs must meet a restriction. That is, the number of newly selecting remote RCN and the number of all selected remote RCNs form a new circle. The distance of each RCN on new circle is larger than $k + m$. Here $k = (n*(n-1))/(2*(n+1))$, n represents the size of RCNs in current community, m represents the times of visiting or indirectly visiting remote RCNs from headstream.

The algorithm of replica location on community layer is described by algorithm 1.

Algorithm 1 (Replica Location on Community Layer)
Step 1 RCN sends replica queries to two neighbors and remote RCNs;
Step 2 If two neighbors found RPA of RLN, return results to RCN and end dissemination, turn to *setp 4*, else continue sending them to their two neighbors;

Step 3 If remote RCNs found the RPA, return results to RCN and end dissemination, turn to *setp 4*, else, generate a new dissemination headstream and send queries to theirs two neighbors and remote RCNs, turn to *setp 2*;

Step 4 If RPA is found, RCN record IP of RCN that maintaining RPA, else turn to replica location on community alliance layer;

Step 5 End.

4 Replica Distribution and Location on Community Alliance Layer

If no desired replicas in community, we need locate replica beyond communities. We distribute and locate replicas based on prefix matching and DHT On this layer.

4.1 Identification and Routing Table of CIN

In order to distribute and locate replica among CINs based on prefix matching and DHT, CINs need be identified with global unique identifiers (LI), represented as k-digit strings of radix-16. RLN can be identified by a logical identifier (RI), represented as 32-digit strings of radix-16. They are achieved by using a secure hashing algorithm like MD5. LI is dynamic and scalable according to the size of current CINs in the system. Suppose N is the size of CINs, $k=\log_N^b+1$, $M=$hash(IP), and LI=PreFix (M, k). If the size of current CINs in system is larger than 16^k, system randomly get a digit, append it to bottom of original LI. If the size of CINs in system is smaller than $16^{K-1}/ (k+1)$, the last digit of strings is cut. We distribute and locate replica by matching LI and RI digit by digit until find a CIN that shares the longest prefix with the RI. This CIN is the root of RI and maintains replica index maps of RI.

To find root digit by digit, each CIN contains links to a set of neighbors that share prefixes with its LI. According to uncertainty of network environment and approach of prefix matching, we design routing table as two-neighbor structure. That is, a CIN with LI α that is a k-digit string of radix-16, its routing table has k columns, and there are 16 entries on each column and 2 neighbors in each entry. Two neighbors of the ith entry on the Lth (here $1<=L<=k$) column share the LI that begins with prefix(α, L-1)+i, and digits begin from $L+1$ are closest with the digits at relative location in α.

4.2 Replica Distribution and Replica Location with Cache Mechanism

We route replica index maps of RI to its root by using prefix matching and DHT to implement replica distribution on community alliance layer. The process of distributing replica index maps can be described by algorithm 2.

Algorithm 2 (Replica Distribution on Community Alliance Layer)

Step 1 CIN gets RI by applying MD5 to RLN;

Step 2 CIN matches RI with neighbors in relative entry of routing table to find root of RI. The replica index maps are stored at the CINs passed by;

Step 3 Stores maps at root;

Step 4 End.

On this layer, replica location is achieved by finding the root of RI based on prefix matching and DHT. In addition, according to the small world characteristics, Cache mechanism is used to cache replica information that is discovered recently by CINs. It can shorten path of dissemination and avoid matching and bandwidth redundancies. After Cache table is filled, the oldest information would be replaced with new one. The process of locating replica can be described by algorithm 3.

Algorithm 3 (Replica Location on Community Alliance Layer)
Step 1 CIN gets RLN and matches it with cached information. If there are desired items, gets RPA and turns to *Step 6*;
Step 2 CIN gets RI and matches it with routing table to find root of RI;
Step 3 RI is matched with replica index maps stored at CINs passed by. If there are desired items, return results. Else CIN continue routing to root and return the results;
Step 4 If results are Null, CIN turns to *Step 7*, else, gets the CINs in results and sends queries to these CINs;
Step 5 CINs in results get RPA and return to CIN;
Step 6 CIN appends RLN and its RPA to Cache table;
Step 7 CIN returns results to RCN;
Step 8 End.

5 Experiment and Performance Evaluation

We focus on two preliminary performances in experiment: the performance of storage and computational load balancing. These measurements were run on four machines (CPU 2.0GHz, memory 256MB) running windows XP. Replica information and replica index maps are stored at tables of SQL server2000 database. Java multithread is used to act as CINs, RCNs and DSNs. They communicate each other by Socket.

Load balancing: Experiment 1 is to test storage load. Fig.2 shows the number of replica index maps on every CIN after 1000, 2000 RLN is hashed to sixteen CINs. From Fig.2 we can see that replica index maps are rough evenly distributed on every CIN by using DHT and prefix matching, which roughly achieve storage load balancing; Experiment 2 is to test computational load. 100 query operations on system that respectively contains 1, 15, 25, 40 CINs are executed. Each CIN maintains 1500 maps. Each community contains 15 RCNs that maintaining 100 replica information. Average query latency is measured

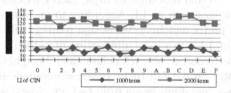

Fig. 2. Number of Maps on CINs

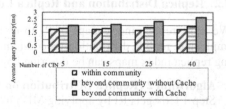

Fig. 3. Query Latency under Different Scenes

under three scenes: 100 query operations within community, 100 query operations beyond community without cached information on CIN, 100 query operations beyond community with cached information on CIN. Results of test are shown in Fig.3: Because of localization of replica location by data community, the query latency with community is not increased with the number of CINs increasing. Besides, the query latency is shortened when there is cached information on CIN comparing to no cached information.

Average Query Latency: Experiment 3 is to test average query latency in *RepliLoc* and compare it with PRLS. We execute the same query operations on the same number of nodes that maintain same replica information with PRLS. The result (shown in Fig.4) tells us that the average query latency of *RepliLoc* is shortened comparing with PRLS.

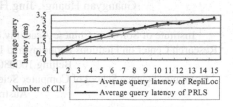

Fig. 4. Average Query Latency of *RepliLoc* and PRLS

6 Conclusion

In *RepliLoc*, we focus on two issues: One is storage load and the other is computational load. So the replica location is classified into two layers. Different replica location approach respectively is used on different layer according to characteristics of each layer. To do so, computational load balancing can be achieved. Replica index maps are rough evenly distributed among CINs by using prefix matching and DHT to accomplish storage load balancing. In addition, using of Cache mechanism can shorten path of dissemination and avoid matching and bandwidth redundancies generated when users repeatedly accessing the same data.

References

1. Chervenak, A., Dedman, E., Foster, I.: Giggle: A framework for constructing scalable replica location services. In: Proceedings of Supercomputing 2002 (SC 2002), pp. 1–17 (November 2002)
2. Chervenak, A., Cai, M., Frank, M.: Applying peer-to-peer technology to grid replica location services. Journal of Grid Computing, 49–69 (2006)
3. You, X., Chang, G., Yang, W., et al.: Replica.location mechanism based on DHT and the small-world theory. In: Jin, H., Pan, Y., Xiao, N., Sun, J. (eds.) GCC 2004. LNCS, vol. 3251, pp. 319–326. Springer, Heidelberg (2004)
4. Zhao, B.Y., Huang, L., Stribling, J., et al.: Tapestry: a resilient global-scale overlay for service deployment. IEEE Journal on Selected Areas in Communications 22(1), 41–53 (2004)
5. IamLItchi, A., Ripeanu, M., Foster, I.: Locating data in (Small-World) peer-to-peer scientific collaborations. In: Workshop on Peer-to-Peer Systems, Cambridge, Massachusetts (2002)
6. Newman, M.E.J., Strogatz, S.H., Watts, D.J.: Random graph with arbitrary degree distribution and their applications. Phys. Rev.e 64 (2001)
7. Raman, S., McCanne, S.: A medel, analysis, and rrotocol rramework for soft state-based communication. Computer Communication Review 29 (1999)

Wireless Video-Based Sensor Networks for Surveillance of Residential Districts

Guangyan Huang[1], Jing He[2,3], and Zhiming Ding[1]

[1] Institute of Software, Chinese Academy of Sciences, Beijing 100080, P.R. China
[2] Research Center on Fictitious Economy and Data Science, Chinese Academy of Sciences, Beijing 100080, P.R. China
[3] Computer Science School of Computer Science and Mathematics, Victoria University, Australia
huanggy@ercist.iscas.ac.cn, hejing@gucas.ac.cn,
zhiming@iscas.ac.cn

Abstract. Compared to traditional wired video sensor networks to supervise a residential district, Wireless Video-based Sensor Networks (WVSN) can provide more detail and precise information while reduce the cost. However, state-of-the-art low cost wireless video-based sensors have very constrained resources such as low bandwidth, small storage, limited processing capability, and limited energy resource. Also, due to the special sensing range of video-based sensors, cluster-based routing is not as effective as it apply to traditional sensor networks. This paper provides a novel real-time change mining algorithm based on an extracted profile model of moving objects learnt from frog's eyes. Example analysis shows the extracted profile would not miss any important semantic images to send to the Base Station for further hazards detection, while efficiently reducing futile video stream data to the degree that nowadays wireless video sensor can realize. Thus it makes WVSN available to surveillance of residential districts.

Keywords: sensor networks, mining video stream, contrast data mining.

1 Introduction

Wireless Sensor Networks (WSN) possess of a large number of low cost sensor nodes capable of sensing, computing, and communication. Nowadays low cost CMOS or CCD cameras and digital signal processors make more capable multimedia nodes available to be applied in WSN. Video information provides significant benefits to many sensor networking applications, such as environmental monitoring, health-care monitoring, and security or surveillance etc. This paper focuses on the Wireless Video-only Sensor Networks (WVSN) for surveillance of residential districts.

Compared with wired camera networks, WVSN has three advantages similar to those in [1]: to reduce costs for network installation, to extend the monitoring target areas from indoors of houses or buildings to the whole residential districts, and to easily adapt upon changes of utilizations. Also, because large number of wireless

Y. Zhang et al. (Eds.): APWeb 2008, LNCS 4976, pp. 154–165, 2008.
© Springer-Verlag Berlin Heidelberg 2008

cameras with constrained resources and low cost replace a few traditional powerful expensive wired camera, more detail and wider range of the region can be monitored.

However, WVSN provides a formidable challenge to the underlying infrastructure, because it generates large amount of video data, which consume orders of magnitude more resources, such as storage, computation, bandwidth and energy resource, than their scalar sensor counterparts [4]. Thus, WVSN should be designed to satisfy limited resource demand in surveillance of residential districts and semantically relevant information should be extracted from the raw video data to send to the Base Station (BS) for further hazards detection, no matter BS with abundant resources can finish the detecting tasks automatically by using the proposed extracted profile or people may detect hazard events by watching the screen about the extracted profile.

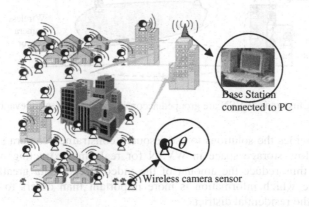

Fig. 1. Wireless camera networks for residential districts

Different from traditional scalar sensor networks, video cameras have the unique feature of capturing object images of a region at arbitrary locations, perhaps distant from the camera [2]. Thus, the sensing range of sensor nodes is replaced with the camera's Field of View (FoV), which is defined as the maximum volume visible from the camera [3]. Video sensors are deployed to make the sensing range of WVSN to cover most of the region in residential districts, such as each road, each corridor, door and stairway of the building, etc, while WVSN also ensures the wireless connection between each camera sensor and BS. An example is shown in Fig. 1. Fig. 1 shows only one BS connected with a PC as a control center, wireless camera sensors, distributed everywhere in the district, route data to BS through one hop or multi-hop.

Based on the unique way that cameras capture data, the authors in [2] analyze how an algorithm designed for traditional WSN, which integrates the 2-D coverage and routing problem, behaves in video-based networks. For application-aware protocols, it uses redundant camera sensors to cover the same region and balances the energy dissipation in WVSN by choosing those cameras that do not capture data as routing nodes. However, this is not suit for the applications without much redundancy of cameras for the same region in surveillance of residential districts.

Cluster-based routing protocols, such as LEACH [5] and BCDCP [6], which are adopted to reduce data by fusing redundant data among sensors with relevance content, are not suit for WVSN. Different from traditional WSN where scalar sensors may evenly randomly deployed, wireless camera sensors are deployed in uneven densities because of their FoV without any obstructs between the objects and them. It's easy to understand the camera sensors are distributed as Fig. 2 shows in a residential district. Also, to fuse video data efficiently, camera sensors forms clusters in a WVSN not according to the Euclid distance but the content relevance of the video data. In Fig. 2, camera sensors that monitor Building 1, Building 2, Road, and Lawn are grouped into different clusters respectively. Clusters are fixed and video data would not be processed further along CHs to the BS. Therefore, cluster-based routing protocols are very simple in WVSN and cannot reduce video stream data greatly.

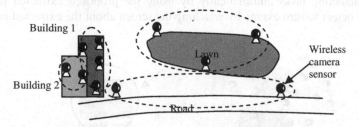

Fig. 2. Clusters in WVSN are grouped according to the content relevance

This paper provides the solutions to use resource constrained camera sensors with low bandwidth, low storage space in WVSN for real-time extracting of important information and thus reduce the amount of the video stream data greatly. The key point is to define which information is more important than others to help detect hazard events in the residential districts.

The rest of the paper is organized as follows. Section 2 introduces related works. Section 3 presents a novel extracted profile model of moving objects learnt from frog's eyes to reduce the amount of video stream. Section 4 gives a real-time change mining algorithm based on three-windows to execute the extracting task. Section 5 provides an example to analyze performance of the proposed change mining methods for surveillance of the residential districts. Section 6 discusses advantages and limitations of the proposed methods. At last Section 7 concludes the paper.

2   Related Works

2.1   Traditional Schemes to Reduce Video Stream Data

Source Coding. Uncompressed raw video streams require excessive bandwidth for routing, thus source coding is useful to compress the raw data. Intra-frame compression reduces redundancy within one frame, while inter-frame compression exploits redundancy among subsequent frames to reduce data to be transmitted and stored [4]. However, existing video source coding algorithm is too complex to be applied on the source nodes in WVSN due to processing and energy constraints.

In-network Processing. Camera sensors in WVSN would collaborate to reduce the transmission of redundant information, merge data originated from multiple views, or even transmit semantically relevant information. In [7], IrisNet (Internet-scale,

Resource-intensive Sensor Network Services) sends only a potentially small amount of processed data instead of transferring raw data. IrisNet uses two main techniques to digest the video data into semantic form: key points corresponding to real object/region and image stitching. The first one is to map the region in the real world to a particular pixel in a camera's image. The second one is to fuse information collected from multiple camera sensors with partially-overlapping FoV and generate panoramic views. Thus, IrisNet only sends the occupancy information of each parking space instead of rich video streams to reduce network bandwidth consumption in a parking space finder application. However, the processing of image data locally on the node will require a significant amount of energy that cannot easily be neglected.

2.2 Change Mining Methods to Reduce Video Stream Data

In [8], it argues that the changes to the patterns may be more critical and informative than other general snapshots of data streams. Also, stream data flows in-and-out dynamically and change rapidly and thus most of stream data may only be examined in a single pass because of limited memory or disk space and limited processing power [9]. In [10], change detection filtering is used to compare a pixel by pixel in an image frame and only record the changed frames to reduce video data. But low level (e.g. pixels) change detection is not efficient to achieve the user-cared information. Instead of bogging down to every detail of data stream, a demanding request is to provide on-line analysis of changes, trends and other patterns at high levels of abstraction [9]. How to abstract video stream with informative form and at the same time to process the video stream online on each resource constrained video-sensor determine whether WVSN can be used widely.

A general model of changes in data stream is given in [11], which use a two-window paradigm instead of storing the full history of the stream. In this detection algorithm, the data is compared in reference window to the data in a current window. The reference window is updated whenever a change is detected, while the current window slides forward with each incoming data point. Also, a meta-algorithm for change detection is given in [11] to reduce the problem from the streaming data scenario to the problem of comparing two (static) sample sets. However, the key to this meta-algorithm is the intelligent choice of the method to detect the difference between two windows. In this model, the changes are difficult to define. But one thing is sure that users are usually not interested in arbitrary change, but rather in change that has a succinct representation that they can understand [11]. Thus, to abstract video stream at high level by online change detection is a good solution for WVSN. The main idea of this paper is to define clearly the change model for extracting profile of moving objects with sufficient information to detect hazards, and to use a three-window algorithm for real-time mining of the defined changes.

3 Proposed Solutions

3.1 Modeling of Extracted Profile

Camera sensors would capture video data all the time and form a video data stream (e.g. 30 frames each second). With limited storage, camera sensors cannot store all the video data locally or send them to the BS, thus the process unit, MCU, must extract important information from the video stream promptly. We can learn from frog's eyes. In [12], *"A frog hunts on land by vision. He escapes enemies mainly by seeing them. But Frog's eyes do not move, to follow prey, attend suspicious events, or search for things of interest."* A frog can handle the image captured by his eyes quickly and response suddenly to the situation around him, because there are four separate detections of sustained contrast, net convexity, moving edge and net dimming on the image in the frog's eye to reduce data for brain to process given in [12]. Analyzing the four operations by frog's eyes, we suppose frog's eyes use three main steps to extract important image based on above change detection. Firstly, frog's eyes separate location-moved or color-changed objects of foreground from background. Secondly, frog's eyes detect the object changes based on object edge or convexity shape, object speed and object size. Thirdly, frog's eyes simplifies the complex situation by considering only single larger object other than a group of smaller ones, thus the movements of background (such as flowers and grass) is neglected. Based on the above steps, frog's eyes sense the whole process of the objects' movements from importing to leaving the field of view. *Thus, the extracted profile is defined as important frames with great changes of moving objects based on object's location and shapes.* Speed changes that can be deduced by locations are neglected.

3.2 Processing Procedure of Video Stream

In Fig. 3, it gives the scheme when no moving objects, no data are stored. Fig. 3 also shows camera unit capture raw image from the physical environment, then MCU runs the change detection algorithm and record extracted profile data (e.g. a group of representative images of moving objects) into the storage, at last radio unit route necessary data (a subset of extracted profile) to the BS. The key point is focused on how to determine changes.

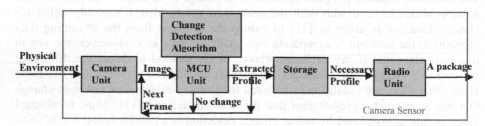

Fig. 3. Data stream and processing procedure of a camera sensor

4 Real Time Mining of Changes from Video Stream

4.1 Modeling of Changes

In this paper, three levels of changes inter-frames in video stream are defined: PiXel Change (PXC), Object Feature Change (OFC) and AcTion Change (ATC).

Pixel Change (PXC). A simple pixel change detection method for comparing two frames of image in video stream is introduced in [10]. A pixel is denoted as P (*color_vector, position_in_X_Y_plane*). Generally, a pair of *color_vector* in different image frames are compared when their values of *position_in_X_Y_plane* are equal. *color_vector* can be the values based on color spaces such as RGB, HSV, and etc. Suppose there always one frame at the very beginning is a background image without any foreground objects. In the first phase, pixel by pixel comparison is used to detect foreground pixels. Then foreground pixels will be clustered into different group to form objects. At the second phase, multi objects that have been recognized are tracked by only searching the change pixels nearby. PXC is the basis for detecting OFC.

Object Feature Change (OFC). Suppose PXC can produce separated objects. We give some general features of each visual foreground object. The spatial positions of objects are a 2D coordinate data denoted by $S\,(x,\,y)$, where $(x,\,y)$ is the centroid of the object on X-Y-Plane. Another character of video stream is its time order, we use *frame ID* to describe it. According to extracted profile learnt from frog's eyes in Section 3.1, we summarize the objects feature in concise representation as follows: object size (the number of pixels), object shape (contour), and object spatial location (centroid) on X-Y-Plane. Object size is used to compute depth for 3D location of object, which is used to adjust the shape change ratio for shape change detection.

Action Change (ATC). Action change is the highest level change that can be understood by user directly. Based on OFC detection, ATC can be detected according to different applications. In surveillance of residential districts, we suppose the situation of intruder detection is similar to the situation of enemies detection by frog's eyes, thus we only define two human actions: moving based on location change, and something separated from the human, or something combinated with the human based on image distorted, including the situation that only part of the object is in FoV. Some constraints in real world determine the action changes we care. To ensure online processing, the changes happen at a time span no greater than three frames. Also, the source data is a serial of 2D images that we can use to detect changes.

4.2 Change Detection Algorithm for Video Stream

Both OFC(Centroid) and OFC(Shape) are used to detect location changes and image distorted changes respectively for extracting representative images from the video stream. OFC(Centroid) is based on centroid on X-Y-Plane of the object and OFC(Shape) is based on both shape and size of the object.

Object centroids at three sequential frames of *frame i-1, frame i* and *frame i+1* are denoted by f_{i-1}, f_i and f_{i+1} respectively. The value of OFC (Centroid) is given by

$$\varpi_c = diff_2(diff_1(f_{i+1}, f_i), diff_1(f_i, f_{i-1}))\,. \tag{1}$$

Given the value of the first level OFC as follows:

$$diff_1(g, h) = \overrightarrow{gh} = (g.x - h.x, g.y - h.y)\,. \tag{2}$$

We define the second level OFC as follows,

$$\varpi_c = \arccos(\frac{u \bullet v}{|u| * |v|}), \varpi_c \in [0,180°] \tag{3}$$

where $u = diff_1(f_{i+1}, f_i) = (x_{i+1} - x_i, y_{i+1} - y_i)$, and $v = diff_2(f_i, f_{i-1})$ $= (x_i - x_{i-1}, y_i - y_{i-1})$.

Fig. 4 shows the mechanism of how to use OFC (Centroid) to detect a key frame of ATC (Moving). A, B and C are three centroids of the moving object at three sequent frames. θ can be computed by Eq. (2). Given a threshold of $\xi = 90°$, the moving pattern change can be detected. Examples are shown in Fig. 4 (a) and Fig. 4 (b). In Fig. 4 (a), $\theta < \xi$, thus, no moving pattern change is detected, while in Fig. 4 (b), $\theta > \xi$, thus moving pattern is determined to be changed.

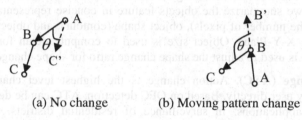

(a) No change   (b) Moving pattern change

Fig. 4. OFC (Centroid) detection mechanism

Object shape is denoted by a serial of difference values of $\Delta Y_i = \{ \Delta y_{1i}, \Delta y_{2i},$ $... \Delta y_{xi}, ..., \Delta y_{mi} \}$, where $\Delta y_{xi} = \max(Y_{xi}) - \min(Y_{xi})$ and m is the maximum width of the image frame. $Y_{xi} = \{ y_{1xi}, y_{2xi}, ..., y_{kxi} \}$ ($x \le m$) at *frame i*, where k is the number of pixels in the set of object pixels, whose X-coordinate value equals x.

$$\varpi_s = diff_1(\Delta Y_i, \Delta Y_{i-1}) = \sum_{j=1}^{m} |\Delta y_{ji} - \Delta y_{j(i-1)}| / N_i \tag{4}$$

where N_i is object size (the number of pixels) at *frame i*.

Fig. 5 shows simplified method to measure the difference between two frames of the object shapes. The curves of S_1S_2 (*Shape1*) and S_3S_4 (*Shape2*) are different projections of two shape contours on X-Coordinate. Let *P1* and *P2* denote the closed area of AS_1S_2C and BS_3S_4D respectively. The number of pixels in hatched area denotes the value of OFC (Shape) between *Shape1* and *Shape2*, given as follows:

$$\sum_{j=1}^{m} |\Delta f_1.yj - \Delta f_2.yj| = P1 \cup P2 - P1 \cap P2 . \tag{5}$$

Eq. (5) can be used to explain Eq. (4), Eq. (4) has eliminated the impact of view depth by dividing value of object size. The curves of both *Shape1* and *Shape2* in Fig. 5 are discrete points.

We can imagine that shape changes smoothly in an action process. Thus ϖ_s computed by Eq. (4) is near to (maybe a little greater, or a little less than) 1. However, if the human changes his action from one kind to another, then $\varpi_s > \psi$, where ψ is the threshold determined by different applications.

OFC (Centroid) is computed by two levels of OFC and OFC (Shape) is computed by only one level of OFC. Fig. 6 shows the levels of OFC. Generally, n levels of OFC can be computed by setting each level of function according to different applications, and the output values of low level OFC are input parameters of its direct high level OFC function. Also OFC (Centroid) can be extended to three-dimension easily.

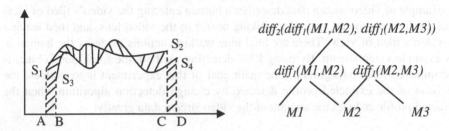

Fig. 5. Difference of two shapes,OFC (Shape)   **Fig. 6.** Levels of OFC

Table 1. Online change detection algorithm based on three windows

Step 1. Compute centroids and shapes in the three sequent windows: the last window (*LW*), the current window (*CW*) and the next window (*NW*). That means three frames must be stores in memory.
Step 2. Compute ϖ_c and ϖ_s according to Eq.(3) and Eq.(4) respectively;
IsChanged(ϖ_c, *centroid*, *CW*); *IsChanged*(ϖ_s, *shape*, *CW*);
Step3. The frame next to NW, named Future Window (*FW*).
if (*FW*!=0) {*NW=FW*; *CW=NW*; *LW=CW*; Go To **Step 1**; } else {Stop;}

Table 2. *IsChanged*() function

1	*IsChanged*(ϖ, *feature*,*CW*)
2	{ Switch (*feature*)
3	{ case *centroid*:
4	if ($\varpi < \xi$) *CW* is marked as changed frame;
5	case *shape*:
6	if ($\varpi > \psi$) *CW* is marked as changed frame;
7	}}

We have developed a novel change detection algorithm based on three windows to detect important frames from the data stream online to reduce the amount of data without losing key points in wireless video-based sensor networks. The main steps of our change detection algorithm are given in Table 1, where *IsChanged()* function in Step 2 is given in Table 2. In Step 2, computing of ϖ_c needs three windows of LW, CW and NW, and computing of ϖ_s only needs two windows of LW and CW.

5   Examples for Performance Evaluation

An example of video stream data describes a human entering the video's filed of view from the far away place, naturally walking nearer to the video lens, and then leaving the video's filed of view. There are total nine walking actions made by the human in this example video stream. By using PXC detecting method, the foreground object is discriminated from background. The main goal of this experiment is to evaluate the precision of the extracted profile detected by change detection algorithm when the extracted profile reduces the amount of the video stream data greatly.

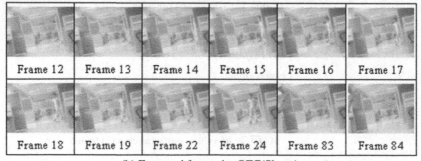

(a) Extracted frames by OFC(Centroid).

(b) Extracted frames by OFC(Shape).

Fig. 7. Sample frames

5.1 Evaluating Change Detection Precision

In this example, both OFC(Centroid) and OFC(Shape) of the change detection algorithm in Table 1 are used to detect important frames from video stream and the results are shown in Fig. 7 (a) and Fig. 7 (b) respectively. There are total of 20 frames are detected from 85 frames in both Fig. 7 (a) and Fig. 7 (b), including two redundant frames of Frame 15 and Frame 22 in both Fig. 7 (a) and Fig. 7 (b). Change frames in Fig. 7 (a) are distributed evenly while change frames in Fig. 7 (b) focalize at the beginning and at the end few frames. The detailed analysis is given as follows.

In Fig. 8, the changed frames detected by using centroid change detection method are plot in Fig. 8 also marked with red pane, which are a set of frames: Cset={11, 15, 22, 30, 36, 41, 48, 52, 60, 71}. The change parameters are computed by Eq. (3) and detection threshold, given in line 4 of Table 2, is set to $\xi = 90°$ in this example. Thus sampling 10 frames from the original 85 frames, the video stream data is reduced to about 11.8% and at the same time the key points about the walking action in the video stream are captured shown in Fig. 7 (a).

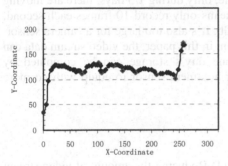 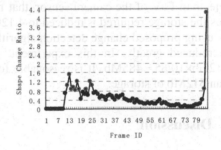

Fig. 8. Trajectory of Moving Object   **Fig. 9.** Shape Change Ratio

In Fig. 8, it's easy to understand the detected frames that capture the image when the human is just during striding actions and the centroid of the human descends naturally. However, how to detect the frame with the special action out by only knowing one last frame and one next frame is more difficult. By using the proposed change detection method based on centroid, ten points are detected but two faults (Frame 11 and Frame 71) among them. Thus, seven walking actions are detected correctly shown in Fig. 7 (a) when the images are captured completely. Fig. 7 (a) also shows walk 3 is detected twice in both Frame 30 and Frame 36, because the human object in the image turns its body left. Two walking actions are missed because of incompletely capturing of the human object by the FoV of the camera. If all the frames are captured completely, then the precision of important frame detection is 100% with redundant frames. That means the proposed change mining algorithm would not miss any important frames.

Shape change ratio computed by Eq. (4) is plot in Fig. 9. The changed frames detected by using shape change detection method shown in Fig. 7 (b) are also marked with red pane in Fig. 9, which are a set of frames: Sset={12, 13, 14, 15, 16, 17, 18, 22, 24, 25, 83, 84}. The change parameters are computed by Eq. (4) and detection threshold, given in line 6 of Table 2, is set to ψ =0.8 in this example. Thus 12 frames are sampled from the original 85 frames.

Thus, extracted profile by change detection algorithms in this paper provides a more semantic content of the video stream to split the whole process of the objects' movements from importing to leaving the field of view into three phases: importing phase, leaving phase and middle phase. In the middle phase, lots of important actions that are detected by OFC(Centroid) will send to the BS for further hazards detection. But extracted points of importing phase and leaving phase detected by OFC(Shape) can be used to split the whole process and not necessarily to send to BS.

5.2 Storage Analysis

Suppose image size is 320×240 pixels, a compressed image (such as jpg) is nearly 10kb. The capture speed of camera is 30 frames each second. If record the entire video stream, the total storage required to record one day's data is 10kb×30×60 ×60×24 ≈ 24.72Gb. It's impossible to equipped camera sensors with so large storage. Also, using the change detection filter in [10], only images with moving objects are recorded, thus it can reduce data. For example, only during 1/3 days, there are moving objects in FoV of the camera sensor, that means only record 10 frames each second, thus the storage required is reduced to 4.12Gb. It's also too large for a camera sensor. By using proposed change detection algorithm in this paper, the video stream data can be reduced greatly. In this experiment, one day's storage can be computed by 4.12Gb×10/85 ≈ 485Mb. It is reasonable for a camera sensor to equip with no more than 512Mb of storage.

6 Discussion

The proposed solution has four advantages: (1) Reducing the amount of video stream data greatly. (2) Keeping sufficient information available to detect hazards such as human intruders. (3) Real time processing image data by using limited storage and CPU resources. (4) Based on extracted frames from video stream, other image processing can be done to reduce the video data further, such as compressing or recognition. The change detection methods are very simple and can be running on video sensors online. Also, the two features of shape and location can resist noise well, and losing a few of pixels cannot affect the values of the two features much.

However, some extensions need be done in the future work: (1) The change detection methods for only one object should be extend to handle multi objects. (2)Detect shape changes other than convexity changes. (3) More actions should be detected other than walking action, and appearing and disappearing processes.

7 Conclusions

Nowadays VWSN node with limited resources has not more power than a frog's eyes. We learn from frog's eyes to simplify the process of handling video stream. Thus, this paper proposes the extracted profile and its mining methods leant from the mechanism of frog's eyes. A novel real-time change mining method based on definition of changes on extracted profile is given. Example analysis shows the extracted profile would not miss the important semantic images to send to the BS for further hazards detection, while efficiently reducing futile video stream to the degree that nowadays wireless video sensor can realize. Although this work uses contrast data mining to detect changes in video stream at very beginning stage, some exciting results have been achieved. In the near future, more statistical data analysis will be used to prove the effectiveness of the proposed change mining methods.

Acknowledgments. The work was partially supported by the National Natural Science Foundation of China under Grant No. 60573164, 70602034, 70531040, 70472074, 70621001, by ARC Discovery Project DP0345710, VU New Research Directions Grant, and SRF for ROCS, SEM. The authors are grateful to the referees for their useful and constructive comments, which have resulted in an improved paper.

References

1. Derbel, F.: Reliable Wireless Communication for Fire Detection Systems in Commercial and Residential Areas. In: Proc. IEEE Wireless Communications and Networking Conference, vol. 1, pp. 654–659 (2003)
2. Soro, S., Heinzelman, W.B.: On the Coverage Problem in Video-Based Wireless Sensor Networks. In: Proc. 2nd International Conference on Broadband Networks, vol. 2, pp. 932–939 (2005)
3. Holman, R., Stanley, J., Ozkan-Haller, T.: Applying Video Sensor Networks to Nearshore Environment Monitoring. IEEE Pervasive Computing 2(4), 14–21 (2003)
4. Akyildiz, I.F., Melodia, T., Chowdhury, K.R.: A Survey on Wireless Multimedia Sensor Networks. Computer Networks 51, 921–960 (2007)
5. Heinzelman, W.B., Chandrakasan, A.P., Balakrishnan, H.: An Application-Specific Protocol Architecture for Wireless Microsensor Networks. IEEE Transactions on Wireless Communications 1, 660–670 (2002)
6. Muruganathan, S.D., Ma, F., Bhasin, D.C., Fapojuwo, R.I., O., A.: A Centralized Energy-Efficient Routing Protocol for Wireless Sensor Networks. IEEE Communications Magazine 43, 8–13 (2005)
7. Campbell, J., Gibbons, P.B., Nath, S., Pillai, P., Seshan, S., Sukthankar, R.: IrisNet: An Internet Scale Architecture for Multimedia Sensors. In: ACM MM 2005, Singapore (2005)
8. Dong, G., Han, J., Lakshmanan, L.V.S., Pei, J., Wang, H., Yu, P.S.: Online Mining of Changes from Data Streams: Research Problems and Preliminary Results. In: Proceedings of the 2003 ACM SIGMOD Workshop on Management and Processing of Data Streams (2003)
9. Chen, Y., Dong, G., Han, J., Pei, J., Wah, B.W., Wang, J.: Online Analytical Processing Stream Data: Is It Feasible? In: ACM DMKD (2002)
10. Feng, W.-C., Kaiser, E., Feng, W.C., Baillif, M.L.: Panoptes: Scalable Low-Power Video Sensor Networking Technologies. ACM Transactions on Multimedia Computing, Communications and Applications 1(2), 151–167 (2005)
11. Kifer, D., Ben-David, S., Gehrke, J.: Detecting Change in Data Streams. In: Proc. of the 30th VLDB Conference, Toronto, Canada, pp. 180–191 (2004)
12. Lettvin, J.Y., Maturana, H.R., McCulloch, W.S., Pitts, W.H.: What the Frog's Eye Tells the Frog's Brain. In: Proc. IRE, vol. 47, pp. 1940–1951. reprinted in Warren S. McCulloch, Embodiments of Mind, MIT Press, Cambridge (1959)

Trust Maintenance Toward Virtual Computing Environment in the Grid Service

Dongbo Wang[1] and Ai-min Wang[2]

[1] School of Computer Science and Engineering, BeiHang University, Beijing 100083, China
[2] School of Computer and Information Engineering, Anyang Normal University, Anyang 455000, China
dongbo_wang@126.com

Abstract. The latest virtual machine technology has provided us a better means of customizing executing environment for jobs in a Grid computing service. However, this is far from enough to guarantee the correctness of the computing outcome, since the executing environments may be compromised, for example some executable contents or configuration files may be tampered or some untrusty packages may be installed in the executing environment. If so, the results of the computing tasks would undoubtedly be affected. In this article, by taking advantage of the virtual machine technology and the relevant ideas in the field of Security and Attestation, two mechanisms are proposed to offer Grid service the function of guaranteeing the trustiness of virtual machine computing environment --- Trust Attestation for virtual software environment based on TPM (Trusted Platform Module) and TKVI (Trusted Kernel of Virtual OS Image), and the Trusted Loading Method for virtual machine based on Xen.

Keywords: trust Attestation, virtual computing environment, trusted loading, virtual machine.

1 Introduction

The key aim of Grid Computing has always been coming up with approaches to coordinate and share heterogeneous resources with high efficiency, so as to carter to the requirement of large-scale computing capacity from some complex applications. Currently, Grid systems mainly choose to encapsulate computing resources in the forms of service to ensure the share and coordination. Surely, this method has proven its effectiveness in many available Grid systems, and carter to the need of many applications to some extend. However, this way of resource encapsulation suffers from some critical drawbacks. Briefly they are:

1. Unable to easily provide jobs with proper executing environments.

Clients might need some customized computing environments for their jobs' execution in many circumstances, for example the job waiting to be executed may rely on some specific software or the support from particular Operating System. For those special requirements, it is probably difficult to locate an existing node in the Grid system that can perfectly match the whole need, even though in the Grid the required resources actually exist.

Y. Zhang et al. (Eds.): APWeb 2008, LNCS 4976, pp. 166–177, 2008.

2. Hard to guarantee the quality of service.

Even thought there is a suitable host for the specific job submitted, in most cases the job can not utilize the resources provided exclusively, but has to share them with many other jobs running in the same host. So it is hard to keep them from affecting each other in executions, and there exists the possibility that the failure of one job may lead to the breakdown of the whole host and therefore cause all jobs running in the host fail.

3. Hard to provide fine-grained resource management.

The current solution cannot configure concretely to what extent the resource can be utilized by a job executing in a node. For example, we cannot customize the utmost limit of CPU utilization for a specific job. The lack of fine-grained resource management may give rise to the waste of resources.

Fortunately, the latest development in the field of Virtual Machine has presented us solutions for those problems listed above. By simulating the underlying hardware, the virtual machine can effectively shield the heterogeneousness of different nodes in the Gird System, and provide better means for the customization of executing environments. At meanwhile, its ability of encapsulation and isolation make it possible for each job to enjoy its own executing environment exclusively with little loss in utilization of resources, so the quality of execution can be guaranteed. Due to these outstanding features of virtual machine, the virtualization of Grid has become prevailing currently.

Nonetheless, merely customizing the suitable and exclusive executing environment for each job is far from enough to guarantee that the outcome of each execution is of full validation and correctness. Despite no negative influence from other execution procedures in the same virtual machine, the running of a job still relies on the specific software context, which may, with high chance, contain some untrusty or tampered executable files, configuration files and development kits. Therefore, we present two mechanisms (mainly aiming at applications under Linux) to ensure the security of a virtual executing environment in Grid:

1. Remote Attestation toward Virtual Machines.

Two methods of the attestation is discussed here, the attestation based on the secure hardware as Trust Base and the attestation based on the Trusted OS kernel as Trust Base. In both scenarios, administrators are able to get the integrity measurements of the operating system and the additional running software from any virtual machine deployed in the Grid System, so that the validation and the credibility of the software environment are certifiable. And based on the integrity information, job-scheduler would be more competent for providing authentic executing environment to guarantee the correctness of the final results.

2. Trusted loading Method for Virtual Executing Environment.

Propose a way to guarantee the reliability of a virtual computing environment by creating the customized virtual machine in certain protocol. Make sure the loading process of the operating system and additional software in specific virtual machines are under the complete supervision for generating the proper and secure executing environments.

2 Background

2.1 Basic Concept

The Remote Attestation is a solution for trusted computing proposed by TCG (Trusted Computing Group). It allows a host to verify the hardware and software running on a remote host, and the host can then used to decide whether or not it trusts that hardware and software configuration. For instance, a host may only trust a remote host running an operating system with the proper patch level or software certified for certain purposes. TCG has proposed a specification for remote attestation that allows a host to remotely prove its hardware and software while protecting its privacy. There are mainly three functions involved in the process of Remote Attestation called trusted measurement, trusted storage and trusted reporting. Trusted measurement consists of taking measurements of platform characteristics. At the booting phase, a secure hardware on the host is responsible for measuring the BIOS and boot loader to ensure the Secure Boot, then when it comes to the startup of the operating system, the integrity measurement tool, which extend the TCG trust measurement concepts to dynamically take measurement in the application layer, would take over and continue the measurement toward software environment, including some modules of the operating system (in the Linux). In the process of trusted measurement, all executable content form BIOS all the way up to the application layer would be measured before they are loaded and running.

In particular, once a piece of executable content is measured, the measurement would be extended into the secure hardware, for it possess the secure storage that can farthest protect the integrity information from being tampered. The secure storage provides a "ratcheting" feature such that each new measurement hash is based on the current hash value and the one to be extended. For instance, if the hash of boot loader is currently contained in secure storage, when the operating system loads, the kernel would be measured and the secure storage would be updated with the hash of (boot loader hash + kernel measurement). Those measurements of the executable contents are also recorded in some specific measurement lists, including the name of executable contents and the values of the hashes.

Trusted reporting is the key component for the attestation and is accomplished by exporting the integrity information and vouching for its authenticity. The host reports trusted measurements by returning the measurement lists and the value of the secure storage digitally signed with the private key generated by secure hardware. Together the trusted measurement (measurement lists and value from the secure storage representing a specific set and sequence of values) and the signature of the key, provide the powerful assurance property that the host is functioning from a specific configuration. All a remote challenger need to do is to verify the signature on the trusted report using the public key get from the certifying authorities of the secure hardware to make sure the report is validate and then compare the measurement values against its set of trusted configurations. If all values in the measurement lists are same as those in trusted set, the testified host would be deemed as authentic.

2.2 Related Work

There have already existed some available implementations of the secure hardware and the integrity measurement tool. For example, TCG has presented the detailed specification of TPM (Trusted Platform Module), a board-embedded chip responsible for the secure boot with the ability of secure storage and encrypting functions; and also the TPM functions can be provided by some external secure coprocessor card that provides maximum security for sensitive data, such as private keys, through a tamper-responsive environment. As for the integrity measurement tool, IBM has presented an effective solution called IMA (Integrity Measurement Architecture), which is designed as the security module of the Linux kernel and can fully extend the trust measurement concepts into the application layer.

The virtual machines are also able to take advantage of the functions of secure hardware. With the assistance of proxy, virtual machines can easily access the service provided by secure hardware, for example, IBM has presented the vTPM (virtual TPM) as a solution for virtual domains of Xen to access the TPM functions through a proxy in VMM (Virtual Machine Monitor) called vTPM manager. This proxy maintains one vTPM instance for each virtual domain, and forward messages between TPM on the hosts and vTPM instances in VMM. So the support of secure hardware is equally available in virtual machines. In fact, as to be discussed next, due to some special traits in the creation and booting process of the virtual machine, even for those hosts without the support of secure hardware, virtual machines in them can also be authenticated based on the trusted OS kernel as the Trust Base.

3 Remote Attestation to Virtual Executing Environments

Remote attestation for virtual machines shares the same basic ideas as for real hosts, consisting of trusted measurement, trusted storage and trusted reporting. However, the virtual machine also enjoys many different traits from the real host, which makes the attesting process quite different from that for real machines. The virtual machine actually depends on the services provided by certain kind of virtual machine management program, and both the management program and a virtual machine itself are running in the host as ordinary executable contents, therefore, if the host is compromised, there is no way to assure the security of a virtual machine running in it. Only when the host is certified to be secure can the credibility of the virtual executing environment in it be discussed. For Instance, all the virtual instances in Xen are actually maintained in the domains created by Xend, the management program of Xen which is in charge of the creation and the booting of all virtual instances. Thus, if the host is compromised, there is high possibility that the Xend has also be compromised, so under such circumstance, we can come up with no measure to guarantee the security of the virtual instances running in the host, for the compromised Xend would involve in all the fundamental processes of a virtual machine's creation. On the other hand, if the host is though to be secure, we can believe that the booting phase of a virtual machine should be trusted.

Furthermore, another special trait of virtual machine is that its operating system exists as an image file, which can be easily customized according to requirements.

Especially for Linux, we even can generate the root file system and the kernel separately. Thereby, it is possible for us to undertake authentication against the kernel for a specific virtual machine, by comparing the measurement value of this kernel to the trusted value of the same version we got in advance. Again take Xen as an instance. The Linux operating system for a virtual instance of Xen is composed by three parts: the root file system, the kernel, and the initrd (Initialization Ram Disk). The root file system can be quite different with each other according to the varying requirements from jobs, but the kernel and initrd are nearly consistent for all computing tasks, except for the different versions in need. So we can create kernel and initrd files for all available versions in a secure server and get their trusted measurement values as their fingerprints. When deploying a new virtual environment, the kernel and initrid files can only be selected from those available in the server. Then at the specific node, we can authenticate the validation of the kernel and initrd files before the deployment, so as to make sure that they are not compromised. In doing so, the loading of kernel is absolutely secure and the kernel can be deemed as a Trust Base for the attestation toward this virtual machine.

As analyzed above, due to the special features of the virtual machine, the attestation against virtual executing environment in Grid can be implemented in two different ways, based on the secure hardware as Trust Base and the verified OS kernel as the Trust Base respectively.

3.1 The Attestation Based on Secure Hardware

With the secure hardware as Trust Base, the process of attestation is quite similar to the process toward real hosts. Whereas with the prerequisite that the host of virtual machines need to be reliable, the main purpose of secure hardware is not to guarantee the secure boot, but to take measurements of the loading process of the OS kernel for the virtual machine. As previously discussed, if the host is not compromised, we can believe that the BIOS and boot loader part is reliable, so what we seek from the secure hardware here is its measurement to the loading of OS kernel, its secure storage and its signature in the trusted reporting process.

Here we suppose TPM as the underlying secure hardware of our host platform. The integrity measurement tool for the software environment is certainly integrated into the Linux kernel as the security module, and when the kernel is loaded, the security module would take over the measurement work to record the integrity information of executable contents before they are loaded. The measurement values would be extended into a PCR of TPM (TPM uses SHA-1 as the algorithm to get the digest from data, and the extension process is like this: PCR[n]•SHA-1(PCR[n] + measured data)), and when it comes to the trusted reporting part, it is TPM's responsibility to sign the final digest value in the PCR for the validation later. The architecture of the attestation based on TPM is showed as Fig 1 (a). The TPM proxy works in VMM to forward TPM messages between the hardware and virtual instances. After the integrity measurement tool starts up, it takes measurements, records them into measurement lists and extends them into the TPM.

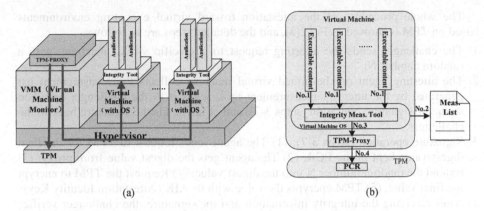

Fig. 1. Figure (a) describes the architecture of attestation based on TPM, and figure (b) reflects the flow of the attestation based on TPM

The detailed flow of attestation goes as depicted in Fig 1(b):

1. The executable contents would be measured by the integrity measurement tool before they are loaded;
2. The integrity measurement tool keeps names of those executable contents and their measurement values in a list;
3. The extension request is sent to the TPM proxy;
4. The proxy utilizes the function of TPM to perform the extension and keeps it in the local storage of the proxy.

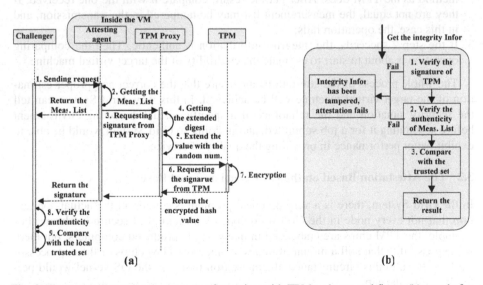

Fig. 2. Figure (a) illustrates the process of attesting with TPM as base, and figure (b) stands for the verification going in the side of challenger

The whole procedure of the attestation toward virtual executing environments based on TPM is showed as Fig 2(a), and the detailed steps are as follows.

1. The challenger sends the attesting request to a specific virtual machine, with a random number N;
2. The attesting agent on the aimed virtual machine collects the measurement list generated by the integrity measurement tool, sends it to the challenger and then starts the signature operation (steps 3-7), and at last sends the signature back to the challenger;
3. Signature operation (steps 3-7): (1) The agent sends request to TPM for the final digest value kept in the PCR; (2) The agent gets the digest value from proxy; (3) Extend the random number N into the digest value; (4) Request the TPM to encrypt the final value; (5) TPM encrypts the value with its AIK (Attestation Identity Key);
4. After receiving the integrity information and the signature, the challenger verifies the authenticity of the information to see whether it can be used to estimate the credibility of the remote virtual executing environment;
5. If the verification is successful, the measurements are compared with the trusted set, or we can call it the fingerprint list. Then the final estimation is drawn according to the result of the comparison.

The flow of verification going in a challenger is showed in Fig 2(b), and the detailed steps are as follows:

1. Decrypt the encrypted digest value received from the target virtual machine, if the decryption is not successful, the value may have been tampered, so operation fails;
2. If step 1 succeeds, the correct value of extension is obtained, and then the challenger extends the hash values in the measurement list in sequence using the same method as the TPM does. After get the result, compare it with the one received. If they are not equal, the measurement list may be tampered in the transmission, and in this case, the operation fails;
3. If the step 2 succeeds, the integrity information is authentic. Then the comparing operation is about to start to estimate the credibility of the target virtual machine.

The whole process of the attestation can ensure that the correct and proper estimation of the target virtual machine will be achieved. In this way, Grid System can tell the security situation and the reliability of a specific virtual executing environment before scheduling it for a job submitted, and in doing so Grid System would be able to exhibit better performance in providing the quality of service.

3.2 The Attestation Based on the OS Kernel as Trust Base

In the Grid System, there is a very practical problem occurring in the actual application that not every node in the Grid is equipped with a kind of secure hardware. For example, the TPM chips are embedded in many newly produced computers, but there is the possibility that still a mount of nodes do not have TPM chips or the coprocessor card on them. In this circumstance, the attestation based on the OS kernel would perfectly bridge the gap.

As has been discussed above, the operating system can be easily customized for virtual machines, especially for Linux, which can even be prepared in three separate parts: the root file system, the kernel and the initrid. And we also know that the loading of Linux consists of two phases. In the first phase, boot loader would load the initrid into the memory. And then the kernel would firstly access to this initrid to load driver modules that are necessary for the further access to the root file system. Then it comes to the operations toward the root file system --- the phase two. The loading of the security module is completed in phase one, which means the integrity measurement tool can undertake its duty after the loading process in phase one. As pointed out at the beginning of this section, we are able to drag the appropriate kernel and initrid files from the server and authenticate the reliability of them to make sure that they are not compromised by comparing their measurement values with fingerprints we got in advance. So once the kernel and initrd files are proven to be reliable, the first loading phase of the operation system would be undoubtedly trusted. After the first phase, integrity measurement tool would start and take measurements of the coming up executable contents. So, this method of attestation actually heightens the Trust Base from the hardware up to the software by guaranteeing the reliability of the OS kernel.

Compared to the attestation using secure hardware as Trust Base, the process here is quite different. Firstly, the kernel and initrid files need to be authenticated at the creation time of the specific virtual machine. Secondly, since no secure hardware is involved in this process, the integrity measurement tool needs to be altered to provide some similar functions of the secure hardware, such as to extend hash values, to store the extended digests in memory and to encrypt the digest at the trusted reporting part. To some extent, we can say that the integrity measurement tool has replaced the secure hardware in this condition, but it is proper because the first phase of loading the operating system is reliable, and also the integrity measurement tool loaded at that time definitely operates without being compromised.

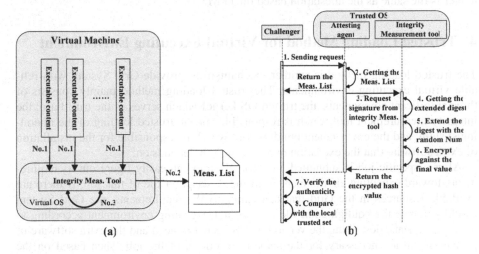

Fig. 3. Figure (a) illustrates the flow of the attestation based on OS, and figure (b) stands for the process of attesting with OS as base

The flow of the attestation based on OS kernel is depicted as Fig 3(a), and the detailed steps are as follows:

1. The integrity measurement tool gets the digest of each executable file by using the SHA-1 algorithm before it is loaded;
2. The names of those executable contents and their measurement values are recorded in a list;
3. The integrity measurement tool automatically completes the extension of the new measurement value.

The whole process of the attestation is showed in Fig 3(b) and the detailed steps are as follows:

1. The challenger sends the attesting request to a specific virtual machine, with a random number N;
2. The attesting agent on the aimed virtual machine collects the measurement list generated by the integrity measurement tool, sends it to the challenger and then starts the signature operation (steps 3-6), and at last sends the signature back to the challenger;
3. signature operation (step 3-6): (1)The agent sends request to integrity measurement tool for the final digest kept in memory; (2) The integrity measurement tool gets the final digest; (3) Extend the digest with the random number N; (4) The integrity measurement tool encrypts the final value;
4. After receiving the integrity information and the signature, the challenger verifies the authenticity of the information to see whether it can be used to estimate the credibility of the remote virtual executing environment;
5. If the verification is successful, the measurements are compared with the trusted set. Then the final estimation is drawn according to the result of the comparison.

The flow of verification of the received integrity information at the side of challenger is the same as the attestation based on TPM.

4   Trusted Loading Method for Virtual Executing Environment

The trusted loading method is another mechanism to provide Grid System with reliable virtual executing environments. The trusted loading method mainly consists of three significant components: the trusted OS kernel which serves as the trust base; the integrity measurement tool which is responsible for the trusted loading of the operating system; and the environment building tool which is responsible for the installation of extra software that the execution of a specific job depends on.

After a specific job is submitted, the Grid System can easily figure out its requirements toward executing environment from the description file of this job. If there are available resources in the Grid that can carter to the requirements, the Grid System would generate the configuration file of a virtual executing environment according to the requirements, describing the version of OS kernel in need and the extra software or packages that are necessary for the smooth execution of the job. Then based on the configuration file, Grid System would choose the suitable Linux kernel and the smallest root file system for the job. The reliability of the chosen kernel should be verified in the same way explained previously. The Integrity measurement tool still is integrated in the kernel and the environment building tool, which is added into the auto-booting list, is contained in the root file system. The root file system also contains a list of extra software that needs to be installed and the signed trusted set for all the executable contents that will be loaded during the startup of an operating system using such a root file system. In fact, when constructing the virtual executing environment with the trusted loading method, the root file system for every computing task is the same, so we can make a template and get the trusted set by measuring the startup of an OS using such template as the root file system.

The kernel is verified, so the loading process of OS is trusted until the integrity measurement tool starts. Then the integrity measurement tool would take responsibility to guarantee the second phase of loading process. Firstly, the integrity measurement tool would try to verify the signed trusted set contained in the root file system. If the verification fails, the trusted set may have been tampered, so the loading process fails and error message is returned. If the verification is successful, the integrity measurement tool would take measurement of every executable file loaded during the OS's second phase of startup, and compare the measurement value with the corresponding item in the trusted set. If there exists any value that does not match with the existed item in the trusted set, or can not find the corresponding item of this executable content, the integrity measurement tool would deem the root file system as being tampered and call off the creation of current virtual environment. The integrity measurement tool continues that operation till the environment building tool starts. Then the control of the creation would be taken over by environment building tool. The environment building tool firstly would parse the list of extra software, and installs them in sequence.

The environment building tool would generate a queue of installation tasks according to the list. For each task in its queue, the building tool would perform protocol as follows:

1. The environment building tool sends a request containing the detailed information of the software package it needs to a particular server to ask for the corresponding software package and the signature;
2. The server locates the software package requested, signs it and sends the signed package back to the environment building tool;
3. The environment building tool verifies the authenticity of the signed software package it received, to tell whether they are trusted;
4. If the verification succeeds, the environment building tool would invoke the command of the Linux to install the software package. If not, repeat this protocol. If the verification fails for three times, the environment building tool will call off the operation, and the loading fails.

The trusted loading of operating system is guaranteed by the integrity measurement tool and the trusted installation of extra software is assured by the environment building tool. The security and correctness of the final result of the execution can be ensured by creating executing environment in this way.

5 Experiments and Analysis

Experiments are conducted to evaluate those two mechanisms. For the attestation, since the aim of this mechanism is to truly reveal the executable files or codes in the system that may have been compromised, in the experiment we deliberately changed the content of an executable file in a virtual machine to see if this alteration can be reflected by the mechanism of attestation we discussed here. As illustrated in Fig 4, the modification was clearly reflected.

Auth_HashCode	Fingerprint_HashCode	Executive_Code	Match?	
2f59f0fb4a0...	2f59f0fb4a090aa3b5...	/usr/bin/mesg	Matched	
776dcd05762...	776dcd05762d9f95f5...	/bin/ls	Matched	
2c1c3fa599c...	2c1c3fa599c33477d5...	/bin/cp	Matched	
b3ac843ab86...	83713e35062223850a...	/vtpm/test_case	UnMatched	

Fig. 4. The result of the attestation in experiment, the file */vtpm/test_case* is our test case

As for the trusted loading method, it involves the whole creation and booting process of a virtual machine, so in the experiments, we manually customized the configuration of the virtual machine to check whether the trusted loading method is capable of creating the virtual environment we wanted with desirable security. Results indicate that the trusted loading can successfully create and boot the virtual machine as being customized or stop the booting process when we deliberately compromise the executable file that would be loaded at the booting phase.

Generally, the installation of extra software is usually undertaken during the process of creating the image file of the operating system. After the proper root file system is chosen, the additional software would be installed in that file system by using the Linux command "chroot". However, in order to guarantee the reliability of extra software, the trusted loading mechanism holds the operation of installation until the startup of a specific virtual machine. Although the installation of extra software can be tightly supervised in this way, the latency time of the startup process will be greatly increased since the communication and transmission can cause extra time loss. Besides, the verification of each executable file would also bring additional time cost.

In order to clearly estimate the time efficiency of the trusted loading method, we made three experiments to gain an insight into the time cost in trusted loading method. The *first one* was to compare the startup time between an ordinary virtual machine and one with integrity measurement tool, when both are equipped with the same OS image file and no other additional software. The *second one* was to record the startup time of an ordinary virtual machine and one adopting the trusted loading mechanism when they both are customized with the extra software. The *last one* was to record the respective time consumption of installing JDK 1.5.0 through environment building tool and by directly using the "chroot" command.

For the second experiment, the basic configuration is as follows: *OS Debian etch/ kernel 2.6.18/ extra software JDK 1.5.0 and SSH Server/ network DHCP*. For the third experiment, the software package to be installed is sun java Development Kit (JDK) 5.0.

The results are shown in Fig 5. As we can see from the results, when booting the pure OS (without additional software packages), the trusted loading would only cost

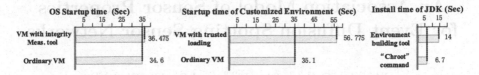

Fig. 5. The comparison of time for different situations

about 2sec more than the ordinary VM. But when the installation of additional software is required, the startup time would increase by 20sec. So it is clearly that the bottleneck of the time efficiency lies in the installation of extra software in the trusted loading process. In fact, since the purpose of this loading method is to generate a secure virtual executing environment for a job in Grid System, contrasted to the execution time of the job, the added 20 seconds or even more actually are very limited. So the efficiency of the trusted loading method can totally be acceptable.

6 Conclusion

This paper presents two mechanisms for ensuring the security of a virtual executing environment in Grid System, including the attestation method and the trusted loading method. The protocols and operation flows involved in the attestation and trusted loading are explained in detail, and experiments are also conducted to verify the effectiveness and analyze the efficiency of these two mechanisms. These two mechanisms discussed in this paper have already been successfully implemented in the Virtual Environment Deployment and Runtime Management System developed by BeiHang University, and exhibited good performance in maintaining reliability of the virtual executing environment.

References

1. Sailer, R., Zhang, X., Jaeger, T., van Doorn, L.: Design and Implementation of a TCG-based Integrity Measurement Architecture. IBM T. J. Watson Research Center
2. Berger, S., Caceres, R., Goldman, K.A.: "vTPM: Virtualizing the Trusted Platform Module. IBM T. J. Watson Research Center
3. Haldar, V., Chandra, D., Franz, M.: Semantic Remote Attestation—A Virtual Machine directed approach to Trusted Computing. Department of Computer Science University of California
4. Jin-Peng, H., et al.: Research and Design on Hypervisor Based Virtual Computing Environment. Journal of Software 18(8), 2016–2026 (2007)
5. Smith, J.E., Nair, R.: The Architecture of Virtual Machines. IEEE Computer 38(5), 32–38 (2005)
6. Virtuoso: Resource Management and Prediction for Distributed Computing Using Virtual Machines (2007)
7. TCG. "TPM Design Principles", "TPM Structures", "TPM Commands"

An Association Model of Sensor Properties for Event Diffusion Spotting Sensor Networks

Xiaoning Cui[1,2,3], Qing Li[2,3], and Baohua Zhao[1,2]

[1]Department of Computer Science and Technology,
University of Science & Technology of China, Hefei, China
[2]Joint Research Lab of Excellence,
CityU-USTC Advanced Research Institute, Suzhou, China
[3]Department of Computer Science, City University of Hong Kong, Hong Kong, China
cxning@mail.ustc.edu.cn, itqli@cityu.edu.hk, bhzhao@ustc.edu.cn

Abstract. Recent years of research on sensor networks have resulted in multi-scale processing techniques for sensor data mining able to reflect the dynamic nature of real-world context. However, few of these techniques provide a systematic view of the relationships between sensor data streams and correlated network behaviors. In this paper, an association model of inherent, data and network properties is presented and analyzed for a suite of event diffusion spotting applications. Based on the associated model, window-based in-network cooperation is conducted for sensitive event diffusion spotting. Experimental results verify the performance of our approach, and confirm the scalability of our association perspective of sensor properties in such event diffusion spotting networks.

Keywords: Association model; Diffusing event; Sensor property; Correlation; In-network cooperation.

1 Introduction

A sensor network, from the perspective of pervasive computing, provides an effective tool for interacting with the physical world [1]. Several wide-area phenomena, such as forest fires, typhoons, floods and fumes, cause widespread destruction and distress. These phenomena are usually quite difficult to predict or detect in time. However, the self-organizing and auto-coordinating capabilities of sensor networks make these networks suitable for offering timely information about such phenomena and thus reducing economic losses due to natural calamities.

In view of the terrible consequence of environmental disasters, the sensitivity of energy-constrained sensor networks is highly required in such disaster-monitoring applications, largely depending on distinct sensor data streams [2], application-specific event models and available data processing models [3]. As disasters are differentiated by frequency, intensity, duration, spreading speed, and so on, any of these distinctions would directly affect sensing strategy and

Y. Zhang et al. (Eds.): APWeb 2008, LNCS 4976, pp. 178–189, 2008.

routing protocol of the sensor network. In our work, we monitor a class of diffusing events and make emergency alert via an event diffusion spotting sensor network. Here, the term of "diffusing event" is defined by the influential range of the event, i.e., the range of the event expands as time goes by. Liquid/gas diffusion and sound/noise propagation are representative diffusing events.

In such kind of urgency-triggered sensor networks, the data-centric nature is most essential, since the emergency notification is much more important than energy conservation. In our work, therefore, we aim at capturing urgent diffusing events and sending out timely report. Our approach is based on an association model of sensor properties to detect and notify diffusing events with high efficiency, i.e., trying to fully exploit the relationships of sensor data streams and corresponding environmental events. The main objectives and contributions of this paper are as follows:

- We target at a group of disaster monitoring applications and adopt an event diffusion spotting network to issue useful emergency reports whenever necessary.
- We exploit sensor properties and integrate information into an association model for such a network.
- Based upon the association framework, we devise window-based in-network cooperation strategies for runtime event notification. Experimental results basically verify the performance of our cooperation strategies and the scalability of the association model.

The rest of this paper is organized as follows: In Section 2 we illustrate a "fume diffusion detection" scenario, followed by the event model and network structure. In Section 3, an association model of sensor data and network behaviors is designed and window-based algorithms are presented for sensitive event diffusion spotting. The performance is evaluated in Section 4. Related work is reviewed in Section 5, and finally, conclusions and future work are given in Section 6.

2 Motivating Scenario

Based on the idea of utilizing diffusion information as warning signals for the scale and trend of the upcoming or ongoing event, in this section, we first specify a fume diffusion detection scenario in Section 2.1, and then briefly discuss the event model and the network structure applied to our work in Section 2.2.

2.1 Fume Diffusion Detection Scenario

Fume diffusion is a serious environmental problem, especially in dense populated places. As shown in Fig. 1, the black hole in the center denotes the source of the fume. The fume spreads in all directions, shown as the dotted circles around the hole. The region covered by these circles grows larger and larger, implying more and more sufferers involved in the danger of poisonous fume.

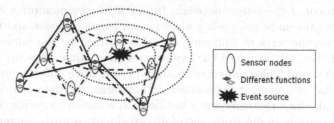

Fig. 1. Fume diffusion detection scenario

Apparently, the diffusion hints may contain potentially interdependent aspects, e.g., wind force, temperature and gas concentration. Correspondingly, different function-based sensor nodes (ref. oval nodes with different-colored diamonds in Fig. 1) conduct in-network information sharing and cooperation, eventually producing a complete view of the diffusing event. In view of such heterogeneous functions, parallel logical tiers are divided accordingly, shown as the solid and dashed lines in Fig. 1. Each tier reflects a certain parameter of the event. Information integration on all of the logical function tiers results in an extensive knowledge of the physical phenomenon, i.e., fume diffusion in this scenario.

Motivated by the sensitivity requirement, i.e., quick response of diffusing events, in such a network, in our work we attempt to exploit the association of the raw data collected by sensor nodes and the required behaviors for in-network cooperation, so that emergency events can be detected and notified before widespread fume diffusion.

2.2 Event Model and Network Structure

A large number of environmental factors affects fume diffusion, including temperature, humidity, gas concentration, etc. In our current work, two major interrelated factors, temperature and gas concentration, are considered for simple access to network sensitivity. Temperature is measured by a threshold list, marked with ranks of possible danger. Gas concentration, on the other hand, is expressed by gas diffusion rate, defined by Formula (1).

$$F_i = -D_i \cdot dc_i/dx \,. \tag{1}$$

Suppose the fume is comprised of $m(m \geq 1)$ kinds of gases $\{g_i\}_{i=1}^m$. In Formula(1), F_i is the diffusion rate of g_i, dc_i/dx is the gradient of g_i concentration, and D_i is its diffusion coefficient, available in statistical environmental data. The minus sign in Formula (1) means that the gas spreads from dense place towards sparse place. In our work, for the sake of simplicity, F_i is assumed to increase approximately with a constant $r_i = kr$, where k is a positive integer and r is the sensing radius of each sensor node, i.e., the event area grows $(2n-1)\pi(kr)^2$ larger during the nth time unit.

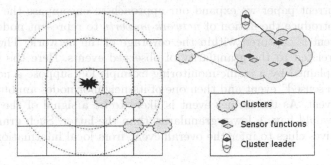

Fig. 2. Backbone topology of the fume diffusion spotting sensor network

In view of the diffusing effect and the predictable flexibility of network operation, Fig. 2 gives the basic network structure. Sensor nodes are spontaneously organized as grid-based clusters (cf. clouds in Fig. 2). Every cluster is guaranteed to involve several function tiers and share local information among heterogeneous members. The cluster-leader is in charge of internal information aggregation and external communication. Cluster-leaders may be elected according to various criteria, e.g., residual energy, possessive function points, and so on.

3   Association Modeling

Following our previous work [4], in this section, we extract the most significant sensor properties for association modeling. Specifically, the properties are classified in Section 3.1. An association model is established in Section 3.2, bridging various properties and integrating comprehensive in-network information. Window-based cooperation strategies are devised accordingly in Section 3.3.

3.1   Property Sorting

A sensor data item is essentially a representation of some physical meaning, measured by elements such as the ones listed in Table 1.

Table 1. Inherent property and data properties

Inherent property	Data properties		
Sensor function	Time	Distance	Value

Here the *inherent property* concerns the physical meaning of a data item, e.g., temperature or concentration, corresponding to a function tier mentioned in Section 2.1. Meanwhile, as listed in Table 1, measurements of a certain physical meaning are grouped into *data properties*.

In the current paper we expand our approach to encompass the entire network. We introduce the notion of *network property* to represent node behaviors when an event is captured within the coverage of the network. The behaviors are highly related to the granularity of observed events. Here the granularity could be explained by a traffic-monitoring example [5]: suppose a node reports a "vibration sensed" event and then one of its neighbor nodes announces a "car detected" event. As the former event is likely to be a signal of the latter one, the former would have a lower granularity than the latter. Such granularity relationship gives clues to infer the overall event from local information pieces.

3.2 Model Establishment

To cover the three types of properties mentioned in Section 3.1, our association model embodies two parts, a 3-D correlation model for inherent and data properties and a hierarchical multi-path model for network property. The two parts are connected with bridging properties, chosen as "sensor function" in inherent property and "data value" in data properties in our work.

The correlation model is sketched on the left hand of Fig. 3. It is displayed in a 3-D Cartesian coordinate system, with three axes denoting data properties. Here "time" records when the data item is collected, "distance" describes how far the sensor is from its neighbors, and "value" is the numerical quantity of the data item. In addition, the sensor function, representation of the inherent property, is a hidden parameter differentiated by line style of data streams.

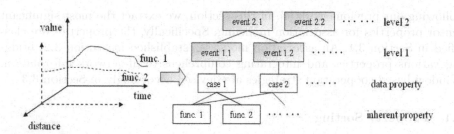

Fig. 3. Association model of inherent, data and network properties

Since nodes' behaviors are highly related to the available event granularity, events are graded into different levels accordingly, as shown on the right hand of Fig. 3. Note that bridging properties constitute the foundation of event levels. Specifically, varied combination of sensor functions and relevant data values produce different cases as triggers for different events. All of them form a hierarchical multi-path structure, functioning as guidance of nodes' behaviors.

3.3 Window-Based In-Network Cooperation Strategies

Employing the proposed association model, we introduce some window-based in-network cooperation strategies for efficient event detection and notification.

The basic event format is as defined in Table 2, where "case" involves thresholds of correlation values of sensor functions. "Level.No." is set by a case or case combination and matched with association template in Fig. 3. During the matching process, we attempt to find out the *highest available level* (HAL) of events within sliding window. Such an HAL is utilized for timely event notification. Basically, sensor data streams would be first processed on each sensor node, and then aggregated by cluster-leaders based on the association model. For the sake of energy efficiency, only the cluster-leaders on internal boundary of the event area are responsible for HAL notification. Details are listed in Algorithms $1 \sim 3$.

Table 2. Basic event format

Level.No.	Time	Location	Case(set)

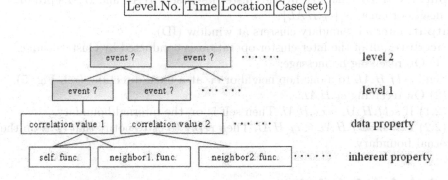

Fig. 4. From association template to instance

Algorithm 1. Adaptive linear regression on a single node
Pre-configuration: window size m, basic threshold T, offset bound ε.
Input: sensing data sample sequence $(q_1, q_2, q_3 \ldots)$.
Output: event data points(marked by window(ID)).
Procedure: between window (i) and window $(i+1)$.
   (1) Read m samples $q_{i+1} \sim q_{i+m}$.
   (2) Forecast the following m data items $q^*_{i+m+1} \sim q^*_{i+2m}$ by linear regression and then cache them.
   (3) Read next m samples $q_{i+m+1} \sim q_{i+2m}$.
   (4) Compare: $T^* = \frac{1}{m} \cdot \sum_{j=1}^{m} |q^*_{(i+1)m+j} - q_{(i+1)m+j}|$.
   (5) If $|T^* - T| > \varepsilon$ Then output $p_{i+1} = q_{(i+1)m+1}$ as a representative event point of window $(i+1)$ and mark window $(i+1)$; Else $T^* = T$.

Algorithm 2. Internal integration on cluster leader
Pre-configuration: a constant-set for value adjustment $\{\delta_{f(i)}\}_{f(i)=1}^{f}$, where f is the total number of sensor functions.
Input: all of the event points reported by cluster members $(p_1, p_2, p_3 \ldots)$, where $p_i = <f_i, t_i, d_i, v_i>$ (ref. Table 1); association template (ref. Fig. 3).
Output: HAL of the cluster at window (ID).

Procedure: at the end of window i.

(1) Function check: calculate distance-based average of data values of each function. For function $f(j)$, $\bar{v}_{f(j)} = [t_i \in window(ID)] \sum_{i=1}^{k} (v_i - \delta_{f(j)} \cdot d_i)/k$.

(2) Case check: refer to the association template to see whether the value combination of functions, $\{\bar{v}_{f(j)}\}$, satisfy certain cases.

(3) Find HAL: record $Level.No.$ list when satisfying certain cases, and match the association model level by level in a bottom-up way until reach HAL (cf. Fig. 3 and Fig. 4).

Algorithm 3. Internal-boundary-cluster determination

Pre-configuration: according to the network structure, a cluster is supposed to have at most four one-hop neighbor-clusters. Each cluster is identified by a number c_k.

Input: the HAL generated by the cluster itself $self.HAL$ and $HALs$ produced by neighbor clusters $\{c_k.HAL\}$.

Output: internal boundary clusters at window (ID).

Procedure: all of the inter-cluster operation is conducted by cluster-leader.

(1) On receiving no message:

Send $self.HAL$ to a one-hop neighbor c_k at a fixed direction. (ref. Fig. 5)

(2) On receiving $c_k.HAL$:

(2.1) If $self.HAL > c_k.HAL$ Then self is on the internal boundary.

(2.2) Else if $self.HAL < c_k.HAL$ Then reply to c_k to notify that c_k is on the internal boundary.

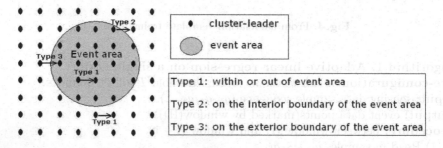

Fig. 5. An example of fixed transmission-direction for internal boundary determination

Once validated as an internal boundary cluster, the cluster-leader should send out event notification to external clusters. However, for the sake of energy efficiency, not all of the cluster-leaders on the internal boundary have to do so, as it is assumed that an event diffusion region naturally expands in all directions and thus no concave boundary exists. Therefore, it can be proved that there are higher probabilities of internal boundary leaders to have one or two external boundary neighbors as the event region extends. In densely populated sensor networks, the internal boundary nodes tend to have the highest chance to possess two external boundary neighbors. To this end, we exploit two methods for event notification.

- **Method 1.** Each internal boundary leader randomly selects one of its external neighbors and transmits its higher HAL to the neighbor.
- **Method 2.** Only the convex points, i.e., internal boundary leaders that have two external neighbors, are responsible for HAL transmission.

4 Experimental Results and Performance Evaluation

The performance is mainly evaluated depending on the sensitivity of such an association model. Section 4.1 gives detailed sensitivity analysis on notification cooperation strategies. In Section 4.2, we consider the scalability of the available notification region for our proposed methods.

4.1 Sensitivity on Notification Cooperation Strategies

Recall the fume diffusion detection scenario in Section 2.1. In our experiment, we configure three types of sensors (temperature sensor t, concentration sensor of Gas g_1 and Gas g_2) in a grid network. Each sensor has two possible states, normal(N) and abnormal(A); specifically, $t(N), t(A); g_1(N), g_1(A); g_2(N), g_2(A)$. Events are consequently classified into three levels, with level 0 representing states of a single type, level 1 combining states of the three types, and level 2 distinguishing danger from safety. Table 3 gives the basic configuration of our experiment, where states at level 1 are combinations of (t, g_1, g_2), and level 2 classifies events by the number of abnormal states. In addition, the two gases are supposed to diffuse at different speed, i.e., following Formula(1) with different c_i and D_i. The experiment is conducted to verify the sensitivity at each level. We simulate a 17×17 grid network to carry on the experiment. In Table 3, the state information at level 0 is produced according to Algorithm 1. The results of levels 1 and 2 are obtained following Algorithm 2. Algorithm 3 and Method 1 or 2 are in charge of event notification.

Table 3. Basic configuration of event levels

Level	State
2	$3A, 2A1N, 1A2N, 3N$
1	$NNN, NNA, NAN, NAA, ANN, ANA, AAN, AAA$
0	$t(N), t(A), g_1(N), g_1(A), g_2(N), g_2(A)$

Algorithm 1 is essentially implemented on each sensor node to measure a certain type of data. We adopt real-life sensor data [9] to evaluate its performance in Fig. 6. From Fig. 6 we can see that the result accurately fits the source data. Moreover, the communication cost can be sharply reduced with an appropriate ε and window-size. For example, when $\varepsilon = 2$ and window-size=16, only 21 event data points are picked up for HAL analysis, with the capability of reflecting the diffusion trend of the entire 400 source data items.

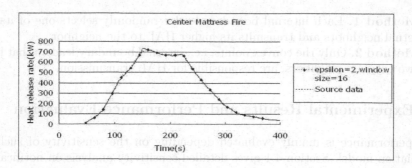

Fig. 6. Adaptive linear regression on each node

In order to verify the performance of Algorithms 2 and 3 and Methods 1 and 2, as shown in Fig. 7, the total number of witnesses (nodes that detect events) and announcers (nodes that make event notification) reflects the coverage of the detection and notification range in the sensor network. In comparison with the source HAL and the HALs produced by our cooperation strategies, it is notable that both methods exhibit a larger coverage than that of the event source, demonstrating the sensitivity of our window-based in-network cooperation strategies.

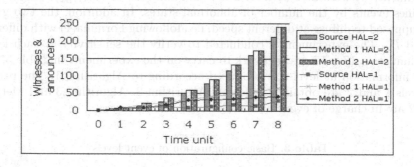

Fig. 7. HAL comparison of two cooperation methods

4.2   Scalability of the Available Notification Region

As it is very important to guarantee the energy efficiency when the sensor network grows large, we consider the scalability of our model mainly from the energy perspective. Fig. 8 shows that the communication amount in both methods would not increase too much compared with the spreading of event source.

In addition, Method 2 increases much slower than Method 1 with respect to communication cost. Here, the communication cost is measured by the amount of message exchange. Therefore, Method 2 tends to have better scalability and

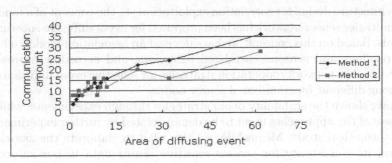

Fig. 8. Scalability evaluation of the two methods

energy-efficiency in emergency notification. Such an evaluation also sheds light on and confirms the scalability of our window-based in-network cooperation strategies with respect to our association model.

5 Related Work

A number of approaches have been proposed for event/outlier detection such as spurious measurement isolation [6], R-tree-based event region detection [7] and distributed deviation detection [8]. Such approaches address various ways of event detection. In our work, however, a distinct perspective is adopted to focus on a suite of diffusing events, i.e., considering the characteristics of diffusing effects and thereby devising an association model for sensitive detection/notification via sensor data mining techniques. There are traditionally five categories of major techniques for sensor data mining [10], from which linear forecasting is employed in our approach, as well as three kinds of popular data processing models [3], where sliding-window is utilized as basis of our strategies.

In addition, as it is anticipated that diverse information could be integrated to detect the diffusing event, stream analysis of heterogeneous data correlation is crucial for such information integration. Some methods for correlation processing have been proposed, including pipe-lined framework of spatial-temporal correlation processing [11], structure-based spatial sampling [12], cost-aware gathering strategies for correlated data [13], and etc. These approaches provide some hints on our design of window-based in-network cooperation strategies. Furthermore, the strategies proposed in this paper are, in principle, in accordance with the general association model of sensor properties for event diffusion spotting task.

6 Conclusions and Future Work

By leveraging traditional techniques for sensor data mining, in this paper we have focused on sensitive detection/notification of diffusing events through the alliance of heterogeneous information streams. Following our previous work on node architecture design for correlation exploitation [4], the concept of a function-based multi-tier sensor network has been proposed for event diffusion spotting applications. Based on this concept, we have devised an association model to bridge sensor readings and sensors' correlated behaviors, and accordingly presented window-based in-network cooperation strategies for sensitive event diffusion spotting among different function-based sensor nodes.

We have shown the scalability of our strategies through experiments, while the robustness of the approaches is yet to be demonstrated by further experiments as well as theoretical study. Meanwhile, we also plan to elaborate the association model on its consistency for context-sensitive sensor data in our subsequent research.

Acknowledgments. The authors would like to thanks Prof.s Elisa Bertino and Yanchun Zhang for their useful and constructive comments on an earlier version of this paper. The work of this paper has been supported, in part, by the National Basic Research Fund of China ("973" Program) under Grant No.2003CB317006.

References

1. Ganesan, D., Estrin, D., Heidemann, J.: Why do we need a new Data Handling architecture for Sensor Networks? In: ACM SIGCOMM Computer Communications Review, pp. 143–148 (2003)
2. Babcok, B., Babu, S., et al.: Models and Issues in Data Stream Systems. In: Proceedings of the 21st ACM SIGACT-SIGMOD-SIGART Symposium on Principles of Database Systems, Madison, Wisconsin, USA, pp. 1–16 (2002)
3. Jiang, N., Gruenwald, L.: Research Issues in Data Stream Association Rule Mining. SIGMOD Record 35(1), 14–19 (2006)
4. Cui, X.N., Zhao, B.H., Li, Q.: Exploiting Data Correlation for Multi-Scale Processing in Sensor Networks. In: 2nd International Conference on Scalable Information Systems, Suzhou, China (2007)
5. Chu, D., Tavakoli, A., Popa, L., Hellerstein, J.: Entirely Declarative Sensor Network System. In: 32nd International Conference on Very Large Data Bases, Seoul, Korea, pp. 1203–1206 (2006)
6. Kotidis, Y., Deligiannakis, A., Stoumpos, V., Vassalos, V., Delis, A.: Robust Management of Outliers in Sensor Network Aggregate Queries. In: 6th International ACM Workshop on Data Engineering for Wireless and Mobile Access, Beijing, China, pp. 17–24 (2007)
7. Jiang, C.Y., Dong, G.Z., Wang, B.: Detection and Tracking of Region-Based Evolving Targets in Sensor Networks. In: 14th International Conference on Computer Communications and Networks, San Diego, California, USA (2005)
8. Subramaniam, S., Palpanas, T., Papadopoulos, D., Kalogeraki, V., Gunopulos, D.: Online Outlier Detection in Sensor Data Using Non-Parametric Models. In: 32nd International Conference on Very Large Data Bases, Seoul, Korea, pp. 187–198 (2006)
9. http://science.fire.ustc.edu.cn/
10. Faloutsos, C.: Stream and Sensor data mining. In: 9th International Conference on Extending DataBase Technology, Heraklion-Crete, Greece (2004)

11. Jeffery, S.R., Alonso, G., Franklin, M.: J., Hong, W., Widom, J.: A Pipelined Framework for Online Cleaning of Sensor Data Streams. In: Proceedings of the 22nd International Conference on Data Engineering, Atlanta, GA, USA, IEEE Computer Society Press, Los Alamitos (2006)
12. Quan, Z., Kaiser, W.J., Sayed, A.H.: A Spatial Sampling Scheme Based on Innovations Diffusion in Sensor Networks. In: Proceedings of the 6th International Conference on Information Proceeding in Sensor Networks, Cambridge, Massachusetts, USA, pp. 323–330. ACM Press, New York (2007)
13. Liu, J.N., Adler, M., Towsley, D., Zhang, C.: On Optimal Communication Cost for Gathering Correlated Data through Wireless Sensor Networks. In: 12th Annual International Conference on Mobile Computing and Networking, Los Angeles, California, USA, pp. 310–321 (2006)

CROWNBench:
A Grid Performance Testing System
Using Customizable Synthetic Workload

Xing Yang, Xiang Li, Yipeng Ji, and Mo Sha

School of Computer Science, Beihang University, Beijing, China
{yangxing, lixiang, jiyipeng, shamo}@act.buaa.edu.cn

Abstract. The Grid middleware must be developed iteratively and incrementally, so Grid performance testing is critical for middleware developers of Grid system. Considering the special characters of Grid system, in order to gain meaningful and comprehensive results of performance testing, it is necessary to implement testing on real Grid environment with various types of workload. CROWN-Bench, as described in this paper, is a system for helping Grid middleware developers to evaluate middleware design and implement using customizable synthetic workload. Middleware developers can customize testing workload basing on the model of Grid workload derived from real workload traces, including its structure and parameters, and then workload is synthesized automatically and contained jobs will be submitted by CROWNBench in a distributed manner. CROWNBench defines several metrics for measuring Grid performance as automatic testing results. The experiment, which used CROWNBench to test the performance of Grid system with CROWN Grid middleware, shows that the system already finished have accomplished its prospective goal. It can implement Grid performance testing in an efficient, flexible, controllable, replayable and automatic way to help middleware developers evaluate and improve their products effectively.

Keywords: Grid computing, performance testing, synthetic workload.

1 Introduction

Grid is a new computing infrastructure, and is developing quickly since proposed in 1990s. Performance testing is needed by people including Grid system designers, developers of Grid middleware, application developers and system users. Comparing to traditional distributed system, such as cluster, the content and implication of Grid system performance is different because of its characteristics, for instance, heterogeneity, dynamics and the special way for sharing resources. Accordingly, performance testing methods and tools especial for Grid system are necessary.

The greatest motivation of our work is to provide a performance testing tool for Grid middleware developers, firstly for CROWN [1] middleware developers, to validate function and performance of their designing and implementing. To gain meaningful results, the Grid system, including physical resources, system architecture and workload,

Y. Zhang et al. (Eds.): APWeb 2008, LNCS 4976, pp. 190–201, 2008.
© Springer-Verlag Berlin Heidelberg 2008

used in test should be close to the target system. Because testing we discuss and implement here is real testing which is executed on real circumstance, physical resources and system architecture are exactly the same as the target system. Thus, what we are focusing on in our system is workload used in testing.

Workload impacts Grid system performance greatly. Usually, there are three kinds of workloads can be used in testing to assess the performance of Grid systems: real Grid workload, synthetic Grid workload and benchmark workload. Synthetic workload, which is basing on workload model derived from real workload traces, is regarded to be a more appropriate candidate in our performance testing system.

In this paper, we present CROWNBench, a system for Grid performance testing using customizable synthetic workload. In order to gain flexibility and universality, CROWNBench allows testers to customize their workload model, including application and job submission rules, basing on workload statistical model extracted from real workload trace. In this way, Grid middleware developers can acquire comprehensive performance data of Grid system by running various kinds of workload. What is more, testing procedure can be controlled and replayed easily.

2 Related Works

Even though workload in CROWNBench system could be customized, synthetic workload should still be basing on workload model and its contained elements and parameters. So real workload traces of some certain Grid systems should be well studied and concluded to a model, which could depict general characteristics of real workload. Workload models of kinds of computer systems have been well studied and researched [2] [3]. Even Grid system workload model has been started to research on some Grid system testbeds to evaluate Gird system performance [4] [5]. Base general rules in these models are extracted. And for testing performance capacity of Grid system, amount and distribution of workload should also be tunable to simulate real workload in different time and on different Grid testbeds.

GRASP [6] is a project for testing performance and validating dependency of Grid system by using probes, some low level benchmarks; NGB[7] is developed for Grid from NPB. It contains a suite of benchmark the structure of which is described by data flow graph; the benchmarks of GridBench [8] cover multiple levels benchmarks, including Micro–benchmark and Micro–kernel. GRENCHMARK [9] is a Grid performance testing framework with synthetic workload. But only four settled applications are supported to synthesize workload, and customizable workload is not supported.

Comparing to projects discussed above, CROWNBench provides tools to analyze real workload trace and maintain testing environment, which are both not covered in these projects. Analyzing real workload trace could find out approximate workload statistical model, and testers could use it to customize the synthetic workload close to real workload and acquire meaningful result of Grid performance testing by running it. Maintaining an isolated testing environment for a Grid performance testing could eliminate the influence among different performance testing and minimize the influence which is exerted on working Grid by performance testing.

3 Grid Performance Testing System

3.1 Grid Performance

When discussing performance of distributed systems, we are usually focusing on the quality and quantity of resources they providing. Grid is regarded as a virtual computer in [10], and performance testing of Grid system is to test performance of this virtual computer constituted by each layers of Grid system while it is running.

Compared to traditional distributed system, performance of physical resources is not the only content of Grid system performance. Grid middleware performance is a very important part in the whole Grid system, especially when we consider requirements of CROWNBench. Throughout development, the middleware must be developed iteratively and incrementally. On each milestone, performance of middleware should be tested on real circumstance. After that, results of performance metrics, which defined before, are required to validate the design and implement of Grid middleware.

3.2 Grid Performance Testing System

Usually, there are three ways of testing Grid performance, model analysis, emulating and real testing respectively. Comparing to real testing, former two have difficulties in emulating dynamic behaviors of the Grid system. Thus, we choose the way, implementing Grid performance testing on real Grid environment to gain performance results in CROWNBench.

In order to fulfill our system goal, helping middleware developers to evaluate and compare performance of different middleware versions, Grid system performance testing should be implemented on same physical resources and Grid environment structure deployed with different middleware version. As a result, performance data could be acquired to compare the performance of different middleware version, to testify whether the new one is better than the old ones.

As a Grid performance testing system, CROWNBench faces challenges in three aspects, and now our contributions also locate in them: generating meaningful and comprehensive workload for Grid performance testing. They will be described in detail in 5.4.

4 Grid Workload Model

Generally, Grid workload includes all jobs submitted to Grid within a period of time. A given real workload on some certain period of time and certain condition is not necessarily representative of workloads on other times and conditions. Thus, Gird workload model should be derived from traces of history workloads, and then be studied as the basis of workload synthesizing.

When we study Grid workload model from a higher view, there are two aspects discussed in our system. Firstly, arrival process and submission process of a certain kind of jobs in workload would be stable in a long term. Distribution of interarrival times and submission nodes are derived from this stable rule, and are used to model Grid workload and customize synthetic workload. Secondly, job running time is also an aspect in Grid workload model as well as in traditional parallel and distributed system workload. In our system, job running time will be measure after testing running as turnaround time. So it will not be settled down before testing.

5 CROWNBench

5.1 System Architecture

CROWNBench is a system for Grid performance testing with customizable synthetic workload, which has been used on Grid deployed with CROWN Grid middleware. Figure 1 shows its architecture.

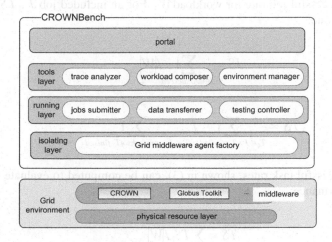

Fig. 1. CROWNBench system architecture

CROWNBench is built upon physical resources and Grid middleware. It is isolated from certain Grid middleware instance by **Grid middleware agent factory**. Layers in CROWNBench above isolating layer do not know which type of Grid middleware was deployed above physical resource layer in Grid environment. In running layer, there are three components. When testing starts running, **job submitter** will submit jobs in synthetic workload automatically. **Data transferrer** takes the responsibility to transfer related data, including testing running files and results data, among nodes in Grid environment. **Testing controller** provides a general testing running platform in heterogeneous Grid environment. In tools layer, trace analyzer, workload composer and environment manager work before testing starts. **Trace analyzer** extracts statistical model of real workload by analyzing workload trace. **Workload composer** synthesizes workload for testing through definition of testers. **Environment manager** maintains a testing environment for a performance testing by building up a testing environment before testing starts and clearing it after testing finishes. Testers could use CROWNBench through system **portal**, including edit testing environment, view workload analyzing results and customize synthetic workload.

5.2 Grid Performance Metrics

W stands for workload during a period of time. The number of applications included in workload is $|W|$. Job J_i includes $|J_i|$ tasks, $T_1...T_{|J_i|}$. For a task T_i, it could have 0 to $(|J_i|-1)$ pre-tasks. It can run when these tasks finished. Similarly, T_i could have 0 to $(|J_i|-1)$ post tasks, which could run simultaneously only when T_i finished.

Because of characters of Grid system, a ratio of successful jobs to whole jobs can be looked as a metric of Grid system performance, especially in Grid system where testers look job correctness more important than processing speed.

JS is the successful job rate for workload W. For an included job J, TS is the successful tasks rate:

$$JS = \sum_{J \in W \wedge Jsucceed} 1 \ /|W| \tag{1}$$

$$TS = \sum_{T_i \in J \wedge T_i succeed} 1 \ / \sum_{T_i \in J \wedge pretasksofT_i finised} 1 \tag{2}$$

Average successful task rates, shown in (3), can be computed to evaluate whole Grid system performance:

$$\overline{TS} = \sum TS_i \ /|W| \tag{3}$$

In Grid system which pays attention to quality of service, turn-around time TT of a job (time from submission to finishing) and each component processing time could be chosen as metrics. In a special scenario, workload contains only one kind of application. (4) is a metric of high level Grid system performance:

$$\overline{TT} = \sum_{J_i \in W} TT_i \ /|W| \tag{4}$$

Processing time of a certain component can be looked as a metric of that component's performance. For example, (5) can be used for evaluating performance of scheduling component in Grid middleware.

$$\overline{ST} = \sum_{J_i \in W} ST_i \ /|W| \tag{5}$$

5.3   CROWNBench Testing Process

A typical testing process with CROWNBench is described below as Figure 2.

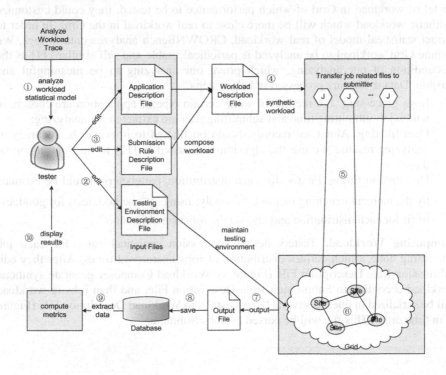

Fig. 2. CROWNBench system testing process

As shown above, workload statistical model could be extracted after analyzing workload trace①; Testers could define testing environment structure in topology description file②, and then testing environment will be built by environment manager; Basing on workload trace analyzing result, testers could edit Application Description File, which contains detailed tasks and working flow and dependency among them; The other file testers should edit is Submission Rule Description File, in which submission statistic rule is set③; Workload Composer composes workload into Workload Description File for testing from Application Description File and Submission Rule Description File④; Jobs Submitter parses workload and submits each job from appointed node and in appointed time⑤; Testing Controller and script which describe the workflow of an application will control the running order of tasks ⑥; When testing is finished, output file was generated⑦; After Data Transferrer transfers result files to the testing control node, result will be extracted and saved into database⑧; After computing test results of metrics⑨; Final results will be fed back to testers⑩.

5.4 Key Techniques of CROWNBench

Analyzing Workload Trace. If testers could acquire some knowledge about statistical model of workload in Grid of which performance to be tested, they could customize synthetic workload which will be more close to real workload in the Grid. In order to extract statistical model of real workload, CROWNBench analyzes data in logs. We assume Grid workload to be analyzed is periodical, stable and self similar [11] as the precondition of our analyzing, which prove our analyzing to be meaningful and feasible. Data in logs are processed in below steps.

1. Logs processing. Job records of some certain type of application are chosen, in which job submitting time and submitting user are extracted for analyzing;
2. Data filtering. Abnormal records should be filtered to improve the accuracy of analyzing results. We use the algorithm discussed in [12] to find out abnormal data;
3. Distribution fitting. Firstly, for each distribution, parameters should be estimate by the moment matching method. Secondly, using χ^2 method, tests for goodness of fit for each distribution and choose the most suitable one.

Composing Workload. Testers define submission statistical rules for each job submitting node, which defines distribution of jobs interarrival times. After they edit Submission Rule Description File (Figure 3), Workload Composer generate synthetic workload according to Submission Rule Description File, and then jobs in workload will be serialized in lines as several "<request>"s in Workload Description File (Figure 4) in time order. This file will be parsed by Jobs Submitter.

```
<test starttime="2006-12-27 22:49:00" lasttime="30">
...
     <workload statistics="Poison">
        <parameter name="request-number" value="1" /><!-- default
value -->
        <parameter name="lamda" value="4" />
        <parameter
name="targetselector"value="certainSelect"applicationid="1234" />
        <parameter name="submit-node" value="192.168.1.54" />
     </workload>
     <workload statistics="Periodical">
        <parameter name="request-number" value="1" /><!-- default
value -->
        <parameter name="period" value="5" />
        <parameter name="target-selector" value="pollSelect" />
        <parameter name="submit-node" value="192.168.1.199" />
     </workload>
...
</test>
```

Fig. 3. Submission rule description file

```
<?xml version="1.0" encoding="UTF-8"?>
<requests starttime="2007-05-15 19:56:30">
        <request id="0" submiter="192.168.1.36" timeoffset="0">
                <application-submit applicationid="1234" />
        </request>
        <request id="1" submiter="192.168.1.25" timeoffset="2">
                <application-submit applicationid="2205" />
        </request>
...
</requests>
```

Fig. 4. Workload description file

Maintaining Testing Environment. To eliminate the influence among different performance testing and minimize the influence which is exerted on Grid environment by performance testing, CROWNBench maintains a testing environment for a performance testing by building an isolated testing environment before testing starts and clearing it after testing finishes. Concretely, a suit of Grid middleware was deployed and configured on each node in basing on its status in Grid environment according to Testing Environment Topology File (Figure 5), and middleware and related files and database will be deleted after performance testing.

Generally, these tasks are finished manually. In CROWNBench, we developed a tool, WSAnt, to implement automatically. Building and clearing of testing environment

```
<topo>
  <regionregistry ip="192.168.1.40" >
    <regionswitch name="cnt_region1"
      ip="192.168.1.40" >
      <rlds name="cnt_rlds1" ip="192.168.1.40" >
        <rlds name="cnt_rlds1" ip="192.168.1.41" >
          <nodeserver ip="192.168.1.41" >
            <gims ip="192.168.1.41"/>
            <schedule ip="192.168.1.41" />
          </nodeserver>
          <nodeserver  ip="192.168.1.42" />
        </rlds>
      </rlds>
      <rlds name="cnt_rlds2" ip="192.168.1.42" >
        <nodeserver  ip="192.168.1.42" />
      </rlds>
    </regionswitch>
  </regionregistry>
</topo>
```

Fig. 5. Testing environment topology file

process are implemented in ANT script, and WSAnt controls its running in remote node by using web service technology. In this way, the whole process is implemented in a more automatic, accelerated and simple manner.

Application's Workflow Control. To control application workflow, we chose a flexible script tool, Ant [13], to avoid a large mount of developing works. Ant is a free open source tool of Apache Software Foundation. CROWNBench uses it to control workflow in applications for following reasons: at first, it is developed by JAVA, so it is platform independent and testing applications written by it can be integrated with Grid services frame easily; Ant and its additional project Ant-contrib [14] provide tasks and dependencies which can be described by XML. So testers' description of tasks' dependencies and workflow can easily mapping to relations between tasks of Ant. We can extend any Grid task to Ant task, and construct them to Grid application. Parameters of those tasks can be defined to change the amount of workload.

In order to support complex workflow of Grid application, CROWNBench extended Ant, and added some new tasks and conditions. These tasks have been collected in workflow control dictionary of CROWNBench. Figure 6 is a fraction of a sample workflow control script.

```
<target name="first">
  <timer name="timer.compute">
    <compute countamount="3" />
  </timer>
  <parallel>
    <sequencial>
      <schedule return="matched.node1" policy="random" count="1"
nodes="http://localhost:8080/wsrf/services/WSAntService" />
      <wsant return="io.result" path="lib" target="second"
url="${matched.node1}" datachannel="${wsantftp}" antfile="build.xml"
daemon="true" />
    </sequencial>
    <sequencial>
      <schedule return="matched.node2" policy="random" count="1"
nodes="http://localhost:8080/wsrf/services/WSAntService" />
      <wsant return="transfer.result" path="lib" target="second"
url="${matched.node2}" datachannel="${wsantftp}" antfile="build.xml"
daemon="true" />
    </sequencial>
  </parallel>
</target>
```

Fig. 6. Workload control script

6 CROWNBench Performance Testing Experiments

6.1 Experiment Environment

Two suites performance testing experiments with CROWNBench are implemented on CROWN testbed, which is deployed on 30 tree-structured nodes in our research institute. The first experiment is done to study the different influence on Grid system performance caused by different Grid workload model instances. We did the second one to find whether our system could gain meaningful performance results with synthetic workload comparing to real workload.

6.2 Experiment I

In experiment we compose three applications (①②③) with different structure by using three basic tasks, which have been provided by CROWNBench, including file transferring, float computing and I/O operation. ①② all contain sequential and parallel workflow, ③ is pure chain of tasks. In this term, CROWNBench is used to study influence on Grid system performance given by job submission frequency. Three kind of workload described by matrix and testing results showed below.

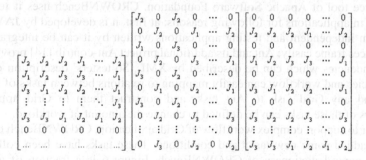

Fig. 7. Workload① workload② workload③

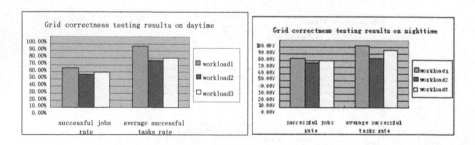

Fig. 8. Correctness testing results(daytime) **Fig. 9.** Correctness testing results (nighttime)

In daytime, besides testing workload, normal running workloads on Grid system are heavy. This situation will be eased during nighttime. Correspondingly, performance of Grid system should also be different in common sense, and it is also revealed by experiments. We can find in both of two experiments that long lasting and continuous workload will give Grid system a more great impact.

In this experiment, workload are same as ones used in first experiment, but they all contain only one application (③) for simplifying testing workload model and computing metrics of Grid system Qos. The testing results are showed as Figure 10.

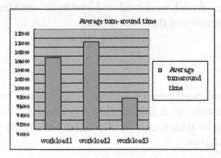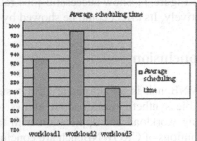

Fig. 10. Grid system Qos testing results (nighttime)

From testing results listed above, we can make some probable conclusions. Jobs submission frequency and lasting time exert influence on both running correctness and Qos of Grid system. Those workloads with higher jobs submission frequency and longer lasting time will spend longer average turnaround time and scheduling time, and also generate greater performance jitter (variance).

6.3 Experiment II

We had learned from jobs observing component in CROWN Grid middleware that the arrival process of AREM jobs, an application used to predict a day's weather in the future, appears to be Poison distribution in a long term, and parameters of distribution are different among periods in a day. Such observing and analyzing processes are finished by component out of CROWNBench system, so it is beyond discussion of this

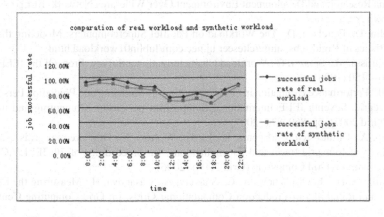

Fig. 11. Successful jobs rate of real workload and synthetic workload

paper. In this application, data firstly is fetched through I/O operation and then is processed. We used CROWNBench to compose synthetic workload using basic tasks of float computing and I/O operation and tuning parameters of Poison to emulate real AREM workload in each two hours period in a day. Then we chose a day to do performance testing every 2 hours. And we compared the average jobs rate of synthetic workload to real workload to evaluate the effectiveness of our system.

The Figure 11 show successful jobs rate of real workload and synthetic workload respectively. Its effect could be showed by results of this experiment.

7   Conclusion

CROWNBench is an automatic Grid system performance testing system using customizable synthetic workload. It allows testers to define testing workload by constructing workload and application with tasks which can be defined by them. Main contributions of CROWNBench are concluded as:

1.   Testers can customize their Grid applications running for Grid performance testing. Grid performance testing is not limited to scenario running certain kinds of applications;
2.   Testers can define jobs submission statistical laws to compose testing workload, and then load them to Grid system in a usual manner;
3.   Running testing workload can be controlled and replayed. Because user can define their applications and tasks with probes, low level performance data of a certain component of system could also be collected to study and improve the performance of the component;
4.   The whole procedure of testing is automatic, because testers only need to take part in works of editing Application Description File, Submission Rule Description File and Testing Environment Topology File.

References

1. China Research and Development Environment Over Wide-area Network, http://www.crown.org.cn/
2. Lublin, U., Feitelson, D.: The Workload on Parallel Supercomputers: Modeling the Characteristics of Rigid Jobs. http://citeseer.nj.nec.com/lublin01workload.html
3. Calzarossa, M., Serazzi, G.: Workload Characterization a Survey. Proc. Of the IEEE 81(8), 1136–1150 (1993)
4. Li, H.: Performance Evaluation in Grid Computing: A Modeling and Prediction Perspective. In: ccgrid. Seventh IEEE International Symposium on Cluster Computing and the Grid (CCGrid 2007), pp. 869–874 (2007)
5. Iosup, A., Dumitrescu, C., Epema, D., Li, H., Wolters, L.: How are real grids used? The analysis of four grid traces and its implications. In: proceedings of 7th IEEE/ACM Intl. Conference on Grid Computing (Grid 2006) (2006)
6. Khalili, O., He, J., Olschanowsky, C., Snavely, A., Casanova, H.: Measuring the Performance and Reliability of Production Computational Grids. In: Grid Computing Conference (2006)

7. Frumkin, M., Van der Wijngaart, R.F.: NAS Grid Benchmarks: A Tool for Grid Space Exploration. Cluster Computing 5(3), 247–255 (2002)

8. Tsouloupas, G., Dikaiakos, M.: GridBench: A Tool for Benchmarking Grids. In: Proceedings of the 4th International Workshop on Grid Computing (Grid 2003), pp. 60–67 (2003)

9. Iosup, A., Epema, D.H.J.: GrenchMark: A Framework for Analyzing, Testing, and Comparing Grids. In: Proc. of the 6th IEEE/ACM Int'l Symposium on Cluster Computing and the Grid (CCGrid 2006), May 2006, pp. 313–320. IEEE Computer Society Press, Los Alamitos (2006)

10. Németh, Z., Gombás, G., Balaton, Z.: Performance Evaluation on Grids: Directions, Issues, and Open Problems. In: Parallel, Distributed and Network-Based Processing (2004)

11. Lublin, U., Feitelson, D.G.: The workload on parallel supercomputers: modeling the characteristics of rigid jobs. J. Parallel & Distributed Comput. 63(11), 1105–1122 (2003)

12. Calzarossa, M., Serazzi, G.: Workload characterization: a survey. Proc. IEEE 81(8), 1136–1150 (1993)

13. The Apache, A.N.T.: Project, http://ant.apache.org

14. Ant-Contrib Tasks, http://ant-contrib.sourceforge.net

A Decision Procedure for XPath Satisfiability in the Presence of DTD Containing Choice

Yu Zhang[1,2], Yihua Cao[1], and Xunhao Li[1]

[1] Department of Computer Science & Technology,
University of Science & Technology of China, Hefei, 230027, China
[2] Anhui Province Key Lab of Software in Computing and Communication, Hefei, China
yuzhang@ustc.edu.cn

Abstract. XPath satisfiability is one of the most basic problems of XML query optimization. A satisfiability decision framework, named *SAT-DTD*, is proposed to determine, given a set of XPath queries P and a DTD τ, which subset of P are satisfiable by an XML tree conforming to DTD τ. In the framework, an indexed NFA is constructed from the set of XPath queries P, and then the NFA is driven by simple API for DTD (SAD, something like SAX) events, derived from DTD τ, to evaluate the predicates in P and to decide the satisfiability of P. Especially, DTD choice (*i.e.* '|' operator) is taken into consideration, and an algorithm, named *SAT-DTD_C*, which bases on *SAT-DTD*, is put forward to determine the unsatisfiability caused by DTD choice. At last, the complexity of the algorithms is analyzed, and the correctness of the algorithms is tested by experiments.

Keywords: DTD choice; XPath satisfiability; automaton.

1 Introduction

XPath [1] is a query language used to navigate the node or node set in XML files. As a sub-language of XSLT, XQuery *et al.*, XPath has been widely used in XML query, transformation and update.

To improve the efficiency on XML access control and query transformation, the query containment problem received a great attention recently [2-6]. The general formulation of this problem is as follows: given two XPath queries p and q, check whether for any tree t, the results from evaluating p always contain those of q, and if so we call that p contains q, denoted as $q \subseteq p$. In the literature, much attention has been paid to analyzing the complexity [2,3], to simplifying the NP-complete containment problem or to converting it into the homomorphism problem among XPath trees in order to exploit EXPTime or PTime approximate algorithms [2,4-6].

XPath satisfiability refers to the state that given an XPath query p, if there exists an XML document d, so that the results from evaluating p are nonempty, then p is satisfiable, denoted as $SAT(p)$. Furthermore, XPath satisfiability can be considered together with an XML specification definition (*i.e.* DTD or XML Schema) τ, that is, if there exists an XML document d such that d conforms to τ and the answer of p is

Y. Zhang et al. (Eds.): APWeb 2008, LNCS 4976, pp. 202–213, 2008.

nonempty, then we say (p, τ) is satisfiable, denoted as $SAT(p, \tau)$. As a new hot spot of research, XPath satisfiability is subsumed by the complement of the containment problem for XPath, and is considered to be far from tight than the latter [8].

Hidders first discussed XPath satisfiability in [10]. He introduced TDG (Tree Description Graph) to describe XPath expression and analyzed the complexity of deciding TDG's satisfiability, however, he did not consider any XML specification languages. Later Lakshmanan *et al.* began to study the XPath satisfiability in the presence of XML Schema [11], the Node Identity Constraint (NIC), which is common in XPath 2.0, was discussed and a method for deciding the satisfiability of TPQ (Tree Pattern Query) was presented, but they did not consider the choice element in XML Schema Definition (XSD). Benedikt and Fan *et al.* [8] analyzed a variety of factors contributing to the complexity of XPath satisfiability in the round, such as with or without DTD, with or without data values, with or without predicates etc. Other related researches include: Marx studied conditional axes in XPath 2.0 [12]; Geerts and Fan discussed sibling axes in the presence of XML Schema [13].

All the papers above mainly studied the complexity of XPath satisfiability under various factors, most of which did not provide with verifying algorithms on the satisfiability. Particularly, deciding the satisfiability of TPQ will make the complexity ascending from PTime to NP-complete, if DTD contains choice (*i.e.* '|' operator). It is of great significance to propose and optimize suitable algorithms for identifying and checking the unsatisfiability for a main class of XPaths under this common situation.

In this paper, we intend to design algorithms on deciding the satisfiability for a set of XPath queries P in the presence of DTD τ, in which a subset of XPath 1.0, denoted as $XP^{\{//,*,[\,]\}}$ (those include descendent axes, wildcards and predicates), and DTD with or without choice are considered. Our major contributions are as follows:

- A framework based on automaton techniques, named *SAT-DTD*, is proposed to decide which subset of P are satisfiable in the presence of τ without choice. In the framework, an indexed NFA constructed from P is driven by a sequence of SAD (simple API for DTD) events derived from τ, to decide the satisfiability for P.
- Based on the framework *SAT-DTD*, an algorithm, named *SAT-DTD_C*, is proposed to identify direct and indirect conflict of XML elements caused by DTD choice and to decide the unsatisfiability of XPath queries caused by the conflict.
- The complexities of the above algorithms are analyzed, and the experimental results demonstrate the correctness and the efficiency of our techniques.

The rest of this paper is organized as follows: Section 2 presents some basic concepts. Section 3 and 4 describe *SAT-DTD* and *SAT-DTD_C* respectively. Section 5 analyzes the complexities of our techniques, shows the experimental results and indicates potential optimization points. Section 6 concludes this paper.

2   Basic Concepts

2.1   DTD

A subset of DTD which only contains element declarations is considered here. Algorithms based on the subset could be easily extended on DTD that further contains attribute-list declarations.

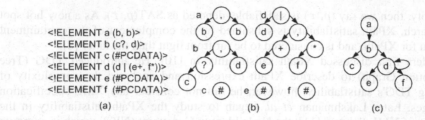

```
<!ELEMENT a (b, b)>
<!ELEMENT b (c?, d)>
<!ELEMENT c (#PCDATA)>
<!ELEMENT d (d | (e+, f*))>
<!ELEMENT e (#PCDATA)>
<!ELEMENT f (#PCDATA)>
```

(a)   (b)   (c)

Fig. 1. (a) A DTD document; (b) the corresponding content model trees of (a) in Xerces; (c) the corresponding DTD graph of (a)

DTD document (e.g. Fig.1(a)) declares the content model for each XML element. The existence of operator '?', '*', '+', especially '|' in an element declaration makes the content model complex. There are many representations of content model in practice. Xerces [14] represents each element's content model as a binary tree, called *content model tree* (e.g. Fig.1(b)) in order to assist in validating XML document. [4] proposed a kind of DTD graph (e.g. Fig.1(c)) to simplify the DTD. We can see that DTD graph gives a monolithic view of DTD, while content model tree similar to Xerces gives a partial but detailed view. So we combine these two models to represent DTD.

Definition 1. Given a DTD document τ, the corresponding **DTD Graph** is a directed graph $G=(N, C)$, where each node $n \in N$ corresponds to one element in τ, each edge $< n_1, n_2 > \in C$ represents that the element corresponding to n_2 is the sub-element of the element corresponding to n_1 in τ.

Definition 2. Given a DTD document τ, for any element e declared in τ, the corresponding **Content Model Tree** (CM Tree) of its declaration $decl_e$ is a binary tree $T_e^{(CM)} = (N, C)$, where $N = N_b \cup N_f$ is the node set, and C is the edge set. Each branch node $n_b \in N_b$ corresponds to an operator in $decl_e$ (such as '|'), each leaf node $n_f \in N_f$ corresponds to a sub-element of e. For each edge $< n_1, n_2 > \in C$, where $n_1 \in N_b$, $n_2 \in N$, if $n_2 \in N_b$, then n_2 is the next level operator of n_1 in $decl_e$; if $n_2 \in N_f$, then n_2 is the operand of n_1 in $decl_e$.

2.2 XPath

Fig.2 gives the grammar of $XP^{\{//,*,[ ]\}}$. Generally this kind of XPath can be represented in tree pattern [2] or in automaton [7,9,15]. We choose the latter to describe a set of *to-be-decided* XPath queries P, and DTD τ will be converted into a sequence of events to drive the automaton. Now we briefly introduce some related concepts.

```
[1] P ::= / E | // E
[2] E ::= E/E | E//E | E[Q] | label | text( ) | * | @* | . | @label
[3] Q ::= E | E Op Const | Q and Q | Q or Q |
          not(Q) | func(Q*)
[4] Op ::= < | ≤ | > | ≥ | = | ≠ | * | div | + | -
```

Fig. 2. XPath fragment supported   **Fig. 3.** Corresponding automaton of /a//*[c]//e

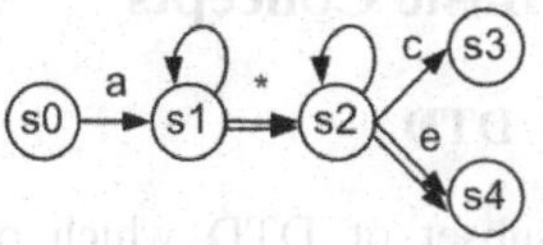

Definition 3. Given an XPath query p, the **main path expression** of p is the remaining expression of p in which all predicates are removed. The **nested path expression** is the XPath expression appearing in one of the top level predicates of p.

For example, if p is /a//*[c]//e, then /a//*//e is the main path expression of p, and c is the nested path expression of p.

Definition 4. Given a set of XPath queries P, the corresponding **Nondeterministic Finite Automaton** (NFA) A_P is incrementally constructed from the *main* and *nested path expressions* in P by path sharing technique in [9]. States in A_P can be further labeled with the following kinds: a **result state** is the state matching at least one *main path expression* in P, and the path from the initial state to one of the result states is called the **main path** in A_P; a **leaf state** is the state matching at least one *nested path expression* in P; a **branch state** is the state in a *main path* of A_P which branches to one or more leaf states; an **APS** (After-Predicate State) is the state in a *main path* of A_P whose parental state is a branch state; an **FPS** (First-location-step-of-Predicate State) is the state matching the first location step of some *nested path expression* in P.

Fig.3 shows the corresponding automaton of XPath /a//*[c]//e. Circles in the figure represent states, where s_4 is a result state and also an APS state, s_3 is a leaf state and also an FPS state, s_2 is a branch state. The child axes and descendant axes are represented by line with arrow and crewel with arrow, respectively.

Suppose that set *Preds* contains all top level predicates appearing in a set of XPath queries P, and set $P' \supseteq P$ contains all nested path expressions in *Preds*. We can construct the corresponding NFA and label the state kinds according to definition 4, the set of NFA states is denoted as *States*. In order to accelerate the running of NFA, the following indices are further added to some kinds of the NFA states.

- Add index table LR(s) to result state s, where LR(s) $\subseteq P$. $\forall p \in P$, if state s matches the main path expression of p, then $p \in$ LR(s);
- Add index table LP(s) to leaf state s, where LP(s) $\subseteq P' \times Preds$. $\forall p \in P$, for each predicate *pred* in p, if state s matches nested path expression q in *pred*, then <q, *pred*>$\in$ LP(s);
- Add index table LB(s) to branch state s, where LB(s) $\subseteq Preds$. $\forall p \in P$, for each location step *ls* of p and each predicate *pred* in *ls*, if s is the state matching *ls*, then *pred*$\in$ LB(s).

Particularly, in order to decide the XPath unsatisfiability caused by DTD choice (details are to be discussed in section 4), more indices are added to the NFA:
- Add index table LS(s) to result state or leaf state s, where LS(s) $\subseteq$ *States*. $\forall p \in$ LR(s) or <q, *pred*>$\in$ LP(s), if state t is the APS state of p or the FPS state of *pred*, then $t \in$ LS(s);
- Add parental state index *parent* and table LAPS(s) to APS state s, where *parent*$\in$ *States* and LAPS(s) $\subseteq P$. $\forall p \in P$, if there exists at least one predicate of p on the location step corresponding to *parent*, then $p \in$ LAPS(s);
- Add parental state index *parent* and table LFPS(s) to FPS state s, where *parent*$\in$ *States* and LFPS(s) $\subseteq P' \times Preds$. $\forall p \in P$, if *pred* is the top level predicate of p, and s is the state matching the first location step of a nested path expression q in *pred*, then <q, *pred*>$\in$ LFPS(s).

3 Deciding XPath Satisfiability in the Presence of DTD

Similar to the approach of converting XML into SAX events [16], DTD is firstly converted into SAD (simple API for DTD) events, and then be input to an NFA constructed from the *to-be-decided* XPath set to decide the XPath satisfiability.

3.1 SAD Events

Currently four kinds of SAD events have been defined: startDTDDocument( ), startElementDecl(*a*), endElementDecl(*a*), endDTDDocument( ), where *a* is the element name. First the DTD document is read and parsed into a DTD graph, and then the graph is converted into a tree for the convenience of issuing SAD events.

Definition 5. Given a DTD document τ and the corresponding DTD graph G, we can create the corresponding *DTD tree* T from G:

1) if G is a tree, then $T = G$;
2) if G is a DAG, then T is the expanded tree of G constructed as below:
   i)   construct the root r' of T from the root r of G, and add r' to a queue Q.
   ii) if Q is empty, then the construction completes. Else, take a node v' out of the Q and get its corresponding node v in G, then construct the corresponding edges and child nodes of node v' in T according to the outgoing edges of node v, add all the constructed child nodes of node v' to Q.
   iii) repeat ii).
3) if G contains a circle, then we can deduce from the set of actual XML documents that for each ring $ring_i$ ($0 \le i \le k$, k is the number of rings in G), its possible maximum expanded depth $depth(ring_i)$ satisfies that the actual maximum expanded depth of $ring_i$ in G is not larger than $depth(ring_i)$.

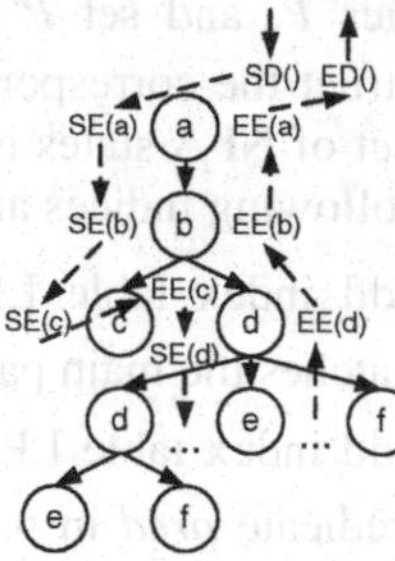

Fig. 4. Generating SAD events according to DTD tree

After converting a DTD graph to a DTD tree, we can depth-first traverse the DTD tree to issue SAD events. For instance, Fig.4 shows the corresponding DTD tree expanded from the DTD graph in Fig.1(c) and its SAD events propagated. In Fig.1(c) the self loop of element d is expanded by 2 levels. SD, SE, EE, ED in Fig.4 is short for startDTDDocument, startElementDecl, endElementDecl, endDTDDocument respectively, the parameter *a* followed by SE and EE is the element name. This figure omits the content model information.

3.2 SAD Events Handling: Deciding Satisfiability

The execution process of NFA is defined by the handling rules on SAD events. This process completes the decision of XPath satisfiability. We first discuss the decision procedure without considering DTD choice.

3.2.1 State Transition
Function *trans(stack, a)* does transition from the top *state set* of *stack* on the input *a* by the following rule: all states reached from any state in the top *state set* along edge

labeled '/a', '/*', '//a' or '//*' are matched states. Meanwhile, to handle multiple match, *trans* records the match layer *matchlayer* for each runtime state s, see section 3.2.4.

Algorithm.1 SAT-DTD

s_0: the initial state; *stack*: the NFA runtime stack, whose frame is a set of NFA states.

startDTDDocument() //briefly, SD
(1) *push({s_0}, stack)*;

endElementDecl(a) //briefly, EE
(1) *sset = pop(stack)*;
(2) for each s in *sset*{
(3)   for each *pred* in LB(s)
(4)   *resetPredRes(pred)*;
     }

endDTDDocument() //briefly, ED
(1) stop or wait another DTD document;

startElementDecl(a) //briefly, SE
(1) *sset = trans(stack, a)*;
(2) *push(sset, stack)*;
(3) for each s in *sset*{
(4)   for each $<q, pred>$ in LP(s)
(5)   *evaluatePred(pred, q)*;
(6)   for each p in LR(s)
(7)   *setRes(p)*;
     }

3.2.2 Deciding the Satisfiability of Main Path Expression

Given an XPath p, consider its main path expression p_m first. If after *trans(stack, a)*, a result state s is reached, and p is in LR(s), then p_m is satisfiable according to the former definitions. So we use *setRes(p)* to record that p_m is matchable. (see line (6)~(7) in SE event handler of algorithm 1)

Predicates may contain XPath expressions, *i.e.* nested path expressions. Suppose p's predicate *pred* contains an XPath q, namely, *pred* looks like [...q...]. Denote q's main path expression as q_m. If after a transition, result state s is reached and $<q_m, pred>$ is in LP(s), then q_m is satisfiable. Next, we need to decide the satisfiability of predicates. (see line (4)~(5) in SE event handler of algorithm 1)

3.2.3 Predicates Evaluation

The satisfiability decision on predicates is mainly done through *evaluatePred(pred,q)*.

Definition 6. Given a DTD τ and an XPath p, *pred* is a predicate of p, e is the general expression in *pred*. Suppose that D is the set of XML documents conforming to τ. e is satisfiable if there exists $d \in D$ so that the evaluation of e in d is true.

evaluatePred(pred, q) uses a bottom-up approach to evaluate the validity of *pred*. In line (4) of algorithm 2, e's satisfiability is decided by evaluating *func(...)*, e.g. *not(Q)*, its satisfiability is opposite to the parameter Q.

In EE event handler, if current NFA backtracking state s is a branch state, then get each predicate *pred* from LB(s), and reset the

Algorithm.2 evaluatePred

evaluatePred(pred, e)
(1) switch (*e.type*) {// see Fig.2
(2)   case E: *e.valid = true*;
(3)   case *E* Op *Const*: *e.valid = E.valid*;
(4)   case *func(Q_1, ..., Q_n)*: //include not(Q)
             e.valid=func(Q_1.valid,..., Q_n.valid);
(5)   case Q_1 or/and Q_2:
             e.valid = Q_1.valid or/and Q_2.valid;
         }
(6) if (*e.parent* != null)
(7)   return *evaluatePred(pred, e.parent)*;
(8) else *setPredRes(p, e.valid)*;
(9) return *e.valid*;

evaluating status of *pred* (see line (2)~(4) in EE handler of algorithm 1) to prepare for next predicate evaluation.

3.2.4 Multiple Match

Multiple match is difficult for XSIEQ [15], an XPath query engine on XML stream. It also occurs in deciding XPath satisfiability.

Definition 7. Given a DTD τ and an XPath p, when deciding p's satisfiability using τ, one location step ls of p may match more than one DTD elements in different depths. This occurrence is called **multiple match**, ls is called **multiple match occurrence point**.

The necessary condition of multiple match occurrences is that the location steps contain descendent axes. A serious problem caused by multiple match is that either the evaluation among predicates on multiple match occurrence point or the evaluation between the predicate and the main path expression are out of sync. Let's take the latter case for example. Suppose that multiple match occurrence point ls matches two DTD elements e_1 and e_2, and there is a predicate $pred$ on ls, the XPath expression on the right side of ls is denoted as p_r. If the evaluation result is that $pred$ is satisfied on input e_1, but p_r is satisfied on input e_2, then the evaluation between $pred$ and p_r are out of sync, and may falsely evaluate the satisfiability of the whole XPath expression p.

Theorem 1. Given a DTD τ and an XPath p, assume that location step ls_i of p contains n predicates and can match m elements, $i.e.$ $e_1,...,e_m$, in the same path in τ. p can be represented as: $...//ls_i[pred_{i1}][...][pred_{in}]/ls_{i+1}/....$ Then the XPath fragment $//ls_i[pred_{i1}][...][pred_{in}]/ls_{i+1}$ can be satisfied if the following condition is met: when ls_i matches e_j ($1 \leq j \leq m$), $pred_{i1}, ..., pred_{in}$ and XPath fragment $/ls_{i+1}$ can also be satisfied.

Proof (briefly). It can be inferred from the definition of XPath satisfiability.

To handle multiple match occurrences, according to theorem 1, for any multiple match occurrence point ls of p, there should be at least one matched element to make all predicates on ls and the right side XPath fragment of ls satisfied simultaneously. For example, if we use DTD in Fig.1(a) to decide the satisfiability for XPath set:{/a//*[c]//e,/a//*[c]/e}, then the satisfied XPath is /a//*[c]//e, and the reason why /a//*[c]/e is not satisfied is that on multiple match occurrence point //*, predicate [c] and main path expression fragment /e cannot be satisfied simultaneously.

4 XPath Unsatisfiability Caused by DTD Choice

DTD element declaration may contain operator '|', which states that its operands cannot occur in the same XML document simultaneously. To simplify the statement, in this section we use the term *node* (in DTD tree) instead of *element* (in DTD).

4.1 Conflict and XPath Unsatisfiability

Definition 8. If DTD formulates that certain two nodes cannot occur in the same XML document simultaneously, then there is a conflict between the two nodes, we call it **node conflict**. The conflict between sibling nodes is **direct conflict**; conflict between other nodes is **indirect conflict**.

Apparently, DTD can only declare conflicts between sibling nodes using operator '|', so all indirect conflicts are caused by direct conflicts. The way to find direct conflicts from indirect conflicts is as follows: given two nodes n_1 and n_2, if n_1 and n_2

conflict indirectly (they cannot be sibling nodes), the layers of n_1 and n_2 are $layer_1$ and $layer_2$ respectively, assume $layer_1 \leq layer_2$, their least recent common ancestor n_a ($layer_a < layer_1$), and n_a's two child nodes n_{ac1}, n_{ac2} are **AOS** (Ancestor-Or-Self) nodes of n_1, n_2 respectively, then the indirect conflict between n_1 and n_2 is caused by the direct conflict between n_{ac1} and n_{ac2}. The following theorem gives the necessary and sufficient condition of direct conflict between nodes.

Theorem 2. Assume $E=(e_1,\ldots,e_n)$ is the sub-element tuple of DTD element e, $T_e^{(CM)}$ is e's content model tree. The necessary and sufficient conditions of e_i and e_j conflict directly are: 1) their least recent common ancestor in $T_e^{(CM)}$ is operator '|', denoted as n_{a1}; and 2) the least recent ancestor of n_{a1} (except operator '|'), denoted as n_{a2}, is not '+' or '*'; and 3) if n_{a2} is '?', then there are no '+' or '*' in the path from n_{a2} to its least recent binary operator ancestor, such as ','.

Proof. If e_i and e_j conflict directly, then condition 1) must be true, consider n_{a1}'s least recent ancestor (except '|') (*i.e* condition 2)): if n_{a2} is ',', then there is definitely direct conflict between e_i and e_j; n_{a2} cannot be '+' or '*', this means that e_i and e_j can occur more than once without resulting conflicts; n_{a2} may be '?', if so, it needs more consideration (*i.e.* condition 3)). When n_{a2} is '?', assume its least recent ancestor n_{a3}, no matter ',' or '|', e_i and e_j conflict or not is unrelated to n_{a3}, we do not need to consider further about n_{a3}'s ancestors. The existence of '+' or '*' in the path between n_{a2} and n_{a3} also indicates that e_i and e_j can occur more than once, without resulting conflicts.

According to theorem 2, we can conclude a further condition on node conflict influencing XPath satisfiability.

Definition 9. Given two node sets A and B, if for each node n_a in A, n_a conflicts with each node n_b in B, then there is conflict between A and B, we call it **node set conflict**.

Theorem 3. Assume XPath p is represented as $\ldots//ls_i[pred_{i1}][\ldots][pred_{in}]/ls_{i+1}/\ldots$, and node set N_i contains nodes in the input DTD tree which match the location step ls_i, any conflict among the following node sets (*i.e. node set conflict*) will result in the unsatisfiability of p: 1) the node set N_{ik} whose elements are from the *topmost layer* and further selected from N_i by $pred_{ik}$ ($1 \leq k \leq n$), where nodes in *topmost layer* are those *closest* to root node in the DTD tree; 2) the node set N_{i+1} whose elements are from the *topmost layer* and further selected from N_i by the next location step $/ls_{i+1}$.

Proof. Just take the node sets N_{ik} selected by $pred_{ik}$ ($1 \leq k \leq n$) and N_{i+1} selected by the next location step $/ls_{i+1}$ for example: if the two node sets conflict each other, then according to definition 9, $pred_{ik}$ and $/ls_{i+1}$ cannot be satisfied simultaneously; if there is no conflict between the two node sets, then $pred_{ik}$ and $/ls_{i+1}$ can always select certain nodes to avoid conflicts, thus they are both satisfied. So, if there are node set conflicts in any two of the above node sets 1) and 2), p is unsatisfiable.

4.2 Deciding XPath Unsatisfiability Caused by Conflict

According to theorem 2, direct conflict can be statically decided. Assume $E=(e_1,\ldots,e_n)$ is the sub-element tuple of DTD element e, then for each n-bit bitset

$BS_i=(b_{i1},\ldots, b_{in})$ of every node e_i: $b_{ij}= 0$ indicates that there is no conflict between e_j and e_i; else vice versa. Node conflicts can be detected by traversing content model trees.

Algorithm 3 gives the decision procedure for XPath satisfiability in the presence of DTD containing choice, which bases on the framework *SAT-DTD*. In the algorithm, the node conflicts caused by DTD choice are identified to decide the unsatisfiabilty of XPath expressions according to theorem 3.

Algorithm.3 SAT-DTD_C (*see Algorithm 1 for other event handlers*)

s_0: the initial state; *stack*: the NFA runtime stack, whose frame is a set of NFA states.

startElementDecl(*a*) //briefly, **SE**	**endElementDecl**(*a*) //briefly, **EE**
// see algorithm 1	// see algorithm 1
(3) for each *s* in *sset*{	(2) for each *s* in *sset*{
...	...
//(8~11) add matched element(ME for short)	// clear matched element in APS or FPS.
(8) for each <*q,pred*> in LFPS(*s*)	(5) if (*s.isFirstMatch*() and (*s.isAPS*() or *s.isFPS*()))
(9) *s.addME*(*q, a*);	(6) *s.clearAllMEs*();
(10) for each *p* in LAPS(*s*)	//check conflicts while reaching branch state
(11) *s.addME*(*p,a*);	(7) if (*s.isBranchState*())
// promote the matched elements.	(8) for each *p* which has predicates
(12) for each *st* in LS(*s*){	*pred₁*, ..., *predₙ* corresponding to *s*{
(13) for each <*q,pred*> in LFPS(*st*)	(9) for any p_1, p_2 in {*pred₁,...., predₙ, p*}
(14) *st.parent.addMEs*(*q, st.getME*(*q*));	(10) *checkConflict*(*s.getMEs*(p_1), *s.getMEs*(p_2));
(15) for each *p* in LAPS(*st*)	(11) if (*p.isLeftMostBranchState*(*s*) and
(16) *st.parent.addMEs*(*p, st.getME*(*p*));	*p.noConflicts*( ))
}	(12) *p.satisfied* = true;
}	}
	// clear matched elements in branch state
	(13) *s.clearAllMEs*();
	}

The operations in algorithm 3 are as follows:

In **SE** event handler,

1) *add* operation: when reaching FPS or APS states, record currently matched nodes for each corresponding XPath expression beforehand. (line(8)~(11));

2) *promote* operation: when reaching result states or leaf states, the corresponding XPath expression is matched. Find all the corresponding APS and FPS states according to LS(*s*) index (line (12)), then promote the matched nodes recorded previously on those states to the parental branch state, in order to decide satisfiability (line (13)~(16)). A same XPath can be matched more than once, so it can be promoted multiple times.

In **EE** event handler,

1) *clear* operation: when backtracking to APS, FPS or branch states, clean matched nodes (line(5)~(6), (13)) to prepare for the next decision.

2) *check* operation: when backtracking to branch state *s* (line(7)), for each XPath *p* that contains predicate(s) on location step corresponding to *s*, decide the conflicts and record them (line(8)~(10)). Finally, if there is no conflict occurrence, *p* is satisfiable (line(11)~(12)).

5 Complexity Analysis, Optimization and Experimental

In this section we first analyze the complexity of *SAT-DTD* and *SAT-DTD_C*, then propose optimization approaches on *SAT-DTD_C*, and finally check the correctness through experiments.

5.1 Analysis on the Complexity of SAT-DTD Algorithms

According to the conclusion in [8], for the XPath with descendent axes and predicates, the complexity of its satisfiability decision is PTIME only if DTD does not contain choice; and the complexity is NP-complete for arbitrary DTD.

5.1.1 Complexity of SAT-DTD Algorithm without Considering DTD Choice

The time complexity of building an automaton from XPath queries is polynomial-time depending on the size of XPath queries. Therefore we focus on the runtime complexity of *SAT-DTD* (see algorithm 1).

Theorem 4. Assume the length of DTD event sequence is $2m$, there are x XPath queries, n states in automaton, q predicates, average x_q nested path expressions and op operators in each predicate respectively, average q_l predicates in each location step with predicates, average u multiple match occurrences in each multiple match occurrence point. Then the time complexity is $O(m\cdot(n+x+x_q\cdot op\cdot q+(x_q\cdot op\cdot q_l+x)\cdot u\cdot q/q_l))$.

Proof. Assume every time all states participate in state transition, then the maximum times of state transition is $O(m\cdot n)$ (for tree NFA, on average each state will transform once to another state). If every time the element start events reach all the result states, the complexity upper bound of main path expression's satisfiability decision is $O(m\cdot x)$, so the complexity upper bound of the satisfiability decision of nested path expressions is $O(m\cdot x_q\cdot q)$. And, each predicate expression needs op computations, the upper bound of predicate computation complexity is $O(m\cdot x_q\cdot op\cdot q)$.

On each multiple match occurrence point, every predicate and the main path expression need additional satisfiability decision at the same time, the time complexity is $O((m\cdot x_q\cdot op\cdot q_l+m\cdot x)\cdot u)$.

Finally the result is $O(m\cdot n+m\cdot x+m\cdot x_q\cdot op\cdot q+ (m\cdot x_q\cdot op\cdot q_l+m\cdot x)\cdot u\cdot q/q_l)$.

From theorem 4 we can see that multiple match has a great impact on the efficiency of *SAT-DTD* algorithm. But the probability of multiple match occurrence is low except that an XPath expression contains "//*".

5.1.2 Complexity of SAT-DTD_C Algorithm

Suppose a DTD contains w elements, the average length of an element's content model is l, then the complexity of deciding node conflict is $O(w\cdot l^2)$.

Theorem 5 gives the extra complexity of runtime *SAT-DTD_C* compared with *SAT-DTD* without considering DTD choice.

Theorem 5. Suppose the number of all predicates is q, each contains x_q XPaths on average; each location step that contains predicate(s) has q_l predicates on average; each FPS state or APS state has v matches on average. The *SAT-DTD_C* algorithm has an extra complexity of $O(v^{q_l\cdot x_q+l}\cdot q/q_l)$ than *SAT-DTD* without considering DTD choice.

Proof. The time complexity of detecting node set conflict in each location step with predicate(s) is $O(v^{q_l \cdot x_q + l})$, and there are q/q_l such location steps.

5.2 Optimization on SAT-DTD_C Algorithm

Theorem 5 points out that *SAT-DTD_C* is exponential as it has parameter q_l and x_q in the exponent. It can be optimized in the following aspects.

- Add functions to determine whether the content models contain operator '|'. For those exclude '|', the complexity of *SAT-DTD_C* will not exceed *SAT-DTD* without considering DTD choice.
- From the observation we see that the complexity of *SAT-DTD_C* is mainly caused by detecting node set conflicts, thus buffer with a lookup table can be used to reduce the complexity of *SAT-DTD_C*.
- Filter out those XPath expressions with high q_l and x_q, ignore them.

5.3 Correctness Checking Experiments

Table 1. The decision results of XPath satisfiability using *SAT-DTD*, where SAT represents the actual satisfiability, T represents *satisfiable*, and F represents *unsatisfiable*

XPath query	SAT	SAT-DTD	Remark
//category[description]/name	T	T	
//text[bold]/keyword	T	T	see condition 2) in theorem 2, there is no conflict between *bold* and *keyword*
//*[text="sth"]/parlist	F	F	there is conflict between *text* and *parlist*
//category[.//listitem]/text	F	F	there is conflict between the parent of *listitem*, *parlist*, and *text*
//category[.//listitem]//text	T	T	there is no conflict between the parent of *listitem, parlist*, and some *text*

Ordinary DTD documents contain only a small portion of operator '|', making the experiments focused on it more difficult to carry out. The experiments discussed in this sub-section focus on the correctness, that is, whether *SAT-DTD* can present a correct decision of XPath satisfiability. Table.1 shows some evaluation results in common circumstances. The DTD document used in this experiment is a fragment in XMark [17]

```
<!ELEMENT site      (categories)>
<!ELEMENT categories (category+)>
<!ELEMENT category  (name, description)>
<!ELEMENT name      (#PCDATA)>
<!ELEMENT description (text | parlist)>
<!ELEMENT text      (#PCDATA | bold)*>
<!ELEMENT bold      (#PCDATA | bold)*>
<!ELEMENT parlist   (listitem)*>
<!ELEMENT listitem  (text | parlist)*>
```

Fig. 5. Fragment of xmark.dtd.

described in Fig.5. The experiment result shows that *SAT-DTD* can correctly decide the XPath satisfiability in typical circumstances with wildcards and descendent axes.

6 Conclusion

The proposed *SAT-DTD* algorithm can decide XPath satisfiability correctly. *SAT-DTD* without considering DTD choice has been applied to our XML access control system to optimize access control rules and queries. Experiments have proved the correctness of the algorithm. Our next task is to apply optimized *SAT-DTD_C* to practical systems, in the meantime we will consider more factors that influence XPath satisfiability, such as: predicates of various properties, operators '?','*', '+' in DTD, etc.

Acknowledgments. This work has been supported by the National Natural Science Foundation of China under Grant No. 60673126.

References

[1] Clark, J., DeRose, S.: XPath Version 1.0. W3C Recommendation (1999), http://www.w3.org/TR/xpath

[2] Miklau, G., Suciu, D.: Containment and equivalence for a fragment of XPath. Journal of the ACM 51(1), 2–45 (2004)

[3] Wood, P.: Containment of XPath Fragments under DTD Constraints. In: Proceedings of Int. Conference on Database Theory (2003)

[4] Böttcher, S., Steinmetz, R.: A DTD Graph Based XPath Query Subsumption Test. In: Proceedings of XML Database Symposium at VLDB, Berlin, Germany (2003)

[5] Liao, Y., Feng, J., Zhang, Y., Zhou, L.: Hidden Conditioned Homomorphism for XPath Fragment Containment. In: Li Lee, M., Tan, K.-L., Wuwongse, V. (eds.) DASFAA 2006. LNCS, vol. 3882, pp. 454–467. Springer, Heidelberg (2006)

[6] Yoo, S., Son, J.H., Kim, M.H.: Maintaining Homomorphism Information of XPath Patterns. In: IASTED-DBA, pp. 192–197 (2005)

[7] Fu, M., Zhang, Y.: Homomorphism Resolving of XPath Trees Based on Automata. In: Dong, G., Lin, X., Wang, W., Yang, Y., Yu, J.X. (eds.) APWeb/WAIM 2007. LNCS, vol. 4505, pp. 821–828. Springer, Heidelberg (2007)

[8] Benedikt, M., Fan, W., Geerts, F.: XPath Satisfiability in the Presence of DTDs. In: Proceedings of the 24th ACM SIGMOD-SIGACT-SIGART Symposium on Principles of Database Systems, pp. 25–36 (2005)

[9] Diao, Y., Altinel, M., Franklin, M.J., et al.: Path Sharing and Predicate Evaluation for High-Performance XML Filtering. ACM TODS 28(4), 467–516 (2003)

[10] Hidders, J.: Satisfiability of XPath expressions. In: Proceedings of the 9th International Workshop on Database Programming Languages, pp. 21–36 (2003)

[11] Lakshmanan, L., Ramesh, G., Wang, H., et al.: On testing astisfiablity of the tree pattern queries. In: VLDB (2004)

[12] Marx, M.: XPath with conditional axis relations. In: EDBT (2004)

[13] Geerts, F., Fan, W.: Satisfiability of XPath Queries with Sibling Axis. In: Bierman, G., Koch, C. (eds.) DBPL 2005. LNCS, vol. 3774, pp. 122–137. Springer, Heidelberg (2005)

[14] Xerces2 Java Parser 2.6.0., http://xerces.apache.org/xerces2-j/

[15] Zhang, Y., Wu, N.: XSIEQ - An XML Stream Query System with immediate Evaluation. Mini-Micro Systems 27(8), 1514–1518 (2006)

[16] Sax Project Organization, SAX: Simple API for XML. (2001), http://www.saxproject.org

[17] Schmidt, A., Waas, F., et al.: A Benchmark for XML Data Management. In: Proceedings of the 28th VLDB Conference, Hongkong, China (2002)

Performance Analysis and Improvement for Transformation Operators in XML Data Integration

Jiashen Tian[1], Jixue Liu[1], Weidong Pan[1], Millist Vincent[1], and Chengfei Liu[2]

[1] School of Computer and Information Science, University of South Australia
[2] Swinburne University of Technology
Jiashen.Tian@postgrads.unisa.edu.au,
{jixue.liu,weidong.pan,millist.vincent}@unisa.edu.au,
cliu@swin.edu.au

Abstract. Data transformation from a source schema to a global schema is an important issue in data integration. For this purpose, a set of transformation operators have been defined to transform both DTDs and the conforming XML documents. In this paper we analyze the performance of individual operators both theoretically and experimentally. The performance of a sequence of operators is also investigated via experiment. Some algorithms and structures are proposed to improve the performance of operators so that linear efficiency can be achieved. The result shows that most of the operators can be implemented in linear time to the size of data to be transformed. The performance of sequence of operators is also close to linear. Java and XQuery language are used respectively to implement the operators and their efficiency is compared.

Keywords: Data Integration, Transformation,Performance Analysis and Improvement, XML, DTD.

1 Introduction

Creating mapping between source schema and global schema is indispensable for establishing a data integration system. Mapping can be built by using query languages or other specifically defined mapping languages. In XML data integration, where the source data is modeled in XML[3], it can also be built by defining some fundamental transformation operators[1], like the relational algebra operators for the relational database. These operators form a mapping when they are applied in sequence. Following the sequence, one schema can be transformed to another. For example, Fig.1 presents two DTDs. By applying the operators *expand*, *collapse*, *shift*, *nest* and *unnest*, defined in[1], DTD (a) can be transformed to DTD (b). This example will be presented in detail later in this paper.

The advantage of the transformation approach is that the operators are fundamental in terms of semantic and data structure of XML model. Complex

Y. Zhang et al. (Eds.): APWeb 2008, LNCS 4976, pp. 214–226, 2008.

transformation can always be represented by these fundamental operators. With these operators, users need only to focus on the transformation behavior rather than have to learn a query language or a programming language to build a mapping.

(a)
```
<!ELEMENT root (actor)*>
<!ELEMENT actor (name,movie*)>
<!ELEMENT movie (title)>
<!ELEMENT name (#PCDATA)>
<!ELEMENT title (#PCDATA)>
```

(b)
```
<!ELEMENT root (movie)*>
<!ELEMENT movie (title,actor*)>
<!ELEMENT actor (name)>
<!ELEMENT name (#PCDATA)>
<!ELEMENT title (#PCDATA)>
```

Fig. 1. Source and Target DTDs

A set of operators have been defined In[1] to transform one DTD to another, with its conforming XML document also transformed. The multiplicities, nested brackets, and disjunction expressions are all considered. In another work, these operators are further proved complete, which means any DTD can be transformed to its *multiplicity compatible DTD* by these operators [2]. In this paper we not only foresee the performance problem when the current operators are used in long sequences and against large XML documents, but also try to improve the performance of expensive operators.

Different sets of transformation operators have been proposed in many papers. However, those operators either defined in relational data model[6], in specified XML schema model[9], or with no consideration of DTD syntaxes like multiplicities and nested brackets[7,8]. None of above work has analyzed the performance of their operators. Our work differs from other work since it is based on a set of complete transformation operators with fully consideration of DTD syntax, and the analysis is conducted theoretically and practically, with the implementation in both XQuery and Java languages, which enables our work unique.

We make the following contributions in this paper. Firstly, we propose some fundamental procedures that are commonly used in the implementation of the operators. By doing so, we reduce the problem of analyzing the performance of each operator to the problem of analyzing the performance of the common procedures. This method can be generally applied to other possible operators as well. Secondly, we propose some algorithms and structures to improve the performance of some operators so that linear efficiency can be achieved. Thirdly, for choosing a more efficient implementation language, we compare the performances of XQuery and Java in implementing the operators and choose Java as the implementation language based on the comparison results. Finally, we study the performance of applying a sequence of operators to transform one XML document to another XML document. The results show that the transformation can be performed in close to linear time to the size of the XML document, which enables us to apply the operators for building a materialized scalable integration system and for XML data exchange.

2 Preliminaries

Definition 1 (XML DTD). An XML Document Type Definition (**DTD**) is defined to be $D = \langle\!\langle\ EN, G, \beta, root\ \rangle\!\rangle$ where:

   (a) EN is a finite set of element names;

   (b) the set of element structures G is defined by $g \in G$ if any of the following holds:

      $g = Str$ where Str is a symbol denoting #PCDATA (text)[3];

      $g = e$ where $e \in EN$;

      $g = \epsilon$, indicating the EMPTY type;

      $g = g_1, g_2$ or $g_1|g_2$ or g_1^c or $[g_1]$ where

         ',' and '|' are conjunction and disjunction respectively, $g_1 \neq Str$, $g_2 \neq Str$, $g_1 = g$ and $g_2 = g$ are recursively defined; g_1 and g_2 are called sub-structures of g and '[ ]' is the sub-structure delimiter; $c \in \{?, 1, +, *\}$;

   (c) β is a set of functions from EN to G as $\beta(e) = [g]^c$ defining the type of e;

   (d) $root$ is the root of the DTD. □

Example 1. The DTD in Fig.1(a) can be formulated as follows.

   $D = \langle\!\langle\ EN, G, \beta, root\ \rangle\!\rangle$ where

   $EN = \{root, actor, movie, name, title\}$,

   $G = \{Str, actor*, [name, movie*], title\}$,

   $\beta(root) = [actor]^*$, $\beta(actor) = [name, movie*]$,

   $\beta(movie) = [title]$, $\beta(name) = Str$, $\beta(title) = Str$. □

Note that DTD defined here has the restrictions of no recursion and of no consideration of duplicated elements.

Definition 2 (XML tree). Let EN be a finite set of element names, V a finite set of node identifiers, VAL an infinite set of text strings, $\perp$ a special value. An **XML tree** T is defined to be $T = \phi(\text{empty})$ or $T = (v:e:val, T_1, T_2, \cdots, T_f)$ where $v \in V$, $e \in EN$, ($val \in VAL$ or $val = \perp$), the triple $v:e:val$ is called a node, and

   (i) if $val = \perp$, then $f \geq 0$, and $T_1, T_2, \cdots, T_f$ are recursively defined trees denoted by $Ch(v)$. The triple $v:e:val$ is often simplified to "e" or "$v:e$" when the context is clear;

   (ii) if $val \in VAL$, then $f = 0$. □

Example 2. The following shows how an XML document is represented in our notation, which is also graphically represented in Fig. 2.

$T_r = (v_r : root, T_1, T_7)$, $T_1 = (v_1 : actor, T_2, T_3, T_4)$, $T_2 = (v_2 : name : Brad\ Pitt)$,
$T_3 = (v_3 : movie, T_5)$, $T_4 = (v_4 : movie, T_6)$, $T_5 = (v_5 : title : Troy)$
$T_6 = (v_6 : title : Babel)$, $T_7 = (v_7 : actor, \cdots)$ □

Definition 3 (hedge). A hedge H is a sequence of consecutive trees under one parent node that conforms to the definition of a specific DTD structure. □

Hedges are defined to capture occurrences of a DTD structure in a document. A hedge relates only to one structure, and a structure may have multiple conforming hedges. A hedge may contain one tree and an XML document is a hedge

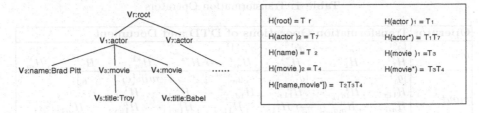

Fig. 2. XML tree   **Fig. 3.** Corresponding Hedges

that conforms to the element *root*. We use $H(g)$ to denote the hedge conforming to the structure g. The concatenation of two hedges H_1 and H_2 is denoted by H_1H_2.

In Example 2, T_3 and T_4 are two hedges each conforming to the structure $g = movie$. When the context structure is changed to $g = movie*$(notice the asterisk at the end), the only hedge conforming to g is $H(g) = T_3T_4$. Similarly, as to $g = [name, movie*]$, its conforming hedge is the sequence $T_2T_3T_4$. More details are shown in Fig. 3. Hedge is the key concept in our performance analysis.

Definition 4 (Multiplicity Operators). The operators we define for manipulating multiplicities rely on the intervals of multiplicities. The intervals of the multiplicities $'?', '1', '+', '*'$ are $[0,1]$, $[1,1]$, $[1,n]$, $[0,n]$ respectively where n represents 'more than one'. Let c_1 and c_2 be two multiplicities. $c_1 \oplus c_2(= c_1c_2)$ has the semantics of the multiplicity whose interval encloses the intervals of c_1 and c_2. Thus, $+? = *$ and $1? =?$. $c_1 \ominus c_2$ is the multiplicity whose interval equals to the interval of c_1 taking that of c_2 and adding that of 1. Thus, $?\ominus? = 1$ and $*\ominus+ =?$.   □

We now list the operator definitions in Table 1 where g means a list of structures, e an element name, H^k a short hand notation of $H(g_k)$, H^{*k} a short hand notation for $H(g_k^c)$ where the $'*'$ in the superscript indicates multiple hedges each conforming to g. Note that $H^{*k} = H_1^k \cdots H_n^k$, and $min(c_1, \cdots, c_n)$ returns the multiplicity the interval of which is the minimal among $c_1, \cdots, c_n$. Each operator is defined in two parts. One transforms DTD structures and the other transforms the conforming hedges. The symbol '$\rightarrow$' indicates the transformation from left to right. Detailed definitions and examples can be found in [1].

Example 3. Continuing our examples, assume that $\beta(actor) = [name, movie*]$, $\beta(movie) = title$ and $H(actor) = (v_1 : actor, (v_2 : name : Brad\ Pitt), (v_3 : movie, (v_5 : title : Troy)), (v_4 : movie, (v_6 : title : Babel)))$ is a hedge in terms of the structure *actor*. Then after applying the operator **collapse** against the structure *movie*, i.e. $collapse(movie) \longrightarrow \beta_1(actor) = [name, title*]$, and $H_1(actor) = (v_1 : actor, (v_2 : name : Brad\ Pitt), (v_5 : title : Troy), (v_6 : title : Babel))$. This transformation can be reversed by using **expand**$(title, movie)$.

Table 1. Transformation Operators

Operator	Transformation Operations of DTD and Document
min	$[g_1^{c_1}, \cdots, g_n^{c_n}]^c \longrightarrow [g_1^{c_1 c}, \cdots, g_n^{c_n c}]^1$; $H_1^{*1}, \cdots, H_1^{*n}, \cdots, H_m^{*1}, \cdots, H_m^{*n} \longrightarrow H_1^{*1}, \cdots, H_m^{*1}, \cdots, H_1^{*n}, \cdots, H_m^{*n}$
mout	$[g_1^{c_1}, \cdots, g_n^{c_n}]^c \longrightarrow [g_1^{c_1 \ominus c_c}, \cdots, g_n^{c_n \ominus c_c}]^{c c_c}$ $c_c = min\{c_1, \cdots, c_n\}$; $H_{11}^1, \cdots, H_{1d_{11}}^1, \cdots, H_{11}^n, \cdots, H_{1d_{1n}}^n, \cdots, H_{m1}^1, \cdots, H_{md_{m1}}^1, \cdots,$ $H_{m1}^n, \cdots, H_{md_{mn}}^n \longrightarrow H_{11}^1, \cdots, H_{11}^n, \cdots, H_{1w_1}^1, \cdots, H_{1w_1}^n, \cdots, H_{m1}^1, \cdots,$ $H_{m1}^n, \cdots, H_{mw_m}^1, \cdots, H_{mw_m}^n$ $w_i = min(d_{i1}, \cdots, d_{in})(i = 1, \cdots, m)$
rename	$(v_1{:}e, H_0) \longrightarrow (v_1{:}e_1, H_0)$
shift	$\cdots, g_{i-1}, g_i, g_{i+1}, \cdots, g_{i+j}, \cdots, g_n \longrightarrow \cdots, g_{i-1}, g_{i+1}, \cdots, g_{i+j}, g_i, \cdots, g_n$; $H^{*i}, H^{*i+1}, \cdots, H^{*i+j}, \cdots, \longrightarrow H^{*i-1}, H^{*i+1}, \cdots, H^{*i+j}, H^{*i}, \cdots$
group	$g_1, \cdots, g_n \longrightarrow [g_1, \cdots, g_n]^1$ hedge unchanged
ungroup	$[g_1, \cdots, g_n]^1 \longrightarrow g_1, \cdots, g_n$ hedge unchanged
expand	$g = g_e \longrightarrow g = e \wedge \beta(e) = g_e$; $H^e \longrightarrow (v{:}e, H^e)$
collapse	$g = e \wedge \beta(e) = g_e \longrightarrow g = g_e$; $(v{:}e, H^e) \longrightarrow H^e$
unnest	$[g_r, g_o^c]^{c_1} \longrightarrow [g_r, g_o^1]^{c_1 +}$; $H_1^r, H_{1_1}^o, \cdots, H_{1_{f_1}}^o, \cdots, H_h^r, H_{h_1}^o, \cdots, H_{h_{f_h}}^o \longrightarrow$ $H_1^r, H_{1_1}^o, \cdots, H_1^r, H_{1_{f_1}}^o, \cdots, H_h^r, H_{h_1}^o, \cdots, H_h^r, H_{h_{f_1}}^o$
nest	$[g_r, g_o^c]^{c_1} \longrightarrow [g_r, g_o^{c+}]^{c_1}$; $H_1^r, H_1^{*o}, \cdots, H_n^r, H_n^{*o} \longrightarrow H_1^r, H_{11}^{*o}, \cdots, H_{1_{f_1}}^{*o}, \cdots, H_h^r, H_{h_1}^{*o}, \cdots, H_{h_{f_h}}^{*o}$ $H_1^r, \cdots, H_h^r$ are distinct, *and* $H_{i_1}^{*o}, \cdots, H_{i_{f_h}}^{*o}$ are paired with the same H^r
fact	$[e_1^1 | \cdots | e_h^1]^c \wedge \beta(e_i) = [g_0, g_{ir}]^1 \longrightarrow [g_0, [e_1^1 | \cdots | e_h^1]]^c \wedge \beta(e_i) = [g_{ir}]^1$; $(v_i{:}e_i, H^0, H^{ir}) \wedge i \in [1, \cdots, h] \longrightarrow H^0, (v_i{:}e_i, H^{ir})$
defact	$[g_0, [e_1^1 | \cdots | e_h^1]]^c \wedge \beta(e_i) = [g_{ir}]^1 \longrightarrow [e_1^1 | \cdots | e_h^1]^c \wedge \beta(e_i) = [g_0, g_{ir}]^1$; $H^0, (v_i{:}e_i, H^{ir}) \wedge i \in [1, \cdots, h] \longrightarrow (v_i{:}e_i, H^0, H^{ir})$
unite	$\beta(e_a) = [g_0, g_a]^1 \wedge \beta(e_b) = [g_0, g_b]^1 \longrightarrow \beta(e) = [g_0, g_a, g_b]^1$; $(H^0, H^a) \wedge (H^0, H^b) \longrightarrow (H^0, H^{*a}, H^b)$
split	$\beta(e) = [g_0, g_a, g_b]^1 \longrightarrow \beta(e_a) = [g_0, g_a]^1 \wedge \beta(e_b) = [g_0, g_b]^1$; $(H^0, H^{*a}, H^b) \longrightarrow (H^0, H^a) \wedge (H^{*0}, H^{*b})$

3   Theoretical Analysis on Operators

In this section, we conduct theoretical study on the performance of the operators. We analyze the operators by decomposing each of them into some basic consecutively-executed procedures. The idea is that if the execution time of each procedure for an operator can be proved linear to the size of data, then the performance of the operator is also linear. As to the operators that can not be analyzed in this way, we will present some algorithms and do practical experiments for them in the next section.

3.1   Identification of Procedures

On analyzing the operator definitions given in [1], it can be found that most of the operators share the following three procedures to transform XML documents.

(i) **locate**: identifies the starting and ending indexes of $H(g')$ in $H(g)$ where g' is the sub-structure of g. We note that more than one hedge might be identified if multiplicities are applied;

(ii) **copy**: duplicates specific hedges relevant to the transformation in the source document to the target document. These specific hedges might be copied for multiple times as required by an operator definition;

(iii) **create**: creates a new node in the target document. Note that this new node does not exist in the source document so that it cannot be derived by procedure **copy**;

As a document is transformed to a target document by a sequence of operators, and most of the operators are the aggregation of one or more procedures, so we show how the procedures are used in the operators by a few examples.

For the operator **min**, it requires the use of the *locate* procedure to locate each group $H_i^{*1} \cdots H_i^{*n}$, denoted by H_i, in $H_1^{*1} \cdots H_1^{*n}$, $\cdots$, $H_m^{*1} \cdots H_m^{*n}$ where $i = 1, \cdots, m$. Then for each H_i, the *locate* procedure is further used to find each H_i^{*j} where $j = 1, \cdots, n$. After all the indexes of the hedges are obtained, the procedure *copy* is used to copy the hedges to the target document in the order of $H_1^{*1} \cdots H_m^{*1}$, $\cdots$, $H_1^{*n} \cdots H_m^{*n}$.

The **expand** operator transforms $H(g)$ to $H(e)$ where $H(e) = (e,\ H(g))$. In this operation, *locate* is needed to find $H(g)$; after that the *create* procedure is used to create a new node labeled with e; Then the *copy* procedure is used to copy $H(g)$ under the node of e.

Note that if the performance of each procedure is linear to the size of the hedges to be transformed, and the number of the procedure applications in an operator is a constant, then the performance of the operator is linear too to the size of the hedges to be transformed.

We summarize the use of procedures in the operators in Table 2. Note that the operators **group** and **ungroup** do not change the source document, and operator **rename** only modifies the label of one specific node, which can be executed in a constant time. Thus, they are not included in the table. Operator **nest** and **unnest** are also not shown in the table due to their complexity and will be analyzed later in this paper. Next we will analyze the performance of each operator by analyzing the performance of the procedures.

Table 2. Procedures of operators

	min	mout	shift	expand	collapse	fact	defact	unite	split
locate	✓	✓	✓	✓		✓	✓	✓	✓
copy	✓	✓	✓	✓	✓	✓	✓	✓	✓
create								✓	✓

3.2   Performance of Procedures

For the procedure *create*, the cost of creating a new node is constant time, and there are no more than two new nodes to be created within any operator in Table 2. As a result, its execution time in any operator is a constant.

For the procedure *copy*, the time is linear to the size of the hedge. Even if a hedge is duplicated in a limited number of times in the target document, the execution time is also linear.

Finally, for the procedure *locate*, we need to propose an algorithm and then analyze the performance of the algorithm. The algorithm computes the start and end indexes of a hedge within another hedge based on a structure definition.

The algorithm assumes structures g and g' and $g' \in g$. H denotes the hedge $H(g)$ in which $H(g')$ will be located. *beginIndex* denotes the position of $H(g)$ from which the *locate* procedure begins. There are two main functions in the algorithm. The function *parse* parses $H(g)$ for $H(g')$ according to g'. Once the label of a node in $H(g)$ matches the first element of g', the start index is found. The last index is corresponds to the node whose label is not in g'. The start and the end indexes will be used by other procedures to process the found $H(g')$. *parse* is defined recursively to cater for possible nested '[ ]' in g.

The function *getMul* is used to get the multiplicity of a structure, *lab* returns the root element name of a tree, and *getStruIn* is used to get structure g_1 within structure g where $g = [g_1]^c$. H_i represents the ith tree in hedge H.

The algorithm can deal with the constructs of disjunction, multiplicities, nested brackets, and empty hedges of DTD syntax which make its design non-trivial and challenging.

Algorithm 1

```
Procedure locate(H,g,g',beginIndex){
//initialing the global variables.
start=end=0; i=beginIndex; located=false;
parse(g); //calling recursive function
return(start, end-1);}
//definition of recursive functon
bool parse(Struture s){
  matchNo=0; matched=false; c= getMul(s);
  if (s∈g' && start==0) then start=i;
  if (s∉g' && start>0) then {end=i;located=true; return false;}
  if s ∈ EN{
     do{if(lab(Hᵢ)==s) then {i++; matchNo++; matched=true;}
          if(i>H.size) then {located=true; return(matchTime>0);}
        }while(matched && (c=='*' or '+'));}
  else if s is a group{ //begins with '['
     let t=getStruIn(s);
     do{ if(parse(t)) then matchNo++;
          if(located) then return(matchNo>0);
        }while(matched&&(c=='*' or '+'));}
  else for each part t in s{
     if(parse(t)) then matchNo++;
     if(located) then return(matchNo>0);
     else if s is disjunction structure then
             { if(matchNo>0) then return(matchNo>0);}}}
  return (matchNo>0);}
```

The core of Algorithm 1 is to keep comparing element names in g and labels on the nodes in $H(g)$ in order. Although the algorithm uses recursion to deal with nested brackets, it scans $H(g)$ only once. Therefore, it executes in linear time to the size of $H(g)$. We assume that N_g is the number of element names in g and N_h is the number of trees in $H(g)$. In the worst case, the number of comparisons will be less than $N_g * N_h$. As the time for comparing is constant and N_g could be also considered as constant since the number of elements included in g is normally small, we can say that the complexity of the algorithm is time-linear to the number of trees in $H(g)$.

We note that procedure *locate* might also be called for multiple times within one operator. For example, as to operator **min**, let $g_a = [g_1^{c_1}, \cdots, g_n^{c_n}]$ and $g = g_a^c$. Firstly the procedure is called to locate several hedges conforming to g_a, each denoted by $H_i(g_a)$, in $H(g)$. There is no intersection between each $H_i(g_a)$. This step costs linear time according to the above analysis. Then the procedure will be called for multiple times to locate each $H(g_j^{c_j})(j = 1, \cdots, n)$ in each $H_i(g_a)$. However, this can be done by scanning each $H_i(g_a)$ only once, thus the time complexity of locating all $H(g_j^{c_j})s$ is equal to that of the case where procedure *locate* is applied only once to scan the whole $H(g)$. Finally, since such multiple-calls within any operator in Table 2 appears at most once, we can say that the whole performance of *locate* within one operator is still linear.

Operator **unite** needs to be mentioned since besides calling the three procedures, it costs extra time to identify the common element names in $\beta(e_a)$ and $\beta(e_b)$. In this case, the number of times for comparisons is no more than the number of element names either in $\beta(e_a)$ or $\beta(e_b)$. Provided that either number of elements in $\beta(e_a)$ or $\beta(e_b)$ is small(e.g. less than 100), this extra cost is time constant. As to the part of transformation, only one new node is created by procedure *create*, and no hedges are duplicated during applying procedure *copy*, and procedure *locate* is applied only each in the hedge $H([g_0, g_a]^1)$ and $H([g_0, g_b])^1$ respectively. Therefore, the performance of operator **unite** is linear.

Based on the theoretical analysis above, we can say that the performance of all the operators in Table 2 is linear in time. Such analysis that applies the concepts of hedge and procedure can be easily applied to other possible sets of operators to see if the operators are suitable for the document transformation and data exchange.

3.3   Performance of Transforming a Document Using One Operator

Each operator only transforms one or more related subtrees in the source document. Each subtree is identified by its root node specified by the parameter of the operator. The document to be transformed is traversed by applying the depth-first search algorithm. When the root node of the subtree to be transformed is identified, the subtree is transformed. Otherwise the root node will be copied directly to the target document.

The time complexity of depth-first search algorithm is proportional to the number of vertices plus the number of edges they traverse $(O(|V| + |E|))$[10]. When all the nodes are traversed, $|E| = |V| - 1$, thus the cost will be $O(|V|)$.

Therefore if the performance of the operator is linear, we can conclude that the performance of transforming a document by one operator is linear .

4   Experimental Analysis on Operators

In this section, we take performance comparison between operators **nest** and **unnest** through an experiment. After that, we propose a novel algorithm to improve the performance of the operator **nest** to be close to linear. Finally, we conduct a typical experiment on transformation. This shows how a sequence of operators is used and how the whole performance of transformation is.

4.1   Complexity of Nest and Unnest

Although the operators **nest** and **unnest** use the procedures defined above, but their uses are in a nested loop structure and the operator *nest* also involves an expensive comparison operation. In what follows, we see that these two operators have a performance that is squared to the size of hedge.

In the case of **nest**, a nested loop is used to compare the values of hedges. These hedges have the same structure and the similar size that can be considered as constant. Therefore, the performance of each comparison is also constant. While the number of times to do the comparison increases greatly as the hedges need to be compared increase. If there are S hedges that need to be compared, then the overall number of comparisons will be given by $S(S+1)/2$.

As to **unnest**, hedge $H(g_r)$ may need to be duplicated for multiple times. For example, we assume the size of $H([g_r, g_0^*])$ is S, the size of $H(g_r)$ is $S/2$, and the size of each $H(g_0)$ is 1. Then $H(g_r)$ will be duplicated for $S/2 * S/2$ times.

Under these circumstances, the performance of both operators is not linear.

4.2   Algorithms for Nest and Unnest

Due to the space limitation, we only present the principle of the algorithms for these two operators.

According to the definition of operator **nest**, it firstly locates $H([g_r, g_0])$ in $H([g_r, g_0]^c)$. Note that there might be only one hedge in extreme situation, and nothing will be changed after the transformation in this case. The next step is to locate $H(g_r)$ in each $H([g_r, g_0])$ with $H(g_0)$ being located at the same time. As a result, We split every $H([g_r, g_0])$ into two parts(hedges): *compareH* ($=H(g_r)$) and *nestH* ($=H(g_0)$). Two lists named *compareL* and *nestL* are also created to save those two parts. For each pair *compareH* and *nestH* in $H([g_r, g_0]^1)$, if there are no hedges in *compareL* equivalent to *compareH*(same structure and same node values), the *compareH* and *nestH* will be added into *compareL* and *nestL* with the same index respectively. Otherwise, the *nestH* will be appended (denoted by ⊎) to *nestL*. After such an operation of comparing and nesting, every *compareH* and *nestH* in the two Lists that have the same index will be re-combined into one group, and all of the groups combined together create a new transformed hedge.

As both of *compareH* and *nestH* are possible to be empty, we assume that the value of any two empty hedges are the same. In addition, if *nestH* is empty, nothing will be appended into *nestL*.

The operator *unnest* is much simpler than *nest*. It needs to locate H^r first. Then each H_i^{*0} is located and is appended to H^r. At last, each group of $[H^r, H_i^{*0}]$ will be appended together as the transformed hedge.

4.3 Preparation for Experiment

Environment of Experiment All the experiments in this paper are conducted on a computer with a 2.21GHz AMD Athlon(tm) 64 Processor and the memory size is 512MB. All XML documents with different sizes and structures are created randomly.

Selection of Implementation Language. XQuery[4] and XSLT[5], developed and recommended by W3C, are two languages for XML querying and transformation, and are increasingly used in many applications related to XML. Due to the extra verbosity of XSLT and its equivalent expressive power compared with XQuery [5][4], we choose to use XQuery1.0 to implement the operators. In addition, we use Java language to process XML documents without applying any features of XQuery, and implement the same function of an operator. As an implementation of the XQuery 1.0 specifications, *Saxon8.7.3* is used to execute an XQuery file, in which user-defined functions, conditional expressions, and FLWOR expressions are all applied. The version of Java is 1.4.1.

We apply the two languages to implement **nest** operator . The result is shown in Table 3. It is very clear that the processing time of using XQuery is much longer than that of using Java. When the size of hedge to be transformed reaches to 10,000, the difference has already been over ten times.

Table 3. Comparison between Two languages

Size(k)	2	4	8	10	20
Java(sec)	0.34	0.43	0.771	1.002	3.074
Xquery(sec)	1.74	3.77	10.54	15.63	53.17

In fact, XQuery does have some limitations. For example, it does not allow explicit logic termination conditions, and variables can be evaluated only once etc. These limitations, to some extent, affect its performance and effectiveness in processing some commands beyond the XML Query. Moreover, too many inevitable recursion nestings are more likely to lead to the over stack of memory. However, we should remember the fact that XQuery really has powerful functionality in dealing with some complex operations besides the basic query function over the XML documents.

As a result, we choose to use Java language in the rest of the implementation.

4.4   Performance Comparison between Nest and Unnest

As the two operators have opposite operations, we transform two DTDs to each other where **nest(movie)** transforms $\beta(root) = [actor, movie]^*$ to $\beta(root) = [actor, movie^+]^*$, while **unnest(movie)** transforms in reverse. The DTDs are defined as simple as possible so that we can focus on the transformation-related hedges since the time of copying irrelative hedges is linear. The number of the elements under node *root* can be considered as the size of the document. Noting that there might exist more complicated structures, we increase the text length of the nodes to simulate possible complex child trees under these nodes. The XML documents are created such that the size of hedge to be transformed is from 2k to 10k, and the value of each node is a string of 100 characters.

The results of the performance of both **nest** and **unnest** are shown in Fig. 4. Although the previous theoretical analysis indicates that the performance of **unnest** is squared to the size of the document, but the experiment results show a close-to-linear performance. In contrast, the performance of **nest** is obviously worse and becomes the major concern among all operators. It is for this reason that we will put special effort to improve the performance of the **nest** operator.

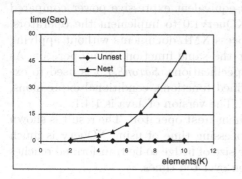
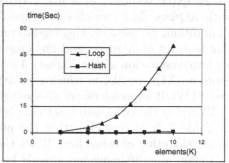

Fig. 4. Performance of Nest and Unnest   **Fig. 5.** Improved Performance of Nest

4.5   Improved Performance of Nest

Now we present an algorithm to improve the execution time of the nest operator using hashing. The main idea of hashing is to store the hedges $H^{g_r} H^{g_o}$ of $[g_r, g_o]$ in a hash table with H^{g_r} being the hash key and H^{g_o} the hash value. Via the process of hashing, all H^{g_o}s with the same H^{g_r} are grouped together to achieve the goal of nesting. We employed Java's *LinkedHashMap* for the implementation. Due to the space limitation, the algorithm is omitted here.

In *LinkedHashMap*, to avoid the collision problem , two or more objects are allowed to have the same hash code, and they are linked one by one in a data structure like List and each List would be put into hash table as one element whose position in hash table is decided by the hash code and the current size of hash table. The only way to distinguish them is by comparing them according to the definition of the objects.

Fig. 5 shows the experiment results of the **nest** operator using two methods - nested loop and hash. It is easy to see that the performance has been improved greatly and is close to linear to the size of data. In fact, it is the usage of hash table that greatly saves the time of searching and comparing between *comparedHedges*, which is the most important factor to the performance.

4.6 Performance of a Sequence of Operators

After analyzing the performance of all the operators separately, we conduct another experiment by using a sequence of operators to implement a typical transformation. This will further show how these operators are used in transformation and how the overall performance is.

We use DTDs in Fig.1 as source and target DTDs. According to their definitions, the transformation can be implemented by the sequence of operators shown in Fig.6.

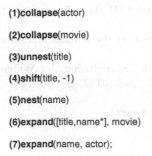

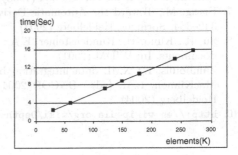

Fig. 6. Sequence of Transformations   **Fig. 7.** Performance of Operators

In the experiment, source documents with different number of elements from 30K to 270K are transformed to the respective target documents automatically by using these operators. According to the experiment result in Fig. 7, we can say that the performance of whole transformation is close to linear. Note that the performance of operator **unnest** is linear in this case because the duplicated hedges are only half of the hedges to be transformed.

5   Conclusions and Future Work

Basing on the thorough theoretical analysis and the experimental results, we can draw the conclusion that except operator **unnest** all the operators can be performed in linear time to the size of the XML document. A sequence of operators also has close to linear performance. This will provide valuable information for the possible application of materialized integration system where these systematical operators are used. Our next work is to study the automated or semi-automated transformation and query rewriting in virtual integration system where these operators are used.

References

1. Liu, J., Park, H.-H., Vincent, M., Liu, C.: A formalism of XML Restructuring Operations. In: Mizoguchi, R., Shi, Z.-Z., Giunchiglia, F. (eds.) ASWC 2006. LNCS, vol. 4185, pp. 126–132. Springer, Heidelberg (2006)
2. Liu, J., Vincent, M., Liu, C., Park, H.-H.: Transformation of XML Data. (To be published)
3. Bray, T., Paoli, J., Sperberg-McQueen, C.M.: Extensible markup language(xml)1.0 (1998), http://www.w3.org/TR/1998/REC-xml-19980210
4. Boag, S., Chamberlin, D., Fernndez, M.F., Florescu, D., Robie, J., Simon, J.: Xquery 1.0 An xml query language (2007), http://www.w3.org/TR/xquery
5. Kay, M.: Xsl transformations (xslt) version 2.0 (2004), http://www.w3.org/TR/xslt20/
6. McBrien, P., Poulovassilis, A.: A Formalisation of Semantic Schema Integration. Information Systems 23, 307–334 (1998)
7. Erwig, M.: Toward the Automatic Derivation of XML Transformations. ER, pp. 342–354. Springer, Heidelberg (2003)
8. Su, H., Kuno, H., Rundensteiner, E.A.: Automating the transformation of XML documents. In: WIDM (2001)
9. Zamboulis, L.: XML data integration by graph restructuring. In: Williams, H., MacKinnon, L.M. (eds.) BNCOD 2004. LNCS, vol. 3112, pp. 57–71. Springer, Heidelberg (2004)
10. http://en.wikipedia.org/wiki/Depth-first_search

Similarity Computation for XML Documents by XML Element Sequence Patterns

Haiwei Zhang, Xiaojie Yuan, Na Yang, and Zhongqi Liu

Department of Computer Science and Technology,
Nankai University, 300071 Tianjin, China PR
zhhaiwei@mail.nankai.edu.cn, yuanxj@nankai.edu.cn

Abstract. Measuring the similarity between XML documents is the fundamental task of finding clusters in XML documents collection. In this paper, XML document is modeled as XML Element Sequence Pattern (XESP) and XESP can be extracted using less time and space than extracing other models such as tree model and frequent paths model. Similarity between XML documents will be measured based on XESPs. In view of the deficiencies encountered by ignoring the hierarchical information in frequent paths pattern models and semantic information in tree models, semantics of the elements and the hierarchical structure of the document will be taken into account when computing the similarity between XML documents by XESPs. Experimental results show that perfect clustering will be obtained with proper threshold of similarity computed by XESPs.

Keywords: XML, XESPs, Similarity, Clustering.

1 Introduction

XML is de facto a new standard of data exchange and representation on the web. Documents and data represented by XML have been increasing rapidly. As a result, the processing and management of XML data are popular research issues. Dealing with XML document collections is a particularly challenging task for Machine Learning (ML) and Information Retrieval (IR)[1].

Clustering XML documents based on their structure along with semantic tags is considered effective for XML data management and retrieval. Usually, clustering XML document includes three stages: modeling XML documents, similarity measuring between models of XML documents and clustering models of XML documents. The clustering stage is relatively straightforward because there are many effective methods for clustering data, such as k-means, hierarchical agglomerative clustering algorithms, etc.

Researchers pay more attention to similarity measuring based on the models of XML documents extracted in the modeling stage. Modeling XML documents as rooted ordered labeled trees (named XML Document Tree) is an ordinary representation [2][3][4][5]. Measuring similarity by tree edit distance is in common use based on tree models [2][3]. Frequent patterns are important models transformed from tree models, such as frequent paths model and frequent sub-tree model [4][5].

Y. Zhang et al. (Eds.): APWeb 2008, LNCS 4976, pp. 227–232, 2008.

In this paper, we take XML document similarity into account by determining the similarity of XML elements because elements similarity can be used to measure the semantic and structural heterogeneity between XML documents. And we make the following contribution:

- We propose a model named XML Element Sequence Pattern (XESP) for XML documents to reduce time and space cost of model extraction in previous work.
- We propose a method for measuring structural similarity along with semantic of elements based on XESP model.
- We use k-means algorithm in experiment to cluster XML documents based on the similarity measured by XESP model.

2   Extraction of XML Element Sequence Patterns

XML Document Tree (XDT) is a common logical model to represent XML documents. There are a number of models transformed from XDT for XML documents. XESP model presented in this paper is also based on XDT.

Definition 1 (XML Element Sequence Pattern (XESP)). *Sequence of XML elements can be described as XML Element Sequence Pattern (XESP) $X=(E, \vdash, \prec, \models, Fe, Fx, level_x, min_e, max_e)$, where*

E is a set of ordered elements corresponding to nodes in V of XDT.

$\vdash$ defines parent-child relationship between elements. If u is the parent node of v, note $u \vdash v$.

$\prec$ defines brother relationship between sequential elements.

$\models$ defines brother relationship between alternative elements.

Fe is one or more elements of E in a certain level identified by $level_x$.

Fx is a set of XESPs that compose of X if X is merged by other XESPs.

$level_x$ denotes the level of elements in Fe.

$\exists e \in E$, min_e and max_e is the minimum and maximum times that e occurs following its parent element.

X can be represented by a sequential form:

$$X_{v_i} = e_0^{<min,max>} \star e_1^{<min,max>} \star ... \star e_i^{<min,max>} \tag{1}$$

by joining all relationship defined by $\vdash$, $\prec$, $\models$ and min_e, max_e. Operator $\star \in \{\cdot, |\}$, where "." is corresponding to logical operator "AND" and identifies relationship "$\vdash$" and "$\prec$", "|" is corresponding to logical operator "OR" and identifies relationship "$\models$". $min=min_{e_i}, max=max_{e_i} (i=0,1,...)$

Path of leaf node v_i in XDT can be represented by XESP X_{v_i}, where Fe is the element related to v_i, Fx is X_{v_i} itself, $level_x$ is the level of v_i, min_e and max_e of each element is set to 1.

Merging all XESPs representing paths of leaf nodes in XDT will obtain a new XESP to represent XML document. There are two operations for merging XESPs: joining and factoring. These operations are used to reduce repeated elements in different XESPs.

Definition 2 (Joining Operation). *Joining XESPs: X_1 and X_2 will get a new XESP: $X=(E, \vdash, \prec, \models, Fe, Fx, level_x, min_e, max_e)$,where:*

$E=E_1 \cup E_2$, $\vdash=\vdash_1 \cup \vdash_2$, $\prec=\prec_1 \cup \prec_2 \cup(X_1.Fe \prec X_2.Fe)$, $\models=\models_1 \cup \models_2$, $Fx=X_1.Fx \cup X_2.Fx$, $Fe=X_1.Fe \cup X_2.Fe$, $level_x$, min_e and max_e are the same as original values.

Definition 3 (Factoring Operation). *Factoring XESPs: X_1 and X_2 will get a new XESP: $X=(E, \vdash, \prec, \models, Fe, Fx, level_x, min_e, max_e)$,where:*

$E=E_1 \cup E_2$, $\vdash=\vdash_1 \cup \vdash_2$, $\prec=\prec_1 \cup \prec_2$, $Fx=X_1.Fx \cup X_2.Fx$, $\models=\models_1 \cup \models_2$ $\cup(X_1.Fe \models X_2.Fe)$, $Fe=X_1.Fe$, $min_e=\min(X_1.min_e,X_2.min_e)$,
$max_e=\max(X_1.max_e,X_2.max_e)$, $level_x$ is the same as original values.

For example, given two element sequential form of XESP: $X_1=C^{<1,1>} \cdot P^{<1,1>} \cdot B^{<1,1>}$ and $X_2=C^{<1,1>} \cdot P^{<1,1>} \cdot S^{<1,1>}$. Joining X_1 and X_2 will get a new XESP whose element sequential form is: $X=C^{<1,1>} \cdot P^{<1,1>} \cdot B^{<1,1>} \cdot S^{<1,1>}$. Factoring X_1 and X_2 will get: $X=C^{<1,1>} \cdot P^{<1,1>} \cdot (B^{<0,1>} \mid S^{<0,1>})$.

3 Similarity Computation

This section computes similarity between XML documents based on the element sequential form of XESP model. Similarity between XESPs is based on similarities of elements in XESPs by considering semantic, structural and constraint information.

XML element often can be a combination of lexemes (e.g. SigmodRecord, Act_Number), a single letter word (e.g. P for person), a preposition or a verb (e.g. related, from, to) that make them syntactically different.[6]. Therefore, element names should be changed into a token set.

Semantic similarity of elements is composed of token similarities in the corresponding token sets. Tokens similarity is computed in two ways, one is by synonym set for semantics based on WordNet (e.g. movie $\rightarrow$ film), and the other is based on string edit distance [7] for syntax. Semantic similarity along with syntactic between XML elements pre-processed is defined as:

$$SeSim(e_1, e_2) = \frac{\sum_{W_1} \sum_{W_2} WSim(w_1, w_2) + \sum_{W_2} \sum_{W_1} WSim(w_2, w_1)}{\mid W_1 \mid + \mid W_2 \mid} \quad (2)$$

where W_1 and W_2 is token set of element e_1 and e_2 respectively, $WSim(w_1,w_2)$ is similarity between tokens w_1 and w_2, and

$$WSim(w_1, w_2) = \begin{cases} 0.8 & \text{if } w_1 \text{ and } w_2 \text{ are synonym,} \\ 1 - \left[\frac{SED(w_1,w_2)}{\max(\mid w_1 \mid, \mid w_2 \mid)}\right] & \text{else.} \end{cases} \quad (3)$$

where $\mid w_1 \mid$ and $\mid w_2 \mid$ are the length of token w_1 and w_2, respectively. Function $SED(w_1,w_2)$ measuring string edit distance of w_1 and w_2.

The structural similarity between XML elements of XESPs is measured by hierarchical and contextual information. Given an element e belongs to XESP X. Note that a sub XESP $SubX$ starts from the first element of X and ends at the previous element of e. Structural measure firstly finds hierarchy similarity between two elements and then finds the similarity of sub XESPs $SubX_1$ and $SubX_2$ as contextual similarity. As a result, structural similarity is defined as:

$$StSim(e_1, e_2) = \alpha * HSim(e_1, e_2) + \beta * ContextualSim(e_1, e_2) \qquad (4)$$

where $\alpha + \beta = 1$, $HSim(e_1, e_2)$ is the hierarchy similarity and defined as:

$$HSim(e_1, e_2) = 1 - \left[\frac{|e_1.Level - e_2.Level|}{\max(|e_1|, |e_2|)} \right] \qquad (5)$$

where $e_1.Level$ and $e_2.Level$ is the level of element e_1 and e_2 in XML document. $ContextualSim(e_1, e_2)$ is the contextual similarity and defined as:

$$ContextualSim(e_1, e_2) = XSim(SubX_1, SubX_2) \qquad (6)$$

where $XSim(SubX_1, SubX_2)$ is the similarity of sub XESP $SubX_1$ and $SubX_2$. This measure will be introduced in section 3.4 for measuring similarity of XESPs.

According to definition 1, occurrence times and relationship is important information of elements that consist of XESPs. Hence, such information also makes a small contribution in determining element similarity.

Occurrence times of elements are represented by min and max value in XESPs. Occur Similarity Coefficient (OSC) is used to measure similarity of occurrence times and defined as:

$$OSC = 1 - \frac{|Occurs_1 - Occurs_2|}{\max(Occurs_1, Occurs_2)} \qquad (7)$$

where $Occurs_i (i=1,2)$ is min or max value of elements in XESPs.

Each element in XESPs has logical relationship to its neighboring elements. In the sequential form of XESPs, the relationship is expressed by logical operator "." and "|", which express "AND" and "OR" relationship respectively. Logic Similarity Coefficient (LSC) is used to describe logic similarity between elements. Similarity between different logical relationship is set to 0.8.

Constraint similarity between elements is defined as:

$$CSim(e_1, e_2) = \omega_1 * OSC + \omega_2 * LSC \qquad (8)$$

where weights $\omega_1 + \omega_2 = 1$. The weight determines the importance of measures in constraint similarity of elements. The default value is set to 0.5 for each weight because the importance of each weight for constraint similarity is equivalent.

Measuring similarity of XESPs will summarize similarity of each pair of elements consist of XESPs. Similarity between a pair of elements is defined as:

$$ESim(e_1, e_2) = \lambda_1 * SeSim(e_1, e_2) + \lambda_2 * StSim(e_1, e_2) + \lambda_3 * CSim(e_1, e_2) \qquad (9)$$

where weights $\lambda_1 + \lambda_2 + \lambda_3 = 1$.

Similarity of XESPs is defined as:

$$XSim(X_1, X_2) = \begin{cases} 0 & X_1=\text{NULL OR } X_2=\text{NULL} \\ \frac{\sum_{i=1}^{n}\sum_{j=1}^{m} ESim(e_1,e_2)}{\max(|X_1|,|X_2|)} & \text{ELSE} \end{cases} \qquad (10)$$

where n and m is the number of elements in XESP X_1 and X_2. The degree of similarity between XESPs is monitored by threshold . If the similarity exceeds the threshold, the XESPs are similar.

4   Experimental Results

Experiments are conducted with XML collection from [8]. All XML documents are derived from the same and different domains and that each has its distinct structure. To show the efficacy of the measure presented in this paper, XESPs will be extracted from XML documents firstly, and then similarity will be computed between pairs of XESPs. Finally, k-means algorithm will be used to cluster XESPs based on the similarity between XESPs and the clustering results will be compared with other two traditional methods: Tree Edit Distance (TED) and Frequent Paths Pattern (FPP).

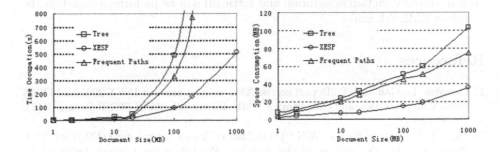

Fig. 1. Comparison of time occupation and space consumption

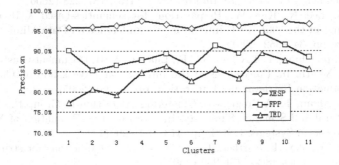

Fig. 2. Precision of Each Cluster

It is shown in Fig. 1 that extraction of XESPs occupies less time and space than extracting the other two models. With the size of a document growing, this advantage becomes more and more apparent.

Similarity between XESPs is usually used for clustering. Fig. 2 shows the precision of clustering based on the similarities measured by XESPs is better than methods based on tree edit distance and frequent paths.

5   Conclusion and Future Work

In this paper, we proposed a method for determining similarity between XML documents based on XML Element Sequence Patterns (XESPs). It makes use of XESPs to represent XML documents and it has the ability to include the elements' semantic and hierarchical structure information of document in the computation of similarity between XML documents. Similarity measured by XESPs brings better clustering result than methods based on tree edit distance and frequent paths.

Future works focus on applications of clustering results obtained by the method presented in this paper. One being included is the extraction of common XML Schema for each cluster after clustering. Toward this direction, other applications such as data extraction integration and retrieval will be performed based on the common XML Schema.

References

1. Denoyer, L., Gallinari, P.: Report on the XML mining track at INEX 2005 and INEX 2006: Categorization and clustering of XML documents. ACM SIGIR Forum 41(1), 79–90 (2007)
2. Moh, C.H., Lim, E.P., Ng, W.K.: DTD-Miner: A tool for mining DTD from XML documents. In: Proceedings of the 2nd Int. Workshop on Advance Issues of E-Commerce and Web-Based Information Systems, Milpitas, California, pp. 144–151 (2000)
3. Dalamagas, T., Cheng, T., Winkel, K., Sellis, T.K.: A Methodology for Clustering XML Documents by Structure. Inform. Syst. 31(3), 187–228 (2006)
4. Lee, J.W., Lee, K., Kim, W.: Preparations for semantics-based XML mining. In: Proceedings of the 2001 IEEE International Conference on Data Mining, San Jose, California, pp. 345–352 (2001)
5. Leung, H.-p., Chung, F.-l., Chan, S.C.F., Luk, R.: XML Document Clustering Using Common XPath. In: Proceedings of the International Workshop on Challenges in Web Information Retrieval and Integration, pp. 91–96 (2005)
6. Nayak, R., Tran, T.: A Progressive Clustering Algorithm to Group the XML Data by Structural and Semantic Similarity. In: Comparative Evaluation of XML Information Retrieval Systems. LNCS, pp. 473–484 (2007)
7. Rice, S.V., Bunke, H., Nartker, T.A.: Classes of cost functions for string edit distance. Algorithmica 18(2), 271–280 (1997)
8. XML Repository, At (2006), http://xmlmining.lip6.Fr/Home

Evolving Weighting Functions for Query Expansion Based on Relevance Feedback

A. Borji[1] and M.Z. Jahromi[2]

[1] School of Cognitive Sciences, Institute for Studies in Theoretical Physics and Mathematics,
Niavaran Bldg. P.O.Box 19395-5746, Tehran, Iran
[2] Department of Computer Science and Engineering, Eng. no. 2 Building,
Molla Sadra St., Shiraz University, Shiraz, Iran
borji@ipm.ir, zjahromi@shirazu.ac.ir

Abstract. A new method for query expansion using genetic programming (GP) is proposed in this paper to enhance the retrieval performance of text information retrieval systems. Using a set of queries and retrieved relevant and non-relevant documents corresponding to each query, GP tries to evolve a criteria for selecting terms which when added to the original query improve the next retrieved set of documents. Two experiments are conducted to evaluate the proposed method over three standard datasets: Cranfield, Lisa and Medline. In first experiment a formula is evolved using GP over a training set and is then evaluated over a test query set of the same dataset. In the second experiment, evolved expansion formula over a dataset is evaluated over a different dataset. We compared our method against the base probabilistic method in literature. Results show a higher performance in comparison with original and probabilistically expanded method.

Keywords: Query expansion, relevance feedback, weighting function, GP, IR.

1 Introduction

There has been a huge raise in digital content production during the last decade. Due to this rapid increase and large amount of data, finding relevant information has turned to become a very challenging task for web users.

A large number of studies have tackled this problem in information retrieval (IR) systems to enhance the performance of existing systems. One category of such methods utilizes user's judgment (relevance feedback) over the retrieved documents to build a revised query with more chance to retrieve more relevant documents at first items. This method has proved to be very efficient and has been widely used in many operational search engines.

Query reformulation techniques proposed in the literature fall into two categories. The idea behind the first category is to use information derived from the context of retrieved document to reformulate the original query. Two main strategies have been proposed: local clustering [1], [2] and global analysis [3], [4] or a combination of both local and global context [5].

In the second category, judgment of user on the retrieved documents is utilized for query expansion and term reweighing [6], [7], [8]. Query expansion techniques consider

Y. Zhang et al. (Eds.): APWeb 2008, LNCS 4976, pp. 233–238, 2008.

distribution of terms in both relevant and non-relevant retrieved documents and also over entire corpus to assemble a revised new query.

Applications of genetic techniques for processing query optimization have been proposed by several authors. Gordon [9] adopted a GA to derive better descriptions of documents. Each document is assigned N descriptions represented by a set of indexing terms. Genetic operators and relevance judgment are applied to the descriptions in order to build the best document descriptions. The author showed that the GA produces better document descriptions than the ones generated by the probabilistic model. Redescription improved the relative density of co-relevant documents by 39.74% after twenty generations and 56.61% after forty generations.

Yang et al. [10] proposed a GA for query optimization by reweighing the query term indexing without query expansion. They used a selection operator based on a stochastic sample, a blind crossover at two crossing points, and a classical mutation to renew the population of queries. The experiments showed that the queries converge to their relevant documents after six generations.

Horng et al. [11] propose a novel approach to automatically retrieve keywords and then uses genetic techniques to tune the keywords weights. The effectiveness of the approach is demonstrated by comparing the results obtained to those using a PAT-tree based approach.

One drawback with a large number of previous studies on query expansion is, they use a fixed criteria for choosing expansion terms and mainly ignore the habits, interests and needs of the users which might change over time. In this work we tackle this shortcoming by proposing a new query expansion technique using GP to evolve new criteria which could be used to customize IR systems adaptively according to users varying needs. Specifically Genetic programming is used to find a formula for selecting a set of terms to be added to the original submitted query.

The reminder of the paper is organized as follows. Principles of our proposed approach for query expansion are introduced in section two. It is then followed by experiments and results in section three. Section four brings conclusions.

2   Query Expansion Using GP

In this work, we use genetic programming [12] to evolve a formula for query expansion during evolutionary process. Like probabilistic RF [13], it then adds a number of highly weighted terms to the query.

To start the evolutionary process, a set of queries and tagged relevant documents to each one is needed. Success of GP is dependent on the size of this set. For this purpose, we used Cranfield dataset which is largest of the three datasets we used in terms of number of queries. In all experiments discussed in this paper, weighting functions are evolved over a subset of Cranfiled dataset. This dataset is partitioned to three train, test and validation fractions with 100, 50 and 75 queries in each fraction respectively. Validation set is used to overcome the over-fitting.

For each query in the training set, a fixed number of documents (DCV[1]) are retrieved. As relevant and non-relevant documents for each query are known, statistical

---

[1] Document Cutoff Frequency.

features could be calculated at leaves of trees after query submission. Terminals used by the GP are listed in table 2.

To evaluate a term weighting formula, m (=10) terms with highest weight, were selected and added to the base query. The new expanded query is then applied to remaining set of documents in the dataset. It is important to notice that at this stage, the first set of retrieved documents must be eliminated when evaluating the revised query. The reason is that the IR system is aware of relevancy of this documents which are marked by the user. Effectiveness of this weighting function is evaluated as the mean precision over 11 standard recall points.

This process is repeated for all queries in the dataset and finally average over all queries is considered as the fitness of the formula (genetic program). During the evolutionary process, initial set of formulas gradually produce more fit individuals. Best formula from the final population is considered as the weighting function. Evolved formula in this way is evaluated over a test query set of either the same dataset i.e. Cranfiled or a different one (Lisa and Medline). To avoid falling system into local minima, we stop the GP as soon as a decrease over validation set is observed.

In a typical scenario, DCV documents are first retrieved and presented to the user. Assuming that user will determine all relevant documents to this query, weighting is performed using generated weighting function. Terms of the relevant documents are weighted according to this weighting function and a number of terms with highest weight are added to the query. A new retrieval session is done with this expanded query. This process is repeated for all queries in the train set and average precision over all queries is considered as the fitness of the individual. Applying genetic operators, next generation of formulas are generated and so on until the last generation or observation of a performance decrease on validation set. For the best query in the final population, the same is done for all queries in the test set and the average precision over all queries is reported as the performance.

3 Experiments and Results

In order to evaluate the proposed method, we conducted two experiments over three standard datasets. Statistics of these sets are given in table 1.

Table 1. Statistics of standard datasets we used in our experiments

Dataset	Documents	Queries	Terms
Cranfield	1400	225	4486
Medline	1033	30	5831
Lisa	6004	35	11560

For stop-word elimination we used the smart system [14]. Stemming was performed using porter stemmer [15]. Remaining indexing terms were weighted by Tf-Idf weighting formula. A cosine similarity measure was used. Parameters used for constructing trees are shown in tables 2 and 3.

Table 2. Terminals used in GP implementation

R	Number of relevant retrieved documents containing term i
NR	Number of Non-relevant retrieved documents containing term i
N	Number of all documents in the dataset
DR	Number of relevant retrieved documents in response of query q
DN	Number of non-relevant retrieved documents in response of query q
TAR	Average term frequency in relevant retrieved documents
TMR	Maximum term frequency in relevant retrieved documents
TAN	Average term frequency in non-relevant retrieved documents
TMN	Maximum term frequency in non-relevant retrieved documents
L	*Document length*
C	Constant

Table 3. Other parameters of the genetic programming

Functions (operators)	+, -, ×, /, sqrt, log
Fitness function	Average precision over all queries
Parse tree initial depth	3
Number of retrieved documents for a query	25
Terms added to each query (m)	10
Population size (pop)	30
Generations (maxgen)	20
Cross-over	0.6 (tree replacement)
Mutation rate	0.1 (Single node mutation)
Copy	0.1

3.1 Experiment I

In this experiment, a subset of 125 queries was randomly selected from Cranfield dataset. Best evolved formula over this set was then evaluated over the test queries of the Cranfield. Due to stochastic nature of GP, we averaged the results over ten runs of different train-test-validation shuffles. Table 4 shows that results of experiment one. Reported results in both experiments are compared against the bared original query and query enhanced by PRF. Improvements are calculated as percentage of the difference between GP and PRF.

Table 4. Average precision over Cranfield dataset

Test Queries	Initial Query	Query Expansion with PRF	Query Expansion with GP	Improvement
125	0.1359	0.1666	0.2158	**29.53 %**

Best individual derived after evolution is given in following equation.

$$\frac{(2N-3DR+\log(L))+N/R+1.4\times\log(DN+0.21)}{(DR+DN)+1/TMN}-L+\log(\frac{\log(N)(NR+N+2.1)}{N+0.56L})+TAN/DN \quad (1)$$

3.2 Experiment II

In this experiment, the best genetic formula derived in experiment one is applied to other two datasets. An interesting result in table 5 is that the best function derived over a subset of queries of Cranfield performed better than PRF over both datasets. Lower performance than of Cranfield using a genetically evolved weighting function suggests datasets have different properties might be due to their different context. It also supports the idea that there might be more promising weighting functions over all datasets which work better than previously suggested by experts. Expanded query using GP showed a higher performance over both datasets than the original query.

Table 5. Average precision over Lisa and Medline datasets

Dataset	Initial Query	Query Expansion with PRF	Query Expansion with GP	Improvement
Lisa	0.0836	0.1180	0.1265	7.2 %
Medline	0.2001	0.3250	0.3537	8.8 %

4 Conclusions

In this paper we proposed a new query expansion method using GP based on relevance feedback. In the first experiment an evolved formula over a training query set was evaluated over a distinct query test set randomly picked from the same dataset. In the second, evolved classifier over the first dataset was applied to the second and third datasets.

In both of experiments, we observed that very small and very large values for m do not enhance results significantly maybe because small number of terms are not enough and large numbers are ineffective as they cause great overlap among queries. From the computational complexity point of view, PRF and GP perform the same as both propose weighting functions. It should be noted that GP calculations are done offline and then a function is then used for query expansion in online search.

The gain in retrieval performance we achieved in this study, approach the five percent mark suggested by savoy and vrajitoru [16] as an indicator of significant performance improvement. Our results are in accordance with this finding. These results indicate the merits of further investigation of our method.

References

1. Attar, A., Franenckel, S.: Local Feedback in Full Text Retrieval Systems. Journal of the ACM, 397–417 (1980)
2. Xu, J., Croft, W.B.: Query Expansion Using Local and Global Document Analysis. In: Proc. ACM SIGIR Annual Conference on Research and Development, Zurich (1996)

3. Qiu, Y., Frei, H.P.: Concept Based Query Expansion. In: Proceedings of the 16th ACM SIGIR Conference on Research and Development in Information Retrieval, Pittsburg, USA, pp. 160–169 (1993)

4. Schutze, H., Pedersen, J.: A Cooccurrence- Based Thesaurus and two Applications to Information Retrieval. Information Processing & Management 33(3), 307–318 (1997)

5. Mandala, R., Tokunaga, T., Takana, H.: Combining multiple evidence from different types of thesaurus for query expansion. In: Proceedings of the 22 the Annual International ACM SIGIR. Conference on research and development in information retrieval, August, Buckley, USA (1999)

6. Rocchio Jr., J.J.: The SMART Retrieval System Experiments in Automatic Document Processing. In: Relevance Feedback in Information Retrieval, pp. 313–323. Prentice Hall, Englewood Cliffs (1971)

7. Salton, G., Buckley, C.: Improving Retrieval Performance By Relevance Feedback, Journal of The American Society for Information Science 41(4), 288–297 (1990)

8. Robertson, S., Walker, S., Hanckock, M.M.: Large test collection experiments on an operational interactive system: Okapi at TREC. Information Processing and Management (IPM) journal, 260–345 (1995)

9. Gordon, M.: Probabilistic and genetic algorithms for document retrieval. Communications of the ACM, 1208–1218 (1988)

10. Yang, J.J., Korfhage, R.R.: Query optimisation in information retrieval using genetic Algorithms. In: Proceedings of the fifth International Conference on Genetic Algorithms (ICGA), pp. 603–611. The University of Illinois Press, Urbana, IL (1993)

11. Horng, J.T., Yeh, C.C.: Applying genetic algorithms to query optimisation in document retrieval. Information Processing and Management, 737–759 (2000)

12. Koza, J.R.: Genetic Programming: On the Programming of Computers by Means of Natural Selection. MIT Press, Cambridge, MA (1992)

13. Crestani, F.: Comparing Neural and Probabilistic Relevance Feedback in an Interactive Information Retrieval System. In: Proceedings of the IEEE International Conference on Neural Networks, Orlando, Florida, USA, June 1994, pp. 426–430 (1994)

14. Salton, G.: The SMART Retrieval System: Experiments in Automatic Document Processing. Prentice Hall, New Jersey (1971)

15. Porter, M.F.: An algorithm for suffix stripping. Program 14(3), 130–137 (1980)

16. Savoy, J., Vrajitoru, D.: Evaluation of learning schemes used in information retrieval. Technical Report CR-I-95-02, Université de Neuchâtel, Faculté de droit et des Sciences Économiques (1996)

Ontology-Based Mobile Information Service Platform

Hongwei Qi, Qiangze Feng, Bangyong Liang, and Toshikazu Fukushima

NEC Laboratories, China
14/F, Bldg. A, Innovation Plaza, Tsinghua Science Park, Beijing 100084, China
{qihongwei, fengqiangze, liangbangyong,
fukushima}@research.nec.com.cn

Abstract. This paper presents a Mobile Information Service Platform (MISP) based on multi-domain & multilingual ontology, which enables mobile users to obtain information by making natural language queries via short messaging, and the paper also describes the platform's evaluation results. Information services via mobile phones, closely linked to people's daily lives, are now booming in the world market. The Short Message Service (SMS) is very popular in China, and it is more convenient for Chinese users to access an information service in natural language style via SMS. However, it is difficult for general natural language processing methods to achieve high processing accuracy. In this paper, a natural language processing method based on a Multi-domain & Multilingual Ontology (MMO) and Multi-domain & Multilingual Query Language (MMQL) is proposed. The MMO is used to provide domain-specific semantic knowledge, and the MMQL is used to provide domain-specific linguistic knowledge. Four categories of trial services, namely a traffic congestion information service, location-related information service, driving route planning service, and bus trip planning service, show that this MISP can provide considerable commercial-level services to mobile users.

Keywords: Short Message Service, Mobile Information Service, Multi-domain & Multilingual Ontology, Multi-domain & Multilingual Query Language.

1 Introduction

With the proliferation of mobile communication networks, especially with the current 2.9 billion mobile subscriber numbers worldwide (490 million in China), information services via mobile phones, closely linked to people's daily lives, will become a big market in the near future [1]. Eyeing the big potential market, mobile operators are also beginning to shift their revenue strategies from just mobile phone calling services to include information services [2].

Basically, users can access mobile information services via their mobile phone's embedded micro-browser or by using text/multimedia messaging (Short Message Service, SMS). In China, almost all users have the experience of sending short messages, and 1.1 billion messages are sent every day [3]. People are accustomed to writing short messages in natural language, so we developed natural language-based approach for accessing mobile information services.

Y. Zhang et al. (Eds.): APWeb 2008, LNCS 4976, pp. 239–250, 2008.
© Springer-Verlag Berlin Heidelberg 2008

Although it is convenient for users to access information using natural language, it is difficult for a general natural language processing method [4] to achieve high accuracy, especially when allowing users to query information from multiple domains in multiple languages.

Firstly, it does not have enough domain-specific semantic knowledge, and therefore it is difficult for the general method to understand a domain and process the related queries with high accuracy. For example, when a user queries "How is the traffic near Silver Plaza?", it is difficult for the system to understand that the user wants to query the traffic situation at Baofusi Bridge, if the system does not know that Baofusi Bridge is the nearest congestion location to Silver Plaza.

Secondly, it does not have enough domain-specific linguistic knowledge. Therefore it is difficult for it to understand flexible natural language queries from different users. Moreover, multilingual queries make the understanding even more difficult.

This paper proposes a natural language processing method based on a Multi-domain & Multilingual Ontology (MMO) and Multi-domain & Multilingual Query Language (MMQL) to solve the above problems.

To tackle the first problem, the MMO, constructed primarily in the spirit of Gruber's view of ontologies [5], is developed to provide domain-specific semantic knowledge by conceptualizing multiple domains. The MMO consists of a group of domain ontologies to conceptualize various domains and a mapping ontology to connect the domain ontologies together organically.

To tackle the second problem, the MMQL, a hierarchical query language based on the MMO, is developed to provide domain-specific linguistic knowledge. It can cover many natural language query forms by defining limited syntax, and parallel lingual syntax is also defined to process multilingual queries.

Finally, with the MMO- and MMQL-based natural language processing method as the core technology, a Mobile Information Service Platform (MISP) is realized. The MISP aims to enable mobile users to access accurate multi-domain information anytime and anywhere, just by making multilingual natural language queries via short messaging. Four categories of trial services, namely a traffic congestion information service [6], location-related information service, driving route planning service, and bus trip planning service, show that this MISP can provide considerate commercial-level services to mobile users.

The rest of this paper is organized as follows: An overview of related work is given in Section 2. Section 3 describes the architecture of the MISP. Section 4 describes the Multi-domain & Multilingual Ontology (MMO). Section 5 explains the Multi-domain & Multilingual Query Language (MMQL). Section 6 presents the natural language query processing engine. Section 7 outlines the implementation and evaluation of the MISP. Section 8 concludes the paper.

2 Related Work

There have been many efforts in system development for mobile information services.

Keyword-based mobile search engines are commonly used by current systems, such as Google mobile search [7], Yahoo mobile search [8], and MSN mobile search [9], among others. Mobile devices have small displays, and mobile user interfaces are often less than optimally usable, because the above systems often directly impose the desktop web search paradigm on the mobile user.

There are also some natural language-based mobile search engines, such as Ask-MeNow [10], Any Question Answered [11], and mInfo [12], among others. They can provide a more usable interface to users.

However, looking through the related natural language-based mobile search engines, we find it is difficult to achieve a high rate of accurate understanding of flexible natural language queries, without the help of human beings, because:

- Most of them do not have enough domain-specific semantic knowledge. They often gather many terms from various domains but limited semantic relations among terms. Therefore it is difficult for them to understand some complex queries and irrelevant answers or no answers are often provided to the user.
- Most of them do not have enough domain-specific linguistic knowledge. Many of them only used general lexical and syntactical features that can be used for any domain, and even if some of them considered domain-specific linguistic features, semantic features were seldom considered. So they often support limited natural language query forms. Moreover, most of them cannot support multilingual query.

3 Architecture of the MISP

We designed the MISP (Mobile Information Service Platform, shown as Fig. 1, to provide natural language-based services. Furthermore, this platform was also designed in order to solve the two problems described in the previous sections.

In the lower level of Fig. 1, it can be seen that domain-specific semantic knowledge and linguistic knowledge for processing users' natural language queries through two methods: the Multi-domain & Multilingual Ontology (MMO), and the Multi-domain & Multilingual Query Language (MMQL), respectively.

The MMO is necessary to provide semantic knowledge for natural language processing. Firstly, semantic knowledge includes that from multiple domains, while different sources may contain the knowledge in different languages, such as Chinese, English, and Japanese, etc. Therefore the MMO is needed to represent multi-domain and multilingual knowledge. Secondly, knowledge of different domains is not separate, and nor is the knowledge in different languages. There are some relevancies among different domains and languages, and the MMO is needed to map them together in order to provide more effective semantic knowledge. So MMO is designed as a group of domain ontologies to represent multi-domain and multilingual knowledge and a mapping ontology to represent the relevancies among domain ontologies.

The MMQL is necessary for flexible natural language query processing. Firstly, users often have different query forms for different domains, and the MMQL is needed to summarize the query forms of each domain. Secondly, there are some common query features of different domains, and the MMQL is needed to extract them in order to reduce repeated definitions. So MMQL is designed as a group of domain query languages to summarize the query forms of each domain and a common query language to extract the common query features of multi-domains, and such a hierarchical style can reduce repeated definitions. Furthermore, to process multilingual queries, parallel lingual syntax is also defined by the MMQL. It comprises Chinese syntax and the corresponding syntax in English, Japanese, etc.

In the middle level of Fig .1, a natural language query processing engine is depicted. Based on the MMO and MMQL, the engine can process users' flexible natural language queries and generate relative answers. Firstly, as the user can make a multilingual query, the query language identification module is used to identify the language type of the user query. Secondly, the query analysis module can understand flexible natural language queries. Thirdly, the information search and answer generation module can find the appropriate information and generate answers for users, according to the above query analysis result.

The top level of Fig .1 shows commercial-level services for mobile users. Using the natural language query processing engine provided by the middle level, an effective information service can be provided, such as a traffic congestion information service, location-related information service, driving route planning service, and bus trip planning service.

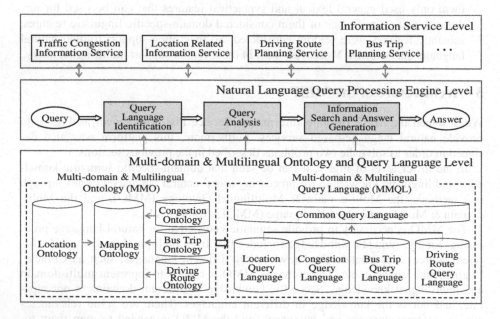

Fig. 1. Architecture of the Mobile Information Service Platform (MISP)

4   Multi-domain and Multilingual Ontology (MMO)

4.1   MMO Representation

The MMO is used to provide semantic knowledge for natural language processing. It consists of a group of domain ontologies and a mapping ontology. The domain ontology is used to record semantic knowledge for each domain, and the mapping ontology is used to record the relations among the domain ontologies. The domain ontology and mapping ontology will now be explained in detail.

4.1.1 Domain Ontology

The domain ontology is created separately for each domain, such as the location ontology, congestion ontology, bus trip ontology, driving route ontology, etc. At the same time, the domain ontologies are defined as a parallel union of multiple languages. The domain ontology is defined in the spirit of Gruber's view of ontologies [5], and it is composed of a set of concepts that are connected with relations. In other words, each concept is described with attributes and its relations to other concepts. Axioms are also considered in order to keep constraining the interpretation and detecting errors in these concepts and their relations. Axioms can also be used to make inferences. The axioms are described by horn rules [13].

```
DefClass Road
{
    Relation: SectionOf
        : lang English
        : type SectionOf(RoadX, RoadY)
        ...
}
DefClass Bridge
{
    Relation: Between
        : lang English
        : type Between(BridgeX, BridgeY, BridgeZ)
    Axiom: Between(Z, X, Y) → Between(Z, Y, X)
        ...
}
DefClass Point-of-Interest
{
    Attribute: Address
        : lang English
        : type String
        ...
}
DefInstance North Fourth Ring Road: Road
{
    SectionOf(North Fourth Ring Road, Fourth Ring Road)
}
DefInstance Baofusi Bridge: Bridge
{
    Between(Baofusi Bridge, Xueyuan Bridge, Jianxiang Bridge)
}
DefInstance Silver Plaza: Point-Of-Interest
{
    Address: No.9, North Fourth Ring West Road
}
```

```
DefClass CongestionLocation
{
    Attribute: ChineseName
        : lang Chinese
        : type string
    Attribute: RushHour
        : lang English
        : type string
        ...
}
DefClass Direction
{
    Relation: PartOf
        : lang English
        : type PartOf(DirectionX, DirectionY)
        ...
}
DefClass TrafficStatus
{
        ...
}
DefInstance Baofusi Bridge: CongestionLocation
{
    ChineseName: 保福寺桥
    RushHour: 7:30~8:30, 17:30~18:30
}
DefInstance West to East: Direction
{
    PartOf(West to East, East-West)
}
DefInstance Jammed: TrafficStatus
{
        ...
}
```

Fig. 2. Examples of Location Ontology and Congestion Ontology

Fig. 2 presents examples of location ontology and congestion ontology, respectively. Location ontology comprises all roads (e.g. North Fourth Ring Road), bridges (e.g. Baofusi Bridge), points of interest (e.g. Silver Plaza) and their geospatial relations (e.g. "Between (Baofusi Bridge, Xueyuan Bridge, Jianxiang Bridge)" means Baofusi Bridge is located between Xueyuan Bridge and Jianxiang Bridge). Similarly, congestion ontology comprises all congestion locations (e.g. Baofusi Bridge), directions (e.g. west to east) and traffic status descriptions (e.g. jammed).

4.1.2 Mapping Ontology

Integrated information services require the domain ontologies to be connected together. So the mapping ontology, which saves the relation information among domain ontologies, is used to integrate the domain ontologies. As shown in Fig. 3, there are three kinds of relations used to make the mapping:

(1) Expression mapping. The mapping is built among synonymous or abbreviated words. For example, "Silver Plaza" and "Silver Tower" are mapped because they are synonymous.
(2) Language mapping. The mapping is built among the same words that are described in different languages. For example, "North Fourth Ring Road" and "北四环路" are mapped because "北四环路" is the Chinese translation of "North Fourth Ring Road." The language mapping enables the MMO to provide multilingual semantic knowledge for multilingual query processing.
(3) Geospatial mapping. The mapping is built among geospatial-related words. For example, "Silver Plaza" and "Baofusi Bridge" are mapped because the congestion location "Baofusi Bridge" is spatially near to the point of interest "Silver Plaza."

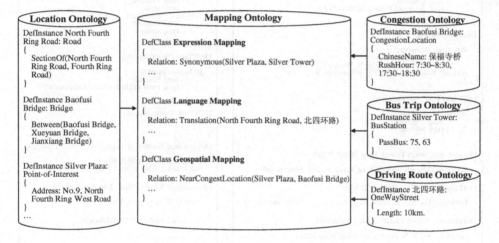

Fig. 3. Architecture of Mapping Ontology

4.2 MMO Creation

Domain ontology is created as a union of multiple languages from its information source. Location ontology, bus trip ontology and driving route ontology are automatically created from the electronic map, based on the functions of the Geographic Information System (GIS). Congestion ontology is semi-automatically extracted from traffic reports, which comprise the traffic status descriptions of all places.

Mapping ontology is created automatically. Firstly, we build the following data libraries to be used for the creation of the mapping ontology: (1) synonymy dictionary. This comprises the words from each domain and corresponding synonymous expressions; (2) abbreviation rules. Those words that have long names often have many abbreviated expressions, especially in the Chinese language and the English language; (3) multilingual dictionary. This comprises the words from each domain and corresponding translations. Secondly, the mapping ontology is created automatically based on the above data libraries: (1) expression mapping is performed based on the synonymy dictionary and abbreviation rules; (2) language mapping is performed based on the multilingual dictionary, and the mapping among the words that have the same meanings but are in different languages is built automatically; (3) geospatial mapping is performed based on the GIS functions, and the mapping among geospatial-related words is built automatically.

5 Multi-domain and Multilingual Query Languages (MMQL)

5.1 MMQL Representation

The MMQL is used to provide domain-specific linguistic knowledge. It can cover many natural language query forms by defining limited syntaxes. The MMQL consists of a group of domain query languages and a common query language. The domain query languages are built to summarize the query forms of each domain. The common query language is summarized to extract the common query features of multi-domains. It can be then inherited by various domain query languages. Such a hierarchical style can reduce repeated definitions through the inherit strategy. Furthermore, both the domain query language and common query language are defined as a parallel union of multiple languages.

5.1.1 Domain Query Language
The domain query language is designed for each domain accordingly, such as the congestion query language, bus trip query language, etc. The domain query language is defined based on the corresponding domain ontology, and it consists of the following components:

(1) Syntax is a grammatical definition system, which attempts to record all possible natural language query forms. Parallel lingual syntax, including English syntax, Chinese syntax, Japanese syntax, etc., is also defined. There are several components in the syntax, and each component has its own symbol definition:

- ■ <X> means X is a constant, and <?X> means X is a variable.
- ■ <?X(Constraint)> means X should satisfy the constraint 'Constraint', which is a set of classes that can be found in the corresponding domain ontology.
- ■ '|' means 'or' logic operator. '[ ]' means the string in '[' and ']' is optional.

(2) The agent is represented as "Syntax(name)→QueryType(keywords)", which means if the user query conforms to the syntax of "name", then the query semantic structure "QueryType(keywords)" is generated.

Fig. 4(a) and Fig. 4(b) present an example of congestion query and bus trip query language, respectively. If the query conforms to the syntax of "Traffic Query Between Points", then the query semantic structure "Traffic(?Start, ?End)" is generated, which means it obtains the traffic status at the points that are located between "?Start" and "?End". If the query conforms to the syntax of "Bus Trip Planning", then the query semantic structure "BusPlanning(?Start, ?End)" is generated, which means it obtains the bus route from "?Start" to "?End".

```
Defsyntax
{
    Syntax: Traffic Query Between Points
    {
        EN: <traffic>[<from>]<?Start(CongestionLocation)>[<to>]<?End(CongestionLocation)>
        CN: [<从>]<?Start(CongestionLocation)>[<到|去>]<?End(CongestionLocation)><路况拥堵|畅通>
        JP: [<から>]<?Start(CongestionLocation)>[<に>]<?End(CongestionLocation)><混雑>
    }
}
Defagent
{
    Agent: Syntax(Traffic Query Between Points) → Traffic(?Start, ?End)
}
```

Fig. 4(a). Example of Congestion Query Language

```
Defsyntax
{
    Syntax: Bus Trip Planning
    {
        EN: <bus>[<from>]<?Start(BusStation)>[<to>]<?End(BusStation)>
        CN: [<从>]<?Start(BusStation)>[<到|去>]<?End(BusStation)><公交汽车>
        JP: [<から>]<?Start(BusStation)>[<に>]<?End(BusStation)><汽車>
    }
}
Defagent
{
    Agent: Syntax(Bus Trip Planning) → BusPlanning(?Start, ?End)
}
```

Fig. 4 (b). Example of Bus Trip Query Language

5.1.2 Common Query Language

The common query language is designed by extracting the common query features of various domain query languages, and the domain query languages are redefined by inheriting from the common query language. The inherit strategy is implemented by instantiating appropriate parameters.

As an example (see Fig. 5), we can extract the common query language from Fig. 4(a) and Fig. 4(b). That is to say, the common query features of "Traffic Query Between Points" in Fig. 4(a) and "Bus Trip Planning" in Fig. 4(b) are summarized as "BetweenPoints" in Fig. 5. Then, the syntax of the congestion query language can be redefined as "inherit BetweenPoints(?para1=traffic, ?para2=CongestionLocation, ?para3=路况|拥堵|畅通)", and the syntax of bus trip query language can be redefined as "inherit BetweenPoints (?para1=bus, ?para2=BusStation, ?para3=公交|汽车)".

```
Defsyntax
{
    Syntax: BetweenPoints
    {
        EN: <?para1>[<from>]<?Start(?para2)>[<to>]<?End(?para2)>
        CN: [<从>]<?Start(?para2)>[<到|去>]<?End(?para2)><?para3>
        JP: [<から>]<?Start(?para2)>[<に>]<?End(?para2)><?para3>
    }
}
```

Fig. 5. Example of Common Query Language

5.2 MMQL Creation

Currently, the MMQL is mainly built manually. Firstly, we analyze the common query features of various domains, and then build a hierarchical structure to show the top-down relations of the query features. Secondly, based on the hierarchical structure, we define lingual syntaxes and agents in parallel for each node by the inheritance between higher-level syntax and lower-level syntax. The higher-level syntax is taken as the common query language, and the lower-level syntax is taken as the domain query language. As the common query language depends on the given set of domain query languages, when a new domain is added, the above manual process must be executed again to extend the MMQL.

6 Natural Language Query Processing Engine

The MISP can take Chinese, English and Japanese queries within the domains of congestion information, location-related information, driving route planning and bus trip planning. Furthermore, in China, as per SMS limitations, almost all of the queries are under 70 characters. The natural language query processing engine can process these flexible queries and generate appropriate answers. It is developed based on the MMO and MMQL, and it includes the following three steps (as shown in Fig. 1):

Firstly, as users may query in multi-languages, such as Chinese, English, Japanese, etc., the query language identification module identifies the language type of the user query by matching the query with MMO and MMQL.

Secondly, the query analysis module finds the matched syntax in MMQL by a relaxed keyword matching method. This means the method can transform the query keywords that cannot satisfy the constraints of the matched syntax, based on the expression mapping, language mapping and geospatial mapping in the mapping ontology of MMO. Through this method, the query analysis module can handle a flexible query from a mobile user. Finally, the module generates the corresponding query semantic structure by the agent definition of the matched syntax.

Thirdly, the information search and answer generation module search for appropriate information and generate an answer, according to the query semantic structure.

For example, a user queries "How is the traffic from Jianxiang Bridge to Silver Plaza?" Firstly, the query is identified as English language, because "traffic", "from" and "to" are comprised in English syntax of MMQL and "Jianxiang Bridge" and "Silver Plaza" are denoted as English language by MMO. Secondly, the query is matched with the syntax of "Traffic Query Between Points" that inherits from "Between-Points", where "?Start" and "?End" are "Jianxiang Bridge" and "Silver Plaza" respectively. Because "Silver Plaza" cannot satisfy the corresponding constraint "CongestionLocation", it is transformed into "Baofusi Bridge" by the mapping information "NearCongestLocation(Silver Plaza, Baofusi Bridge)". Then generates the corresponding query semantic structure "Traffic(Jianxiang Bridge, Baofusi Bridge)". Thirdly, search traffic reports and generate the answer "9:59 Jianxiang Bridge to Baofusi Bridge, west to east, traffic slow".

7 Implementation and Evaluation

7.1 Implementation

As shown in Fig. 6, the implementation of the MISP consists of 5 components:

(1) Information Sources, including traffic reports, WebPages, etc. They provide the MISP with multi-domain & multilingual information.
(2) Content Providers, who provide basic information for the service, such as a congestion database, location database, etc.
(3) Mobile Users, who query information via SMS.
(4) Mobile Operators such as China Mobile and China Unicom. They transmit queries and system responses between Mobile Users and Service Providers.
(5) Service Providers, which are the key component of the MISP. They work between Content Providers and Mobile Operators.

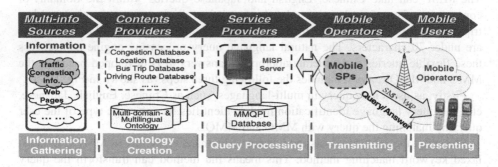

Fig. 6. Implementation Architecture of the MISP

Currently, we have developed:

(1) A Multi-domain & Multilingual Ontology (MMO), which comprises four domain ontologies: the congestion ontology, location ontology, driving route ontology and bus trip ontology. They comprise seven cities (Beijing, Tianjin, Shanghai, Guangzhou, Heilongjiang, Jiangsu, Jilin) of China, about 150 classes, 300 attributes and relations, 120 axioms, and 2,000,000 instances.
(2) A Multi-domain & Multilingual Query Language (MMQL), which comprises the above 2300 syntaxes.
(3) Four trial services: the traffic congestion information service, location-related information service, driving route planning service, and bus trip planning service.

7.2 Experiment 1: Accuracy Evaluation

Because the inputs to MISP are natural language queries, the first experiment was used to examine the accuracy of the MISP to process them, in two categories: (1) Accuracy of query processing, and this was used to evaluate the ability to understand various queries. (2) Accuracy of answer generation, and this was used to evaluate the ability to generate correct answers. We correspondingly defined two types of accuracies as follows:

$$\text{Query Accuracy} = \frac{\text{Number of queries whose query semantic structures are correctly extracted}}{\text{Number of all queries}}$$

$$\text{Answer Accuracy} = \frac{\text{Number of queries whose answers are correct and have no irrelevant information}}{\text{Number of queries whose query semantic structures are correctly extracted}}$$

We selected a test set that comprises 1,500 queries (500 for Chinese, 500 for English, 500 for Japanese) collected for the traffic congestion information service, location-related information service, driving route planning service, and bus trip planning service, respectively. In other words, the test set comprises 2000 English queries, 2000 Chinese queries and 2000 Japanese queries. As these services had already been launched in China, we were able to gather these different types of queries from real-life users.

Twenty users were required to check the query semantic structure and answer for each query and give their evaluations ('satisfied' or 'unsatisfied'), and then the above accuracies were calculated based on the users' evaluations.

Table 1(a) gives the multi-domain query processing evaluation results, which show 93.6~94.1% query accuracy and 94.7~96.8% answer accuracy achieved for the four domains. In examining those queries that were not answered correctly, two types of errors were identified: (1) User errors, which were generated in the query by the user himself, e.g., incomplete query, semantic error or spelling error in the query; and (2) System errors, in which the user's query is correct, but the system cannot answer it appropriately, e.g., due to an unrecognized word or unrecognized syntax in the query.

Table 1(b) gives the multilingual query processing evaluation results, which show 92.4~95.2% query accuracy and 94.7~95.9% answer accuracy achieved for the three languages. In examining the queries that were not answered appropriately, besides being due to user errors and system errors as discussed in the above paragraph, the accuracy of English query processing is higher than Chinese and Japanese, because Chinese and Japanese queries are often more flexible.

Table 1(a). Multi-domain Evaluation

Domains \ Accuracy	Query Accuracy	Answer Accuracy
Traffic Congestion	94.1%	94.7%
Location-Related	93.9%	96.8%
Driving Route Planning	93.7%	96.1%
Bus Trip Planning	93.6%	95.5%

Table 1(b). Multilingual Evaluation

Languages \ Accuracy	Query Accuracy	Answer Accuracy
Chinese	93.5%	95.1%
English	95.2%	95.9%
Japanese	92.4%	94.7%

7.3 Experiment 2: Performance Evaluation

The second experiment was used to examine the performance of the MISP. We deployed the MISP on a HP DL580 server with 2200MHz CPU×4 and 4GB memory. Then we used the 6000 queries from Experiment 1, repeated 10 times (60,000 queries in all), as a performance test. The scalability and response time evaluations are shown in Table 2. It can be seen that the driving route planning service performed the worst because it requires geographical computing using the GIS function, which is time-consuming. The MISP had already been applied to a real service in China, and its performance, shown in Table 2, was good enough for the real service.

Table 2. Performance Evaluation of MISP

Service ⟍ Performance	Scalability	Response Time
Traffic Congestion	18x15 (Thread) Queries/Second	55 Milliseconds/Query
Location-Related	9x15 (Thread) Queries/Second	110 Milliseconds/Query
Driving Route Planning	5x15 (Thread) Queries/Second	200 Milliseconds/Query
Bus Trip Planning	8x15 (Thread) Queries/Second	125 Milliseconds/Query
Average	10x15 (Thread) Queries/Second	100 Milliseconds/Query

8 Conclusion

In this paper, we present a Mobile Information Service Platform (MISP) based on a Multi-domain & Multilingual Ontology (MMO) and Multi-domain & Multilingual Query Language (MMQL). The MMO can provide domain-specific semantic knowledge, and the MMQL can provide domain-specific linguistic knowledge. With the help of the MMO and MMQL, the MISP can process users' flexible natural language queries with higher accuracy. Experiments on a traffic congestion information service, location-related information service, driving route planning service, and bus trip planning service show that the MISP can provide considerate commercial-level services to mobile users. Furthermore, these services have been already launched in China.

References

1. Portio Research Ltd.: Worldwide Mobile Market Forecasts 2006-2011: Global Analysis and Forecasts of Mobile Markets, Technology and Subscriber Growth (2006)
2. Paulson, L.D.: Search Technology Goes Mobile. IEEE Computer 38(8), 19–22 (2005)
3. iResearch Co., Ltd.: Market Size of China SMS Market 2003-2008. 2005 China WVAS Market Research Report (2006)
4. Dekleva, S.M.: Is Natural Language Querying Practical. J. Database, 24–36 (1994)
5. Gruber, T.R.: A Translation Approach to Portable Ontology Specification. Knowledge Acquisition 5(2), 199–220 (1993)
6. Qi, H., et al.: A Map Ontology Driven Approach to Natural Language Traffic Information Processing and Services. In: The 1st Asian Semantic Web Conference, pp. 696–710. Springer, Beijing (2006)
7. Google Mobile. http://www.google.com/intl/zh-EN/mobile/
8. Yahoo Mobile. http://mobile.yahoo.com/search
9. MSN Mobile. http://mobile.msn.com/
10. AskMeNow. http://www.askmenow.com/
11. Any Question Answered. http://www.issuebits.com/
12. mInfo. http://www.minfo.com/
13. Chandra, A., Harel, D.: Horn Clause Queries and Generalizations. J. Logic Programming. 2, 1–15 (1985)

On Safety, Computability and Local Property of Web Queries
[Extended Abstract]

Hong-Cheu Liu[1] and Jeffrey Xu Yu[2]

[1] Department of Computer Science and Information Engineering, Diwan University,
Madou, Tainan County, Taiwan 72153
hcliu@dwu.edu.tw
[2] Department of System Engineering and Engineering Management,
The Chinese University of Hong Kong,
Shatin, New Territories, Hong Kong
yu@se.cuhk.edu.hk

Abstract. In this paper, we study the safety of formulas, computability and local property of Web queries. The limited access capability and loosely structured information make querying the Web significantly different from querying a conventional database. When evaluating queries on the Web, it is not feasible to exhaustively examine all objects on the Web. The local property, studied in this paper, can be informally described as the reverse scenario: in order to check if an object participated in the result of a Web query, is it sufficient to examine a bounded portion of the Web? We start our investigation by using the Web machine proposed by Mendelzon and Milo [1]. We review the genericity, domain independence and computability of Web queries. We present a syntactic class of local Web queries and a sound algorithm to check if a query belongs to this class. We then examine the notion of locality of two popular XML query languages, namely XPath and XQuery, and show that they are not local.

1 Introduction

The connection of the Web and databases bring many challenges and new opportunities for creating advanced database applications. Web sites are also increasingly powered by accessing their databases directly. However, the limited access capability, the lack of concurrency control and loosely structured information make querying the Web different from querying a conventional database. These essential aspects have significantly effects on the *computability* of queries [1,2].

The theory of database queries has grown into a rich research area, which includes the expressiveness and complexity of query languages, the domain independence and the safety of queries, query translation and optimization. However, the development of Internet technology has occurred very rapidly and much of the above traditional framework of database theory needs to be re-investigated

Y. Zhang et al. (Eds.): APWeb 2008, LNCS 4976, pp. 251–262, 2008.
© Springer-Verlag Berlin Heidelberg 2008

in the Web scenario [3]. This paper will focus on the study of the issues of safety, computability and local property of Web queries.

The most common technology currently used for searching the Web depends on sending information retrieval requests to "index servers" that provide associative access to a large collection of pre-computed data. However, these index servers and similar search engines do not guarantee they have indexed every Web object, nor do they give any guarantee of currency of the index [1]. One problem with these index servers is that queries cannot exploit the structure and topology of the Web objects network. We argue that designing feasible Web query languages and investigating fundamental aspects of querying the Web are still demanding and important. There are some interesting and challenging problems for these languages.

Expressiveness of database query languages is one of the major research topics in finite model theory. Most database query languages have limited power and some tools have been developed for analyzing these query languages. However, most of those tools are only applicable to first-order logic and some of its extensions but they do not apply to languages querying the Web. We are especially interested in computability and local property of query languages that are applied to the Web. By locality, we mean that the result of a query can be determined by looking at a certain predetermined portion of the input. The conventional approach in database theory is to estimate query evaluation time as a function of the size of the database. The classical complexity and expressivity is irrelevant in the Web context. For a query to be practical, it should not attempt to access too much of the network. The goal of this paper is to give a general study of local property of queries in the context of Web scenario, which goes beyond the pure first-order case, and then analyze expressive power of Web queries including XQuery.

In the relational database framework, queries are defined as generic and computable mapping from database instances to well-typed results. Three query paradigms have been developed, and these are *Calculus, Algebra, Datalog*. Ideally, we expect that every formula of a query language can serve as a query. If the formula has free variables, $x_1, ..., x_k$, then the answer to the query is the set of k-tuples which satisfy the formula in the intended structure. In practice, not every formula can be accepted as a query as *finite* and *effectively computable* conditions required by database systems.

The notions of "domain independence" and "safety" have been developed to capture intuitive properties related to the finite query results criterion. Due to the different scale of volume of the Web databases, the limited data access capabilities, the lack of concurrency control, and the dynamic nature of the Web, the notions of the above computability and safety are quite different from the finite relational model. For example, a query such as "find nodes with no incoming links" is easily expressible as $\{x \mid Node(x, ...) \land \forall y (Node(y, ...) \rightarrow \neg Link(y, x, ...))\}$, which is not computable in the context of Web. The work of [1,3] has dealt this issue in the context of the Web.

Unlike relational calculus, playing the role of *relational completeness*, there are no well-accepted yardsticks for expressiveness of query languages in the context of the Web. The query languages for the Web attract much attention. Some languages are targeted at querying the Web as a whole, while others are aimed at semi-structured data and XML [4,5,6,7,8,9]. We consider complex value query languages as Web information is easily modeled in the complex value data model.

The main contributions of this paper are as follows:

- We generalize the notions of *local query* to the Web context and show that not all eventually computable queries are local.
- An algorithm for providing a sufficient condition for a calculus query being finitely computable has been developed. Then we show that every finitely computable query is local in the context of Web.
- We define the notion of "continuous" for Web queries and investigate its relationship with local property. This result will be helpful in checking query computability.
- We analyze the expressive power of Core XPath and XQuery fragments.

The benefit of this study is that the results enhance our understanding of some fundamental aspects of querying the Web and could be used in practice for designing appropriate Web query languages.

The remainder of this paper is organized as follows. Section 2 briefly reviews basic concepts. Section 3 investigates the local property of Web queries. Section 4 analyzes the expressive power of XPath and XQuery. Finally, we provide some conclusions in Section 5.

2   Queries on the Web

The World Wide Web is a large, heterogeneous, distributed collection of Web objects connected by hyperlinks. It can be viewed as a graph whose nodes are Web objects that are identified by a uniform resource locator (URL). We use the complex value data model as a model of the Web that captures the Web's graph nature, and the semi-structured information it holds. Intuitively, the global Web database can be viewed as a finite *complex value* data structure over the fixed schema S which contains DB, $Node$ and $Link$ schemas.

$$DB = \{R_1, ..., R_n\}$$
$$Node = \{id, title, subject, date, length, ...\}$$
$$Link = \{source, Ref(label, destination, parameters)\}$$

DB is a finite set of base relations containing the relevant data such as bookmarks, local files, etc. It also contains semantic predicates that a Web query may apply. Relation $Node$ specifies the set of Web objects. The Web objects may refer to Web pages, Web sites, or other objects on the Web. Complex value relation $Link$ specifies, for each node, a finite set of links to other nodes and the labels associated with them.

In relational databases, it is usually required that the query be *generic, domain independent* and *computable*. Let **I** be a Web instance. A Web query q over schema $\mathcal{S}$ is a generic mapping which associates to each pair $(o, \mathbf{I})$, where o is a particular object in **I**, a subset of the Web node structure. The object o is the starting point (or source) of the query. The semantics for Web query follows the calculus semantics. We write $\varphi(x_1, ..., x_n)$ to indicate that $x_1, ..., x_n$ is a list of the variables occurring free in φ. A formula $\varphi(x_1, ..., x_n)$ over $\mathcal{S}$ defines a Web query q which is expressed as follows.

$$q(o, \mathbf{I}) = \{v(\boldsymbol{x}) \mid \mathbf{I} \models_{\mathbf{d}} \varphi[v], v \text{ is a valuation over free variables of } \varphi\},$$

where **d** is the domain of a Web instance **I**, i.e., the set of valid values in **I**. An *isomorphism* ρ is a mapping that is one-to-one, onto mapping from domain **d** to **d**. Each isomorphism is extended to Web instance in the obvious way.

Definition 1. *A Web query q is generic iff for any Web instance (o, I) and each isomorphism ρ, $q(\rho(o), \rho(I)) = \rho(q(o, \boldsymbol{I}))$.*

The above definition captures the data independence principle in databases. It means that the result only depends on the information in **I** and is independent of any particular encoding scheme.

Definition 2. *A Web query q is domain independent iff any Web instance $I = (\boldsymbol{d}; DB, Node, Link)$ and each pair of domains $\boldsymbol{d}_1$, $\boldsymbol{d}_2$, $q(\boldsymbol{d}_1; DB, Node, Link) = q(\boldsymbol{d}_2; DB, Node, Link)$.*

We adopt the Web query model of [1]. As stated in that article, a *Web machine* is a Turing machine augmented with an oracle. It has two input tapes, an ordinary tape on which the input DB base relations are encoded, and an oracle tape on which the $Node$ and $Link$ relations are encoded. It also has an ordinary working tape and an oracle working tape. The result of the computation is written to the output tape in a standard append only manner. In the dynamic Web environment, we view a Web as an *infinite* sequence of Node and Link relation pairs. We review the notion of *computable query* proposed by Abiteboul and Vianu [2].

Definition 3. *A Web query q is finitely computable if there exists a Web machine that computes on input tape enc($\boldsymbol{I}$), halts and produces enc($q(\boldsymbol{I})$) on the output tape.*

Definition 4. *A query q is eventually computable if there exists a Web machine whose computation on input enc($\boldsymbol{I}$) has the following properties:*

- *The content of the output tape at each computation step forms a prefix of enc($q(\boldsymbol{I})$), and*
- *For each tuple in enc($q(\boldsymbol{I})$), its encoding occurs on the output tape after a finite number of computation steps.*

We study queries on the Web structures. Informally, by locality of query in the finite relational model, we mean that in order to check if a tuple belongs to the result of a query, one has to look at a certain predetermined portion of the input. Now we would like to define local query in the Web scenario. Given the schema S introduced earlier, its graph $\mathcal{G}(S)$ is defined as $< \mathbf{d}, E >$ where (a, b) is in E iff there is a tuple $\mathbf{t} \in Link$ such that both a and b are in $\mathbf{t}$ and $a = \mathbf{t}.source$, b is in $\mathbf{t}.Ref.destination$. The distance $dist(a, b)$ is defined as the length of the shortest path from a to b in $\mathcal{G}(S)$. Given $a \in \mathbf{d}$, and $r \geq 0$, the r-*sphere* of a, denoted by $S_r(a)$, is $\{b \mid dist(a, b) \leq r, b \in \mathbf{d}\}$. For a tuple $\mathbf{t}$, $S_r(\mathbf{t}) = \bigcup_{0 \leq i < k} S_r(\mathbf{t}[i])$, where $\mathbf{t}$ is a k-ary tuple.

Given a tuple $\mathbf{t}$, its r-*neighborhood* $N_r(\mathbf{t})$ is defined as a Web structure $< S_r(\mathbf{t}), R \mid_{S_r(\mathbf{t})}>$, where $R \in \{(R_i \cup Node \cup Link)\}$ and for each relation R, $R \mid_{S_r(\mathbf{t})}$ refers to that the domain of R is restricted to $S_r(\mathbf{t})$. Now we give the definition of locality which is adopted from [10].

Definition 5. *Given a query φ, its locality rank is the minimum number $r \in N$ such that, for every structure S and for every two vectors t_1, t_2 of elements of $\mathbf{d}$, it is the case that $N_r(t_1) \cong N_r(t_2)$ implies $S \models \varphi(t_1)$ iff $S \models \varphi(t_2)$. A query is local if it has a finite locality rank.*

Intuitively, the above definition states that the result of a query φ can be determined by looking at a neighborhood of its argument $N_r(t_1)$ or $N_r(t_2)$. According to the definition, r is the minimum number for which the case holds. It implies that we only need to look at a *small* neighborhood of the arguments of the query.

As stated in [10], every first-order (relational calculus) query is local in the finite model setting.

A Complex Value Calculus
In the complex value data model, the calculus is a many-sorted calculus. Calculus variables may denote sets so the calculus will permit quantification over sets. The calculus, denoted $CALC^{cv}$, is a strongly sorted extension of first order logic [11]. The vocabulary of the calculus language is defined as follows.

1. parentheses (,);
2. logical connectors $\wedge$, $\vee$, $\neg$, $\rightarrow$;
3. quantifiers $\exists$, $\forall$;
4. equality $=$, membership $\in$, and containment $\subseteq$ symbols;
5. sorted predicate symbols;
6. sorted tuple functions $<>$, and sorted set functions $\{\}$.

Definition 6. *Terms of the complex value calculus language are defined as follows:*

1. complex value constants of some sort τ;
2. variables whose sorts can be inferred from the context, and
3. if x is a tuple variable and C is an attribute of x, then $x.C$ is a term.

Definition 7. *Atomic formulas (positive literals) are sorted expressions of the form*

$$R(t_1, ..., t_n), \quad t = t', \quad t \in t', \quad or \quad t \subseteq t',$$

where R is a relation predicate and t_i, t, and t' are terms or function symbols with the obvious sort compatibility restrictions.

Formulas are defined from atomic formulas using standard connectives and quantifiers.

Example 1. The replacement of the second component of tuples of relation R:$\{<A, B : \{\} >\}$ by its cardinality is represented by the formula: $\{x \mid \exists y(R(y) \wedge x.A = y.A \wedge x.B = count(y.B))\}$, where *count* is a function which computes cardinality of a set.

3 Computability and Local Property of Web Queries

The goal of this section is to investigate whether or not (eventually) computable Web queries possess local property. Informally, our results show that not every eventually computable query has local property even if it can be specified as a positive fragment of first order logic expression. We also show that if a query is not eventually computable, then it is not local, i.e., we cannot just look at "small neighborhoods" of its arguments in order to check if a tuple belongs to the result of a query.

We will develop an algorithm for providing a sufficient condition for a calculus query being finitely computable. Then we show that every finitely computable query is local in the context of Web.

3.1 Safe Calculus Queries

We adopt a syntactic condition, called range-restricted, that ensures calculus queries are finitely computable, given an input Web instance [2]. Based on the notion of range-restricted condition, we define safe formulas. A formula is *safe* if and only if each variable is range-restricted. That is, the set of values assigned to each variable have a finite proof.

As in complex value databases, we define the set of range-restricted variables of a formula using the following procedure, which returns either the symbol ℧ (which indicates that a quantified variable is not finitely computable) or the set of free variables that are range-restricted (i.e., each computation value of these variables has a finite proof; each variable may have finite set of computation values).

In the following algorithm, if several rules are applicable, the one which returns the largest set of safe-range variables is chosen.

procedure range-restricted (*rr*)

Input: a calculus formula φ
Output: a subset of the free variables of φ or ℧

begin
(*pred* is a predicate in $\{=, \neq, \in, \subseteq\}$)
if for some parameterized query $\{x \mid \psi\}$ occurring as a term in φ, $x \notin rr(\psi)$
then return ℧

case φ **of**

$R(t)$	: **if** $R \in DB$ **then** $rr(\varphi) = free(t)$
$Node(a, x_1, ..., x_n)$	: $rr(\varphi) = free(Node)$
$Link(a, Ref(y_1, ..., y_n))$	: $rr(\varphi) = free(Link)$
$\phi(x_1, ..., x_n)$	
$\wedge Link(x_i, Ref(y_1, ..., y_k))$	: **if** ϕ is safe **then**
	$\quad rr(\varphi) = rr(\phi)$
	$\quad \cup free(Link(x_i, Ref(y_1, ..., y_k)))$
$\phi(x_1, ..., x_n)$	
$\wedge Node(x_i, y_1, ..., y_k)$	: **if** ϕ is safe **then**
	$\quad rr(\varphi) = rr(\phi)$
	$\quad \cup free(Node(x_i, y_1, ..., y_k))$
t $pred$ $t' \wedge \psi$	: **if** ψ is safe and $free(t') \subseteq free(\psi)$
	$\quad$ **then** $rr(\varphi) = free(t) \cup free(\psi)$
t $pred$ t'	: **if** $free(t') = rr(t')$
	$\quad$ **then** $rr(\varphi) = free(t') \cup free(t)$
	$\quad$ **else** $rr(\varphi) = \emptyset$
$\psi_1 \wedge \psi_2$	: **if** ψ_1 and ψ_2 are safe,
	$\quad$ **then** $rr(\varphi) = rr(\psi_1) \cup rr(\psi_2)$
$\psi_1(u) \wedge \neg \psi_2(v)$	: **if** ψ_1 and $\psi_2(v)$ are safe , $v \subseteq u$
	$\quad$ **then** $rr(\varphi) = free(\psi_1)$
	$\quad$ **else return** $\emptyset$
$\exists x \psi_1$	: **if** $x \in rr(\psi_1)$
	$\quad$ **then** $rr(\psi) = rr(\psi_1) - \{x\}$
	$\quad$ **else return** $\mho$
$\neg \psi$	: $rr(\varphi) = \emptyset$

end

Note that the disjunction of the two formulas $\varphi_1 \vee \varphi_2$ is replaced by the equivalent $\neg \varphi_1 \wedge \neg \varphi_2$ before the above algorithm is applied.

We say that a formula is *safe* if $rr(\varphi) = free(\varphi)$, and a query is safe if its associated formula is safe. Every safe Web query is finitely computable. However, characterizing the (largest) class of Web queries that is finitely computable is an open problem.

Example 2. The query: Find the set of nodes which are directly linked by a node o, can be expressed by $\{t \mid \exists o \in Node \wedge t \in o.Ref.destination\}$ is a safe formula. So it is finitely computable.

We first investigate the relationship between the notion of locality and computability. The positive first order language FO^+ is eventually computable [2]. We show that local property of FO^+that was conjectured to hold, fails in the context of Web scenario.

As stated in [2], FO^+ is eventually computable. There is a query in FO^+ which cannot be terminated in finite time. So we cannot just look a predetermined portion of the input to check a tuple whether or not belongs to the result of the query.

Fact: In the context of Web scenario, not every query expressed in FO^+ is local.

Now we consider safe calculus queries. A complex value calculus formula is *strongly safe range*, denoted $CALC^{CV-}$, if it is safe and the inclusion predicate does not occur in it.

Theorem 1. *Every safe query expressed in first-order (or $CALC^{CV-}$) language is local.*

PROOF SKETCH. The answer of finitely computable query is always finite and computable by a halting Web machine. For each such query, it reads a finite input and produces a finite output. This aspect is equivalent to the first order query language in the finite model. So it is local.   □

Example 3. Consider query: Find all node objects referencing the node object o. This query could be expressed as the following $CALC$ calculus formula.

$$\{t \mid \exists x(Link(x) \land o \in x.Ref.destination \land t = x.source)\}$$

It is eventually computable. However, it cannot be effectively computed in finite time. So it is not a local query.

Note that safe $CALC^{cv}$ may not be local, for example, transitive closure query can be expressed in $CALC^{cv}$ but it is not a local query.

In the finite model, not every local query is generic or domain independent. For example, every first-order query is local [10] but not every firs-order query is domain independent.

For a graph G, its degree $deg(G)$ is the cardinality of the set of all possible in- and out-degrees that are realized in G. A query q is said to have the *bounded degree property* [12], if there is function $f_q : N \to N$ such that $deg(q(\mathcal{S})) \le f_q(k)$ for every structure $\mathcal{S}$, where k is the maximum degree in G.

Corollary 1. *Every finitely computable query has bounded degree property and all queries expressed in safe FO are local and domain independent in the context of the Web.*

Theorem 2. *If a query is not eventually computable, then it is not local.*

PROOF SKETCH. If q is a query which is not eventually computable. Suppose that q is local, then for each tuple we only need to look at a predetermined portion of the input to check whether it belongs to the result. So this query would be eventually computable.   □

Example 4. The query: Find all objects that do not belong to a cycle, is not eventually computable. So it is not local.

We now explore the local property in the dynamic Web environment. As described in [1], we view a Web as an infinite sequence of Node and Link relation pairs. A dynamic Web machine is exactly like a Web machine, except that the oracle can non-deterministically switch from one Node, Link pair to the next one in the sequence. There are three classes of queries in the dynamic Web environment: dynamic computable Web queries, dynamic semi-computable Web queries and eventually computable dynamic Web queries. Dynamic computable Web queries can be evaluated by a terminating program. Dynamic semi-computable Web queries could be evaluated by a terminating computation if the computation were sufficiently faster than the update rate. The following theorem shows that the first two classes are local.

Theorem 3. *Dynamic computable and dynamic semi-computable queries are local.*

PROOF SKETCH This class of Web queries consists of queries that could be evaluated by a terminating computation if the computation were sufficiently faster than the update rate. This class queries can be specified by first-order or complex value logic. So it is local. $\square$

Now we define the notion of "continuous" for Web queries. First we define an "approximation" of a database instance. Suppose that $I_1 = (d_1; DB_1, Node_1, Link_1)$, $I_2 = (d_2; DB_2, Node_2, Link_2)$, we say that I_1 approximates I_2, written $I_1 \sqsubseteq I_2$, iff $d_1 \subseteq d_2$, $DB_1 \subseteq DB_2$, $Node_1 \subseteq Node_2$, $Link_1 \subseteq Link_2$. We then define the least upper bound of a directed family of Web databases. Let $\mathcal{I}$ be a finite set of Web database instances. We say that $\mathcal{I}$ is *directed* iff it is nonempty and for any I_1, I_2, there exists $I \in \mathcal{I}$ such that $I_1 \sqsubseteq I$ and $I_2 \sqsubseteq I$. For a directed family of database instances $\mathcal{I} = \{I_1, I_2, ..., I_k\}$, $I_i = (d_i; DB_i, Node_i, Link_i)$, we define its *least upper bound* $\bigcup \mathcal{I} \overset{\text{def}}{=} (d; DB, Node, Link)$, where d is the union of the domains of all Web databases in $\mathcal{I}$, $DB = \bigcup_i DB_i$, $Node = \bigcup_i Node_i$, $Link = \bigcup_i Link_i$.

Definition 8. *A query q is called continuous iff for any directed set of Web database instances $\mathcal{I}$, $\bigcup\{q(I) \mid I \in \mathcal{I}\}$ exists and $q(\bigcup \mathcal{I}) = \bigcup\{q(I) \mid I \in \mathcal{I}\}$.*

Here $\bigcup\{q(I) \mid I \in \mathcal{I}\}$ denotes the least upper bound of the results $q(I)$ with $I \in \mathcal{I}$.

Intuitively, the concept of continuity captures the property that the result of a query depends only on a finite approximation of the database. During the computation we can inspect only a finite fragment of a potentially infinite input.

Observation that not each local query is continuous.

Example 5. Consider a first-order query, $\{t \mid \exists x(p(x) \wedge \neg q(x)) \wedge Node(t) \wedge t = x\}$, where p and q are relation predicates in DB. It is local. But it is not continuous as it is not monotonic.

Example 6. Consider the query: Find all objects which are not referenced by any other object. We can express this query as follows.

$$\{t \mid Node(t) \wedge \neg \exists x(Node\text{-}Link(x) \wedge t \in x.Ref.destination)\}$$

Let $\mathcal{I} = \{I_1, I_2\}$. $I_1 = \{d, Node = \{A, B, C\}, Node\text{-}Link = \emptyset\}$; $I_2 = \{d, Node = \{B, C\}, Node\text{-}Link = \{B, \{label\text{-}1, C\}\}\}$. Then, $q(\bigcup \mathcal{I}) = \{A, B\}$. $\bigcup\{q(I) \mid I \in \mathcal{I}\}$ $= q(I_1) \cup q(I_2) = \{A, B, C\} \cup \{B\} = \{A, B, C\}$. So $q(\bigcup \mathcal{I}) \neq \bigcup\{q(I) \mid I \in \mathcal{I}\}$. The query is not continuous. By definition, this query is not eventually computable. So it is not local by Theorem 2.

We end this section with a brief comments on safety, locality, computability and continuity, i.e., safety and computability are closely related. Every safe query expressed in FO or $CALC^{cv-}$ is finitely computable and domain independent. Locality and continuity are independent concepts but relevant. These two notions help us to inspect only a finite fragment of a potentially infinite input during any computation of a query.

4   A Case Study: FO^{tree} and XQuery

In this section, we check whether some languages for data exchange on the web, namely XPath and XQuery exhibit the locality property. XPath 1.0 is a variable free language designed to specify paths between nodes in XML documents. Core XPath [13] is the logical core of XPath 1.0 language that contains just what is needed to define paths in, and select nodes from an XML document tree. It is the fact that Core XPath without descendant axis relation (i.e., $//$) possesses local property.

The central expression in XPath is the location path, axis::$node\_label$[filter]. When evaluated at node n, it yields an answer set consisting of nodes n' such that the axis relation goes from n to n', the node tag of n' is $node\_label$, and the expression filter evaluates to true at n'.

Core XPath is not powerful enough to express every first-order definable path. We describe the language FO^{tree} [14] which captures the class of first-order definable paths.

Definition 9. *The language FO^{tree} is defined as follows:*

- *the atomic formulas are $x = y$, $P_i(x)$, $xR_{\Downarrow}y$, and $xR_{\Rightarrow}y$, for any variables x, y, for any unary predicate symbol P_i;*
- *if ϕ, ψ are formulas, then $\neg\phi$, $\phi \wedge \psi$, and $\exists x\phi$ are formulas.*

where $R_{\Downarrow}$ and $R_{\Rightarrow}$ denote the descendant and the following sibling relation respectively.

There exists a natural expansion of Core XPath, called Conditional XPath, which is expressively complete for first-order definable paths. Unlike relational calculus, the first-order tree language FO^{tree} does not possess the local property. We establish that XQuery and FO^{tree} are not local, as follows.

Theorem 4. *Conditional XPath, FO^{tree} and XQuery are not local.*

PROOF SKETCH. Conditional XPath is expressively complete for first-order definable paths [14]. Conditional XPath can express the reflexive and transitive closure of the relation expressed by (child :: *[F])*, where F is a filter expression. By [10], it is clear that Conditional XPath is not local as one cannot only look at a certain predetermined portion of the input in order to check if a node belongs to the result of the above reflexive and transitive closure of the relation expressed by (child :: *[F])*. As stated in [14], for every FO^{tree} formula, there exists an equivalent Conditional XPath path expression. So FO^{tree} is not local either. Note that XQuery supports (possibly recursive) user-defined functions and is capable of expressing transitive closure, which is known to be non-local. □

Our framework considers two kinds of query languages. One is targeted at querying the Web as a whole, while the other is aimed at semi-structured data and XML. However, our opinion is that feasible Web query languages should combine these two functions and build the latter technology on top of the framework proposed in Section 3.

5 Conclusion

We studied the local property of computable query in the context of Web. We have shown that all queries expressed in safe first order logic (or $CALC^{CV-}$) are local. The problem of which fragments of FO^{tree} and XQuery possessing local property is worthy of investigating and the expected results could be helpful for updating XML views and query processing in XQuery query engine.

Acknowledgments

This work was supported by a grant (No. 418205) from the Research Grants Council of the Hong Kong Special Administrative Region, China.

The authors would like to thank Byron Koon Kau Choi for his fruitful discussions and helpful comments. Part of the work performed while the first author was visiting The Chinese University of Hong Kong.

References

1. Mendelzon, A., Milo, T.: Formal models of web queries. Information Systems 23, 615–637 (1998)
2. Abiteboul, S., Vianu, V.: Queries and computation on the web. Theoretical Computer Science 239, 231–255 (2000)
3. Vianu, V.: A web odyssey: from codd to **XML**. In: Proceedings of ACM Symposium on Principles of Database Systems, pp. 1–15 (2001)
4. Buneman, P., Davidson, S., Hillebrand, G., Suciu, D.: A query language and optimization techniques for unstructured data. In: Proceedings of the ACM SIGMOD International Conference on Management of Data, pp. 505–516 (1996)
5. Mihaila, G., Mendelzon, A., Milo, T.: Querying the world wide web. Journal of Digital Libraries 1, 68–88 (1997)
6. Abiteboul, S., Quass, D., McHugh, J., Widom, J., Wiener, J.: The lorel query language for semi-structured data. Journal of Digital Libraries 1 (1997)
7. Deutsch, A., Fernandez, M., Florescuall, D., Levy, A., Suciu, D.: A query language for **XML**. In: Proceedings of WWW8, pp. 11–16 (1999)

8. Chamberlin, D., Robie, J., Florescu, D.: Quilt: An **XML** query language for heterogeneous data sources. In: Suciu, D., Vossen, G. (eds.) WebDB 2000. LNCS, vol. 1997, pp. 53–62. Springer, Heidelberg (2001)
9. Abiteboul, S., Buneman, P., Suciu, D.: Data on the Web. Morgan Kaufmann, San Francisco (2000)
10. Dong, G., Libkin, L., Wong, L.: Local properties of query languages. Theoretical Computer Science 239, 277–308 (2000)
11. Abiteboul, S., Beeri, C.: The power of languages for the manipulation of complex values. The Very Large Data Bases Journal 4, 727–794 (1995)
12. Libkin, L., Wong, L.: Query languages for bags and aggregate functions. Journal of Computer and System Sciences 55, 241–272 (1997)
13. Gottlob, G., Koch, C.: Monadic queries over tree-structured data. In: Proceedings of the 17th Annual IEEE Symposium on Logic in Computer Science (2002)
14. Marx, M.: Conditional xpath. ACM Transactions on Database Systems 30, 929–959 (2005)

Privacy Inference Attacking and Prevention on Multiple Relative K-Anonymized Microdata Sets

Yalong Dong[1], Zude Li[2], and Xiaojun Ye[1]

[1] Key Laboratory for Information System Security, Ministry of Education,
School of Software, Tsinghua University, 100084 Beijing, China
dongyl05@mails.tsinghua.edu.cn, yexj@tsinghua.edu.cn
[2] Computer Science Dept., University Of Western Ontario
London, Ontario, Canada, N6A 5B7
zli263@uwo.ca

Abstract. In k-anonymity modeling process, it is widely assumed that a relational table of microdata is published with a single sensitive attribute. This assumption is too simple and unreasonable. We observe that multiple sensitive attributes in one or more tables may incur privacy inference violations that are not visible under the single sensitive attribute assumption. In this paper, a new (k, ℓ)-anonymity model is introduced beyond the existed ℓ-diversity mechanism, which is an improved microdata publication model that can effectively prevent these multiple-attributed privacy violations. The (k, ℓ)-anonymity process consists of two phases: k-anonymization on identifying attributes and ℓ-diversity on sensitive attributes. The related (k, ℓ)-anonymity algorithms are proposed and the data generalization metric is provided for minimizing the anonymization cost. A running example illustrates this technique in detail, which also convinces its effectiveness.

1 Introduction

As a microdata publication model, *k-anonymity* aims to preventing potentially existed privacy violations (e.g., individual re-identification) beyond published microdata sets [11]. Many instances illustrating such attacks are listed in literature such as [11,12,5,6,9,10,2,15]. In general, a *k-anonymized* microdata set requires that each record is indistinguishable from at least *k-1* other records with respect to (w.r.t.) certain *identifying* attributes [13,1].

It is observed that privacy inference violations may exist in a *k-anonymized* dataset when there is little *diversity* in those *sensitive* attributes [9]. The precise privacy inference attack analysis is studied in our previous paper [15] based on the privacy inference logic theory. Through further research, we observe that multiple sensitive attributes together in a (or more) relational table of microdata may incur privacy inference violations, even the table is handled with proper k-anonymity techniques under the ad hoc assumption that only one sensitive attribute is contained in the table. So in this paper, we focus on the privacy discovery and prevention solutions against such multi-attribute privacy violations.

Y. Zhang et al. (Eds.): APWeb 2008, LNCS 4976, pp. 263–274, 2008.

Example for Motivation: Table 1 shows a set of medical records from a fictitious hospital. We divide the attributes into two groups, (Allergy, Disease) as the *sensitive* attributes and (BirthDate, Sex, Zipcode) as the *identifying* attributes.

Table 1. A table of health data

	Bi.Da.	S.	Zip.	Allergy	Disease
1	03-24-39	F	13053	Penicillin	Cancer
2	02-02-39	F	13068	Sulfur	Cancer
3	11-12-39	F	13068	Sulfur	Cancer
4	08-02-57	M	14890	No	Polio
5	08-02-57	M	13053	Penicillin	Polio
6	11-22-42	M	13092	No	Stoke
7	07-25-42	M	13092	Sulfur	Stoke

The naive ℓ-*diversity* mechanism is proposed [9] to provide each value cluster of a sensitive attribute that contains at least ℓ well-represented value elements. A table with multiple sensitive attributes is ℓ-*diverse* if the table on any one of them is ℓ-*diverse* when others are all treated as identifying attributes [9]. By taking the ℓ-*diversity* mechanism on Table 1, we can get two anonymized tables (in Table 2 & 3) that can be published since each of them is 2-*diverse* ($\ell = 2$).

Table 2 & 3. Two 2-diversity tables on 2 sensitive attributes

	Bi.Da.	S.	Zip.	Disease	Allergy
1	*-*-39	F	130**	Cancer	Penicillin
2	*-*-39	F	130**	Cancer	Sulfur
3	*-*-39	F	130**	Cancer	Sulfur
4	08-02-57	M	1****	Polio	No
5	08-02-57	M	1****	Polio	Penicillin
6	*-*-42	M	13092	Stoke	No
7	*-*-42	M	13092	Stoke	Sulfur

	Bi.Da.	S.	Zip.	Allergy	Disease
1	*-*-*	*	13053	Penicillin	Cancer
5	*-*-*	*	13053	Penicillin	Polio
2	*-*-*	*	130**	Sulfur	Cancer
3	*-*-*	*	130**	Sulfur	Cancer
7	*-*-*	*	130**	Sulfur	Stoke
4	*-*-*	M	1****	No	Polio
6	*-*-*	M	1****	No	Stoke

According to the naive ℓ-*diversity* mechanism, the first table is 2-*diverse* on sole sensitive attribute Allergy (Disease is taken as an identifying attribute) while the second one is 2-*diverse* on Disease (Allergy is taken as an identifying attribute). While both of them are 2-*anonymous*. In each of them, there are no privacy violations that can increase the probability of individual re-identification. But if both of them are published, we can infer some useful information that are just provided by their combination. We take query-answer form to illustrate this kind of inference attacks.

- Q: $\pi_{\{BirthDate,\ Sex,\ Zipcode\}} \sigma_{\{Allergy\ =\ Penicillin\ and\ Disease\ =\ Cancer\}}$
- A: $<*-*-39,\ F,\ 13053>$ (*inferred narrowly*)
- Q: $\pi_{\{BirthDate,\ Sex,\ Zipcode\}} \sigma_{\{Allergy\ =\ Penicillin\ and\ Disease\ =\ Polio\}}$

- A: <08-02-57, M, 13053> (*inferred exactly*)
- Q: $\pi_{\{Allergy,\ Disease\}}\sigma_{\{BirthDate\ =\ 08-02-57\ and\ Sex\ =\ M\ and\ Zipcode\ =\ 14890\}}$
- A: <No, Polio> (*inferred exactly*)

The above scenario illustrates a requirement for the succeeding *k-anonymity* models, i.e. preventing inference violation over published microdata sets with multiple sensitive attributes. That is the main motivation of this paper work.

We observe that such kind of inference violations is caused by lacking enough relativity analysis and corresponding measures between multiple sensitive attributes. This is not a marginal effect on the initial dataset. In real practice, we can find many public microdata sets have relativity on identifying attributes. For instance, two microdata sets published by two hospitals have large probability to share same attribute (e.g., BirthDate, Sex, Zipcode) but different sensitive attributes. Such a situation can be simulated as a single microdata set with multiple sensitive attributes.

Consequently, we propose the $(k,\ \ell)$-*anonymity* model towards preventing the inference violations on a microdata table with multiple sensitive attributes. The $(k,\ \ell)$-*anonymity* process consists of two phases:

1. Phase 1: *k-anonymization*, which is to reach the general k-anonymity requirement through generalizing identifying attributes, and
2. Phase 2: ℓ-diversity, which ensures the multiple sensitive attributes satisfy the ℓ-diversity requirement, which is different from the ℓ-*diversity* mechanism [9], since it can avoid the multiple-attribute inference violations, which cannot be avoided by the ℓ-*diversity* mechanism.

Further, we propose the optimal and approximation $(k,\ \ell)$-*anonymity* algorithms. They address the following three objectives: (I) avoiding inference violations among multiple sensitive attributes as well as on single one; (II) guaranteeing *k-anonymity* satisfied on identifying attributes and ℓ-*diversity* required on all sensitive attributes; and (III) making the data generalization cost for the process on identifying attributes as minimal as possible. Built based on previous work, we can easily observe that the advantages of this model will make it more robust on anti-inference and practicable in real microdata publishing applications.

The remainder of the paper is arranged as follows. We interpret some basic concepts and notations in Section *2* and discuss the theoretical foundation for multiple attribute inference analysis in Section *3*. Then, we define the data generalization metric and the precise and approximation $(k,\ \ell)$-*anonymity* algorithms in Section *4*. After it, we compare our model with some related work in Section *5*. Finally, we give a short conclusion in Section *6*.

2   Concepts and Notations

In this section, we introduce some basic concepts and notations for usage convenience, as well as some preliminary concepts for our further discussion later on.

We suppose individual data are recorded in a relational table, mainly including identifying attributes and sensitive attributes (here we ignore the unique identity attributes and others). The schema of a table in database refers to m-tuples $<A_1$, $A_2, \cdots, A_m>$, where A_i $(1 \leq i \leq m)$ is an attribute, and the value domain of A_i is denoted as D_i including the values that may appear in A_i. In it, an attribute is named as a *sensitive* (information) attribute ($\mathcal{SI}$) whose value for any particular individual must be kept secret from others. Beside, an instance of such a relational table is a set of m-tuples $<a_1, a_2, \cdots, a_m>$, where $a_i \in D_i$. We assume that each record refers to an individual, and a table instance is just a subset of some larger population Ω. Generally, we use table as the meaning of a table instance in this paper.

In a table, a set of nonsensitive attributes is called a *Quasi-Identifier attribute set* ($\mathcal{QI}$), if they can be joined with external information to uniquely re-identify at least one individual in Ω (with sufficiently high probability) [11,6,5]. A table $\mathcal{T}$ satisfies k-anonymity requirement if for every tuple t ($t \in \mathcal{T}$), there exist k-1 other tuples $ti_1, \cdots, ti_{k-1} \in \mathcal{T}$, such that $t[\mathcal{QI}_i] = ti_1[\mathcal{QI}_i] = \cdots = ti_{k-1}[\mathcal{QI}_i]$, where $t[\mathcal{QI}_i]$ denotes the projection of t onto the attributes in $\mathcal{QI}_i$, $\mathcal{QI}_i \in \mathcal{QI}$ (similar to $ti_j[\mathcal{QI}_i]$, $1 \leq j \leq k$-1).

A table satisfies the ℓ-*diversity* requirement (different from that in [9]) if any cluster mapping to a $\mathcal{QI}$ attributes tuple includes at least ℓ different values for all $\mathcal{SI}$ attributes. We say two $\mathcal{SI}$ attributes tuples are different if and only if on any $\mathcal{SI}$ attribute, their corresponding values are different.

Definition 1 ($\mathcal{K}$Cluster, $\mathcal{L}$Cluster). In a k-*anonymized* table, the cluster mapping to a $\mathcal{SI}$ attribute tuple is called a $\mathcal{K}$*Cluster*. The cluster mapping to a $\mathcal{QI}$ attribute tuple is called a $\mathcal{L}$*Cluster*.

For example, in Table 2, we can derive: $\mathcal{L}Cluster^1 = \{<$Penicillin, Cancer$>$, $<$Sulfur, Cancer$>$, $<$Sulfur, Cancer$>\}$. $\mathcal{K}Cluster^1 = \{<$*-*-39, F, 130**$>$, $<$*-*-39, F, 130**$>$, $<$*-*-39, F, 130**$>\}$.

Given an attribute domain, it is possible to construct a more "general" domain in a variety of ways, such as in [11,12]. We use $\preceq$ to denote the value generalization relation. For two value v_i and v_j, we take $v_i \preceq v_j$ to denote v_j is an *anonymity* (i.e. *generalization*) *form* of v_i, written as $\mathcal{AF}(v_i)$. For instances, 03-24-39 $\preceq$ 03-*-39 $\preceq$ *-*-39 holds on BirthDate, $F \preceq *$ holds on Sex. We define the *height* between two values is the distance between them in a generalization hierarchy, denoted as $\mathcal{H}(v_i, v_j)$. And the *generalization range* ($\mathcal{R}$) from v_i to v_j, written $\mathcal{R}(v_i, v_j)$, means the probability to ascertain v_i according to v_j, as $\mathcal{R}(03\text{-}24\text{-}39, 03\text{-}*\text{-}39) = 30$, $\mathcal{R}(F, *) = 2$.

Actually, it is hard to arrange generalization hierarchies with propositional generalization. For example, on (a) of Figure 1, the *generalization range* ($\mathcal{R}$) from B0 to B1 is about 30, denoted as $\mathcal{R}(B0, B1) = 30$, which means the probability to ascertain the exact value (belonged to B0) according to B1 is just about $\frac{1}{30}$, but it from B1 to B2 is 12, as $\mathcal{R}(B1, B2) = 12$, where the probability is $\frac{1}{12}$.

It is very obvious among two generalization hierarchies, such as on BirthDate and Sex. The probability to ascertain the right sex is $\frac{1}{2}$, much larger than that on the BirthDate generalization hierarchy.

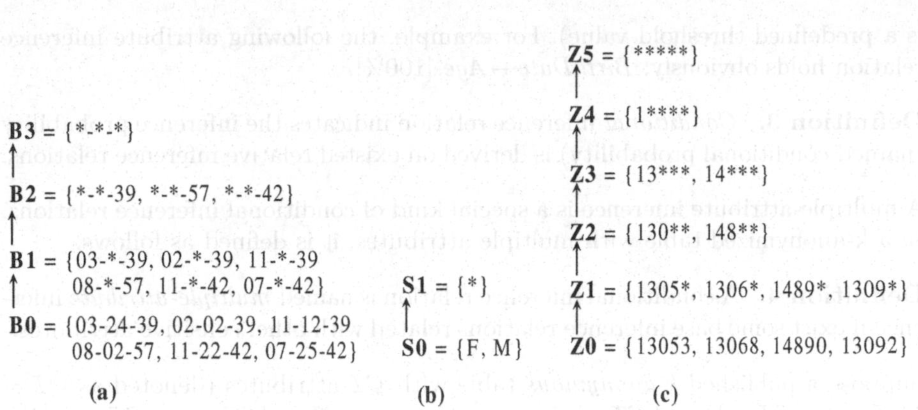

B3 = {*-*-*}
↑
B2 = {*-*-39, *-*-57, *-*-42}
↑
B1 = {03-*-39, 02-*-39, 11-*-39
              08-*-57, 11-*-42, 07-*-42}
↑
B0 = {03-24-39, 02-02-39, 11-12-39
              08-02-57, 11-22-42, 07-25-42}

(a)

S1 = {*}
↑
S0 = {F, M}

(b)

Z5 = {*****}
↑
Z4 = {1****}
↑
Z3 = {13***, 14***}
↑
Z2 = {130**, 148**}
↑
Z1 = {1305*, 1306*, 1489*, 1309*}
↑
Z0 = {13053, 13068, 14890, 13092}

(c)

Fig. 1. Domain generalization hierarchies for BirthDate (a), Sex (b), and Zipcode (c)

3 Multiple-Attribute Inference Analysis

Considering the example in Section 1, it is a special kind of inference violations if we can ascertain a narrower range of attributes values on Table 2 and 3 together. For instance, the result of the first query above can be derived as: <*-*-39, F, 130**> ∩ <*-*-*, *, 13053> = <*-*-39, F, 13053>, which is obviously narrower than any one from each single table. Such a situation can be analyzed with the Set theory and the Probability theory.

Definition 2. An *inference relation* from value v_i to v_j holds, written as $v_i \hookrightarrow v_j[p]$, if v_j can be ascertained w.r.t. v_i with *inference probability* p within $[0, 1]$.

Specifically, the inference probability from $\mathcal{AF}(v)$ to v in the *population* Ω_v, as $\mathcal{P}_{\Omega_v}(\mathcal{AF}(v) \hookrightarrow v)$, indicates the likelihood of correctly ascertaining v in Ω_v according to $\mathcal{AF}(v)$, as:

$$\mathcal{P}_{\Omega_v}(\mathcal{AF}(v) \hookrightarrow v) = \frac{|\mathcal{V}_{\Omega_v}(v)|}{|\mathcal{V}_{\Omega_v}(\mathcal{AF}(v))|} \tag{1}$$

Where $\mathcal{V}_{\Omega_v}(v) = \{v'|v' \in \Omega_v, v' \preceq^* v\}$, $\mathcal{V}_{\Omega_v}(\mathcal{AF}(v)) = \{v'|v' \in \Omega_v, v' \preceq^* \mathcal{AF}(v)\}$. Similarly, we define the inference probability within domain of v, $\mathcal{D}_v$ ($v \in \mathcal{D}_v \subseteq \Omega_v$), as: $\mathcal{P}_{\mathcal{D}_v}(\mathcal{AF}(v) \hookrightarrow v) = \frac{|\mathcal{V}_{\mathcal{D}_v}(v)|}{|\mathcal{V}_{\mathcal{D}_v}(\mathcal{AF}(v))|}$. Obviously, the following formula holds:

$$\mathcal{P}_{\Omega_v}(\mathcal{AF}(v) \hookrightarrow v) \leq \mathcal{P}_{\mathcal{D}_v}(\mathcal{AF}(v) \hookrightarrow v) \tag{2}$$

Beside, it obviously holds: $v \preceq \mathcal{AF}(v) \preceq \mathcal{AF}(\mathcal{AF}(v)) \Longrightarrow \mathcal{P}_{\Omega_v|\mathcal{D}_v}(\mathcal{AF}(\mathcal{AF}(v)) \hookrightarrow v) \leq \mathcal{P}_{\Omega_v|\mathcal{D}_v}(\mathcal{AF}(v)) \hookrightarrow v)$, where $\Omega_v|\mathcal{D}_v$ refers to any one of Ω_v and $\mathcal{D}_v$. It indicates that the more information loss on v (i.e. the less available of v), the less possible to correctly ascertain v.

In $\mathcal{T}$, we define an *attribute inference* relation from attribute $\mathcal{A}_i$ to $\mathcal{A}_j$, if for any value v_j on $\mathcal{A}_j$, there exists a value v_i on $\mathcal{A}_i$, satisfying $\mathcal{P}_{\mathcal{T}}(v_i \hookrightarrow v_j) > \alpha$ (α

is a predefined threshold value). For example, the following attribute inference relation holds obviously: $BirthDate \hookrightarrow Age$ [100%].

Definition 3. *Conditional* inference relation indicates the inference probability (named conditional probability) is derived on existed relative inference relations.

A multiple-attribute inference is a special kind of conditional inference relations. In a k-anonymized table with multiple attributes, it is defined as follows.

Definition 4. A conditional inference relation is named *multiple-attribute* inference if exist some base inference relations related with other sensitive attributes.

Suppose a published *k-anonymous* table with $\mathcal{QI}$ attributes (denoted as $\mathcal{QI} = \{\mathcal{QI}_1, \cdots, \mathcal{QI}_{m1}\}$) and $\mathcal{SI}$ attributes (denotes as $\mathcal{SI} = \{\mathcal{SI}_1, \cdots, \mathcal{SI}_{m2}\}$). We define $\Sigma(\mathcal{SI}) = \{\emptyset, \mathcal{SI}_1, \{\mathcal{SI}_1, \mathcal{SI}_2\}, \cdots\}$, which refers to the complete closure of the $\mathcal{SI}$ set in the table. $\Sigma^*(\mathcal{SI}) = \Sigma(\mathcal{SI})\text{-}\{\emptyset\}$. Now we can formally describe the multi-attribute inference on $\mathcal{QI}$ attributes as follows:

Suppose in tuple t, $\forall$ $\imath$, $\wp$, $\hbar \in \Sigma^*(\mathcal{SI})$ and the corresponding $\mathcal{KCluster}^t_\imath$, $\mathcal{KCluster}^t_\wp$, $\mathcal{KCluster}^t_\hbar$ on $\mathcal{QI}$ attributes tuples. The inference probability to ascertain an element in $\mathcal{KCluster}^t_\hbar$ (denoted as $Pr^t(\hbar)$) is defined as follows (if there exists a common subset in $\mathcal{KCluster}^t_\imath$, $\mathcal{KCluster}^t_\wp$, $\mathcal{KCluster}^t_\hbar$):

$$Pr^t(\hbar) = \frac{|\mathcal{V}_T(\hbar)|}{|\mathcal{V}_T(\cap\{\mathcal{KCluster}^t_\imath, \mathcal{KCluster}^t_\wp, \mathcal{KCluster}^t_\hbar\})|} \tag{3}$$

Multi-attribute inferences may happen if $|\cap\{\mathcal{KCluster}^t_\imath, \mathcal{KCluster}^t_\wp, \mathcal{KCluster}^t_\hbar\}|$ is enough small. In detail, if $\mathcal{KCluster}^t_\imath$ and $\mathcal{KCluster}^t_\wp$ are not published, the inference probability $Pr^t_0(\hbar)$ is $\frac{\mathcal{V}_T(\hbar)}{|\mathcal{V}_T(\mathcal{KCluster}^t_\hbar)|}$. But after they are published, the inference probability is $Pr^t(\hbar)$ as above, which is larger than $Pr^t_0(\hbar)$ if $\exists$ $e \in \mathcal{KCluster}^t_\hbar$ and $e \notin \cap\{\mathcal{KCluster}^t_\imath, \mathcal{KCluster}^t_\wp\}$. In such a view, any $\mathcal{SI}$ attribute is *relative with* other $\mathcal{SI}$ attributes.

The multi-attribute inference on $\mathcal{SI}$ attributes can be describe similarly to the above on $\mathcal{QI}$ attributes. What we need to do is to make the inference probability to ascertain an element in any $\mathcal{KCluseter}$ or $\mathcal{LCluseter}$ enough small against multi-attribute inference after publishing the table with multiple sensitive attributes. A feasible approach is to make the data (for publishing) more robust against multi-attribute inference attack while satisfy *k-anonymization* on $\mathcal{QI}$ attributes and *ℓ-diversity* on $\mathcal{SI}$ attributes. This is the main idea of our *(k, ℓ)-anonymity* model.

Definition 5. A table with multiple sensitive attributes satisfies the *(k, ℓ)-anonymity* requirement, if for each tuple t in the table, $|\mathcal{KCluster}^t| \geq k$, $Diverse(\mathcal{LCluster}^t) \geq \ell$, where *Diverse* returns the diversity degree on the cluster, simply, it returns the amount of different tuples in the cluster.

The *(k, ℓ)-anonymity* model is to implement *k-anonymization* on $\mathcal{KClusters}$ for $\mathcal{SI}$ attributes while keep $\mathcal{LClusters}$ for $\mathcal{QI}$ attributes *ℓ-diverse*. The *Diverse* measurement is on the whole $\mathcal{SI}$ tuple but not a single attribute, which effectively avoid multi-attribute inference violations.

From the perspective of using the services on published microdata tables with (k, ℓ)-*anonymity*, k and ℓ can be seen as two thresholds for deciding whether users' *query* on the table results in privacy disclosure. Still, they are not enough precise, since the different generalization ranges may exist on $\mathcal{QI}$ tuples . For example, to ascertain an exact element in two $\mathcal{KClusters}$, {<08-02-57, M, 1****, Polio>, <08-02-57, M, 1****, Polio>} and {<*-*-42, M, 13092, Stoke>, <*-*-42, M, 13092, Stoke>} are different, since $\mathcal{R}$(<08-02-57, M, 14890, Polio>, <08-02-57, M, 1****, Polio>) = $\frac{1}{10^4}$, $\mathcal{R}$(<07-25-42, M, 13092, Stoke>, <*-*-42, M, 13092, Stoke>) = $\frac{1}{30 \times 12}$. But the two clusters are treated same in the general k-anonymization process.

For more precise privacy violation discovery and prevention, we define the inference *threshold* value ∂^k and ∂^ℓ, which are used for judging whether querying $\mathcal{QI}$ attributes values w.r.t. $\mathcal{SI}$ attribute values and querying $\mathcal{SI}$ attributes values w.r.t. $\mathcal{QI}$ attribute values incur privacy disclosure respectively, i.e. on any tuple t, we should guarantee the following two conditions for (k, ℓ)-*anonymization*.

$$Pr(\Sigma^*(\mathcal{SI}^t) \rightarrow \mathcal{KCluster}^t) \leq \partial^k; \tag{4}$$

$$Pr(\mathcal{QI}^t \rightarrow \mathcal{LCluster}^t) \leq \partial^\ell. \tag{5}$$

Definition 6. In tuple t, the query on attribute tuple $\hbar$ is *anti-inference* if the following formula holds ($\partial_\hbar = \partial^k$ if $\hbar \in \mathcal{QI}$, or $\partial_\hbar = \partial^\ell$ if $\hbar \in \Sigma^*(\mathcal{SI})$):

$$Pr^t(\hbar) \leq \partial_\hbar \tag{6}$$

Suppose query on $\hbar$ in tuple t is *anti-inference*, the lower bound of $\partial_\hbar$ is:

$$Lower\_Bound^t(\partial_\hbar) = \frac{\mathcal{V}_T(\hbar)}{|\mathcal{V}_T(Cluster_\hbar^t)|}. \tag{7}$$

If query on the whole table $\mathcal{T}$ is *anti-inference*, we define the lower bounds of ∂^ℓ and ∂^k as:

$$Lower\_Bound(\partial^\ell) = Max_{\substack{\hbar \in \Sigma^*(\mathcal{SI}) \\ t \in \mathcal{T}}}(Lower\_Bound^t(\partial_\hbar)) \tag{8}$$

$$Lower\_Bound(\partial^k) = Max_{t \in \mathcal{T}}(Lower\_Bound^t(\partial_{\mathcal{QI}})) \tag{9}$$

So the inference probability on $\hbar$ in t should satisfy:

$$Pr^t(\hbar) = \frac{\mathcal{V}_T(\hbar)}{|\mathcal{V}_T(\cap_{i=1}^{sz} Cluster_{\hbar_i}^t)|} \geq Lower\_Bound^t(\partial_\hbar) \tag{10}$$

where $Cluster$ is $\mathcal{KCluster}$ if $\hbar \in \mathcal{QI}$, otherwise it is $\mathcal{LCluster}$.

The (k, ℓ)-*anonymity* model need to guarantee any queries on the published (k, ℓ)-*anonymized* microdata tables (with multiple sensitive attributes) are *anti-inference*, which is mainly achieved by the (k, ℓ)-*anonymity* algorithm (process).

4   (K, ℓ)-Anonymity Algorithm

In this section, we propose the precise and approximation algorithms and some necessary conditions to establish the (k, ℓ)-*anonymity* model. Both two algorithms are based on the *anonymization cost* notion. It is relative with the *generalization range*.

For example, the cost on 13053 $\preceq^+$ 130∗∗ for Zipcode generalization is much more than that on $F \preceq$ ∗ for Sex generalization. Another it is relative with the *generalization height*, as the cost on 13053 $\preceq^+$ 130∗∗ is more than that on 13053 $\preceq^+$ 1305∗ for Zipcode generalization, since the ratio of the *height* on the former to the whole height is $\frac{2}{5}$, but the ratio is $\frac{1}{5}$ on the latter.

Suppose the cluster size is sz, each tuple element in the cluster contains m attributes: $A_1, \cdots, A_m$. $\mathcal{H}_{A_i}$ denote the total *height* of the generalization hierarchy for attribute A_i. The corresponding attribute values in the derived cluster satisfying k-*anonymization* is denoted as $A_i^{'}$ ($i=1,\cdots,m$), which is the generalization of all A_i^j ($j=1,\cdots,sz$). The total cost on a cluster can be calculated as follows:

$$S = \sum_{j=1}^{sz}(\sum_{i=1}^{m}(c_i^j \times log(\mathcal{R}(A_i^j, A_i^{'})) \times \frac{\mathcal{H}(A_i^j, A_i^{'})}{\mathcal{H}_{A_i}})) \tag{11}$$

where factor c_i^j is used to balance the influence of *height* and *range* on the result.

We should note that the cost simply based on the amount of ∗ in the derived table is not enough meaningful (it is often used in many algorithms, such as in [2,3], discussed more in our paper [7]), since this calculation cannot present the real generalization degree and the effect after generalization.

For example, the generalization cost on <∗, 13053> (derived from <F, 13053> and <M, 13053>) and on <F, 1305∗> (derived from <F, 13053> and <F, 13054>) are equivalent according to the simple calculation on the amount of ∗. But in effect, the exact attribute value tuple can be inferred with $\frac{1}{2}$ probability, which is larger than that on the latter, which is $\frac{1}{10}$. If we just require 2-*anonymization*, then the latter is over-generalized.

Overall, the main idea of two algorithms is make the whole cost of all $\mathcal{KClusters}$ as small as possible while guarantee the corresponding $\mathcal{LClusters}$ are ℓ-*diverse*. To achieve this goal, we adopt the graph representation method in the algorithm (similar to the algorithm in [2] but ours satisfy the ℓ-*diversity* requirement for each cluster). The graph representation method takes each tuple in the table as a node in the graph. We define the weight on each edge between two nodes as the S value evaluated on the two tuples' $\mathcal{QI}$ attributes. Beside, we denote n' as the amount of clusters in the graph.

Based on the graph representation, we can calculate the temporal complexity of the precise algorithm for the optimal (k, ℓ)-anonymity solution is $\mathcal{O}(C_{n'}^{\ell} \times (\frac{n}{n'})^{\ell})$, where $C_{n'}^{\ell}$ denote the combination amount of choosing ℓ numbers from n' numbers.

We can briefly describe the algorithm for this optimal solution.

1. Building the completely equivalent set w.r.t. all $\mathcal{SI}$ attributes, suppose there are n' clusters;
2. From these n' clusters, choose k-clusters combination, and pick up just one element from each cluster, guaranteeing they are ℓ-*diverse*;
3. From the combination, choose the cluster set covering the whole table with the minimal cost.

To improve the efficiency, we propose an approximation algorithm. The process of this algorithm is (1) establish a graph with n nodes; (2) get all $\mathcal{KClusters}$ with at least ℓ totally different values on all $\mathcal{SI}$ attributes and the whole S on the graph is as small as possible.

It should be noted that the necessary conditions for successfully running this algorithm to build the (k, ℓ)-*anonymity* model are:

(1) $k \geq \ell$; and (2) $Diverse(<\mathcal{SI}_1, \mathcal{SI}_2, \cdots, \mathcal{SI}_{m2}>) \geq \ell$ $(m2 = |\mathcal{SI}|)$.

A sufficient condition is that exist $\lceil \frac{n}{\ell} \rceil$ ℓ-*diverse* clusters on $\mathcal{SI}$ tuples.

Simply, we just publish one table derived from the initial table, which includes all $\mathcal{SI}$ attributes as the maximal combination. To prevent the multi-attribute inference, the algorithm guarantees there are at least ℓ completely diverse $\mathcal{SI}$ attributes elements in any $\mathcal{LCluster}$. It is also very convenient to extend this algorithm for multiple tables publishing. What only need to do is to guarantee at least ℓ elements co-existed in all $\mathcal{LClusters}$ for different $\mathcal{SI}$ attributes or combinations mapping to the same $\mathcal{QI}$ tuple, at the same time, make the inference probability on $\mathcal{QI}$ attributes less than the corresponding threshold.

(k, ℓ)-anonymity approximation algorithm
Input: An initial table $\mathcal{T}$
Output: The (k, ℓ)-*anonymized* table for publishing.
Body:
(1)   Transform the table into a graph with isolated n nodes;
(2)   Calculate the cost S on the edges between any two nodes, where all
(3)   factors are 1 in default;
(4)   For $i = 2$ to k do
(5)   Establish clusters with at least i and at most $(2 \times k\text{-}1)$ nodes
(6)   in each of complete ℓ'-*diverse* elements $(\ell' \leq i, \ell' = \ell$ if $i = k)^1$,
(7)   while S on each cluster is as small as possible;
(8)   End For
(9)   Establish the table for publishing w.r.t to the clusters in the
(10)   graph through data generalization on $\mathcal{QI}$ attributes according
(11)   to the corresponding generalization hierarchies;

[PROOF OF COMPLETENESS AND SOUNDNESS] As this algorithm is on the graph representation, which is discussed in [2], as well as its completeness and soundness. Beside, similarly to the algorithm in [2], the temporal complexity of this algorithm is $\mathcal{O}(kn^2)$. Further, it can be easily revised to implement more comprehensive mechanisms, such as Entropy ℓ-*diversity* in [9].

---

[1] For instance, complete *2-diversity* in a cluster indicates existing a node, in which the $\mathcal{SI}$ attributes values are completely different from another one.

Suppose all factors c_i^j ($j=1,\cdots,7$, $i=1,2,3$) are 1 in default, the cost S on any edge in the graph can be calculated, as illustrated in Table 4, where the right-upper data items are the result calculated according to Formula (11), the left-lower data items are the result simply considering the amount of $*$. From Table 4, we can also observe that the S metric method is more precise than the previous one (i.e. the amount of $*$ [2,3]).

Table 4. Cost S on the edge between any two nodes

	1	2	3	4	5	6	7
1	\	5.76	5.76	18.55	11.18	13.02	13.02
2	4	\	3.92	18.55	13.02	13.02	13.02
3	4	2	\	18.55	13.02	13.02	13.02
4	8	8	8	\	7.36	17.85	17.85
5	4	6	6	4	\	12.33	12.33
6	6	6	6	8	6	\	3.92
7	6	6	6	8	6	2	\

Then we can build the optimal $(2, 2)$-*anonymized* table (in Table 5) and the approximation $(2, 2)$-*anonymized* table (in Table 6) according to the precise and approximation algorithms respectively.

Table 5 (left) & **Table 6 (right).** Optimal&approximate $(2, 2)$-*anonymized* table

	Bi.Da.	S.	Zip.	Allergy	Disease
1	*-*-*	*	130**	Penicillin	Cancer
6	*-*-*	*	130**	No	Stoke
2	*-*-*	*	130**	Sulfur	Cancer
3	*-*-*	*	130**	Sulfur	Cancer
5	*-*-*	*	130**	Penicillin	Polio
4	*-*-*	M	1****	No	Polio
7	*-*-*	M	1****	Penicillin	Stoke

	Bi.Da.	S.	Zip.	Allergy	Disease
1	*-*-*	*	1****	Penicillin	Cancer
4	*-*-*	*	1****	No	Polio
7	*-*-*	*	1****	Sulfur	Stoke
2	*-*-*	*	130**	Sulfur	Cancer
5	*-*-*	*	130**	Penicillin	Polio
3	*-*-*	*	130**	Sulfur	Cancer
6	*-*-*	*	130**	No	Stoke

5 Compare with Related Work

Publishing data about individuals without revealing sensitive information about them is an important problem. Instead of simple data suppression, k-*anonymity* is proposed for more usage convenience.

It is proven that the k-*anonymity* problem is $\mathcal{NP}$-*hard* even when the attribute values are ternary [10,2,1]. An algorithm with $\mathcal{O}(k)$-*approximation* is proposed for this problem [2]. Individual-specific k mechanism is discussed in [8], which is an improvement of the current table-level k mechanism. Violation of k-anonymity occurs when a particular attribute value of an entity can be determined to be among less than k possibilities by using the views together with the schema information of the private table [14]. It is proven that whether a set of views occur privacy disclosure for *k-anonymity* is also a computationally *hard* problem [14]. Formal research on data utility is studied in [4], towards increasing the utility through injecting additional information on k-anonymized data. Through observation on the original dataset, [3] gives a way to clustering records more useful information, through deleting or non-releasing some records (using a ϵ fraction). The approach to precisely measure the privacy inference violations based on probability theory is discussed in our previous paper [15].

The *ℓ-diversity* [9] mechanism is the work on how to prevent the inference on $\mathcal{SI}$ attributes values when the corresponding $\mathcal{QI}$ attributes values are known. Its main work just focus on single sensitive attribute. In many situations, multiple sensitive attributes are always existed in the same table. So our (k, ℓ)-*anonymity* model and the algorithms are motivated to solve such a kind of challenges, with advantages to some extent over the work in [9]. Further, our algorithm is based on graph representation, which is similar to the mechanism in [2], in which, the algorithm to implement *k-anonymity* is also based on graph. In [2], the algorithm needs $\mathcal{O}(kn^2)$. In our algorithm, the components are directly accomplished by the aggregation iteration process from 2 to k while guarantee the corresponding $\mathcal{LClusters}$ are *ℓ-diverse* after iteration. The complexity of the whole process is $\mathcal{O}(kn^2)$. Beside, as above discussed, the mean complexity of our precise algorithm is $\mathcal{O}(C_{n'}^{\ell} \times (\frac{n}{n'})^{\ell})$. The worst situation is about $\mathcal{O}(n^{\ell})$. If ℓ is 2 or 3, it is very easy to find the optimal solution for data publishing with the equivalent time cost comparing with the approximation algorithm and even others.

From the above discussion, we can conclude that they are three privacy attacks that can be prevented by our model, which can be describe as follows (1) Inferring $\mathcal{QI}$ attributes when single $\mathcal{SI}$ are known; (2) Inferring $\mathcal{SI}$ attributes when $\mathcal{QI}$ are known; (3) Inferring $\mathcal{QI}$ attributes when multiple $\mathcal{SI}$ (in one or several tables) are known. As we know, the (k, ℓ)-*anonymity* model is the first study on inference prevention among multiple sensitive attributes for privacy protection in microdata publication.

6 Conclusion

In conclusion, the primary work in the paper is the (k, ℓ)-*anonymity* model. It consists of two processes: *k-anonymization* on $\mathcal{QI}$ attributes and *ℓ-diversity* on $\mathcal{SI}$ attributes. With the algorithm, we explicitly achieve the *ℓ-diversity* characteristic while implicitly capture *k-anonymization* on $\mathcal{QI}$ attributes. Overall, the advantages of our (k, ℓ)-*anonymity* model include: (1) It is available for more general dataset publishing, especially when the dataset contains multiple sensitive attributes that should be protected against inference; (2) It supports multi-attribute inference violation prevention, which is not achieved yet by previous related techniques; (3) The cost metric on data generalization is more reasonable comparing to some existed measurements.

In future, we will continue study the model on more efficient algorithms supporting more individuals' flexibility, as our preliminary research in [8].

References

1. Aggarwal, G., Feder, T., et al.: Anonymizing Tables for Privacy Protection (2004), http://theory.standford.edu/~rajeev/privacy.html
2. Aggarwal, G., Feder, T., et al.: Approximation Algorithms for K-Anonymity. Journal of Privacy Technology (November 2005)
3. Aggarwal, G., Feder, T., et al.: Injecting Utility into Anonymized Datasets. In: Proc. of PODS 2006, June 2006, pp. 153–163 (2006)
4. Kifer, D., Gehrke, J.: Injecting Utility into Anonymized Datasets. In: Proc. of SIGMOD 2006, June 2006, pp. 217–229 (2006)
5. LeFevre, K., DeWitt, D.J., Ramakrishnan, R.: Multidimensional K-Anonymity. Technical Report (2005), www.cs.wisc.edu/techreports/2005/
6. Lefevre, K., DeWitt, D.J., Ramakrishnan, R.: Incognito: Efficient Full-Domain K-Anonymity. In: Proc. of SIGMOD 2005 (June 2005)
7. Li, Z., Zhan, G., Ye, X.: Towards a More Reasonable Generalization Cost Metric For K-Anonymization. In: Bell, D.A., Hong, J. (eds.) BNCOD 2006. LNCS, vol. 4042, Springer, Heidelberg (2006)
8. Li, Z., Zhan, G., Ye, X.: Towards a Microdata Anonymization Model with Individual-Defined Ks. In: Bressan, S., Küng, J., Wagner, R. (eds.) DEXA 2006. LNCS, vol. 4080, pp. 883–893. Springer, Heidelberg (2006)
9. Machanavajjhala, A., Gehrke, J., Kifer, D.: ℓ-Diversity: Privacy Beyond K-Anonymity. In: Proc. of ICDE 2006 (2006)
10. Meyerson, A., Williams, R.: On the Complexity of Optimal K-Anonymity. In: Proc. of PODS 2004 (2004)
11. Samarati, P., Sweeney, L.: Protecting Privacy when Disclosing Information: K-Anonymity and Its Enforcement Through Generalization and Suppression, Technical Report, SRI Computer Science Lab. (1998)
12. Sweeney, L.: Achieving K-Anonymity Privacy Protection Using Generalization and Suppression. Intl. Journal on Uncertainty, Fuzziness and Knowledge-based Systems 10(5), 571–588 (2002)
13. Sweeney, L.: K-Anonymity: A Model For Protecting Privacy. Intl. Journal on Uncertainty, Fuzziness and Knowledge-based Systems 10(5), 557–570 (2002)
14. Yao, C., Wang, X.S., Jajodia, S.: Checking for K-Anonymity Violation by Views. In: Proc. of VLDB 2005 (2005)
15. Ye, X., Li, Z., Li, Y.: Capture Inference Attacks for K-Anonymity with Privacy Inference Logic. In: Kotagiri, R., Radha Krishna, P., Mohania, M., Nantajeewarawat, E. (eds.) DASFAA 2007. LNCS, vol. 4443, Springer, Heidelberg (2007)

Exposing Homograph Obfuscation Intentions by Coloring Unicode Strings

Liu Wenyin, Anthony Y. Fu, and Xiaotie Deng

Department of Computer Science, City University of Hong Kong, Hong Kong SAR., China
{csliuwy, anthony, csdeng}@cityu.edu.hk
http://antiphishing.cs.cityu.edu.hk

Abstract. Unicode has become a useful tool for information internationalization, particularly for applications in web links, web pages, and emails. However, many Unicode glyphs look so similar that malicious guys may utilize this feature to trick people's eyes. In this paper, we propose to use Unicode string coloring as a promising countermeasure to this emerging threat. A coloring algorithm is designed and prototyped to assign colors to a set of required languages/scripts such that each language/script is displayed uniquely in color, while the color difference among different languages is maximized. Based on that, we proposed both fixed and adaptive coloring schemes to render Unicode strings in weblinks and documents so as to distinguish mixed Unicode characters from different language/script groups and vividly illustrate potential Homograph Obfuscation intentions. Our user study shows that it is helpful to remind end users of weirdly displayed strings.

1 Introduction

Universal Character Set (UCS) is a union of characters/symbols of most languages in the world. It is more and more popular and important in our daily life. We use it to compose web links, web pages, and emails. However, there are many similar characters in UCS, as shown in Figure 1. This could cause a severe web security problem. Malicious people can use various similar characters from different languages to mimic "citibank", as shown in Figure 2. A real case is that another "paypal.com" (in which the second 'a' is U-0430 in Unicode) was successfully registered in 2005. Unwary users (and even expert users) could be easily spoofed by such phishing scam to expose their security sensitive information, such as credit card number, password, etc. This attack is referred to as homograph attack [4] and can also be expanded to webpages and emails, in which similar characters can be used to generate content with the same reading effect to human users but escapable of content based phishing detection and spam filtering. In this paper, we address this problem and propose a method based on coloring to differentiate "abnormal" weblinks, webpages, and emails from relatively "normal" ones.

We have ever briefly mentioned in [3] to color the Unicode strings to help end users find out these weird strings, and therefore, to relieve the threat of this kind of phishing attack to end users. In this paper, we fully explore this idea and propose a fixed coloring scheme and an adaptive coloring scheme. Both schemes are based on

Y. Zhang et al. (Eds.): APWeb 2008, LNCS 4976, pp. 275–286, 2008.
© Springer-Verlag Berlin Heidelberg 2008

the same coloring algorithm we proposed, which selects colors from a set of available colors and assigns them to a set of required languages/scripts such that each language/script is displayed uniquely in color, while the color difference among different languages is maximized. As the characters in Unicode belong to different language regions [6], a straightforward idea is to assign the characters in each language region with an identical color, as used in Quero Toolbar [10]. However, our study shows that there are actually similar characters from the same language, as shown in Figure 1(a). Nevertheless, some languages are sharing with the same set of characters, e.g., basic Latin characters are used by French, German, and Dutch, etc. Hence, we analyze the UCS first and design a grouping mechanism to classify the characters, and assign each group of characters with one specific color. The additional color property could provide users with more information for understanding Internationalized Resource Identifiers (IRIs) or document context semantics. According to our proposed coloring scheme, we prototyped this idea and conducted user study on this Unicode string coloring scheme. Our user study shows that well designed coloring algorithms can warn the users better against homograph obfuscation.

(a) From Latin characters	A	Ā	Ă	Ą
	0041	0100	0102	0104
(b) From CJK characters	行	行		
	884C	FA08		
(c) From CJK characters	银	銀	铌	銀
	94F6	9280	9512	92C3

Fig. 1. Examples of visually similar characters in Unicode

The rest of this paper is organized as follows. Section 2 introduces related work. In Section 3, we discuss the Unicode character coloring issues, which include character grouping, coloring palette construction, and the two coloring schemes. Section 4 demonstrates our prototype tool, Unicode String Illustrator. Section 5 presents our user study, which shows the effectiveness of the proposed coloring schemes. Finally, we conclude this paper in Section 6 and discuss future work in Section 7.

2 Related Work

2.1 Anti-phishing

The Web facilitates our daily lives, but also causes a lot of security problems, including DoS, worms, DNS poisoning, and router attacks. In recent years, phishing has emerged very quickly as a new but serious threat. It was reported that U.S. consumers lost 630 million US dollars over the past two years to fraudulent phishing e-mail scams, c.f. Consumer Reports [12]. The amount of money lost to online banking fraud in the UK increased 55 percent to £22.5m in the first half of 2006, according to figures from banking industry body, c.f. Apacs [11]. As a consequence, phishing becomes a hot topic to discuss in both the computer security society and the law enforcement society.

Most frequently used anti-phishing strategies focus on toolbars or extensions of Web browsers (e.g., Firefox and MS IE7). These can be classified into 5 main categories by their methodologies.

(1) *Black/white list.* The representative black/white list based systems include Phish-Tank SiteChecker, Google Safe Browsing, FirePhish, and CallingID Link Advisor, etc. The biggest challenge of this method is to maintain the black/white list. To maintain the black list, we need the help from the user community. A typical way is to collect the phishing reports from users and then process them by anti-phishing analysts (employed or volunteered). The drawback is that not all phishing weblinks will be reported and not all anti-phishing analysts are reliable and professional.

(2) *Reputation scoring.* This technique can use reputation scores either reported from the anti-phishing community or computed from the given webpages, e.g., WOT and iTrustPage. Therefore the reliability and reputation scoring algorithm is crucial for this technique.

(3) *Malware detection.* Malware is not phishing but it can be used to assist phishing. With the development of the anti-phishing technologies, ordinary and old phishing methods fail to work and thereby more phishers could turn to use malware to assist. The representative product is Finjan.

(4) *Relevant domain name suggestion.* This technique suggests users the most relevant legitimate domain names when they are accessing the web, e.g. SpoofStick. The biggest challenge is to recognize the accessing webpage and make reasonable suggestions.

(5) *Personalized visual indicator.* This technique is like posting a personalized stick note to the legitimate websites, such that we can recognize it when we access it again. The representative product is TrustBar. Such system assumes and relies on that the users can assign indicators correctly to real websites and always keep in mind that webpages without such indicator have phishing potentials.

These tool bar based strategies are end-user level solutions only and heavily rely on the end-users' education level, experience, and vigilance. However some researchers have shown that security tool bars do not effectively prevent phishing attacks [14]. In addition, they also bring inconvenience to the end-users, including especially, too many false alarms. In [1], Liu et al proposed an active and more comprehensive anti-phishing strategy, which determines suspicious links from emails (at end-users' email readers/senders, or enterprise email servers, e.g., bounced-backed emails) and all possible cousin domain names. The webpage pages at the suspicious links are taken and compared with the protected webpages. Visually similar webpages are reported. Especially, enterprise users can use this strategy for early detection of possible phishing Web pages and prepare themselves to clear potential security threats in their e-commerce environments without inconveniencing any end user.

2.2 Homograph Obfuscation

Gabrilovich et al demonstrated homograph attacks [4] that visually identical weblinks can be created in International Domain Name (IDN). Phishers can simply find similar characters from UCS to replace certain ones in the legitimate web link to carry out the attack. Major Web browsers, such as Microsoft IE7 and Firefox, can display an IDN in Punycode (a translation of Unicode into the ASCII plane) in the address bar. However, many users prefer to display their IDNs in their own language scripts. Because Punycode is an ASCII based encoding form rather than a user understandable form. It is to make IDN compatible to ASCII based DNSes. However, it is very difficult for a human to understand the meanings of the Punycode strings. Homograph attack is kind of Unicode attack [3], however, Unicode attack refers to more than visually similar web links but also semantically similar web links, as shown in [13]. There are also two tools, IRI/IDN SecuChecker [3] and REGAP [2], developed to fight against Unicode attack. The Unicode attack detection tools can help ICANN [5] and DNS registrars to detect malicious registration applications.

2.3 Coloring Techniques

The standard color space is represented with the RGB system. Other color systems (such as Non-linear R'G'B', HSV, CIE $L^*a^*b^*$ and CIE $L^*u^*v^*$, YCbCr, YIQ, and YUV etc.) have direct mappings to the RGB system. Each color system has its advantage for a specific representation. Although Non-linear R'G'B', YUV, and CIE $L^*u^*v^*$ are considered as better for differentiating colors, Riemersma's investigation [8] shows that all of them have disadvantages and an approximated color distance assessment formula based on the RGB system is proposed, as shown in Eq. (1). Since about one out 12 people has color deficiency problem in the world, we need to consider using some color safe strategy in our Unicode coloring scheme and we found the web safe color palette proposed by BT (British Telecommunications) [7] is a quite good match for this requirement.

$$\bar{r} = \frac{C_{1,R} + C_{2,R}}{2}$$

$$\Delta R = C_{1,R} - C_{2,R}$$

$$\Delta G = C_{1,G} - C_{2,G}$$

$$\Delta B = C_{1,B} - C_{2,B}$$

$$\Delta C(C_1, C_2) = \sqrt{(2 + \frac{\bar{r}}{256}) \times \Delta R^2 + 4 \times \Delta G^2 + (2 + \frac{255 - \bar{r}}{256}) \times \Delta B^2}$$

(1)

3 Unicode String Coloring

3.1 Criteria for Coloring Scheme Design

Coloring Unicode strings is not as straightforward as assigning colors randomly to each language/scripts type. A good design should have the following features.

Differentiability

Visually similar characters could be identifiable from one another in the same context. A suitable color distance is necessary to measure the human's perception to differentiate colors. We can use such measure to generate a reasonable color palette to use. By considering color-blindness users, we can use a "color-blindness safe palette" as the candidate color set. In addition, we should construct a color palette with as many colors as possible.

Scalability

UCS is a developing and growing repertoire, which means the Unicode consortium [6] is continuously adding more scripts to UCS to make it more "complete" in each new version. Hence, the scalability of the coloring scheme is important. We indeed hope the proposed coloring scheme is still workable after the Unicode consortium publishes new versions of UCS.

Readability/Usability

The coloring scheme is to vividly show the mixture of different languages in certain context. We need to make the coloring result comfortable for users. As we know, too colorful or contrastive images could make readers feel uneasy. Therefore, one big concern of the coloring scheme is the readability/usability problem.

3.2 Unicode Character Grouping

We first should divide all the Unicode characters into different language/script groups. Phishers may carry out Unicode attack by employing different language scripts into the same context. "www.citibank.com" is an example combining both Latin Basic and Cyrillic symbols. The highlighted "c" in this IDN is Cyrillic while the others are Latin Basic. Hence, the motivation of Unicode grouping is to list all of the elementary groups that we should assign colors to, such that we can finally distinguish how many and what languages/scripts are used in a particular Unicode string.

It is reasonable to display characters of different languages/scripts in different colors. The Unicode consortium classified the Unicode characters into 11 region based groups and 122 subgroups based on different language scripts. We can assign one color to each subgroup. The complete list is available at [13]. However, these subgroups do not represent the language / script difference sufficiently. Particularly, for instance, when we want to differentiate Simplified Chinese, Traditional Chinese, Japanese, and Korean in a Unicode string, we found that some of these characters are actually used by two or more languages. According to [6], the Unicode consortium merges all of the Chinese style characters into CJK Ideographs. If we simply present CJK Ideographs into an identical color, we cannot differentiate these four scripts. Therefore we need more specific level subgroups to replace CJK Ideographs. We denote the four types of scripts as S, T, J, and K, and we can generate 15 combinations to form subgroups from CJK Ideographs: S, T, J, K, ST, SJ, SK, TJ, TK, JK, STJ, STK, SJK, TJK, STJK, e.g., ST denotes the group containing only Simplified Chinese and Traditional Chinese but no Japanese and Korean.

3.3 Coloring Palette Construction

Before we start the coloring process, we need to know how many and what colors are available for us to use. Such set of colors is referred to as the color palette. It should contain as many differentiable colors as possible. We also design the palette as color-blindness safe; because 5% to 8% of men and 0.5% of women in the world are born colorblind (We limit the discussion to protans (red weak) and deutans (green weak) because they make up 99% of this group). The department for older and disabled customers of British Telecom (BT) proposed a list of safe web colors for color-deficient vision in [7], which consists of 216 colors. We can simply choose colors from these colors to generate our coloring palette. In order to provide the best differentiability (as mentioned in Section 1.4), we have to rank the available colors in the palette from the most distinguishable to the least and we can use the ones having inter-distances greater than a threshold. This problem is equal to the Maximum Clique Problem [9], which is NP-Complete. To simplify the computation, we use a greedy algorithm (it is worth to note that greedy algorithm does not find the global optimal solution in most times, however, our experiment with approximated optimal algorithm [15] shows similar result to the greedy algorithm but with worse computational performance). In the greedy algorithm, we first choose the color which is the most distant to the background and foreground, and then recursively choose the one which is the most distant to the chosen ones. Finally, we obtain the color palette with the colors ranked in the descendant order of preference. Both the algorithm and the coloring palette are available at [13]. Figure 3 (a) shows some samples from the coloring palette. The text in each color (e.g., in the nth row) shows its RGB value and the minimum color distance among all of the first n colors. E.g., "0 255 51 MinDistance=516" (green color in the 4th row) means the minimum distance among the first four colors is 516. The first color in the list does not have minimum distance to any previous color and is therefore considered as infinite.

3.4 Unicode String Coloring

Given a color palette, the Unicode string coloring algorithm assigns colors in the palette to each language/script subgroup of Unicode where we should follow the criteria in Section 1.4 for best of user experience. We propose two coloring schemes: Fixed Coloring Scheme and Adaptive Coloring Scheme.

Fixed Coloring Scheme

In the fixed coloring scheme, each language/script subgroup is assigned to a fixed color. Therefore, users will have a straightforward association between language and its dominant color, e.g., red to Chinese. The fixed coloring scheme is suitable for coloring IDNs and IRIs. Besides, we can also color the ASCII letters using the main foreground color (e.g., black) such that the highly-reputable websites, such as www.hsbc.com and www.citibank.com, are displayed as they are. It is also acceptable if we maintain a white list of domain names and not color (or just color using the main foreground color) the URLs in the white list. Figure 2 shows some examples of colored "citibank" using the fixed coloring scheme. In these examples, we use

UC-SimList_v0.9 [3], which is now referred to as Similar Unicode Character Index (SUCI), to find the similar characters to "citibank". UC-SimList_v0.9 contains all of the visually similar character groups with 90% or higher similarity using the pixel-overlapping assessment method.

Language / Script Subgroup	"c"	"i"	"t"	"i"	"b"	"a"	"n"	"k"
Basic Latin	c 0063	i 0069	t 0074	i 0069	b 0062	a 0061	n 006E	k 006B
Latin-1 Supplement		¡ 00A1		¡ 00A1				
Latin Extended-A								ķ 0137
Latin Extended-B					b 0185			ƙ 0199
IPA Extensions					ɓ 0253			
Greek and Coptic	c 03F2	ί 03AF		ί 03AF				
Cyrillic	с 0441	і 0456		і 0456		а 0430	п 043F	
Latin Extended Additional					ḅ 1E05	ạ 1EA1	ṇ 1E47	ḳ 1E33
Number Forms	c 217D	i 2170		i 2170				
Halfwidth and Fullwidth Forms	c FF43	i FF49	t FF54	i FF49	b FF42	a FF41	n FF4E	k FF4B
Mixed Example	c FF43	i 2170	t 0074	ί 03AF	ɓ 0253	а 0430	ṇ 1E47	ƙ 0199

Fig. 2. Examples of the fixed coloring scheme's results

Adaptive Coloring Scheme

Sometimes, we may want to be aware of the diversity of language/script usage in a given basic background and foreground, and do not really care about the association between language/scripts and colors. Hence, we can use the given background and foreground colors as a start to calculate the remaining colors to generate an adaptive coloring palette in real-time with the algorithm in [13] and select as many colors as needed from the palette.

We can use it to illustrate the diversity of language/script usage in webpage blocks or document context. Figure 3 demonstrates some examples colored by the adaptive coloring scheme. Obviously, we can understand the diversity of language/script usage after coloring. Suppose we have one Unicode string composed by three languages, as shown in Figure 3(a) (white as the background color and black as the foreground color; the most frequently used language/script is assigned the foreground color). The result after rendering is shown in Figure 3(b). The adaptive coloring palette is in the column on the right side. Figure 3(c) and Figure 3(d) show more results using different color palettes.

(a) Original context	**Phishing is Harmful!** 网钓很危险！ フィシングはあぶないです！	
(b) BG=White, FG=Black	**Phishing is Harmful!** 网钓很危险！ フィシングはあぶないです！	255 255 255 0 0 0 MinDistance=764 204 0 255 MinDistance=517 0 255 51 MinDistance=516 255 102 0 MinDistance=431 0 153 255 MinDistance=407
(c) BG=Black, FG=Yellow	Phishing is Harmful! 网钓很危险 フィシングはあぶないです	0 0 0 255 255 0 MinDistance=649 0 204 255 MinDistance=579 255 0 255 MinDistance=569 0 204 0 MinDistance=408 0 0 255 MinDistance=402
(d) BG=LightYellow, FG=Red	**Phishing is Harmful!** 网钓很危险！ フィシングはあぶないです！	255 255 204 255 0 0 MinDistance=585 0 51 255 MinDistance=579 0 204 0 MinDistance=526 0 255 255 MinDistance=408 0 0 0 MinDistance=402

Fig. 3. Example results of the adaptive coloring scheme

4   Unicode String Coloring Tool

We prototyped the Unicode string coloring scheme and call it Unicode String Illustrator. It contains two parts, IRI Illustrator (the fixed coloring scheme) and Context Illustrator (the adaptive coloring scheme).

4.1   IRI Illustrator

IRI Illustrator is a web browser plug-in. It replaces the function of the address bar. When users are accessing (by clicking or copy-and-pasting) phishing IRI(s), the language/scripts' property will be illustrated by different colors.

4.2   Context Illustrator

Context Illustrator is a tool to present the language/script's diversity in a context. The background color and foreground color are configurable. The coloring palette is automatically generated using the greedy algorithm. Figure 4 shows the coloring

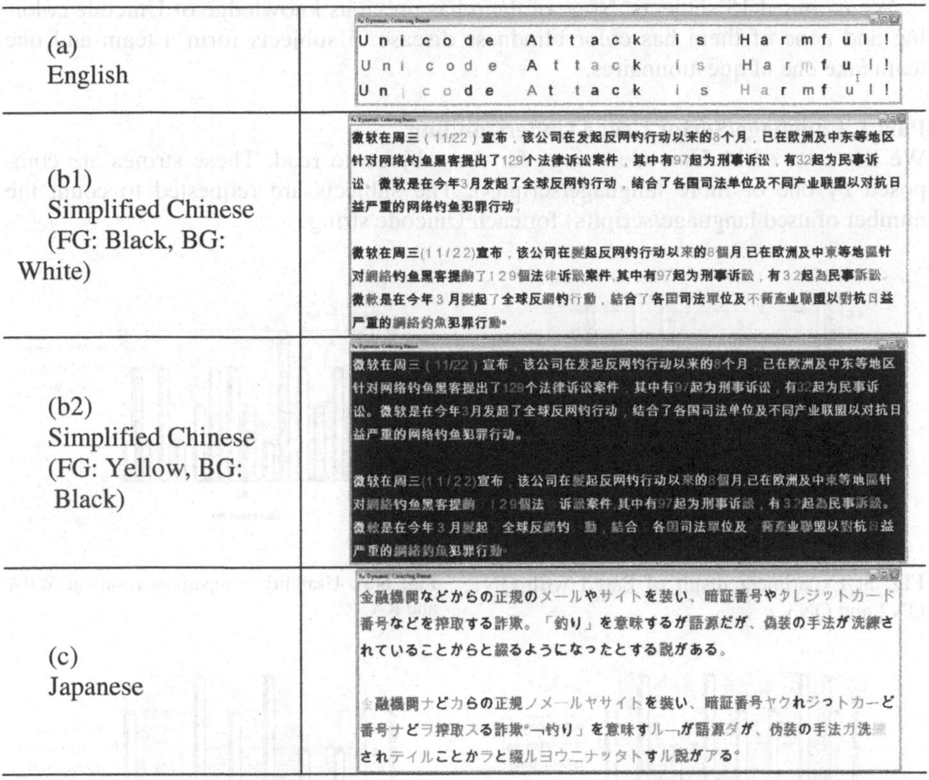

(a) English	
(b1) Simplified Chinese (FG: Black, BG: White)	
(b2) Simplified Chinese (FG: Yellow, BG: Black)	
(c) Japanese	

Fig. 4. Demo of the Context Illustrator

results of the Context Illustrator processing English, Chinese, and Japanese context with different background and foreground configurations.

We can see that both the IRI Illustrator and the Context Illustrator can render Unicode strings into a style that is effective to observe character-based obfuscation.

5 User Study

We have done a user study to show the effectiveness of our coloring scheme through user study, which consists of two parts: Part I is to test the effect of Unicode coloring on the subjects' ability to recognized different languages; and Part II is to compare the usability of the two coloring approaches. Three questionnaires (QN1, QN2, and QN3) are used for both parts, and all questionnaires are available at [13]. QN1 tests the white/black coloring approach (WBA). In this approach, characters are always in black and the background is always white. QN2 tests the random coloring approach (RA). In this approach we assign a random color to each language/script group and display the characters with their assigned colors. QN3 tests our fixed and adaptive coloring approach (FAA, which is the coloring approach we presented in Section 3 and demonstrated in Section 4).

We recruited 15 subjects. None of them has previous knowledge of Unicode coloring and none of them has color blindness disease. 5 subjects form a team and one team take one of questionnaires.

Part I, Language/Script Usage Understanding
We list a set of 11 Unicode strings for the subjects to read. These strings are composed by one or more language/script(s). The subjects are requested to count the number of used language/script(s) for each Unicode string.

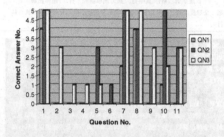

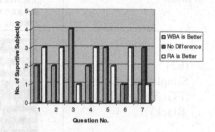

Fig. 5. Experiment result of Part I with QN1, QN2 and QN3

Fig. 6(a). Usability comparison result of WBA and RA

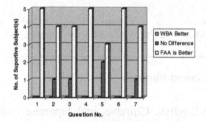

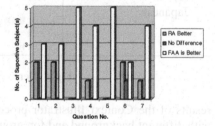

Fig. 6(b). Usability comparison result of WBA and FAA

Fig. 6(c). Usability comparison result of RA and FAA

Figure 5 shows the experiment result. Among the 11 (number of Unicode strings) * 5 (number of subjects) = 55 answers, there are in total 11 questions correctly answered by the subjects of QN1, and 27 for QN2, and 29 for QN3. We can see that in all questions, subjects from QN2 and QN3 can do much better than the ones in QN1, and subjects from QN3 can do slightly better than the ones from QN2. Hence, we can conclude that coloring (RA and FAA) can do much better than none-coloring (WBA) on users' language/script(s) usage understanding. We also conclude that FAA can do very similar to RA but slightly better. Our explanation is that we did not use many colors for all the 11 questions, so even random selection of colors can make a good result. Hence, FAA cannot show much advantage to RA. However, when a Unicode string contains many colors, FAA should perform much better than RA.

Part II, Usability Comparison for Different Coloring Approaches

We request the subjects to compare the usability of different coloring approaches. QN1 compares WBA with RA, QN2 compares WBA with FAA, and QN3 compares RA with FAA.

Figure 6 shows the experiment result. Figure 6(a) shows that WBA is relatively similar to RA. The reason could be that RA does not consider the background and the used colors. Hence RA could assign similar color(s) to other language/script(s) or background. Figure 6(b) shows that FAA is much better than WBA, and Figure 6(c) shows that FAA is better than RA. Hence, FAA is better than both WBA and RA. Therefore we conclude that a well designed coloring approach can help end users better understand language/script(s) usage in a Unicode string.

6 Conclusion

In this paper, we proposed to use Unicode string coloring as a solution to character-level obfuscation. We proposed two coloring schemes: Fixed coloring and Adaptive coloring. The fixed coloring scheme assigns a specific color to each language/script as its basic color to satisfy the security requirement and is therefore a good match for web address coloring. The adaptive coloring scheme calculates a coloring palette based on the given background, foreground colors, and the language/script composition in real-time. The purpose of this coloring scheme is to provide an easy way for users to understand the language/script diversity in a Unicode string context. We prototyped the two coloring schemes in the Unicode String Illustrator, which contains the IRI Illustrator and the Context Illustrator. Our user study shows that even though coloring is a generally useful method to show the mixture of different languages/scripts and is useful for revealing the obfuscation intentions, only well designed coloring schemes can keep high usability and effectiveness for reading and understanding the colored Unicode strings.

7 Future Work

We have demonstrated that Unicode string coloring can help illustrate language/script(s) usage in IRIs and text paragraphs. However, it cannot illustrate the phishing intention of strings such as "www.citi-bank.com". We plan to add this coloring scheme into other phishing detection systems, such as REGAP [2], to make detection of such kind of phishing intentions possible. In Unicode String Illustrator, Ver. 1.0, we do not use all of the 15 subgroups of CJK. We only considered two subgroups: Simplified Chinese and the rest CJK characters. We leave the more thorough grouping work in our future work. In fact, languages/scripts that do not have similar characters can also share the same color. It is also of our future interest to save the number of colors by doing so.

Acknowledgement

The work described in this paper was fully supported by a grant from the Research Grants Council of the Hong Kong Special Administrative Region, China [Project No.

CityU 117907] and the National Grand Fundamental Research 973 Program of China under Grant No. 2003CB317002. We would like to thank Yeung Wan Hang, Chau Kin Man, and Mak Sheung Man for their help in the experiments and user studies in this project.

References

1. Liu, W., Deng, X., Huang, G., Fu, A.Y.: An Anti-Phishing Strategy based on Visual Similarity Assessment. IEEE Internet Computing 10(2), 58–65 (2006)
2. Fu, A.Y., Deng, X., Liu, W.: REGAP: A Tool for Unicode-based Web Identity Fraud Detection Journal of Digital Forensic Practice 1(2), 83–97.(Special Edition on Anti-phishing and Online Fraud) (2006)
3. Fu, A.Y., Deng, X., Liu, W., Little, G.: The Methodology and an Application to Fight against Unicode Attacks. In: Proceedings of SOUPS 2006, CMU, Pittsburgh, USA (July 2006)
4. Gabrilovich, E., Gontmakher, A.: The Homograph Attack. Communications of the ACM 45(2), 128 (2002)
5. ICANN, http://www.icann.org
6. Unicode Consortium, The Unicode Character Code Charts By Script, http://www.unicode.org/charts
7. BTPLC, Safe Web Colours for Colour-Deficient Vision, http://www.btplc.com/age_disability/technology/RandD/colours/colours1.htm
8. Riemersma, T.: Colour Metric, http://www.compuphase.com/cmetric.htm
9. Karp, R.: Reducibility among Combinatorial Problems. In: Proceedings of Symposium on the Complexity of Computer Computations (1972)
10. Krammer, V.: Phishing Defense against IDN Address Spoofing Attacks. In: Proceedings of the 4th Annual Privacy Security Trust Conference 2006 (PST 2006), October 2006, pp. 275–284. ACM Press, New York (2006)
11. The UK Payment Association, http://www.apacs.org.uk
12. Computer Times, $8 Billion Lost to Online Scams, http://www.computertimes.com/oct06Articfle8BillionLostToOnlineScams.htm
13. CityU Coloring Palette, http://antiphishing.cs.cityu.edu.hk/ColoringScheme
14. Wu, M., Miller, R.C., Garfinkel, S.L.: Do Security Toolbars Actually Prevent Phishing Attacks? In: Proceedings of SIGCHI 2006, pp. 601–610 (2006), http://groups.csail.mit.edu/uid/projects/phishing/chi-security-toolbar.pdf
15. Macambira, E.M.: An Application of Tabu Search Heuristic for the Maximum Edge-Weighted Subgraph Problem. Annals of Operations Research 117, 175–190 (2002)

On the Complexity of Restricted k-anonymity Problem*

Xiaoxun Sun[1], Hua Wang[1], and Jiuyong Li[2]

[1] Department of Mathematics & Computing
University of Southern Queensland, QLD, Australia
{sunx, wang}@usq.edu.au
[2] School of Computer and Information Science
University of South Australia, Adelaide, Australia
jiuyong.li@unisa.edu.au

Abstract. One of the emerging concepts in microdata protection is k-anonymity, introduced by Samarati and Sweeney. k-anonymity provides a simple and efficient approach to protect private individual information and is gaining increasing popularity. k-anonymity requires that every tuple(record) in the microdata table released be indistinguishably related to no fewer than k respondents. In this paper, we introduce two new variants of the k-anonymity problem, namely, the *Restricted k-anonymity problem* and *Restricted k-anonymity problem on attribute* (where suppressing the entire attribute is allowed). We prove that both problems are $\mathcal{NP}$-hard for $k \geq 3$. The results imply the main results obtained by Meyerson and Williams. On the positive side, we develop a polynomial time algorithm for the *Restricted 2-anonymity problem* by giving a graphical representation of the microdata table.

1 Introduction

Today's globally networked society places great demand on the sharing of information. However, the use of data containing personal information has to be restricted in order to protect individual privacy. To ensure the anonymity of the entities to which the sensitive data undergoing public or semipublic release refer, data holders often remove or encrypt explicit identifiers such as names, medical care card numbers(MCN) and addresses. The process is called de-identifying the data.

However, such a de-identification procedure does not guarantee the privacy of individuals in the data. Released information often contains other data, such as race, date of birth, gender and Zip code, which can be linked to publicly available information to re-identify respondents and to infer information that was not intended for release. Sweeney reported that 87 percent of the population of the United States can be uniquely identified by the combinations of attributes: gender, date of birth, and 5-digit zip code [11].

* This research was funded by Australian Research Council (ARC) grant DP0774450 titled *"Privacy Preserving Data Sharing in Data Mining Environments"*.

Y. Zhang et al. (Eds.): APWeb 2008, LNCS 4976, pp. 287–296, 2008.
© Springer-Verlag Berlin Heidelberg 2008

Table 1. De-identified Private Table

MCN	Gender	Age	Zip	Diseases
*	Male	25	4350	Hypertension
*	Male	23	4351	Hypertension
*	Male	22	4352	Depression
*	Female	28	4353	Chest Pain
*	Female	34	4352	Obesity
*	Female	31	4350	Flu

Table 2. A 3-anonymous view of Table 1

MCN	Gender	Age	Zip	Diseases
*	Male	22-25	435*	Hypertension
*	Male	22-25	435*	Hypertension
*	Male	22-25	435*	Depression
*	Female	28-34	435*	Chest Pain
*	Female	28-34	435*	Obesity
*	Female	28-34	435*	Flu

Besides de-identification, an alternative approach is to restrict the release of information in some way. In this paper, we focus on the strategy of k-anonymity, which was first proposed by Samarati and Sweeney [10]. A microdata table satisfies k-anonymity if every record in the table is identical to at least $(k-1)$ other records with respect to the set of quasi-identifier attributes.[1] Such a data set is called k-anonymous. As a result, an individual is indistinguishable from at least $(k-1)$ individuals in a k-anonymous data set.

Among the techniques proposed for providing anonymity in the release of microdata, the k-anonymity proposal focuses on two techniques in particular: generalization and suppression, which unlike other existing techniques, such as de-identification, preserve the truthfulness of the information. Generalization consists in substituting the values of a given attribute with more general values. We use * to denote the more general value. For instance, we could generalize two different Zip code 4350 and 4373 to 435*. The other technique, referred to as data suppression, removes the part (cell suppression) or entire value (attribute suppression) of attributes from the microdata table. Note that suppressing an attribute to reach k-anonymity can equivalently be modeled via a generalization of all the attribute values to *.[2]

To illustrate the concept, consider the data in Table 1, which exemplifies medical data to be released after de-identification. This table does not contain personal identification attributes, such as name, address, and medical care card number(MCN). However, values of other released attributes, such as age, gender and Zip may appear in some external table jointly with the individual identity, and can therefore allow tracking. For instance, age, gender and Zip can be linked within Table 3 to reveal Name, Address, and City. In Table 1, for example, the first record is unique in these three attributes, and this combination, if unique in the external world as well, uniquely identifies the corresponding tuple as pertaining to "Lee, 10 Collard Court, Toowoomba", thus revealing that he has reported Hypertension.

To avoid breaching privacy, Table 1 can be modified to Table 2. In Table 2, age is grouped into intervals, and Zips are clustered into large areas (the symbol

---

[1] The set of attributes included in the microdata table, also externally available and therefore exploitable for linking is called *quasi-identifier*.

[2] This observation holds assuming that attribute suppression removes only the values and not the attribute (column) itself. This assumption is reasonable since removal of the attribute (column) is not needed for k-anonymity.

Table 3. Non de-identified Publicly available table

Name	Address	City	Age	Zip	Gender
......	......	......	......	......	......
......	......	......	......	......	......
Lee	10 Collard Court	Toowoomba	25	4350	Male
......	......	......	......	......	......

$*$ denotes any digit). A (tuple) record in the quasi-identifier is identical to at least three other records in Table 2, and therefore, no individual is identifiable.

A k-anonymous table protects individual privacy in the sense that, even if an adversary has access to all the quasi-identifier attributes of all the individuals represented in the table, he would not be able to track down an individual's record further than a set of at least k records. Thus, releasing a k-anonymous table prevents definitive record linkages with publicly available databases and keeps each individual hidden in a crowd of $k - 1$ other people.

In recent years, numerous algorithms have been proposed for implementing k-anonymity via generalization and suppression. Samarati [9] presents an algorithm that exploits a binary search on the domain generalization hierarchy to find minimal k-anonymous table. We recently improve his algorithm by integrating the hash-based technique [12]. Bayardo and Agrawal [3] presents an optimal algorithm that starts from a fully generalized table and specializes the dataset in a minimal k-anonymous table, exploiting ad hoc pruning techniques. LeFevre, DeWitt and Ramakrishnan [7] describes an algorithm that uses a bottom-up technique and a priori computation. Fung, Wang and Yu [4] present a top-down heuristic to make a table to be released k-anonymous. The approach applies to both continuous and categorical attributes. As far as k-anonymity problem is concerned, fewer theoretical results were obtained. The exceptions are Meyerson and Williams [8] and Aggarwal et al. [1,2] proved the optimal k-anonymity is $\mathcal{NP}$-hard (based on the number of cells and number of attributes that are generalized and suppressed) and describe approximation algorithms for optimal k-anonymity.

2   Paper Organization and Contributions

In Section 3, we introduce two new variants of the k-anonymity problem, namely, the *Restricted k-anonymity problem* and *Restricted k-anonymity problem on attribute* and we discusses the connection between *Restricted k-anonymity problem* and general k-anonymity problem which stresses the significance of investigating this new class anonymity problem.

Our first contribution is the $\mathcal{NP}$-hardness proof the *Restricted k-anonymity problem* and the *Restricted k-anonymity problem on attribute*, and are presented in Section 4 and 5. The theoretical results for *Restricted k-anonymity problem* also provide an alternative $\mathcal{NP}$-hardness proof of general k-anonymity problem, which imply the main results obtained in [1,2,8].

$v_1 = (1,0,1,0)$	$t(v_1) = (*,0,*,0)$
$v_2 = (1,0,0,0) \overset{t}{\Longrightarrow}$	$t(v_2) = (*,0,*,0)$
$v_3 = (0,0,1,0)$	$t(v_3) = (*,0,*,0)$

$v_1 = (1,0,1,0)$	$v_1 = (1,0,1,0)$
	$v_2 = (1,0,0,0)$
$v_2 = (1,0,0,0) \overset{t}{\Longrightarrow}$	$v_3 = (0,0,1,0)$
	$v_4 = (0,1,0,1)$
$v_3 = (0,0,1,0)$	$v_5 = (1,1,0,1)$
	$v_6 = (0,1,1,1)$

Fig. 1. Suppressing a dataset by t **Fig. 2.** Constructing a restricted instance

The second contribution is presented in Section 6. Through a graphical representation of the microdata table, we develop a polynomial time algorithm for the *Restricted 2-anonymity problem*. Considering the connection between *Restricted k-anonymity problem* and general k-anonymity problem, we could develop another efficient algorithm for general k-anonymity problem as well. We will include this application part in a separate paper. Finally, conclusions and future work are given in Section 7.

3   Restricted k-anonymity Problem

We consider degree-m tuples in the private database to be m-dimensional vectors v_i, drawn from $\sum^m$, where $\sum$ is a (finite) alphabet of possible values for attributes(columns). Thus, the private databases under consideration are formally represented as subsets $V \subseteq \sum^m$. Let $*$ be a symbol that is not in $\sum$.

Definition 1. *Let t be a map from V to $(\sum \cup \{*\})^m$. We say that t is a suppressor on V if for all $v \in V$ and $j = 1, 2, \cdots, m$, it is the case that $t(v)[j] \in \{v[j], *\}$.*[3]

Intuitively speaking, a suppressor defines some kind of anonymous vector $t(v) = v'$ in an anonymous set $V' \subseteq (\sum \cup \{*\})^m$. The coordinates of V' are identical to the coordinates of V, except some may be suppressed by $*$. Consider following example. Let $V = \{1010, 1000, 0010\}$, with suppressor $t(a_1 a_2 a_3 a_4) = *a_2 * a_4$ (each $a_i \in \{0,1\}$), then the resulting $t(V) = \{*0*0, *0*0, *0*0\}$ (See Fig.1).

Now, we can extend the definition of a suppressor t to a set of vectors V. Here, we regard $t(V)$ as a multiset [4], when two or more vectors in V map to the same suppressed vector. (i.e. $v \neq v' \in V$, but $t(v) = t(v')$). Following, we define k-anonymity.

---

[3] We consider a special case of suppressions, i.e. each entry is either included in the output, or omitted entirely, with a $*$ character taking its place.

[4] A multiset is a set in which elements can appear more than once. Notice that, given a multiset M and an element e, we may have that $e \in M$ match more than once; i.e., $\{e|e \in M\}$ is a multiset and its cardinality can be larger than 1. The usual set operations are extended to multisets accordingly.

Definition 2. *Let t be a suppressor on the set $V = \{v_1, v_2, \cdots, v_n\} \subseteq \sum^m$. Then $t(V)$ is k-anonymous if and only if for all $v_i \in V$, there exists $k-1$ indices $i_1, i_2, \cdots, i_{k-1} \in \{1, 2, \cdots, n\}$, such that $t(v_{i_1}) = t(v_{i_2}) = \cdots = t(v_{i_{k-1}}) = t(v_{i_k})$.*

In other words, when a suppressor makes the database k-anonymous, it means that every anonymous vector is a member of a multiset of (at least) k identical vectors. For example, the left dataset in Fig.1 becomes 3-anonymous after suppressing by t.

Restricted k-anonymity problem. Given $V \subseteq \sum^m$ (where $\sum = \{0, 1\}$) such that the number of zeroes in each attribute (column) is exactly k; Is there a suppressor t, such that $t(V)$ is k-anonymous and suppresses the minimum number of vector coordinates?

EXAMPLE: The left dataset in Fig.2 is an instance of general 3-anonymity problem and the right dataset is an instance of *Restricted 3-anonymity problem*.

Another version of the *Restricted k-anonymity problem* is where we choose whether or not to suppress various attributes from the database. We say that attribute(column) j is suppressed by t if for all $v \in V$, $v[j] = *$. Formally, we define the *Restricted k-anonymity problem on attribute* as follows:

Restricted k-anonymity problem on attribute. Given $V \subseteq \sum^m$ (where $\sum = \{0, 1\}$) such that the number of zeroes in each attribute (column) is exactly k; Is there a suppressor t, such that $t(V)$ is k-anonymous and suppresses the minimum number of attributes (columns)?

The reason why we introduce the definition of *Restricted k-anonymity problem* is its close connection with general k-anonymity problem. Given an instance of general k-anonymity problem, we could construct an instance of the *Restricted k-anonymity problem* (take Fig.2 as an example, by adding three vectors v_4, v_5, v_6, we can make the left dataset an instance of the *Restricted 3-anonymity problem*), which could provide an alternative approach to solve the general k-anonymity problem. Currently, we are working on developing algorithms to find solutions of the restricted problem and with some post-processing, the solution can be a good solution for the general problem.

4 Cell Suppression Is Hard

In this section, we prove that the *Restricted k-anonymity problem* is $\mathcal{NP}$-hard for $k \geq 3$. First, recall the *Restricted 3-anonymity problem*.
Restricted 3-anonymity problem: Given $V = \{v_1, v_2, \cdots, v_n\} \subseteq \sum^m$ (where $\sum = \{0, 1\}$) such that the number of zeroes in each attribute (column) is 3, and $l \in N$: Is there a suppressor t, such that $t(V)$ is 3-anonymous, and the total number of suppressed vector coordinates in $t(V)$ is at most l?

Theorem 3. *The Restricted 3-anonymity problem is $\mathcal{NP}$-hard.*

Proof. The reduction is from *Exact cover by 3-sets (X3C)* [5]: Given a finite set X with $|X| = 3q$ and a collection C of 3-element subsets of X, does C contain an exact cover for X; that is, a sub collection $C' \subseteq C$ such that every element of X occurs in exactly one member of C'?

Let $X = (x_1, x_2, \cdots, x_{3q})$ and $C = (C_1, C_2, \cdots, C_m)$, where $|C_i| = 3$ for $i = 1, 2, \cdots, m$. We construct a database V as follows. For each x_i, define an m-dimensional vector $v_i \in \sum^m$:

$$v_i[j] = \begin{cases} 0 & \text{if } x_i \in C_j \\ 1 & \text{otherwise} \end{cases}$$

Set $V = (v_1, v_2, \cdots, v_{3q})$, $l = 3q(m-1)$, and the number of zeroes in each attribute(column) is 3 because of $|C_j| = 3$. Then V is an instance of the *Restricted 3-anonymity problem*.

Assume t is the optimal suppressor on V (i.e. suppresses the minimum number of vector coordinates and maintains 3-anonymity). We claim that the total number of coordinates suppressed by t is at most $3q(m-1)$ if and only if there is an $X3C$ in C.

Sufficiency. Suppose that there is an $X3C$ $C' \subseteq C$. For $i = 1, 2, \cdots, n$, let $j(i)$ be such that $C'_{j(i)}$ is the unique set in C' that contains x_i. Define a suppressor t by:

$$t(v_i)[j'] = \begin{cases} 0 & \text{if } j' = j(i) \\ * & \text{otherwise} \end{cases}$$

Since $x_i \in C'_{j(i)}$, the $v_i[j(i)] = 0$, and all the other are $*$. Therefore, t is a suppressor on V.

Now consider any $t(v_i)$. There are three elements x_i, $x_{i'}$, $x_{i''}$ in the set $C'_{j(i)}$, and each element has identical anonymous vectors; i.e. $t(v_i) = t(v_{i'}) = t(v_{i''})$. Hence there are two vectors in $t(V)$ which are identical to $t(v_i)$. This shows that $t(V)$ is 3-anonymous and t is feasible. Since in our solution, every $t(v) \in t(V)$ has exactly one non-$*$ coordinate, the number of $*$'s is exactly $3q(m-1)$. Therefore the optimal 3-anonymous solution has at most $3q(m-1)$ $*$'s in its vectors.

Necessity: Suppose that t suppresses at most $3q(m-1)$ coordinates and there does not exist $X3C$. We draw a contradiction as follows. Consider any 3-anonymous solution t for V. First, we answer the question: can there exist a vector with two non-$*$'s in its anonymous form? Suppose that v_i is such a vector. Since $t(V)$ is 3-anonymous, there must exist two other vectors $v_{i'}$ and $v_{i''}$, which have the same value as $t(v_i)$ in the anonymous form, say, $t(v_{i'})$ and $t(v_{i''})$. Since the non-$*$ coordinates have the same values as in the original v_i vectors, we must have v_i, $v_{i'}$ and $v_{i''}$ identical in two different coordinates, j and j'. By construction, any two vectors in V can match only in coordinates where they are 0, and $v_i[j] = 0$ only if the element x_i is in the set C_j. Hence v_i, $v_{i'}$ and $v_{i''}$ are in the two different sets, C_j and $C_{j'}$ of C. However, this means two different sets are identical in C, which is not possible. So for any feasible 3-anonymous suppressor t for V, every vector $v_i \in V$ has at most one non-$*$ coordinate in its 3-anonymous form $t(v_i)$. Hence at least $3q(m-1)$ coordinates in $t(V)$ are suppressed.

Therefore, if we have a $t(V)$ with at most $3q(m-1)$ suppressed coordinates, it must be that every vector in $t(V)$ has exactly one non-$*$ coordinate. Given this fact, we can construct an $X3C$ C' for C in the following way. For each $i = 1, 2, \cdots, n$, consider the non-$*$ coordinate in $t(v_i)$. This coordinate must have value 0 (otherwise there can be no identical vectors). If this corresponds to the coordinate j, we add the set C_j to a cover C'. Clearly we produce a collection of sets such that each element in X is in at least one set. Since there are 3 identical vectors for every vector $v \in V$(including v), it follows that there are at most q sets in C'. Since we need at least q sets to cover every element, there must be exactly q sets in C', which is exactly an $X3C$. This contradicts our assumption, so it follows that there is an $X3C$ in C if and only if the optimal 3-anonymous solution has at most $3q(m-1)*$'s.

Corollary 4. *The Restricted k-anonymity problem is $\mathcal{NP}$-hard for $k \geq 3$.*

Corollary 5. *The k-anonymity problem is $\mathcal{NP}$-hard for $k \geq 3$.*

Corollary 5 was first obtained by Meyerson and Williams [8].

5   Attribute Suppression Is Hard

In this section, we consider the situation whether or not to suppress various attributes from the database and we prove it is hard as well.

Suppose that $X = (x_1, x_2, \cdots, x_{3q})$ and $C = (C_1, C_2, \cdots, C_m)$, where $|C_i| = 3$, for $i = 1, 2, \cdots, m$, and let $\sum = \{0, 1\}$. We build the database $V = (v_1, v_2, \cdots, v_{3q})$ where each v_i represents an element in X. Assume t suppresses the least number of attributes and is defined as in Theorem 3.

Theorem 6. *The Restricted 3-anonymity problem on attribute is $\mathcal{NP}$-hard.*

Proof. (proof sketch) We claim that there exists a suppressor that suppresses at most $m - q$ attributes and maintains 3-anonymous if and only if C has an $X3C$. If C has an $X3C$, then by suppressing those $m - q$ attributes not in the cover, each remaining attributes has 3 vectors that has the same value, which is 3-anonymous. Conversely, if we have a suppressor t as above, then for every j, since the anonymous table is 3-anonymous and the number of zeroes is 3, there are exactly 3 vectors v_k such that $v_k[j] = 0$. It follows that if an attribute is not suppressed, then there exists 3 vectors with the same value under this attribute. Since the two attributes i and j are not suppressed in a 3-anonymous table if and only if $C_i \cap C_j = \emptyset$, at least $m - q$ attributes must be suppressed in any 3-anonymous table. Therefore, if we obtain a $t(V)$ with at most $m - q$ suppressed attributes, it must be that exactly $m - q$ attributes are suppressed in a 3-anonymous table. Then we can obtain an $X3C$ as in Theorem 3.

Corollary 7. *The Restricted k-anonymity problem on attribute is $\mathcal{NP}$-hard for $k \geq 3$.*

Corollary 8. *The k-anonymity problem on attribute is $\mathcal{NP}$-hard for $k \geq 3$.*

Corollary 8 implies the result obtained by Aggarwal *et al.*[1,2].

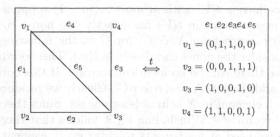

Fig. 3. Dataset(right)and its graphical representation(left)

6   Algorithm for Restricted 2-anonymity Problem

In this section, we present a graphic representation of the *Restricted 2-anonymity problem*, which produces a polynomial time algorithm with running time in $O(n^2m)$. First, recall the *Restricted 2-anonymity problem*:

PROBLEM: *Restricted 2-anonymity problem*

INSTANCE: Dataset $V = \{v_1, v_2, \cdots, v_n\} \subseteq \sum^m$, where $\sum = \{0, 1\}$ and the number of zeroes in each attribute (column) is 2, and $l \in N$.

QUESTION: Is there a suppressor t such that $t(V)$ is 2-anonymous and the total number of suppressed vector coordinates in $t(V)$ is at most l?

The transformation is made from the *perfect matching problem* in a simple graph. We include its definition here for completeness.

Perfect matching problem: Given a graph $G = (U, E)$ with $|U| = n$ and $|E| = m$, is there a subset $S \in E$ of $n/2$ edges such that each vertex of U is contained in exactly one edge of S?

Without loss of generality, assume that no two columns in the dataset have the same values. (If not, we could simplify the dataset by deleting the repeated one, which has no effect on the anonymity process) Also assume that $V = \{v_1, v_2, \cdots, v_n\} \in \sum^m$. Now construct a graph as follows:

Let $U = (v_1, v_2, \cdots, v_n)$ and $E = \{e_{ik}(j)\}$ where for each $j = 1, 2, \cdots, m$, $e_{ik} = (v_i(j), v_k(j))$ with $v_i(j) = v_k(j) = 0$ according to the assumption. Then we get the simple graph $G = (U, E)$ with $|U| = n$ and $|E| = m$. (See Fig.3 as an example.) On the contrary, if we have the simple graph $G = (U, E)$ with $U = (v_1, v_2, \cdots, v_n)$ and $E = (e_1, e_2, \cdots, e_m)$, then construct a database V as follows:

For each v_i, define an m-dimensional vector $v_i \in \sum^m$ as $v_i[j] = 0$ if $v_i \in e_j$; Otherwise, $v_i[j] = 1$; Set $V = \{v_1, v_2, \cdots, v_n\}$. Then because the graph G is simple, obviously, the number of zeroes in each attribute (column) is 2, which is an instance of the *Restricted 2-anonymity problem*.

Theorem 9. *Given an instance of the Restricted 2-anonymity problem, the optimal Restricted 2-anonymous solution has at most $n(m - 1)$ *'s suppressed by t if and only if there is a perfect matching in the corresponding constructed graph G.*

	$e_1 e_2 e_3 e_4 e_5$		$e_1\ e_2\ e_3 e_4\ e_5$
$v_1 = (0, 1, 1, 0, 0)$		$t(v_1) = (0, *, *, *, *)$
$v_2 = (0, 0, 1, 1, 1)$	t	$t(v_2) = (0, *, *, *, *)$
$v_3 = (1, 0, 0, 1, 0)$	$\Longleftrightarrow$	$t(v_3) = (*, *, 0, *, *)$
$v_4 = (1, 1, 0, 0, 1)$		$t(v_4) = (*, *, 0, *, *)$

	$e_1 e_2 e_3 e_4 e_5$		$e_1\ e_2\ e_3 e_4\ e_5$
$v_1 = (0, 1, 1, 0, 0)$		$t(v_1) = (*, *, *, 0, *)$
$v_2 = (0, 0, 1, 1, 1)$	t	$t(v_2) = (*, 0, *, *, *)$
$v_3 = (1, 0, 0, 1, 0)$	$\Longleftrightarrow$	$t(v_3) = (*, 0, *, *, *)$
$v_4 = (1, 1, 0, 0, 1)$		$t(v_4) = (*, *, *, 0, *)$

Fig. 4. Dataset and its two 2-anonymous tables

Proof. The proof is similar to Theorem 3, we omit it here due to the page limit.

Corollary 10. *The Restricted 2-anonymity problem can be solved in polynomial time.*

Algorithm 1. Polynomial time algorithm for the *Restricted 2-anonymity problem*

Input :   A dataset $V = (v_1, v_2, \cdots, v_n) \subseteq \sum^m$

Output: The 2-anonymous dataset $t(V)$ (where t is a suppressor)

1.   Construct the graph $G = (U, E)$ where $U = (v_1, v_2, \cdots, v_n)$ and $E = \{e_{ik}(j)\}$
     and for each $j = 1, 2, \cdots, m$, $e_{ik} = (v_i(j), v_k(j))$, with $v_i(j) = v_k(j) = 0$
2.   Find one perfect matching M in G.
3.   If found, let $M(i)$ denote the unique edge in M containing node i and let
     $t(M(i)) = 0$ and $t(j) = *$, if $j \neq M(i)$. Output $t(V)$.
4.   If not found. Output $t(V)$ with each value replaced by $*$ in V.

Running Time. The running time of Algorithm 1 depends on Step 2, which can be solved in $O(n^2 m)$ [6]. Since the transformation could be done in at most $O(n^2)$, the algorithm time complexity for *Restricted 2-anonymity problem* is in $O(n^2 m)$. Also, since the graph can be specified by its vertex adjacency matrix A, which could be described by at most nm bits of input, so the space (memory) complexity of the algorithm is $O(nm)$. Note that if we find out all the perfect matchings M in G, then we could find all the possible 2-anonymous tables.

EXAMPLE: We use Fig.4 as an example to illustrate how Algorithm 1 works. Our objective is to make the left dataset in Fig.4 2-anonymous. The left graph in Fig.3 is the graphic representation of the dataset in Fig.4. In that graph, we could find all perfect matchings $\{e_1, e_3\}$ and $\{e_2, e_4\}$ and according to Algorithm 1, all the 2-anonymous tables are shown in Fig.4.

7   Conclusions and Future Work

In this paper, we introduce two new variants of the k-anonymity problem, namely, the *Restricted k-anonymity problem* and *Restricted k-anonymity problem on attribute*. We prove that both problems are $\mathcal{NP}$-hard for $k \geq 3$. The results imply the main results obtained by Meyerson and Williams. We have also developed a polynomial time algorithm for the *Restricted 2-anonymity problem* by giving a graphical representation of the microdata table.

Our future work is to develop applicable algorithms for general k-anonymity problem based on the theoretical results obtained in this paper. More specifically, it involves developing a new efficient exact algorithm and providing better approximate algorithm scheme for general k-anonymity problem based on the connection and transformation between the *Restricted k-anonymity problem* and general k-anonymity problem.

Acknowledgements

We would like to thank Professor Jeffrey Yu for his useful comments on the paper.

References

1. Aggarwal, G., Feder, T., Kenthapadi, K., Motwani, R., Panigrahy, R., Thomas, D., Zhu, A.: Anonymizing tables. In: Eiter, T., Libkin, L. (eds.) ICDT 2005. LNCS, vol. 3363, pp. 246–258. Springer, Heidelberg (2004)
2. Aggarwal G, Feder T, Kenthapadi K, Motwani R, Panigrahy R, Thomas D, Zhu A.: Approximation algorithms for k-anonymity. Journal of Privacy Technology, (paper number 20051120001)
3. Bayardo, R.J., Agrawal, R.: Data privacy through optimal k-anonymization. In: Proc. of the 21st International Conference on Data Engineering (ICDE 2005), Tokyo, Japan, pp. 217–228 (2005)
4. Fung, B., Wang, K., Yu, P.: Top-down specialization for information and privacy preservation. In: Proc. of the 21st International Conference on Data Engineering (ICDE 2005), Tokyo, Japan (2005)
5. Garey, M.R., Johnson, D.S.: Computers and Intractability: A Guide to the Theory of $\mathcal{NP}$-Completeness. Freeman, San Francisco (1979)
6. Lawler, E.L.: Combinatorial Optimization: Networks and matroids. Holt. Rinehart and Winston, New York (1976)
7. LeFevre, K., DeWitt, D.J., Ramakrishnan, R.: Incognito: Efficient fulldomain k-anonymity. In: Proc. of the 24th ACM SIGMOD International Conference on Management of Data, Baltimore, Maryland, USA, pp. 49–60 (2005)
8. Meyerson, A., Williams, R.: On the complexity of optimal k-anonymity. In: In Proc. of the 23rd ACM-SIGMOD-SIGACT-SIGART Symposium on the Principles of Database Systems, Paris, France, pp. 223–228 (2004)
9. Samarati, P.: Protecting respondents' identities in microdata release. IEEE Transactions on Knowledge and Data Engineering 13(6), 1010–1027 (2001)
10. Samarati, P., Sweeney, L.: Generalizing Data to Provide Anonymity when Disclosing Information (Abstract). In: Proc. of ACM Symposium on Principles of Database Systems, p. 188 (1998)
11. Sweeney, L.: k-anonymity: a model for protecting privacy. International journal on uncertainty, Fuzziness and knowledge based systems 10(5), 557–570 (2002)
12. Sun, X., Li, M., Wang, H., Plank, A.: An efficient hash-based algorithm for minimal k-anonymity problem. In: Thirty-First Australasian Computer Science Conference (ACSC2008), Wollongong, Australia (to appear 2008)

An Efficient Electronic Marketplace Bidding Auction Protocol with Bid Privacy

Wenbo Shi, Injoo Jang, and Hyeong Seon Yoo

School of Computer Science and Engineering, Inha University
Incheon, 402-751, Korea
swb319@hotmail.com

Abstract. We modified the multi-agent negotiation test-bed which was proposed by Collins et al. In 2004, Jaiswal et al. have modified Collins's scheme, but Jaiswal's scheme still has some security weaknesses: such as replay data attack and DOS (denial-of-service) attack, anonymity disclosure, collision between customers and a certain supplier. So the proposed protocol tries to reduce DOS attack and avoids replay data attack by providing ticket token and deal sequence number to the supplier. It utilizes efficient LPN-based authentication method to accomplish lightweight authentication. And it publishes an interpolating poly-nomial for sharing the determination process data and avoids collusion between a customer and a certain supplier. Also the proposed scheme relaxes the trust as-sumptions for three-party in Jaiswal's scheme. According to comparison and analysis with other protocols, our proposed protocol shows good security and less computation cost.

1 Introduction

In 2002, Collins et al. presented a multi-agent marketplace, MAGNET (Multi-Agent Negotiation Test-bed) for electronic business-to-business market [1]. Jaiswal et al. proposed security protocol and put it into real-world networks and analyze security problem to improve MAGNET in 2004 [2]. The proposed major modification is the use of a publish/subscribe system by the market to notify the agents about the auctions. Also they adopted time-release cryptography to guarantee non-disclosure of the bids and anonymous communication to hide the identities of the bidders. According to this, the MAGNET is improved in security. But the improved protocol still has some weaknesses: vulnerable to the replay data attack, DOS (denial-of-service) attack, anonymity disclosure weakness, collusion between a customer and a certain supplier.

The proposed protocol utilizes ticket token to restrict download, also market gen-erates deal sequence number (dsn) and random number (r) for suppliers who have download requests for quotes (RFQ). It utilizes efficient LPN-based authentication method to accomplish lightweight authentication. When auction is closed, market constructs a simple interpolating polynomial for sharing the determination process data in supplier group who have taken part in this auction. Sharing the determination process data can totally avoid collusion between a customer and a certain supplier.

Y. Zhang et al. (Eds.): APWeb 2008, LNCS 4976, pp. 297–308, 2008.

In 2003, Chang et al. proposed an anonymous auction protocol, they applied a simple method for ensuring anonymity of bidders, and it also provided some important properties of auction protocol [3]. However, Jiang et al. found there were still some weaknesses in initial phase of Chang et al.'s protocol, so they improved it and proved its security in 2005 [4]. Because computation cost is not taken into account in their improvement, Chang et al. proposed the enhancement with the alias in their protocol and analyzed the computation cost in 2006 [5]. Another protocol which provides auction properties was proposed by Liaw et al. in 2006. By comparison with Hwang et al.'s protocol [6], they indicate that their protocol has strong security and more efficient [7]. By comparative analysis with those protocols, the proposed protocol shows good security and less computation cost.

2   Jaiswal's Protocol and Its Vulnerabilities

2.1   Jaiswal's Protocol

We introduce Jaiswal et al.'s protocol briefly. It contains planning, bidding, auction close and winner determination phase.

2.1.1   Planning

The customer sends a signed *RFQ* message $S_{S_{kc}}(RFQ)$ to the market for publishing. S_{kc} is secret key of the customer.

2.1.2   Bidding

The supplier receives the *RFQ* through the publish/subscribe system. If interested, the supplier generates a bid-message comprising of three parts: the *RFQ* number and a random number (*RFQ#*, *r*), auction-session key: a symmetric session key K_a, bid data.

It then signs and encrypts the message and sends it to the market: $M = [(RFQ\#, r), TE\{E_{pk_c}(k_a)\}, E_{k_a}\{S_{sk_s}(bid)\}]$.*TE* means a time-release encryption method. For all the bid-messages received by the market, it publishes (*RFQ#*, (*r, hash(M)*)) on the publish/subscribe system, where *M* is the bid-message sent by a supplier to the market. The supplier can also check the publish/subscribe system and verify that whether its bid was actually received and displayed by the market. The customer can then retrieve the bids from the publish/subscribe system. However, it cannot access the bid data unless it decrypts the time-release cryptography [8], the supplier agent would construct a puzzle that would take the customer longer than the auction deadline to solve.

2.1.3   Auction Close

Once the auction is closed, the suppliers would release k_a to the market in an encrypted form, along with the customer's copy: $E_{pk_m}(S_{sk_s}(k_a, r)), E_{pk_c}(S_{sk_s}(k_a, r))$ .The market would then pass on the customer's portion of the encrypted key: $E_{pk_c}(S_{sk_s}(k_a, r))$.

2.1.4 Winner Determination

The customer agent uses various algorithms to determine the winner from the bids it has received. Once the winner has been determined, it would use the market's white-board to notify suppliers $S_{sk_c}(RFQ\#, r_{winner})$ .Once market publishes this result on the publish/subscribe system any supplier can check to see if it is the winner.

2.2 Vulnerabilities

2.2.1 Replay Attack

The Jaiswal et al.'s protocol is vulnerable to the replay data attack. We suppose that an attacker has eavesdropped and intercepted messages sent by suppliers for a period of time, collected lots of data pairs $M=[(RFQ\#,r),TE\{E_{pkc}(k_a)\},E_{ka}\{S_{sks}(bid)\}]$ and $E_{pkm}(S_{sks}(k_a,r))$, $E_{pkc}(S_{sks}(k_a,r))$. When the attacker gets right opportunity, he can utilize the appropriate replay data for current bid process. In the message $M=[(RFQ\#,r),TE\{E_{pkc}(k_a)\},E_{ka}\{S_{sks}(bid)\}]$, RFQ number $(RFQ\#)$ and random number (r) are not put under protection, that gives the attacker an opportunity to cheat by using replay data to forge a fake message. If a legitimate supplier doesn't take part in current bid process or the attacker shuts down the communication between market and the legitimate supplier. Then the attacker utilizes and constructs a bid-message $M'=[(RFQ\#,r'),TE'\{E_{pkc}(k_a)\},E'_{ka}\{S_{sks}(bid)\}]$. In message M', $RFQ\#$ is the current RFQ number and other data blocks are old data blocks. The market just publish $(RFQ\#, (r', hash(M')))$, and the supplier can check the hash value of M'. Because the legitimate supplier doesn't take part in this bid process, he will not pay attention to the publish message. Moreover, because of the time-release encryption of message, the market cannot know the identities of the suppliers, also he cannot verify the integrity and freshness of this message. When auction is closed, the attacker sends message $E'_{pkm}(S_{sks}(k_a,r))$, $E'_{pkc}(S_{sks}(k_a,r))$ to market. Because r and k_a are uniform with the former replay messages, market cannot verify whether this message is a replay data or not, and the case of the customer is similar with market. Only after the customer de-crypts data and checks the bid data carefully, he might identify whether this message is replay data or not.

2.2.2 Customer Collusion

The customer determines a winner from messages he has received. We assume that customer take sides with a certain supplier. After customer knew all suppliers' identities, customer only choose his partner supplier, without considering other suppliers, which is unfair to the others in process. So we should have a mechanism to inform the other suppliers of the bid determination process data to avoid collusions.

3 Security Assumptions

In Jaiswal et al.'s protocol they proposed several trust assumptions for market as: conveys RFQs from customer to supplier agents, communicates the bids from supplier agent back to customer agent, keep records of all the transactions of RFQs and bids, aggregates statistical data to ensure non-repudiation. Also they proposed trust assumptions for three-party as: customer agent will not collude with any of the supplier agents, customer agent communicates with the supplier agents only through market. But in our proposed scheme, we do not need to follow those trust assumptions of Jaiswal et al.'s protocol for three-party. In our proposed protocol, it is impossible if a customer wants to collude with a supplier. Also we allow the customer agent communicating with the supplier agents to exchange information. We will introduce how we can relax these trust assumptions for three-party in later section.

4 Proposed Architecture

4.1 Key Techniques

4.1.1 Ticket Token

For our auction scheme, we need a secure authentication system for group communication. Not only authentication, we also need a mechanism to protect identities of participant in supplier group. So we import the concept of Conference key distribution into our auction scheme. The concept of conference key distribution was first proposed by Ingemarrsson et al. [9]. In 2005, Ryu et al. showed incompleteness of another scheme and improved it by adding verification phase [10]. Because of adopting publish/subscribe system we are going for anonymity primarily, not just for the mutual authentication. In our situations, it is more important that the identities of the suppliers participating in an auction procedure are hidden from other attending suppliers. Furthermore, the participant suppliers should be anonymous to market and customer. Also ticket token can help avoid DOS attack to some extent.

4.1.2 LPN-Based Authentication

At first, we introduce the Learning parity with noise (LPN) problem. The LPN problem with security parameters q, k, η. Let A be a random $q*k$ binary matrix, let x be a random k-bit vector, let η be a constant noise parameter, where $\eta \in [0, 1/2]$, and let v be a q-bit vector such that $|v| \leq \eta q$. Given A, η, and $z = A \cdot x \oplus v$, find a k-bit vector x' such that $|A \cdot x' \oplus z| \leq \eta q$. The hardness of the computational LPN problem has been shown to be NP-complete [11]. Several authentication protocols based on LPN problem were proposed, and most representative of them were HB protocol and HB^+ protocol [12]. Also Gilbert proposed that HB^+ is not secure against a man-in-the-middle attack [13].

When we import the LPN-based authentication into our auction scheme, we are going to show that the man-in-the-middle attack will not happen during the communication of market and suppliers. The two k-bit vectors x and y are secret keys shared by the market and a supplier S, where S denotes identity of the supplier. Figure 1 illustrates one round of our LPN-based authentication.

1. S chooses a blinding vector a_1 randomly, and sends $\{dsn, S, a_1, T_1\}$ to market, where T_1 denotes a valid timestamp.
2. After receiving $\{dsn, S, a_1, T_1\}$, market checks T_1 and chooses challenge b_1 randomly, publishes (dsn, S, a_1, b_1, T_2), where T_2 denotes a valid timestamp.

3. S gets b_1 from publish board, then computes response $z_1 = a_1 \cdot x \oplus b_1 \cdot y \oplus v$, and sends $\{dsn, S, z_1, T_3\}$, where v denotes a noise bit, T_3 denotes a valid timestamp.

4. After receiving $\{dsn, S, z_1, T_3\}$, market checks T_3 and publishes (dsn, S, z_1) and accepts the round if $a_1 \cdot x \oplus b_1 \cdot y = z_1$.

The condition when HB$^+$ protocol attack happen is that the attacker can manipulate challenges sent by the market to a supplier during the authentication message exchanges. In our LPN-based authentication process, market utilizes publish system to show the message which suppliers have sent, so suppliers can check and verify whether their own data message was actually received and displayed by market. We assume that there is an attacker in the middle, and he can manipulate the message sent by a supplier to the market. Because of the publish system, the attacker cannot manipulate the challenges published by the market. According to what is mentioned above, the attacker doesn't have any opportunity to disturb the authentication process.

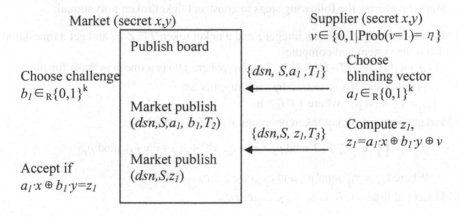

Fig. 1. One round of LPN-based authentication

4.2 Proposed Scheme

The proposed protocol has the following phases: planning, bidding, auction close and winner determination, figure 2 illustrates the proposed protocol, as explained in the following:

For convenience, we assume there are n_1 customers and n_2 suppliers and one market in our auction scheme. In order to guarantee the reliability and safety among market, customer group and supplier group, we need a certification authority (CA) in key pre-distribution long-term key process.

CA chooses and publishes large prime number p_1, p_2 such that p_1-1 and p_2-1 have large prime factors. Let q_1 be a prime divisor of p_1-1 and g_1 be a generator with order q_1 in GF(p_1), q_2 be a prime divisor of p_2-1 and g_2 be a generator with order q_2 in GF(p_2). Let S be the identity of the supplier, C be the identity of the customer. By using the Diffie-Hellman scheme, CA assigns a secret key $x_{i_1} \in Z^*q_1$ and computes public key

$y_{i_1} = g_1{}^{x_{i_1}} \bmod p_1$ for each customer and market, where $1 \leq i_1 \leq n_1 + 1$. Then, gives the secret key x_{i_1} to each customer and market in a secure way; CA assigns a secret key $x_{i_2} \in Z^* q_2$ and computes public key $y_{i_2} = g_2{}^{x_{i_2}} \bmod p_2$ for each supplier and market, where $1 \leq i_2 \leq n_2 + 1$. Then, CA gives the secret key x_{i_2} to each supplier and market in a secure way. CA assigns two symmetric key $x_{i_3} \in_R \{0,1\}^k$, $y_{i_3} \in_R \{0,1\}^k$, for each supplier and market, where $1 \leq i_3 \leq n_2 + 1$. Then, CA gives the symmetric secret key x_{i_3} and y_{i_3} to each supplier and market in a secure way.

4.2.1 Planning

A customer sends a *RFQ* message which is signed by his secret key S_{kc} to market for publishing. After receiving the *RFQ* message, market verify and publish it. After that,
Market performs the following steps to construct ticket token polynomial:

1 Market randomly choose an integer r and a ticket token $T \in Z^* q_2$ and get a timestamp t from the system and compute
$A = g^r \bmod p$, $B = r*T + H(t\|A)*x_{m_2} \bmod q_2$, where H() is a one-way hash function.
2 Compute the secret key shared by each supplier as:
$k_{mi_2} = y_{i_2}{}^r \bmod p_2$,where $1 \leq i_2 \leq n_2$
3 Market constructs ticket token polynomial $f(x)$ for publishing.

$$f(x) = \prod_{i_2=1}^{n_2} (x - k_{mi_2}) + T \bmod p_2 = x^n + c_{n-1}x^{n-1} + \ldots\ldots + c_1 x + c_0 \bmod p_2$$

Where $k_{mi_2} = y_{i_2}{}^r \bmod p_2$, and $c_{n-1}, c_{n-2}, \ldots\ldots c_1, c_0 \in Z^* q_2$
4 Market publishes A, B, t, $c_{n-1}, c_{n-2}, \ldots\ldots c_1, c_0.$

4.2.2 Bidding

If a supplier S is interested in this auction session, S will

1 Get A, B, t, $c_{n-1}, c_{n-2}, \ldots\ldots c_1, c_0$ from market's publish board.
2 Check the timestamp t is a valid data or not.
3 Compute the secret key shared with market as:
$k_{i_2m} = A^{x_{i_2}} \bmod p_2$
4 Get ticket token by computing $f(k_{i_2m})$

$$f(k_{i_2m}) = (k_{i_2m})^n + c_{n-1}(k_{i_2m})^{n-1} + \ldots\ldots + c_1(k_{i_2m}) + c_0 \bmod p_2 = T \bmod p_2$$

5 Verify ticket token T by computing $H(t\|A)$ and check the equation
$g^B \equiv A^T * y_{m_2}{}^{H(t\|A)} \bmod p_2$

After getting T, the supplier use the ticket token T to download *RFQ*. When a legitimate supplier downloads the *RFQ*, market generates *dsn* and r to the supplier. After getting *dsn* and r from market, the supplier generates a bid-message comprising of *RFQ* number (*RFQ#*), *dsn* and r, symmetric auction-session key (k_a), bid data (*bid*). Then he signs, hashes and encrypts the message (M) and sends M to market. Where

$M=[RFQ\#,dsn,hash(r,dsn), E_{ka}(S_{Sks}(bid)), hash(r,E_{ka}(S_{Sks}(bid)))], S_{Sks}(bid)$ is a bid data block signed by supplier's secret key S_{ks}. After receiving messages came from suppliers, Market publishes all $(RFQ\#, (dsn, M, hash(M)))$ on publish board. Suppliers can check and verify whether their own bid-message was actually received and displayed by market.

4.2.3 Auction Close

When the auction is closed, market authenticate supplier's agent by using LPN authentication method. After that, market publishes notice which announced whether the supplier's agent passes the authentication or not. If passed, supplier's agent send $S_{Sks}(dsn, r, x_{i_3} \oplus y_{i_3} \oplus K_a)$, after receiving it, market publishes $dsn, r, x_{i_3} \oplus y_{i_3} \oplus K_a$. Suppliers can check and verify whether their messages were actually received and displayed by market. Market publish $RFQ\#, E_{p_{kc}}((dsn_1,r_1,k_{a_1}),......(dsn_i,r_i,k_{a_i}), ...$ $(dsn_n,r_n,k_{a_n}))$, where $E_{p_{kc}}()$ is encrypted by customer's public key. According to the message market published, customer download corresponding M from the publish/subscribe system and decrypt data block and get all valid bid information. Having authenticated supplier group twice, market can make sure that he would publish valid supplier's information.

4.2.4 Winner Determination

The customer determines a winner and sends message $S_{Skc}(RFQ\#, r_{winder})$ to inform market winner information. After market receiving the customer's message, market checks whether the winner contained in legal supplier list $(dsn_1,r_1,k_{a_1}),......(dsn_i,r_i,k_{a_i})$, $......(dsn_n,r_n,k_{a_n})$. If the winner is contained in the list, market publishes the winner information on board for notifying suppliers. If there is a controversy about winner, market will choose a large prime number p and a primitive element g for GF(p), where GF(p) is the set of integers $\{0,1, ... ,p-1\}$ with arithmetic operations defined modulo p. And market generate a symmetric session key K, $K \in Z^*q$. Then market uses the signature datum of suppliers who have taken part in current auction to construct a derivation function $F(x)$ and conceals K in it. After that, market publish the coefficient $c_{n-1},c_{n-2},......c_1,c_0$ of the $F(x)$, where $S_{Sks_i}(k_{a_i},r_i)$ are supplier's signature data ($i=1,2,....n$). Market encrypts bid determination process data and publish $RFQ\#, E_k(bid\ determination\ process\ data)$.

$$F(x) = \prod_{i=1}^{n}(x-h_i) + K \bmod p = x^n + c_{n-1}x^{n-1} + ... + c_1x + c_0 \bmod p$$

where $h_i = g^{S_{Sksi}(k_{ai},r_i) \bmod p-1}$ and $c_{n-1},c_{n-2},......c_1,c_0 \in Z^*q$

5 Security Analysis

5.1 Replaying Attack

In planning phase, market published ticket token polynomial, while each legitimate supplier can get valid dsn and r from the market, then the suppliers will embed dsn and

r into message (M) which the suppliers sends to market in bidding phase. Furthermore, dsn and r are always fresh in current auction, so we don't worry that an attacker can use a replay data to cheat market. In bidding phase, the market utilizes publish/subscribe system to publish ($RFQ\#, (dsn, hash(M))$), it also avoids replaying attack to some extent. When auction is closed, the supplier will be authenticated by market using LPN-based method first. During the LPN-based authentication, the market will choose challenge $b_i \in_R \{0,1\}^k$ randomly each round and each auction time, also the suppliers' message are published on board each time. So even if there is an attacker use replay data to send market, he will not pass the authentication. According to what is mentioned above, we do not worry about the replay data.

5.2 Man-in-the-Middle Attacks

We assume that there is a rogue S' between market and a supplier, and he tries to impersonates a legitimate supplier by intercepting exchanged message.In bidding phase, even if S' intercepted message $M=[RFQ\#,dsn,hash(r,dsn), E_{ka}(S_{Sks}(bid))$ $,hash(r,E_{ka}(S_{Sks}(bid)))]$, he cannot modify it, because the legitimate supplier will check the hash(M) on the publish board.When auction is closed, S' tries to disturb LPN-base authentication process. But he also cannot succeed because he cannot solve the LPN problem, also S' cannot modify the message which a supplier sent to market, because market will publish the data on board, every legitimate supplier can check the information. After authenticating, the legitimate supplier will send $S_{Sks}(dsn, r, x_{i_3} \oplus y_{i_3} \oplus K_a)$ to market, and the legitimate supplier can verify it on publish board. So there is not any opportunity for attacker to cheating.

So It is impossible to S' to disturb auction process by intercepting or modifying data block. According to this, the mechanism between the communication parties can avoid Man-in-the-middle attack.

5.3 DOS (Denial-Of-Service) Attack

D. Kesdogan and L. Buttyan discussed some question about the anonymity in group communication protocols. The study showed broadcast networks can be safe using some method under some situation, the ticket token mechanism is a good solution for our auction protocol [14, 15]. Each legitimate supplier can get ticket token from ticket token polynomial, interested suppliers utilize ticket tokens to download the RFQ. This method can efficiently avoid some attackers taking part in bid-process. And the data dsn and r will be embedded in the message (M) for checking. When a supplier sends M to market, data dsn and $E_{ka}\{S_{Sks}(bid)\}$ have identified data block $hash(r,dsn)$ and $hash(r, E_{ka}\{S_{Sks}(bid)\})$ separately. This improvement will make market identify junk data efficiently and relieve DOS attack on market to some extent.

5.4 Public Verifiability and Avoid Collusion

After market published the winner information on board, if there is a controversy about winner, market constructs key interpolating polynomial for sharing the bid determination process data. All participants can use their own $S_{Sks}(k_a, r)$ to work out derivation

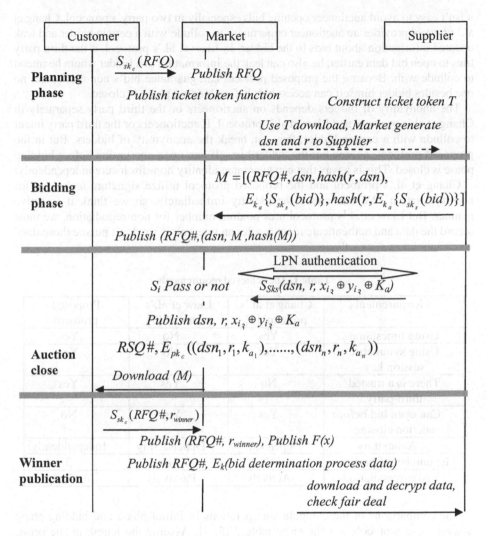

Fig. 2. The proposed protocol

function and get the key to decrypt the data, each participant in this bid process can analyze the fairness of the bid process. It will supervise whether the customer is fair to every supplier or not.

6 Properties Achieved and Discussion

In this section, we discuss the proposed protocol with other protocols. The first comparison is summarized in table 1 [5, 7].

There isn't a deliberate mechanism which can avoid opening bid before bidding phase is closed in Chang et al.'s protocol. And sealed-bid security is inherent weakness, because it isn't easy to avoid auctioneer opening bids especially in two party's protocol. Chang et al.' protocol provides an auctioneer opportunity to collude with a certain bidder and leak updated information about bids to the bidder. In Liaw et al.'s protocol, if the third party tries to open bid data earlier, he also can leak the information to a bidder whom he intend to collude with. Because the proposed protocol can guarantee bid's non-disclosure, no one besides bidder himself can access bids before bidding phase is closed.

The anonymity of bidders depends on auctioneers or the third party separately in Chang et al.'s protocol and Liaw et al.'s protocol. If auctioneers or the third party intent to collude with a certain bidder, they can break the anonymity of bidders. But in the proposed protocol, no one but the bidder himself can open encryption before bidding phase is closed. This is a way that can guarantee identity non-disclosure independently.

Chang et al.'s protocol and the proposed protocol utilize signature technique for non-repudiation. We can check the identity immediately, so we think it is active manner. But Liaw et al.'s protocol uses random number for non-repudiation, we must record the data and authenticate identity after an interval. If we don't pursue those data, we cannot know it exactly.

Table 1. The achieved requirements

Requirements	Chang et al.'s protocol [5]	Liaw et al.'s protocol [7]	Proposed protocol
Using timestamp	Yes	No	Yes
Using symmetric session key	Yes	No	Yes
There is a trusted third-party	No	Yes	Yes
Can open bid before auction closing	Yes	Yes	No
Anonymity	Dependently	Dependently	Independently
The number of parties	2	4	3
Non-repudiation	Actively	Passively	Actively

The comparisons of the computation operations of initial phase and bidding phase among these protocols are shown in table 2 [5, 7]. Assume the length of the prime number p is 1024 bits in Diffie-Hellman and public key encryption, symmetric key length is 128 bits (for AES), hash function digest is 160 bits (for SHA-1), public key certification is 1024 bits, signature length is 320 bits (for DSA). Because RSA's computation operation can be summarized as a modular exponentiation operation, and the computation operation of a modular exponentiation is about $O(|n|)$ times that of a modular multiplication, where $|n|$ denotes the bit length of n. So compared with a modular multiplication computation in Zn^*, the computation time consumed by hashing operations, symmetric encryptions or decryptions can be neglected [16]. And symmetric cryptosystem is faster 1000 times than asymmetric cryptosystem and hash function is faster 10 times than symmetric cryptosystem [17].

Even if liaw's protocol excludes all the computation operations of bank party, it is still more inefficient than Chang's protocol. Obviously, the proposed protocol is more efficient than Chang's protocol and Liaw's protocol in total.

Table 2. The numbers of different computation operations

Phase	Chang et al.'s protocol [5]	Liaw et al.'s protocol [7]	Proposed protocol
The initiation phase	4 HF 2 SKE 2 SKD 2 PKE 2 PKD 4 ME 2 R	5 HF 0 0 3 PKE 3 PKD 0 5 R	2HF 0 0 1 PKE 1 PKD 4 ME 3 R
The bidding phase	0 2 SKE 2 SKD 1 PKE 1 PKD 0	0 0 0 5 PKE 5 PKD 0	4 HF 1 SKE 0 1 PKE 0 0

PKE: Public Key Encryption; PKD: Public Key Decryption;
SKE: Symmetric Key Encryption; SKD: Symmetric key decryption;
HF: Hash Function; ME: Modular Exponentiation; R: generate a random number

7 Conclusions

As mentioned above, the current paper demonstrated that the proposed electronic marketplace bidding auction protocol relieves DOS attack on market to some extent, protects supplier anonymity, provides sealed bid with independent private values and avoids collusion among parties. Also it satisfies the security requirements of an electronic auction, such as anonymity, non-repudiation, verifiability etc. And the proposed scheme relaxes some trust assumptions for three-party in Jaiswal's scheme. According to the discussion and analysis with Chang et al.'s protocol and Liaw et al.'s protocol, our proposed protocol needs less computation cost in initiation phase and bidding phase, and has better security in anonymity, fairness and reliance on the third party. Therefore, the advantages of our proposed bidding auction protocol are that collusion will be difficult and computation cost is low.

Reference

1. Collins, J., Ketter, W., Gini, M.: A multi-agent negotiation testbed for contracting tasks with temporal and precedence constraints. International Journal of Electronic Commerce 7(1), 35–57 (2002)
2. Jaiswal, A., kim, Y., Gini, M.: design and implementation of a secure multi-agent marketplace. Electronic Commerce Research and Applications 3(4), 355–368 (2004)
3. Chang, C.C., Chang, Y.F.: Efficient anonymous auction protocols with freewheeling bids. computer & security 22(8), 728–734 (2003)
4. Jiang, R., Pan, L., Li, J.H.: An improvement on efficient anonymous auction protocols. computer & security 24(2), 169–174 (2005)

5. Chang, Y.F., Chang, C.C.: Enhanced anonymous auction protocols with freewheeling bids. In: Proceedings of 20th International Conference on Advanced Information Networking and Applications(AINA), vol. 1, pp. 353–358 (2006)
6. Hwang, M.S., Lu, E.J.L., Lin, I.C.: Adding timestamps to the electronic auction protocol. Data & Knowledge Engineering 40(2), 155–162 (2002)
7. Liaw, H.-T., Juang, W.-S., Lin, C.-K.: An electronic online bidding auction protocol with both security and efficiency. Applied Mathematics and Computation 174(2), 1487–1497 (2006)
8. Rivest, R.L., Shamir, A., Wagner, D.A.: Time-lock puzzles and timed-release crypto, Technical Report MIT/LCS/TR-684. MIT Press, Cambridge (1996)
9. Ingemaresson, I., Tang, T.D., Wong, C.K.: A conference key distribution system. IEEE Trans. Info. Theory 28(5), 714–720 (1982)
10. Tang, Q., Mitchell, C.J.: Comments on two anonymous conference key distribution systems. Computer Standards & Interfaces 27(4), 397–400 (2005)
11. Regev, O.: On Lattices, Learning with errors, Random Linear Codes. and Cryptography. In: 37th ACM symposium on theory of computing, pp. 84–93 (2005)
12. Juels, A., Weis, S.: Authenticating pervasive devices with human protocols. In: Shoup, V. (ed.) CRYPTO 2005. LNCS, vol. 3621, pp. 293–308. Springer, Heidelberg (2005)
13. Gilbert, H., Robshaw, M., Silbert, H.: An active attack against HB+ − a provable secure lightweight authentication protocol. Cryptology ePrint Archive, Report 2005/237 (2005)
14. Kesdogan, D., Palmer, C.: Technical challenges of network anonymity. Computer Communications 29(3), 306–324 (2006)
15. Buttyan, L., Ben Salem, N.: A Payment Scheme for Broadcast Multimedia Streams. In: Proceedings of the 6th IEEE Symposium on Computers and Communications, pp. 668–673 (2001)
16. Fan, C.-I., Chan, Y.-C., Zhang, Z.-K.: Robust remote authentication scheme with smart cards. Computers & Security 24(8), 619–628 (2005)
17. Chang, C.C., Lail, C.S., Harn, L.: Contemporary Cryptography and its Applications, 2nd edn. Unalis Co (2001)

A Provable Secure Authentication Protocol Given Forward Secure Session Key

Wenbo Shi, Injoo Jang, and Hyeong Seon Yoo

School of Computer Science and Engineering, Inha University
Incheon, 402-751, Korea
swb319@hotmail.com

Abstract. This paper proposes a key distribution and authentication protocol between user, service provider and key distribution center (KDC). This protocol is based on symmetric cryptosystem, challenge-response, Diffie-Hellman component and hash function. In proposed protocol, user and server update the session key under token-update operation, and user can process repeated efficient authentications by using updated session keys. Another merit is that KDC needs not to totally control the session key between user and server in proposed protocol. Even if an attacker steals the parameters from the KDC, the attacker still can not calculate session key. We use BAN logic to proof these merits of our proposed protocol. Also according to the comparison and analysis with other protocols, our proposed protocol provides good efficiency and forward secure session key.

Keywords: key distribution, authentication, forward secure session key.

1 Introduction

As the explosive growth of internet, many applications need to provide security service to more and more users, so user authentication and key distribution have become an important part of network security especially over untrustworthy networks.

The Kerberos authentication system is based on the well-known Needham-Schroeder authentication protocol [1, 2]. The Kerberos is a fairly mature standard and has good security properties, but it still has a number of limitations and weaknesses. For examples, it is not effective against password guessing attacks and replay attacks, exposure of session keys and so on [3, 4]. So some proposed authentication protocols want to improve Kerberos using public key cryptography [5, 6]. They distributed the authentication workload form the centralized KDC to the individual principals, but the implementation cost is still high.

Another protocol providing authentication and key distribution services, KryptoKnight was proposed in 1992 [7, 8]. They implement a family of novel authentication and key distribution protocols different from Kerberos. Those protocols offer several advantages over Kerberos and are more flexible than Kerberos. Paul

Y. Zhang et al. (Eds.): APWeb 2008, LNCS 4976, pp. 309–318, 2008.

Syverson provided an enhancement on KSL protocol and NS protocol in 1993 [9]. And AUTHMAC_DH protocol used the message authentication codes to exchange the Diffie-Hellman components to overcome some Kerberos's drawbacks [10]. Although most of discussed protocol provided the authentication and key distribution services, they did not establish the true session key between user and server [7, 8, 9, 10].

Chien proposed a protocol based on public key, challenge-response and hash chaining, with several practical merits [11]. It alleviates the burden of client side in computational cost, but the total computation cost is still more than those protocols based on symmetric key [11, 12, 13]. Hwang's protocol is more efficient than Chien's protocol and the communication cost is lower than it [11, 13].

In this paper, an authentication and key distribution protocol given forward secrecy is proposed. It updates session key every time by token-update operation. In addition, the proposed protocol provides another merit is that KDC need not to totally control the session key between a user and a server, it relieves the computation cost burden of KDC and prevents the attacker calculating session key even after stealing the information from KDC.

2   Proposed Protocol

The proposed protocol has two parts: initial phase and subsequent phase (Fig. 1). The KDC maintains a secret key K_c to calculate the symmetric key for his users. K_{uc} is the long-term shared key between the user and the KDC, while K_{sc} is the long-term shared key between the server and the KDC, where $K_{uc}=f(K_c,U)$, $K_{sc}=f(K_c,S)$, U and S denote their identities, f(•) is a kind of hash function.

2.1   Initial Phase

In order to get service from a server (S), a user (U) performs the initial phase to establish the session key with the server as following steps.

(I.1)   U->S $\{U, a^x \bmod p, h(a^x \bmod p, K_{uc})\}$
   Where p is a large prime and a is a generator with order $p-1$ in $GF(p)$, h(•) denotes a one-way hash function.
   1 U randomly selects a secret number x to compute $a^x \bmod p$.
   2 U calculates $h(a^x \bmod p, K_{uc})$.
   3 U sends his identity (U), $a^x \bmod p$, $h(a^x \bmod p, K_{uc})$ to server.

(I.2)   S->KDC $\{U, a^x \bmod p, h(a^x \bmod p, K_{uc}), S, a^y \bmod p, h(a^y \bmod p, K_{sc})\}$
   1 S stores $a^x \bmod p$.
   2 S randomly selects a secret number y to compute $a^y \bmod p$.
   3 S calculates $h(a^y \bmod p, K_{sc})$.
   4 S sends $U, a^x \bmod p, h(a^x \bmod p, K_{uc})$ and his identity (S), $a^y \bmod p$,
      $h(a^y \bmod p, K_{sc})$ to KDC.

(I.3) KDC->S { E_{Ksc} (a^y mod p, n, r, E_{Kuc} (a^x mod p, a^y mod p ,n, r))}
     1 KDC authenticates user and sever by checking the hash value
       $h(a^x$ mod p, $K_{uc})$ = $h(a^x$ mod p, $K_{uc})'$ and $h(a^y$ mod p, $K_{sc})$ = $h(a^y$ mod p, $K_{sc})'$
       separately.
     2 KDC chooses n and generates a random number r. Where n is the number of
       times that U can communicate with S.
     3 KDC sends E_{Ksc} (a^y mod p, n, r, E_{Kuc} (a^x mod p, a^y mod p, n, r)) to S.

(I.4) S->U { E_{Kuc} (a^x mod p, a^y mod p, n, r), E_{Kn} (a^x mod p|| a^y mod p)}
     1 S decrypts message E_{Ksc} (a^y mod p, n, r, E_{Kuc} (a^x mod p, a^y mod p, n, r)) using K_{sc}.
     2 S authenticates KDC by checking the value of a^y mod p.
     3 S computes the a^{xy} mod p to get K_n and encrypts a^x mod p,a^y mod p using K_n.
     4 S sends E_{Kuc} (a^x mod p, a^y mod p, n, r), E_{Kn} (a^x mod p|| a^y mod p) to U.
     5 S keeps U, K_n, n, r.

(I.5) U receives { E_{Kuc} (a^x mod p, a^y mod p,n, r), E_{Kn} (a^x mod p|| a^y mod p)}
     1 U decrypts message E_{Kuc} (a^x mod p, a^y mod p,n, r) using K_{uc}.
     2 U authenticates KDC by checking the a^x mod p.
     3 U computes the a^{xy} mod p to get K_n.
     4 U decrypts (a^x mod p || a^y mod p) using K_n.
     5 U authenticates server by checking the value of a^x mod p and a^y mod p.
     6 U keeps K_n,n, r.

2.2 Subsequent Phase

In subsequent phase, without losing the generality, we assume the user is requesting the
ith service now.

(S.1) U->S {U, E_{Kn-i+1} (r_{n-i+1})}
     1 U encrypts r_{n-i+1} using K_{n-i+1}.
     2 U sends U, E_{Kn-i+1} (r_{n-i+1}) to S.

(S.2) S->U { E_{Kn-i} (r_{n-i+1}|| r_{n-i})}
     1 S decrypts the message E_{Kn-i+1} (r_{n-i+1}) using K_{n-i+1}.
     2 S authenticates U by checking r_{n-i+1} ?= r'_{n-i+1}.
     3 If r_{n-i+1} == r'_{n-i+1}, S computes K_{n-i}= hash(K_{n-i+1}|| r_{n-i+1}).
     4 S generates a new random number r_{n-i}.
     5 S encrypts(r_{n-i+1}|| r_{n-i}) using K_{n-i}.
     6 S stores K_{n-i} and r_{n-i}, updates i.

(S.3) U receives { E_{Kn-i} (r_{n-i+1}|| r_{n-i})}
     1 U computes K_{n-i}= hash(K_{n-i+1}|| r_{n-i+1}).
     2 U decrypts E_{Kn-i} (r_{n-i+1}|| r_{n-i}), and authenticates U by checking r_{n-i+1} ?= r'_{n-i+1}.
     3 If r_{n-i+1} == r'_{n-i+1}, U stores K_{n-i} and r_{n-i}, updates i.

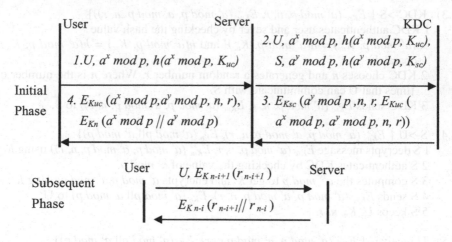

Fig. 1. The proposed protocol

3 Security Analysis

3.1 Replaying Attack

In initial phase, user and server generate nonce x and y, calculate $a^x \bmod p$ and $a^y \bmod p$. After checking the freshness of those nonce numbers, KDC encrypts $a^x \bmod p$ and $a^y \bmod p$ separately using their symmetric key, only user and server themselves can decrypt the number and check the validity of their own number. If an attacker wants to cheat others by replaying old messages, it can be prevented easily.

In subsequent phase, the user used the K_{n-i+1} and the random number r_{n-i+1} to constructed K_{n-i} =hash(K_{n-i+1}|| r_{n-i+1}), and encrypt the r_{n-i} to login server. Server can validate the data by decrypting it, and judge the data is from the legal user or not. In addition, the session key will be updated every login, we do not worry about the replay data.

3.2 Know Key Attack and Forward Secrecy

In proposed protocol, even if an attacker compromises an old session key K_{n-i+1} shared by user and server, the known key attack still fails, because the session key is updated every login by the random number generated by server. In addition, the random number is updated on every login and only shared by user and server, because the attacker can not get the updated random number to calculate the new session key, so the known key attack fails.

To prove the forward secrecy in the proposed protocol, we also give three assumptions.

We assume that an attacker knows the shared key K_{uc}. Because the attacker does not know the user's secret number x, even if the attacker can get $g^y \bmod p$, n and r, he does not have any opportunity to obtain K_n. Even if he can launch impersonate user attack to cheat the server and the KDC in initial phase, but the session key is still secure.

We assume that an attacker knows the shared key K_{sc}. Because the attacker does not know the server's secret number y, so the situation is almost same as the attacker knows the shared key K_{uc}.

We assume that an attacker knows the shared key K_{uc} and K_{sc}. Because the attacker does not know the user's secret number x or the server's secret number y, even if he can get $g^x \bmod p$, $g^y \bmod p, n$ and r, he has not any opportunity to obtain K_n. Even if he can launch impersonating attack to cheat the server or the KDC in initial phase, but the session key is still secure.

3.3 Man-in-the-Middle Attacks

We assume that there is a rogue S', he impersonate S by intercepting message1 and constructing message2', but it is impossible to him to forge a message2' can cheat KDC, because the attacker can not get the long-term keys K_{uc} and K_{sc}, message3 and message4 are also protected by the key K_{uc}, K_{sc} and session key K_n. According to this, the authentication between the communication parties can avoid Man-in-the-middle attack.

3.4 Authentication Proof Based on BAN Logic

In this section, we will prove the proposed protocol can achieve the goals of authentication and key distribution using BAN logic [14]. The BAN logic can analyze the required initial assumptions of the participants and their final beliefs.

The symbols U, S and C denote the user, service provider and key distribution center. $\{X\}_K$ represents the formula X encrypted under the key K. $<X>_Y$ represents X combined with the formula Y. $Y \mid\Rightarrow X$ represents a principal Y has jurisdiction over X. $Y \mid\equiv X$ represents a principal Y believes X. $\#(X)$ represents the formula X is fresh [14].

According to BAN logic, we give some initial assumptions [14]:

1 $U \mid\equiv U \xleftrightarrow{K_{uc}} C$ 6 $S \mid\equiv (C \mid\Rightarrow S \xleftrightarrow{K_n} U)$

2 $S \mid\equiv S \xleftrightarrow{K_{sc}} C$ 7 $U \mid\equiv (C \mid\Rightarrow S \xleftrightarrow{K_n} U)$

3 $C \mid\equiv U \xleftrightarrow{K_{uc}} C$ 8 $U \mid\equiv \#(a^x \bmod p)$

4 $C \mid\equiv S \xleftrightarrow{K_{sc}} C$ 9 $S \mid\equiv \#(a^y \bmod p)$

5 $C \mid\equiv S \xleftrightarrow{K_n} U$

Assumptions 1-5 are about the shared keys among U, S and C. Assumptions 6 and 7 indicate that U and S trust in C, they believed that C provide good protection on the exchanged messages and forward the messages to the other side honestly. Assumptions 8 and 9 show that two nonces have been generated by various principals who consider them to be fresh.

According to the BAN logic [14], the goals of authentication and key distribution are defined as follows:

$1\ U \models S \xleftrightarrow{K_n} U$ $5\ U \models C \mid \sim (r)$

$2\ S \models S \xleftrightarrow{K_n} U$ $6\ S \models C \mid \sim (r)$

$3\ U \models S \models S \xleftrightarrow{K_n} U$ $7\ C \models U \models (r)$

$4\ S \models U \models S \xleftrightarrow{K_n} U$ $8\ C \models S \models (r)$

Our goals include two aspects. Goals 1-4 are about that U believes that S trusts in the distributed key K_n, and S believes that U trusts in the distributed key K_n. Goals 5-8 are about that C believes that U and S trust in the freshness of the random number generated by C, and U and S trust in the random number.

The idealized protocol is as follows:

Step 1 $U \rightarrow S$ $< a^x \bmod p > k_{uc}$

Step 2 $S \rightarrow C$ $< a^x \bmod p > k_{uc}, < a^y \bmod p > k_{sc}$

Step 3 $C \rightarrow S$ $\{ a^y \bmod p, S \xleftrightarrow{K_n} U, r, \{a^x \bmod p, a^y \bmod p, S \xleftrightarrow{K_n} U, r\} k_{uc} \} k_{sc}$

Step 4 $S \rightarrow U$ $\{a^x \bmod p, a^y \bmod p, S \xleftrightarrow{K_n} U, r\} k_{uc}, \{a^x \bmod p, a^y \bmod p, S \xleftrightarrow{K_n} U\} k_n$

The proof may be briefly outlined as follows:
Since step 3 and assumption 2, message-meaning rule applies and yields:

$$\text{"} S \models C \mid \sim \left(a^y \bmod p, S \xleftrightarrow{K_n} U, r, \{a^x \bmod p, a^y \bmod p, S \xleftrightarrow{K_n} U, r\} k_{uc} \right) \text{"}$$

Inference (1)

Since Inference (1), belief conjuncatenation rule applies and yields:

$$\text{"} S \models C \mid \sim \left(a^y \bmod p, S \xleftrightarrow{K_n} U, r \right) \text{"}$$

Inference (2)

Since Inference (2), belief conjuncatenation rule applies and yields:

$$\text{"} S \models C \mid \sim \left(r \right) \text{"}$$

Inference (3)

Since Inference (3) and assumption 4, yields:

$$\text{"} C \models S \models \left(r \right) \text{"}$$

Inference (4)

Since assumption 9, freshness conjuncatenation rule applies and yields:

$$\text{"} S \models \# \left(a^y \bmod p, S \xleftrightarrow{K_n} U, r \right) \text{"}$$

Inference (5)

Since Inference (2) and (5), nonce-verification rule applies and yields:

$$\text{"} S \models C \models \left(a^y \bmod p, S \xleftrightarrow{K_n} U, r \right) \text{"}$$

Inference (6)

Since Inference (6), belief conjuncatenation rule applies and yields:

$$\text{"}S \models C \models \left(S \xleftrightarrow{K_n} U \right)\text{"}$$ Inference (7)

Since Inference (7) and assumption 6, jurisdiction rule applies and yields:

$$\text{"}S \models \left(S \xleftrightarrow{K_n} U \right)\text{"}$$ Inference (8)

Since step 4 and assumption 1, message-meaning rule applies and yields:

$$\text{"}U \models C \mid \sim \left(a^x \bmod p, a^y \bmod p, S \xleftrightarrow{K_n} U, r \right)\text{"}$$ Inference (9)

Since Inference (9), belief conjuncatenation rule applies and yields:

$$\text{"}U \models C \mid \sim (r)\text{"}$$ Inference (10)

Since Inference (10) and assumption 3, yields:

$$\text{"}C \models U \models (r)\text{"}$$ Inference (11)

Since assumption 8 and Inference (9), freshness conjuncatenation and nonce-verification rules apply and yield:

$$\text{"}U \models C \models \left(a^x \bmod p, a^y \bmod p, S \xleftrightarrow{K_n} U, r \right)\text{"}$$ Inference (12)

Since assumption 7 and Inference (12), jurisdiction rule applies and yields:

$$\text{"}U \models \left(S \xleftrightarrow{K_n} U \right)\text{"}$$ Inference (13)

Since step 4 and Inference (13), message-meaning rule applies and yields:

$$\text{"}U \models S \mid \sim \left(a^x \bmod p, a^y \bmod p, S \xleftrightarrow{K_n} U \right)\text{"}$$ Inference (14)

Since assumption 8 and Inference (14), freshness conjuncatenation and nonce-verification rules apply and yield:

$$\text{"}U \models S \models \left(a^x \bmod p, a^y \bmod p, S \xleftrightarrow{K_n} U \right)\text{"}$$ Inference (15)

Since Inference (15), belief conjuncatenation applies and yields:

$$\text{"}U \models S \models \left(S \xleftrightarrow{K_n} U \right)\text{"}$$ Inference (16)

If U uses the session key K_n to encrypt and send the message to S, then we yields:

$$\text{"}S \models U \models \left(S \xleftrightarrow{K_n} U \right)\text{"}$$ Inference (17)

By Inference (3),(4),(8),(10),(11),(13),(16) and (17), we prove that the proposed protocol achieves the goals of authentication and key distribution.

4 Discussions

In this section, we compare the proposed protocol with other protocols. The first comparison is summarized in table1.

Kerberos V5 protocol and Shieh's protocol send their symmetric key during key distribution [1, 12]. But Chie's protocol, Huang's protocol and our proposed protocol do not, that means an improvement at security aspect [11, 13].

We assume that KDC is controlled by an attacker, trojan horse is planted. If the attacker steal the session key information (a,n) in Chie's protocol and Huang' protocol, he can know very well about every session key between the user and the server [11, 13]. In proposed protocol, even if the attacker knows all the parameters from KDC, he can not calculate the session key. It also improves in secure aspect.

Table 1. Comparisons with some other protocols

	Kerberos V5 [1]	Shieh [12]	Chie [11]	Huang [13]	Proposed protocol
Cryptographic primitives	SK	SK, Ce	PK	PK	SK, DH
Initial/Subsequent	4/2	5/3	4/2	4/2	4/2
Nonce/clock	timestamp	nonce	nonce	nonce	nonce
send symmetric key	yes	yes	no	no	no
Know key attack	Insecure [11,13]	Insecure [11,13]	Secure [11,13]	Secure [13]	secure
Dictionary attack	Insecure [11]	Insecure [11]	Secure [11]	secure	secure
KDC calculate	yes	yes	yes	yes	no

Message run of Initial and Subsequent phase: Initial/Subsequent;
KDC can calculate the session key: KDC calculate;
SK: Symmetric Key; Ce: certificate; PK: Public Key; DH: Diffie-Hellman;

The comparisons of the computation cost of initial phase among these protocols are shown in table 2 [11, 12, 13].

Assume the length of the prime number p is 1024 bits in Diffie-Hellman, and public key encryption is 1024 bits (for RSA), symmetric key length is 128 bits (for AES), hash function digest is 160 bits (for SHA-1), public key certification is 1024 bits, signature length is 320 bits (for DSA).

Because RSA's computation cost can be summarized as a modular exponentiation operation, and the computation cost of a modular exponentiation is about $O(|n|)$ times that of a modular multiplication where $|n|$ denotes the bit length of n. So compared with a modular multiplication computation in Zn^*, the computation time consumed by hashing operations, symmetric encryptions or decryptions can be neglected [15]. And symmetric cryptosystem is faster 1000 times than asymmetric cryptosystem and hash function is faster 10 times than symmetric cryptosystem [16, 17].

Shieh's protocol needs five message runs in initial phase, it is more inefficient than others [12]. According to the table2, the proposed protocol is more efficient than Chie's protocol [11]. Huang's protocol is more efficient than Chie's protocol and proposed protocol in initial phase [11, 13]. But the proposed protocol can adopt pre-computing to shorten the communication time.

Table 2. The comparisons of the computation cost of initial phase

	Shieh [12]			Chie [11]			Huang [13]			Proposed protocol		
	U	S	C	U	S	C	U	S	C	U	S	C
PKE	0	0	0	1	1	2	0	0	0	0	0	0
PKD	0	0	0	1	1	2	0	0	0	0	0	0
SKE	2	1	2	0	0	0	0	1	2	0	1	2
SKD	1	2	0	0	0	0	2	1	0	2	1	0
HF	0	0	0	1	1	n	n+2	2	n+4	0	0	0
others	0	0	0	1Si, 1Ce	1Si, 1Ce	1Si, 2Ce	0	0	0	2 ME	2 ME	0

PKE: Public Key Encryption; PKD: Public Key Decryption;
SKE: Symmetric Key Encryption; SKD: Symmetric key decryption;
HF: Hash Function; Si: signature; Ce: certificate; ME: Modular Exponentiation

The comparisons of the computation cost of subsequent phase among these protocols are shown in table 3 [11, 12, 13]. Shieh's protocol needs three message runs in subsequent phase, it is more inefficient than others [12]. Huang's protocol is obviously more efficient than Chie's protocol [11, 13]. The proposed protocol is more efficient than Huang's protocol in subsequent phase [13].

Table 3. The comparisons of the computation cost of subsequent phase

	Shieh [12]		Chie [11]		Huang [13]		Proposed protocol	
	U	S	U	S	U	S	U	S
SKE	n	n	n	n	n	n	n	n
SKD	n	n	n	n	n	n	n	n
HF	0	0	n(n+2)/2	3n	n(n+2)/2	3n	2n	2n
Others	0	0	n PKE	n PKD, n Si	0	0	0	0

PKE: Public Key Encryption; PKD: Public Key Decryption;
SKE: Symmetric Key Encryption; SKD: Symmetric key decryption;
HF: Hash Function; Si: signature;

5 Conclusions

In this paper, a more secure authentication and key distribution protocol given forward secrecy is presented. In addition, the protocol changed the situation that KDC needs to totally control the session key between user and server. The session key is efficiently decided by a user, a server and a KDC. Even if a attacker steals all the parameters about the session key from the KDC, the attacker still can not calculate the session key. The user and the server update session key and authenticate each other every time. According to the analyses in section 3 and section 4, the proposed protocol shows improved security and provides good efficiency.

References

1. Kohl and Neuman, The Kerberos network authentication service (v5). In: Internet Request for Comments RFC-1510 (1993)
2. Needham, R.M., Schroeder, M.D.: Using encryption for authentication in large networks of computers. Communication of the ACM 21(12), 993–999 (1978)
3. Neuman, B.C., Ts'o, T.: Kerberos: An authentication service for computer networks. IEEE Communications 32(9), 33–38 (1994)
4. Bellovin, S.M., Merrit, M.: Limitations of the Kerberos authentication system. Computer Communication Review 20(5), 119–132 (1990)
5. Ganesan, R.: Yaksha: augmenting Kerberos with public key cryptography. In: Ganesan, R. (ed.) SNDSS'95: Proceedings of the 1995 Symposium on Network and Distributed System Security, pp. 132–143. IEEE Computer Society Press, Los Alamitos (1995)
6. Sirbu, M.A., Chuang, J.C.I.: Distributed authentication in Kerberos using public key cryptography. In: Proceedings of the 1997 Symposium on Network and Distributed System Security, IEEE Computer Society, pp. 134–141. IEEE Computer Society, Los Alamitos (1997)
7. Molva, R., Tsudik, G., van Herreweghen, E., Zatti, S.: KryptoKnight authentication and key Distribution System. In: Deswarte, Y., Quisquater, J.-J., Eizenberg, G. (eds.) ESORICS 1992. LNCS, vol. 648, pp. 155–174. Springer, Heidelberg (1992)
8. Bird, R., Gopal, I., Herzberg, A., Janson, P., Kutten, S., Molva, R., Yung, M.: The KryptoKnight family of Light-weight protocols for authentication and key distribution. IEEE Transactions on Networking 3(1), 31–42 (1995)
9. Syverson, P.: On key distribution protocols for repeated authentication. ACM SIGOPS Operating Systems Review 27(4), 24–30 (1993)
10. Aslan, H.K.: Logical analysis of AUTHMAC_DH: A new protocol for authentication and key distribution. Computers & Security 23(4), 290–299 (2004)
11. Chien, H.Y., Jan, J.K.: A hybrid authentication protocol for large mobile network. Journal of Systems and Software 67(2), 123–130 (2003)
12. Shieh, S.P., Ho, F.S., Huang, Y.L.: An efficient authentication protocol for mobile networks. Information Science and Engineering 15(4), 505–520 (1999)
13. Hwang, R.-J., Su, F.-F.: A new efficient authentication protocol for mobile networks. Computer Standards & Interfaces 28(2), 241–252 (2005)
14. Burrows, M., Abadi, M., Needham, R.: A logic of authentication. ACM Transactions on Computer Systems 8(1), 18–36 (1990)
15. Fan, C.-I., Chan, Y.-C., Zhang, Z.-K.: Robust remote authentication scheme with smart cards. Computers & Security 24(8), 619–628 (2005)
16. Hwang, M.S., Lin, I.C., Li, L.H.: A simple micro-payment scheme. Journal of Systems and Software 55(3), 221–229 (2001)
17. Chang, C.C., Lail, C.S., Harn, L.: Contemporary Cryptography and its Applications, 2nd edn. Unalis Co (2001)

A Secure Multi-dimensional Partition Based Index in DAS

Jieping Wang[1,2] and Xiaoyong Du[1,2]

[1] School of Information, Renmin University of China
[2] Key Laboratory of Data Engineering and Knowledge Engineering, MOE
{wangjieping,duyong}@ruc.edu.cn

Abstract. Database-as-a-Service is an emerging data management paradigm wherein data owners outsource their data to external untrusted server. To keep server from unauthorized access, sensitive data are encrypted before outsource. Although partition based index is an effective way of querying encrypted data, several such indexes would lead to extra information leakage because of un-uniform multi-dimensional data distribution. In this paper we first introduce two security constraints based on minimal confidential interval and minimal occupation number, then propose a multi-dimensional partition which satisfies previous security constraints. Since optimal multi-dimensional partition is a NP-hard problem, we propose a heuristic based greedy algorithm which is simple and efficient. Experiments show that our index could achieve a stable trade-off between security and efficiency compared to multiple single-dimensional indexes.

Keywords: Database security, Multi-dimensional partition, DAS.

1 Introduction

Database-as-a-Service (DAS) [1] is an emerging data management paradigm, in which enterprises outsource their data to external service provider instead of install and maintain DBMS in-house. DAS could give enterprises higher data availability and more effective disaster protection than in-house operations. Besides, it could reduce maintenance cost by sharing expertise of database professionals. DBMS is comprised of two major components: the client and the server. According to whether they are trusted or not there are four different types of models which are illustrated in table 1.

Table 1. Four database models

	Trusted client	Untrusted client
Trusted server	single-user model	multi-user model
Untrusted server	DAS	data publishing

In DAS scenario, untrusted server poses many significant security challenges, foremost of which is the issue of data privacy. In multi-user scenario encryption [2,3] is an effective way of protecting data privacy. The encrypted data is also called cipher.

Y. Zhang et al. (Eds.): APWeb 2008, LNCS 4976, pp. 319–330, 2008.

However, encryption is often supported for sole purpose of protecting the data in storage and assumes trust in the server, which decrypts data for query execution, because major challenges come from outside attackers. While in DAS scenario, challenges not only come from outside attackers, but also come from the untursted server itself. Since the server is untrusted, all encryption related work has to be delegated to the trusted client. The optimization principle is to push as much work as possible to the server. The major challenge is how to build secure indexes on encrypted data. Various techniques are proposed [5,6,7,8,9,10], among which partition is an effective one. However, current partition only focuses on single-dimension. We will show how multiple such indexes would lead to extra information leakage in the following.

1.1 Attacks on Multiple Single-Dimensional Partition Based Index

We will show how multiple single-dimensional partition based indexes lead to extra information leakage by the following example.

Example 1: Consider the plaintext table patient in table 2(a), attribute age and disease are sensitive. Tuple level(referenced [1] about encryption granularity) encrypted table along with two partitions on sensitive attributes is illustrated in table 2(b). The partition scheme is plotted in table 2(c).

Table 2(a). Plaintext table

Name	Age	Disease
Alice	23	Flu
Jack	25	Pneumonia
Jone	28	High pressure
Linda	31	Diabetes
Susan	33	Diabetes
Tom	36	Cancer
Mike	42	Flu
Stone	45	Aids
Rose	49	Cancer

Table 2(b). Cipher table

ETuple	Age	Disease
&*(jk&*!@#	1	A
Sdfsd$%#@$	1	A
*jkk4566#$%	1	B
Jklkl)^%%$$	2	B
Sdf^}Y@#$o	2	B
Asdf@##$$%	2	C
Jklk$&*^^%$	3	A
Y%^@#$$%	3	C
^&jkl*(#$$%	3	C

Table 2(c). Partition scheme

Attribute	Partition interval	Identifier	percentage
	[20,30]	1	33%
Age	[30,40]	2	33%
	[40,50]	3	34%
	{Flu, Pneumonia }	A	33%
Disease	{Diabetes, High pressure}	B	33%
	{Cancer, Aids}	C	34%

Here we use equi-depth partition scheme, the depth of which is 3. In partition scheme, the tuples within the same bucket are indistinguishable from each other. So for any single attribute range query, there is a security constraint which ensures that the server will return at least 3 tuples if match. However, the security constraint is broken for multiple attribute range query, i.e., there is no lower limit (positive integer) on the number of returned tuples. For example, for predicate such as where age=1 and disease=B, the server will return only one tuple. The security broken comes from the un-uniform data distribution in multi-dimensional space. A graph representation is illustrated in figure 1, where each asterisk represents data along two dimensions: age and disease. Although the number of tuples within each bucket along single dimension is at least 3, the number of tuples in 2-dimensional cell has no lower limit (positive integer).

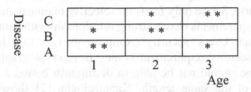

Fig. 1. 2-dimensional representation of tuples in table 2(b)

Observation 1: Although data distribution along single dimension is uniform, in most cases the data distribution along multiple dimensions is un-uniform. The more the number of dimensions, the higher un-uniform the distribution is.

1.2 Contributions and Paper Outline

From previous example we see that the security constraint, which holds on single-dimensional partition, would be broken when there are multiple such partitions. To solve above problem, we propose a secure multi-dimensional partition. Specifically, our major contributions are as follows:

1. We propose two security constraints based on minimal confidential interval and minimal occupation number, and one efficiency evaluator: normalized squared variance.
2. Single-dimensional partition considers one dimension at a time, while our multi-dimensional partition considers multiple dimensions at a same.
3. Since optimal multi-dimensional partition is a NP-hard problem, we propose a heuristic based greedy algorithm which is simple and efficient.

The rest of the paper is organized as follows. We first introduce security and efficiency evaluators in Section 2. Section 3 has an overview of related work in single-dimensional partition and multi-dimensional histograms. Section 4 describes how to construct a secure multi-dimensional partition. In section 4.2 we propose a heuristic based greedy algorithm for approximately optimal partition. Experiments in section 5 compare our partition with single-dimensional one under various scenarios. Section 6 concludes the paper and gives potential directions for future work.

2 Security and Efficiency Evaluation

2.1 Security Evaluation

To keep the server from unauthorized access, sensitive data are encrypted before outsource. To keep cipher secret, what we should do is to keep the attacker from inferring corresponding plaintext or keys. Security depends on the attacker's prior knowledge, users' security requirements and plaintext distribution. According to attacker's prior knowledge, there are three classes of attacks: known plaintext attack, selective plaintext attack and cipher only attack, among which selective plaintext is the most common scenario. So we only focus on selective plaintext attack.

Users' security requirements have a strong relationship with data distribution. So we discuss them together. Users' security requirements vary to a large extent, among which the strictest security requirement comes from cryptography community. The main idea is that one should not be able to distinguish between encryptions of two different messages of the same length. Kantarcioghu [7] showed that any secure database system, which meets the strictest security definition, will require a linear database scan to execute queries and thus will be inefficient and impractical. Fortunately in most cases users' security requirements are not so strict. In the following we will introduce two practical security requirements. The first one is based on minimal confidential interval and interval density. The formal definition is as follows.

Definition 1: *Density based minimal confidential interval.* Domain P is divided into s intervals such that $n_1/w_1=n_2/w_2=...n_s/w_s$, and $\min\{w_i|1<=i<=s\} >= w_{min}$, where w_{min} is minimal confidential interval, n_i and w_i represent the number of tuples and width of the ith interval respectively. Here we call n_i/w_i the density of the ith interval.

Under minimal confidential interval an attacker need not determine the exact plaintext p corresponding to a cipher value c. Instead, what an attacker need is to obtain a tight enough estimate of p. Definition 1 is only for numeric data. For non-numeric data, such as categorical data, preprocessing is needed to transform them to numeric data. When the data distribution is uniform, above partition can be reduced to equi-width partition.

The second security requirement is based on minimal occupation number. The formal definition is as follows.

Definition 2: *Minimal occupation number.* Minimal occupation number specifies the least number of tuples contained in any cell under multi-dimensional space. Assume minimal occupation number is k, then $\min\{cell_i|1<=i<= n)) >= k$., where n means the number of cells in multi-dimensional space, $cell_i$ means the number of tuples in the ith cell.

Minimal occupation number has some similarity with k-Anonymity in data publishing in that they are all used for de-identification of individual tuple. However, the solutions to them are quite different. K-Anonymity [18] is mainly achieved by data suppression and generalization. Suppression is the process of deleting cell values or entire tuples, mainly used for outliers. Generalization involves replacing specific values with more general ones. Data publishing is mainly used for statistical information query. Cell values can be updated under the condition that the result of statistical query is acceptable. However, in DAS scenario, there are various queries including statistical queries and entity queries. Any Update of cell values would lead to error of entity queries. So cell values must be kept untouched. In this paper we achieve minimal occupation number through cells merge, which will be discussed in section 4.2.

Above two security definitions are complementary. If only one is satisfied, the partition is still insecure. For example, for table 2(b) we assume that minimal confidential intervals of attribute age and disease are 10 and 2 respectively and data distributions of them are uniform. So the partition in fig 1 is density based minimal confidential interval satisfied. However, from example 1 we can see that the partitioning is insecure.

2.2 Efficiency Evaluation

From security's perspective, the number of tuples in each cell is as many as best. While from efficiency's perspective, the number of tuples in each cell is as little as best. To evaluate the efficiency we propose a new evaluator, which is formally defined as follows:

Definition 3: *Average squared variance of cell.* Assume minimal occupation number is k, then average squared variance of cell is $(\sum_{i=1}^{s}(n_i - k)^2 / s)^{1/2}$, where s is the number of cells, n_i is the number of tuples in the ith cell.

It's intuitive that the value of above evaluator depends on the value of k. To make partition based on different k comparable, we introduce another evaluator:

Definition 4: *Average cell size.* Assume minimal occupation number is k, then average cell size is $(\sum_{i=1}^{s} n_i / s) / k$, where s is the number of cells, n_i is the number of tuples in the ith cell.

3 Problem Statement and Related Work

3.1 Problem Statement

In this paper we mainly address secure and efficient multi-dimensional partition problem, which holds the following characteristics:

1. The cells after multi-dimensional partition are security guaranteed. Specifically, after partition each cell is both minimal confidential interval guaranteed and minimal occupancy number guaranteed.
2. Our multi-dimensional partition is efficient, which means that average squared variance of cell is much less than that of multiple single-dimensional partitions under various circumstances.

3.2 Related Work

There is much previous research on querying over encrypted data [4,5,6,8,9,10]. An algebraic framework [4,10] is developed for query rewriting over encrypted representation to minimize the computation at the client site. For range queries, there are three solutions: node level encrypted B+-tree [5], order-preserving encryption [9], partition based index [4,8]. For equality queries, partition [4,8] and hash [5] based indexes are proposed. For aggregate queries, additional partition level statistical information [6] is proposed. There are also various security and efficiency evaluators, including equivalent class and graph automorphism [5], average squared error of estimation and entropy [8]. However, as far as I know all above research only focus on single attribute. Besides efficient querying, there is still some work on secure and efficient storage of encrypted data [11]. [12] proposes secure SQL evaluation on XML database.

To address extra information leakage induced by multiple single-dimensional partitions, we propose a secure and efficient multi-dimensional partition. One related research area is multi-dimensional histograms. There are many multi-dimensional histograms including equi-depth [13], MHist [14], GenHist [15], STGrid [16], STHoles [17]. According to whether the partitions are overlapped or not, there are two classes of histograms. Our multi-dimensional partition shares some intriguing features with non-overlapping histograms. First, they all use rectangles as partitions (in 2-dimension space). Second, they use the same partition schemes, which are illustrated in Fig. 2. We say that each partition scheme is Fig. 2. is more flexible than those to its left, since for example we can simulate recursive partition with arbitrary partition using at most the same number of buckets, but not vice-versa. While from the simplicity of partition each scheme is more difficult than those to its left. In spite of these connections, there are fundamental differences between multi-dimensional histogram and our partition. First, the main objective of the histogram is to form buckets enclosing close-to-uniform tuple density whenever possible. On the other hand, the goal of our partition is to form partitions which are security constrained and efficiency optimized. Second, histograms are used for estimation of queries. Bounded estimation error is acceptable. However, our partition is used for index. Any error is unacceptable.

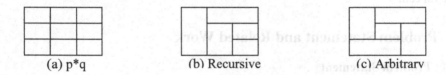

(a) p*q   (b) Recursive   (c) Arbitrary

Fig. 2. Three classes of partition schemes for rectangle buckets

In our partition we require that each cell must be security constrained and efficient optimized. We select arbitrary scheme for it's the most flexible (efficient) of the three schemes.

4   Secure and Sub-optimal Multi-dimensional Partition

To solve security degradation caused by multiple single-dimensional partitions, we propose a multi-dimensional partition. For simplicity of illustration, we show how to build partition in 2-dimensional space. The extension to high dimensions is straightforward. The cells in 2-dimensional space are rectangles. In the following discussion we will use cells and rectangles alternatively. The procedure of multi-dimensional partition includes two steps: cells partition and cells merge. In the first step, each dimension is partitioned to p(q) intervals according to domain range and data distribution. After the first step, each cell produced should be minimal confidential interval constrained and the whole 2-dimensional space is partitioned to p*q cells. In the second step, adjacent cells are merged together to ensure the number of tuples in each cell is at least k and as minimal as possible. Since optimal cells merge is a NP-hard problem, we propose a sub-optimal heuristic based greedy algorithm. In the following we will discuss these two steps in detail.

4.1 Cells Partition

Given a 2-dimensional space, we first partition it to p*q rectangles. Specifically we first represent every tuple using a point in 2-dimensional space. Then every dimension is partitioned to p(q) intervals according to domain range and data distribution respectively. Each interval is minimal confidential interval constrained. We will show this procedure by an example.

Example 2. Again we use table in example 1. Assume there are 100 tuples in it. Domain range of attribute age is [20, 50]. The type of attribute disease is categorical. So we transform it to numeric type first. To simplify problem, the transformation is straightforward. Six diseases are mapped to 10, 20, 30, 40, 50 and 60 according to their lexicographic order. So domain range of attribute disease is [10, 60]. Data distribution of two attributes are uniform. We partition each dimension into 3 intervals uniformly. The whole 2-dimensional space is divided into 3*3 cells, which are illustrated in Fig. 3.

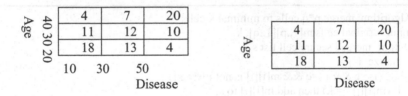

Fig. 3. Cells partition   Fig. 4. Cells merge

In Fig. 3. the number in each cell demonstrates the number of tuples in it. Since the number of tuples in each cell is distinct, we would use that number to represent corresponding cell.

4.2 Cells Merge

After cells partition, we have p*q cells, any of which is minimal confidential interval constrained. However the number of tuples in each cell varies to a large extent. Assume the minimal occupation number is k. For those cells the number of tuples in which is less than k, we will apply cells merge to ensure the number of tuples in each cell is at least k. While from efficiency's perspective, the number of tuples in each cell should be as little as possible. So cells merge problem can be defined as achieving minimal k problem, which we call optimal cells merge problem.

Cells merge can be done iteratively until minimal k is satisfied. In each iteration cells merge has two limitations: first, merge can only be applied to adjacent cells, so in p*q context, a cell has at most 4 adjacent cells; second, merged cells must be rectangles. For example, in Fig. 3 assume minimal occupation number is 15, so cell {4, 7} can only be merged either with cell {20} or with cell {11, 12} in one iteration. In order to find the optimal cells merge, we must enumerate all possible cells merge. However, it is a NP-hard problem. The proof can be found in [18].

Since optimal cells merge is a NP-hard problem, we propose an approximately optimal algorithm, which is a heuristic based greedy algorithm. For easy understand of heuristics, we will introduce the heuristics through an example using the cells in figure 3. In the example we assume minimal occupation number is 15.

1.   Cells merge is processed by scanning cells linearly, so a cell has at most two cells to be merged instead of four. For example, assume the scan order is first from up to down, then from left to right. So center cell {12} has two cells {10} or {13} to merge in one iterative.

2.   Once a cell which has exact 15 tuples occurs, terminate any iterative containing that cell and get a merged cell. For example, cell {4,11} has 15 tuples, so we terminate any iterative containing cell {4,11} and output cell {4,11} as the merged cell.

3.   A cell can has several secure merged cells. We call each one candidate cell. Among candidate cells, select the cell which has the least number of tuples. If there are many such cells, select one randomly. For example, cell {7} can be merged with cell {20} or cell {12}. We select cell {7, 12} for it has less tuples.

```
Algorithm: merge p*q cells to minimal k cells
Input: original cells m[1..p][1..q], k
Output: merged secure cell list s
Initialize: i=j=1
while (i<=p && j<=q && m[i][j] is not merged) {
   if (m[i][j] >= k) then add m[i][j] to s;
   else {
         form all possible mergers,
         if (max(merger) >= k) then {
               select the merger which is minimal and not less than k;
               add the merger to s;
            } else select a cell from s to merge;
            remove the cells enclosed by s from m;
    }
    increment i,j;
}
return s;
```

Fig. 5. A heuristic based greedy algorithm

4. For cells which can not be merged to form secure cell even after exhaustive try,
    merge them to appropriate previously formed secure cells. For example in Fig. 3.,
    after the cells are merged to secure cells {4,11}, {7,12}, {20}, it's turn for cell
    {10}. Because it could not form a secure cell, so we merge it to an already formed
    secure cell {20}. The reason that we select cell {20} but not cell {7, 12} is that the
    merged cell will have less tuples.

After cells scanning, the cells in Fig. 3 are merged to secure ones illustrated in Fig. 4.
Cells merge algorithm is illustrated in Fig. 5. Assume after cells partition there are n
cells in total. The space complexity of our greedy algorithm is $O(n)$. In the worst case,
to merge $n^{1/2}$ cells we need scan n cells. In average to merge one cell we need scan $n^{1/2}$
cells. Since there are n cells in total, so the time complexity is $O(n^{3/2})$.

5 Experimental Evaluation

5.1 Experimental Setup

We carried out experiments on a 1.7G Hz Pentium machine, with 512M RAM. The
algorithm of cells partition and merge was implemented by Java 1.4.1.

Our experiments were based on Adults database from the UCI Irvine Machine
Learning Repository [19], which is comprised of data from US census. The table has 14
attributes, 8 of which are continuous and 6 are categorical. We only encrypted two
attributes in our experiments: age and occupation. Age is continuous and has 74 distinct
values in total. Occupation is categorical and has 14 distinct values in total. We deleted
the tuples with missing values in our experiments.

In our experiments we compared the efficiency of our multi-dimensional partition
with that of single-dimensional partition using averaged squared variance of cell and
average cell size introduced in Section 2.2.

5.2 Results Analysis

In our experiments we explored the impacts of the following dimensions on partition
efficiency:

1. The impact of minimal occupation number.
2. The impact of data density
3. The evaluator used: average squared variance of cell (denoted as asvc), average
    size of cell (denoted as asc).

First, we experimented on 1000 tuples with minimal occupation number varied. The
result is plotted in Fig. 6(a), which show that efficiency of multi-dimensional partition
(denoted as MD) is much better than that of single-dimensional partition (denoted as
SD) independent of minimal occupation number. With the increase of minimal
occupation number, efficiency of two partitions all decreased, which means that the
efficiency evaluator (average squared variance of cell) is negatively related to minimal
occupation number. The decrease degree of SD is quicker than MD, which means that
SD is more sensitive to the value of minimal occupation number.

To explore the impact of data density on efficiency, we made the second experiment on a much denser dataset, which includes 10000 tuples with the same domain range as that in the first experiment. The result was plotted in Fig. 6(b). An interesting counter-intuition is the efficiency degradation when minimal occupation number is least (here 10). The unexpected degradation is due to too little minimal occupation number compared to data density, which leads to many big original cells compared to minimal occupation number and almost no cells merge in the end. So it's recommended to increase minimal occupation number when data density increases.

We adjusted the value of minimal occupation number according to data density in the third experiment. The result was plotted in Fig. 6(c). The values of average squared variance of cell are much larger than those in the first experiment, which means that this evaluator is dependent of minimal occupation number. So we did the fourth experiment using another evaluator: average cell size. The result in Fig. 6(d) illustrated that this evaluator is independent of minimal occupation number.

Whatever minimal occupation number, data density and efficiency evaluator, the performance of our multi-dimensional partition is stable compared to that of single-dimensional partition.

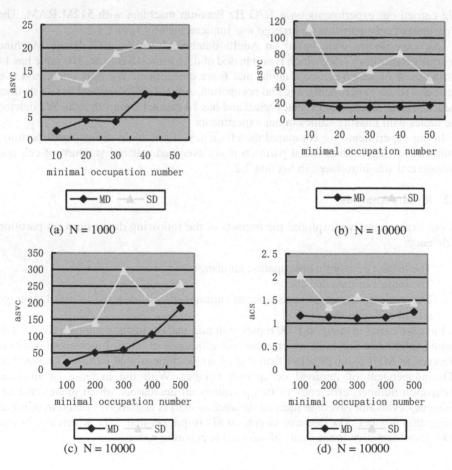

(a) N = 1000

(b) N = 10000

(c) N = 10000

(d) N = 10000

6 Conclusions and Future Work

In this paper we first illustrate multiple single-dimension partition indexes are insecure because of un-uniform data distribution in multi-dimensional space, then propose a multi-dimensional partition based index which includes two steps: cells partition and cells merge. Since optimal cells merge is a NP-hard problem, we propose heuristic based greedy algorithm. Experiments show that our greedy algorithm is secure and effective.

One desirable feature of multi-dimensional partition is dynamic maintenance of partition. In above sections we only build a secure and efficient multi-dimensional partition for static scenarios. While in dynamic scenario, the cells may become insecure (too few tuples) or inefficient (too many tuples). A straightforward solution is to split and merge. Besides, in above discussion we assume that query workload is uniform. However, it is not the case in some real applications. So it's expected to develop a workload aware partition. The solution to above two problems will be our future work.

Acknowledgements. This paper is in part supported by the National Natural Science Foundation of China (60573092).

References

1. Hacigumus, H., Iyer, B., Mehrotra, S.: Providing Database as a Service. In: Proc. of 18th International Conference on Data Engineering, San Jose, CA, USA (February 2002)
2. DES. Data encryption standard. FIPS PUB 46, Federal Information Processing Standards Publication, (1977)
3. Oracle Corporation: Database Encryption in Oracle9i (2001), http://otn.oracle.com/deploy/security/oracle9i
4. Hacigumus, H., Iyer, B., Li, C., Mehrotra, S.: Executing SQL over Encrypted Data in the Database Service Provider Model. In: SIGMOD, Madison, Wisconsin, USA, June 4-6 (2002)
5. Damiani, E., Vimercati, S.D.C., Jajodia, S., Paraboschi, S., Samarati, P.: Balancing Confidentiality and Efficiency in Untrusted Relational DBMSs. In: 10th ACM CCS, Washington (2003)
6. Mykletun, E., Tsudik, G.: Aggregation Queries in the Database-As-a-Service. In: IFIP. WG 11.3 Working Conference on Data and Applications Security (DBSec) (July 2006)
7. Kantarcioglu, M., et al.: Security Issues in Querying Encrypted Data. Technical Report, Purdue University, (2004)
8. Hore, B., Mehrotra, S., Tsudik, G.: A Privacy-Preserving Index for Range Queries. In: Proc. of the 30th VLDB Conference, Toronto, Canada (2004)
9. Agrawal, R., Kierman, J., Srikant, R., Xu, Y.: Order Preserving Encryption for Numeric Data. In: Proc. of ACM SIGMOD 2004, Paris, France (June 2004)
10. Bouganim, L., Pucheral, L.: Chip-Secured Data Access: Confidential Data on Untrusted Servers. In: Proc. of the 28th VLDB Conference (2002)
11. Iyer, B., Mehrotra, S., Mykletun, E., Tsudik, G., Wu, Y.: A Framework for Efficient Storage Security in RDBMS. In: Bertino, E., Christodoulakis, S., Plexousakis, D., Christophides, V., Koubarakis, M., Böhm, K., Ferrari, E. (eds.) EDBT 2004. LNCS, vol. 2992, pp. 147–164. Springer, Heidelberg (2004)

12. Wang, H., Lakshmanan, L.V.S.: Efficient Secure Query Evaluation over Encrypted XML Databases. In: VLDB 2006 (September 2006)
13. Muralikrishna, M., DeWitt, D.J.: Equi-Depth Histograms for Estimating Selectivity Factors for Multi-dimensional Queries. SIGMOD Record 17(3), 28–36 (1988)
14. Poosala, V., Ioannidis, Y.E.: Selectivity Estimation without the Attribute Value Independence Assumption. In: Proceedings of the Twenty-third International Conference on Vary Large Databases (VLDB 1997) (1997)
15. Gunopulos, D., Kollios, G., Tsotras, V.J., Domeniconi, C.: Approximating Multi-dimensional Aggregate Range Queries over Real Attributes. In: Proceedings of the 2000 ACM International Conference on Management of Data (SIGMOD 2000) (2000)
16. Aboulnage, A., Chaudhuri, S.: Self-tuning histograms: Building histograms without looking at data. In: proceedings of the 1999 ACM International Conference on Management of Data (SIGMOD 1999) (1999)
17. Bruno, N., Chauhuri, S., Gravano, L.: STHoles: A Multidimensional Workload-Aware Histogram. In: proceedings of the 2001 ACM International Conference on Management of Data (SIGMOD 2001) (2001)
18. LeFevre, K., Dewitt, D.J., Ramakrishnan, R.: Mondrain Multidimensional K-Anonymity. In: ICDE (2006)
19. Blake, C., Keogh, E., Merz, C.: UCI repository of machine learning databases. University of Califonia, Ivrine, Dept. of Informaiton and Computer Science,
http://mlearn.ics.uci.edu/MLRepository.html

Multilateral Approaches to the Mobile RFID Security Problem Using Web Service

Namje Park[1], Youjin Song[2,*], Dongho Won[3], and Howon Kim[1]

[1] Information Security Research Division, ETRI,
161 Gajeong-dong, Yuseong-gu, Daejeon, 305-350, Korea
{namjepark,khw}@etri.re.kr
[2] Department of Information Management, Dongguk Univertsity,
707 Seokjang-dong, Gyeongju, Gyeongsangbuk-do, 780-714, Korea
song@dongguk.ac.kr
[3] Information Security Group, Sungkyunkwan University,
300 Cheoncheon-dong, Jangan-gu, Suwon, Gyeonggi-do, 440-746, Korea
dhwon@security.re.kr
http://www.security.re.kr

Abstract. The mobile RFID (Radio Frequency Identification) is a new application to use mobile phone as RFID reader with a wireless technology and provides new valuable services to user by integrating RFID and ubiquitous sensor network infrastructure with mobile communication and wireless internet. However, there are an increasing number of concerns, and even some resistances, related to consumer tracking and profiling using RFID technology. Therefore, in this paper, we describe the secure application portal service framework leveraging globally mobile RFID based on web service which complies with the Korea's mobile RFID forum standard.

1 Introduction

Though the RFID technology is being developed actively and lots of efforts made to generate its market throughout the world, it also is raising fears of its role as a 'Big Brother'. So, it is needed to develop technologies for information and privacy protection as well as promotion of markets (e.g., technologies of tag, reader, middleware, OIS, ODS, OTS, etc.). The current excessive limitations to RFID tags and readers make it impossible to apply present codes and protocols. The technology for information and privacy protection should be developed in terms of general interconnection among elements and their characteristics of RFID in order to such technology that meets the RFID circumstances.

---

* Youjin Song is the corresponding author for this paper. This research was financially supported by the Ministry of Commerce, Industry and Energy (MOCIE) and Korea Industrial Technology Foundation(KOTEF) through the Human Resource Training Project for Regional Innovation, and the IT R&D program of MIC/IITA [2005-S088-04, Development of Security Technology for Secure RFID/USN Service].

Y. Zhang et al. (Eds.): APWeb 2008, LNCS 4976, pp. 331–341, 2008.
© Springer-Verlag Berlin Heidelberg 2008

While common RFID technologies are used in B2B models like supply channels, distribution, logistics management, mobile RFID technologies are used in the RFID reader attached to an individual owner's cellular phone through which the owner can collect and use information of objects by reading their RFID tags; in case of corporations, it has been applied mainly for B2C models aiming at their marketing. Though most of current RFID application services are used in fields like the search of movies posters and provision of information in galleries where less security needs are required, they will be expanded to and used more frequently in such fields like purchase, medical cares, electrical drafts, and so on where security and privacy protection are indispensible. Therefore, in this paper, we describe the multilateral approaches to the mobile RFID security problem using web service which complies with the Korea's mobile RFID forum standard. This is new technology to RFID will provide a solution to protecting absolute confidentiality from basic tags to user's privacy information.

2   UHF Mobile RFID Network

2.1   Networked Mobile RFID Technology

RFID is expected to be the base technology for ubiquitous network or computing, and to be associated with other technology such as telemetric, and sensors. The mobile phone integrated with RFID can activate new markets and end-user services, and can be considered as an exemplary technology fusion. Furthermore, it may evolve its functions as end-user terminal device, or 'u-device (ubiquitous device)', in the world of ubiquitous information technology.

Networked RFID means an expanded RFID network and communication scope to communicate with a series of networks, inter-networks and globally distributed application systems. So it makes global communication relationships triggered by RFID, for such applications as B2B, B2C, B2B2C, G2C, etc. Networked RFID loads a compact RFID reader in a cellular phone, providing diverse services through mobile telecommunications networks when reading RFID tags through a cellular phone. Internet-enabled mobile phone which equips RFID reader will bring new service concepts to mobile telecommunication.

Networked RFID technology is focusing on the UHF range (860~960MHz), since UHF range may enable longer reading range and moderate data rates as well as relatively small tag size and cost. Then, as a kind of handheld RFID reader, in the selected service domain the UHF RFID phone device can be used for providing object information directly to the end-user using the same UHF RFID tags which have widely spread. The service area of networked RFID is expected to be unlimited, and its services, diverse; currently, however, the service scenarios using RFID tags chiefly as offline hypertext owing to the constraints of cellular phone performance and business models are still at the proposal stage.

2.2   Mobile RFID's Wireless Specifications

For a mobile terminal with an RFID reader embedded, the configuration of reader chip and adjacent circuitry can be illustrated as shown in [Fig. 1]. Inside the reader

chip are two components: the digital component, which processes Host/RFID protocols, and the analog component, which processes base band signals and 900MHz RF
signals.

[Passive RFID Tags]   **[Mobile RFID Reader SOC]**

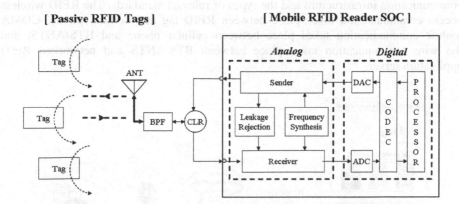

Fig. 1. Mobile RFID's SoC

To eliminate the analog component from the design, it is necessary to prepare corresponding wireless specifications for it. This paper prepared the domestic wireless
specifications for mobile RFID in Korea, in reference to applicable RFID frequency,
cell radius, channel allocation, and relevant radio regulation acts (ordinances), technical standards, etc.

In general, the minimum power to be delivered to a passive RFID tag is -10dBm
(100uW). On the contrary, since a mobile RFID terminal has good accessibility to
tags, it can fully meet the requirements for application in mobile RFID services, if any
tag can be recognized within a 1 mile radius. Accordingly, it was estimated that the
signal output of a mobile RFID should be at least 20dBm or higher, in overall consideration of link loss = -315dB, tag antenna gain ≤ 2dBi, built-in reader antenna gain ≤
0dBi, and more. However, a mobile RFID cannot emit as much output as a fixed type
reader because it works only with power supplied from a mobile phone battery. Thus,
our wireless specifications determined the sender output by allowing for minimum
power based on link analysis, limitations of CMOS power amplifier, and the mobile
phone's battery power.

On the other hand, it is not necessary for mobile RFID to recognize a massive
number of tags at once since it is designed primarily for the reader's portability. The
Mobile RFID has to just request and send information on several tag recognition
codes, so it can make any application service, if necessary, at a data rate of about
40Kbps without difficulty. In Korea, the frequency band allocated for RFID ranges
from 908.5MHz to 914MHz. The RFID device supports data rates as high as 640Kbps
at this band and can communicate with other terminals only if a wide channel bandwidth is available in the restricted area. It is not appropriate for terminals like mobile
RFID that may be used by an uncertain number of multiple users. Therefore, the mobile RFID was based on 200 KHz channel bandwidth at data rate of about 40Kbps.

2.3 Network Architecture of the Mobile RFID

Networked RFID services are provided by using mobile telecommunication network and typical Internet. Figure 2 shows the interface structure for MRFID service's communication infrastructure and the types of relevant standards. The RFID wireless access communication takes place between RFID tag and cellular phone, CDMA mobile communication takes place between cellular phone and BTS/ANTS, and the wire communication takes place between BTS/ANTS and networked RFID application server.

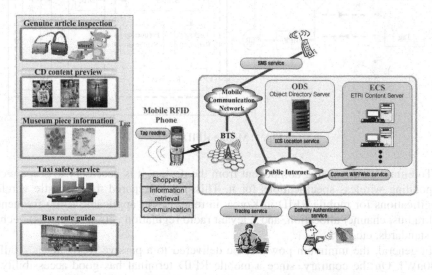

Fig. 2. Conceptual Network Architecture of Networked RFID

Networked RFID service structure is defined to support ISO/IEC 18000-6 A/B/C through the wireless access communication between RFID tag and the RFID reader, but there is no RFID reader chip supporting all three wireless connection access specifications yet that the communication specification for the cellular phone will be determined by the mobile communication companies. It will be also possible to mount the RF wireless communication function to the reader chip using SDR technology and develop ISO/IEC 18000-6 A/B/C communication protocol in software to choose from protocols when needed.

Networked terminal's function is concerned with the recognition distance to the RFID reader chip built into the cellular phone, transmission power, frequency, interface, technological standard, PIN specification, UART communication interface, WIPI API and WIPI-HAL API extended specification to control reader chip. RFID reader chip middleware functions are provided to the application program in the form of WIPI API. Here, mobile RFID device driver is the device driver software provided by the reader chip manufacturer.

Networked network function is concerned with the communication protocols such as the ODS communication for code interpretation, the message transmission for the transmission and reception of contents between the cellular phone terminal and the application server, contents negotiation that supports networked RFID service environment and ensures the optimum contents transfer between the cellular phone terminal and the application server, and session management that enables the application to create and mange required status information while transmitting the message and the WIPI extended specification which supports these communication services.

Cellular phone requires a common control interface between various RFID readers and the application or the middleware, and EPCglobal Inc. and ISO are defining the functions that RFID reader has to commonly support as well as various common command and standardizing message types. Mobile RFID functions will be extended continuously into standard cellular phone RFID reader, and the RFID supported WIPI extension model using WIPI which is the wireless internet standard platform will define the API required in using the reader suitable for mobile environment as the API extension of WIPI which is the wireless internet standard platform while maintaining compatibility among various devices.

First, mobile RFID phone reads RFID tags on object and fetches code stored in it. Second, mobile RFID phone should do code resolution with which mobile RFID phone obtains the location of remote server that provides information of the product or adequate mobile service. Code resolution protocol is identical with DNS protocol. ODS (Object Directory Service) server in figure 3 acts DNS server and is like EPCglobal's ONS (Object Name Service) server. Mobile RFID phone queries the location of server with code to ODS server then ODS server replies with the location of the server. Finally, mobile RFID phone requests contents or service to the designated server acquired from ODS server.

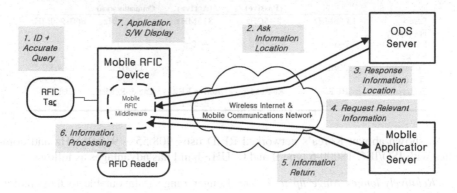

Fig. 3. Block Basic Communication Scenario for Networked RFID

ODS server plays a role of DNS server which informs the mobile RFID phone of the contents/service server's location as explained above. ODS server may be organized in hierarchical structure similar to DNS server. The OTS (Object Traceability Service) server keeps history of tag reading in RFID readers throughout the lifecycle of objects. Its main purpose is for tracking object in SCM. The OIS (Object Information Service) records the reading RFID tag event to OTS server and may provide additional information of an object in detail such as manufactured time, manufacturer name, expiration time, etc. The RPS (RFID Privacy management Server) controls access to information of the object in accordance with privacy profile which owner of the object makes up. WAP and Web server are content servers providing wireless Internet contents such as news, game, music, video, stock trading, lottery, images and so on.

2.4 Previous Works

For realization of networked RFID services, it is required that RFID devices such as RFID tag or RFID reader should be installed to mobile phones. The Nokia phone support RFID technology based on 13.56MHz, and ISO 14443A. This Nokia phone is based on NFC protocol that uses 13.56MHz complying with ISO/IEC 18092. Mobile phone users can automate and initiate tasks, such as browsing service instructions or logging time-stamped data like meter readings. KDDI in Japan developed slide-in RFID readers that can be easily attached to the back side of a mobile phone. There are two types of RFID reader according to frequency band. One is 2.45GHz passive type and the other is 315MHz active type. Some pilot tests were scheduled in March 2005. Location-based services with RFID equipped mobile devices are being tested e.g. in Japan. RFID tags are placed in public venues; pedestrians may then use handheld devices to obtain additional information on their surroundings. Table 1 shows the summary of mobile RFID implementations.

Table. 1. Class of Mobile RFID

	Nokia's Mobile RFID	KDDI's Mobile RFID (Passive)	KDDI's Mobile RFID (Active)	NFC (Near Field Communication)	Korea's Mobile RFID
Radio Frequency	13.56MHz	2.45GHz	315MHz	13.56MHz	860~960MHz
Reading Range	2~3cm	~5cm	~10m		
Compliant Standards	ISO/IEC 14443 A			ISO/IEC 18092	ISO/IEC 18000-6 B/C
Feature	HF RF Reader	RF Reader	Active RFID Reader	Tag & Reader	UHF RF Reader

As shown above, Korea's networked RFID uses 908.55 ~ 913.95MHz and complies with ISO/IEC 18000-6 type B and C. UHF band has advantages as follows:

- *Relatively longer range up to 100cm*: Longer range is favourable to most mobile RFID services for convenience
- *Short range available up to 2 or 3cm*: If needed in case of payment system, short range may be supported by reducing RF strength by application
- *Avoiding duplicate investment for RFID tag*: Most RFID tags in SCM work in 900MHz-range, i.e., ISO 18000-6 type A/B/C and EPCglobal. This means that both SCM and mobile RFID application can share a RFID tag. That is, a single RFID tag can provide different contents according to its application

3 Combination of Web Service and RFID

Web service, even though it has been commonly available for quite some time, it has not yet reached the level of popularity expected. Despite this reality, however, web service has evolved steadily and occupies its own position as a very critical technology for the next generation.

Web service is a basic technology for system interlocking among different platforms distributed on a network by means of existing technologies used on the web. Its compatibility or scalability is based on XML. Actually, there is not any one technology called web service, but web service is realized through arranging basic technologies for different purposes, such as SOAP, WDSL and UDDI.

The connection between RFID and web service has recently been highlighted because the web service may provide a solution for development problems associated with the RFID system. Indeed, RFID has nothing firsthand to do with web service. However, web service is indispensable to RFID service. It is not easy to develop the RFID system, fundamentally because of the following reasons, although those problems can all be resolved with the help of web service:

1) Reader has too low level control of API.

   The wrapping reader driver supplied by reader producers with web service can make it possible to overcome the inevitable challenges in developing the RFID system. Web service itself is based on distributed object technology, so it can be said that high level API is implemented only with an interface based on web service.
2) There is not yet any proven way to control the reader via network.

   No matter how upgraded the API setup is in a system having access to a reader within the possible range of conventional technology, controlling the reader via network still remains unresolved. But an interface equipped with web service allows you to directly control the reader via a network, which solves the challenge of control.
3) There is not yet any established method to manage and monitor the reader when it is widely distributed on a network.

If the RFID system is operated without any knowledge of reader failure, the stock amount or the like may not be indicated on the system correctly, possibly resulting in excessive orders. The web service itself runs on a web server so there are lots of tools available on the market for checking the functions of web servers. It is possible that these tools may be applied to management of the RFID reader. However, no matter how high the performance of a reader, if it must involve any independent tool for management, it has little use in applications.

The advantages of using web service for RFID system go beyond this simple management of the reader. For example, using web service for RFID system makes it very easy to take security measures such as encoding communicated data or restricting access. For web service, SSL makes it easy to encode what is communicated and even authenticate IP address through a firewall as well. It is not necessary to develop independent password (cipher) processing or set up an access control module. In addition, load sharing process for access congestion, analysis of transactions based on timeline and more can be easily carried out through conventional software or hardware. Above all, it is important that there is no limitation of accessibility from all platforms. Only one service can make it possible to deal with any request from all platforms, without preparing different drivers, etc. for a variety of platforms.

Summing up, web service can solve the inevitable challenges in developing the RFID system, which cannot be resolved with conventional technologies, and contribute greatly to improving development efficiency and reducing costs.

4   Globally Reliable Mobile RFID Framework Based on Web Service

4.1   Multilateral Approaches to the Mobile RFID Security Problem

This technology aims at RFID application services like authentication of tag, reader, and owner, privacy protection, and non-trackable payment system where more strict security is needed.

1) *Approach of Platform Level*
   This technology for information portal service security in offering various mobile RFID applications consists of application portal gateway, information service server, terminal security application, payment server, and privacy protection server and provides a combined environment to build a mobile RFID security application service easily.

2) *Approach of Protocol Level*
   - It assists Write and Kill passwords provided by EPC Class1 Gen2 for mobile RFID tag/reader and uses a recording technology preventing the tag tracking.
   - It employs information protection technology solving the security vulnerability in mobile RFID terminals that accept WIPI as middleware in the mobile RFID reader/application part and provides end-to-end security solutions from the RFID reader to its applications through WIPI based mobile RFID terminal security/code treatment modules.

3) *Approach of Privacy Level*
   This technology is intended to solve the infringement of privacy, or random acquisition of personal information by those with RFID reader from those with RFID attached objects in the mobile RFID circumstance except that taking place in companies or retail shops which try to collect personal information. Main assumptions are as follows:
   - Privacy in the mobile RFID circumstance comes into force when a person holds a tag attached thing and both information on his/her personal identity (reference number, name, etc.) and the tag (of commodity) are connected to each other.
   - Privacy protection information are concerned with the tag attached object (its name, value, etc.) and the personal identity (of its owner or reference).
   - When it comes to the level of access authority, the owner can have access to any personal information on his/her object, an authorized person (e.g., pharmacist or doctor in medical care) only to access permitted information, and an unauthorized person to nowhere.

4.2 Secure Mobile Application Portal in Web Service Environment

The mobile RFID is a technology for developing a RFID reader embedded in a mobile terminal and providing various application services over wireless networks. Various security issues - Interdomain security, privacy, authentication, E2E (End-to-End) security, and untraceability etc. - need to be addressed before the widespread use of mobile RFID. Model of mobile RFID service as shown in figure 4 defines additional three entities and two relationships compared to that defined in RFID tag, RFID access network, RFID reader, relation between RFID tag and RFID reader, relation between RFID reader and application server.

Generally, in mobile RFID application such as smart poster, Application Service Provider (ASP) has the ownership of RFID tags. Thus, mobile RFID users have to subscribe to both the ASP for these kinds of RFID services and mobile network operator for mobile communication service. Namely, there exist three potentially distrusted parties: user owned RFID reader, mobile network operator, and ASP. Accordingly, trust relationship among three parties must be established to realize secure mobile RFID service. Especially, when a RFID reader tries to read or change RFID service data stored in tag, the reader needs to get a tag access rights. Additionally, it is important that new tag access rights whenever some readers access a same tag must be different from the already accessed old one.

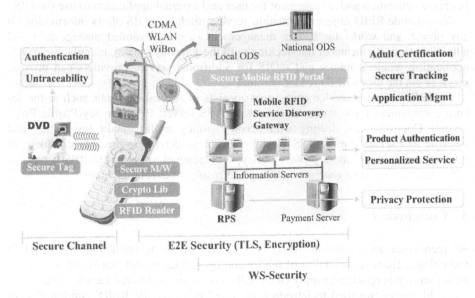

Fig. 4. Conceptual for Secure RFID over Mobile Networks

Secure mobile RFID application portal is a secure service portal for various mobile RFID application services. The service provider using SMAP (Secure Mobile Application Portal) can easily deploy several mobile RFID applications guaranteed with security and privacy protection.

Secure portal discovery service gateway is a system that classifies and defines the mobile OIS system as an individual element system in a mobile RFID security application service network, and also supports internal functions comprising each element service system. This gateway system manages the locations and interface of servers and services as registered from applicable product and service providers that provide product information and contents corresponding to each EPC. Moreover, this system works as a gateway system that seeks appropriate services with the capacity to provide information on any product equipped with a tag containing applicable EPC or contents, etc. related to such EPC, and then provide such services by way of wired and wireless device (particularly, mobile RFID reader terminals) of users who access mobile RFID service network.

This portal allows you to find out off-line product information on an EPC, on-line additional service information, information on authentication for product family or product related to this EPC, and more. This gateway system also has another advantage - applying a privacy policy and security to personal information for inquiry about information in the interest of creating a secured and reliable environment for making information available. Additional features of this system include automatically generating and managing OIS for the mobile RFID. This function allows you to create and use a variety of useful functions, such as database for managing information corresponding to each EPC by simply setting; security for access control, authentication, encoding, etc; data exchange for sharing necessary information among OISs; and interface generation and management for user and external application to use the OIS.

As a mobile RFID object information service module, mOIS offers information on any object, and works for system management or commissioned management and administration. By means of interlocking with the gateway system, manufacturer OIS, application retailer's mOIS, and mOIS for additional service providers can provide services via the user's RFID reader terminal.

The RFID privacy service component consists of several elements, such as the secure communicator for communication with RPS server; PPP (Privacy Profile Processing) Controller for dealing with privacy policy profile, audit log policy and whether to execute duties; RPS Decision Point to determine possible coverage of information on the basis of privacy policy; OIS Schema to be sent to RPS server; or RPS Data Generator to generate audit log and results of executing duties.

5 Conclusion

As mentioned above, mobile RFID is a newly promising application using RFID technology. However, mobility of reader and its service model that is different from RFID service in retail and supply chain will cause some additional security threats.

In this paper, we tried to introduce the concept of mobile RFID, combination of Web service and RFID, and expose some additional security threats caused by it. The frequency band to support the air protocol is allocated at 908.5MHz to 914MHz by TTA (Telecommunication Technology Association) in Korea to comply with ISO 18000-6 for air interface communications at 860MHz to 960MHz. And we describe a way to incorporate its new technology to work with cell phones in particular as an external security reading device (replacing 900MHz) and same time as an added security service based on web service to manage all RFID mobile device mediums. With this purpose, the application areas of this service platform are also briefly presented. By doing so, the customized security and privacy protection can be achieved in web service environment. In this regard, the suggested technique is an effective solution for security and privacy protection in a networked mobile RFID system based on web service.

References

1. Tsuji, T., Kouno, S., Noguchi, J., Iguchi, M., Misu, N., Kawamura, M.: Asset management solution based on RFID. NEC Journal of Advanced Technology 1(3), 188–193 (2004)
2. Chae, J., Oh, S.: Information Report on Mobile RFID in Korea. ISO/IEC JTC 1/SC 31/WG 4 N 0922, Information paper, ISO/IEC JTC 1 SC 31 WG4 SG 5 (2005)
3. Jin, S., et al.: Cluster-based Trust Evaluation Scheme in Ad Hoc Network. ETRI Journal 27(4), 465–468 (2005)
4. Sarma, S.E., Weis, S.A., Engels, D.W.: RFID systems, security and privacy implications. Technical Report MIT-AUTOID-WH-014, AutoID Center, MIT Press, Cambridge (2002)
5. Choi, W., et al.: An RFID Tag Using a Planar Inverted-F Antenna Capable of Being Stuck to Metallic Objects. ETRI Journal 28(2), 216–218 (2006)
6. Weis, S., et al.: Security and Privacy Aspects of Low-Cost Radio Frequency identification Systems. In: First International Conference on Security in Pervasive Computing (SPC) (2003)
7. Ohkubo, M., Suzuki, K., Kinoshita, S.: Cryptographic Approach to Privacy-Friendly Tags. In: RFID Privacy Workshop (2003)
8. W. Park, B. Lee: Proposal for participating in the Correspondence Group on RFID in ITU-T. Information Paper. ASTAP Forum (2004)
9. Park, N., Kwak, J., Kim, S., Won, D., Kim, H.: WIPI Mobile Platform with Secure Service for Mobile RFID Network Environment. In: Shen, H.T., Li, J., Li, M., Ni, J., Wang, W. (eds.) APWeb Workshops 2006. LNCS, vol. 3842, pp. 741–748. Springer, Heidelberg (2006)
10. MRF Forum: WIPI C API Standard for Mobile RFID Reader (2005)
11. MRF Forum: WIPI Network APIs for Mobile RFID Services (2005)
12. Park, N., Kim, S., Won, D., Kim, H.: Security Analysis and Implementation leveraging Globally Networked Mobile RFIDs. In: Cuenca, P., Orozco-Barbosa, L. (eds.) PWC 2006. LNCS, vol. 4217, pp. 494–505. Springer, Heidelberg (2006)
13. Park, N., Kim, H., Kim, S., Won, D.: Open Location-based Service using Secure Middleware Infrastructure in Web Services. In: Gervasi, O., Gavrilova, M.L., Kumar, V., Laganá, A., Lee, H.P., Mun, Y., Taniar, D., Tan, C.J.K. (eds.) ICCSA 2005. LNCS, vol. 3481, pp. 1146–1155. Springer, Heidelberg (2005)
14. Lee, J., Kim, H.: RFID Code Structure and Tag Data Structure for Mobile RFID Services in Korea. In: Proceedings of ICACT (2006)
15. B. Chug, et al.: Proposal for the study on a security framework for mobile RFID applications as a new work item on mobile security. ITU-T, COM17D116E, Q9/17, Contribution 116, Geneva. (2005)
16. Park, N., Kim, S., Won, D.: Privacy Preserving Enhanced Service Mechanism in Mobile RFID Network. In: ASC (Advances in Soft Computing), vol. 43, pp. 151–156. Springer, Heidelberg (2007)
17. Kim, Y., Lee, J., Yoo, S., Kim, H.: A Network Reference Model for B2C RFID Applications. In: Proceedings of ICACT (2006)

Classifying Security Patterns

Eduardo B. Fernandez[1], Hironori Washizaki[2], Nobukazu Yoshioka[2],
Atsuto Kubo[3], and Yoshiaki Fukazawa[3]

[1] Department of Computer Science and Engineering, Florida Atlantic University,
Boca Raton, FL, USA
ed@cse.fau.edu
[2] National Institute of Informatics,
2-1-2 Hitotsubashi, Chiyoda-ku, Tokyo 101-8430, Japan
{washizaki, nobukazu}@nii.ac.jp
[3] Department of Computer Science and Engineering, Waseda University,
3-4-1 Ohkubo, Shinjuku-ku, Tokyo 169-8555, Japan
{a.kubo, fukazawa}@fuka.info.waseda.ac.jp

Abstract. Patterns combine experience and good practices to develop basic
models that can be used for new designs. Security patterns join the extensive
knowledge accumulated about security with the structure provided by patterns
to provide guidelines for secure system design and evaluation. In addition to
their value for new system design, security patterns are useful to evaluate
existing systems. They are also useful to compare security standards and to
verify that products comply with some standard. A variety of security patterns
has been developed for the construction of secure systems and catalogs of them
are appearing. However, catalogs of patterns are not enough because the
designer does not know when and where to apply them, especially in a large
complex system. We discuss here several ways to classify patterns. We show a
way to use these classifications through pattern diagrams where a designer can
navigate to perform her pattern selection.

Keywords: Pattern classification, security patterns, secure system development,
system architecture.

1 Introduction

A *pattern* is a packaged reusable solution to a recurrent problem. Design patterns
embody the experience and knowledge of many designers and when properly
catalogued, they provide a repository of useful solutions for different types of
problems [8]. They have shown their value in many projects and have been adopted
by many institutions. Design patterns have been extended to other aspects of software,
first to architectural aspects of design, then to the analysis stage. Analysis and design
patterns are now well established as a convenient and reusable way to build high-
quality object-oriented software.

Security has become an important concern in current systems. The main objectives
of security are to protect the confidentiality, integrity, and availability of data. Data is

Y. Zhang et al. (Eds.): APWeb 2008, LNCS 4976, pp. 342–347, 2008.
© Springer-Verlag Berlin Heidelberg 2008

a valuable resource and it has become the target of many attacks by people who hope to gain monetary advantages, make political statements, or just vandalize. Security countermeasures are usually classified into five groups: Identification and Authentication, Access Control and Authorization, Logging, Cryptography, and Intrusion Detection. Security patterns describe mechanisms that fall into these categories (or combinations thereof) to stop or mitigate attacks as well as the abstract models that guide the design of these mechanisms. *Security patterns* join the extensive knowledge accumulated about security with the structure provided by patterns to provide guidelines for secure system construction and evaluation. A good number of security patterns have been described in the literature and several catalogs have appeared [3, 9, 11, 12, 13, 14].

However, it is not enough to have a catalog of security patterns, we also need a guidance for the designers about how to select appropriate patterns. We are seeing the appearance of more and more complex systems, incorporating web access and all kinds of distribution structures. Security should be applied in the whole life cycle of applications and at each stage patterns can be applied to provide specific security controls. A designer who is not a security expert would be lost trying to apply them into her design. Designers need guidance about how to select appropriate patterns. A step in this direction is a good classification of security patterns. We present here some possible classifications based on architectural concerns, architectural layers, and some relationships between patterns. We show the use of these classifications through a pattern diagram, where the designer can navigate in order to select appropriate patterns.

Sections 2 and 3 show several classifications based on different viewpoints for security patterns. Section 4 introduces a diagram for describing relationships based on those classifications. The last section presents some conclusions and future work.

2   Classifications Based on Architectural Concerns and Architectural Layers

Because we consider security patterns to be architectural patterns, we should look at software architecture classifications. [1] classifies architectural patterns using the type of concerns they address, e.g. Layered Structure, Data Flow, Adaptation, User Interaction, Distribution. This means we should classify security patterns according to their concerns, e.g. patterns for access control, cryptography, file control, identity, etc. For example, authentication in distributed systems is considered in: **Authenticator, Remote Authenticator /Authorizer**, and **Credential** (see [3] for references to these patterns). Chapters 7 and 8 of [12] are organized in this way.

We can think of a computer system as a hierarchy of layers, where the application layer uses the services of the database and operating system layers, which in turn, execute on a hardware layer. These layers provide another dimension for classification. We can define patterns at all levels. This allows a designer to make sure that all levels are secured, and also makes easier propagating down the high-level constraints. At the highest level we have patterns that describe the use of security models to define access control to the application objects: **Authorization (Access Matrix)**, **Role-Based**

Access Control (RBAC), Reference Monitor, Multilevel Security, Attribute-Based Access Control (ABAC). A recent paper defines Session-Based versions of those patterns [5]. At the operating system level we have patterns such as Secure Process, Controlled Virtual Address Space, and others. Patterns for web services security include: Application Firewall, XML Firewall, XACML Authorization, XACML Access Control Evaluation, and WSPL [2] (References for all these are found in [3]). Fig. 1 shows the combination of concerns and levels where patterns with the same concern, e.g. Authentication, may appear in multiple levels.

The life cycle stage where the pattern is applied is another dimension. For example, security models are selected in the analysis stage while distribution approaches are selected at the design stage.

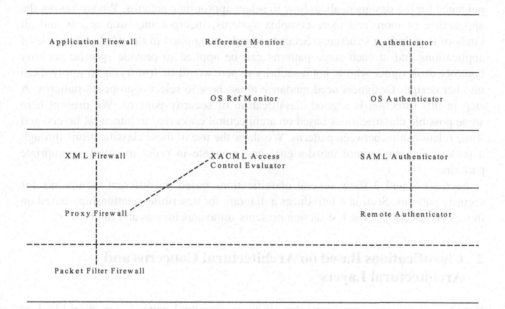

Fig. 1. Types of patterns and levels

3   Automatic Relationships between Patterns

Kubo et al. [10] have proposed an automated relationship analysis technique for patterns. This technique utilizes existing text processing techniques (such as TF-IDF and vector space model) to extract patterns from documents and to calculate the strength of pattern relationships based on document similarity.

We apply now this technique to security patterns belonging to four catalogs: Fernandez' firewall patterns [12], legal cases pattern [6], and other patterns, together with Yoder's architectural patterns [15]. Fig. 2 visualizes some of the results. The resulting graph is a specific classification, which suggests several useful relationships including some which are not explicitly specified in the original documents. For example:

- The pattern **Secure Handling of Legal Cases** [6] is recognized as a sub-pattern included in the resulting context of another pattern **Packet Filter Firewall** [12] denoted by a *Sub In Resulting* relation. This relation is not specified in [6, 12]; however it is reasonable because the domain model of Secure Handling of Legal Cases assumes the use of reference monitors such as the packet filter firewall. In other words, the use of a Packet Filter Firewall leads to a resulting context in which legal cases (such as when clients are sued by another party) can be handled in a secure way.
- As another example, there is a *Similar Forces* relation (i.e. relation in which forces of two patterns are similar) between **Roles** [15] and Secure Handling of Legal Cases. This cross-cutting relation over different catalogs is reasonable because those patterns share similar constraints such as the existence of a large number of users for the target system and the inconvenience of customizing security for each person.

These newly found relations are useful for selecting security patterns belonging to several catalogs. This technique can extract appropriate relationships in a catalog and across different catalogs. Since this technique is automated, this technique is more scalable than other manual classification techniques. However, there is a tradeoff between scalability and validity; manual classification techniques are superior in terms of result's validity because they are based on experts' considerations.

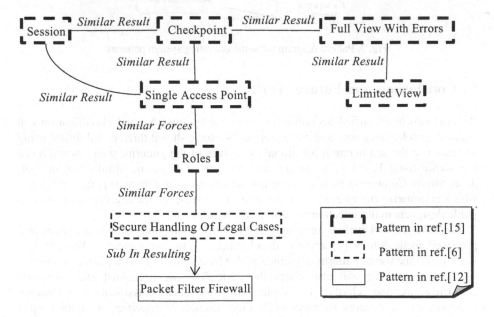

Fig. 2. Automated relation analysis result for security patterns

4 Pattern Diagrams

Fig. 3 shows a pattern diagram that relates some operating system security patterns
[12]. A pattern diagram uses these classifications to help the designer navigate in the
design space. For example, an operating system designer can start from a **Secure
Process** and use a **Controlled Process Creator** to create new processes in a secure
way (controlling their initial rights). These processes can then execute in a
Controlled Virtual Address Space (with controlled rights). The general structure of
the virtual address space is defined through a **Virtual Address Space Structure
Selection.**

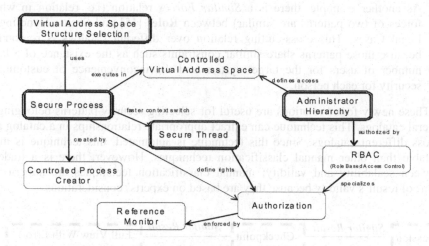

Fig. 3. Pattern diagram for some operating system patterns

5 Conclusions and Future Work

Patterns can be classified according to many viewpoints. A good classification can
make their selection easier and more precise. We have shown three possibilities: using
as reference the architectural/functional objectives of the patterns, using as reference
the architectural layers of a system, and looking at linguistic similarities in their
descriptions. Combining these classifications and expressing them as pattern diagrams
which summarize the relevant patterns at a given stage or for a given concern, can
guide designers in the selection of appropriate patterns.

Future work will include refining these classifications and incorporate them in a
proposed methodology [7] and the development of further patterns. We are also
working on the use of patterns combined with Model-Driven Development to produce
secure systems. We will also check the tradeoff between manual and automatic
classifications and clarify the applicability of each classification technique.
Composition of patterns to accomplish more complex properties is another open
problem. Finally, how well the patterns cover the space of requirements needs to be
evaluated.

References

1. Avgeriou, P., Zdun, U.: Architectural patterns revisited—A pattern language. In: Procs. EuroPLoP 2005, pp. 1–39 (2005)
2. Delessy, N., Fernandez, E.B., Larrondo-Petrie, M.M.: A pattern language for identity management. In: Procs. of the 2nd IEEE Int. Multiconference on Computing in the Global Information Technology (ICCGI 2007), Guadeloupe, French Caribbean, March 4-9 (2007)
3. Fernandez, E.B.: Security patterns. In: Procs. of the Eigth International Symposium on System and Information Security - SSI 2006 (Keynote talk) Sao Jose dos Campos, Brazil, November 08-10 (2006)
4. Fernandez, E.B., Larrondo-Petrie, M.M., Sorgente, T., VanHilst, M.: A methodology to develop secure systems using patterns. In: Mouratidis, H., Giorgini, P. (eds.) Integrating security and software engineering: Advances and future vision, ch. 5, pp. 107–126. IDEA Press (2006)
5. Fernandez, E.B., Pernul, G.: Patterns for session-based access control. In: Procs. of the Pattern Languages of Programming Conference (PLoP 2006) (2006)
6. Fernandez, E.B., la Red Martinez, D.L., Forneron, J., Uribe, V.E., Rodriguez, G.: A secure analysis pattern for handling legal cases. In: Proc. 6th Latin American Conference on Pattern Languages of Programming (2007)
7. Fernandez, E.B., Yoshioka, N., Washizaki, H., Jurjens, J.: Using security patterns to 'build secure systems. In: Procs. 1st Int. Workshop on Software Patterns and Quality (SPAQu 2007), Nagoya, Japan, December 3 (2007)
8. Gamma, E., Helm, R., Johnson, R., Vlissides, J.: Design patterns –Elements of reusable object-oriented software. Addison-Wesley, Reading (1994)
9. Kienzle, D.M., Elder, M.C., Tyree, D., Edwards-Hewitt, J.: Security patterns repository, Version 1.0., http://www.modsecurity.org/archive/securitypatterns/dmdj_repository.pdf
10. Kubo, A., Washizaki, H., Takasu, A., Fukazawa, Y.: Extracting relations among embedded software design patterns. Journal of Integrated Design and Process Science 9(3), 39–52 (2005)
11. The Open Group, Security Design Patterns Technical Guide, http://www.opengroup.org/security/gsp.htm
12. Schumacher, M., Fernandez, E.B., Hybertson, D., Buschmann, F., Sommerlad, P.: Security Patterns: Integrating systems and software engineering. J. Wiley, Chichester (2006)
13. Schumacher, M.: The Security Patterns page, http://www.securitypatterns.org
14. Steel, C., Nagappan, R., Lai, R.: Core Security Patterns: Best Strategies for J2EE, Web Services, and Identity Management. Prentice Hall, New Jersey (2005)
15. Yoder, J., Barcalow, J.: Architectural Patterns for Enabling Application Security. In: 4th Conference on Pattern Languages of Programs (PLoP 1997) (1997)

Enhanced Mutual Authentication and Key Exchange Protocol for Wireless Communications

Yi-jun He[1], Moon-Chuen Lee[1], and Jie Li[2]

[1] Dept of Computer Science & Engineering, The Chinese University of Hong Kong, Shatin, NT, Hong Kong
{yjhe, mclee}@cse.cuhk.edu.hk
[2] School of Information Science and Engineering, Central South University, Changsha 410075, Hunan, China
jli@csu.edu.cn

Abstract. We analyze several mutual authentication and key exchange protocols (MAKEPs) in this paper, and present a number of desirable properties of such protocols for secure wireless communications. To address the security problems of existing protocols, we propose an improved version of MAKEP known as EC-MAKEP. Apart from providing the desired security features supported by the existing protocols, the proposed protocol supports user anonymity, and forward secrecy which are not supported by many of the previously proposed protocols. Further, the proposed protocol outperforms ES-MAKEP, the latest improved version of MAKEP, in terms of computation cost and communication bandwidth.

Keywords: MAKEP; Elliptic Curve; forward secrecy; user anonymity.

1 Introduction

The mutual authentication and key exchange protocol (MAKEP) presented in [1] aimed to provide secure authentication between a user and a server, and to enable them to determine jointly a session key. This session key can then be used to establish a secure communication channel between the user and the server. In general, a good mutual authentication and key exchange protocol should possess the following properties [2,3]:

User Anonymity: In mobile communications, most users require their identity and private information being kept confidential.

Forward secrecy: If the long-term private keys of one or more of the entities are compromised, the secrecy of previously established session keys should not be affected.

Data Integrity: A system with this property implies that it can verify if any data received from the client has been modified during transmission.

Known-key security: If one session key has been obtained by an adversary, the protocol should still be able to achieve its goal; this means that neither the private keys nor other session keys (past or future) would be compromised as a result.

Y. Zhang et al. (Eds.): APWeb 2008, LNCS 4976, pp. 348–358, 2008.

Key control: The secret session key between any two entities should be jointly determined; neither entity can predetermine any portion of the session key.

Key-compromise impersonation resilience: If the long-term private key of an entity *A* is compromised, the protocol would allow the adversary to impersonate *A*; but it should not allow the adversary to impersonate other entities to *A*.

Unknown key-share resilience: An entity *A* cannot be coerced into sharing a key with any entity *C* when in fact *A* thinks that it is sharing the key with another entity *B*.

In addition, based on the considerations for a mobile communication environment, the list below could be seen as performance measurement criteria [4].

1. *Minimum number of passes*: To reduce latency time, the number of message exchanges required between entities should be minimal.
2. *Efficient usage of bandwidth*: Due to the low bandwidth in the mobile communication, the total number of bits transmitted should be kept as small as possible.
3. *Limited computational capability*: Since the mobile device computation capability is generally limited, the protocol should reduce the number of cryptographic operations, and employ more offline computations than online computations as much as possible.

So far, different types of mutual authentication and key exchange protocols have been proposed. In general, such protocols could be grouped into two categories: public key based protocols and symmetric key based protocols.

In the public key based protocols [5], each party holds a pair of private and public keys. The private key is kept by the owner and used either for decryption (confidentiality) or encryption (signature) of messages. The public key is published to be used for the reverse operation. They provide arbitrarily high levels of security and do not require an initial private key exchange. However, these operations can be very inefficient when implemented on low-power wireless devices. Further, it requires the support of the Public-Key Infrastructure to process the certification, their computational complexities are usually so high that prevent the public key cryptosystems being widely deployed in most of the applications running on low-power wireless devices.

In the symmetric key protocols [6], a common key is used by both communication partners for encryption and decryption. The symmetric key algorithms are much faster than the public key algorithms when implemented in wireless devices. However, symmetric key based protocols need the two communicating entities to share a long-lived key. So, how to securely distribute the long-lived key to each communicating pair is a major issue. Moreover, to communicate with different partners, each entity needs to possess a set of distinct long-lived keys. Hence key management is another problem when deploying symmetric key based protocols.

The organization of this paper is as follows. In section 2, we first introduce briefly the original MAKEP proposed by D. S. Wong and A. H. Chan [7]; then we present several improved MAKEPs proposed in the literature [8,9,10]. Section 3 presents a brief review and the cryptanalysis of ES-MAKEP [10] which is the latest MAKEP relatively more secure and efficient than those previous improved MAKEPs. In section 4, we present our proposed improved MAKEP known as EC-MAKEP which has addressed the weaknesses of the contemporary MAKEPs [7,8,9,10]. In section 5, security and performance analysis of EC-MAKEP are presented. Finally, Section 6 concludes the paper.

2   Related Work

In this section, we analyzed several existing MAKEPs which are relatively more efficient than those public-key based protocols, and more secure than the symmetric key protocols.

In 2001, Duncan S. Wong et al. proposed a Server-specific MAKEP [7] (say, Ss-MAKEP) for wireless communications between a low-power wireless device (client) and a powerful base station (server). This protocol eliminates any usage of public-key cryptographic operations at the client side. Instead, it uses efficient symmetric key based operations. Furthermore, in a conventional symmetric key based scheme, the server has to maintain a secure database of all its clients' long-lived keys, but the cost of maintaining and searching through such a database is eliminated in the proposed scheme. Instead, each client keeps a certified long-lived key and sends it securely to the server whenever the protocol is executed. Further, client can change its key anytime by obtaining a new certificate from CA without involving server. Therefore there is no key synchronization problem between client and server. However, this protocol requires the client to possess a certificate specific to the server before communicating with the server. In other words, each certificate is server-specific. A client has to keep a lot of distinct certificates in order to communicate with different servers. As a mobile device with limited storage, it would affect the implementation of this protocol. Also, server knows the long-lived symmetric key of client, which makes malicious server able to impersonate client.

Duncan S. Wong et al. also presented another protocol called Linear MAKEP [7] (say, L-MAKEP) which is an improved protocol of Ss-MAKEP. It allows each client to communicate with as many servers as it wants without keeping numerous distinct certificates. In addition, it prevents any server to impersonate its own clients. But the client has to keep many pairs of public and private keys in order to construct a certificate when it wants to communicate with the specific server.

However, in the Ss-MAKEP and the L-MAKEP schemes, the encrypted messages of the protocol do not include some information of the sender and the recipient. Kyungah Shim [8] broke these two schemes by using unknown key-share attack successfully and proposed an Improved Ss-MAKEP (say, ISs-MAKEP) and an Improved L-MAKEP (say, IL-MAKEP) by including the identities of the sender and recipient in the messages being encrypted to address weakness of the original protocols. Further, this improvement did not incur additional overhead on the proposed protocols.

However, Jim-Ke Jan et al. [9] pointed out that the man-in-the-middle attack can still attack protocol L-MAKEP and IL-MAKEP [8]. They proposed an improved MAKEP (I-MAKEP) based on Girault's method to resist the attacks of unknown key-share and man-in-the-middle. I-MAKEP adopted the pre-computation technique to reduce the overhead of the client just like the IL-MAKEP. At the same time, besides the information of pre-computation and some system parameters, the scheme just needs to keep the user's secret key in client's memory. It is much better than the IL-MAKEP which needs to keep the certificates.

The savings in computations, bandwidth, and number of message exchanges are valuable for key exchange protocols, especially in the environment of wireless communications with low power mobile devices. Fuw-Yi Yang et al.[10] proposed a protocol known as ES-MAKEP for mobile communications, which required only 0.1 modular multiplication for online computation. This online computation is ten times faster than that of conventional protocols. The number of message exchanges and message size are smaller than those of the previous protocols Ss-MAKEP, L-MAKEP, ISs-MAKEP, IL-MAKEP and the I-MAKEP. In addition, the proposed protocol can resist not only the unknown key-share attack mentioned above but also the man-in-the-middle attack which Ss-MAKEP and ISs-MAKEP could not resist.

However, we find that all the protocols Ss-MAKEP, L-MAKEP, ISs-MAKEP, IL-MAKEP and ES-MAKEP do not have the property of perfect forward secrecy, and could not provide user anonymity as well. There is thus a need to develop a new mutual authentication and key exchange protocol to support both perfect forward secrecy and user anonymity. In this paper, we only describe ES-MAKEP in detail, since it is a MAKEP which possesses all advantages of all previously improved protocols. Then we introduce our mutual authentication and key exchange protocol known as EC-MAKEP having the properties of perfect forward secrecy, user anonymity and all the merits of ES-MAKEP.

2.1   Brief Review of ES-MAKEP

Before presenting the details of ES-MAKEP, we first introduce the notations and symbols used in [10]:

SK_S: a private key of server
PK_S: a public key of server
K: a secret key of symmetric encryption/decryption function
ε_{PK_S} (): an asymmetric encryption function
δ_{SK_S} (): an asymmetric decryption function
E_K(): a symmetric encryption function
D_K(): a symmetric decryption function
h(): a hash function
ID_U: the identification of a client
ID_S: the identification of a server
σ: session key
p, q: a private key pair of client
g, n: a public key pair of client
$x \parallel y$: string x concatenates string y
$|n|$: bit length of n
r_{UK}, r_{UF}, r_{UR}: three random numbers selected by client
r_{SK}: a random number selected by server
l: the length of session keys

The message flows for ES-MAKEP are illustrated in Figure1.

Step 1: As an offline initialization step, two large prime numbers p and $q \in \{0,1\}^{k/2}$ are randomly chosen such that $p = 2 p' + 1$ and $q = 2 q' + 1$. Then, selecting a random value g of order $\lambda(n)$ from the multiplicative group $g \in Z_n^*$, where n=pq and $\lambda(n)=lcm(p\text{-}1,\ q\text{-}1)=2p'q'$. Then client announces the public key (n,g) to the public, but keeps its private key (p,q) as a secret.

Step 2: Client encrypts the random number r_{UK} using server's public key, and calculates $CMT=g^{r_{UF}\|r_{UR}} \bmod n$. Then, client sends server the message M_1 including C_1, CMT, and client's identity information ID_U to ask for initiating a new session.

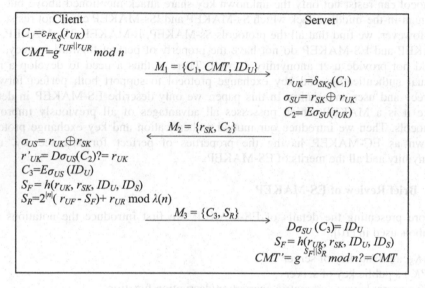

Client	Server
$C_1=\varepsilon_{PK_S}(r_{UK})$	
$CMT=g^{r_{UF}\|r_{UR}} \bmod n$	
$\xrightarrow{\quad M_1 = \{C_1,\ CMT,\ ID_U\}\quad}$	
	$r_{UK}=\delta_{SK_S}(C_1)$
	$\sigma_{SU}= r_{SK}\oplus r_{UK}$
	$C_2= E\sigma_{SU}(r_{UK})$
$\xleftarrow{\quad M_2 = \{r_{SK},\ C_2\}\quad}$	
$\sigma_{US}= r_{UK}\oplus r_{SK}$	
$r'_{UK}= D\sigma_{US}(C_2)?= r_{UK}$	
$C_3=E\sigma_{US}(ID_U)$	
$S_F= h(r_{UK},\ r_{SK},\ ID_U,\ ID_S)$	
$S_R=2^{\|n\|}(\ r_{UF} - S_F)+ r_{UR} \bmod \lambda(n)$	
$\xrightarrow{\quad M_3 = \{C_3,\ S_R\}\quad}$	
	$D\sigma_{SU}(C_3)= ID_U$
	$S_F= h(r_{UK},\ r_{SK},\ ID_U,\ ID_S)$
	$CMT'= g^{S_F\|S_R} \bmod n?=CMT$

Fig. 1. Original ES-MAKEP

Step 3: On receiving M_1, server decrypts the ciphertext C_1 to obtain r_{UK}, and calculates the session key σ_{SU} using r_{UK} and the random number r_{SK} it selects. Server also encrypts the random number r_{UK} using this session key. Then, server sends M_2 which includes r_{SK} and C_2 to client.

Step 4: Upon receiving the message M_2, client calculates the session key $\sigma_{US}= r_{UK}$ r_{SK}, and decrypts the ciphertext C_2 to obtain r'_{UK}. Client authenticates server by checking if r_{UK} equals to r'_{UK}. Because only server can calculate σ_{SU}, thus, if messages M_1 and M_2 are successfully transmitted, σ_{SU} and σ_{US} should have the same value. Accordingly, r_{UK} and r'_{UK} should be equal. After authenticating server, client computes the quantities $S_F= h(r_{UK},\ r_{SK},\ ID_U,\ ID_S)$ and $C_3=E\sigma_{US}(ID_U)$. Then it solves S_R in equation (1).

$$2^{\|n\|}r_{UF} + r_{UR} = 2^{\|n\|}S_F + S_R \bmod \lambda(n)$$
$$S_R=2^{\|n\|}(r_{UF} - S_F)+ r_{UR} \bmod \lambda(n) \quad (1)$$

At last, client sends the response message $M_3 = \{C_3,\ S_R\}$ to server.

Step 5: Server computes the quantities $S_F = h(r_{UK}, r_{SK}, ID_U, ID_S)$ and $CMT' = g^{S_F \| S_R}$ $mod\ n$. Then, server compares the value of CMT with CMT'. Based on Adi Shamir and Yael Tauman's scheme[11], CMT and CMT' should be equal if all the messages are correctly transmitted.

At this point, ES-MAKEP should have completed the mutual authentication and key exchange process. But from the security perspective, it is not safe enough, as it could not support the property of forward secrecy and user anonymity. We give the details of these two problems below.

2.2 Lacking Forward Secrecy Property

Park, *et. al.* gave a definition of the property of forward secrecy in [12] as the following. Even if a long-term private key has been disclosed to an adversary, the session keys established via the protocol runs using the long-term key would not be compromised. However, ES-MAKEP does not support forward secrecy, since the session key could be computed if the server secret key is revealed. The details are as described below.

Assume an adversary E is listening to the session of the ES-MAKEP, and the adversary could obtain C_1. Then it can compute $r_{UK} = \delta_{SK_S}(C_1)$. If the server secret key SK_S has been disclosed, E could obtain r_{SK} from M_2; thus the session key σ_{SU} would then be computed as $r_{SK}\ r_{UK}$. Therefore, ES-MAKEP does not satisfy the requirement for perfect forward secrecy, since the disclosure of the server secret key SK_S would enable an adversary to compute the session key σ_{SU}.

When the above mentioned scenario occurs, the previous session key would be exposed to the attacker. With this session key and those messages previously intercepted during the transmission, the attacker can easily get useful information from those messages encrypted using the session key.

2.3 Lacking User Anonymity Property

Any system supporting user anonymity means that it keeps user secrets confidential or avoids disclosing any confidential user information. Especially in e-business application, user anonymity is a very important issue since online business transactions could incur many security problems if user secrets are disclosed during the process. For instance, an attacker could make use of the user identity to impersonate the user to perform online shopping.

ES-MAKEP suffers from lacking the user anonymity property mentioned above. In the protocol, ID_U is transmitted in M_1 without any encryption. So, an attacker could obtain M_1 to figure out the ID_U. Thus user identity could be exposed to an attacker. The attacker could make use of this ID_U to pretend to be a legitimate user and initiate a session with server.

3   The Proposed Protocol EC-MAKEP

We improve the ES-MAKEP by using ECDH (Elliptic Curve Diffie-Hellman) [13] algorithm in the key exchange process. We employ two computed values R_C and R_S as the exchange data to replace the original random values r_{UK} and r_{SK} respectively, and to make the protocol equipped with the forward secrecy property. Further, we remove ID_U from M_1 in order to prevent user identity from being exposed to an attacker. The improved protocol EC-MAKEP is as outlined in Figure 2 below.

1. As an offline initialization step, the ECDH algorithm has been preset to use a big prime p_1 and two other parameters a and b satisfying the equation $y^2=(x^3+ax+b)$ mod p_1, to form an elliptic group E_{p1} (a,b). Then, it chooses the basic point $G=(x,y)$ with order q_1, where q_1 is the minimum integer satisfying q_1 $\cdot G = O$, O being a point at infinity.

2. In order to communicate with the server, the client chooses an integer $r_C < q_1$, $r_{UF} \in {}_R\{0,1\}^l$ and $r_{UR} \in {}_RZ_{\lambda(n)}$. Then it computes $R_C=r_C \cdot G(mod)p_1$ through an Elliptic Curve Cryptography (ECC) point multiplication operation. According to the ECC property, R_C is an ECC point. Then client computes $CMT=g^{r_{UF}\|r_{UR}}$ mod n, and sends R_C and CMT to server.

3. Server chooses an integer $r_S < q_1$, and computes $R_S=r_S \cdot B(mod)$ p_1. R_S is an ECC point.

4. After receiving R_C, server computes session key $\sigma_{SU}= r_S \cdot R_C$ and uses the symmetric encryption algorithm $E_K()$ to encrypt R_C with the encryption key σ_{SU}. Then server sends the encrypted value C_1 and R_S to client.

5. Client computes session key $\sigma_{US}=r_C \cdot R_S$, and applies the symmetric decryption algorithm $D_K()$ to decrypt C_1 with the decryption key σ_{US} to obtain $R_{C'}$. If $R_{C'}$ is equal to R_C, σ_{SU} and σ_{US} must be equal; otherwise, the values sent by server to client or the values sent by client to server could have been changed by an attacker.

6. Client encrypts ID_U using session key as the encryption key, and computes the quantities $S_F= h(R_C, R_S, ID_U, ID_S)$, and $S_R =2^{|n|}(r_{UF} - S_F) + r_{UR}$ mod $\lambda(n)$. Then client sends C_2 and S_R to server.

7. Server computes the quantities $S_F= h(R_C, R_S, ID_U, ID_S)$ and $CMT'= g^{S_F\|S_R}$ mod n. Server authenticates client by verifying if CMT equals CMT'.

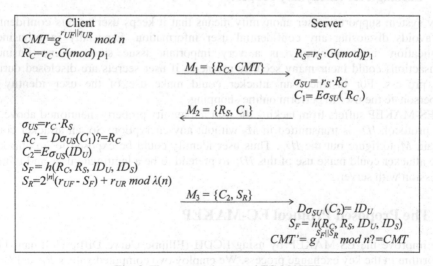

Fig. 2. The proposed protocol EC-MAKEP

4 Performance Analysis

4.1 Security Properties

User Anonymity. In EC-MAKEP, ID_U is not included in M_1 anymore; it is sent in M_3 in an encrypted form $E_{\sigma_{US}}(ID_U)$. As ID_U is encrypted using the session key σ_{US} computed as $r_C \cdot R_S$, only the server can calculate the session key and decrypt C_2 to get ID_U. The server can authenticate client by checking ID_U without worrying about if an attacker could intercept ID_U during the message transmission. Thus, the user identity can be kept confidential, and the requirement for user anonymity can be met.

Forward Secrecy. In EC-MAKEP, the client sends a pre-computed ECC point R_C to server. Accordingly, the server sends a pre-computed ECC point R_S to client. The session key would then be computed as $r_S \cdot R_C$ by the server, and as $r_C \cdot R_S$ by the client. Now, using $\varepsilon_{PK_S}(r_{UK})$ instead R_C in message M_1 from client to server, and using r_{SK} instead of R_S in message M_2 from server to client can help both parties to establish a secret session key satisfying the requirement for forward secrecy. Suppose the attacker could get R_C, R_S and the private key of the server; it still cannot successfully obtain the previous session key, because according to the property of the ECC algorithm, the attacker cannot compute r_S or r_C using R_C and R_S. Therefore, the compromise of the long-term private key of the server does not lead to the disclosure of the previous session key. Thus, EC-MAKEP can support forward secrecy.

Data Integrity. A system supporting data integrity implies that it can check if the data received from the client are correct; that is, it can check if the data transmitted to the receiver have been modified. In EC-MAKEP, the client first computes $S_F = h(R_C, R_S, ID_U, ID_S)$; it then computes S_R using S_F, which is subsequently sent to the server. Since the parameters R_C, R_S, ID_U, ID_S used in computing S_F have all been transmitted in the network, if any one of them has been modified during the transmission, the S_F calculated by client would not be equal to the one calculated by server. As a result, CMT would not be equal to CMT'. So all the data transmitted during the execution of EC-MAKEP can be verified. So, it facilitates the validation of data integrity.

Known-key security. With the proposed EC-MAKEP, if the current session key has been compromised, the other session keys (past and future), and the private keys of the client and the server could still be safe. Since session key computation uses a random value of client (server) and a pre-computed value of server (client), and the random value is different in each session, so the past or future session keys have no relation with the current one. Further, as the private keys of the client and server are not involved in the computation of the session key, they would not be compromised even if the session key has been disclosed.

Key control. In the protocol EC-MAKEP, both client and server cannot predetermine the session key being established, because the establishment of the session key involves both a random value and a pre-computed value. Each of the values comes from a different entity, so neither the client nor the server can determine the session key before the communication.

Key-compromise impersonation resilience. With the protocol EC-MAKEP, even if the private key of the client has been exposed, an attacker can impersonate neither the client nor the server because the protocol does not use the private keys of the two entities in the key exchange and authentication process; it uses the pre-computed value R_C and R_S instead.

4.2 Computation Cost and Bandwidth

To simplify the estimation of the computation cost, we divide the computation cost of the protocol into two parts: offline computation and online computation. Moreover, the costs of additions, hash operation $h()$, symmetric encryption $E_K()$ and decryption $D_K()$ would not be included since the costs of these operations are much smaller than the cost of the elliptic curve point multiplication operation. Fuw-Yi Yang and Jinn-Ke Jan [10] presented their analysis of computation cost and bandwidth of ES-MAKEP. By following their approach, we have carried out a similar analysis on our proposed EC-MAKEP and the results are as presented below. Based on the findings of [14,15], an ECC with 160-bit key length could offer roughly the same level of security as RSA with 1024-bit modulus. As to the modulus exponent function $g^x \mod n$, we set the length of modulus n equal to 1024bits, where x is an 160-bit random integer. The cost of computing such a modulus exponent function is estimated to be about $1.5|x|$ modular multiplications [16], equal to around 240 modular multiplications ($|x|$ indicates the length of x). When the length of p_1 is 160 bits in the elliptic curve point multiplication function, it would be 8 times faster than modulus exponent computation. So it can be deduced that the computation cost of elliptic curve point multiplication is equivalent to around 29 modular multiplications [16]. Tables1 and 2 present the comparisons of computation costs of EC-MAKEP and ES-MAKEP on the client side and the server side respectively.

Table 1. Comparison of computation costs on client side (/MMs)

	EC-MAKEP		ES-MAKEP	
	Online	Offline	Online	Offline
Message M_1	0	1805 [a]	0	1778
Message M_2	0	0	0	0
Message M_3	0.1 [b]	0	0.1	0
Total	0.1	1805	0.1	1778

Table 2. Comparison of computation costs on server side (/MMs)

	EC-MAKEP		ES-MAKEP	
	Online	Offline	Online	Offline
Message M_1	0	29 [c]	1536	0
Message M_2	0	0	0	0
Message M_3	1776 [d]	0	1776	0
Total	1776	29	3312	0

Note: MMs denotes the computation cost of a modular multiplication $a*b$ mod n, where a, b, and n are all set to be 1024 bits.

  a. Computing R_C requires one ECC point multiplication, with a cost of 29MMs. Computing $CMT = g^{r_{UF} \| r_{UR}}$ mod n needs $1.5*(160+1024) = 1776$MMs[17].
  b. Fuw-Yi Yang et al. [10] show that to compute $S_R = 2^{|n|}(r_{UF} - S_F) + r_{UR}$ mod $\lambda(n)$ requires 0.1MMs.
  c. Computing R_S requires one ECC point multiplication, with a cost of 29MMs.
  d. Computing $CMT' = g^{S_F \| S_R}$ mod n requires $1.5*(160+1024) = 1776$MMs.

Table 3 shows the bandwidth overheads of EC-MAKEP and ES-MAKEP. As suggested in [18], for practical cryptographic operations, we set $|l| = |ID_U| = |r_{UF}| = |S_F| = 160$ bits.

Table 3. Bandwidth overheads in EC-MAKEP and ES-MAKEP (/bits)

	EC-MAKEP	ES-MAKEP
Message M_1	1184	2208
Message M_2	480	320
Message M_3	1184	1184
Total	2848	3712

When compared with ES-MAKEP, the proposed protocol EC-MAKEP requires a similar online computation cost on the client side; however, it requires a much smaller online computation cost on the server side. Although it requires a bigger offline computation cost on client side, the small additional cost for computing R_C and R_S are justifiable since the underlying operations could help to provide forward secrecy, and enhance the security of the protocol. Moreover, the offline computation would be performed only once; so the computation cost can be considered insignificant. When compared with ES-MAKEP in terms of bandwidth requirement, EC-MAKEP reduces 864bits of bandwidth. It is certainly an obvious advantage for low bandwidth wireless communication.

5 Conclusions

We studied the early mutual authentication and key exchange protocol (MAKEP), and the improved MAKEPs; we also identified their weaknesses. The latest improved protocol known as ES-MAKEP addresses many of the security problems of the previous MAKEPs. This paper proposes another protocol EC-MAKEP which has improved on ES-MAKEP. Our security analysis of the proposed protocol shows that EC-MAKEP satisfies all of the major security requirements for a secured MAKEP. It compares favorably with ES-MAKEP in terms of a recognized list of security requirements. For instance, ES-MAKEP does not support forward secrecy and user anonymity; whereas EC-MAKEP supports both of the two features. Moreover, when compared with ES-MAKEP, EC-MAKEP has a smaller online computation cost, and requires a smaller bandwidth.

Acknowledgement. The work reported in this article has been supported in part by CUHK under RGC Direct Grant with Project ID 2050347.

References

1. Jakobsson, M., Pointcheval, D.: Mutual Authentication and Key Exchange Protocol for Low Power Devices. In: Financial Cryptography pp. 178–195. Springer, Heidelberg (2001)
2. Chen, L., Cheng, Z., Smart, N.P.: Identity based authentication key agreement protocols from pairings. In: Proceedings of the 16th IEEE Computer Society Foundations Workshop (CSFW 2003) (2003)
3. Certicom's Bulletin of Security and Cryptography. Code and cipher vol.1(2), http://www.certicom.com/codeandcipher
4. Lee, K.-H., Moon, S.-J., Jeong, W.-Y., Kim, T.-G.: A 2-pass authentication and key agreement protocol for mobile communications. In: Song, J.S. (ed.) ICISC 1999. LNCS, vol. 1787, pp. 156–168. Springer, Heidelberg (2000)
5. Harbitter, A.H., Menase, D.A.: Performance of public-key-enabled Kerberos authentication in large networks. In: Proceedings of 2001 IEEE Symposium on Security and Privacy, pp. 170–183 (2001)
6. Needham, R.M., Schroeder, M.D.: Using Encryption for Authentication in Large Networks of Computers. Commun. of the ACM 21(12), 993–999 (1978)
7. Wong, D.S., Chan, A.H.: Mutual authentication and key exchange for low power wireless communications. In: Military Communications Conference, 2001. MILCOM 2001. Communications for Network-Centric Operations: Creating the Information Force, vol. 1, pp. 39–43. IEEE, Los Alamitos (2001)
8. Shim, K.: Cryptanalysis of mutual authentication and key exchange for low-power wireless communications. IEEE Communications Letters 7(5), 248–250 (2003)
9. Jan, J.K., Chen, Y.H.: A new efficient MAKEP for wireless communications. In: Proceedings of the 18th International Conference on Advanced Information Networking and Application (AINA 2004), vol. 2, pp. 347–350. IEEE, Los Alamitos (2004)
10. Yang, F.-Y., Jan, J.-K.: "A Secure and Efficient Key Exchange Protocol for Mobile Communications" Cryptology ePrint Archive 2004/167, (July 2004), http://eprint.iacr.org
11. Shamir, A., Tauman, Y.: Improved online/offline signature schemes. In: Kilian, J. (ed.) CRYPTO 2001. LNCS, vol. 2139, pp. 355–367. Springer, Heidelberg (2001)
12. Park, D., Boyd, C., Moon, S.-J.: Forward Secrecy and Its Application to Future Mobile Communications Security. In: Imai, H., Zheng, Y. (eds.) PKC 2000. LNCS, vol. 1751, pp. 433–445. Springer, Heidelberg (2000)
13. Certicom Research, Standards for efficient cryptography, SEC 1: Elliptic Curve Cryptography, Version 1.0, (September 20, 2000)
14. Zhuxing, L., Zhenglong, L.: Elliptic-Curve Undeniable Signature Schemes. In: The 11th information security conference, pp. 331–338 (2001)
15. Jurisic, A., Menezes, A.J.: ECC Whitepapers: Elliptic Curves and Cryptography. Certicom corp. http://www.certicom.com/research/weccrypt.html
16. Koblitz, N.: Elliptic Curve Cryptosystems. Mathematics of Computation 48(17), 203–209 (1987)
17. Fegruson, N., Schneier, B.: Practical Cryptography. John Wiley & Sons, Chichester (2003)
18. Lenstra, A., Verheul, E.: Selecting Cryptographic Key Size. In: Imai, H., Zheng, Y. (eds.) PKC 2000. LNCS, vol. 1751, Springer, Heidelberg (2000)

Verification of the Security Against Inference Attacks on XML Databases

Kenji Hashimoto, Fumikazu Takasuka, Kimihide Sakano, Yasunori Ishihara, and Toru Fujiwara

Graduate School of Information Science and Technology, Osaka University
1-5 Yamadaoka, Suita, Osaka, 565-0871 Japan

Abstract. A verification framework for the security against inference attacks on XML databases is proposed. The framework treats a concept called k-secrecy with an integer $k > 1$ (or $k = \infty$), which means that attackers cannot narrow down the candidates for the value of the sensitive information to $k - 1$ (or finite), using the results of given authorized queries and schema information. The security verification methods proposed in the framework are applicable to practical cases where authorized queries extract some nodes according to any of their neighboring nodes such as ancestors, descendants, and siblings. Also, using experimental results of prototype systems, we discuss time efficiency of the proposed verification methods.

Keywords: XML database, inference attack, security, verification.

1 Introduction

Nowadays, many people and organizations have a growing interest in data security. For a database system to be secure, secrecy, integrity, and availability of data must be achieved appropriately with respect to a given security policy. View mechanisms has played a principal role in achieving the secrecy of database systems. Views can avoid direct accesses to sensitive information in the database. However, there still is a possibility of indirect accesses, i.e., the sensitive information can be inferred using authorized views as well as general domain knowledge. Inference attacks mean that a user infers the sensitive information from the information to which accesses are permitted. It is important for database managers to know beforehand the possibility that the inference attacks succeed, because databases are often used as the core of the systems requiring high-level security (e.g., e-business, Web services).

Example 1. We show an example of inference attacks on XML databases, which are getting used actively in many organizations recently. Fig. 1 illustrates an XML document D to be attacked. The document D is valid against the following schema A:

Y. Zhang et al. (Eds.): APWeb 2008, LNCS 4976, pp. 359–370, 2008.

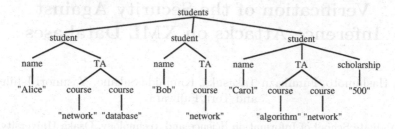

Fig. 1. An XML document D to be attacked

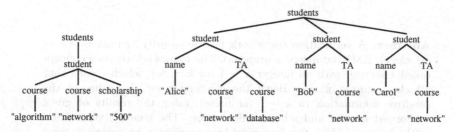

Fig. 2. The result $T_1(D)$ **Fig. 3.** The result $T_2(D)$

```
<!ELEMENT students (student*)>
<!ELEMENT student (name, TA?, scholarship?)>
<!ELEMENT name (#PCDATA)>
<!ELEMENT TA (course+)>
<!ELEMENT course (#PCDATA)>
<!ELEMENT scholarship (#PCDATA)>
```

We consider the following two authorized queries T_1 and T_2 on D: Query T_1 extracts, for each student receiving a scholarship, the courses of which the student is a teaching assistant (TA) and the amount of the scholarship. Fig. 2 shows the result $T_1(D)$. Query T_2 extracts, for each student who is a TA of "network" and/or "database", the student's name and a TA node with the courses "network" and/or "databases" of which the student is a TA. Fig. 3 shows the result $T_2(D)$. Given the queries T_1 and T_2, and the results $T_1(D)$ and $T_2(D)$, it can be inferred that the student named "Alice" does not receive a scholarship, because T_2 and $T_2(D)$ indicate that "Alice" is a TA of both "network" and "database", and T_1 and $T_1(D)$ indicate that no student who is a TA of both courses receives a scholarship.

Inference attacks on relational databases have been studied for a few decades. The structure of relational databases (i.e., sets of tuples) is fixed and simple, so the structural information is useless to infer the sensitive information in many situations. On the other hand, the structure of XML databases, which is modeled as labeled ordered trees, is flexible and complicated. Attackers may use such structural information as well as the equalities of values. Therefore, it is desirable that verification frameworks for the security against inference attacks on XML databases take account of inference using structural information directly.

In this paper, we adopt a concept of k-secrecy against inference attacks on XML databases. The underlying idea is similar to ℓ-diversity in [1]: It is considered to be insecure that the set of candidates for the value of the sensitive information can be narrowed down by inference. More formally, let D be an XML document to be attacked. Suppose that the following information is available to the attacker:

- the schema A of D,
- authorized queries (i.e., view definitions) $T_1, \ldots, T_n$,
- results of the authorized queries on D (i.e., views) $T_1(D), \ldots, T_n(D)$, and
- a query T_S to retrieve the sensitive information in D.

The set of candidates for the value of the sensitive information inferred by the attacker is

$$C = \{T_S(D') \mid D' \in TL(A), T_1(D') = T_1(D), \ldots, T_n(D') = T_n(D)\},$$

where $TL(A)$ is the set of the XML documents valid against A. That is, C is the set of all $T_S(D')$ such that D' behaves the same as D with respect to queries $T_1, \ldots, T_n$. Note that C is never empty because C always contains $T_S(D)$. Then, D is k-secret with respect to T_S under the authorized queries $T_1, \ldots, T_n$ if $|C| \geq k$ for a given integer $k > 1$. In particular, D is ∞-secret if $|C|$ is infinite. Note that this formalization is general and not specific to XML databases; it is easily applicable to other kinds of databases.

The main contribution of the paper is to propose a verification framework with respect to the above security formalization. In the framework, an XML document is modeled as a labeled ordered tree as usual. A schema is represented by a tree automaton, which is the theoretical foundation of Relax NG [2]. A query is represented by a composition of *relabeling* and/or *deleting tree transducers*. Our queries are a subclass of XSLT but still practical because filtering nodes according to any of their neighboring nodes such as ancestors, descendants, and siblings is expressible.

Then, a verification method based on the above framework is straightforwardly derived. The method consists of four steps and simulate the attacker's inference as follows: In the first step, for each pair of an authorized query T_i and its result $T_i(D)$, compute a tree automaton representing the set of candidates for D determined by T_i and $T_i(D)$. In the second step, compute the intersection of the tree automata. In the third step, then, compute a tree automaton representing the set C of candidates for $T_S(D)$. In the last step, it is examined whether $|C|$ is infinite, and if not, whether $|C| \geq k$ for a given integer $k > 1$.

However, it is expected that the straightforward method spends a large time in the first step because large tree automata will be produced as intermediate results. The large tree automata can contain many states and transition rules to represent candidate documents which will not be valid against the schema A in the next step. We improve the first step so that intermediate results which contain less such "unnecessary" states and rules is produced by using the information of the schema. We have found by experiment that time efficiency is improved by the modification.

The rest of the paper is organized as follows. Section 2 reviews related works on inference attacks. Section 3 introduces models of XML documents, schemas, and queries. Section 4 proposes a verification method and its modification for the security against inference attacks on XML databases. Section 5 mentions the prototype systems and discusses a direction of further improvement of the verification methods. Section 6 summarizes the paper.

2   Related Works

As stated in Section 1, inference attacks on relational databases have been studied for a few decades. Inference attacks using aggregate functions are one of the most famous threats in statistical relational databases [3]. Disclosure Monitor [4] is a part of a database management system that monitors information disclosure by inference attacks. [5] incorporates into Disclosure Monitor a mechanism that accomplishes maximum data availability as long as given sensitive information is secure against inference attacks. Also, several sophisticated formalizations of the security have also been proposed, e.g., the one based on information theory [6,7], the one based on rough set theory [8]. Moreover, [9,10] proposes stronger security definitions where the probability distribution of possible secrets does not change before and after authorized views and the answers of them are given.

Recently, the security against inference attacks on relational databases is often discussed in the context of privacy protection. Reference [11] proposes a formalization of the security, called k-anonymity, which means that for each person, there are at least k tuples in the view that can be associated with the person. After that, some weakness of k-anonymity has been pointed out. For example, [1] demonstrates that even if the database is k-anonymous, the k or more tuples for a person may have the same sensitive value, so the person's sensitive information may be identified. Then, [1] proposes the notion of ℓ-diversity, which means that for each person, there are at least ℓ possible values (or ℓ bits in entropy, etc.) for the person's sensitive information. Reference [12] states that a person's privacy should be customizable by the person, and proposes the notion of personalized anonymity. This paper adopts a concept of k-secrecy similar to ℓ-diversity.

As for XML databases, the notion of security views is proposed in [13]. A security view consists of a view XML document and its DTD, and contains no information that may be useful to infer the sensitive information. However, the formalization of the attacker's inference is unclear. In [14], attackers are supposed to use schema information and functional dependencies as well as the view XML document. Then, an algorithm for finding a maximal view document without allowing successful inference attacks is proposed. Both of these researches assume the "single-view" environment, that is, to each user only one view is given. In relational databases, [9] and [10] study the security against multiple-view inference attacks, which can provide more availability in general. This paper assumes the "multiple-view" environment in XML databases.

Fig. 4. Simulating a text node with value "500"

3 Models

In this section, first, ordinary models of XML documents and schemas are introduced. Then, a model of queries is presented, which is one of the originality of this paper.

3.1 XML Documents

An XML document is represented by an *unranked* labeled ordered tree, i.e., a labeled ordered tree where the number of the children of a node is not bounded. Let Σ be an alphabet including special symbols \$ and #, which will be explained in Definition 5 of Section 3.3. Let T_Σ denote the set of the unranked labeled ordered trees over Σ, i.e., every node of the trees in T_Σ is labeled with a symbol in Σ. We denote a tree by $a(t_1 \cdots t_n)$ where $t_1, \ldots, t_n$ $(n \geq 0)$ are trees. It denotes the tree which has a root labeled with a and its immediate subtrees $t_1, \cdots, t_n$ from the left in the order.

Definition 1. T_Σ *is defined as the smallest set satisfying the following condition:* $a(t_1 \cdots t_n) \in T_\Sigma$ *if* $a \in \Sigma$ *and* $t_1, \ldots, t_n \in T_\Sigma$ $(n \geq 0)$.

We write a instead of $a(\epsilon)$.

As many theoretical studies do, we would like to avoid distinguishing text nodes (i.e., nodes with string values) with element nodes (i.e., nodes with tag names). However, regarding string values as tag names is not suitable for our purpose because to deal with ∞-secrecy properly, the set of possible string values should be infinite while the set of tag names defined in a schema is only finite. Therefore, we simulate a text node by an element node named **string** and its children which are also element nodes labeled by the constituent characters. For example, the text node with value "500" in Fig. 1 is simulated by a **string** node and its three children labeled with "5", "0", and "0", respectively (see Fig. 4).

3.2 Sets of XML Documents

In this paper, we use a *non-deterministic tree automaton* (NTA) to represent schemas or sets of candidates for the value of the sensitive information.

Definition 2. *An NTA A is a 4-tuple* $(Q, \Sigma, \hat{q}, R)$*, where*

- Q *is a finite set of states,*
- Σ *is an alphabet,*
- $\hat{q} \in Q$ *is the initial state, and*

- R is a set of transition rules in the form of (q, a, e), where $q \in Q$, $a \in \Sigma$, and e is a non-deterministic string automaton (NA) over Q.

Let $t = a(t_1 \cdots t_n)$, where $a \in \Sigma$ and $t_1, \ldots, t_n \in T_\Sigma$. Let $q(t)$ denote the configuration representing that the root of t is associated with a state q. If $(q, a, e) \in R$ and $q_1 \cdots q_n \in L(e)$, where $L(e)$ is the string language recognized by e, then A can move from $q(t)$ to $a(q_1(t_1) \cdots q_n(t_n))$, which means that the root of each t_i in t is associated with q_i. A tree $t \in T_\Sigma$ is *accepted* by A if it can move from $\hat{q}(t)$ eventually to t. Let $TL(A)$ denote the tree language recognized by A.

3.3   Queries

Each query is represented by a composition of some relabeling tree transducers and a deleting tree transducer introduced below.

Definition 3. *A (deterministic) bottom-up relabeling tree transducer (BRTT) T^B is a 4-tuple $(Q, \Sigma, \hat{q}, P^B)$, where*

- Q is a finite set of states,
- Σ is an alphabet,
- $\hat{q} \in Q$ is the final state, and
- P^B is a set of transformation rules in the form of $a(e) \rightarrow q'(a')$, where $q' \in Q$, $a, a' \in \Sigma$, and e is an NA over Q. If P^B contains two distinct rules $a(e_1) \rightarrow q'_1(a'_1)$ and $a(e_2) \rightarrow q'_2(a'_2)$, then $L(e_1) \cap L(e_2)$ must be the empty language.

A BRTT relabels nodes dependently on their descendants. Formally, let $t_p = a(q_1(t_1) \cdots q_n(t_n))$, where $a \in \Sigma$, $t_1, \ldots, t_n \in T_\Sigma$ and $q_1, \ldots, q_n \in Q$. T^B can move from the configuration t_p to $q'(a'(t_1 \cdots t_n))$ in one step if $(a(e) \rightarrow q'(a')) \in P^B$ and $q_1 \cdots q_n \in L(e)$. Let $T^B(t)$ ($t \in T_\Sigma$) denote $t' \in T_\Sigma$ such that T^B can move from t eventually to $\hat{q}(t')$.

Example 2. The following BRTT $T^B = (Q^B, \Sigma, \hat{q}, P^B)$ relabels every **student** which has no **scholarship** to $\$$, given a document valid against the schema A in Example 1 as an input tree of T^B: $Q^B = \{\hat{q}, q_1, q_2\}$, Σ contains $\{$string, name, course, TA, scholarship, student, students, #, $\$\}$ as well as alphabetic characters and symbols, and P^B consists of the following rules:

- /* Each leaf a is associated with state q_1. */
  $a(\varepsilon) \rightarrow q_1(a)$ for each $a \in \Sigma$,
- /* Each of string, name, course, and TA is not relabeled. */
  $\text{string}(q_1^*) \rightarrow q_1(\text{string})$, $\text{name}(q_1) \rightarrow q_1(\text{name})$,
  $\text{course}(q_1) \rightarrow q_1(\text{course})$, $\text{TA}(q_1^+) \rightarrow q_1(\text{TA})$,
- /* scholarship is not relabeled but the existence of scholarship in this subtree is indicated by q_2. */
  $\text{scholarship}(q_1) \rightarrow q_2(\text{scholarship})$,

- /* student is relabeled with $ if and only if it has no q_2 as a child
  (i.e., no scholarship in its subtrees). */
  student($q_1q_1?q_2$) → q_1(student), student($q_1q_1?$) → q_1($), and
- /* students is not relabeled and the final state $\hat{q}$ is reached
  if students has q_1 as a child. */
  students(q_1) → $\hat{q}$(students).

Definition 4. *A (deterministic) top-down relabeling tree transducer (TRTT)*
T^{T} *is a 4-tuple* $(Q, \Sigma, \hat{q}, P^{\mathrm{T}})$, *where*

- *Q is a finite set of states,*
- *Σ is an alphabet,*
- *$\hat{q} \in Q$ is the initial state, and*
- *P^{T} is a set of transformation rules in the form of $q(a) \to a'(q')$, where
  $q, q' \in Q$ and $a, a' \in \Sigma$. For each pair of q and a, there must be at most one
  rule in P^{T} whose left-hand side is $q(a)$.*

A TRTT relabels nodes dependently on their ancestors. T^{T} can move from configuration $q(a(t_1 \cdots t_n))$ to $a'(q'(t_1) \cdots q'(t_n))$ in one step if $(q(a) \to a'(q')) \in P^{\mathrm{T}}$.
Note that by definition a state associated with a leaf node will disappear after
the next transformation step because a leaf node is a node with the empty se-
quence of child subtrees. Let $T^{\mathrm{T}}(t)$ $(t \in \mathcal{T}_\Sigma)$ denote $t' \in \mathcal{T}_\Sigma$ such that T^{T} can
move from $\hat{q}(t)$ eventually to $t' \in \mathcal{T}_\Sigma$.

Example 3. The following TRTT T^{T} = $(Q^{\mathrm{T}}, \Sigma, \hat{q}, P^{\mathrm{T}})$ relabels every TA
to # and every name to $, given a document valid against the schema
A in Example 1 as an input tree of T^{T}: Q^{T} = $\{\hat{q}, q_1\}$, Σ contains
{string, name, course, TA, scholarship, student, students, #, $} as well as al-
phabetic characters and symbols, and P^{T} consists of the following rules:

- /* students is not relabeled and its children are associated with q_1. */
  $\hat{q}$(students) → students(q_1),
- /* name is relabeled with $.*/
  q_1(name) → $($q_1$),
- /* TA is relabeled with #. */
  q_1(TA) → #(q_1), and
- /* The others are not relabeled. */
  $q_1(a) \to a(q_1)$ for each $a \in \Sigma - \{$students, name, TA$\}$.

Definition 5. *A (deterministic) deleting tree transducer (DTT) T^{D} is a 4-tuple*
$(\{q_0, q, q_d\}, \Sigma, q_0, P^{\mathrm{D}})$, *where*

$$P^{\mathrm{D}} = \{q_0(a) \to a(q), \ q(a) \to a(q), \ q_d(a) \to q_d \mid a \in \Sigma\}$$
$$\cup \{q(\#) \to q, \ q($) \to q_d\}.$$

A DTT deletes #-labeled nodes of the input tree and subtrees rooted by $-
labeled nodes of the tree, traversing in a top-down manner. Exceptionally, the
root of the input tree is never deleted to ensure that the output is a tree. Note

that we have a unique DTT when Σ is fixed. So, we simply write T^D as a unique DTT when Σ is clear from context. For example, $t = a(b\#(de)b(\#f\$)\$(de))$ is transformed to $t' = a(bdeb(f))$ by a DTT.

In this paper, a query is assumed to be a composition of some of BRTTs and/or TRTTs, followed by a DTT. In our model, a query can filter and/or relabel nodes according to any of their neighboring nodes such as ancestors, descendants, and siblings, but cannot reconstruct the input tree with its document order changed or add new nodes to the input tree. From a practical point of view, this class of queries seems sufficient because many queries to databases essentially extracts some parts of its content. We also assume that no constituent relabeling tree transducers of a query T_S to retrieve the sensitive information relabels a node to $\#$. Therefore, T_S can delete some subtrees of the input tree, but cannot delete only internal nodes. Without this assumption, candidates for the value of the sensitive information cannot be represented by an NTA in general. Moreover, we assume that each authorized query T_i is total to $TL(A)$ where A is the schema of the XML document D to be attacked, i.e., for each $t \in TL(A)$, there is $T_i(t) \in \mathcal{T}_\Sigma$.

Example 4. T_1 in Example 1 can be represented by $T^D \circ T^T \circ T^B$ where T^B and T^T are given in Example 2 and 3 respectively.

4   Verification Methods

In this section, a verification method and its modification for the security against inference attacks on XML databases are presented. The input and output of these methods are as follows:

Input. The schema A of an XML document D to be attacked, authorized queries $T_1, \ldots, T_n$ which are compositions of some BRTTs and/or TRTTs followed by a DTT, results $T_1(D), \ldots, T_n(D)$ of the authorized queries on D, and a query T_S to retrieve the sensitive information in D.

Output. "∞-secret" if D is ∞-secret with respect to T_S under the authorized queries $T_1, \ldots, T_n$, "k-secret" if D is not ∞-secret but k-secret for a given constant k, and "not k-secret" otherwise.

4.1   A Straightforward Method: M1

A straightforward method M1 consists of the following four steps:

1. For each i $(1 \le i \le n)$, construct an NTA A_i which accepts D' such that $T_i(D') = T_i(D)$ from T_i and $T_i(D)$ as follows. Let $T_i = T^D \circ T_i^{R,m} \circ T_i^{R,m-1} \circ \cdots \circ T_i^{R,1}$ where each $T_i^{R,j}$ is a TRTT or a BRTT. To construct A_i, first construct NTA A_i^0 which accepts D^0 such that $T^D(D^0) = T_i(D)$, second construct NTA A_i^1 which accepts D^1 such that $T^{R,m}(D^1) \in TL(A_i^0)$, and so on. Finally construct $A_i^m(=A_i)$ which accepts D^m such that $T^{R,1}(D^m) \in TL(A_i^{m-1})$.

2. Construct an NTA A_D which recognizes the set of the candidates for D, i.e.,

$$TL(A_D) = \{D' \mid D' \in TL(A), T_1(D') = T_1(D), \ldots, T_n(D') = T_n(D)\}$$

by computing the intersection of A_is and A.

3. Construct an NTA A_S which recognizes the set of the candidates for the value of the sensitive information $T_S(D)$, i.e.,

$$TL(A_S) = \{T_S(D') \mid D' \in TL(A_D)\}$$

from A_D and T_S. Let $T_S = T^D \circ T_S^{R,l} \circ T_S^{R,l-1} \circ \cdots \circ T_S^{R,1}$ where each $T_S^{R,j}$ is a TRTT or a BRTT. To construct A_S, first construct NTA A_S^0 which accepts $T_S^{R,1}(D^0)$ where $D^0 \in TL(A_D)$, second construct NTA A_S^1 which accepts $T_S^{R,2}(D^1)$ where $D^1 \in TL(A_S^0)$, and so on. Finally construct $A_S^l(=A_S)$ which accepts $T^D(D^l)$ where $D^l \in TL(A_S^{l-1})$.

4. Decide whether $|TL(A_S)|$ is infinite (∞-secrecy), and if not, decide whether $|TL(A_S)| \geq k$ (k-secrecy) for a given positive integer k.

In Step 1, each A_i^j can be constructed from $T_i^{R,j}$ or T^D and A_i^{j-1} using a technique called *inverse type inference* [15]. Roughly speaking, A_i^j can be constructed by composing each transition rule of A_i^{j-1} and each transformation rule of the transducer. Likewise, in Step 3, A_S^j can be constructed from $T_S^{R,j}$ or T^D and A_S^{j-1} using a technique called *type inference*. In Step 4, ∞-secrecy can be decided using the following property: $|TL(A_S)|$ is infinite if and only if non-useless transition rules of A_S form a cycle or the string automaton of some non-useless transition rule of A_S recognizes an infinite language. Also, whether $|A_S| \geq k$ can be decided by counting up to k the number of the distinct trees accepted by A_S.

4.2   A Modified Method: M2

In Step 1 of M1, an NTA A_i^0 such that $TL(A_i^0) = \{D_0 \mid T^D(D_0) = T_i(D)\}$ is constructed as an intermediate result. Since T^D is a DTT which deletes all the #-nodes (and \$-subtrees), $TL(A_i^0)$ consists of all the trees obtained by arbitrarily embedding #-nodes into $T_i(D)$. However, such arbitrary embedding is often wasteful because it causes many "inappropriate" candidates in $TL(A_i^0)$. More precisely, there will be many $D_0 \in TL(A_i^0)$ such that $D_0 \neq T_i^{R,m} \circ T_i^{R,m-1} \circ \cdots \circ T_i^{R,1}(D_A)$ for any $D_A \in TL(A)$.

A modified method M2, obtained by improving Step 1 of M1, uses the information of A in order to avoid arbitrary embedding of #-nodes. First, in Step 1 of M2, an NTA A_F such that

$$TL(A_F) = \{T_i^{R,m} \circ T_i^{R,m-1} \circ \cdots \circ T_i^{R,1}(D_A) \mid D_A \in TL(A)\}$$

is constructed from A by type inference. Next, $T_i(D)$ and the transition rules of A_F are compared in a top-down manner. Roughly speaking, for a node and its children of $T_i(D)$, the positions where deleted #-nodes never exist are identified by the comparison. The positions may be between some children and/or between the parent and some children. Then, the output NTA of Step 1 is constructed by embedding #-nodes into the positions other than the identified ones of $T_i(D)$.

5   Experimental Results

Based on the two verification methods in Section 4, we have implemented prototype systems. The systems are written in Java and are running on a Windows XP machine with Intel Xeon X5355 2.66GHz CPU and 1.4GB Java VM heap memory.

We used the following schema:

```
<!ELEMENT students (student*)>
<!ELEMENT student (name, gender, scholarship)>
<!ELEMENT name (#PCDATA)>
<!ELEMENT gender (#PCDATA)>
<!ELEMENT scholarship (#PCDATA)>
```

where amount of scholarship was either 0 or 500. Also, we used the authorized queries T_1 and T_2, and the secret query T_S as follows: T_1 extracts, for each student receiving a scholarship, i.e., amount of scholarship is not 0, the gender of the student and the amount of the scholarship; T_2 extracts, for each student, the name and the gender of the student; T_S extracts the name and the scholarship of a particular student. In addition, to make the intermediate results the largest, in each of the XML documents half of all students receive scholarships. We executed the prototypes to decide 2-secrecy of such XML documents.

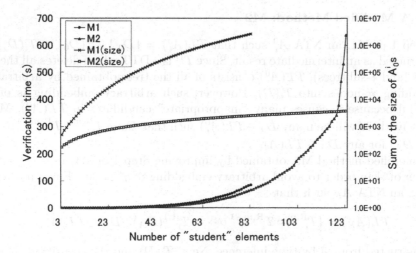

Fig. 5. Verification time

Fig. 5 shows verification times and the sum of the size of A_i^0s in Step 1 of M1 and M2. Here, the size of an NTA is defined as $|Q| + \sum_{(q,a,e) \in R} |e|$, where Q is the set of the states of the NTA, R is the set of the transition rules of the NTA, and the size $|e|$ of a string automaton e is the sum of the number of the states of e and the number of the tuples (consisting of a source state, an input symbol, and a destination state) satisfying the transition relation of e. As shown in Fig. 5, the size of A_i^0 in M2 was much smaller than that in M1. While M1 only verified at up to eighty-three because it exhausted heap memory in Step 1 at eighty-four, M2 did at up to a hundred and twenty-three and it exhausted memory in not Step 1 but Step 3. Also, M2 verified about twenty seconds faster than M1 at eighty-three.

6   Conclusion

We have proposed a verification framework for the security against inference attacks on XML databases. The methods proposed in the framework are applicable to practical cases where the class of authorized queries can express filtering nodes according to any of their neighboring nodes such as ancestors, descendants, and siblings. We have presented the straightforward method and its modification, and confirmed by experiment that the modification has effect in time efficiency.

One of our future works is to confirm the effect of the modification in time efficiency by experiment with many other instances. Another future work is to take account of attacks using functional dependencies and extend the proposed verification method against such attacks. Also, we would like to propose a static verification method for the security, i.e., given authorized queries and schema information, deciding whether every document valid against the schema is k-secret. Since the running time of static verification does not depend on the results of authorized queries on a database instance, the static verification can work even for a large database instance.

Acknowledgment

This research was supported in part by "Global COE (Centers of Excellence) Program" of the Ministry of Education, Culture, Sports, Science and Technology, Japan, and by the Telecommunications Advancement Foundation, Japan. Also, the authors thank Mr. Shinichi Minamimoto of Osaka University for implementing the prototype system.

References

1. Machanavajjhala, A., Gehrke, J., Kifer, D., Venkitasubramaniam, M.: ℓ-diversity: Privacy beyond k-anonymity. In: Proceedings of the 22nd International Conference on Data Engineering, p. 24 (2006)
2. Clark,   J.,   Murata,   M.:   RELAX   NG   (2001), http://www.oasis-open.org/committees/relax-ng/spec-20011203.html

3. Denning, D.E.R.: Cryptography and Data Security. Addison-Wesley, Reading (1982)
4. Brodsky, A., Farkas, C., Jajodia, S.: Secure databases: Constraints, inference channels, and monitoring disclosures. IEEE Transactions on Knowledge and Data Engineering 12(6), 900–919 (2000)
5. Farkas, C., Toland, T., Eastman, C.: The inference problem and updates in relational databases. In: Databases and Application Security, vol. XV, pp. 181–194. Kluwer, Dordrecht (2002)
6. Morgenstern, M.: Security and inference in multilevel database and knowledge-base systems. In: Proceedings of the 1987 ACM SIGMOD International Conference on Management of Data, pp. 357–373 (1987)
7. Moskowitz, I., Chang, L.: An entropy-based framework for database inference. In: Pfitzmann, A. (ed.) IH 1999. LNCS, vol. 1768, pp. 405–418. Springer, Heidelberg (2000)
8. Zhang, K.: IRI: A quantitative approach to inference analysis in relational databases. In: Database Security, vol. XI, pp. 279–290 (1998)
9. Deutsch, A., Papakonstantinou, Y.: Privacy in database publishing. In: Proceedings of the Tenth International Conference on Database Theory, pp. 230–245 (2005)
10. Miklau, G., Suciu, D.: A formal analysis of information disclosure in data exchange. Journal of Computer and System Sciences 73(3), 507–534 (2007)
11. Sweeney, L.: k-anonymity: A model for protecting privacy. International Journal on Uncertainty, Fuzziness and Knowledge-based Systems 10(5), 557–570 (2002)
12. Xiao, X., Tao, Y.: Personalized privacy preservation. In: Proceedings of the 2006 ACM SIGMOD International Conference on Management of Data, pp. 229–240 (2006)
13. Fan, W., Chan, C.Y., Garofalakis, M.N.: Secure XML querying with security views. In: Proceedings of the 2004 ACM SIGMOD International Conference on Management of Data, pp. 587–598 (2004)
14. Yang, X., Li, C.: Secure XML publishing without information leakage in the presence of data inference. In: Proceedings of the Thirtieth International Conference on Very Large Data Bases, pp. 96–107 (2004)
15. Suciu, D.: The XML typechecking problem. SIGMOD Record 31(1), 89–96 (2002)

Mining, Ranking, and Using Acronym Patterns

Xiaonan Ji[1,*], Gu Xu[2], James Bailey[1], and Hang Li[2]

[1] NICTA Victoria Laboratory, Department of CSSE, University of Melbourne,
Australia
{xji,jbailey}@csse.unimelb.edu.au
[2] Microsoft Research Asia, 4F, Sigma Center, No. 49, Zhichun Road, Haidian
District, Beijing, 100080, China
{guxu,hangli}@microsoft.com

Abstract. Techniques for being able to automatically identify acronym patterns are very important for enhancing a multitude of applications that rely upon search. This task is challenging, due to the many ways that acronyms and their expansions can be embedded in text. Methods for ranking and exploiting acronym patterns are another related, yet mostly untouched area. In this paper we present a new and extensible approach to discover acronym patterns. Furthermore, we present a new approach that can also be used for both ranking the patterns, as well as utilizing them within search queries. In our pattern discovery system, we are able to achieve a clear separation between higher and lower level functionalities. This enables great flexibility and allows users to easily configure and tune the system for different target domains. We evaluate our system and show how it is able to offer new capabilities, compared to existing work in the area.

1 Introduction

An acronym is a word formed from the parts of a full name and is used to stand for that name. For example, CPU can be used to stand for "Central Processing Unit". Recognizing acronyms and their full names is useful in many document processing applications. Alternatively, acronyms are often used by users as terms within search queries. By being able to replace acronyms with their most appropriate expansions, a search engine can potentially deliver better search results.

Most of the previous work in this area concentrates on identifying acronym patterns. The drawbacks are either a lack of extensibility (such as [10,6,9,8]) or heavy reliance on the availability of a remarked training corpus (such as [12,2]). In contrast, our new acronym mining system clearly separates higher-level mapping strategies from lower-level mapping rules and is available to achieve a high degree of flexibility. Users can easily add new rules or turn on/off existing ones.

A new and interesting related problem is the ability to rank and deploy acronym patterns for online search. The purpose of ranking method is to properly appraise different patterns, by taking into account the popularity of acronym

---

* Part of the work has been done at Microsoft Research Asia.

Y. Zhang et al. (Eds.): APWeb 2008, LNCS 4976, pp. 371–382, 2008.
© Springer-Verlag Berlin Heidelberg 2008

patterns in conjunction with standard confidence measures in their correctness. The ability to rank acronym patterns can be particularly attractive for online search applications, where the ambiguity of acronyms can cause problems for keyword-matching based IR systems. If the IR system can recognize acronyms and their most popular expansions, the retrieved results can be ranked more appropriately.

Related work. Some algorithms discover acronym patterns by finding a best alignment from acronym letters to letters of words in the full name. Some of them use predefined rules [10,14,11,9], while others use machine learning methods [2,12]. An interesting idea that aims to identify syntax-based relationships for word phrases is studied in [1,15]. These methods can be used to discover various types of patterns such as acronyms and expansions or books and authors. Acrophile is an online acronym dictionary based on work in [6]. It uses a very simple method of acronym ranking, which is just based on counting acronym frequencies. We are not aware of any research which has addressed the problems of how to rank acronym patterns and how to use them for query extension.

Our contributions. In this paper, we introduce a new acronym pattern mining framework called `AcroMiner`. `AcroMiner` uses an architecture built on mapping rules, but allows users to flexibly configure predefined rules or add new ones. Experiments show that `AcroMiner` is able to handle large data sets and can achieve promising results. Furthermore, we also address the novel problems of how to rank acronym patterns and extend acronym queries using their expansions.

2   Preliminary Definitions

The full name of an acronym is called its *expansion*. The way of coming up with the acronym from the expansion is called the *mapping* from the former to the latter. This mapping can be broken into lower-level mappings from individual letters of the acronym to positions in the expansion. In this paper, letters of acronyms and the corresponding positions of the expansions are underlined, in order to show how acronyms are mapped to expansions. An acronym pattern is the tuple of an acronym word (A) and its expansion (E), denoted as <A, E>.

An acronym may stand for multiple expansions. For example, WTO stands for "World Trade Organization" as well as "World Tourism Organization" and sometimes even "World Toilet Organization". This property is called *the ambiguity of acronyms*. On the other hand, an expansion is usually abbreviated by only one acronym. We call this property *the unambiguity of expansions*.

This paper discusses the following problems. *The mining problem:* identifying acronym patterns from documents. *The ranking problem:* recognizing the acronym patterns and then ordering them by a popularity measure. *The acronym query extension problem:* using ranked acronym pattern set to improve the capability of an IR system to handle queries containing acronyms.

Given a set of documents $D = \{s_1, s_2, ..., s_n\}$, the *occurrence frequency (tf)* of an object (e.g. a word or an acronym pattern) p is the total number of occurrences of p in documents of D , that is, $tf(p) = \sum_{i=1}^{n} |\{o \in s_i | o = p\}|$. The *document frequency* (df) of p is the number of documents containing p, that is, $df(p) = |\{s_i | o \in s_i \text{ and } o = p\}|$.

Letters contained in documents[1] are categorized as follows. *Invisible letters*: tab (ASCII 09), line feed (ASCII 10), carriage return (ASCII 13) and space (ASCII 32). *Punctuation letters*: exclamation mark (!), question mark (?), brackets ((,), [,], <, >, {, }), hyphen (-), underscore (_), colon (:), semi-colon (;), comma (,), period (.), slash (/), apostrophe ('), quotation marks(" ", ' '). *Alphabetical letters*: [A-Z] and [a-z]. *Numerical letters*: [0-9]. *Symbolic letters*: @, ~, #, \$, %, ^, \, &, *, +, =, |.

Letters that are not allowed to be contained in acronyms and expansions are marked as delimiters. Delimiters indicate the boundaries of potential expansions.

3   The Framework of AcroMiner

AcroMiner consists of four components: document preprocessing, identification of acronyms, identification of acronym patterns and postprocessing. The four components operate sequentially: the outputs of earlier components are input into the later ones.

Document Preprocessing. The input documents are reformatted in the first component. The preprocessing removes meta data, marks some punctuation and symbolic letters as delimiters and concatenates sentences from a document into a single long sequence.

Identifying Acronym Words. The second component identifies acronyms from input sequences. A regular expression $\mathcal{R}$ is used to identify acronyms. During the scan of word sequences, each word w is checked to see whether it satisfies $\mathcal{R}$ or not. The regular expression currently used in AcroMiner is:

$$\mathcal{R} = \{num\}?((U\{sep\}?)^+([\{num\}, L, \{sym\}])?)?(U\{sep\}?)^+([\{num\}, S])?$$

In $\mathcal{R}$, $\{num\}$ stands for numerical letters; U means alphabetical letters written in uppercase ([A-Z]); $\{sep\}$ stands for period (.); L stands for alphabetical letters written in lowercase ([a-z]); $\{sym\}$ stands for '&', '/' or '-'; S stands for lowercase letter 's'; (...) is used to group subexpressions; [...] is used to wrap alternative subexpressions; '?' means that the preceding subexpression may appear at most once; '+' means that the preceding subexpression appears at least once. $\mathcal{R}$ covers the format of the majority of acronyms, such as "U.S.A.", "SVMs", "3D", "P2P", "DoD", "AT&T" and "TCP/IP".

It could be the case that there are acronyms written in different ways to that specified by $\mathcal{R}$. For example, the acronym of "Tool Command Language" is written as Tcl rather than TCL[2]. In order to discover these acronyms, two writing formats are additionally considered. The first writing format is: a single

---

[1] In this paper we only consider documents written in English and encoded by ASCII.
[2] According to Tcl Wikipedia: http://en.wikipedia.org/wiki/Tcl.

word embedded in a pair of parentheses, following closely after a sequence of words. It is denoted as ... *a sequence of words* (w).... The other writing format is: a sequence of words embedded in a pair of parentheses, preceded closely by a word. It is denoted as ... w (*a sequence of words*).... According to [15], acronyms are usually written at the position w and the expansions are usually contained in the sequence of words next to it.

Identifying Acronym Patterns. Once an acronym is found, the next task is to identify the expansion it stands for. We make the assumption that *an acronym's expansion lies close to the acronym.* In other words, we only try to identify expansions in regions appearing W words before or after the acronym occurrence. We call these expansion regions *context windows* or CWs.

This problem can be divided into two levels. (a) The lower-level subproblem deals with how to map individual letters in the acronym to individual positions in the CW. Several letter-to-word mapping rules are used as constraints to define such mappings. Mapping rules are weighted with different mapping scores according to a confidence measure of correctness. (b) The higher-level subproblem deals with the strategy of mapping multiple letters to multiple positions. A backtracking algorithm is used to discover all letter-to-word mappings. The quality of the mapping from acronym to an expansion is measured by the sum of mapping scores of the applied rules. We next discuss these subproblems in detail.

(a) Lower-level letter-to-word mapping rules. The letter-to-word mapping rules are categorized into three types: *fixed*, *shiftable* and *neglectable*. These types are considered in the higher-level mapping strategy and their meanings will be explained later. We now list the mapping rules.

1. A letter can be mapped to the 1st position of a non-neglectable word[3], if it is the same as the letter in that position. This is a *fixed* mapping rule. For example, in <CMU, "Carnegie Mellon University">, each letter in the acronym is mapped to the 1st position of each word.
2. A letter can be mapped to the 2nd or 3rd position of a non-neglectable word if it is the same as the letter in that position and its preceding letter is mapped to the preceding position. For example, in <DASFAA, "Database Systems for Advanced Applications">, the first 'A' in the acronym is mapped to the 2nd position of the first word in the expansion. This is a *shiftable* mapping rule.
3. A letter can be mapped to the leading position of a neglectable word if it is the same as the initial letter of this word. For example, in <WOW, "World of Warcraft">, the letter 'O' in the acronym is mapped to the 1st position of the neglectable word "of". This is a *shiftable* mapping rule.
4. A letter can be mapped to a hyphen-connected word if it is the same as the first letter of that word. A hyphen-connected word is a non-leading word in a

---

[3] Neglectable words currently used are: &, after, an, and, are, as, at, de, en, for, from, in, is, la, of, on, or, the, to, up, with. New neglectable words can be added easily.

hyphenated phrase. For example, in the hyphenated phrase "Peer-to-Peer", "to" and the second "peer" are hyphen-connected words. By applying this rule to <XWC, "X-Windows Commander">, the letter 'W' in the acronym is mapped to the 1st position of "Windows". This is a *shiftable* mapping rule.

5. For words containing English prefixes[4], it is common to map two letters from the acronym to the prefix and suffix separately. A letter can be mapped to the first letter of the suffix by applying rule 1 given that its preceding letter is mapped to the first letter of the prefix. For example, in <MSDN, "Microsoft Developer Network">, 'S' is mapped to the first letter of "-soft" given its preceding letter 'M' is mapped to the first letter of "micro-". This is a *shiftable* rule.

6. Letters of acronyms can be mapped to sub-words of some compound words[5]. A letter can be mapped to the first letter of the latter sub-word according to rule 1 given that its preceding letter is mapped to the initial letter of the former word. For example, in <DBA, "Database Administrator">, 'B' can be mapped to the sub-word of "database" given that its preceding letter 'D' is mapped to the first letter of the former sub-word. The compound word list can be extended by users. This is a *shiftable* rule.

7. A letter from the acronym can be mapped to the 4th, 5th or 6th position of a word if that word contains at least 8, 9 or 10 letters and its preceding letter is mapped to the first letter of this word by applying rule 1. It is *shiftable*.

8. 'X' can be mapped to any word initialized with prefix "ex". This is a *fixed* rule.

9. The ending letter 'S' (or 's') of the acronym can be neglected if no proper mapping can be found. The reason is that the ending 'S' (or 's') can represent the plural format of the acronym thus have no position to be mapped to. For example, in <SVMS, "Support Vector Machine(s)">, the ending 'S' is mapped to nothing. This is a *neglectable* mapping rule.

10. Numerical letters [0-9] can be mapped to their English names, e.g. in <3D, "Three Dimension">, '3' is mapped to "three". This is a *fixed* mapping rule.

11. Some special letters can be mapped to English words. In the current implementation, '&' is mapped to "and", '2' is mapped to "to" and '4' is mapped to "for". For example, in <AT&T, "American Telephone and Telegraph">, '&' is mapped to "and". It is a *fixed* mapping rule.

12. If the acronym contains numerical letters, its preceding letter (if there is any) or its following letter is repeated that many times to create a new acronym. This newly-created acronym is also used to find possible expansions by using other rules. For example, "W3C" can be changed to "WWWC" in order to discover <W3C, "World Wide Web Consortium">.

---

[4] The English prefix currently used are: anti-, auto-, bi-, bio-, cent-, centi-, chem-, circum-, contra-, counter-, deci-, dis-, euro-, ex-, extra-, fore-, inter-, kilo-, mega-, micro-, mini-, multi-, out-, over-, post-, pre-, pro-, quad-, semi-, sub-, super-, tele-, trans-, tri-, manu-, ultra-. New prefixes can be added easily.

[5] Compound words currently used are: data-base, play-station, on-line, world-wide, north-west. New compound words can be added easily.

(b) Higher-level mapping strategy. The higher-level mapping strategy tells how to map multiple letters to multiple positions. When more than one mapping way is found from the acronym to a substring in the CW, the higher-level mapping strategy picks one expansion with the highest mapping score.

We use a backtracking algorithm to explore all possible ways of mapping. During the backtracking, the types of mapping rules decide which mappings are not able to be changed and which ones are able to be shifted after they are created.

If a letter is mapped to some position by a fixed mapping rule, it cannot be changed to map to other positions. If a fixed mapping cannot be applied, backtracking is required until a previous mapping established by a shiftable mapping rule is found. If a letter is mapped to some position by a shiftable mapping rule, this mapping can be removed and the letter can be mapped to some other position to the right. The neglectable mapping rules allow a letter to be neglected.

Each mapping rule is assigned a mapping score. These mapping scores are used to measure the quality (the confidence of correctness) of an acronym-to-expansion mapping. The general principles of the score setting are: longer expansions are preferred to shorter ones and regular mappings are preferred to less regular ones. Writing formats can also give hints about correct mappings, such as letters written in uppercase. Mapping scores can be set approximately and still achieve good performance, as long as they reflect these principles.

Algorithm 1. AcroMiner(A, CW)

Require: A: the acronym word. CW: the context window.
Ensure: $\mathbb{P}$: the set of mined acronym patterns whose acronym is A.
1: $j = 1$;
2: $\mathbb{P}' = \varnothing$;
3: **while** $j < |CW|$ **do**
4:     FindAE(A, CW, 1, j); /* $\mathbb{P}'$ is used to store discovered acronym patterns whose expansions start from the j-th position in CW. */
5:     $<A, E> = \arg\max_{ms(<A,E_i>)}\{<A, E_i> \in \mathbb{P}'\}$; /* $ms(<A, E>)$ is the mapping score from the acronym A to the expansion E. */
6:     add $<A, E>$ to $\mathbb{P}$;
7:     $j = $ 1st position of the word following the last word of E;
8:     $\mathbb{P}' = \varnothing$;
9: **end while**

Given the acronym word and a CW, the mapping strategy works according to Algorithm 1. `AcroMiner` was designed to explicitly separate lower-level mapping rules (the mapping constraints) from higher-level mapping strategy. Line 2 of Algorithm 2 deals with choosing the applicable lower-level mapping rule for the specific letter and the rest of the code deals with the higher-level mapping strategy. If this line is treated as a black box, the higher-level mapping strategy is clearly separated from lower-level mapping rules.

Acronym Pattern Postprocessing. A postprocessing step is necessary for better acronym pattern ranking, as well as better acronym query extension. This component merges acronym patterns containing duplicate expansions that have

---

Algorithm 2. FindAE(A, *CW*, i, j)

Require: i: the starting position of A. j: the starting position of *CW*. ℙ′: store patterns whose
   expansions start from position *j* in *CW*. '#': the delimiter.
Ensure: Map letters A[i... |A|] to *CW*[j... k] according to certain k. If new pattern is found, add
   it to ℙ.
1: $k = \arg\min_{j \leq k \leq |CW|}\{k | \exists r$ related to the position of $k\}$; /* *r* is lower-level mapping rule */
2: **if** $k == \varnothing$ **then**
3:    **return**
4: **end if**
5: **if** '#'∈ *CW*[j... k] **then**
6:    **return** /*   */Expansions containing delimiters are illegal.
7: **end if**
8: **if** *r* can be applied to the mapping from A[i] to *CW*[k] **then**
9:    **if** i==|A| **then**
10:       add the newly discovered acronym pattern to ℙ′;
11:    **else**
12:       FindAE(A, *CW*, i+1, k+1);
13:       **if** i==1 **then**
14:          **return** /*   */The mapping from A[1] to other positions than j is handled in AcroMiner()
15:       **else if** *r* is shiftable **then**
16:          FindAE(A, *CW*, i, k+1); /* Try to find all the possible mappings */
17:       **end if**
18:    **end if**
19: **else**
20:    **if** *r* is shiftable **then**
21:       FindAE(A, *CW*, i, k+1); /* On failure, shift the letter to the next applicable position */
22:    **else**
23:       **return** /*   */If *r* is fixed, the failure is not shiftable, backtrack to earlier shiftable mappings
24:    **end if**
25: **end if**

---

the same meaning, but are written in different expressions. Once a duplication
is detected between two acronym patterns, the one having smaller document
frequency is removed and its document frequency is added to the one retained.

4   Ranking Acronym Patterns

Acronym patterns are not all equally useful. Some patterns are not popular and
used by few people. `AcroMiner` may mistakenly discover false patterns, where
the acronyms do not stand for mapped expansions. It is not desirable to treat the
less popular or false acronym patterns in an equal fashion to the more popular
or correct ones. A method for ranking, based on scores for the acronym patterns,
is described next.

The qualities of acronym patterns can be quantitatively measured by *ranking
scores*. The ranking score is controlled by three factors: (I) *pattern popularity*, (II) *gap between the acronym and its expansion* for every occurrence of the
pattern and (III) *mapping score from the acronym to the expansion* for every
occurrence of the pattern.

Acronym pattern popularity is measured by document frequency. The gap
between the acronym and the expansion is measured by the number of words
in-between the acronym and the expansion. The larger the gap is, the weaker
the relevance is between the acronym and the expansion and thus there is less
confidence to say the mapping is correct. The mapping score of an acronym
pattern (more accurately, from the acronym to the expansion) is the sum of the

scores of the mapping rules that were applied to map letters of the acronym to positions of the expansion.

The *rating score* (*rs*) of an occurrence of an acronym pattern is measured by combining the gap and the mapping score and is given by:

$$rs(o) = \frac{\sum_i^{|A|} ms(A[i])}{f \times |A| \times g}. \tag{1}$$

There, o is an occurrence of the acronym pattern $<A, E>$. $|A|$ is the length of the acronym. $ms(A[i])$ is the score of the mapping rule applied to map the i-th letter of A to certain position of E. $\sum_i^{|A|} ms(A[i])$ is the mapping score from A to E. f is the maximum score among all the mapping rules. $f \times |A|$ can be thought as the "highest achievable" mapping score obtainable for A.

The *ranking score* is calculated by multiplying the average rating score among all occurrences of the acronym pattern with the popularity (df) of the pattern:

$$rank(p) = \frac{\sum_{o \in s_i, o=p} rs(o)}{tf(p)} \times df(p). \tag{2}$$

As we can see, an acronym pattern is ranked by considering the "fitness" of mapping the acronym word to the expansion and the frequency of seeing this pattern in the data set. Patterns with higher ranking scores should be placed at higher positions in the result list.

5   Acronym Query Extension Using Acronym Patterns

We now discuss the following questions about using acronym patterns (more precisely, the expansions) for query extension. If a user query is submitted that contains a word not found in a dictionary, should the system consider it as an acronym? Should every acronym be extended by its expansions for the retrieval task? Should the system consider all of the acronym's possible expansions, or only a subset of them?

Addressing these questions requires us to estimate the probability that a query term T is an acronym corresponding to an expansion E, i.e. $P(T \text{ is } A, E)$. Now $P(T \text{ is } A, E) = P(E|T \text{ is } A) \times P(T \text{ is } A)$, where on the right hand side, the former term is the probability that T stands for E, if T is definitely an acronym. The latter term is the probability that T is used as acronym. $P(E|T \text{ is } A)$ can be computed by:

$$P(E|T \text{ is } A) = \frac{rank(<A, E>)}{\sum_{E_i, <A, E_i> \in \mathbb{P}} rank(<A, E_i>)}. \tag{3}$$

Equation (3) says that the probability for acronym A to stand for E, is the ratio of the ranking score of acronym pattern $<A, E>$, to the sum of the ranking scores of all patterns having acronyms as A. $P(T \text{ is } A)$ can be computed by:

$$P(T \text{ is } A) = \frac{df(T \text{ is } A)}{df(T)}, \tag{4}$$

where $df(T$ is $A)$ is the number of documents containing acronym pattern $<$A, E$>$ and $df(T)$ is the number of documents containing the word T. If T is used without any associated expansion (i.e. no expansion is discovered) in many documents, it is likely that T is not an acronym and is instead just a normal English word.

Finally, $P(T$ is $A, E)$ is derived by combining equations (3) and (4):

$$P(T \text{ is } A, E) = \frac{rank(<A, E>)}{\sum_{E_i, <A, E_i> \in \mathbb{P}} rank(<A, E_i>)} \times \frac{df(T \text{ is } A)}{df(T)}. \tag{5}$$

For a query term T, all expansions whose probabilities (as calculated by Equation (5)) are larger than a predefined threshold, will be selected and submitted along with T for the IR system to process with the query.

6 Experiments

Our experiments have been designed to evaluate the accuracy of AcroMiner, the efficiency of the ranking method and the usefulness of using acronym patterns to extend acronym queries.

Experiments on AcroMiner. The V.E.R.A[6] acronym dictionary was used to evaluate the performance of AcroMiner. This data set contains 11201 computer and IT related acronyms and their intended expansions. AcroMiner discovered 9134 acronym patterns, of which 8763 ones were correct. The recall is 78.2% and the precision is 95.9%. Statistics for the misidentified 371 patterns are given in Table 1. The table shows that it is more difficult to identify acronyms with shorter lengths. The fewer letters are in the acronym, the easier it is to mismatch them.

Table 1. Statistical information about misidentified patterns from V.E.R.A. data set

Acronym length	# ground truth	# misidentified	percentage(%)
2	700	207	30
3	5537	153	3
4	3255	11	0.3
5 and above	1709	0	0

The performance of the online acronym extraction system Acrophile[7] can be compared with AcroMiner on the same data set. It discovered 8570 acronym patterns, among which 8058 ones were correct. The recall is 72% and the precision is 94%. This is not a precise comparison, because there is no means to adjust parameters, such as the window size W, for Acrophile. A large number of mistakes made by Acrophile were because it rigidly tried to map a letter to the 4th, 5th or 6th position in a word.

---

[6] Virtual Entity of Relevant Acronyms: http://cgi.snafu.de/ohei/user-cgi-bin/veramain-e.cgi

[7] Acrophile demo: http://ciir.cs.umass.edu/irdemo/acronym/getacros.html

`AcroMiner` was run on two real-world data sets in order to generate comprehensive acronym dictionaries for querying and other applications. The Web database is a collection of around 20 million crawled web pages. The Wiki data set[8] contains around 4 million Wikipedia articles written in XML. Only the plain text content was used for pattern mining and lower-level mapping rules 2 and 7 were switched off. It took roughly 12 hours to discover 563440 acronym patterns from the Web database and 2.5 hours to discover 118028 acronym patterns from the Wiki database. This time includes both the mining and ranking processes. On average, almost three new acronym patterns were discovered from every 100 Wikipedia articles or web pages. For the Web database, each acronym was mapped to 4.2 expansions on average. Interestingly, the acronym "ACE" was mapped to 493 expansions (not all of which may be valid).

It is difficult to evaluate the performance on these databases since no ground truth is available, but precision based on random sampling can be computed in the following way: 200 acronym patterns were randomly selected at a time and checked in Google for their correctness. This process was repeated for four times and the average precision measured in this way was 81%. If only the acronym patterns that ranked among the top five were selected, the precision was 91% and for only the top one, the precision was 98%.

Experiments on Ranking Acronym Patterns. Table 2 lists the top ranked acronym patterns for (I) `AcroMiner` using the Web database, (II) `AcroMiner` using the Wiki database and (III) The Acrophile [6] system. The Acrophile online acronym dictionary was mined using military and government documents. Many of its top ranked patterns are less well-known. For the ranking result for `AcroMiner` using the Web database, we compared with the ranks from AcronymFinder[9]. This web site is used for human assisted collection and ranking of acronym patterns using 5 levels, based on the popularity of use. The number next to each expansion indicates the ranking level given by that web site. As shown, most of the acronym patterns were ranked similarly by both `AcroMiner` and AcronymFinder, but in `AcroMiner` the ranking was done automatically, to a smaller level of granularity.

Experiments on Acronym Query Extension. 161 acronym queries were picked out from a query set[10] used as the benchmark data to evaluate algorithms of query extension.

The normalized discounted cumulative gain scoring measure (NDCG), for each of the first-page search results from a Web search engine was used to compare the qualities of the query results before and after being extended by acronym patterns. We use a special operation[11] to embed acronyms and expansions. The op-

---

[8] Hyperlink to download Wikipedia database:
    `http://en.wikipedia.org/wiki/Wikipedia:Database_download`
[9] AcronymFinder online dictionary: `http://www.acronymfinder.com`
[10] This data set is product-related and its information is hidden due to privacy issues.
[11] The operation is product-related and is hidden due to privacy issues. We denote it as OP.

Table 2. Ranking results of some acronym patterns returned from `AcroMiner` and Acrophile

AcroMiner using Web database			
CS	AI	CSU	ACA
1.Counter Strike(1)	1.Artificial Intelligence(1)	1.Channel Service Unit(1)	1.American Correctional Association(1)
2.Computer Science(1)	2.Amnesty International(1)	2.California State University(1)	2.American Camp Association(1)
3.Customer Service(1)	3.American Idol(1)	3.Colorado State University(1)	3.Australian Communications Authority(1)
4.Creative Suite(1)	4.Adobe Illustrator(1)	4.Charles Sturt University(1)	4.American Chiropractic Association(1)
5.Community Server(-)	5.All Inclusive(1)	5.Christian Social Union(1)	5.American Counseling Association(1)
AcroMiner using Wiki database			
CS	AI	CSU	ACA
1.Counter Strike	1.Artificial Intelligence	1.Colorado State University	1.American Camp Association
2.Club Sport	2.Amnesty International	2.Christian Social Union	2.American Chiropractic Association
3.Computer Science	3.Artificial Insemination	3.California State University	3.American Counseling Association
4.Credit Suisse	4.Appenzell Innerrhoden	4.Cleveland State University	4.Amputee Coalition of America
5.Chief of Staff	5.Air India	5.Charles Sturt University	5.Agile Combat Aircraft
Acrophile			
CS	AI	CSU	ACA
1.Combat Support	1.Artifical Intelligence	1.California State University	1.Associate Contractor Agreements
2.Containment Spray	2.Amnesty International	2.Colorado State University	2.American Counseling Association
3.Congenital Syphilis	3.Active Ingredient	3.Computer Software Unit	3.Airspace Control Authority
4.Computer Science	4.Assignment Instruction	4.Computer Software Units	4.Airspace Coordination Area
5.Core Spray	5.Action Items	5.Conservation System Units	5.Administrative Cost Allowance

eration tells the search engine that phrases embedded should be treated equally and interchangeably but documents containing more of the phrases are not necessarily ranked higher than the ones containing fewer of the phrases. The results of using and not using this operation are compared.

45 acronym queries, after being extended to include their expansions, had improvements of their NDCG scores. 86 extended queries had no NDCG score change, while another 30 extensions led to decreases of NDCG scores. While these results indicate that acronym extension is indeed promising, it is hard to explain the 86 unchanged cases. The search engine we used considers many factors, which give alternative clues about how to properly rank results, regardless of any use of acronym patterns. It is therefore perhaps not surprising if the operation does little to affect the overall results. Also, the data sets we used for mining by `AcroMiner` to create its acronym dictionaries are still small compared to the entire Web. So this data limitation may cause `AcroMiner` to miss some popular acronym patterns. Acronym patterns are likely to provide more obvious improvements for other types of IR systems, where keyword-matching plays a more important role in overall ranking function.

7 Conclusion

In this paper we have studied the problem of mining acronym patterns from unstructured documents. We have developed AcroMiner, a highly open and flexible mining system that can handle large-scale data sets with high accuracy. We also presented a study on how to rank acronym patterns and use them for acronym query extension, a new problem in the area.

Acknowledgement. This work was partially supported by Microsoft Research Asia and National ICT Australia. We thanks Jiafeng Guo and Zhichao Zhou for their helpful suggestion.

References

1. Brin, S.: Extracting Patterns and Relations from the World Wide Web. In: WebDB, pp. 172–183 (1998)
2. Chang, J.T., Schütze, H., Altman, R.B.: Creating an Online Dictionary of Abbreviations from MEDLINE. Journal of the American Medical Informatics Association 9, 612–620 (2003)
3. Hawking, D., Craswell, N., Bailey, P., Griffihs, K.: Measuring Search Engine Quality. Information Retrieval 4(1), 33–59 (2001)
4. Pustejovsky, J., Castano, J., Kotecki, M., Morrell, M.: Automatic Extraction of Acronym-Meaning Pairs from Medline Databases. Medinfo 10, 371–375 (2001)
5. Järvelin, K., Kekäläinen, J.: IR evaluation methods for retrieving highly relevant documents. SIGIR, 41–48 (2000)
6. Larkey, L.S., Ogilvie, P., Price, M.A., Tamilio, B.: Acrophile: An Automated Acronym Extractor and Server. ACM DL, 205–214 (2000)
7. Vladimir, I.: Levenshtein: Binary Codes Capable of Correcting Deletions, Insertions and Reversals. Doklady Akademii Nauk SSSR 163(4), 845–848 (1965)
8. Rimer, M., O'Connell, M.: BioABACUS: A Database of Abbreviations and Acronyms in Biotechnology and Computer Science. Bioinformatics 14(10), 888–889 (1998)
9. Schwartz, A.S., Hearst, M.A.: A Simple Algorithm for Identifying Abbreviation Definitions in Biomedical Texts. In: Pacific Symposium on Biocomputing (2003)
10. Taghva, K., Gilbreth, J.: Recognizing Acronyms and Their Definitions. IJDAR 1(4), 191–198 (1999)
11. Jonathan, D., Wren, H.R.: Garner: Heuristics for Identification of Acronym-Definition Patterns Within Text: Toward an Automatic Construction of Comprehensive Acronym-Definition Dictionaries. Methods of Information in Medicine 41(5), 426–434 (2002)
12. Xu, J., Huang, Y.: Using SVM to Extract Acronyms from Text. Soft Comput. 11(4), 369–373 (2007)
13. Baeza-Yates, R.A., Berthier, A.: Ribeiro-Neto: Modern Information Retrieval. ACM Press, New York (1999)
14. Yeates, S.: Automatic Extraction of Acronyms from Text. In: New Zealand Computer Science Research Students Conference, pp. 117–124 (1999)
15. Yi, J., Sundaresan, N.: Mining the Web for Acronyms Using the Duality of Patterns and Relations. In: Workshop on Web Information and Data Management, pp. 48–52 (1999)

A Method for Web Information Extraction

Man I. Lam[1], Zhiguo Gong[1], and Maybin Muyeba[2]

[1] Faculty of Science and Technology
University of Macau, Macao, PRC
{ma46522, fstzgg}@umac.mo
[2] School of Computing
Liverpool Hope University, Liverpool, L16 9JD, UK
muyebaM@hope.ac.uk

Abstract. The Word Wide Web has become one of the most important information repositories. However, information in web pages is free from standards in presentation and lacks being organized in a good format. It is a challenging work to extract appropriate and useful information from Web pages. Currently, many web extraction systems called web wrappers, either semi-automatic or fully-automatic, have been developed. In this paper, some existing techniques are investigated, then our current work on web information extraction is presented. In our design, we have classified the patterns of information into static and non-static structures and use different technique to extract the relevant information. In our implementation, patterns are represented with XSL files, and all the extracted information is packaged into a machine-readable format of XML.

1 Introduction

During the past decade, information extraction has been extensively studied with many research results as well as systems developed. Since the late 1980's, through the message understanding conference (MUC) sponsored by defense advances research project agency, many information extraction systems have been successfully developed and quantitatively evaluated [1].

The information source can be classified into three main types, including free text, structured text and semi-structured text. Originally, the extraction system focuses on free text extraction. Natural Language Processing (NLP) techniques are developed to extract this type of unrestricted, unregulated information, which employs the syntactic and semantic characteristics of the language to generate the extraction rules. The structured information usually comes from databases, which provide rigid or well defined formats of information, therefore, it is easy to extract through some query language such as Structured Query Language (SQL). The other type is the semi-structured information, which falls between free text and structured information. Web pages are a typical example of semi-structured information. In this paper, we will focus on extracting text information from web pages.

According to the statistical results by Miniwatts Marking Group [URL1], the growth of web users during this decade is over 200% and there are more than 1 billion

Y. Zhang et al. (Eds.): APWeb 2008, LNCS 4976, pp. 383–394, 2008.

Internet users from over 233 countries and world regions. At the same time, public information and virtual places are increasing accordingly, which almost covers any kind of information needs. Thus this attracts much attention on how to extract the useful information from the Web.

Currently, the targeted web documents can easily be obtained by inputting some keywords with a web search engine. But the drawback is that the system may not necessarily provide relevant data rich pages and it is not easy for the computer to automatically extract or fully understand the information contained. The reason is due to the fact that web pages are designed for human browsing, rather than machine interpretation. Most of the pages are in Hypertext Markup Language (HTML) format, which is a semi-structured language, and the data are not given in a particular format and change frequently [1].

There are several challenges in extracting information from a semi-structured web page such as the lack of a schema, ill formatting, high update frequency and semantic heterogeneity of the information. In order to overcome these challenges, our system design transforms the page into a format called Extensible Hypertext Mark-up Language (XHTML) [URL2]. Then, we make use of the DOM tree hierarchy of a web page and regular expressions are extracted out using the Extensible Style sheet Language (XSL) [URL3, URL4] technique, with a human training process. The relevant information is extracted and transformed into another structured format—Extensible Mark-up Language (XML) [URL5].

The remainder of the paper is organized as follows: some related works are illustrated in section 2, which involve a brief overview of the current web information extraction systems; then detail techniques in our approach are addressed in section 3; experimental results are explained in section 4; finally, the conclusion and future work are mentioned in the last section.

2 Related Work

From time to time, many extraction systems have been developed. In the very beginning, a wrapper is constructed to manually extract a particular format of information. However, the wrapper is not adaptive to change, it should be reconstructed accordingly to different types of information. In addition, it is complicated and knowledge intensive to construct the extraction rules used in a wrapper for a specific domain. Therefore only experts may have knowledge to do that. No doubt, the inflexibility and the development cost for construction are the main disadvantages of using wrappers.

Due to the extensive work in manually constructing a wrapper, many wrapper generation techniques have been developed. Those techniques could be classified into several classes, including language development based, HTML tree processing based, natural language processing based, wrapper induction based, modeling based and ontology based [2].

In order to assist the user to accomplish the extraction task, a new language was developed for a language development based system. The famous systems for this type include TSIMMIS [3] and Web-OQL [4]. One of the drawbacks of such a model is that not all users are familiar with the new query language, so the performance of the system may not be as expected. Then, as most of the web pages are in HTML format, another type of extraction system, HTML tree processing based system, was proposed. By parsing the tree structure of a web page, a system is able to locate useful pieces of information. XWRAP [5] and RoadRunner [6] are examples in this respect. In this solution, web pages need to be transformed into XHMTL or XML format due to limitations of the HTML format.

For some pages which are mainly composed of grammatical text or paragraphs, Natural Language Processing (NLP) systems can be used. NLP is popularly used to extract free text information, and makes use of filtering, part-of-speech tagging and lexical semantic tagging technology to build up the extraction rules. SRV [7], WHISH [8] and KnowItAll [9] are examples of this technique. However, for some pages which are composed of the tabular or list format, NLP based tools may not be effective since the internal structure of the page can not be fully exploited.

The wrapper induction based systems can induce the contextual rules for delimiting the information based on a set of training samples. SoftMealy [10] and STALKER [11] are typical examples. In modeling based systems, according to a set of modeling primitives, for example tables or lists, the data are conformed to the pre-given structure. Then the system tries to locate the information against given structures. NoDoSe [12] is an example of this type of systems. The last type is ontology based systems. Ontology techniques can be used to decompose a domain into objects, and further to describe these objects [13]. This type of system does not rely on the structures of web pages or the grammars of texts but instead an object is constructed for a specific type of data. WebDax [13] is a typical example in this respect.

Besides classifying by the main techniques used, the wrapper can also be grouped into semi-automatic wrapper or fully-automatic wrapper. For the semi-automatic wrapper, human involvements are necessary. Most of the systems belong to this type, such as TSIMMIS [3] and XWRAP [5]. For the fully-automatic wrapper, no human intervention is needed, examples include Omini [14] and STAVIES [15], which make use of tree structures, or the visual structures of pages to perform the extraction task.

Our proposed system belongs to the type of semi-automatic wrappers. Through the training process, our system learns rules for extraction. We suppose that with training, the system can be more adaptive to different type of pages if the training samples are broad enough. In addition, for different types of information, we make use of different techniques for extraction. For most existing systems, usually, only one main methodology is applied for extraction. The benefit of a multi extraction methodology is that the extraction can produce higher performance.

3   System Design

The system works in two phases, pre-processing phase and the extraction phase as shown in figure 1. In the Pre-processing phase, in order to overcome the ill representations of HTML documents, all the pages are transformed into XHTML format. Then, the training process is performed, which gathers patterns or rules for the extraction phase. In the extraction phase, based on the human training results, the system chooses suitable extraction methods for different information fields.

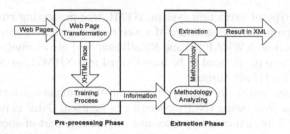

Fig. 1. System Architecture

3.1 Patterns and Rules of the Information

All of the web pages are transformed from HTML into a W3C [3] recommended XHTML format in the preprocessing phase. Though the current web browsers can present correctly the ill formulated HTML documents, it is difficult to identify the hierarchical structures of web pages. For example, the tag <P> or <BR> may be used alone for HTML elements, without the corresponding closing tags associated. However, such usage is not allowed in XHTML format. Therefore the closing tag should be added accordingly during the preprocessing phase.

An open source library called Tidy [URL6], provided by W3C organization, is used to transform the web pages. Tidy is able to fix up a broad range of ill formed HTML. After pages are transformed into XHTML format, the training process is performed. Users need to highlight the extracted word in sample pages. Figure 2 shows the interface of the training process in our system.

The target information is modeled in a schema as $r(f_1, f_2, \ldots, f_n)$, where f_i is the field of information. Let $PSet=\{p_1, p_2, \ldots, p_m\}$ be a set of sample training web pages and we suppose that target records can be extracted at least partly from each of those training pages. Each web page p in $PSet$ is annotated with a vector $(f_1:l_1, f_2: l_2, \ldots, f_n:l_n)$, where l_i is the location of the field f_i in page p.

The objective of the training process is to mine out (extract) patterns and rules for target information. For each field f_i, a pattern set, annotated as PS_i, is constructed from those sample pages. And a pattern pn_{ij} in PS_i is the characteristic of context of f_i. In our system, we represent pn_{ij} into the format $\{PW_{ij}, EW_i,\}$, where PW_{ij} is the words which occur just before the instance of f_i in page p_j, and EW_i is the formulation rule of f_i's value, which is described using regular expression in our work and given by the users. And we further suppose EW_i is irrelevant to the individual training pages. As we know, a web page is represented with a DOM tree. And all the fields for the same record are embedded in different tag nodes of the tree. We further use directed graph to describe the organizational constraint of the fields in the web page p_j. Let $F=\{f_1, f_2, \ldots, f_n\}$, then the constraint is defined as a directed graph $CG_j=<F, E_j>$, where E_j is the set of directed edges such that $f_i \rightarrow f_k$ in E_j if and only if the tag node of f_i is embedded in the tag node of f_k, $f_i \leftrightarrow f_k$ if and only if the node of f_i and node of f_k are sibling elements in the DOM tree. Then, $rl_j=(PS_j, CG_j)$ is called an extraction rule for target record $r(f_1, f_2, \ldots, f_n)$ with respect to training page p_j. In fact, with the training process, multiple rules can be derived from those potential training web pages. We use RS to denote the set of all the possible rules for the given schema. That is, $RS=\{rl_j\}$.

The number of rules in the original *RS* can be as many as the number of training pages in *PSet*. However, many rules may show redundant information or contradictory information. To reduce the size of rules, some reduction algorithm is performed on the original set of *RS*. To do so, we suppose that patterns of fields are orthogonal to constraints of fields. In fact, the former give the local context of fields, and latter describe their occurrence relationships in web pages. With such assumptions, we merge the patterns and the constraints independently. For the patterns of field f_i, the merged pattern is defined as: $s\text{-}PS_i = \{\{t_{ij}:tf(t_{ij})\}, EW_i\}$, where t_{ij} is a pre-word extracted from sample pages with respect to field f_i, $tf(t_{ij})$ is the number of occurrences of t_{ij} for field f_i. In one extreme situation, all the pre-words extracted for f_i is the same, denoted as t, then $tf(t)=m$ (the total number of training pages). In another extreme, all the pre-words are different from each other, then $tf(t_{ij})=1$. Therefore, the weight of t_{ij} indicates its significance degree as a pre-word for field f_i. To normalize the weight, we replace $tf(t_{ij})$ with $ntf(t_{ij})=tf(t_{ij})/m$. Then, $\sum_{\{j\}} ntf(t_{ij})=1$. Thus, the overall merged pattern for schema $r(f_1, f_2, \dots, f_n)$ is then described as $SP=(s\text{-}PS_1:EW_1, s\text{-}PS_2:EW_2,\dots, s\text{-}PS_n:EW_n)$.

As we know, each constraint of the fields gives the organizational structure of those fields in one training page. The constraint is represented as a directed graph, with fields as nodes, and embedded relationship as the directed edges. We concatenate all those constraint graphs into one graph, with nodes as those fields, and weight directed edge from f_i to f_k with weight $n(i,k)$, where $n(i,k)$ is the number of constraints having the edge from f_i to f_k. The merged constraint graph is denoted as *CG*. Then, (SP,CG) is called the extraction rule for schema $r(f_1, f_2, \dots, f_n)$.

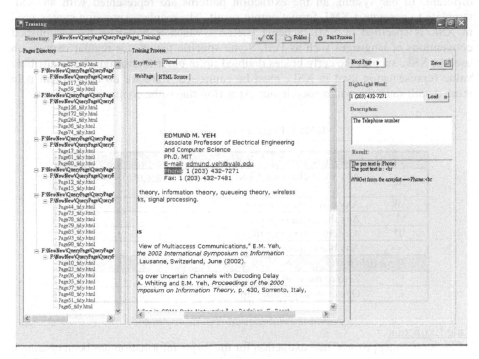

Fig. 2. Screen Shot for training process

3.2 Information Extraction Processing

According to the rule (*SP,CG*) of the target information, our system classifies each extraction field into static field (*SF*) or non-static field (*NSF*) based on the EW_i in *SP*. It is easy to classify if the extracted word contain any special character, i.e., the character other than the alphabetical and numerical character. If most of the *EWi* in one extraction field contains rather similar special characters, then this field is classified as *STF*. For example, all the e-mail addresses contain "@" and ".", and the structure is stable, which can be easily represented by a regular expression.

In the following section, the extraction method for *SF* will be discussed first, followed by the *NSF*.

3.2.1 Methodology for Static Structure Information

As the extracted information is in a static structure, the system makes use of this feature to generate an extraction rule using a regular expression. Before going into detail, a brief introduction to regular expressions will be provided first. Then the rule generation process will be explained in detail.

Introduction to Regular Expression

Regular expression is the value formulating pattern, which is defined by using regular expression syntax, to represent some information requirements. The regular expression syntax which has been used in our system is shown in table 1.

For different types of languages used, the regular expression syntax is slightly different. In our system, all the extraction patterns are represented with an XSL template to generate a XML format output result. The regular expression syntax used in XSL is specified in XML Schema [URL5], which is based on the established conventions of languages such as Perl. Some specific feature for regular expression used in XSL, such as the curly braces doubled up in order to distinguish the attribute value representation used in XSL, must be cautioned. Here we do not explain it in more detail, since the syntax used in our extraction rule is the commonly used one.

Table 1. Regular expression syntax

Character Classes	
\w	Any word character which composed of a-z or A-Z or 0-9 or _
\W	Any non-word character
\d	Any digit, that is 0-9
\D	Any character other than a digit
\s	Any blank space character
Repetition	
*	Zero or more relationship
+	One or more relationship

Rule Extraction

After the training process, rule extraction for the target information has been obtained. As shown in the following equations, a particular processing extraction for a field f_i is shown in equation (1), which is composed of several rule patterns rl_{ij} with "or" relationship. For each rule pattern rl_{ij}, it is composed of the pre-word PW_{ij} and the regular expression of the extracted word EW_i, as shown in equation (2), where δ function is used to transform the content into a regular expression. In order to avoid the confusion when more than one f_i have the similar structure, such as phone number and fax number, therefore, the $s\text{-}PW_{ij}$ is added in the rule pattern.

$$s - PW_i = \{PW_{i1}, PW_{i2}, ... PW_{im}\} \tag{1}$$

$$ps_{ij} = s - PW_{ij} : \delta(EW_{ij}) \tag{2}$$

For the δ function, the main syntaxes we have used are shown in table 1. The main idea is to transform all the space character into "\s*" and all the text or number patterns into [\w]*. For simplicity, figure 3 shows the step of generating the extraction rule for email. Assume that the PW_{ij} is "E-mail:", and the EW_{ij} is "profa@umac.mo". As shown in the following steps, the process of δ function is to replace all the word character into [\w]* and then replace the space character into \s*.

$$\begin{aligned} rp_{ji} = PW_i \cup \delta(EW_i) &= "E - mail :" \cup \delta(" profa @ umac.mo") \\ &= "E - mail :" \cup "[\backslash w] * @ [\backslash w] * . [\backslash w] *" \\ &= E - mail : \backslash s * ([\backslash w] * @ [\backslash w] * . [\backslash w] *) \backslash s * \end{aligned}$$

Fig. 3. Steps of generating the extraction rule for email

Target Information Extraction

After generating the extraction rule PS_i for an extraction field f_j, the rule is then represented with a XSL template in order to generate the XML output result. Figure 4 shows the XSL template used in our system for single information extraction. In order to take the advantages of the regular expression, the newest version of XSL, i.e. v.2.0, can provide some new instructions for this purpose. After the regular expression is passed into the XSL template, the xsl:anayze-string instruction will start to test if the regular expression "regexp" matches the content in the input string, here the selected input string is ".", that is the whole page. After that, each of the matched parts will be processed by xsl:matching-substring child instruction, here we only use the most simplest instruction xsl:value-of to get out the value of the matched part.

XSL [URL4] can help us to find the element nodes in XML document. If the XSL file is set probably, it is able to capture out the most meaningful data in the transformed web page. In our system, XSL 2.0 is used, which is not popularly used, and not all the software systems support this version, however, it is worth using it. The XSL processor used in our system is Saxon-B 8.7, which is a limited but free version of XSL processor [URL7].

```
<Xsl:stylesheet xmlns:xsl="http://www.w3.org/1999/XSL/Transform" version="2.0">
<xsl:param name="regexp" />
<xsl:template match="*">
        <xsl:analyze-string select="." regex="{$regexp}">
                <xsl:matching-substring>
                        <xsl:value-of select="." />
                </xsl:matching-substring>
        </xsl:analyze-string>
</xsl:template>
</xsl:stylesheet>
```

Fig. 4. XSL Template

3.2.2 Methodology for Non-static Information Extraction

For the information without static structure, it is not suitable to use the regular expression rule. Therefore, our system tries to exploit the special features of web page, the organizational structure of the fields in the DOM tree architecture.

As mentioned before, all the pages are transformed into XHTML format before the extraction process can be performed. So the DOM tree should be valid, well formatted and correct. These requirements are essential and a pre-requisite for our system, because the whole extraction algorithm relies on the HTML tags.

$$PTageSet = \{tag_1, tag_2, ..., tag_n\} \subset p$$
$$KWSet_i = \{KW_1, KW_2, ..., KW_l\}$$

For each $KW_i \in KWSet_i$ {

  If $KW_i \subset p$ {

    $\exists ktag_i \in PTagSet$, where $ktag_i = nearest(ot)$
    $KTagSet = KTagSet + ktag_i$

  }

}

For each $ktag_{k(k=1..n)} \in KTagSet$ {

  $WP_l = Split(p, ktag_k)$

For each WP_l {

  If $\exists KW_i \in WP_l$ {

    $Output = Output + WP_l$

    $p = p - WP_l$

  }

}

}

Fig. 5. Extraction Steps

Let *p* be a web page, and *PTagSet* be the set of tags in *p*, that is *PTagSet* = {*tag₁*, *tag₂*, *tag₃*, ... , *tagₙ*}, and each of the tag is either open or close tag. From the result training process, keywords exist in the schema information. After analyzing, a set of keywords, *KWSetᵢ*={*KW₁*, *KW₂*, ... , *KWₗ*} for extraction field *fⱼ* is formed.

For each *KWᵢ* in *KWSetᵢ*, the location of the *KWᵢ* will be identified in the page *p*. Then the first most nearest open tag against *KWᵢ* the keyword *KTagSet* will be put into a set. After all *KWᵢ* in *KWSetᵢ* are applied, *KTagSet*= {*KTage₁*, *KTage₂*, ... , *KTageₙ*} will be formed.

For each *KTagᵢ* in *KTagSet*, the page *p* will be split up to several parts by using *KTagᵢ*. For every part of the page, if any *KWᵢ* is located in between, then the whole part will be added to the output result set.

The above methodology will be illustrated in figures 6 and 7. As an example, if keyword *Wᵢ* appears between the first pair of the <LI> as shown in figure 6, then the *KTag* will be <UL>, since it is the first open tag for the keyword *Wᵢ.* Then the whole page is split by the tag <UL>. After that, for each separated part, if the *Wᵢ* exists, the part will add to the result set.

In example 7, a more complicated example is presented. As shown in figure 7, keyword *Wᵢ* appears more than one time in the page. In this example, the key tag list is {<UL>, <LI>}. Then the page is split by <UL> first and any part that contains *Wᵢ* will be output to the result set. After that, the remaining parts of the page will be further split by <LI> and the process continues.

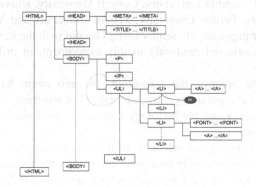

Fig. 6. DOM tree structure of web page example 1

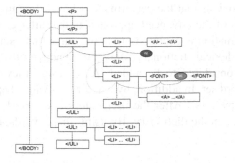

Fig. 7. DOM tree structure of web page Example 2

The methodology mentioned above is only the basic one. In order to enhance the extraction result, instead of collecting the keyword *KW* from adifferent training sample pages, some equivalent and synonymous keywords will be added. In our system, the additional keywords added to the *KWSet* are selected from a lexical dictionary called WordNet [17], which provides semantic relations among words. For example, the "Teach" in WordNet 2.1 has the synonymous words "instruct" and "learn". The performance of the basic methodology and the extended one will be discussed later in the evaluation section.

4   Evaluation

In order to evaluate our system in more scientific way, the evaluation metrics recall and precision measurements are used. Recall measures the amount of the relevant information that the system correctly extracts, and the precision measures the reliability of the information extracted [18].

In our experiment, the extraction record contains 4 fields.

$$FSet = \{Telephone, Fax, TeachingCo\,urse, Re\,searchInte\,rest\}.$$

There are totally 450 pages for the training process and 2100 pages for testing. All of them come from the staff information page of different universities, including Boston University, Columbia University, Cornell University, University of Maryland, New York University, Perdue University, Macau University and Yale University. In addition, some other pages are chosen by using Google with the keyword "professor". All these pages are selected randomly in this experiment, in order to evaluate the normal performance.

After the training process, the *telephone*, *fax* and *email* have a rather static structure, therefore the regular expression is used. For *teaching course* and *research interest*, DOM tree analyzing methodology is used. Table 2 shows the results of our experiments on average.

Since all the web pages are selected from different sources, the structures and content of them are comparatively divergent. By using only the regular expression, it is fair that the recall rate for the *telephone* and *fax* are around 0.7, while the precision rate can reach more than 0.9. To extract the information on *teaching course* and *research interest*, by comparing the two different methodologies, basic and the advanced one, it is obvious that the performances have some improvement. However, the major advantages of the advanced approach is that after enhancing the key word set, most of the information is captured out at the same time, since those of the added words come from a lexical dictionary, only the synonymous words with the same word type will be considered to be added. As a result, they will not cause much "noise" to the key word set. In addition, by using the advanced approach, it is possible to omit the training process. After the user enters a keyword, a set of synonymous words can be found from the dictionary. By using those words, it is enough to extract some information.

Table 2. Resulting figure

	Methodology					
	Regular Expression		DOM Tree Analyzing (Basic)		DOM Tree Analyzing (Advanced)	
	Telephone	Fax	Teaching Course	Research Interest	Teaching Course	Research Interest
Recall	0.721622	0.696703	0.77593	0.875255	0.82098	0.878205
Precision	0.9821	0.952381	0.831667	0.559836	0.807409	0.562824

5 Conclusion and Future Work

In this paper, we have firstly discussed the history and current developments in web information extraction. A further detailed analysis of our approach shows that it relies on the human training process. In contrast to most existing extraction systems, our system uses different methodologies to extract the information, either through the regular expressions or through analysis of the DOM tree. Furthermore, we have also proposed an advanced approach for DOM tree analysis by enhancing the keyword set. After the extraction process, the result is output to an XML file format. However, some limitations also exist. In the current system, the extraction task is only individual page based. It means that all the fields for the same record are supposed to be contained in the same pages. However, in many other situations, the fields may be located in different relevant pages, such as several linked web pages. In future work, we are going to extend our system to handle multi-page extractions.

Acknowledgement. This Work was supported in part by the University Research Committee under Grant No. RG069/05-06S/07R/GZG/FST and by the Science and Technology Development Found of Macao Government under Grant No. 044/2006/A.

References

1. Eikvil, L.: Information Extraction from World Wide Web – A Survey, Technical Report 945, Norweigan Computing Center, Oslo, Norway (July 1999)
2. Laender, A.H.F., Ribeiro-Neto, B.A., da Silva, A.S., Teixeira, J.S.: A brief survey of Web data extraction tools. ACM Sigmod Record 31(2), 84–93 (2002)
3. Hammer, J., McHugh, J., Garcia-Molina, H.: Semistructured Data: The TSIMMIS Experience. In: Hammer, J., McHugh, J., Garcia-Molina, H. (eds.) Proc. I East-European Workshop on Advances in Database and Information Systems - ADBIS 1997, Petersburg, Russia (1997)
4. Arocena, G., Mendelzon, A.: WebOQL: Restructuring Documents, Databases, and Webs. In: Proc. IEEE Intl. Conf. Data Engineering 1998, Orlando (February 1998)
5. Liu, L., Pu, C., Han, W.: XWRAP: An XML-enabled wrapper construction system for web information sources. In: Proceedings of the international conference on data engineering (ICDE), pp. 611–621 (2000)

6. Crescenzi, V., Mecca, G., Merialdo, P.: Roadrunner: Towards automatic data extraction from large web sites. In: Proc 27th Very Large Databases Conference, VLDB 2001, pp. 109–118 (2001)
7. Freitag, D.: Information Extraction from HTML: Application of a General Learning Approach. In: Proceedings of the 15th National Conference on Artificial Intelligence (AAAI 1998) (1998)
8. Solderland, S.: Learning Information Extraction Rules for Semi-structured and Free Text. Machine Learning 34, 233–272 (1999)
9. Etzioni, O., Cafarella, M., Downey, D., Kok, S., Popescu, A.-M., Shaked, T., Soderland, S., Weld, D.S., Yates, A.: Web-scale information extraction in KnowItAll (preliminary results). In: Proceedings of the 13th World Wide Web Conference, pp. 100–109 (2004)
10. Hsu, C.-N., Dung, M.-T.: Generating finite-state transducers for semi-structured data extraction from the web. Information Systems 23(8), 521–538 (1998)
11. Muslea, I., Minton, S., Knoblock, C.: Hierarchical wrapper induction for semistructured information sources. Autonomous Agents and Multi-Agent Systems 4(1/2), 93–114 (2001)
12. Adelberg, B.: NoDoSE—A Tool for Semi-Automatically Extracting Structured and Semistructured Data from Text Documents. In: Proceedings of the 1998 ACM SIGMOD International Conference on Management of Data, Seattle, Washington, June 1998, pp. 283–294 (1998)
13. Snoussi, H., Magnin, L., Nie, J.-Y.: Toward an Ontology-based Web Data Extraction (2002)
14. Buttler, D., Liu, L., Pu, C.: A Fully Automated Object Extraction System for the World Wide Web. In: Proceedings of the 21th International Conference on Distributed Computing Systems, pp. 361–370 (2001)
15. Papadakis, N.K., Skoutas, D., Raftopoulos, K.: IEEE Computer Society. In: Varvarigou, T.A. (ed.) STAVIES: A System for Information Extraction from Unknown Web Data Sources through Automatic Web Wrapper Generation Using Clustering Techniques, IEEE Transactions on Knowledge and Data Engineering, vol. 17(12), pp. 1638–1652 (December 2005)
16. Xiao, L., Wissmann, D.: Information Extraction from the Web: System and Techniques. Applied Intelligence 21, 195–224 (2004)
17. Miller, G.A., Beckwith, R., Fellbaum, C., Gross, D., Miller, K.: Introduction to WordNet: An On-line Lexical Database, Revised (August 1993)
18. Cardie, C.: Empirical methods in information extraction. AI Magazine 18(4), 65–80 (1997)

URL1 http://www.internetworldstats.com/   *Miniwatts Marking Group*
URL2 http://www.w3.org/TR/xhtml1/   *XHTML, W3C Recommendation*
URL3 http://www.zvon.org/xxl/XSLTutorial/Output/contents.html   *XSLT Tutorial*
URL4 http://www.w3.org/TR/xslt.html   *XSL Transformations, W3C Recommendation*
URL5 http://www.w3.org/TR/xmlschema-0/   *XML Schema Primer, W3C Working Draft*
URL6 http://tidy.sourceforge.net/   *HTML Tidy Library Project*
URL7 http://www.saxonica.com/   *Saxon Processor*

Information Presentation on Mobile Devices: Techniques and Practices

Lin Qiao, Ling Feng, and Lizhu Zhou

Dept. of Computer Science & Technology, Tsinghua University, Beijing, China
qiaol-03@163.com, {fengling, dcszlz}@tsinghua.edu.cn

Abstract. The popularity of hand-held mobile devices, such as personal digital assistants (PDA) and smart cell phones, is growing. Compared with traditional desktop computers, these mobile devices have distinct limitations, including tiny displays, scarce computing hardware resources, bandwidth fluctuations, ad-hoc communications, voluntary, and involuntary disconnections, etc., presenting new challenges to human-computer interaction on mobile devices. In this paper, we survey research efforts done on mobile device user interface design. Some recently developed techniques for diverse information presentation via visual, audio, and tactile channels are reported. We also brief on our experiences in virtually presenting database query results on PDAs, including both presentation style and presentation content.

1 Introduction

Mobile devices have gained increasingly popularity due to its portability nature. People use these small mobile devices to manage personal information, do simple work with poor processing requirement, or remotely control PCs and computerized appliances [1]. Nowadays, the use of mobile devices has penetrated into the domains of education, business, military, etc. [1].

Compared with traditional desktop computers, mobile devices have many limitations in terms of 1) small-sized display with poor resolution, fewer colors, and different width/height ratio from the normal 4:3; 2) limited CPU processing and memory capacities; 3) slow connection with fluctuated bandwidth; and 4) non-user-friendly input facilities (ordinarily used keyboard and handwriting demanding lots of screen space, resulting in quite inaccurate results) [2,3]. Table 1 gives some quantitative indicators.

Due to these large differences, the classic desktop solutions have limited applicability to mobile user interface design. In this paper, we survey research efforts on the design of mobile device interfaces, focusing on information presentation through diverse channels including visual, audio, and tactile channels. The remainder of the paper is organized as follows. Section 2 describes the guidelines for mobile device interface design. Following these guidelines, some recently developed techniques and approaches for mobile device interface design are presented in Section 3. We also describe our experiences of virtually presenting database query results on PDAs in Section 4, and conclude the paper in Section 5.

Y. Zhang et al. (Eds.): APWeb 2008, LNCS 4976, pp. 395–406, 2008.

Table 1. The comparison between HP rx5965 PDA and ChaoYang A550 desktop

	Screen Size	Resolution	Speeds	Memory Size	Input Tools
Desktop	17 Inches	1600×1200	2.80GHz	2.0G	Keyboard&Mouse
PDA	3.5 Inches	240×320	400MHz	64MB RAM	Virtual keyboard
				2.0G Flash	stylus

2 Guidelines for Mobile Device User Interface Design

[4,5] identified several main challenges in mobile human-computer interaction, leading to the following design considerations for mobile device interfaces.

- *Small devices with limited input/output facilities.* Considering the small screen size, poor sound output quality, and tiny keypads, etc., it would be practical to provide word selection instead of requiring text input whenever possible, and allow for single- or no-handed operation.
- *Mobility.* Environments may change drastically as users move. The mobile device interface will inevitably experience multiple and dynamic contexts through various sensors and networks. These unreliable or patchy sensors may bring incomplete and varying context information. Here, the interface should preferably permit users to configure output to their needs and preferences (e.g., text size, brightness), and meanwhile adapt itself automatically to the user's current environment.
- *Widespread population.* Simple user interface should be designed, because users often don't have any formal training in their technologies. Besides, it must allow for personalization, providing users the ability to change settings themselves. Also, the interface should be visually pleasing and fun as well as usable to offer enjoyment.
- *Multi-tasking with limited and split attention.* Due to the mobile environment, frequency of interruptions is likely to be much higher than on desktop. Designers should consider the multi-tasking, allowing applications to be stopped, started, and resumed with little or no effort. In addition, multimodality approach can be adopted to accomplish multi-tasks.

3 Techniques for Mobile Device User Interface Design

Many human-interface researchers are trying new methods to enable and enhance information presentation, utilizing visual, audio, and tactile channels of mobile devices, respectively.

3.1 Visual Channel of Mobile Devices

Different types of information (such as Web pages, texts, images, maps, and structured data) are visually presented in different ways [2].

1) Web Page Presentation

Mobile Web search receives great attention nowadays. Web contents, mostly designed for desktop computers, are badly suited for mobile devices [6,7]. Currently, the majority of commercially available mobile web browsers use single-column viewing mode to avoid horizontal scroll [7]. But this approach tends to have much more vertical scroll and destroys the layout of original view [7].

Based on small- and large-scaled user studies, [8] provided a list of development principles for Web page display. They are: 1) developing phone-based applications that provide direct, simple access to focused valuable contents; 2) trimming the page-to-page navigation down to a minimum; 3) providing more rather than less information for each search result; 4) using simple hierarchies which are similar to the phone menus that users are already familiar with; 5) adapting for vertical scrolling or reducing the amount of vertical scrolling by simplifying the text to be displayed; 6) reducing the number of users' keystrokes; 7) providing a quick way for users to know whether a search result points to a conventional HTML page or a small screen optimized page; 8) pre-processing conventional pages for better usability in small screen contexts; and 9) combining theoretical and empirical evaluation to provide further insights [8].

Google's PDA interface is similar to Google's XHTML interface [9]. However, the diversity of queries in mobile environments was far less than in desktop [9]. This might be due to the enormous amount of efforts (in terms of time and key presses) needed for users to enter query terms, so that each session on mobile devices had significantly fewer queries than sessions initiated on the desktop [9]. Users for the most part tended to search similar content as desktop queries, and the percentage of adult queries was vastly larger [9].

To deliver adaptive Web contents on mobile devices, researchers also considered to re-author web pages, which can be done at *server side, intermediate side*, or *client side* [10].

A. *Re-authoring Web pages at server-side.* Server-side adaptation provides the Web page author maximum control over content delivery for mobile devices [10]. [11] reported a system which used the W3C's Document Object Model (DOM) API to generate an XML tree-like structure, and Extensible Style Sheet Language Transformations (XSLT) to generate Wireless Markup Language (WML) or HTML content for display on mobile devices. This system also adapted to the users' dynamic context. [12] presented another system which could adapt multimedia Web documents to optimally match the capabilities of the client device. In a scheme called InfoPyramid, content items on a Web page were transcoded into multiple resolution and modality versions, so that they could be rendered on different devices [12]. On the other hand, a customer could select the best parameters from the InfoPyramids to meet the resource constraints of the client device while still delivering the most "value" [12].

B. *Re-authoring Web pages at intermediate-side.* Proxies typically apply intermediate adaptations [10]. Today, many of web page visualization efforts fall into this category. Without changing the layout of original web pages, [13] reduced the size of images which were larger than that of mobile screens and removed media which mobile devices did not support. [6] described a scaled-down version to fit the mobile devices screen. Images embedded in a web page and the Internet address bar were removed; and the font size of textual contents was adjusted by the user [6]. The focus + context visualization was also employed in the display of Mobile Web. Users could choose what they are interested in with a large font size, while other information in the surrounding area was displayed in a reduced font size [6].

[14] splited a Web page's structure into smaller but logically related units. A two-level hierarchy was used with a thumbnail representation at the top level to provide a global view and an index to a set of subpages at the bottom level for detailed information. [15] introduced heuristics for structure-aware Web transcoding which considered a Web page's structure and the relative importance of its components. [7] proposed to display a web page as a thumbnail view, but preserving the original page layout, so that users can identify the overall page structure and recognize pages they previously viewed. [16,17] also proposed to show pages in a modified original layout.

C. *Re-authoring Web pages at client-side.* A client device can use style sheets to format contents in a browser [10]. For instance, the font size of textual contents can be adjusted by users [6]. Together with the above intermediate-side approaches, by storing user's operations with the DOM tree in a profile, the system automatically generated a DOM-tree with branches expanded or hidden based on the user's interest [6].

2) Text (Lengthy Document) Presentation

Two popular ways to view lengthy documents on small screens in the literature are *Rapid Serial Visual Presentation* (RSVP) and *Leading Format Presentation* (LFP) [18]. 1) RSVP presents one or more text words at a time at a fixed location on the screen [19]. Two variants of RSVP, i.e., Adaptive RSVP and Sonified RSVP, were detailed in [20]. Adaptive RSVP adjusts each text chunk exposure time with respect to content (e.g., the number of characters and words to be exposed) as well as to context (e.g., the result of content adaptation, the word frequencies of the words in the chunk, and the position of the chunk in sentence being exposed). Sonified RSVP plays appropriate sound when a certain text chunk is displayed. 2) LFP method scrolls the text in one line horizontally/vertically across the screen [18,19,20]. Considering that sentence boundary is important in reading, a sentence-oriented presentation manner was developed for a small window, which presented complete sentences one at a time [19].

In general, sentences are read more accurately and more natural in the RSVP format than in the LFP format [20,21], since when the eye processes information during fixed gazes, it is more comfortable that the text moved successively rather than continuously [20]. However, the experiments of [22] showed that comprehension for smooth scrolling Times Square was at least as high as that for RSVP at presentation rates ranging from 100 to 300 words per minute.

3) Image Presentation

To visualize data-intensive images on mobile devices, an intuitive solution is to compress and transcode images to reduce data transmission and processing.

JPEG 2000 [23] detailed a progressive transmission mechanism which allowed images to be reconstructed by different pixel accuracy or spatial resolution and be delivered for different target devices with different capabilities. [24] introduced an non-uniform resolution presentation method, in which resolution was the highest at the fovea but falls off away from the fovea. [25] classified images into image type and purpose, and transcoded images in order to adapt to the unique characteristics of the devices with a wide range of communication, processing, storage, and display capabilities, thus improving the delivery.

In addition to treating an image as a whole, [26,27] proposed to separate region-of-interest and deliver the most important region to the small screen according to human's attention model. They used RSVP presentation technique to simulate the attention shifting process, and noticed that there was an important psycho physiological activity - visual attention shifting. Image browsing on small devices could be improved by simulating the fixation and shifting process in a way similar to RSVP. An image was decomposed into a set of regions which were displayed serially, each for a brief period of time. [26] further introduced a generic and extensible image attention model based on three attributes (i.e., region of interest, attention value, and minimal perceptible size) associated with each attention object. [27] tried to find an optimal image browsing path based on the image attention model to simulate the human browsing behavior.

4) Map Presentation
Maps play an important role in mobile location-based services. However, they are often too large to be fully displayed on mobile device screens [2]. To this end, [28] used 3D arrows to point towards the objects and by the side of the arrows the information about distance and name of point object was provided with text. The 3D arrows were semi-transparent for comfortable visual. "Halo" [29] represented off-screen locations as abstract "streetlamps" with their lights on the map. The map was overlayed with translucent arcs, indicating the location of off-screen places. Each arc was part of a circular ring that surrounds one of the off-screen locations. The arcs on the map allowed viewers to recognize the missing off-screen parts, and let viewers understand its position in space well enough to know the location of the off-screen targets.

5) Structured Data Presentation
There are also some interesting visualized methods developed for structured data, such as relational databases, 3D objects, and calendar, etc.

Database. [30] designed a graphical database interface for mobile devices. In this method, as soon as a connection was made, the relations in the database were displayed on their interface. Initially, only "top-level" relations were shown, and for the sake of conserving screen space, a nested relation structure was imposed on non-nested database systems. On the interface, users could select any number of relations, and display all the possible join paths between them. The resulting join was displayed on an auxiliary screen, which showed the actual SQL query and the actual answer set for that query [30].

3D Object. To visualize 3D model on mobile devices, Virtual Reality Modeling Language (VRML) and Extensible 3D(X3D) allow a content developer to re-use a large collection of existing Web-Based 3D worlds in the mobile context and develop content for different platforms with the same tools [31]. For location-aware presentation of VRML content on mobile devices, the user interface was divided into two parts: an upper area where the actual 3D world was visualized and a lower area providing status information and tools for users to navigate the 3D world, setting the system and moving the viewpoint [31].

Calendar. [32] showed an interesting fisheye calendar interface called DataLens on PDAs. On the interface, first, users could have an overview of a large time period with a graphical representation of each day's activities. Then, users could tap on any day to expand the area representing that day and reveal the list of appointments in context [32]. The "semantic zooming" approach used in DataLens was utilized to visually represent objects differently depending on how much space is available for displaying. The graphical views were scaled to fit the available space, while the textual views used a constant-sized font, and the text was clipped to fit in the available space [32].

3.2   Audio Channel of Mobile Devices

With the hard handling and limited screens, it is beneficial to make use of the speech channel of mobile devices for the following reasons [33]. First, voice is portrayed as the most naturalistic way to interact with a system, so speech interface is more natural for interaction. Second, speech interface helps increase interaction efficiency, because speech is faster than any other common communication method like typing and writing. Third, voice interaction avoids "hand-busy" and "eyes-busy" operations which happen with the visual interface. Fourth, people think that telephony network is often more trustworthiness than Web. Finally, speech interface can also serve as a good input manner, where speech recognition avoids password input [33]. Ease-of-use and the speed of interaction are the two most important requirements for the voice interface, and voice interface must be an integral part of the whole user interface of the device, but should not be overused due to the misrecognition [34].

[34] designed a multi-lingual speaker-dependent voice dialing user interface, which could support speech recognition and speech synthesis. Users don't need to train the voice tag, and the interface system can generate the tag automatically. [35] offered a speech interface model, where users can use a single personalized speech interface to access all services and applications.

3.3   Tactile Channel of Mobile Devices

Apart from visual and audio channels, tactile sensation can also be explored throughout mobile user interaction. By tactile feedback, we can reduce possible mobile interaction mistakes, since audio feedback is difficult to apply when the environment is noisy, and visual feedback is also difficult as users have to pay much attention to others and the screen is small [36]. Users can feel the vibration with their fingers as they press the screen. [36] did text entry experiments and showed that users with tactile feedback user interface entered significantly more text, made fewer errors, and corrected more errors they did make.

[37] used paper metaphor to design the switching of scrolling and editing operations, where a touch sensor is attached on the PDA. In map or Web browser, when a user does not touch sensor, the screen scrolls according to the movement of the pen when dragging, and when touching, the screen does not scroll and edit while dragging. In the photograph browser, when the user does not touch the sensor, the screen also scrolls the photograph, but when touching, if dragging the pen upward, the photograph is zoomed in; and if downward, the photograph is zoomed out. Dragging the pen left to right invokes clockwise rotation, and right to left invokes counter clockwise rotation [37].

4 Our Practice

In this section, we describe our experience of virtually presenting database query results on a small PDA, focusing on presentation style and presentation content. We use Microsoft SQL server's sample *Northwind* relational database as an example. It has an *employee* table, containing 16 attributes named *employee-ID, LastName, FirstName, BirthDate, Title*, and so on.

4.1 Dynamic Presentation Style

The commonly used visual database interface on desktop PCs is form-based [38], containing all the database query results that satisfy a user's query request. However, for mobile PDA users, it will be too heavy and even unreadable to showing all the result records once only on the small screen. Hence, we follow the dynamic text presentation principle, and adopt the two well-developed methods, i.e., leading format and serializing format [19], for dynamically displaying database query results on small screens.

Leading display of query results. The leading can be either *tuple*-wise or *attribute*-wide. Tuple-wise leading presentation, as its name implies, is to present result tuples one by one from right to the left side of the screen; while attribute-wise leading presentation is to display result attribute values one by one from right to left, as illustrated in Figure 1 and Figure 2 respectively, where symbol "||" is used to separate different attribute values.

Here, two closely related important factors that affect reading are *display speed* and *jump length* . The former measures the amount of words presented per minute; while the latter measures the continuity of leading display characters movement. The speed setting must take human's eye fixation time for reference. Performance study in [21] showed that the slow speed of 171 wpm (Words Per Minute) and the fast speed of 260 wpm makes non-evident difference on reading accuracy. However, with the same speed setting, the shorter the jump length is, the smoother the movement of characters looks on the screen. [21] showed that reading accuracy declines as jump length decreases from 3 characters to 1

Fig. 1. Tuple-wise leading display   **Fig. 2.** Attribute-wise leading display

character. Another study [39] also demonstrated that when the jump length is set to 1 or 2 characters, reading accuracy is very poor, but when jump length increases, the reading accuracy increases. Any jump length setting between 4 and 10 characters has the same effect.

In our experimental setting, considering a query result is multi-lined moving, making the understanding more difficult than single-lined moving, we set the jump length to 5 characters, and the display speed to $260/n$ wpm, where n is the number of lines to be output on the screen, dynamically determined upon each user's query request.

Besides display speed and jump length, the color difference between text and background also influences the reading comprehension [40]. According to [40], for Video Display Terminals, the color difference between text and background colors is preferably not less than a threshold value, and *red-on-white,blue-on-yellow*, and *white-on-black* are all favorable options complying to this requirement [41]. Our test shows that readers are mostly comfortable with their familiar combination of black texts on white background.

Serially display of query results. The above leading display inevitably incurs human's eye movement during reading. To decrease the amount of eye movement, we adopt the serial display method [19] to present a query result page by page, with the latter one covering the previous one. This method simulates human's normal reading experience (like turning pages of a book). It enables to visualize the structure of a database tuple. Figure 3 presents a page containing one complete result tuple. According to [42], we set the display speed to 250 wpm for better reading comprehension. The jump length parameter is not applicable for this method. We compare the effect of leading display and serially display in Table 2.

Some other considerations during our presentation design are as follows.

1) *Pause/continue/stop* functions. During dynamic display, it is desirable to permit mobile users to interrupt information moving at will and to continue or stop later.

2) *Fast page turning.* To be efficient, anytime, the presenter should allow mobile users to quickly turn the page up and down for their interested contents.

3) *Enlarged display.* To enhance readability on small screens, mobile users sometimes may want to have a bigger view about some important contents. The presenter shall provide a larger version of information, selected by users.

Table 2. The comparison between leading display and serial display

	Leading Display	Serial Display
Reading efficiency	slow(260/n wpm)	fast(250 wpm)
Reading comprehension	familiar & acceptable initially[20]	lengthy texts efficiently[20]
Suitable scenario	long time operation	short time operation
	e.g., information gathering	e.g., information confirmation

4.2 Presentation Content

While making efforts to dynamically present query results on small screens, we also try to adapt the result contents, aiming to present the most potential important and useful results on the screen.

Observing that users often need some aspects of query results depending on query context, we propose to select the most potentially useful attributes to be displayed on the screen. To do this, we apply three strategies.

1) Method 1: Considering that users' similar interest usually last for a certain period, we select attributes based on the access history. That is, each time the most frequently used attribute will be selected for display. Let F_i^j denote the probability that the jth attribute is selected for display for the ith query. Assume each database query is independent. According to the Maximum likelihood estimate, the possibility that the jth attribute is interesting for the $k+1$th query is $\frac{\sum_{a=1}^{k} F_a^j}{k}$.

2) Method 2: The classical least recently used (LRU) strategy is also a simple decision making method utilizing the access history. That is, each time the least recently used attribute will not be selected for display.

3) Method 3: Considering that user's interested attributes usually depend on query context, like user's profession, query intention, time and place of the query, query duration, etc., we do context-aware attribute selection. We first let users identify their query context elements such as their *"duty"*, *"month"*, *"place"*, *"query intention"*, and *"time spent on query"*, etc., with which we build a decision tree by training the historic database access log, using the classical decision-tree arithmetic C4.5. We then make selection decision upon every new query request.

4) Method 4: Considering similar query contexts lead to similar user's interested attributes, we cluster similar query contexts. When a new query comes, we choose the most similar cluster according to the distance between the cluster's center and the new query contexts. Then, we vote the user's interested attributes among the chosen cluster. We use k-means arithmetic to cluster the query contexts.

Of course users can always add their interested attributes or delete the suggested ones.

We conducted a small-scaled case study among the undergraduate university students inside the campus to let them point out the commonly interested attributes in the *employee* table under different query situations. Totally we have 2946 such records, 60% of which was used for training, and 40% of which was used for testing. Assume that at most five attributes values will be presented on PDAs at one time. Figure 4 shows the performance result, where the horizontal axis represents the total 16 attributes. As illustrated, the third context-aware attribute selection approach performs consistently the best among the three. This is obvious since it takes into account both the access history and current access context, while the second LRU method makes the least use of the historic knowledge, thus performing the worst, and the first method - the most frequently used approach stays in the middle.

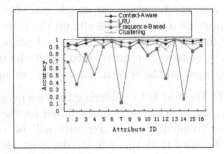

Fig. 3. Serial display   **Fig. 4.** Performance Results

5 Conclusion

The growing popularity of mobile computing devices presents new challenges to mobile device interface designers. In this paper, we surveyed related work on information presentation through the visual, audio, and tactile channels of mobile devices. Our preliminary experiments with virtually presenting database query results on PDAs were also reported.

Acknowledgement

The work is jointly supported by the Ministry of Education of China (Changjiang Scholar Program), Tsinghua University (Hundred-Talent Program), and Faculty of Information Science and Technology (Teaching and Research Starting Fund).

References

1. Myers, B.A., Nichols, J., Wobbrock, J.O., Miller, R.C.: Taking handheld devices to the next level. Computer 37(12), 36–43 (2004)
2. Chittaro, L.: Visualizing information on mobile devices. Computer 39(3), 40–45 (2006)

3. Kärkkäinen, L., Laarni, J.: Designing for small display screens. In: Proc. of the second Nordic conference on Human-computer interaction, ACM, pp. 227–230 (2002)
4. Dunlop, M., Brewster, S.: The challenge of mobile devices for human computer interaction. Personal Ubiquitous Comput. 6(4), 235–236 (2002)
5. Gong, J., Tarasewich, P.: Guidelines for handheld mobile device interface design. In: Proc. the 2004 DSI Annual Meeting (2004)
6. Zhang, D.: Web content adaptation for mobile handheld devices. Commun. ACM 50(2), 75–79 (2007)
7. Lam, H., Baudisch, P.: Summary thumbnails: readable overviews for small screen web browsers. In: CHI 2005: Proc. of the SIGCHI conference on Human factors in computing systems, pp. 681–690. ACM, New York (2005)
8. Buchanan, G., Farrant, S., Jones, M., Thimbleby, H., Marsden, G., Pazzani, M.: Improving mobile internet usability. In: Proc. of the 10th international conference on WWW, pp. 673–680. ACM, New York (2001)
9. Kamvar, M., Baluja, S.: A large scale study of wireless search behavior: Google mobile search. In: Proc. of the SIGCHI conference on Human Factors in computing systems, pp. 701–709. ACM, New York (2006)
10. Laakko, T., Hiltunen, T.: Adapting web content to mobile user agents. IEEE Internet Computing 9(2), 46–53 (2005)
11. Pashtan, A., Kollipara, S., Pearce, M.: Adapting content for wireless web services. IEEE Internet Computing 7(5), 79–85 (2003)
12. Mohan, R., Smith, J.R., Li, C.S.: Adapting multimedia internet content for universal access. IEEE Transactions on Multimedia 1(1), 104–114 (1999)
13. Berhe, G., Brunie, L., Pierson, J.M.: Modeling service-based multimedia content adaptation in pervasive computing. In: Proc. of the first conference on computing frontiers on Computing frontiers, pp. 60–69 (2004)
14. Chen, Y., Xie, X., Ma, W.Y., Zhang, H.J.: Adapting web pages for small-screen devices. IEEE Internet Computing 9(1), 50–56 (2005)
15. Hwang, Y., Kim, J., Seo, E.: Structure-aware web transcoding for mobile devices. IEEE Internet Computing 7(5), 14–21 (2003)
16. Roto, V., Kaikkonen, A.: Perception of narrow web pages on a mobile phone. In: Proc. of Human Factors in Telecommunications (2003)
17. Roto, V., Popescu, A., Koivisto, A., Vartiainen, E.: Minimap: a web page visualization method for mobile phones. In: Proc. of the SIGCHI conference on Human Factors in computing systems, pp. 35–44. ACM, New York (2006)
18. Laarni, J.: Searching for optimal methods of presenting dynamic text on different types of screens. In: Proc. of the second Nordic conference on Human-computer interaction, pp. 219–222. ACM, New York (2002)
19. Rahman, T., Muter, P.: Designing an interface to optimize reading with small display windows. Human Factors 41(1), 106–117 (1999)
20. Öquist, G., Goldstein, M.: Towards an improved readability on mobile devices: Evaluating adaptive rapid serial visual presentation. In: HCI 2002: Proc. of the 4th International Symposium on Mobile Human-Computer Interaction, London, UK, pp. 225–240. Springer, Heidelberg (2002)
21. Juola, J.F., Tiritoglu, A., Pleunis, J.: Reading text presented on a small display. Applied Ergonomics 26(3), 227–229 (1995)
22. Kang, T.J., Muter, P.: Reading dynamically displayed text. Behaviour & Information Technology 8, 33–42 (1989)
23. Christopoulos, C., Skodras, A., Ebrahimi, T.: The jpeg2000 still image coding system: an overview. IEEE Transactions on Consumer Electronics 46(4), 1103–1127 (2000)

24. Chang, E., Mallat, S., Yap, C.: Wavelet foveation. Applied and Computational Harmonic Analysis 9(3), 312–335 (2000)
25. Smith, J.R., Mohan, R., Li, C.S.: Content-based transcoding of images in the internet. In: Proc. 1998 Intl. Conf. on Image Processing, pp. 7–11 (1998)
26. Chen, L.Q., Xie, X., Fan, X., Ma, W.Y.: A visual attention model for adapting images on small displays. Multimedia systems 9(4), 353–364 (2003)
27. Liu, H., Xi, X., Ma, W.Y., Zhang, H.J.: Automatic browsing of large pictures on mobile devices. In: Proc. ACM Multimedia 2003, pp. 148–155 (2003)
28. Chittaro, L., Burigat, S.: 3d location-pointing as a navigation aid in virtual environments. In: AVI 2004: Proceedings of the working conference on Advanced visual interfaces, pp. 267–274. ACM, New York (2004)
29. Baudisch, P., Rosenholtz, R.: Halo: A technique for visualizing off-screen locations. In: Proc. CHI 2003, pp. 481–488 (2003)
30. Alonso, R., Haber, E.M., Korth, H.F.: A database interface for mobile computers. In: Proc. Globecom Workshop on Networking of Personal Communication Applications (1992)
31. Burigat, S., Chittaro, L.: Location-aware visualization of vrml model in gps-based mobile guides. In: Proc. the tenth Intl. Conf. on 3D Web technology, pp. 57–64 (2005)
32. Bederson, B.B., Clamage, A., Czerwinski, M.P., Robertson, G.G.: Datelens: A fisheye calendar interface for pdas. ACM TOCHI 11(1), 90–119 (2004)
33. Fan, Y., Saliba, A., Kendall, E.A., Newmarch, J.: Speech interface: an enhancer to the acceptance of m-commerce applications. In: Proc. ICMB 2005, pp. 445–451 (2005)
34. Iso-Sipila, J., Moberg, M., Acoustics, O.V.: Multi-lingual speaker-independent voice user interface for mobile devices. In: IEEE ICASSP, vol. 1, pp. 1–1 (2006)
35. Pakucs, B.: Butle: A universal speech interface for mobile environments. In: Brewster, S.A., Dunlop, M.D. (eds.) Mobile HCI 2004. LNCS, vol. 3160, Springer, Heidelberg (2004)
36. Brewster, S., Chohan, F., Brown, L.: Tactile feedback for mobile interactions. In: Proc. SIGCHI Conf. on Human factors in computing systems, pp. 159–162 (2007)
37. Siio, I., Tsujita, H.: Mobile interaction using paperweight metaphor. In: Proc. UIST 2006, pp. 111–114 (2006)
38. Jagadish, H.V., Chapman, A., Elkiss, A., Jayapandian, M., Li, Y., Nandi, A., Yu, C.: Making database systems usable. In: Proc. SIGMOD Conf, pp. 13–24 (2007)
39. Granaas, M.M., McKay, T.D., Laham, R.D., Hurt, L.D., Juola, J.F.: Reading moving text on a CRT screen. Human Factors (1984)
40. Shieh, K.K., Chen, M.T.: Effects of screen color combination and visual task characteristics on visual performance and visual fatigue. In: Proc. the National Science Council, vol. 21, pp. 361–368 (1997)
41. Wang, A.H., Fang, J.J., Chen, C.H.: Effects of vdt leading-display design on visual performance of users in handling static and dynamic display information dual-tasks. International Journal of Industrial Ergonomics 32, 93–104 (2003)
42. Bernard, M.L., Chaparro, B.S., Russell, M.: Examining automatic text presentation for small screens. In: Proc. Human Factors and Ergonomics Society 45th Annual Meeting, pp. 637–639 (2001)

Pattern-Based Extraction of Addresses from Web Page Content

Saeid Asadi, Guowei Yang, Xiaofang Zhou, Yuan Shi,
Boxuan Zhai, and Wendy Wen-Rong Jiang

School of Information Technology & Electrical Engineering
The University of Queensland
GP78 South, St. Lucia, Brisbane, Australia
{asadi, guowei, zxf, yuanshi, bxzhai, wjiang}@itee.uq.edu.au

Abstract. Extraction of addresses and location names from Web pages is a challenging task for search engines. Traditional information extraction and natural processing models remain unsuccessful in the context of the Web because of the uncontrolled heterogenous nature of the Web resources as well as the effects of HTML and other markup tags. We describe a new pattern-based approach for extraction of addresses from Web pages. Both HTML and vision-based segmentations are used to increase the quality of address extraction. The proposed system uses several address patterns and a small table of geographic knowledge to hit addresses and then itemize them into smaller components. The experiments show that this model can extract and itemize different addresses effectively without large gazetteers or human supervision.

Keywords: Address Extraction, Web page Analysis, Address Itemization.

1 Introduction

Search engines are the main tools for searching the resources on the World Wide Web. Although general search engines are widely used to retrieve Web pages, they fail to handle many queries according to the users' needs and interests. Specialized search engines instead are dedicated to find either particular types of resources such as sounds, photos and movies or Web pages based on different criteria e.g. language or geographic location.

Geographic or location-based search has recently received attention from both academic and industrial groups. The popularity of the Web originally was because of providing access to the resources regardless of their physical locations. However, parallel to this global accessibility, many services and facilities have become available on the Web which refer much more to specific locations. People use search engines to find Web pages of local services and events around them or in a particular area. They search for restaurants, accommodation, governmental offices, clubs and social activities on the Web. While a significant proportion of the Web queries contain geospatial dimensions and refer to specific locations,

Y. Zhang et al. (Eds.): APWeb 2008, LNCS 4976, pp. 407–418, 2008.

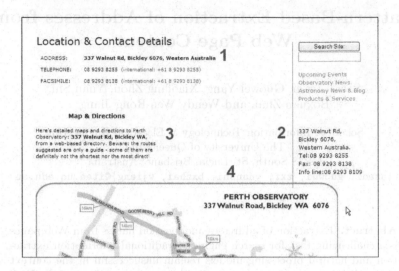

Fig. 1. Different ways of mentioning an address in a Web page

general search engines can not differentiate these queries and handle them properly [15,2]. Few commercial geographic search engines have been commercially developed among them Google Map [1] and Yahoo Local [2] are notable which are geographically limited and they often search commercial databases such as Yellow Pages instead of covering real Web pages. The problem basically originates from the ambiguous dynamic nature of location names, various addressing styles, lack of geographic information, and multiple locations related to a Web resource.

Geographic Web search has its own specifications which makes it different from general search. A brief of the location-based search engine architecture and tasks can be found in [7,11,1]. Basically, a geographic search engine must be able to find related addresses and location names and assign them to Web pages. Current address extraction techniques basically require large gazetteers which are expensive and unavailable for many countries. Also different markup styles e.g. HTML, XML and DOM increase the complexity of address extraction on the Web. As a result, natural language processing models are not able to extract all addresses and location names from Web page contents. Figure 1 shows different styles of mentioning an address in a typical Web page. The equivalent HTML code of these addresses are different and as a result the extracted addresses are not identical.

In this paper, we introduce a pattern-based model which uses HTML and visual segmentations to improve address extraction on Web pages. Our model detects an address and itemizes it to its components using patterns and a small set of triggers and location names. Our contribution to address extraction is trying to divide an address to its semantic components and also eliminate large scale gazetteers. This model is automatic and does not require expensive gazetteers

---

[1] http://maps.google.com/
[2] http://local.yahoo.com/

or much human effort. It also accumulates new location names to its knowledge repository.

In section 2, we review the previous work on information extraction and geo-tagging of Web pages.In sections 3 and 4, the general architecture and the details of the proposed model are described in depth respectively. The experimental setup and results are discussed in sections 5.

2   Background

Information Extraction (IE) systems extract the information about pre-specified types of events, entities or relationships from a text and this information is often added to a database for more applications. IE requires knowledge about proper names. In address extraction, this knowledge comes normally from gazetteers. Several works have focused on the use of supervised learning techniques such as Hidden Markov Models [10,17] and rule learning [18]. These techniques learn a language model or a set of rules from a set of hand-tagged trained documents; then apply the model or rules to new texts. Models learned in this manner are effective on documents similar to the set of training documents, but they are ineffective when applied to documents with a different genre or style. As a result, this approach has problem scaling to the Web due to the diversity of text styles and genres on the Web and the prohibitive cost of creating an equally diverse set of manually tagged documents. The experiments on extraction of location names with language patterns and without gazetteers such as [12] often end in bad results because of the ambiguities in geographic names.

The geography or location of Web resources has been measured as *geographical scope* or the location of Web pages that link to a page [8], *target location* or the location of people who visit a Web page [3], and *geographic focus* or a location which most of the content or addresses on a page refer to [1]. A review of different sources of geographic information in Web pages can be found on [11]. Geographic features e.g postcodes and telephone codes haven been used e.g. in [16,4] for geo-tagging. In [6], the visual structure of addresses, as they are shown to users, has been used to detect and extract address blocks from Web pages.

Gazetteer approach [14] and pattern-based models [2] often lead to weak results for extraction of geographic features on the Web. Web-a-Where [1] uses gazetteers to find and disambiguate location names on Web pages. Their results indicates that having all location names in the database does not insure a full coverage. KnowItAll [9] construct extraction rules based on the instants found in documents. Then the system decides whether a name belongs to a particular class or not using statistical rules. Similarly, Ourioupina [13] reports an algorithm for geographic knowledge acquisition in which a location name is added to different patterns and sent to search engines. Based on the frequency of the results, the best class is selected for the candidate. For example, the phrase *"city of Paris"* seems to be more repeated on the Web than *"cities in Paris"*. As a result, the system decides that Paris is a CITY not a COUNTRY.

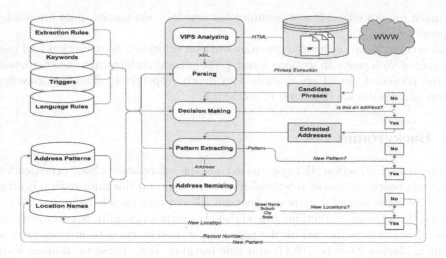

Fig. 2. An overview of the address extraction system described in this paper

3   The System Architecture

The proposed address extraction system consists of five components: HTML Pre-Processor, Parser, Knowledge Searcher, Decision Maker, and Knowledge Accumulator. A general picture of the proposed system is shown in Figure 2.

3.1   HTML Pre-processor

The content of Web pages are embedded in HTML tags. An effective information extraction model should analyze HTML properly. By parsing HTML alone, many pieces of geographic information which are embedded in separate lines (Fig. 1) or tabular structures can be lost. An information extractor parses a file line by line and in this case, it can not find the component of each address properly. To overcome this issue, we convert HTML files to XML in a way that the visual structure of Web pages will be considered. The system first transforms a HTML file into a XML file by employing the VIPS Demo software developed by Microsoft Research Group [3]. Vision-Based Page Segmentation (VIPS) [5,20] is a technique for automatic segmentation of Web page into smaller blocks according to the visual segmentations that a user can see on a Web page. The next step is in-depth analyzing and traversing the XML to obtain content information from every XML leaf node, and sorting them in a linear sequence together with their node numbers. At the same time, a node index is built where children nodes and their parents are connected together with bidirectional pointers which later can support trace of the parent node and then sister nodes to get a whole view of a complete address.

---

[3] For more details see: http://www.ews.uiuc.edu/ dengcai2/VIPS/VIPS.html

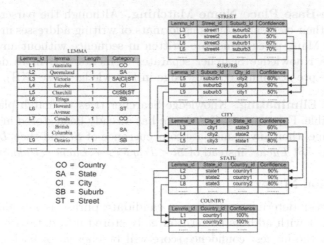

Fig. 3. A view of the Knowledge-Base (KB)

3.2   XML Parser

The XML Parser is responsible for reducing the search scope from a whole file to the areas of interest by defining syntactic filtration. It tries to find all candidate phrases (potential addresses) in a node. Besides, the parser also divides a potential address into its component inferred from content clues such as *comma*, *
* etc. Each word of an address element is coarsely classified into different categories (We have defined these categories: *NUM, Trigger, NOTICE, Cased Word, Preposition, Period Ended, quotes Ended*). Each segment obtained in this step, will be utilized as default searching unit of Database Searcher.

3.3   Knowledge Searcher

Knowledge Searcher itemizes elements of a potential address according to the accumulated knowledge and patterns Knowledge-Base (KB) (Figure 3). It finds all possibilities of a potential address and forms them into a list of possible patterns in three steps:

Standardizing Word Formats. Many words and names have different spells (e.g. *Tehran* and *Teheran*), abbreviations (e.g. *street* and *st.*) and name variations or synonyms (e.g. *Netherlands* and *Holland*). Knowledge Searcher first standardizes or converts different variations to their pre-selected formats. For example, wherever there is *QLD* in an address, the system considers it as Queensland. This is not always correct since variations or abbreviations of some words may also be part of other names. For example, *St* is a common abbreviation for street in addresses. However, it is also a common abbreviation for Saint in many people or place names. The Knowledge Searcher retains the original spelling as well as the standardized format, and both will be looked up in the Knowledge-Base step later.

Knowledge-Base Place Name Matching. Although the parser divides addresses into their elements, confusing formats of writing addresses in HTML files still remain. Many place names are written in sequence without any separation symbol. The Knowledge Searcher separates elements into more delicate level based on the pre-accumulated knowledge in Knowledge-Base (KB).

Ambiguity Eliminating. Knowledge Searcher tries to match place name as long as possible. For example, if *St Lucia* has already been stored in database as a unique name, then any immediate co-occurrence of *St* and then *Lucia* will be considered as *"St Lucia"* instead of *St <Trigger>+ Lucia <Unknown word>*.

3.4   Decision Maker

Decision Maker determines whether a candidate phrase is an address or not by matching it with address patterns already stored in a database. A matched pattern with the highest confidence score will be chosen for each candidate. If the the phrase exceeds a threshold, it will be confirmed as an address, and then it will be itemized to its components. The decision maker basically performs three tasks: 1) Delimitating ambiguities and conflicts of place names; 2) Itemizing each potential address to its elements; and 3) Adding the lost parts to address based on a location tree wherever it is possible. For example, the address *"No. 10, William Street, Toowong, Queensland"* will be modified as *"No. 10, William Street, Toowong, Brisbane, Queensland, Australia"* in it Brisbane and Australia has been added to the address according to a location tree.

3.5   Knowledge Accumulator

Knowledge Accumulator is the last component of the system. Knowledge accumulation exhibits in two aspects:

Location Accumulation. Place names are collected based on the addresses recognized in previous part. Knowledge Accumulator estimates whether an unknown word or phrase (few words) can be considered as a place name or not, and if yes, which type of place it should be classified to, and what confidence score should be assigned to it. Besides, it recomputes (increase/decrease) confidence scores of known place names.

Address Pattern Accumulation. The system gathers new patterns and increases their confidence score with next occurrences of them. This step is similar to [19] which automatically builds a gazetteer by using new locations extracted through patterns.

4   Details of Our Work

4.1   Ambiguity Elimination

There are often two types of ambiguities in addresses: syntactic and semantic. *Syntactic ambiguity* comes from the fact that one word may have different spellings or abbreviations e.g. *"St"* for *street*and *Saint*. The lemma *"St"* has at least two explanations. The second ambiguity, *semantic*, resides in different understandings of a known string. Take lemma *"Sydney"* for example. The famous Sydney is the capital city of Australia; but, it also refers to few other places in USA, Canada etc. Such geo/geo ambiguities happen more pervasively in suburb and street levels. Besides, geo/non-geo ambiguities are also common: *Sydney* is also a common female name in several countries.

Syntactic ambiguity can be avoided by combining a word with its previous and following words and search them in Knowledge-Base (KB). The method is carried out following the heuristic that the longer a word list can be matched in KB, the more chance to be a unique name. For example, having in phrase "... Shopping Center St Lucia..." if *St Lucia* is found in KB, then "St" will be interpreted as "Saint" instead of *"street"*. Semantic ambiguity is solved based on Pattern Base(PB) and Knowledge-Base(KB). All possible permutations of different identities of a known place names in a potential address can be obtained by searching KB. A confidence score is calculated for each permutation by matching them with patterns. The permutation with the highest confidence will be chosen.

The next step is finding all possible paths that connect identities of every two adjacent place names in a potential address. If a place name in a path cannot be connected to its neighbors with any of its known identities, it will be labelled as *Unknown Word*. Pattern-based heuristic is then applied to calculate confidence of each possible path. If there is a matched pattern, its confidence will be used to make a final decision by considering both pattern confidence and identity confidence.

Example: Assume that there are 9 lemmas in KB that cover all the levels of the address in Figure 3. Among them, 3 lemmas have multiple identities. *"Victoria"* is a street name in different cities, a state of Australia and also a city in Canada. *"Churchill"* can be a city, a suburb and also a street. *"Howard Avenue"* refers to few streets of different cities in Australia and other countries.

Following algorithm indicates how place names are detected in Phrases. Let W_i to be the *i*th word in a candidate phrase:

Place Name Detecting Algorithm.
input: PW - A candidate phrase
        W_i - the *i*th word in PW
        f - any syntactic format of W_i
        KB - Knowledge-Base
        C_i - Result Collection

```
1. PW(pre_word, Wᵢ) {
2.    if ((pre_word + f) = a place name found in KB)
3.       add (pre_word + f) to Cᵢ;
4.    if (pre_word + f) = part of a name in KB
5.       pre_word = pre_word + f;
6.       PW(pre_word, Wᵢ+1);//try next word in PW
7. }
```

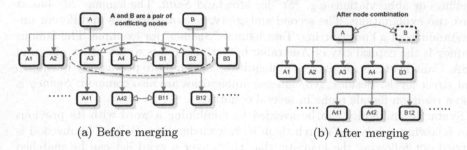

(a) Before merging   (b) After merging

Fig. 4. Merging Conflict Nodes: (a)Before merging (b)After merging

Based on PW(), the syntactic ambiguity elimination function is like the following:

Syntactic Ambiguity Elimination.
1. SyntacticAE(Potential) {
2.   current_word = first word in Potential
3.   C = NULL; //initialize C
4.   While current_word != EOF
5.   {
6.   C = SAE (C, current_word); //add longest result in C
7.   current_word = next new word in Potential;
8.   }
9. }

4.2   Conflict Elimination

Address Conflict refers to inconsistencies between accumulated knowledge in KB and extracted information from the Web. According to our experience, there are two causes of address conflict. One comes from word misspelling and synonymy. *"Sydny"* for example, is a misspelling of *Sydney*. The other conflict comes forth due to incompleteness of KB e.g. *"Newcastle"* is recorded as a city in Australia in KB while it is also a city in UK not covered in KB. Conflict elimination is performed via a self-learning process. Since it is error-prone for us to determine at a early stage which type of conflict it belongs to, and what to do with it, we delay making decision to later. Figure 4 illustrates how the system deals with conflicting nodes.

1. Keeping the Conflict: Conflicting pairs of *conflicting element* and *original element* in a potential address are recorded in a *conflicting table*. Then, a new branch of address for the conflicting element will be built in KB *conflicting element*. All the place names in the result address that belongs to original element will be issued with new identities and stored as new nodes in the new branch with relatively low confidence. If this branch is meaningful, it will appear next time and its confidence will increase accordingly. Otherwise, it will have the same low score.

2. Removing Meaningless Conflict Element: Regular work of conflict checking is carried out after a certain period. First, the confidence of each conflicting node is checked. If the confidence is still low, probability is high for the conflicting element to be a misspelling. Such element has little value to the system and it will be removed from KB. On the contrary, if its confidence has increased to a certain score, it will be taken as a correct place name and kept in KB.

3. Finding Synonymous Sub-tree: After removing all meaningless elements, conflicts still remain in KB between the highly confident pairs because of synonymy. In our model, the similarity between two different potential locations is computed by comparing their offsprings with preassumption that they share a similar parent. According to our experience, it is likely for different places to have sub-areas entitled with a similar names. The heuristics for determining relationship is that the more offsprings they share, the more possible they represent the same place.

4. Merging Synonymous Sub-Tree: After Resemblance is confirmed, the next step is merging two nodes together. The node with less confidence(LN) is merged into the other one(HL) and their relationship is recorded in KB as synonymy. Then, all of LN offsprings are recursively merged with HL ones. During Knowledge accumulation process, there may be some itemized place names with no direct parent due to the incomplete information. For example, in *"Fifth Avenue, Brisbane, QLD"*, the suburb is missing. We store the *"Fifth Ave"* as a new street record, whose parent identity is void. These non-parent places are referred as *dangling places* in KB.

A simple way to eliminate a dangling place is setting it as a sub-region of any place that immediately appears as its direct parent. However, such action may induce mistakes in KB considering misunderstandings by misspellings, homonymy, incorrect itemizations etc. Such mistakes can also be safely banished in KB with dangling place and its potential parent with a low confidence score at first. During the later extraction work, when the two place names appear together again, their confidence score will increase. if the score exceeds a threshold, the affiliation between these two place will be confirmed.

5   Experiments

5.1   Dataset and Experimental Setup

For testing our model, we selected 1100 Web pages related to Australia. The Web pages were manually searched and addresses were extracted. Only addresses with at least street information and a recognizable location name were selected. After discarding undesirable candidates, 2030 address were chosen for the experiments.

After manually control of the address, 10 most repeated patterns of addressing style in Australia were selected and the address patterns were built up with different confidence scores. We added triggers, keywords and a small set of coarse location names (country names, Australia states and major cities) with their abbreviations to our database. The database remained quite small as we basically intended to have a pattern-based approach for address extraction rather than a gazetteer approach.

5.2   Results and Discussion

The accuracy of our system has been tested in different ways. Figure 5(a) shows the recall of the proposed address extraction model. The results indicate that using address patterns alone, our system can extract 65 percent of the addresses. Adding a small table of geographic information, the recall increases to 73 percent. The results show that the pattern-based approach alone can not extract all addresses in Web pages. However, figure 6(a) shows that the precision of address extraction increases from 73 percent to 97 percent if we add a very general table of locations to the system. Figure 5(b) compares the *F-measure* of address extraction with or without using geographic knowledge. The second approach still remains more successful and this indicates that the pattern approach needs geographic knowledge to extract addresses from Web pages properly.

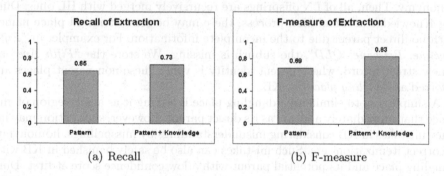

(a) Recall   (b) F-measure

Fig. 5. Address extraction using patterns alone and using patterns and locations

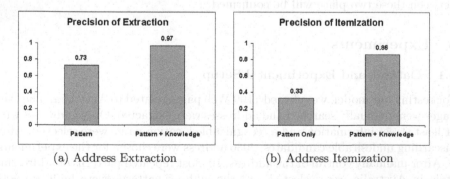

(a) Address Extraction   (b) Address Itemization

Fig. 6. Precision of the system tasks: (a)Address Extraction (b)Address Itemization

For evaluating our address itemizing model, we have measured the precision of itemized addresses. Figure 6 compares the precision of address extraction and address itemization. 6(b) shows that a pure pattern-based model can not handle the itemization task properly and the precision remains poor. However, a small amount of geographic knowledge improves the itemization of addresses dramatically.

From the experimental results it could be concluded that a pure pattern-based address extraction model can not extract and itemize addresses from Web pages properly. This is because of large variation in address patterns on the Web. Finding all address patterns is almost impossible. However, as previously mentioned, adding a small table of general or coarse location names e.g. country and state names, and also using some triggers and keywords, the system provides better results both in extraction and in itemization of addresses. This combined model is cheaper and more flexible than gazetteer-based extraction approach.

6   Conclusion

In this paper, we discussed the importance of finding the addresses mentioned in Web pages as a basic source of geographic information in location-based search engines. We introduced a pattern-based address extraction approach which uses only a small table of geographic names, triggers and keywords. The results show that although a pure pattern approach can not handle all addresses properly, it can extract and itemize a reasonable proportion of addresses when a small gazetteer is used.

References

1. Amitay, E., Har'El, N., Sivan, R., Soffer, A.: Web-a-where: geotagging web content. In: SIGIR, pp. 273–280 (2004)
2. Zhou, X., Asadi, S., Chang, C.-Y., Diederich, J.: Searching the World Wide Web for Local Services and Facilities: A Review on the Patterns of Location-Based Queries. In: Fan, W., Wu, Z., Yang, J. (eds.) WAIM 2005. LNCS, vol. 3739, pp. 91–101. Springer, Heidelberg (2005)
3. Zhou, X., Asadi, S., Diederich, J., Shi, Y., Xu, J.: Calculation of Target Locations for Web Resources. In: Aberer, K., Peng, Z., Rundensteiner, E.A., Zhang, Y., Li, X. (eds.) WISE 2006. LNCS, vol. 4255, pp. 277–288. Springer, Heidelberg (2006)
4. Borkar, V.R., Deshmukh, K., Sarawagi, S.: Automatic segmentation of text into structured records. In: SIGMOD Conference, pp. 175–186 (2001)
5. Cai, D., Yu, S., Wen, J.-R., Ma, W.-Y.: Block-based web search. In: SIGIR, pp. 456–463 (2004)
6. Can, L., Qian, Z., Xiaofeng, M., Wenyin, L.: Postal address detection fromweb documents. In: WIRI '05: Proceedings of the International Workshop on Challenges in Web Information Retrieval and Integration, Washington, DC, USA, 2005, pp. 40–45. IEEE Computer Society Press, Los Alamitos (2005)
7. Chen, Y.-Y., Suel, T., Markowetz, A.: Efficient query processing in geographic web search engines. In: SIGMOD Conference, pp. 277–288 (2006)

8. Ding, J., Gravano, L., Shivakumar, N.: Computing geographical scopes of web resources. In: VLDB, pp. 545–556 (2000)
9. Etzioni, O., Cafarella, M.J., Downey, D., Kok, S., Popescu, A.-M., Shaked, T., Soderland, S., Weld, D.S., Yates, A.: Web-scale information extraction in knowitall (preliminary results). In: WWW, pp. 100–110 (2004)
10. Freitag, D., McCallum, A.K.: Information extraction with hmms and shrinkage. In: AAAI-99 Workshop on Machine Learning for Informatino Extraction (1999)
11. Markowetz, A., Chen, Y.-Y., Suel, T., Long, X., Seeger, B.: Design and implementation of a geographic search engine. In: WebDB, pp. 19–24 (2005)
12. Mikheev, A., Moens, M., Grover, C.: Named entity recognition without gazetteers. In: EACL, pp. 1–8 (1999)
13. Ourioupina, O.: Extracting geographical knowledge from the internet. In: International Workshop on Active Mining, ACDM-AM (2002)
14. Pouliquen, B., Steinberger, R., Ignat, C., Groeve, T.D.: Geographical information recognition and visualization in texts written in various languages. In: Handschuh, H., Hasan, M.A. (eds.) SAC 2004. LNCS, vol. 3357, pp. 1051–1058. Springer, Heidelberg (2004)
15. Sanderson, M., Kohler, J.: Analyzing geographic queries. In: SIGIR Workshop on Geographic Information Retrieval, GIR 2004 (2004)
16. Silva, M.J., Martins, B., Chaves, M., Cardoso, N.: Adding geographic scopes to web resources. In: SIGIR Workshop on Geographic Information Retrieval, GIR 2004 (2004)
17. Skounakis, M., Craven, M., Ray, S.: Hierarchical hidden markov models for information extraction. In: IJCAI, pp. 427–433 (2003)
18. Soderland, S.: Learning information extraction rules for semi-structured and free text. Machine Learning 34(1-3), 233–272 (1999)
19. Uryupina, O.: Semi-supervised learning of geographical gazetteers from the internet. In: HLT-NAACL Workshop on Analysis of Geographic References, pp. 18–25 (2003)
20. Yu, S., Cai, D., Wen, J.-R., Ma, W.-Y.: Improving pseudo-relevance feedback in web information retrieval using web page segmentation. In: WWW, pp. 11–18 (2003)

An Effective Method Supporting Data Extraction and Schema Recognition on Deep Web[*]

Wei Liu, Derong Shen, and Tiezheng Nie

Department. of Computer, Northeastern University, Shenyang, 110004,China
281439157@QQ.com, Shendr@mail.neu.edu.cn

Abstract. With the rapid development of Internet, data sources on deep web store a large number of high-quality structured data, which demands the development of structured data extraction method. But the existing methods focus on data rather than structure, and some of them are difficult to maintain. To resolve these problems, a complete and effective method supporting data extraction and schema recognition is proposed in this paper. To extract data, a novel algorithm based on clustering is adopted, which is also effective when faced complex data and excessive noise. And a simple extraction rule model is defined to resolve the problem of maintenance. In addition, it does deep mining on result schema recognition. At last, experiments show satisfactory results.

1 Introduction

According to the survey[1] released by UIUC in 2004, there are more than 300,000 data sources on deep web available at that time. And CompletePlanet[2] has collected more than 7,000 data sources and classified them into 42 topics. So we can conclude that deep web provides people a great opportunity to get their desired information.

With proliferation of data sources on deep web, the importance of data integration is growing. These data sources contain a large number of high-quality structured data, which is still growing rapidly. However, this kind of data source must be accessed through its query interface, and finally returns a response page containing some result data to the user. So how to extract the result data automatically and accurately is widely discussed.

The existing data extraction approaches on deep web focus on data rather than structure, which don't care result schema. And many methods are not able to process complex data. In addition, the deep Web data sources tend to change, such as the changes of page structure and result schema, which will affect the accuracy of original extraction methods. Clearly, these issues bring a large number of difficulties to data extraction and new challenges to the data integration researchers.

To resolve these problems, a complete and effective method supporting data extraction and schema recognition is proposed in this paper. To extract data, a novel algorithm based on clustering is adopted, which is also effective when faced complex data and noise. And a simple extraction rule model is defined to resolve the problem of maintenance. In addition, it does deep mining on result schema recognition.

---

[*] This research is supported by the National Natural Science Foundation of China under Grant No. 60673139, 60573090.

Y. Zhang et al. (Eds.): APWeb 2008, LNCS 4976, pp. 419–431, 2008.

The paper is organized as follows. Section 2 presents the related work. In section 3 a clustering-based data extraction method is introduced. Section 4 presents an extraction rule generation strategy. In section 5 how to recognize result schema is discussed. The experimental methods and results are reported in Section 6. Section 7 summarizes our contributions and concludes the paper.

2   Related Work

Early approaches to extract data on deep web are mostly semi-automatic or manual, which needs experts to analyze structure and features of the web pages. And the necessary schema information for data extraction is provided by experts too, such as SG – Wrap[3] and DEByE[4].

In recent years, more and more automatic methods have been proposed, which can automatically generate wrapper and extract data without human involvement. Several typical methods are MDR[5], MDRII[6], RoadRunner[7][8], EXALG[9], DeLa[10] and ViDRE[11].

MDR and MDRII are methods based on Tag Tree structure features, which mine structured data by finding similar sub-tree. When the result data structure is simple, it can get very good results. But if the structure is complex, or there is excessive noise in data region, it can not handle well. In another words, it do not support data with nested structure.

ViDRE is an extraction method based on visual characteristics, which simulates how a user understands web layout structure based on his visual perception to some degree and ultimately achieves the purpose of identifying structured data. This approach is very novel, but needs an effective visualization model[12][13]. Therefore, the efficiency is lower than extraction by analyzing HTML documents directly. In addition, extraction fully based on visual information is not reliable. When pages have no apparent visual characteristics, the extraction will become very difficult.

RoadRunner is a wrapper induction system, which generates wrapper by comparing the similarities and mismatches between the sample pages. RoadRunner can identify optional attribute and nested attribute which may repeat many times in a record, but the algorithm has an exponential time complexity. EXALG is similar to RoadRunner, and has resolved the issue of time cost. But both of them try to generate templates for response pages, the structure of which may change when the web site updates. The original template is no longer valid. Therefore, they can not solve the problem of maintenance very good.

DeLa takes token suffix-tree data structure to deal with data extraction problem, and analyzes the semantic of data extracted by assigning a label to a column. But its semantic analysis of the data is not thorough enough, and it can not obtain real schema information.

The contribution of this paper is to propose a complete data extraction solution, including: (1) To extract data, a novel algorithm based on clustering is proposed, which is also effective when faced complex data and excessive noise. (2) And a simple extraction rule model is defined to resolve the problem of maintenance. (3) In addition, it does deep mining on result schema recognition.

3 Data Extraction

Analyzing both the structure and visual information of response pages, a clustering-based data extraction method is proposed, which can be divided into five phases: building HTML DOM tree model, identifying data region node, identifying data record, alignment and attribute separation.

The data extraction flow in dashed rectangle is shown in **Fig.1**, and the details are as follows.

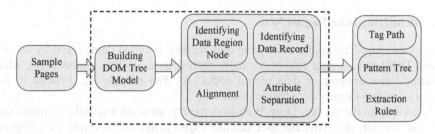

Fig. 1. Data Extraction Flow

Phase 1: According to the sample page document, we firstly build Html DOM Tree Model, which describe the structure and visual characteristics of tags in the document. The whole data extraction process is based on this model.

Phase 2: We use a clustering-based algorithm to identify data region node, which represents the location of records in the HTML DOM model.

Phase 3: To identify data record in the data region we use a cluster combining method, which will find the largest repetition of sub-node under the data region node.

Phase 4: We align all of the nodes in a cluster to find a representative tree structure of the data record.

Phase 5: We analyze the text in the representative tree to separate attributes and labels from the noise.

3.1 HTML DOM Tree Model

In a HTML document, most of tags appear in pair (start tag and end tag), which makes HTML tags representing a nested structure, and we can use a tree structure to describe the nested structure. According to the W3C DOM standard, we construct a Html DOM tree, as shown in **Fig.2**.

Nodes in DOM tree are divided into two kinds: tag node and text node. And the tag and text between the start tag and end tag of a tag node construct their child nodes. In this model, each node contains additional information that is optional attributes and values of the tag, such as CSS, fonts, colors, etc. So the model contains both structure information and visual information, on which the system is based.

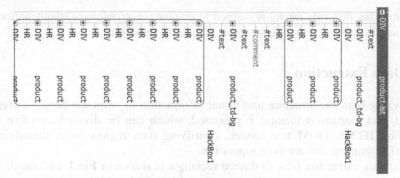

Fig. 2. A HTML DOM Tree

3.2 Identifying Data Region Node

In response page, the region containing data records is called data region by us. And in the DOM tree, a data region normally represents as a data region node, which contains large amounts of data record nodes. As data records are difficult to find directly, we first identify data region node.

According to the survey, Deep Web sites mostly focus on a specific domain, such as book, car, air, etc. After the user submits a query, the web site returns a response page containing several data records, which describe entities of the same type. Therefore, both the external representation and internal code of the page show repetitions.

(a)   (b)

Fig. 3. A response page (a) and its DOM Tree (b)

For instance, www.amazon.com is an e-commerce site that mainly provides electronics, books, music, DVDs, etc. **Fig.3(a)** is a response page from www.amazon. com and the region in the rectangle is the data region. **Fig.3(b)** is the DOM tree of its data region. Obviously, both the external representation and the DOM tree of the three books are similar with each other.

To judge whether a node is a data region node, we need to know whether there are repetitions under the node in the DOM model. We must note that, the repetition may contain one node or several nodes, and repetitions are not always adjacent with each other for there may be noise nodes between them.

Despite the variety of forms, but we can find, most nodes in a data region are similar to their brothers nodes. In **Fig.4**, node Φ and ② form a data record that is a repetition. Node ③ and ④ are noise nodes. No matter how many noise nodes are contained in the data region. As long as there are multiple data records, node like Φ or ② will appears multiple times.

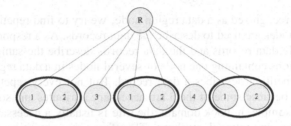

Fig. 4. The DOM Tree of a data region

We use a clustering-based algorithm to identify data region nodes.

Algorithm CBI (N, T)	Algorithm Cluster (N, T)
1 **if** N.depth >= 3 **then**	1 CS = ;
2 Cluster (N, T);	2 **for** each child N.children **do**
3 **for** each child N.children **do**	3 **if** CS = **then**
4 CBI (child, T);	4 CS.newCluster (child);
	5 **else**
	6 **for** each cluster CS **do**
	7 **if** Distance (cluster, child) is minimus **then**
	8 **if** Distance (cluster, child) < T **then**
	9 cluster.insert (child);
	10 **else** CS.newCluster (child);

Fig. 5. Clustering-Based Algorithm

The clustering-based algorithm is shown in **Fig.5**. For any node N, if the tree rooted by N meets the height requirement, we will cluster the child nodes of N, and the structure of the nodes belong to the same cluster are similar with each other. Then we repeat the process to its child nodes.

In the clustering process, we use edit distance between trees rooted by the nodes to compute similarity. The shorter the distance, the higher the similarity. As shown in Fig. 5, N denotes the node to identify, T denotes the distance threshold, CS denotes the cluster set generated after clustering. The clustering process is as follows. For each child node of N, if CS is empty, we create a new cluster, which only contains one node. Otherwise, we find the cluster that has a shortest distance to child. If the distance is smaller than T, we insert the child into the cluster. Otherwise, we also create a new cluster.

Finally, we define the only node in a cluster as isolated node. If the proportion of isolated nodes in a cluster set is smaller than a threshold, we consider the node as a data region node.

3.3 Identifying Data Record

When a node is recognized as a data region node, we try to find repetitions under it to identify which nodes are used to describe the data records. As a response page usually contains multiple data records and the data records describe the same type of entity, there are repetitions containing one node or several nodes in a data region.

Usually, a repetition describes a data record. But not every repetition covers the same attributes, because some of them are necessary attributes and some are optional attributes. For example, in book domain, the title is usually a necessary attribute, and discount may be an optional attribute. Therefore, the repetitions are similar in structure with each other to some degree. To completely describe the record structure, we try to find the largest repetition which contains as much attributes as possible.

Before combining clusters, we define the following concepts:

The distance from node A to node B: the number of node A minus the number of node B.

The distance from node A to cluster C: the number of node A minus the number of node B, which is the nearest node in A's left belonging to C.

The distance from cluster A to cluster B: the average distance from nodes in A to B.

If the distance from every node in the cluster A to cluster B is near the average distance, that the standard deviation of distances is less than a threshold T, we say that A is parallel to B.

Our cluster combining algorithm firstly finds the largest subset of CS, in which clusters are parallel to each other. Then it combines the clusters in the subset according their position.

At first, every node under the data region is replaced by the cluster where it belongs. Supposing that data region node N contains child nodes 123456789 and the numbers denote their positions, we obtain CS={A={1,5,9}B={2,6}C={3,7}D={4,8}} after clustering. So the sequence 123456789 transforms to ABCDABCDA.

Then we discover that B is parallel to A, as well as C and D. So they are able to be combined. Because the distances from BCD to A are 1, 2 and 3, the largest repetition is ABCD. ABCD probably describes a data record.

3.4 Alignment

After we combine the clusters and find the largest repetition, we need to transform repetition from the clustering model into a DOM Tree model, which is called Pattern Tree by us. But a cluster contains several nodes, we should select one of them to ensure that the tree rooted by any other node in the cluster is a sub-tree of it.

As shown in **Fig.6**, there are three trees. The nodes contained in them are ACEF, ABCF and ACDE. First of all, we create a Pattern Sub-Tree only including the root node P. Then we insert nodes into it according their position in the cluster, and record the number of occurrences of each node. Finally, we combine the Pattern Sub-Tree and obtain a complete Pattern Tree.

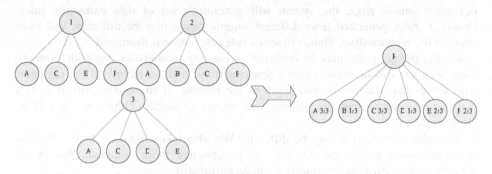

Fig. 6. An example of Alignment

3.5 Attribute Separation

During Attribute Separation phase, we analyze the text in the representative tree to separate attributes and labels from the noise. In response pages that contain multiple records, each record has the same attribute name, but its attribute value may be different. Therefore, if we compare the text in DOM tree of each record, and find out the same text and different text, it should be easy to distinguish the attribute name from the attribute value.

4 Extraction Rule

After analyzing a HTML document, we define a model that contains a set of extraction rules for it to improve extraction efficiency.

4.1 Extraction Rule Model

In our system, an extraction rule model is made up of Tag Path and Pattern Tree. Formally, extraction rule R can be represented as <P, T>. Tag Path P is used to locate the data region node and Pattern Tree T is used to identify data records and attributes.

P can be represented as the regular expression: (/tagname ([tagnum])?)+, where tagname denotes the name of the tag and tagnum denotes the number of tag under its parent tag. The semantic of this expression is defined as usual, + being an iterator and (a)? being a shortcut for (a | ε) which denotes an optional pattern. For instance, in the tag path "/html/body/table/tbody/tr[1]/td[2]", because "body" is the only one body child of tag "html", tagnum is not needed. In addition, "tr[1]" is the first tr child of tag "tbody", and" td[2]" is the second td child of tag "tr[1]".

After alignment and attribute separation we obtain a Pattern Tree T that indicate a DOM Tree structure of a complete record, in the meantime it also contains the statistical information of each attribute.

4.2 Extraction Rule Integration

For each sample page, the system will generate a set of data extraction rules. However, rules generated from different sample pages may be different, and even some of the rules conflict. Thus, extraction rule integration is discussed.

Firstly, the tag paths may be different. According to statistics, tag path from the same web site can be divided into a few situations, and we record all the tag paths with some statistical information, such as the number of successful extraction. If a path success most times, its priority will be set to higher. Otherwise, it will be eliminated.

Secondly, pattern tree may be different. We also take alignment to resolve this problem, which makes the coverage of pattern tree more complete. But if the frequency of an attribute is too low, it will be eliminated.

5 Result Schema Recognition

We have obtained the extraction rules of each data source. But the data extracted has only structure but no semantic. Therefore, we also need to process the data further to understand the meaning of attributes. We assign labels to data to obtain result schema. Firstly, we construct an instance lib for each data source, and then carry on Synchronized Label Assignment to assign labels to each data column. Synchronized Label Assignment consists of Local Label Assignment and Global Label Assignment.

5.1 Instance Lib

An instance lib for each data source constructed is shown in **Table 1**, a row denotes a tuple, which is a data record, and a column denotes an attribute that has no exact semantic information.

Table 1. An Instance Lib

A	B	C	D	E	F
Head First Java, 2nd Edition	Kathy Sierra and Bert Bates	Paperback	Feb 9, 2005	$29.67	$19.94
Java How to Program (7th Edition) (How to Program)	Harvey & Paul Deitel & Associates	Paperback	Dec 27, 2006	$76.07	$58.89
Introduction to Java Programming-Comprehensive Version (6th Edition)	Y Daniel Liang	Paperback	Jul 12, 2006	$85.34	$70.74

5.2 Local Label Assignment

Local Label Assignment is carried on in local data source. And we employ the following five heuristics.

Heuristic 1: match form element labels in query interface to data attributes. In query interface, a query condition is usually consisted of attribute name and keyword. If the query is successful, the keyword will re-present in the results information, by which we could match keywords in instance lib to form element labels in query interface.

Advanced Search Books

Fill in **at least** one field. Fill in more to narrow your search. Or you can browse our most popular titles, subject by subject.

Author: [   ]

○ Exact Name ◉ Last, First Name (or Initials) ○ Start of Last Name

Title: [Java   ]

◉ Title Word(s) ○ Start(s) of Title Word(s)

Subject: [   ]

◉ Subject Word(s) ○ Start(s) of Subject Word(s)

Publisher: [   ]

ISBN: [   ]

Search up to six ISBNs simultaneously, using " | " between each.
Refine Your Search (optional):

Category: [All Subjects   ▾]

Format: [All Formats   ▾]

Reader age: [All Ages   ▾]

Language: [All Languages ▾]

Publication Date: [Before ▾]

[the year   ▾]

[2009]

Sort Results by: [Bestselling   ▾]

(Search now)

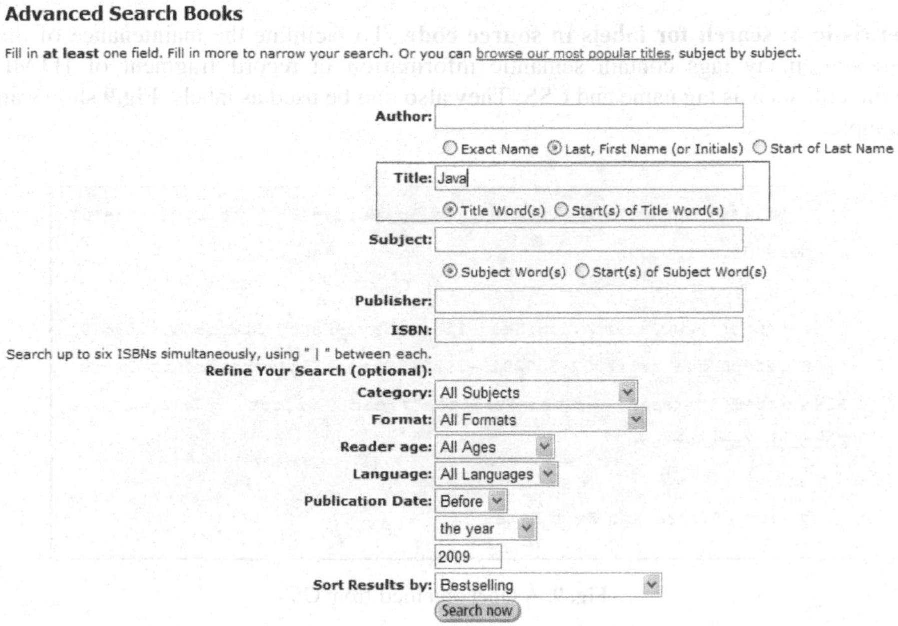

Fig. 7. A query interface

For example, in the query interface in **Fig.7**, a query "title = Java" is submitted. After data extraction, we obtain three records as shown in Table 1. Obviously, all the values in column A contain the keyword "Java". Therefore, we can match "title" to Column A. In other words, we assign a label "title" to column A.

Heuristic 2: search for labels in table headers. Some web sites, such as train and air, usually arrange the results in a table, which has semantic headers, as shown in Figure 7. Every header can be assigned to the column in instance lib as a label. In HTML document, tags such as <TH> and <THEAD> are usually used to design table headers.

Heuristic 3: search for labels in result data. In response pages, sometimes attribute values are encoded together with attribute names. As shown in **Fig.8**, "Buy new" and

"Used & new" are attribute names, which can be used as labels. We try to find the prefix and suffix shared by most of the records to obtain a label.

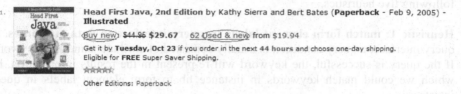

Fig. 8. An example of prefix

Heuristic 4: search for labels in source code. To facilitate the maintenance of the web site, many tags contain semantic information in record fragment of HTML document, such as tag name and CSS. They also can be used as labels. **Fig.9** shows an example.

```
<td    class="dataColumn"><table    cellpadding="0"    cellspacing="0"
border="0"><tr><td>
<a
href="http://www.amazon.com/Head-First-Java-Kathy-Sierra/dp/059600
9208/ref=pd_bbs_sr_1/002-3502018-1923223?ie=UTF8&s=books&qid=11929
32447&sr=1-1"><span    class="srTitle">Head    First    Java,    2nd
Edition</span></a>

  by Kathy Sierra and Bert Bates
```

Fig. 9. A label obtained from CSS

Heuristic 5: some data has conventional format. Some data has conventional format. For example, a date is usually organized as "dd-mm-yy" and "dd/mm/yy". An email usually contain @.

5.3 Global Label Assignment

In a domain, every data source shares not only a similar schema but also similar data record. Thus, we extend the label assignment to the global environment. It is Global Label Assignment that searches for labels in other data sources, as shown in **Fig.10**.

Supposing that there is a record "data mining, Han Jiawei" in the instance lib of source A. We try to send a query with keyword "data mining" to source B. If source B also contains book records on data mining, we will be able to employ the heuristics above. Finally labels from source B is assigned to the attribute of source A.

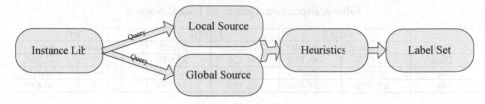

Fig. 10. Label Assignment Flow

As there are many the label sources, we assign a Label Set to each column instead of only one label.

5.4 Schema Dictionary

In order to obtain the exact schema information, a schema dictionary is defined by experts for each domain before schema recognition, as shown in Table 2. A schema dictionary defines the mapping relationship between each attribute in global schema and labels. As long as we compare the label set of each column in instance lib with the labels in schema dictionary, we could obtain the schema information.

Table 2. A schema dictionary

Attribute	Label Set
Title	title, srTitle ,book ,bookname
Author	by, author ,writer
Edition	edition, binding
Price	price, saleprice, $

6 Experiments

In this paper, we use a large data set to evaluate our approach. Our data set is collected from the web site www.completeplanet.com. Completeplanet has collected more than 70,000 web databases and search engines, which is currently the largest directory for deep web. During our selection, duplicates under different topics are not used to maximize the diversity among the selected web sites. For each web site selected, we submit at least ten different queries to get response pages. And only response pages containing at least two data records are used.

Firstly, we experiment on our data extraction method. The selected web sites can be divided into 3 domains. In each domain we extract data in three approaches, ViDRE, MDRII and our approach DE. Then we compare the results with the correct results provided by experts and compute the average precision and recall of each approach, as shown in **Table 3**.

Secondly, we experiment on our schema recognition approach and compare it with DeLa. There are two phases in our approach, Local Label Assignment(LLA) and Global Label Assignment(GLA). We compute precision and recall in each phase, and the result are shown in **Table 4**.

Table 3. Experimental results on Data Extraction

Domain	ViDRE		MDRII		DE	
	Precision	Recall	Precision	Recall	Precision	Recall
Book	92.43%	83.74%	94.62%	85.93%	97.34%	93.78%
Car	90.62%	79.27%	92.34%	80.12%	95.78%	92.41%
Job	87.39%	58.47%	88.45%	76.56%	92.14%	90.47%

Table 4. Experimental results on Schema Recognition

Domain	DeLa		LLA		GLA	
	Precision	Recall	Precision	Recall	Precision	Recall
Book	84.23%	64.63%	86.25%	66.37%	91.29%	79.72%
Car	90.32%	78.45%	92.62%	84.78%	98.76%	91.35%
Job	87.47%	73.62%	89.83%	76.83%	96.34%	86.38%

Obviously, our method is much better in every domain.

7 Conclusion

In this paper, we propose a full solution for Data Extraction and Schema Recognition on deep web. It adopts a clustering-based algorithm, which is effective to extract complex data from response pages containing large number of noise. In addition, a simple extraction rule model is defined to resolve the problem of maintenance. Finally, a novel Synchronized Label Assignment is proposed, which is very helpful for schema recognition. All of these make the data extracted from deep web really available. The experimental result has shown that our proposed approach achieves very satisfactory performance. And we plan to do deeper research on result schema for our future works.

References

1. Chang, C.-C.K., He, B., Li, C., Patel, M., Zhang, Z.: Structured Databases on the Web: Observations and Implications. In: SIGMOD Conference, pp. 61–70 (2004)
2. http://www.completeplanet.com/
3. Meng, X., Lu, H., Wang, H., Gu, M.: SG-WRAP: a schema-guided wrapper generator. In: Proceedings of the 18th International Conference on Data Engineering, pp. 331–332 (2002)
4. Laender, A.H.F., Berthier, A.R., Altigran, S.: DEByE - data extraction by example. Data Knowl. Eng. 121–154 (2002)
5. Liu, B., Grossman, R.L., Zhai, Y.: Mining data records in Web pages. In: Proceedings of the Ninth ACM SIGKDD International Conference on Knowledge Discovery and Data Mining, Washington, pp. 601–606 (2003)
6. Zhai, Y., Liu, B.: Web Data Extraction Based on Partial Tree Alignment. In: WWW, pp. 10–14 (2005)
7. Crescenzi, V., Mecca, G., Merialdo, P.: RoadRunner: towards automatic data extraction from large Web sites. In: Proceedings of the 27th International Conference on Very Large Data Bases, pp. 109–118 (2001)

8. Crescenzi, V., Mecca, G., Merialdo, P.: RoadRunner: automatic data extraction from data-intensive web sites. In: Proceedings of the 21th ACM SIGMOD International Conference on Management of Data, Madison, p. 624 (2002)
9. Arasu, A., Garcia-Molina, H.: Extracting Structured Data from Web Pages. In: SIGMOD Conference, pp. 337–348 (2003)
10. Wang, J., Lochovsky, H.F.: Data Extraction and Label Assignment for Web Databases. In: WWW, pp. 20–24 (2003)
11. Liu, W., Meng, X., Meng, W.: Vision-based Web Data Records Extraction. In: Proc. of the 9th SIGMOD International Workshop on Web and Databases, pp. 20–25. Illinois, Chicago (2006)
12. Cai, D., Yu, S., Wen, J., Ma, W.: Extracting Content Structure for Web Pages Based on Visual Representation. In: Zhou, X., Zhang, Y., Orlowska, M.E. (eds.) APWeb 2003. LNCS, vol. 2642, pp. 406–417. Springer, Heidelberg (2003)
13. Cai, D., He, X., Wen, J., Ma, W.: Block-level link analysis. In: SIGIR, pp. 440–447 (2004)

The Experiments with the Linear Combination Data Fusion Method in Information Retrieval

Shengli Wu[1], Yaxin Bi[1], Xiaoqin Zeng[2], and Lixin Han[2]

[1] School of Computing and Mathematics
University of Ulster, Northern Ireland, UK
{s.wu1, y.bi}@ulster.ac.uk
[2] Department of Computer Science
Hohai University, Nanjing, China
xzeng@hhu.edu.cn, {lixinhan2002}@yahoo.com.cn

Abstract. In data fusion, the linear combination method is a very flexible method since different weights can be assigned to different systems. However, it remains an open question that which weighting schema is good. In many cases, a simple weighting schema was used: for a system, its weight is assigned as its average performance over a group of training queries. In this paper, we empirically investigate the weighting issue. We find that, a series of power functions of average performance, which can be implemented as efficiently as the simple weighting schema, is more effective than the simple weighting schema for data fusion. We also investigate combined weights which concern both performance of component results and dissimilarity among component results. Further performance improvement on data fusion is achievable by using the combined weights.

1 Introduction

In the last couple of years, data fusion has been investigated by quite a few researchers in information retrieval. Several data fusion methods such as CombSum [3], CombMNZ [3], the linear combination method [2,7,8], the probabilistic fusion method [4], Borda fusion [1], Condorcet fusion [5], and the correlation method [10,11], have been proposed, and extensive experiments using TREC data have been reported to evaluate these methods. Experimental results show that, in general, data fusion is an effective technique for effectiveness improvement.

The linear combination data fusion method is a very flexible method since different weights can be assigned to different systems. However, it is unclear that which weighting schema is good. In some previous researches, different search methods such as golden section search [7,8] and conjugate gradient [2] were used to search suitable weights for component systems. One major drawback of using these methods is their very low efficiency. Because of this, data fusion with only two or three component systems were investigated in [2] and [7,8]. In some situations such as the WWW and digital libraries, documents are updated frequently, then each component system's performance may change considerably from time to time. The weights for the systems should be updated accordingly.

Y. Zhang et al. (Eds.): APWeb 2008, LNCS 4976, pp. 432–437, 2008.

In such a situation, it is very difficult or impossible to use those low efficient weighting methods.

In some data fusion experiments, (e.g., in [1,6,9,11]), a simple weighting schema was used: for a system, its weight is set as its average performance over a group of training queries. There is a straightforward relationship between performance and weight. This method can be used in very dynamic situations since weights can be calculated and modified very easily. However, it has not been investigated how good this schema is. We would like to investigate this issue further. We shall demonstrate that, a power function weighting schema, with a power of 2 or 3, is more effective than the simple weighting schema (which is a special case of power function, power = 1) for data fusion, though both of them can be implemented in the same way. In addition, we shall demonstrate that combined weights which concern both performances of component results and dissimilarities among component results can further improve the effectiveness of the fused results.

2   Performance Weights

Suppose we have n information retrieval systems ir_1, ir_2,..., ir_n, and for a given query q, each of them provides a result r_i. w_i is the (performance) weight assigned to system ir_i. Then for any document d in one or all results, the linear combination method uses the equation $M(d, q) = \sum_{i=1}^{n} w_i * s_i(d, q)$ to calculate its score. Here $s_i(d, q)$ is the normalized score of document d in result r_i, $M(d, q)$ is the calculated score of d for q. All the documents can be ranked using their calculated scores.

In this section we only consider performance weights. For each system ir_i, suppose its average performance over a group of training queries is p_i, then p_i is set as ir_i's weight (w_i) in the simple weighting schema, which has been used in previous research (e.g., in [1,6,9,11]). The purpose of our experiment is to try to find some other schemas which are more effective than the simple weighting schema but can be implemented as efficiently as the simple weighting schema. In order to achieve this, we set w_i as a power function of p_i. Besides p_i, we used $p_i^{0.5}$, $p_i^{1.5}$, p_i^{2}, $p_i^{2.5}$ and p_i^{3} as ir_i's weights. If we use a larger power for the weighting schema, then those systems with better performance have a larger impact on the fused results, and those results with poorer performance have a smaller impact on the fused results.

4 groups of TREC data were used for the experiment [1]. Their information is summarized in Table 1. From Table 1 we can see that these 4 groups of submitted results (called runs in TREC) are different in many ways. Therefore, they comprise a good combination for us to evaluate data fusion methods.

The Zero-one linear normalization method was used for score normalization. For a resultant list, the top-scored document was given a score of 1, the

---

[1] Some submitted results include fewer documents. For convenience, those results were not selected.

Table 1. Information summary of four groups of results submitted to TREC

Group	Track	No. of results	No. of queries	No. of retrieved documents
2001	Web	32	50	1000
2003	Robust	62	100	1000
2004	Robust	77	249	1000
2005	Terabyte	41	50	10000

Table 2. Performance (in MAP) of several data fusion methods (In LC(a), the number a denotes the power value used for weight calculation; for every data fusion method, the improvement rate of its MAP value over the average performance of all component results is shown)

Group/ Ave.	Comb- Sum	Comb- MNZ	LC(0.5)	LC(1.0)	LC(1.5)	LC(2.0)	LC(2.5)	LC(3.0)
2001	0.2614	0.2581	0.2620	0.2637	0.2651	0.2664	0.2673	0.2681
0.1861	+40.46%	+38.69%	+40.79%	+41.70%	+42.54%	+43.63%	+43.63%	+44.06%
2003	0.2796	0.2748	0.2841	0.2865	0.2879	0.2890	0.2900	0.2908
0.2256	+23.94%	+21.81%	+25.93%	+26.99%	+27.62%	+28.10%	+28.55%	+28.90%
2004	0.3465	0.3434	0.3482	0.3499	0.3512	0.3522	0.3530	0.3537
0.2824	+22.70%	+21.60%	+23.30%	+23.90%	+24.36%	+24.72%	+25.00%	+25.25%
2005	0.3789	0.3640	0.3857	0.3897	0.3928	0.3952	0.3970	0.3986
0.2991	+26.68%	+21.70%	+28.95%	+30.29%	+31.33%	+32.13%	+32.53%	+33.27%

bottom-scored document was given a score of 0, and any other document was given a score of between 0 and 1 accordingly.

For all the systems involved, we evaluated their average performance in Mean Average Precision (MAP) over a group of queries. Then 0.5, 1.0, 1.5, 2.0, 2.5, and 3.0 were used as the power values to calculate weights for the linear combination method. In a year group, we randomly chose m (m=3, 4, 5, 6, 7, 8, 9, or 10) component results. Then data fusion methods were performed using those chosen results. For a particular setting of m component results in a year group, 200 runs were carried out. Mean Average Precision (MAP) was used to evaluate the fused retrieval results. Besides the linear combination method with different weighting schemas, CombSum and CombMNZ were also involved in the experiment.

Tables 2 shows the performance of the fused result. Each data point is the average of 8*200*q_num measured values. Here 8 is the different number (3, 4,..., 9, 10) of component results used, 200 is the number of runs for each setting, and q_num is the number of queries in each year group. The improvement rate over the average performance of all component results is shown as well.

Comparing CombMNZ with CombSum, CombMNZ is not as good as Comb-Sum in all 4 year groups. With any of the weighting schemas chosen, the linear combination method performs better than the best component result, Comb-Sum, and CombMNZ in all 4 year groups. Comparing with all different weighting schemas used, we can find that the larger the power is used for weighting calculation, the better the linear combination method performs.

Table 3. The improvement rate of the fused result over the best component result Performance (in MAP)

Group/ Best	Comb- Sum	Comb- MNZ	LC(0.5)	LC(1.0)	LC(1.5)	LC(2.0)	LC(2.5)	LC(3.0)
2001	+10.44%	+9.04%	+10.69%	+11.41%	+12.00%	+12.55%	+12.93%	+13.27%
2003	-0.71%	-2.41%	+0.89%	+1.74%	+2.24%	+2.63%	+2.98%	+3.27%
2004	+4.40%	+3.46%	+4.91%	+5.42%	+5.82%	+6.12%	+6.36%	+6.57%
2005	-0.89%	-4.79%	+0.89%	+1.94%	+2.75%	+3.37%	+3.85%	+4.26%

Table 4. Performance (MAP) of several data fusion methods with combined weights, the improvement rate of the fused result over the best component result performance is shown in parentheses. (In LC(a:b), a and b denote the values of the two parameters in Equation 2 in Section 3 for weight calculation.)

Group	LC(3.0:0)	LC(3.0:0.5)	LC(3.0:1)	LC(3.0:1.5)	LC(3.0:2)
2001	0.2681	0.2686	0.2688	0.2687	0.2682
	(+13.27%)	(+13.48%)	(+13.56%)	(+13.52%)	+(13.31%)
2003	0.2908	0.2912	0.2928	0.2935	0.2938
	(+3.27%)	(+3.41%)	(+3.98%)	(+4.23%)	(+4.33%)
2004	0.3537	0.3538	0.3535	0.3528	0.3519
	(+6.57%)	(+6.60%)	(+6.51%)	(+6.30%)	(+6.03%)
2005	0.3986	0.3994	0.4002	0.4009	0.4016
	(+4.26%)	(+4.47%)	(+4.68%)	(+4.87%)	(+5.05%)

Two-tailed tests were carried out to compare the differences between all the data fusion methods involved. The tests show that the differences between any pair of the data fusion methods are statistically significant at a level of .000 ($p < 0.001$, or the probability is over 99.9%). From the worst to the best, the data fusion methods are ranked as follows: CombMNZ, CombSum, LC(0.5), LC(1.0), LC(1.5), LC(2.0), LC(2.5), LC(3.0).

We also compare the fused result with the best component result. The improvement rate of the fused results over the best component results is shown in Table 3. This improvement rate is a hard measure. Both CombMNZ and CombSum do not perform as well as the best component result in some year groups; while they perform better than the best component result in some other year groups. With any of the weighting schemas chosen, the linear combination method performs better than the best component result, CombSum, and CombMNZ in all 4 year groups. Consider the hardness of this measure, the fused result is very good.

In the above experiment, we observed that a power of 3 obtained the best fusion result. Some more experiments are desirable to find out how does a power of bigger than 3 performs.

3 Combined Weights

Now we consider combined weights which combine performance weights and dissimilarity weights. There have been a few different ways of calculating dissimilarity weights. In [10,11], the overlap rate and Spearman rank coefficient of two results were used. In this paper, we use another way to do it. First, we normalize the scores in both component results r_1 and r_2 for a given query q using the Zero-one normalization method. Then we calculate the Euclidean distance $D(r_i, r_j, q)$.

Suppose that we have n component results r_i (i=1, 2, ..., n), then component result r_i' average distance from all other component results can be calculated as

$$dis_{i,q} = \frac{1}{n-1} \sum_{j=1 \wedge j \neq i}^{n} D(r_i, r_j, q) \tag{1}$$

dis_i can be obtained by averaging $dis_{i,q}$ over all the queries. For each system ir_i, its combined weight is defined as

$$w_i = (p_i)^a * (dis_i)^b \tag{2}$$

Here a and b are positive numbers, p_i is the average performance of system ir_i over all the queries.

The experiment was carried out using the same data as in Section 2. Combined weights, calculated using Equation 2, were assigned to each component result. We varied b (0.5, 1, 1.5, and 2) while fixing a to be 3. According to our investigation in Section 2, 3 is a good value for that. Experimental results are shown in Table 4.

Compared with the situation that only performance weights are used, combined weights can bring small but significant improvement (about 0.5%). It suggests that the setting $n = 0.5$ is not as good as the other settings (n=1, 1.5, and 2). However, the three settings (n=1, 1.5, and 2) are very close in performance for data fusion.

Finally, let us compare LC(1.0)(the simple weighting schema) with LC(3.0:1.5) (combined weight, with a performance power of 3 and a dissimilarity power of 1.5). LC(3.0:1.5) performs better than L(1.0) in all 4 year groups by 1% to 3.5%. Although the improvement looks small, it is statistically significant ($p < 0.001$) in all 4 year groups. Since these two schemas require the same relevance judgment information, we consider LC(3.0:1.5) is certainly a better choice than LC(1.0) for data fusion.

4 Conclusions

In this paper we have presented our preliminary work about assigning appropriate weights for the linear combination data fusion method. From the experiments conducted with the TREC data, we observe that for performance weighting, a series of power functions (e.g., a power of 2, 2.5, or 3) are better than the simple weighting schema, in which the performance weight of a system is assigned as its average performance (power equals to 1). The power function schema can be implemented as efficiently as the simple weighting schema. We have also demonstrated that the combined weights, which comprise performance weights and dissimilarity weights, can be used for data fusion to further improve effectiveness. We believe that the investigation outcome in this paper is very useful for practical application of the data fusion technique. As our next stage work, we will investigate the linear combination method with more weighting options.

Acknowledgements. Xiaoqin Zeng and Lixin Han's work was partially supported by the NSFC, China, under Grants 60571048 and 60673186.

References

1. Aslam, J.A., Montague, M.: Models for metasearch. In: Proceedings of the 24th Annual International ACM SIGIR Conference, New Orleans, Louisiana, USA, September 2001, pp. 276–284 (2001)
2. Bartell, B.T., Cottrell, G.W., Belew, R.K.: Automatic combination of multiple ranked retrieval systems. In: Proceedings of ACM SIGIR 1994, Dublin, Ireland, July 1994, pp. 173–184 (1994)
3. Fox, E.A., Koushik, M.P., Shaw, J., Modlin, R., Rao, D.: Combining evidence from multiple searches. In: The First Text REtrieval Conference (TREC-1), Gaitherburg, MD, USA, March 1993, pp. 319–328 (1993)
4. Lillis, D., Toolan, F., Collier, R., Dunnion, J.: Probfuse: a probabilistic approach to data fusion. In: Proceedings of the 29th Annual International ACM SIGIR Conference, Seattle, Washington, USA, August 2006, pp. 139–146 (2006)
5. Montague, M., Aslam, J.A.: Condorcet fusion for improved retrieval. In: Proceedings of ACM CIKM Conference, McLean, VA, USA, November 2002, pp. 538–548 (2002)
6. Thompson, P.: Description of the PRC CEO algorithms for TREC. In: The First Text REtrieval Conference (TREC-1), Gaitherburg, MD, USA, March 1993, pp. 337–342 (1993)
7. Vogt, C.C., Cottrell, G.W.: Predicting the performance of linearly combined IR systems. In: Proceedings of the 21st Annual ACM SIGIR Conference, Melbourne, Australia, August 1998, pp. 190–196 (1998)
8. Vogt, C.C., Cottrell, G.W.: Fusion via a linear combination of scores. Information Retrieval 1(3), 151–173 (1999)
9. Wu, S., Crestani, F.: Data fusion with estimated weights. In: Proceedings of the 2002 ACM CIKM International Conference on Information and Knowledge Management, McLean, VA, USA, November 2002, pp. 648–651 (2002)
10. Wu, S., McClean, S.: Data fusion with correlation weights. In: Proceedings of the 27th European Conference on Information Retrieval, Santiago de Composite, Spain, March 2005, pp. 275–286 (2005)
11. Wu, S., McClean, S.: Improving high accuracy retrieval by eliminating the uneven correlation effect in data fusion. Journal of American Society for Information Science and Technology 57(14), 1962–1973 (2006)

Squeezing Long Sequence Data for Efficient Similarity Search*

Guojie Song[1], Bin Cui[2], Baihua Zheng[3], Kunqing Xie[1], and Dongqing Yang[2]

[1] Key Laboratory of Machine Perception (Peking University),
Ministry of Education, Beijing, China
gjsong@pku.edu.cn, kunqing@cis.pku.edu.cn
[2] School of Electronic Engineering and Computer Science, Peking University, Beijing, China
bin.cui@pku.edu.cn, dqyang@pku.edu.cn
[3] School of Information System, Singapore Management University, Singapore
bhzheng@smu.edu.sg

Abstract. Similarity search over long sequence dataset becomes increasingly popular in many emerging applications. In this paper, a novel index structure, namely *Sequence Embedding Multiset tree(SEM-tree)*, has been proposed to speed up the searching process over long sequences. The SEM-tree is a multi-level structure where each level represents the sequence data with different compression level of multiset, and the length of multiset increases towards the leaf level which contains original sequences. The multisets, obtained using sequence embedding algorithms, have the desirable property that they do not need to keep the character order in the sequence, i.e. shorter representation, but can reserve the majority of distance information of sequences. Each level of the tree serves to prune the search space more efficiently as the multisets utilize the predicability to finish the searching process beforehand and reduce the computational cost greatly. A set of comprehensive experiments are conducted to evaluate the performance of the SEM-tree, and the experimental results show that the proposed method is much more efficient than existing representative methods.

1 Introduction

Recently, similarity search over long sequences attracts more attentions due to some new emerging applications. For example, in computational biology, searching for specific sequences over DNA and protein sequences appears as a fundamental operation for problems such as assembling the DNA chain from the pieces obtained by the experiments, looking for given features in DNA chains and determining how different two genetic sequences were. In such applications, the problem of sequence similarity search is typically based on block edit distance with move[3], which is another kind of distance metric except for character edit distance, and is widely used in computational biology and text processing environments. Due to the great length of sequences, similarity search becomes a challenging problem.

* Supported by the National Natural Science Foundation of China under Grant No. 60703066 and No. 60473051 and supported by the National High-Tech Research and Development Plan of China (863) under Grant No.2006AA12Z217.

Y. Zhang et al. (Eds.): APWeb 2008, LNCS 4976, pp. 438–449, 2008.
© Springer-Verlag Berlin Heidelberg 2008

Many index structures have been proposed to handle similar sequence search problems, such as suffix trees and vector space indexing. Suffix tree based index structures [8] have been gaining favor as the methods for sequence search, but they consume too much memory space. Classical indexing techniques can be used for sequence search, such as X-trees [5]. They typically perform well in low to medium dimensional spaces (up to $20-30$ dimensions), but their performance deteriorates drastically for long sequence match. Differently, embedding based index techniques, such as FastMap [2], have also been proposed to decrease computational cost. However, they have to scan the original sequence dataset in order to construct embedded space. Since the distance computation in the original string space is very expensive, the construction of the embedded space is impractical. Furthermore, the approximate factor keeps increasing as the data size expands, rendering this approach unworkable for large sequence databases.

In this paper, we focus on how similarity query over long sequences dataset can be executed efficiently by using sequence embedding techniques. To reduce the computational cost for long sequences' comparison, we adopted sequence embedding algorithm to convert the sequence to a multiset which has compressed representation. The distance among original sequences can be achieved by computation on multisets with high approximate factor but in linear time. Sequence oriented dimensionality reduction method with predicability, named **SDR**, has been developed based on sequence embedding. As far as we know, this is the first method using embedded sequence for sequence search. A novel index structure, SEM-tree, is proposed by using SDR techniques developed to efficiently shrink the search space and hence speed up the search process. A comprehensive simulation has been conducted to evaluate the performance of SEM-tree. As experimental results indicate, search over proposed index structure shows superior performance over other approaches in terms of efficiency.

The rest of this paper is organized as follows. Section 2 reviews the related work. section 3 presents sequence oriented dimension reduction technique based on multiset, followed by the structure of SEM-tree and detailed algorithms. Experimental results are presented in Section 4. Finally, we conclude this paper in Section 5.

2   Related Work

Many heuristic-based search methods have been developed to conduct sequence search, mainly including two categories: hash-table based methods[12] and suffix-tree-based methods[8]. The former handle short queries well, but become very inefficient, in terms of both time and space, for long queries. The latter manage mismatches inefficiently, and are notorious for their excessive memory usage.

Vector space based indexing is another kind of method, such as vp-tree*[13], which partitions the data space into spherical cuts by selecting random reference points from the data. In the recently proposed reference-based sequence indexing methods [9], reference sequences are selected from dataset by choosing the ones that are far away from each other. All these methods are based on the original sequence space and are still inefficient for long sequence data.

Embedding based index techniques, such as FastMap [2] and MetricMap [10], have also been proposed to decrease the computational cost. However, both approaches ask for a distance-preserving mapping function. Finding a suitable mapping is a tough and time-consuming process, and no such function is available for block edit distance. Additionally, in order to construct embedded space, it has to scan the original sequence ,making it impractical in our case.

Recently proposed indexing techniques, such as M-trees [7], Slim-tree[6] and DBM-tree[1], can be used to support sequence search. The M-tree is a height-balanced tree, where the data elements are stored in leaf nodes. The Slim-Tree is an evolution of the M-Tree, embodying the first published technique to reduce the amount of overlap between tree nodes, which leads to a smaller number of disk accesses to answer similarity queries. These two structures are height balanced and attempt to reduce the height of the tree at the expense of flexibility in reducing overlap between nodes. This constraint was released in the DBM-Tree by reducing the overlap between nodes in high density regions, resulting in an unbalanced tree.

3   The SEM-Tree

3.1   SDR: Sequence Dimensionality Reduction

We use an example to illustrate the embedding process proposed in [3], as shown in Figure 1. The original sequence s, also denoted as $ET_1(s)$ (the sequence before the first embedding iteration), contains $cabagehcadbba$ and it is partitioned into six blocks after first iteration, with each block having 2 or 3 elements. Each block $ET_1(s)[c_s, c_e]$ (c_s and c_e are the start and end character in this block respectively)is thereafter represented by an element based on a hashing function $h(ET(s)_1[*, *])$ and all these new elements form $ET_2(s)$ (the sequence before the second iteration). For example, the first block of $ET_1(s)[ca]$ is hashed to an element k. h is a one-to-one Karp-Miller hash function on sequences of length at most 3, therefore the same sequence must correspond to the same sub-tree in $ET(s)$. The partition and hashing continues until at one level the sequence only contains one element. The time complexity of the whole process is $O(|s| \log^* |s|)$.

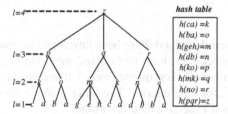

Fig. 1. $ET(s)$ ($s = cabagehcadbba$)

Given an $ET(s)$, all the elements within the entire tree can be represented by a family of multisets, denoted as $T(s)$. $ET_i(s)$ is used to represent the sequence at the i-th level, and the set of the elements within $ET_i(s)$ is denoted by the sub-multiset $T_i(s)$. Via Example 1, it is easy to notice that $ET_i(s)$ and $T_i(s)$ contain the same set of elements, while the order is important in the former but irrelevant in the latter.

The Analysis of Sequence Dimensionality Reduction. The work of [3] only focuses on how to embed sequences into multiset space using a sequence embedding technique, while it does not mine characteristics of (sub-)multisets. Following the traditional dimension reduction ideas, such as PCA, the multisets generated by embedding technique motivate us to exploit the sub-multisets for sequence dimensionality reduction.

Suppose we have two sequences, s_1 and s_2, in the sequence dataset. Let $T(s_1) = \bigcup_{i=1}^{l_1} T_i(s_1)$[1] and $T(s_2) = \bigcup_{j=1}^{l_2} T_j(s_2)$ denote the transformed multisets, where l_1 and l_2 are the heights of trees $ET(s_1)$ and $ET(s_2)$ respectively. Its original distance, $\hat{d}(s_1, s_2)$, is a *edit distance with moves*[3], which can only be approximated by the sequence embedding method.

Since the order of the elements does not affect the multiset, the distance between two multisets is only determined by the number of the different elements σ in the multisets, where they share a common character set $\Sigma = \{a_1, a_2, ..., a_\sigma\}$.

The sub-multisets $T^i(s)$ can be defined as the union of all the sub-multisets in the first i levels of $T(s)$, denoted as $T^j(s) = \bigcup_{k=1}^{j} T_k(s)$, thus the formal definition of distance between two sub-multisets, $T^i(s_1)$ and $T^i(s_2)$, is provided by Equation 1.

$$d_e(T^i(s_1), T^i(s_2)) = card(T^i(s_1) \ominus T^i(s_2)) \tag{1}$$

To facilitate the distance comparison between two sequences with different length, normalized standard distance between two multisets is defined in Equation 2.

$$\hat{d}_e(T^i(s_1), T^i(s_2)) = \frac{d_e(T^i(s_1), T^i(s_2))}{card(T^i(s_1) \cup T^i(s_2))} \tag{2}$$

Lemma 1. *Given two sequences s_1 and s_2, $\forall i, j$ with $i \leq j \leq max(l_1, l_2)$, we have:*

$$\hat{d}_e(T_i(s_1), T_i(s_2)) \leq \hat{d}_e(T_j(s_1), T_j(s_2))$$

Thus, we have contractness property of SDR as following.

Lemma 2. *Contractness: Given any two sequences s_1 and s_2, for any $i \leq j$ we have:*

$$\hat{d}_e(T^i(s_1), T^i(s_2)) \leq \hat{d}_e(T^j(s_1), T^j(s_2)).$$

Lemma 3. *Predictability: For any two sequences s_1 and s_2, upper bound of distance $\hat{d}_e(T^i(s_1), T^i(s_2))$, denoted as $B_u(\hat{d}_e(T^i(s_1), T^i(s_2)))(>\hat{d}_e(T(s_1), T(s_2)))$, can be estimated based on sub-multiset $T^i(s_1)$ and $T^i(s_2))$ as follows:*

$$B_u(\hat{d}_e(T^i(s_1), T^i(s_2))) = \frac{card(T^i(s_1) \ominus T^i(s_2)) + v}{card(T^i(s_1) \cup T^i(s_2)) + v}$$

where $v = card(T_i(s_1) \cup T_i(s_2))$

---

[1] For operations ($\ominus$, $card()$, $\cap$ and $\cup$) on multiset, please refer related references.

3.2   The Index Structure

The basic idea of SEM-tree is based on the characteristic of distance preservation and contractibility. Each node stands for one cluster represented by a sub-multiset which serves as the center and the radius, and the multisets whose sequences located in this cluster form the children nodes. The tree structure based on multisets not only keeps the original clustering characteristics of sequences, but also eases the construction of index and query processing.

Given a sequence dataset SD, we apply sequence embedding process on SD to transform the sequences into a new space, named SEM-Space. Consider a sequence $s \in SD$, we define $\mathbf{SEM}(s, l)$ to be an operator that projects sequence s on its first l-level sub-mulitiset in $ET(s)$ $(1 \le l \le L)$, denoted as $\mathbf{SEM}(s, l) = \bigcup_{i=1}^{l} T_i(s)$ $(i.e.\ T^l(s))$, where l is called an embedding level and L is the maximum level.

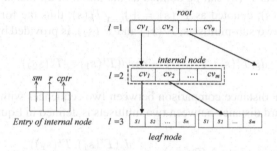

Fig. 2. The Structure of SEM-tree

The SEM-tree is a multi-tier tree and tries to partition the space into clusters and refine the clustering process as tree grows. Figure 2 shows an example. However, the indexing keys at each level of the tree are different—nodes closer to the root use the keys with lower embedding level, i.e. shorter sub-mulitiset, and the keys at the leaves are the original sequence data. At the root node of SEM-tree, only the sub-multisets from the first embedding level of the sequences contribute to the partition. Take the sequence $cabagehcadbba$ in Section 3.1 as an example, we may use T_1 in level 1, $T_1 \cup T_2$ (i.e. T^2) in level 2, and full sequences in leaf level. Note that, we can select different levels of sub-multiset for each level during tree construction. However, the lengths of the sub-multisets are in non-descending order from root to leaf.

The entry in the internal nodes of SEM-tree is a vector, denoted by cluster vector cv, which corresponds to a cluster at the SEM-Space. All the vectors within an internal node form a larger cluster, which refers to an entry in the parent level. We name the vectors within a certain node *vector set* (VS). Each cv is a 3-tuple vector $(sm, r, cptr)$, as shown in Figure 2. For a cv at i-th level, the sub-multiset sm represents the center of the cluster, and r is the radius of this cluster such that any sequence s in this cluster satisfies $\hat{d}_e(sm, T^m(s)) \le r$, where m is the embedding level of multisets of the node. Pointer $cptr$ points to the child node formed by all the immediate child vectors of cv. Different from the internal node, the leaf node simply stores the original sequences within a certain cluster.

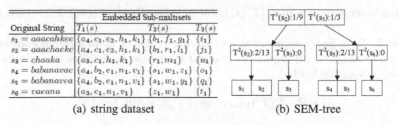

	Embedded Sub-multisets		
Original String	$T_1(s)$	$T_2(s)$	$T_3(s)$
$s_1 = aaacahkee$	$\{a_4, c_1, e_2, h_1, k_1\}$	$\{b_1, f_1, g_1\}$	$\{t_1\}$
$s_2 = aaachaeke$	$\{a_4, c_1, e_2, h_1, k_1\}$	$\{b_1, r_1, l_1\}$	$\{j_1\}$
$s_3 = chaaka$	$\{a_3, c_1, h_1, k_1\}$	$\{r_1, m_1\}$	$\{u_1\}$
$s_4 = babanavae$	$\{a_4, b_2, e_1, n_1, v_1\}$	$\{s_1, w_1, z_1\}$	$\{o_1\}$
$s_5 = babanavea$	$\{a_4, b_2, e_1, n_1, v_1\}$	$\{s_1, w_1, y_1\}$	$\{q_1\}$
$s_6 = vaeana$	$\{a_3, e_1, n_1, v_1\}$	$\{z_1, w_1\}$	$\{t_1\}$

(a) string dataset   (b) SEM-tree

Fig. 3. Construction of SEM-tree

Due to the space constraint, here we only run an example to illustrate the construction algorithm. Suppose each of these sequences has already been partitioned into three sub-multisets, denoted by $T_1(s)$, $T_2(s)$, and $T_3(s)$ in figure 3. In the first level, we got two clusters by using the distance computation based on the first level sub-multiset $T^1(.)$. One cluster take $T_1(s_2)$ as the center and $1/9$ as the radius. The other take $T_1(s_5)$ as cluster centered with radius of $1/3$. Similarly, at the second level, we follow the same principle for each sub-cluster based on the sub-multiset $T^2(.)$. In the leaf level, we index the full sequences if we assume each node can store two sequences.

Note that SEM-tree is a dynamic index structure, which can facilitate the data update based on the properties of multiset. For example, when a new sequence is inserted, we simply transform the original sequence to SEM-Space, and insert it into the appropriate sub-cluster from root.

3.3 Query Processing on SEM-tree

SEM-tree cannot guarantee absolute accuracy, because the search is conducted according to the distance between corresponding multisets rather than real sequences in the internal levels of the tree. In this section, we develop a double-bound theory to make sure that the searching process can cover all candidate sequences and complete query results can be returned.

According to [3], for sequences s_1 and s_2, the real distance can be bounded by the distance on embedding space as follows:

$$\alpha * \hat{d}_e(T(s_1), T(s_2)) \leq \hat{d}(s_1, s_2) \leq \beta * \hat{d}_e(T(s_1), T(s_2)) \qquad (3)$$

where $\alpha = \frac{m}{8nlognlog^*n}$, $\beta = \frac{2m}{n}$, $n = max(|s_1|, |s_2|)$ and $m = max(card(T(s_1)), card(T(s_2)))$.

Based on inequation 3, the real distance between any two sequences can be limited by double bounds $[\alpha * \hat{d}_e(T(s_1), T(s_2)), \beta * \hat{d}_e(T(s_1), T(s_2))]$.

Based on the predictability of multiset, such double-boundary can be applied on the transformed sub-multiset for each internal node of SEM-tree.

Lemma 4. *For sequences s_1 and s_2, the real distance $\hat{d}(s_1, s_2)$ can be approximated by clustering vector at i-th level based on the double boundaries.*

$$\hat{d}(s_1, s_2) \geq \alpha * B_l\left(\hat{d}_e(T^i(s_1), T^i(s_2))\right)$$
$$\hat{d}(s_1, s_2) \leq \beta * B_u\left(\hat{d}_e(T^i(s_1), T^i(s_2))\right)$$

Here, the lower boundary of $\hat{d}_e(T^i(s_1), T^i(s_2))$, i.e., $B_l\left(\hat{d}_e(T^i(s_1), T^i(s_2))\right.$, is itself.

Lemma 5. *For a query sequence q, a clustering vector cv in an internal node at level i and query range γ, all sequences covered by cv are included in the results if the following inequation holds:*

$$\beta_d * B_u(\hat{d}_e(T^i(q), cv.sm)) + \beta_r * rad_u(cv_i) \leq \gamma$$

where β_d is the β value for $\hat{d}_e(T^i(q), cv.sm)$, and so is the β_r for $rad_u(cv_i)$.

This lemma means that if the summation of upper bound distance $\beta_d * B_u(\hat{d}_e(T^i(q), cv.sm))$ and upper bound distance $\beta_r * rad_u(cv_i)$ is less than specified range γ, then all sequences covered by cv must be included in the final results. $B_u(\hat{d}_e(T^i(q), cv.sm))$ has been proved in lemma 4, and we have lemma 6 for $rad_u(cv_i)$.

Lemma 6. *For a clustering vector cv at level i in SEM-tree, if $T^i(s)$ is the center of cv, and r is the radius, then the upper bound of radius cv in multiset can be estimated as:*

$$rad_u(cv_i) = \frac{|T^i(s)| * r + (2 - r) * |T_i(s)|}{|T^i(s)| + (2 - r) * |T_i(s)|}$$

Thus, query processing can be accelerated with the following filtering strategy: At a particular internal node at level i, all irrelevant clustering vectors should be filtered out. First, for each cv, the double bounds of the real distance between $cv.sm$ and query sequence q can be obtained directly based on Lemma 4. Then, a filtering step will prune unqualified cv according to the following criteria.

- If $\beta_d * B_u(\hat{d}_e(T^i(cv.sm), T^i(q))) + \beta_r * rad_u(cv_i) \leq \gamma$, no sequences contained by cluster cv need to be further checked, because they belong to the final results with the guarantee of upper-bound, and only need to be refined from the leaf nodes.
- If $\alpha_d * B_l(\hat{d}_e(T^i(cv.sm), T^i(q))) - \beta_r * rad_u(cv_i) > \gamma$, cv will not be further checked. The sub-tree rooted in cv will be filtered out, because it has no chance of being included in the result set with the guarantee of lower-bound.
- After filtering out above cvs, remaining cv set will be the results for further check at the next level of the tree.

A range query $Q_{Range} = <q, \gamma>$ retrieves all sequences s in the sequence datasets that satisfy the range condition $\hat{d}(q, s) \leq \gamma$, where γ is the query range specified by the user and $\hat{d}(*)$ is the distance measurement in original sequence space. To improve the efficiency of search, (sub-)multisets have been adopted as index keys in internal nodes of the SEM-tree. The algorithm is provided in Algorithm 1.

Firstly, the query sequence q is embedded into a multiset $T(q)$ and a search starting from the root of SEM-tree based on $T(q)$ is thereafter conducted. For each clustering vector in an internal node, those irrelevant clustering vectors will be first filtered out based on proposed filtering principe (in line 3). Otherwise, for each qualified clustering vector cv, it will be determined based on the proposed multiset predictability whether all sequences covered by the sub-tree rooted by cv belong to the final results(in line 4). If so, all these sequences will be appended to the final result RS and this subtree will

---

Algorithm 1. RangeQuery(q,γ)

Input: q: query sequence, γ: searching radius specified by user;
Output: RS: all the sequences that have a distance to q shorter than γ;
Procedure:
1: $RS = \emptyset$;
2: **RangeSearchOnTree**($root, T(q), RS, 0$);
Proc RangeSearchOnTree(Node N_d, multiset $T(q)$, SET RS, level l)
1:   **if** T is not a leaf node **then**
2:     **for** each $cv \in N_d$ **do**
3:       **if** $\alpha_d * \hat{d}_e(T^l(cv.sm), T^l(q)) - \beta_r * rad_u(cv_i) \leq \gamma$ **then**
4:         **if** $\beta_d * B_u(\hat{d}_e(T^l(q), cv.sm)) + \beta_r * rad_u(cv_i) \leq \gamma$ **then**
5:           /*Node filtering based on predictability proposed;*/
6:           $RS = RS\cup \{all\ sequences\ in\ sub\text{-}tree\ rooted\ by\ cv\}$;
7:         **else**
8:           *RangeSearchOnTree*($cv.cptr, T(q), RS, l + 1$);
9:         **end if**
10:      **end if**
11:    **end for**
12: **else**
13:    **for** each $tp \in N_d$ **do**
14:      **if** $\hat{d}(q, tp.seq) \leq \gamma$ **then**
15:        $RS = RS \cup \{tp.seq\}$;/*filtering based on real distance*/
16:      **end if**
17:    **end for**
18: **end if**

---

not be further checked. Otherwise, the left clustering vectors need to be further checked by calling search process **RangeSearchOnTree**(*) recursively(in line 8). At the leaf node, each candidate sequence needs to be examined based on real distance function in original sequence space to determine whether they belong to final result set RS.

4 Experiments

We compared our method with several other existing global sequence indexing methods: 1) the M-Tree[7], 2) the DBM-Tree[6] and 3) the Slim-Tree[1]. All the implementations are running on a PC with Intel P4 CPU 1.5 GHz and 1G MB memory. All the indexes are stored on disk, and the disk page size is set to 4KB. The results are the average performance of 200 range queries.

Dataset simulator SUMATRA [11] is employed to generate the synthetic sequence dataset. It is a popular data simulator in mobile environments, which produces the moving trajectory sequences with various length and scale.

4.1 Scalability in Sequence Length (*sl*)

The resilience to the increase of sequence length is one of the most critical features for an index structure targeting long sequence dataset. The goal of this experiment is to

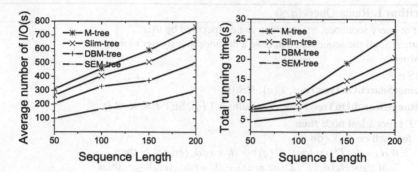

Fig. 4. Average number of I/O(s) and total running time(s) for SEM-tree, M-tree, Slim-tree and DBM-tree for queries with varying sequence length from 50 to 200. The default values are (γ=0.01, s=10000 and τ = 15).

compare the performance of our method with existing methods for increasing lengths of sequences. Figure 4 presents the results with sl(*sequence length*) varying from 50 to 200 under SUMATRA dataset. Query range γ is 0.1.

As sl increases, each sequence has more characters, which results in higher tree and more expensive distance computation. However, the SEM-tree is based on the multiset, which is a compact representation of the real sequence. A multiset only records the distinct characters of sequence and corresponding occurrence times, not the orders. Therefore, an increased sl does not deteriorate the performance of the SEM-tree obviously as depicted in figure 4.

4.2 Scalability in Database Size (s)

We fix the length of sequences at 100 and query range γ at 0.01, and vary the dataset size from 5,000 to 20,000 sequences. The performance of range queries on SUMATRA datasets for SEM-tree, M-tree, DBM-tree and Slim-tree, including an average number of I/O(s) and total running time, has been shown in Figure 5.

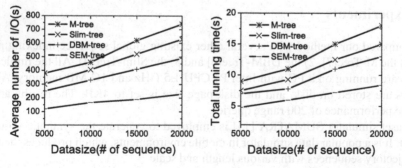

Fig. 5. Average number of I/O(s) and total running time(s) for SEM-tree, M-tree, Slim-tree and DBM-tree for queries with varying data size from 5000 to 20000. The default values are (γ=0.01, sl=100 and τ = 15).

Generally, with the increase of data size, more sequences will be covered for the fixed query range γ, which not only incurs the increment of distance computation but also but also leads to more I/O operations involved. Therefore, the performances of all indexes decrease as depicted in Figure 5. However, the SEM-tree is still about 40-50% better than other methods for range query on SUMATRA dataset. These results clearly show the efficiency of the SEM-tree over the three compared indexes. This is because SEM-tree uses (sub-)multisets rather than the full length in the internal nodes, and the latter can lead to more I/O operations and distance computational cost.

4.3 Impact of Character Set Size (τ)

Here, we compare the performance of SEM-tree with other three indexes by fixing the query range ($\gamma=0.01$) for a varying character set size $\tau = 5, 10, 15, 20$ and 25 under SUMATRA data.

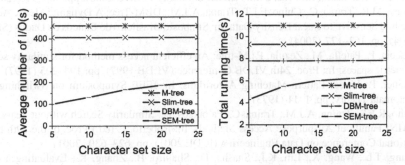

Fig. 6. Average number of I/O(s) and total running time(s) for SEM-tree, M-tree, Slim-tree and DBM-tree for queries with varying character set size from 5 to 25. The default values are ($\gamma=0.01$, $s=10000$ and $sl = 100$).

As shown in Figure 6, as the number of character set size increases, the performance of SEM-tree degrades gradually. The reason is that the effectiveness of sequence embedding is deteriorated, as we need to use longer multiset to represent the original sequence. However, the SEM-tree still performs better than other schemes for a relatively large τ, e.g. 20. Note that, the performances for DBM-tree, Slim-tree and M-tree are kept unchanged because the query processing is unrelated with τ but only influenced by the length of sequence sl.

5   Conclusion

In this paper, we proposed a "dimensionality reduction" like mechanism based on sequence embedding technique to minimize the expensive distance computational cost. As an application of the derived properties, a novel index structure, SEM-tree was presented to index sequences. In a SEM-tree, each level represents the sequence data with different compression level of multisets, whose sizes increase from root to leaf level. As demonstrated by the comprehensive simulations, SEM-tree shows a much better performance than existing schemes, in terms of CPU cost and I/O cost.

References

1. Traina, C., Traina, A.J.M., Seeger, B., Faloutsos, C.: Slim-Trees: High Performance Metric Trees Minimizing Overlap Between Nodes. In: Zaniolo, C., Grust, T., Scholl, M.H., Lockemann, P.C. (eds.) EDBT 2000. LNCS, vol. 1777, pp. 51–65. Springer, Heidelberg (2000)
2. Faloutsos, C., Lin, K.I.: Fast Map: A fast algorithm for indexing, data mining and visualization of traditional and multimedia datasets. In: Proc. of the International Conference on Management of Data (SIGMOD 1995), pp. 163–174 (1995)
3. Cormode, G., Muthukrishnan, S.: The string edit distance matching problem with moves. In: Proc. of the 13th annual ACM-SIAM symposium on Discrete Algorithms, pp. 667–676 (2002)
4. Jagadish, H.V., Ooi, B.C., Tan, K.-L., Yu, C., Zhang, R.: iDistance: An adaptive B^+-tree based indexing method for nearest neighbor search. ACM Trans. on Data Base Systems 30(2), 364–397 (2005)
5. Venkateswaran, J., Lachwani, D., Kahveci, T., Jermaine, C.M.: Reference-based Indexing of Sequence Databases. In: Proc. 24th VLDB Conference (VLDB 2006), pp. 906–917 (2006)
6. Vieira, M.R., Traina, C., Chino, F.J.T., Traina, A.J.M.: DBM-Tree: A Dynamic Metric Access Method Sensitive to Local Density Data. In: Simposio Brasileiro de Bancos de Dados (SBBD 2004), pp. 163–177 (2004)
7. Ciaccia, P., Patella, M., Zezula, P.: M-tree: An efficient access method for similarity search in metric spaces. In: Proc. 24th VLDB Conference (VLDB 1997), pp. 194–205 (1997)
8. Weiner, P.: Linear Pattern Matching Algorithms. In: IEEE Symposium on Switching and Automata Theory, pp. 1–11 (1973)
9. Filho, R.F.S., Traina, A.J.M., Traina, C., Faloutsos, C.: Similarity Search without Tears: The OMNI Family of All-purpose Access Methods. In: Roberto, F. (ed.) Proc. of the 19th International Conference on Data Engineering (ICDE 2001), pp. 623–630 (2001)
10. Wang, T.L., Wang, X., Lin, K.I., Shasha, D., Shapiro, B., Zhang, K.: Evaluating a class of distance-mapping algorithms for data mining and clustering. In: Proc. of the 5th ACM International Conference of Knowledge Discovery and Data Mining (SIGKDD 1999), pp. 307–311 (1999)
11. http://www-db.stanford.edu/pleiades/SUMATRA.html
12. Zhang, Z., Schwartz, S., Wagner, L., Miller, W.: A greedy algorithm for aligning DNA sequences. J. Comput. Biol. 7, 203–214 (2000)
13. Sahinalp, S.C., Tasan, M., Macker, J., Ozsoyoglu, Z.M.: Distance-Based Indexing for String Proximity Search. In: Proceeding of the 19th International Conference on Data Engineering (ICDE 2003), Bangalore, India, March 2003, pp. 125–136 (2003)

Appendix

LEMMA1 PROOF: Suppose $r_1 = T_i(s_1) \cap T_i(s_2)$, $r_2 = T_i(s_1) \ominus T_i(s_2)$, and $r_3 = T_i(s_2) \ominus T_i(s_1)$. According to Equation 2, $\hat{d}_e(T_i(s_1), T_i(s_2)) = \frac{|r_2|+|r_3|}{2|r_1|+|r_2|+|r_3|}$. When $ET_i(s_1)$ and $ET_i(s_2)$ are embedded into $ET_{i+1}(s_1)$ and $ET_{i+1}(s_2)$, let us suppose r' as the common subsequence of $ET_{i+1}(s_1)$ and $ET_{i+1}(s_2)$, i.e., $r' \sqsubseteq ET_{i+1}(s_1)$ and $r' \sqsubseteq ET_{i+1}(s_2)$. Since sequence embedding adopts the hash-function which is deterministic, the common subsequence r'(such as $'k'$) can only be produced by the common subset r_1(such as $'\{a, c\}'$). However, sequences that share the same set of elements r_1 may not be embedded into the same sequences(except for $h(ac) = k$, we can have $h(ca) = q$), i.e., $|r'| \leq \frac{|r_1|}{k}$, with $k = \frac{|ET_i(s_1)|}{|ET_{i+1}(s_2)|}$.

Thus,$\hat{d}_e(T_{i+1}(s_1), T_{i+1}(s_2)) = \frac{(2|r_1|+r_2|+r_3|)/k - |r'|}{(2|r_1|+r_2|+r_3|)/k} \geq \hat{d}_e(T_i(s_1), T_i(s_2))$.

LEMMA2 PROOF: Assume $a_i = d_e(T_i(s_1), T_i(s_2))$ and $b_i = card(T_i(s_1) \cup T_i(s_2))$, according to LEMMA 1, we have:$\frac{a_1}{b_1} \leq \frac{a_2}{b_2} \leq \dots \leq \frac{a_l}{b_l}$, with $\frac{a_i}{b_i} = \hat{d}_e(T_i(s_1), T_i(s_2))$. We have following induction process:

Basic step: Based on LEMMA 1, we have $\frac{a_1}{b_1} < \frac{a_2}{b_2}$. Therefore, $\frac{a_1}{b_1} \leq \frac{a_1+a_2}{b_1+b_2} \leq \frac{a_2}{b_2}$, i.e., $\hat{d}_e(T^1(s_1), T^1(s_2)) \leq \hat{d}_e(T^2(s_1), T^2(s_2))$.

Inductive step: Suppose the Lemma is satisfied when j equals $i+1$ ($i \geq 1$), i.e., $\frac{A_i}{B_i} \leq \frac{A_i+a_{i+1}}{B_i+b_{i+1}}$ with $A_i = \sum_{k=1}^{i} a_k$ and $B_i = \sum_{k=1}^{i} b_k$. Therefore, we have $A_i * b_{i+1} \leq B_i * a_{i+1} \Rightarrow \frac{A_i}{B_i} \leq \frac{A_i+a_{i+1}}{B_i+b_{i+1}} \leq \frac{a_{i+1}}{b_{i+1}}$. Based on LEMMA 1 we have $\frac{a_{i+1}}{b_{i+1}} \leq \frac{a_{i+2}}{b_{i+2}}$. In other words, $\frac{A_i}{B_i} \leq \frac{A_i+a_{i+1}}{B_i+b_{i+1}} \leq \frac{A_i+a_{i+1}+a_{i+2}}{B_i+b_{i+1}+b_{i+2}} \leq \frac{a_{i+2}}{b_{i+2}}$, i.e. $\frac{A_i}{B_i} \leq \frac{A_{i+1}}{B_{i+1}} \leq \frac{A_{i+2}}{B_{i+2}}$.

Therefore, the Lemma is also satisfied for j equals $i+2$, and the proof is completed.

LEMMA3 PROOF: We know, $\hat{d}_e(T^i(s_1), T^i(s_2)) = \frac{card(T^i(s_1) \ominus T^i(s_2))}{card(T^i(s_1) \cup T^i(s_2))}$

Obviously, this distance value is bounded by $\hat{d}_e(T(s_1), T(s_2))$,

$$= \frac{a + card(\cup_{j=i+1}^{l_1} T_j(s_1) \ominus \cup_{j=i+1}^{l_2} T_j(s_2))}{b + card((\cup_{j=i+1}^{l_1} T_j(s_1)) \cup (\cup_{j=i+1}^{l_2} T_j(s_2)))} < \frac{a + card(\cup_{j=i+1}^{l_1} T_j(s_1) \cup (\cup_{j=i+1}^{l_2} T_j(s_2)))}{b + card((\cup_{j=i+1}^{l_1} T_j(s_1)) \cup (\cup_{j=i+1}^{l_2} T_j(s_2)))},$$

where $a = card(T^i(s_1) \ominus T^i(s_2))$ and $b = card(T^i(s_1) \cup T^i(s_2))$.

Since the scale of the union $\cup_{j=i+1}^{l_1} T_j(s_1)$ and $\cup_{j=i+1}^{l_2} T_j(s_2)$ will be smaller than that of $T_i(s_1)$ and $T_i(s_2)$, $card((\cup_{j=i+1}^{l_1} T_j(s_1)) \cup (\cup_{j=i+1}^{l_2} T_j(s_2)))$ can be bounded by $card(T_i(s_1) \cup T_i(s_2))$. In other words, $B_u(\hat{d}_e(T^i(s_1), T^i(s_2)))$ can be approximated by $\frac{card(T^i(s_1) \ominus T^i(s_2)) + v}{card(T^i(s_1) \cup T^i(s_2)) + v}$, where $v = card(T_i(s_1) \cup T_i(s_2))$.

LEMMA6 PROOF: Assume a sequence s' within cv has distance r from s. According to Equation 2, there are many possible choices for sequences within cv satisfying such constraints. To make sure that all sequences within cv are covered by the distance $\hat{d}_e(T(s), T(s'))$, we choose one representative sequence s' satisfying the constraints: $T^i(s') \supset T^i(s)$ with $\hat{d}_e(T^i(s), T^i(s')) = r$.

Obviously, the radius r is the maximal distance among the whole multiset between s and s', which is bounded by $\hat{d}_e(T(s), T(s')) = \frac{\frac{|T^i(s)| * r}{(1-r)} + card(\cup_{j=i+1}^{l_1} T_j(s) \ominus \cup_{j=i+1}^{l_2} T_j(s'))}{\frac{|T^i(s)|}{(1-r)} + card(\cup_{j=i+1}^{l_1} T_j(s) \cup (\cup_{j=i+1}^{l_2} T_j(s')))}$

$< \frac{\frac{|T^i(s)| * r}{(1-r)} + card((\cup_{j=i+1}^{l_1} T_j(s)) \cup (\cup_{j=i+1}^{l_2} T_j(s')))}{\frac{|T^i(s)|}{(1-r)} + card((\cup_{j=i+1}^{l_1} T_j(s)) \cup (\cup_{j=i+1}^{l_2} T_j(s')))}$ where l_1 and l_2 are the height of $ET(s)$ and $ET(s')$ respectively.

Since the scale of the union $\cup_{j=i+1}^{l_1} T_j(s)$ and $\cup_{j=i+1}^{l_2} T_j(s')$ is smaller than that of $T_i(s)$ and $T_i(s')$, $card((\cup_{j=i+1}^{l_1} T_j(s)) \cup (\cup_{j=i+1}^{l_2} T_j(s')))$ can be bounded by $card(T_i(s) \cup T_i(s')) = \frac{(2-r) * T_i(s)}{(1-r)}$. In other words, $rad_u(cv_i)$ can be approximated by $\frac{|T^i(s)| * r + (2-r) * |T_i(s)|}{|T^i(s)| + (2-r) * |T_i(s)|}$.

An Architecture for Distributed Controllable Networks and Manageable Node Based on Network Processor

Tao Liu[1,2,*], Depei Qian[1,2], Huang Yongxiang[1], and Rui Wang[2]

[1] Xi'an Jiao Tong University
710049 Xi An, Shanxi Province, P.R.China
[2] Sino-German Joint Software Institute, Bei Hang University.
100083 Bei Jing, P.R.China
taobell@msn.com

Abstract. We present an architecture for Distributed Controllable Networks (DCN) which provides distributed control and management functions to networks. This control system can collect and correlate network information between several manageable nodes placed at different locations in the network to offer a large scale view to the network operator. Also, this system can deploy management program and commands to the manageable nodes to control the whole network. The system is capable of capturing packets and pre-processing them on the node itself and the more detailed analysis is taken place in a distributed way. The prototype implementation of our system is based on Intel IXP2400 network processor.

Keywords: Distributed Network Management, Network Processor.

1 Introduction

Control is becoming an increasingly import function for computer networks. The increasing complexity of networks in terms of topology, routing behavior, traffic patterns, and throughput performance is due to a rising number of diverse and heterogeneous end-systems and network equipment. The network control can be split to three steps, information gathering, information analysis, and command deployment.

First of all, information including network topology, network events must be gathered by the routers. Then, all information must be analyzed in a central or distributed way. After analysis, a group of commands must be deployed to a number of routers and achieving the control task.

To response to the network events, the control commands must be deployed to specified routers. The result of the information analysis indicates that which control command should be executed on which router, and also the proper parameters are attached to the command. The simplest way to deploy the commands is to deploy the

---

* The work was supported by National Science Foundation of China through the grant No.60673180 and 90612004, the National High Technology Development Program (863 Program) under the grand No.2006AA01A118 and the project of International Cooperation Program of China under the grant No.2006DFA11080.

Y. Zhang et al. (Eds.): APWeb 2008, LNCS 4976, pp. 450–455, 2008.

command one by one. To optimize the deployment, we can use group deployment or other optimization method.

These three steps may be used interoperable. For example, the Information Analyzer detects a kind of worm attacks, but can not find out which worm for lack of detail network information, so the Analyzer deploy new packet gathering modules to get more information, after gathering new packets and new network events, may be the Analyzer can decide which response should be deployed to defense the attack. And finally, new control commands deployed to specified routers, and the network being recovered again.

The distributed aspect of observe, analyze and control the networks aims at correlating packet instances, network metrics, and events across multiple routers placed at different locations in a network. Such kind system expands the view of observation beyond a single node or link, helps the network operator to analyze the network events globally; and finally, to control the networks holistically. The goals of our control architecture by define DCE (Distributed Controllable Environment) architecture.

The design and implementation of such a control system poses a number of challenges in terms of functionality, interoperability, scalability and performance. In our work, we build our information gathering module on a network-processor based prototype implementation, and the information analysis and command deployment modules on a high performance computer. In particular, our contributions are:

A controllable node architecture for observing and collecting network information.

A control framework for bring inter-operation to information gatherer and information analyzer.

A deployment interface for deploy packet capture program, parameters and control command to routers.

A prototype implementation based on Intel IXP2400 network processor.

Network processors are being used as the basic building blocks of network devices. In our prototype system, it is used to build the information collection module. Our information collection module can be configured to observe and collect packets that the network operator desires. That is to say, if the network operator wants to capture new type of packets, he can reconfigure the information collection module or write a new plug-ins which can be added into the module. The network processors are ideal for such tasks since they are optimized for packet processing tasks and also support new program injection and reconfiguration.

Section 2 discusses related works. In Section 3, we introduce the system architecture of controllable networks, and especially discusses information gatherer module. The prototype implementation is presented in Section 4, and Section 5 summarizes and concludes the paper.

2 Related Works

Network monitor can be performed using passive measurement techniques [1]. Passive measurement projects often aim at large scale monitoring. Some of them are NLANR PMA [2], Sprint IPMON [3], and AT&T GigaScope [4]. These projects utilize custom hardware to obtain the performance required for monitoring high speed links. All this projects aim at capture all the packet traces without processing. Our architecture needs to capture and analyze the packet simultaneity, and we have to take responses for specified network events.

A distributed passive measurement infrastructure is presented in [5], which has capture node performs very little post processing of the data and lacks a real-time query capability. Our architecture supports capturing only specified packets and does some quick pre-analyzing to get some basic network status.

ATMEN [6] is an distributed measurement infrastructure uses Gigascopes[4] as its packet capture device. It has a real-time query capability, and the ability to reuse measurement data. However, ATMEN is distributed in nature with its components communicating via the Internet using a communication protocol. Our architecture can be implemented as a router's add-on functions, and supports to reconfigure the capture modules and the routers as controllable network elements.

3   Distributed Controllable Networks

Figure 1 shows the overall system architecture, the primary components of the system are Information Gatherer, Information Analyzer. The Information Gatherer is a DCE architecture based module deployed on network routers that need to be controlled. The Information Gatherer performs the collection, pre-processing and archiving the packet traces. Each Information Gatherer has a number of capture agents which can be extended. The capture agents can pre-process the packet traces and transfer the processing result to a center database.

The Information Analyzer can read and analyze the pre-processing results from the center database. After analyzing, A set of control command may be created and deployed to the controllable routers to fix the network.

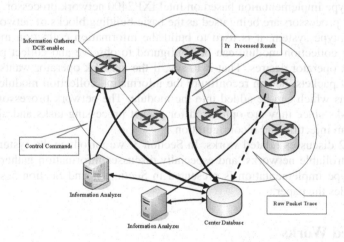

Fig. 1. Distributed Controllable Networks

3.1   DCE Architecture

The Distributed Controllable Environment architecture in Figure 2 is the software architecture of Information Gatherer which can be divided into two parts, 1, fast-path, which classify, capture, dropping, forwarding the packets and do very little processing, 2, low-path, which do some pre-processing. The capture rules will install filters to classifications which indicate the patterns of packets which the capture agents want. And Capture agents do some pre-analysis related to the packets. Additionally, this node also has a Command Processor in order to install new Capture Agents and manage other software and hardware modules.

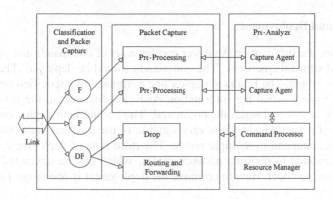

Fig. 2. Distributed Controllable Environment architecture

3.2 Information Gatherer

The Information Gatherer is designed based on DCE. The Information Gatherer includes three modules: packet capture, pre-processing, and capture agent.

The first step of the control scheme is to capture the packets to find out what's happening in the networks. In our design, only the packets which match the capture agents' need will be captured and processed. When capture agent is installed into the node, filters will be installed to the classification module. The filtering ensures that only a subset of observed traffic will be captured and directed to proper pre-processing and capture agent.

In general, the pre-processing did some statistics and/or very small computation to each captured packet. There are a large number of potential types of statistics, for example counter, sliding window, and histogram. The capture agent defines which statistic method will be injected to pre-processing module and configures the statistics parameters dynamically. After the pre-processing, the processed raw packets will be forwarded to the router's original forwarding module. But if the capture agents received a drop command from the operator, for examples drop the P2P traffic of some patterns, the raw packets will be dropped.

The capture agents will collect the pre-processed network events from the pre-processing module. After pre-analysis, the capture agent will get a number of network information. The capture agent will package and transfer the network information to the center database periodically. If an emergency network event is detected on the capture agent, this event will transferred to the center database immediately. Multiple agents can be installed or un-installed on the Information Gatherer by the network operator and Information Analyzer.

3.3 Information Analyzer

Two types of information analyzers are proposed in our design: single analyzer and integrated analyzer. A single analyzer connects only one capture agent. The integrated analyzer can install multiple capture agents on more than one Information Gatherer. So the integrated analyzer can get more network information and has a more clear view of network status.

3.4 Command Deployment

After analysis, the Information Analyzer creates a set of commands and parameters to try to fix and adjust the network. The commands should be deployed .There are three types of deployment. The first is to deploy one command to a specified network node. It is a simple process that the Information Analyzer connects to the network node's Command Processor and sends the command. The second is to deploy one command to a number of network nodes. It is very suitable to use a virtual multicast protocol to send a single command to multiple nodes. The third type of deployment is to deploy multiple commands to multiple network nodes, which can be optimized by a group communication protocol. How to optimize the deployment is one of our future works.

3.5 Capture Agent's Installation

The capture agent should be dynamically installed on the Information Gatherer or un-installed. This is a radical approach to allow high-level applications to download and deploy new code to a network node. This approach raises a different flexibility versus efficiency tradeoff. The ability to download and execute code is less limiting than a pre-defined hardware approach, but we can use network processors to partially solve this problem.

4 Prototype and Performance

We have implemented a prototype with basic filter and gatherer to illustrate the feasibility of the proposed architecture. The Information Gatherer implementation is based on an Intel IXP2400 [7] development board. In our implementation we just add small additional overhead to the router.

We evaluate the time consuming on the IXP2400's development environment software from Intel. The average time consuming of 5 main modules are illustrated as follows. We use 5-tuple classification which classifies IP packets' five parameters (src addr, src port, dest addr, dest port, protocol). The simple pre-processing we used here for test do packet statistics for specified P2P UDP packets.

Table 1. Average time consuming of 5 main modules (ns)

RX	5-Tuple classification	Simple pre-processing	Routing & Forward	Schd & TX
143.3	88.3	75.2	416.7	329.9

In the table, we can see that add a simple pre-processing into the fast path just need a little processing power, and if we limit the number of captured packets, we get lesser impact of the performance of a normal router.

5 Summary

In summary, we have presented our architecture for controllable networks especially the hardware design of the distributed Information Gatherer. The distributed Information Gatherer is the key entity of the control architecture. The ability to dynamically add new filters to the capture module can reduce the bandwidth load for collection database and provides an extensible way to get variable network status data. The pre-analyzer is a key enabler to implement online analyzing which reduces processing load for integrated analyzers. Our prototype system shows the feasibility of this design and did not harm the router's performance.

For future works, a more powerful and extensible classification algorithm should be adopted to support dynamically installed capture agents.

References

1. Bhattacharyya, S., Moon, S.: Network monitoring and measurements: Techniques and experience. In: Tutorial at ACM Sigmetrics 2002, Marina Del Rey (June 2002)
2. National Laboratory for Applied Network Research – Passive Measurement and Analysis. Passive Measurement and Analysis (2003), http://pma.nlanr.net/PMA/
3. Fraleigh, C., Moon, S.B., Lyles, B., Cotton, C., Khan, M., Moll, D., Rockell, R., Seely, T., Diot, C.: Packet-level traffic measurements from the Sprint IP backbone. IEEE Network 17(6), 6–16 (2003)
4. Cranor, C., Gao, Y., Johnson, T., Shkapenyuk, V., Spatscheck, O.: Gigascope: High performance network monitoring with an SQL interface. In: Proc. of the 2002 ACM SIGMOD International Conference on Management of Data, Madison, WI, June 2002, p. 623 (2002)
5. Arlos, P., Fiedler, M., Nilsson, A.: A distributed passive measurement infrastructure. In: Dovrolis, C. (ed.) PAM 2005. LNCS, vol. 3431, pp. 215–227. Springer, Heidelberg (2005)
6. Krishnamurthy, B., Madhyastha, H.V., Spatscheck, O.: ATMEN: A triggered network measurement infrastructure. In: Proceedings of the 14th International World Wide Web Conference (WWW), Chiba, Japan (May 2005)
7. Intel Corp. Intel Second Generation Network Processor (2005), http://www.intel.com/design/network/products/npfamily/

Building a Scalable P2P Network with Small Routing Delay[*]

Shiping Chen[1,2], Yuan Li[2], Kaihua Rao[2], Lei Zhao[2], Tao Li[2], and Shigang Chen[3]

[1] Network Center, University of Shanghai for Science and Technology,
Shanghai, 200093, China
[2] Department of Computer Engineering, University of Shanghai for Science and Technology,
Shanghai, 200093, China
[3] Department of Computer and Information Science and Engineering,
University of Florida, USA
chensp@usst.edu.cn

Abstract. Most existing P2P networks route requests in $O(kN^{1/k})$, $O(log N)$, $O(log N / log k)$ hops, where N is the number of participating nodes and k is an adjustable parameter. Although some can achieve $O(d)$-hop routing for a constant d by tuning the parameter k, the neighbor locations however become a function of N, causing considerable maintenance overhead if the user base is highly dynamic as witnessed by the deployed systems. This paper explores the design space using the simple uniformly-random neighbor selection strategy, and proposes a random peer-to-peer network that is the first of its kind to resolve requests in d hops with a chosen probability of $1-c$, where c is a constant. The number of neighbors per node is within a constant factor from the optimal complexity $O(N^{1/d})$ for any network whose routing paths are bounded by d hops.

Keywords: Peer-to-Peer Networks, Randomized Topology, Routing Delay.

1 Introduction

Peer-to-peer (P2P) systems have many applications in data sharing, notification services, data dissemination, directory lookup, software distribution, and distributed indexes. Because data may be kept at any node, a fundamental problem is to efficiently locate the node that stores a particular data item. Napster uses a centralized directory service. Gnutella [1] and KaZaA [2] rely on flooding-based search mechanisms, which cause tremendous communication overhead for large systems [3,4,5,6].

To solve the scalability problem, many P2P proposals use distributed hash tables (DHT) to uniformly distribute the responsibility of data location management to all nodes. An identifier is associated with each data item, and each node is responsible for storing a certain range of identifiers together with the corresponding data items or

---

[*] This research was Supported by the National Natural Science Foundation of China under Grant No. 60573142, The Shanghai Leading Academic Discipline Project under Grant No. T0502.

Y. Zhang et al. (Eds.): APWeb 2008, LNCS 4976, pp. 456–467, 2008.
© Springer-Verlag Berlin Heidelberg 2008

their locations (addresses). DHT provides a basic function, *lookup(id)*, which maps an arbitrary identifier to the responsible node. To implement such a function, an overlay P2P network is formed among the participating nodes. When a lookup request is issued, the request will be routed to the responsible node via the P2P network. In a highly-dynamic environment where nodes frequently join and depart, the maintenance overhead for the overlay P2P network is a major design concern [7]. A recent survey on different types of P2P networks can be found in [17].

When constructing a P2P network, there exists a fundamental space-time tradeoff between the number of neighbors (i.e., the size of the routing table) and the network diameter (i.e., the length of the routing path) [8]. Many P2P networks have an adjustable parameter (k) that can be tuned for different space-time tradeoffs. For example, if $k = log N$, both time and space complexities of CAN become $O(log N)$, where N is the number of nodes in the system. For all P2P networks, however, the maintenance overhead is minimized when k is a constant — instead of a function of N that changes continuously as nodes join/depart.

PRR [9] and Pastry [10] require $O(k \frac{log N}{log k})$ neighbors per node and route in $O(\frac{log N}{log k})$ hops with high probability. In the following, we shall omit "with high probability" as it is true for most complexities to be described. Tapestry [11] and Chord [12] require $O(log N)$ neighbors and route in $O(log N)$ hops. CAN [13] requires $O(k)$ neighbors and route in $O(k N^{\frac{1}{k}})$ hops.

The first asymptotically-optimal system is Viceroy [14], which requires seven neighbors per node and routes in $O(log N)$ hops. Koorde [15], and Manku [16], achieve asymptotical optimality with $O(k)$ neighbors and $O(\frac{log N}{log k})$ routing hops, where [16] assumes $k = O(polylog(n))$.

In the family of P2P networks, one important member is much less investigated, i.e., one with $O(N^{\frac{1}{d}})$ neighbors per node and d routing hops, where d is a constant. Such a network is appealing in practice because of its small routing delay, which does not grow with respect to the size of the network. Each routing hop in a P2P network requires a message to travel end-to-end from one node to another, likely crossing the Internet. Given the prevalence of inexpensive memory, it is often desirable to trade more neighbors (space) for shorter routing paths (delay). For increased number of neighbors, the main problem is not the space requirement, but the complexity for maintaining the neighboring relationship [7]. This is particularly true for structured networks such as PRR, Pastry, and randomized Chord, where the neighbors of a node x are required to match the top i digits of x and differ at the $(i+1)th$ digit, for $i \in [1...log_k N]$, where k is the base of the digits. By choosing $k = N^{\frac{1}{d}}$, these systems achieve $O(d)$ routing hops with $O(d N^{\frac{1}{d}})$ neighbors. However, the neighbor locations are now a function of N because the base k is related to N. As N changes, the base of the digits ($N^{\frac{1}{d}}$) changes, which can make many existing neighbors no longer valid, causing considerable maintenance overhead.

One solution for reducing maintenance overhead is to use random neighbors, which require little maintenance. A node can take any other nodes as its neighbors based on certain probability distribution. Among the random P2P networks [14,,16], [16] have an adjustable parameter k, which must be a polylog function of N in order for their complexities to hold. For NoN routing [16], $k = O(polylog(N))$. None can achieve constant routing distance by adjusting k.

This paper proposes a new random P2P network that combines arbitrary neighbor selection, typically used only in unstructured P2P networks, with a DHT (distributed hash table) ring. It is the first of its kind to resolve requests in no more than d hops with probability $1-c$, where d and c are two configurable constants. In more conventional terms, choosing a small value (e.g., 10^{-10}) for c, the system resolves an arbitrary request in d hops with high probability (e.g., $1-10^{-10}$). There is a small probability c that a request is not resolved in d hops. When it does happen, a slower routing path will be taken, which guarantees to find the responsible node. The number of neighbors per node is $O((-lnc)^{1/2}dN^{\frac{1}{d}})$. Random neighbors are easy to manage. When nodes join or depart, the random neighbors of all other nodes remain unchanged. Without sacrificing the performance, a node increases (or decreases) its number of random neighbors only when N doubles (or halves). Note that the location of any particular neighbor is independent of N.

In Appendix A we prove that, for routing paths to be bounded by d hops, the lower bound on the number of neighbors is $\Omega(N^{\frac{1}{d}})$. Therefore, the space complexity of the proposed random P2P network is within a constant factor $(-lnc)^{1/2}d$ from the optimal.

The rest of the paper is organized as follows. Section 2 defines the model, notations and performance metrics. Section 3 proposes a random peer-to-peer network. Section 4 presents the simulation results. Section 5 shows the time complexity and the space complexity. Section 6 draws the conclusion.

2 Model, Notations and Performance Metrics

Each data item is mapped to an m-bit identifier by a hash algorithm. The whole ID space can be viewed as a modulo-2^m circle, where the next identifier in the circle after the largest value $\cdot 2^m-1\cdot$ is zero. Consider N participating nodes. Each node is assigned an identifier by hashing its address or domain name. When the node identifiers are marked on the ID space, they split the circle into N segments. A node x is responsible for the segment (denoted as $seg(x)$) that immediately follows its node identifier. The nodes that are responsible for the adjacent preceding (or following) segments are called the predecessors (or successors) of x. The location information about a data item is stored at x if the identifier of the item belongs to $seg(x)$.

When a user queries for a data item whose identifier is id, she submits a lookup request(id), which is routed through an overlay network to the node that is responsible for the identifier, denoted as $node(id)$. The node subsequently returns the data location to the user. The performance/overhead tradeoff achieved by the routing algorithm is fundamentally determined by the structure of the overlay topology.

Table 1. Notations

N	number of nodes in a peer-to-peer network
m	number of bits in an identifier
x, y, z, w	arbitrary nodes in a peer-to-peer network
$seg(x)$	segment of identifiers that x is responsible for
id	arbitrary identifier to be queried
$node(id)$	node that is responsible for id
S_x	set of sequential neighbors of x
R_x	set of random neighbors of x
$sup\_seg(x)$	segment of identifiers that $S_x + \{x\}$ is responsible for
s	number of sequential neighbors
r	number of random neighbors
$1 - P_d$	probability for a request to be resolved in d or less hops
$1 - c$	target probability of resolving a request in d or less hops
$RP2P(d, c)$	random peer-to-peer network that resolves a request in d or less of hops with a probability of at least $(1 - c)$

The notations defined above and later in the paper are listed in Table 1 for quick reference. We evaluate the performance of a peer-to-peer system based on the following metrics.

1. time complexity: the maximum number of hops that a request(id) must travel in the overlay topology before reaching $node(id)$
2. space complexity: the maximum storage that a node is used to keep the neighbor information

The issues of load balancing [18,19], proximity and locality [20], security [21], pricing, etc., are beyond the scope of this paper.

3 Random Peer-to-Peer Network (RP2P)

Given two constants d and c, our goal is to develop a peer-to-peer network whose time and space complexities are $O(d)$ and $O(N^{\frac{1}{d}})$, respectively. We start with an abstract description of the system. We then present some analytical results and discuss the protocols/algorithms that realize the system.

For all above complexities in the forms of $O(N^{\frac{1}{d}})$, we have omitted factors that are functions of d and c. These factors will be shown in the detailed description of the system.

3.1 Overlay Topology

Each node knows a set of neighbors that it will directly communicate with. There are two types of neighbors, as shown in Figure 1, where the circle represents the ID space.

random neighbors: A node x takes a number of randomly selected nodes as its random neighbors, denoted as R_x.

sequential neighbors: A node x takes a number of predecessors and a number of successors as its sequential neighbors, denoted as S_x. The combination of the segments that $S_x + \{x\}$ are responsible for is denoted as $sup\_seg(x)$, which is called the super segment of x.

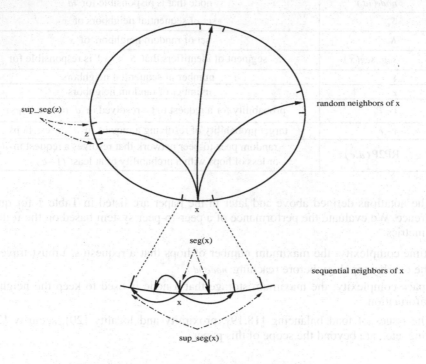

sup_seg(z)

random neighbors of x

z

x

seg(x)

sequential neighbors of x

x

sup_seg(x)

Fig. 1. Random neighbors and sequential neighbors of x

In the example of Figure 1, x has three random neighbors and four sequential neighbors. A node is required to store the following information about its neighbors.

- For each sequential neighbor $y \in S_x$, it uses two integers to store the neighbor's segment, $seg(y)$. Combining all these segments, x also knows its super segment, $sup\_seg(x)$.

- For each random neighbor $z \in R_x$, it uses two integers to store the neighbor's super segment, $sup\_seg(z)$.

The above information is learned from the neighbors. The space complexity for storing the information is equal to the number of neighbors. when x receives a request whose identifier belongs to $sup\_seg(x)$, it knows immediately which node (a sequential neighbor or itself) is responsible for the identifier. On the other hand, if the identifier belongs to the super segment of a random neighbor z, x should forward the request to z.

3.2 Routing Algorithm

When a node x receives a request(id), it processes the request by the following algorithm. Suppose the request carries the address of the node that originates the request.

```
RP2P_Routing( id )
1.   if  id ∈ seg(x)  then
2.            process  request  and  send  result  to
original requester
3.   else if ∃y∈ S_x,id∈ seg(y)  then
4.            forward the request to  y
5.   else if ∃z∈ R_x,id∈ sup_seg(z)  then
6.            forward the request to  z
7.   else
8.            forward  the  request  to  all  random
neighbors
```

A few routing examples are given in Figure 2.

zero-hop case: It takes zero hop to resolve a request if $id \in seg(x)$, as shown by the first plot in the figure and implemented by Lines 1-2 of the algorithm.

one-hop case: It takes one hop if $id \in seg(y), \exists y \in S_x$, as shown by the second plot in the figure and implemented by Lines 3-4 of the algorithm.

two-hop case: It takes two hops if $id \in sup\_seg(z), \exists z \in R_x$, as shown by the third plot in the figure and implemented by Lines 5-6 of the algorithm.

Three-hop case: It takes three or more hops otherwise, as shown by the last plot in the figure and implemented by Lines 7-8 of the algorithm.

For the first three cases, x knows for sure which is the next node to forward the request. For the last case, x has no clue about the next node. Hence, it broadcasts the request to all random neighbors. To restrain the broadcast overhead, we introduce a TTL field in the request message such that the request can only travel d or less hops and allows up to $d-2$ levels of broadcast (to random neighbors). As illustrated in the figure, the last two hops do not require broadcast as the node receiving the request has enough information to determine whether two more hops can reach $node(id)$ and if so, which is the next node to forward the request.

Below we give a basic analytical result. Suppose each node has s sequential neighbors and r random neighbors. To simplify the analysis, assume the nodes are responsible for equal-sized segments of the ID space. We will show that the analytical results with this assumption match very well with the simulation results without this

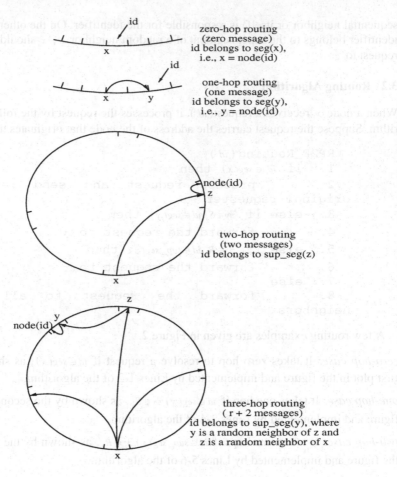

Fig. 2. Routing examples

assumption. Let P_d be the probability for request(id) to NOT reach $node(id)$ in d or less hops. The following upper bound of $P_d, d \geq 2$ is proved in Appendix B.

$$P_d \prec (1 - \frac{s}{N})^{r^{d-1}} \quad (1)$$

We will demonstrate shortly that, by appropriately choosing the values of s and r, a request can be resolved in d or less hops with a chosen probability (e.g., $1-10^{-10}$).

3.3 Determining Appropriate Values for s and r

Consider an integer $d \geq 2$ and a small constant $c \in (0..1)$. We prove that, if $s = r = kN^{\frac{1}{d}}$ where $k = (-lnc)^{\frac{1}{d}}$, then $P_d < c$. By (1), we have

$$P_d < (1 - \frac{kN^{\frac{1}{d}}}{N})^{(kN^{\frac{1}{d}})^{d-1}} \qquad (2)$$

Define the following quantity.

$$q = (\frac{N}{k^{\frac{d}{d-1}}})^{\frac{d-1}{d}}$$

Rewrite (2) as below.

$$P_d < (1 - \frac{1}{q})^{qk^d}$$

$(1 - \frac{1}{q})^q$ is a monotonically-increasing function with respect to q, and $limit_{q \to \infty}(1 - \frac{1}{q})^q = \frac{1}{e}$, where e is the base of natural logarithm. Hence, we have

$$P_d < (\frac{1}{e})^{k^d}$$

$$= (\frac{1}{e})^{(-\ln c)^{\frac{1}{d}d}}$$

$$= c$$

Let RP2P(d,c) be a random peer-to-peer network where each node has $(-\ln c)^{\frac{1}{d}} N^{\frac{1}{d}}$ sequential neighbors and the same number of random neighbors.[1] As an example, when $d = 3$, it becomes RP2P(3, c). Suppose each request carries a TTL field whose initial value is d. We modify the routing algorithm such that the longest routing path has no more than d hops.

```
RP2P_Routing_TTL ( id )
1.    decrease the TTL of the request by one
2.    if  id∈ seg( x ) then
3.    process request and send result to original
requester
4.    else if  ∃y∈ S_x, id ∈ seg( y ) then
5.    forward the request message to  y
6.    else if  ∃z∈ R_x, id ∈ sup_ seg( z ) then
7.    forward the request message to  z
8.    else if TTL of the request ≥2 then
9.    forward the request to all random neighbors
10.   else
11.   discard the request
```

Based on the previous analysis, we have the following theorem.

---

[1] If $c = 10^{-10}$ and $d = 3$, then $(-\ln c)^{\frac{1}{d}} = 2.8$.

Theorem 1. The probability for RP2P(d,c) to resolve a request in d or less hops is larger than $1-c$, where $d \geq 2$ and $c \in (0...1)$.

4 Simulation Results

Our simulation results match very well with the analysis. We simulated RP2P(3, c) on networks of 1000,10000, and 100000 nodes, respectively. The simulation was repeated for different values of c. The results are shown in Table 2. The column of c is the target failure probability. The column of s,r is the number of sequential (random) neighbors. The column of P_3 is the measured probability of NOT resolving a request in 3 or less hops. P_3 is always better (smaller) than the target value c. That is because our analysis made a conservative simplification when using (5) to derive the upper bound of P_d in Appendix B.

Table 2. Simulation results for RP2P(3, c)

	$N = 1,000$		$N = 10,000$		$N = 100,000$	
c	s,r	P_3	s,r	P_3	s,r	P_3
1.0e-1	13	6.8e-2	28	8.8e-2	61	1.1e-1
1.0e-2	16	9.4e-3	35	1.1e-2	77	9.1e-3
1.0e-3	19	4.3e-4	41	7.4e-4	88	8.6e-4
1.0e-4	20	1.4e-4	45	7.1e-5	97	8.1e-5
1.0e-5	22	7.7e-6	48	9.2e-6	104	9.4e-6
1.0e-6	23	1.5e-6	51	8.9e-7	111	8.7e-7
1.0e-7	25	3.5e-8	54	7.0e-8	117	8.4e-8

5 Complexities of RP2P(d,c)

The maximum number of hops that a request will travel in RP2P(d,c) is d, and the time complexity is thus $O(d)$. The number of neighbors per node is $r+s = 2(-lnc)^{\frac{1}{d}}N^{\frac{1}{d}}$, and the space complexity is thus $O((-lnc)^{\frac{1}{d}}N^{\frac{1}{d}})$.

6 Conclusion

This paper designs a random peer-to-peer network with neighbor nodes selected uniformly at random. The network is the first of its kind to resolve requests within a constant number of hops with high probability. A key advantage is the ease of neighbor management when nodes join/depart. The time and space complexities of the proposed network are $O(d)$ and $O((-lnc)^{\frac{1}{d}}N^{\frac{1}{d}})$, respectively. We conduct comprehensive analysis to derive the properties of the systems. Our simulation results match with the analytical results.

References

1 Gnutella: Gnutella, http://gnutella.wego.com
2 KaZaA: KaZaA, http://www.kazaa.com
3 Ritter, J.: Why Gnutella can't Scale. No, Really, http://www.tch.org/gnutella.html
4 Ripeanu, M., Iamnitchi, A., Foster, I.: Mapping the Gnutella Network. IEEE Internet Computing Journal, Special Issue on Peer-to-Peer Networking 6(1) (2002)
5 Sen, S., Wang, J.: Analyzing Peer-to-Peer Traffic across Large Networks. In: ACM SIGCOMM Internet Measurement Workshop (August 2002)
6 Saroiu, S., Gummadi, K.P., Dunn, R.J., Gribble, S.D., Levy, H.M.: An Analysis of Internet Content Delivery Systems. In: Proc. of the 5th Symposium on Operating Systems Design and Implementation (OSDI) (December 2002)
7 Ratnasamy, S., Shenker, S., Stoica, I.: Routing Algorithms for DHTs: Some Open Questions. In: Druschel, P., Kaashoek, M.F., Rowstron, A. (eds.) IPTPS 2002. LNCS, vol. 2429, Springer, Heidelberg (2002)
8 Xu, J.: On the Fundamental Tradeoffs between Routing Table Size and Network Diameter in Peer-to-Peer Networks. In: Xu, J. (ed.) Proc. of IEEE INFOCOM 2003 (April 2003)
9 Plaxton, C., Rajaraman, R., Richa, A.: Accessing Nearby Copies of Replicated Objects in a Distributed Environment. In: Proc. of ACM Symposium on Parallelism in Algorithms and Architectures (SPAA) (June 1997)
10 Druschel, P., Rowstron, A.: Pastry: Scalable, Distributed Object Location and Routing for Large-Scale Peer-to-Peer Systems. In: Proc. of 18th IFIP/ACM International Conference on Distributed Systems Platforms (Middleware 2001) (November 2001)
11 Zhao, B., Kubiatowicz, J., Joseph, A.: Tapestry: An Infrastructure for Fault-Tolerant Wide-Area Location and Routing, Tech. Rep. UCB/CSD-01-1141, University of California at Berkeley, Computer Science Department (2001)
12 Stoica, I., Morris, R., Karger, D., Kaashoek, F., Balakrishnan, H.: Chord: A Scalable Peer-To-Peer Lookup Service for Internet Applications. In: Proc. of ACM SIGCOMM 2001 (August 2001)
13 Ratnasamy, S., Francis, P., Handley, M., Karp, R., Shenker, S.: A Scalable Content-Addressable Network. In: Proc. of ACM SIGCOMM 2001 (August 2001)
14 Malkhi, D., Naor, M., Ratajczak, D.: Viceroy: A Scalable and Dynamic Emulation of the Butterfly. In: Proc. of ACM PODC 2002 (July 2002)
15 Kaashoek, F., Karger, D.R.: Koorde: A Simple Degree-Optimal Hash Table. In: Kaashoek, M.F., Stoica, I. (eds.) IPTPS 2003. LNCS, vol. 2735, Springer, Heidelberg (2003)
16 Manku, G.S.: Routing Networks for Distributed Hash Tables. In: Proc. of 22nd ACM Symposium on Principles of Distributed Computing (PODC) (June 2003)
17 Risson, J., Moorsa, T.: Survey of Research towards Robust Peer-to-Peer Networks: Search Methods. Journal of Computer Networks 55 (2006)
18 Joung, Y.-J., Yang, L.-W., Fang, C.-T.: Keyword search in DHT-based peer-to-peer networks. IEEE Journal on Selected Areas in Communications 25 (2007)
19 Li, Z., Xie, G.: A Distributed Load Balancing Algorithm for Structured P2P Systems. In: Proc. of the 11th IEEE Symposium on Computers and Communications (June 2006)

20 Ferreira, R.A., Jagannathan, S., Grama, A.: Locality in structured peer-to-peer networks. Journal of Parallel and Distributed Computing 66 (2006)
21 Navabpour, S., Nejad, N.F., Abbaspour, M., Behzadi, A.: Secure Routing in Structured Peer to Peer File-Sharing Networks. In: Proc. of International Conference on Communications and Networking in China (ChinaCom 2006) (October 2006)

Appendix A. Number of Neighbors Per Node in Networks of Constant Diameter

Theorem 1: The average nodal degree must be $\Omega(N^{\frac{1}{d}})$ for an N-node network with diameter d.

Proof: For a network with diameter d, starting from an arbitrary node, we can reach all nodes by a breadth-first search tree of d levels in depth. Let x be the average nodal degree. The number of nodes in the tree is $N = O(\sum_{i=0}^{d} x^i) = O(x^d)$. In order for $N = O(x^d)$ to hold, it is required that $x = \Omega(N^{\frac{1}{d}})$.

Appendix B. Upper Bound for P_d in RP2P

We establish an upper bound for P_d, $d \geq 2$, in the following. Consider an arbitrary identifier id and an arbitrary node x. Suppose x issues request(id). Each node has an equal probability of being responsible for id. $sup\_seg(x)$ consists of the segments of $(s+1)$ nodes. Hence, the probability for $id \in sup\_seg(x)$ is

$$P = (s+1)/N \tag{3}$$

It takes zero hop for the request to reach $node(id)$ if $x = node(id)$. Hence, $P_0 = 1 - \frac{1}{N}$. It takes one or less hop if id belongs to $sup\_seg(x)$. Hence, $P_1 \leq 1 - P = 1 - (s+1)/N$. We now derive P_d for $d \geq 2$. The request will not reach $node(id)$ in d or less hops if and only if the following two conditions are satisfied.

Condition 1: $id \notin sup\_seg(x)$

Condition 2: Starting from any random neighbor of x, the request will not reach $node(id)$ in $(d-1)$ or less hops.

The probability for Condition 1 to hold is $1 - P$. The probability for Condition 2 to hold is $(P_{d-1})^r$. Hence,

$$P_d = (1-P)(P_{d-1})^r \tag{4}$$

By induction we have, for $d \geq 2$,

$$P_d = (1-p)\sum_{i=1}^{d-1} r^{i-1} \, (P_1)^{r^{d-1}} \qquad (5)$$

We simplify the formula as follows.

$$P_d < (P_1)^{r^{d-1}}$$

$$< (1 - \frac{s}{N})^{r^{d-1}}$$

ERASP: An Efficient and Robust Adaptive Superpeer Overlay Network*

Wenjun Liu, Jiguo Yu**, Jingjing Song, Xiaoqing Lan, and Baoxiang Cao

School of Computer Science, Qufu Normal University,
Ri-zhao, Shandong, 276826, P.R. China
lwj_job@163.com, jiguoyu@sina.com, jjs54@126.com, bxcao@126.com

Abstract. The concept of superpeer (SP) has been introduced to improve the performance of popular file-sharing applications. However, the current constructing protocols are inefficient. In this paper an efficient and robust adaptive superpeer P2P overlay network ERASP is proposed. It is based on the well-known gossip paradigm and takes capacity and online time into account to estimate the network requirements for constructing and maintaining superpeer network. Additionally, the connectivity from client peers to SPs is strengthened. Experiment results show that ERASP is more efficient and robust than current related protocols.

1 Introduction

Peer to peer (P2P) networks have recently become a popular media through which huge amount of data can be shared. Peers are connected among themselves by some logical links forming an overlay above the physical network. Due to dynamics that take place in the network, maintaining stability of the overlay structure is a major challenge for the P2P network community.

Many typical P2P overlay structures have been described in the literatures [3,7,8]. More recently, some novel P2P systems were proposed [9,10,11]. Currently superpeer (SP) topologies are emerging as most influencing structure among them. By introducing the concept of SP, the P2P topologies are now organized through a two-level hierarchy [4,5,6,8,12]. The SP nodes that are faster and/or more reliable than "normal" nodes connect to each other forming the upper level in the network hierarchy. Each SP works as a server on behalf of the set of client peers who form the lower level of network hierarchy.

The SP paradigm allows decentralized networks to run more efficiently by exploiting heterogeneity and distributing load. However, building and maintaining this topology is not simple. The extreme scale and dynamism call for robust and efficient protocols, capable to self-organize and self-repair a SP overlay in spite of both voluntary and unexpected events like joins, leaves and crashes.

---

* The work is supported by NNSF (10471078) of China and RFDP (20040422004), Promotional Foundation (2005BS01016) for Middle-aged or Young Scientists of Shandong Province, SRI of SPED(J07WH05), DRF and UF(XJ0609)of QFNU.
** The corresponding author.

Y. Zhang et al. (Eds.): APWeb 2008, LNCS 4976, pp. 468–474, 2008.
© Springer-Verlag Berlin Heidelberg 2008

Pyun et al. proposed a scalable unstructured P2P system (SUPS) which is a special kind of SP network [6]. In [4], Montresor proposes a generic mechanism for the construction and maintenance of the SP overlay structures. The approach is that the nodes with smaller capacity are replaced by the bigger in order to receive a minimum number set of SPs. Gnutella protocol 0.6 [1] employs a hybrid architecture combining centralized and decentralized model. The ERSN [12] utilizes peer sampling protocol based on random walks to estimate the network requirements for constructing an efficient SP network and establishes emergence links between leaf peers to get a robust network.

In this paper, we design and implement an efficient and robust adaptive superpeer protocol ERASP on the basis of the existing hierarchical topology management protocols. ERASP takes capacity and online time metric of each participant peer into account. Furthermore, the connection intension from client to SP is enhanced. Experiment results show that ERASP is an efficient and robust adaptive protocol for building P2P overlay network.

The remainder of the paper is organized as follows. In Section 2 we describe the ERASP in detail. Experimental results that validate our mechanism are presented in Section 3. Finally, we conclude our paper and draw directions for future work in Section 4.

2 ERASP

We start out by giving some notations. Let n_i be a peer in an $N-$nodes network. Each SP maintains two neighbor sets: the client peer set $G_{cp}(n_i)$ and the SP set $G_{sp}(n_i)$. In order to distinguish peers that are capable to act as SPs from peers that can join just as clients, we associate each peer n_i with a parameter c_{n_i} representing its capacity. Parameter t_{n_i} denote the online time factor of n_i. Additionally, S denotes candidate SP set and $load_{n_i}$ is the real load of n_i.

Informally speaking, a SP overlay topology is characterized as follows. Peers must be assigned one of two mutually-exclusive roles: SP or CP. Our goal is to produce a SP topology characterized by minimum number of SPs. Peers with higher capacity and longer online time are considered better candidates as SPs. At each time, the target topology is the one composed by the minimum set of peers whose total capacity is sufficient to cover all other client peers and these SPs have longer online time.

2.1 Selecting Clients and Superpeers

To build a topology with desirable characteristics, we propose a mechanism based on the well-known gossip paradigm [1,2]. Topology information such as identifier, capacity, online time, current role and neighborhood of participating peers are disseminated through periodic gossip messages between randomly selected peers. Based on the received information, peers update their neighborhoods in order to obtain a better approximation of the target topology.

Algorithm 1. Outline of the adaptive distributed evolutionary algorithm.
Notations: p is local peer, q and r are identifiers of the remote peer.

```
 1:   q ← NULL;
 2:   S ← { r is underloaded ∧ c_r ≥ c_p ∧ t_r ≥ t_p};
 3:   while (S ! =NULL ∧ q == NULL)
 4:       r ← RANDOM (S);
 5:       S ← S-{r};
 6:       if (load_r < c_r ∧ (c_p < c_r ∨ load_r > load_p))
 7:           q ← r;
 8:   if (p is SP)
 9:       if (q ! = NULL)
10:           < transfer the client peers of p to q; >
11:           < p becomes client peer of q;>
12:       else
13:           < do nothing;>
14:   else
15:       if (q ! = NULL)
16:           < add p into set G_{cp}(q); >
17:       else
18:           < p becomes a SP itself; >
```

Algorithm 1 shows the ERASP evolutionary framework. The rationale behind the process is that all SPs try to push clients towards more powerful and the longer online time peers that are willing to accept more load. In lines 2-7, the protocol performs a random selection among those SPs that are underloaded and whose capacity is larger or equal than the capacity of the local peer. After that each peer judges its role. If the current peer p is SP and the selected SP q is not empty, then p tries to transfer all its clients to the selected SP q and itself becomes p's a client. If q is empty, just do nothing. In line 15, if the role of p is CP and the selected SP q is not empty, then p becomes a client of q. In line 17, a clint peer who can not join any SP will become a SP itself. The result of select is that each client peer is connected to its SPs.

2.2   The Connectivity Enhancement

In order to increase the robustness of our protocol, the connectivity from client peers to SPs is strengthened in ERASP. That is, each client peer maintains d super peers. The client peer n_i maintains different peer sets in different cases:

- If $G_{sp}(n_i)$ is not empty and $|G_{sp}(n_i)| \geq d$, then select d SPs that have maximum capacity and the longest online time performance from SP set $G_{sp}(n_i)$.
- If $G_{sp}(n_i)$ is not empty and $|G_{sp}(n_i)| < d$, then check if S is empty. If not, select $|G_{sp}(n_i)|$ SPs from $G_{sp}(n_i)$ and $d - |G_{sp}(n_i)|$ preliminary SPs with maximum capacity and the longest online time factor from candidate set S.

- If $G_{sp}(n_i)$ is empty but S is not empty, then select d candidate SPs from S as its enhancement links. Meantime, add these d candidate SPs into $G_{sp}(n_i)$.
- If both $G_{sp}(n_i)$ and S are empty, then n_i is connected to other client peers and becomes uncovered peer itself.

Note that the capacity and the online time parameters may be incompatible. Which parameter should be selected depend on particular application occasion. It is easy to see that in the absence of failures, this mechanism eventually produces the target topology. In fact, all SPs continually try to discover peers with larger capacity and longer online time that are not completely utilized. These are selected from the union of the underloaded sets, that progressively shrinks until it becomes empty or reaches its minimum size.

3   Evaluation

To validate ERASP, we have performed numerous experiments based on simulation. We were interested in three main questions: (i) what is the behavior of the protocol with respect to its parameters; (ii) how robust the protocol is; and (iii) what is the performance advantage compared to related protocols.

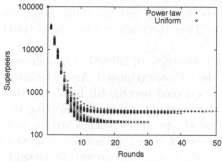 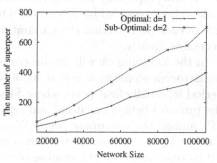

Fig. 1(a). SPs change with rounds   **Fig. 1(b).** SPs change with network size

Fig.1(a) shows the behavior of our protocol over time. As initial configuration, we selected a topology that is the farthest from the target: a random topology where all peers behave as SPs. The curves represent the number of SPs contained in the network after the specified number of rounds, averaged over 20 experiments. The algorithm proves to be extremely fast, independently from the distribution considered: after a few rounds the resulting topologies approximate extremely well the target.

The construction of an efficient SP topology should decrease the number of peers firstly. The theoretical minimum number of super peers is the minimum value of super peers required. Fig.1(b) shows the comparison of optimal ($d = 1$) and suboptimal ($d = 2$) cases. We can see that when $d = 2$ the topology has smaller SPs, but receives the better robustness (see later).

Fig.2 shows relation between time complexity and network size. Three different points in time are considered. Interestingly, the time needed to reach

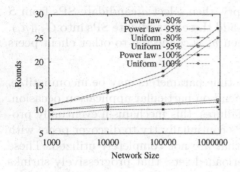
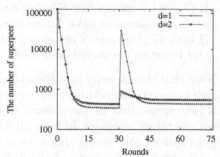

Fig. 2. Number of rounds to obtain different utilization thresholds

Fig. 3. Catastrophic failure scenario at round 30 30% of the SPs are removed

such utilization thresholds is independent from network size. While the target configuration requires a larger number of rounds, that grows (approximately) logarithmically with respect to the size of the network.

From the point of topology stability even if all of the SPs failed, ERASP can self-organized as a new topology quickly. Of course, the probability of this case is very low, especially the selected SPs have the longest online time factor. This makes clients covered by SPs stably. Once the topology reaches stable state, the role of each peer will not change continually. Thus the traffic within the network decreases greatly.

In the following we will demonstrate the robustness of ERASP. Fig.3 shows a catastrophic scenario: at round 30, 30% of the SPs are removed. After the initial period when all client peers whose SPs have crashed become SP by themselves, the protocol behaves as usual and repairs the overlay topology by selecting new SPs among the remaining peers. We can see the protocol reaches steady state quickly again. Comparing the different value of d, $d = 2$ makes the number of SP increase little but the number of uncovered client peer decreases remarkably.

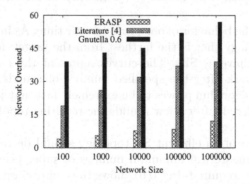

Fig. 4. Network overhead comparison: ERASP, Gnutella 0.6 and Literature [4]

Finally, we compare the efficiency of ERASP with that of Gnutella 0.6 and literature [4] in terms of network overhead and operations performed by each peer to form such SP overlays. The comparison of total network overhead is illustrated in Fig.4. We find that the overhead spent by ERASP to construct SP topology is far less Gnutella 0.6 and literature [4] with the network size increasing. The reason is that ERASP chooses stable peer with longer online time and higher capacity, this results in less network traffic necessarily.

4 Conclusions

In this paper we propose an efficient and robust adaptive protocol ERASP for constructing superpeer P2P overlay network. The peer with higher capacity and longer online time will be selected. The simulation results show that comparing with the existing approaches it decreases network traffic evidently. The future work includes to incorporate ERASP with the ability of handling locality heterogeneity and to explore the behavior of such hierarchical topology under churn.

References

1. Lime Wire LLC, Rfc-Gnutella 0.6,
   http://rfcgnutella.sourceforce.net/development
2. Jelasity, M., Kowalczyk, W., van Steen, M.: Newscast computing, Technical Report IR-CS-006, Vrije Universiteit Amsterdam, Dept. of Computer Science (Nov 2003)
3. Kurmanowytsch, R., Jazayeri, M., Kirda, E.: Towards a Hierarchical, Semantic Peer-to-Peer Topology. In: Proceedings of the Second International Conference on Peer-to-Peer Computing, Sweden (2002)
4. Montresor, A.: A robust protocol for building superpeer overlay topologies. In: Proc. P2P 2004. 43(4), pp. 202–209. IEEE Computer Society Press, Los Alamitos (2004)
5. Niu, C.Y., Wang, J., Shen, R.M.: A Topology Adaptation Protocol for Structured Super-peer Overlay Construction. In: Zhuge, H., Fox, G.C. (eds.) GCC 2005. LNCS, vol. 3795, pp. 953–958. Springer, Heidelberg (2005)
6. Pyun, Y.J., Reeves, D.S.: Constructing a balanced, $(\log(n))/\log\log(n))$-diameter super-peer topology for scalable P2P system. In: Proc. P2P 2004, pp. 210–218. IEEE Computer Society Press, Los Alamitos (2004)
7. Small, T., Li, B., Liang, B.: On Optimal Peer-to-Peer Topology Construction with Maximum Peer Bandwidth Contributions. In: Proceedings of the 23rd Queen's Biennial Symposium on Communications, Kinston, Ontario, Canada, May 29-June 1, pp. 157–160 (2006)
8. Yang, B., Garcia-Molina, H.: Designing a Super-Peer Networks. In: Proceedings of the International Conference on Data Engineering, Los Alamitos, CA (March 2003)
9. Yu, J.G., Song, J.J., Liu, W.J., Zhao, L., Cao, B.X.: KZCAN: A Kautz Based Content-Addressable Network. In: Proc. 8th ACIS International Conference on Software, Engineering, Artificial Intelligence, Networking, and Parallel/Distributed Computing (SNPD 2007), IEEE Computer Society, Los Alamitos (2007)

10. Yu, J.G., Song, J.J., Liu, W.J., Cao, B.X.: BGKR: A novel P2P network based on generalized Kautz and ring with constant congestion. In: Proc. 2th International Conference in Communications and Networking in China(ChinaCom 2007), IEEE, Los Alamitos (2007)

11. Yu, J.G., Song, J.J., Cao, B.X.: REIK: A Novel P2P Overlay Network with Byzantine Fault Tolerance. In: The Third International Conference on Semantics, Knowledge and Grid (SKG 2007), pp. 146–151. IEEE Computer Sciety, Los Alamitos (2007)

12. Zhang, Q.B., Peng, W., Lu, X.C.: ERSN: An efficient and Robust Super-Peer P2P Network. Journal of Computer Research and Development, 607–612 (2006)

Traceable P2P Record Exchange
Based on Database Technologies

Fengrong Li[1] and Yoshiharu Ishikawa[2]

[1] Graduate School of Information Science, Nagoya University
[2] Information Technology Center, Nagoya University
Furo-cho, Chikusa-ku, Nagoya 464-8601, Japan
lifr@db.itc.nagoya-u.ac.jp, ishikawa@itc.nagoya-u.ac.jp

Abstract. Information exchanges in P2P networks have become very popular in recent years. However, tracing how data circulates between peers and how data modifications are performed during the circulation before reaching the destination are not easy because data replications and modifications are performed independently by peers. This creates a lack of reliability among the records exchanged. To provide reliable and flexible information exchange facilities in P2P networks, we propose a framework for a record exchange system based on database technologies. The system consists of three layers: a user layer, a logical layer and a physical layer. Its tracing operations are executed as distributed recursive queries among cooperating peers in a P2P network. This paper describes the concept and overviews the framework.

1 Introduction

A *peer-to-peer* (*P2P*) network which consists of a large number of autonomous computers (*peers*) and is not dependent on a specific server is widely used in various applications such as file exchange, user communication, and content distribution. During information exchange in a P2P network, since duplications and changes to data may be performed by every peer without central control, it is difficult to determine the origin of data and to determine the movement of data between peers. This causes a lack of reliability in the data exchanged. As an example, when searching for images of beautiful scenes in a P2P file exchange service, reliability may not be significant, but when researchers exchange and share scientific information such as genome data with other researchers, the lack of reliability would be a critical concern. If a researcher is not sufficiently confident that the data was obtained from reliable sources, he will hesitate to use it for research purposes.

In this context, we propose a framework for *reliable record exchange* in P2P networks, where a *record* means a tuple-structured data item that obeys a predefined schema globally shared in the network. Records are exchanged between peers and peers can modify, store, and delete their records independently. The architecture of our P2P record exchange framework consists of three layers: the *user layer*, the *logical layer*, and the *physical layer*. The user layer provides a user interface for the record exchange system and the logical and physical layers

Y. Zhang et al. (Eds.): APWeb 2008, LNCS 4976, pp. 475–486, 2008.

support its internal representations. The two underlying layers are based on the relational data model, which is used for representing records in the user layer. In addition, they maintain record exchange and modification histories, facilitating traceability. In the physical layer, each peer in the P2P network maintains its own relations for storing information, and the logical layer provides virtual views by integrating the distributed relations. The abstraction in the logical layer provides a comprehensive framework for representing traceability requirements as database queries. Tracing queries are expressed as recursive *datalog* queries and executed over the distributed peers in the network.

The remainder of this paper is organized as follows. In Section 2, we describe the fundamental concept of P2P record exchange. In Section 3, we introduce the logical layer of the system. In Section 4, we describe the physical layer and present the concept underlying query processing. Section 5 reviews related work. Finally, in Section 6 we conclude the paper and outline future work.

2   P2P Record Exchange

2.1   Motivating Example

In this paper, we propose the concept of *traceable record exchange* in a P2P network and present the system architecture and its query processing framework. We assume that each peer corresponds to a user and maintains a set of *records* owned by the user. Each record has the same structure, which is defined by a predetermined schema which is globally shared within the network.

As an example, consider the sharing of information regarding novels among peers in a P2P network. Each peer maintains its own records and wishes to incorporate appropriate records from other peers to enhance its own record set. Figure 1 shows an example record set `Novel` owned by a peer that consists of four attributes: `title`, `author`, `language`, and `year`. Other peers also maintain their `Novel` records with the same structure, but their contents are not the same.

title	author	language	year
Pride and Prejudice	Jane Austen	English	1813
Madame Bovary	Gustave Flaubert	French	1857
War and Peace	Leo Tolstoy	Russian	1865

Fig. 1. Example Record Set `Novel`

In our record exchange framework, every peer can act as a provider of information. In this example, suppose that a user is interested in the novels written by `Jane Austen`. The user finds the desired records from other peers in the network by issuing a query. The user then examines the retrieved records which include (`Persuasion`, `Jane Austen`, `English`, 1818), and the selected records are registered in the local record management system as additional records. Of course, the user can modify and/or delete the obtained records in the local record set. In addition, the user can allow access to the record set from the other peers in the network.

A *traceability problem* occurs, for example, when the owner of the record set shown in Fig. 1 has the question: "Was 'Pride and Prejudice' actually published in 1813?" The user might check that the record (Pride and Prejudice, Jane Austen, English, 1813) in Fig. 1 is correct and try to find evidence supporting its validity. If the record was obtained from a well-known and credible peer, the user may suppose that the record is probably correct. Alternatively, if the original creator of the record is reliable, the record would also be reliable. However, finding such evidence from a P2P network is quite difficult. Peers are highly distributed and there is no central server that can answer traceability queries using the complete histories of all the records in the network.

To solve the traceability problem, we propose a framework for record exchange with a traceability facility. All the information required for tracing is maintained in distributed peers. When a tracing query is given, it is executed with the cooperation of the peers in a distributed manner. The next subsection describes the overall framework of the record exchange system.

2.2 System Framework

The *record exchange system* consists of the following three layers:

1. *User layer*: offers a user interface for the record management system.
2. *Logical layer*: provides a virtual view containing whole records in the P2P network including information for tracing.
3. *Physical layer*: implements the logical virtual views based on the cooperation of autonomous peers.

We now briefly explain the function of the user layer. The user layer provides the following functions to the user:

- *Search*: The system executes given search queries using distributed peers. The details are beyond the scope of this paper.
- *Registration*: After registration, records are under the control of the system and are potential targets for tracing.
- *Deletion/Update*: A user can delete and modify the records in his local system.
- *Tracing*: The system provides tracing facilities to the user. The details are described below.

In the following discussion, we omit the details of the user layer since it is not the main topic of this paper. The two underlying layers, the logical layer and the physical layer, are described in detail in Sections 3 and 4.

2.3 Requirements for Traceability

If there is no support for traceability in record exchange, the following problems may occur:

1. *The source of a record cannot be determined*: For example, it is not clear whether a record which exists at peer A was created at peer A, or was obtained from other peers. Moreover, it is difficult to know which peer initially

created the record. Since the record is insufficiently reliable, the advantages of record exchanging will be lost.

2. *Duplicate detection is not possible*: For example, when two or more records with the same values exist in a local record set, it is difficult to judge whether they were obtained from one single source or from different sources.
3. *The destination of a record cannot be identified*: Suppose that peer A discovers an error in one of its own records which is made available to other peers. If the peer wants to make error notifications to the peers that have incorporated the record, the identification of such peers is difficult.
4. *Updates cannot be traced*: Suppose that peer B has obtained a record from peer A and modified it, and that peer C has subsequently copied the modified record from peer B. Even if peer C wants to know the original source of the record, it is not possible because a simple search does not match the original value of the record at peer A.

The proposed record exchange framework copes with these problems using database technologies. The next section describes the logical layer, which represents information for tracing based on the relational model.

3   The Logical Layer

3.1   Data Representation

In the logical layer, virtual *views* are constructed by unifying the record sets (relations) maintained by distributed peers. By providing virtual integrated views, the users in the logical layer (e.g., the system administrator) can formulate queries for tracing more easily, as described in the following example. We simplify the example shown in Fig. 1 and assume that each peer maintains a `Novel` record set that has two attributes `title` and `author`. Figure 2 shows three record sets in the user layer maintained by peers A, B and C.

Peer A

title	author
t1	a1
t5	a5

Peer B

title	author
t1	a1
t2	a3

Peer C

title	author
t1	a1

Fig. 2. Record Sets among Three Peers

In the logical layer, records in the user layer are managed based on the relational data model. In each peer, three relational views are constructed and maintained. First, relation `Data[Novel]` in Fig. 3 expresses a view that unifies all the novel records held by peers A, B and C shown in Fig. 2. The view also contains old record values that were deleted or modified by users. They are hidden from the user layer but used for tracing histories. Attributes `title` and `author` are visible from the user layer, but the other two attributes `peer` and `id` are used for management. Attribute `peer` represents the logical name of the

title	author	peer	id
t1	a1	A	#A011
t5	a5	A	#A028
t1	a1	B	#B032
t2	a2	B	#B040
t2	a3	B	#B051
t1	a1	C	#C005
t6	a6	C	#C077

from_id	to_id	time
–	#A011	5/2/07
–	#A028	8/18/07
–	#B032	4/10/07
–	#B040	4/20/07
#B040	#B051	6/10/07
–	#C005	3/20/07
#C077	–	10/06/07

Fig. 3. View `Data[Novel]`   **Fig. 4.** View `Change[Novel]`

peer used for sending a message to other peers in the P2P network. Attribute id is used to identify records during query processing.

Second, the relation `Change[Novel]` shown in Fig. 4 is a global view containing the insertion, modification, and deletion histories. Attributes `from_id` and `to_id` express the record ids before/after a modification. Attribute `time` represents the timestamp of the modification. When the value of the `from_id` attribute is the null value ($-$), it signifies that the record has been inserted. Similarly, when the value of the `to_id` attribute is the null value, it means that the record has been deleted.

Finally, view `Exchange[Novel]` shown in Fig. 5 stores information regarding record exchange among peers. Attributes `from_peer` and `to_peer` express the origin and the destination of record exchanges, respectively. Attributes `from_id` and `to_id` contain the logical ids of the exchanged record in both peers. Attribute `time` stores a timestamp expressing the time when the record was copied to the peer. For example, the first tuple shows that peer A copied the record from peer B, where it had the id value `#B032`, and peer A assigned a new id `#A011` for the record when it was registered at peer A. Record exchanges among peers can be traced using the three views.

from_peer	to_peer	from_id	to_id	time
B	A	#B032	#A011	5/2/07
C	B	#C005	#B032	4/10/07

Fig. 5. View `Exchange[Novel]`

3.2 Representation of Queries in the Logical Layer

In this subsection, we describe the representation of queries in the logical layer. Since recursive processing is needed in order to trace information in a network, queries are written using *datalog* [2,9]. Datalog has been used for network-oriented query processing in fields such as in *declarative networking* [8]. We now present some example tracing queries.

Query 1: Suppose that peer A holds a record with title `t1` and author `a1` and that peer A wants to know which peer originally created the record. This query may be described as follows. Strings such as P and I1 represent variables and '_' indicates an anonymous variable. The last rule represents the final result expected by the user.

```
BReach(P, I1) :- Data[Novel]('t1', 'a1', 'A', I2),
                 Exchange[Novel](P, 'A', I1, I2, _)
BReach(P1, I1) :- BReach(P2, I2), Exchange[Novel](P1, P2, I1, I2, _)
Origin(P) :- BReach(P, I), NOT Exchange[Novel](_, P, _, I)
Query(P) :- Origin(P)
```

Relation BReach defined by the first two rules means "Backward Reachable". In BReach(P, I), the symbol P represents the name of the peer which was in the path from the originator of the record to peer A and I is the id of the record (t1, a1) when it was at peer P. In the first rule, the name of the peer which handed peer A the record directly is sought using the information in Data[Novel] and Exchange[Novel]. The second rule is for recursive processing. Thus, BReach collects information regarding all the peers which are in the path from the originator to peer A for the record in question. The third rule is used for selecting the peer at which the record originated. The originating peer should be reachable from peer A and should not have received the record from any other peer. If this query is processed according to the example views shown in Fig. 3, C is returned as the originating peer.

Query 2: Suppose that peer A wishes to know which of its own records were obtained via peer B. This query may be expressed as follows:

```
BReach2(P, I1, T, A) :- Data[Novel](T, A, 'A', I2),
                        Exchange[Novel](P, 'A', I1, I2, _)
BReach2(P1, I1, T, A) :- BReach2(P2, I2, T, A),
                         Exchange[Novel](P1, P2, I1, I2, _)
ViaB(T, A) :- BReach2('B', _, T, A)
Query(T, A) :- ViaB(T, A)
```

BReach2 has a similar structure to BReach in Query 1. The difference is that BReach2 holds additional information regarding novel titles and author names. BReach2(P, I, T, A) means that a record at peer A was a copy of the record with title T and author A held by peer P with id I. In the third rule, the titles and the authors which satisfy the constraints are extracted.

Query 3: In the process of P2P record exchange, there is a chance that a record (t1, a1) obtained from another peer in the past may be obtained again at some point. Suppose that peer A wishes to verify whether the recently obtained record is the same as the record already registered in its local system. It is easy to discover whether there is a record at peer A which has the same value with the given record (t1, a1) using a search query. However, even if the value is same, there is a possibility that some other peer has independently created a record with the same value. Suppose that peer A received a record (t1, a1) from peer D in which the record has the id #D051. The following query checks whether the originator of the given record (t1, a1) is the same as that of the record (t1, a1) at peer A with id #A011.

```
BReachA(P, I1) :- Data[Novel]('t1', 'a1', 'A', '#A011'),
                  Exchange[Novel](P, 'A', I1, I2, _)
BReachA(P1, I1) :- BReachA(P2, I2), Exchange[Novel](P1, P2, I1, I2, _)
BReachD(P, I1) :- Data[Novel]('t1', 'a1', 'D', '#D051'),
                  Exchange[Novel](P, 'D', I1, I2, _)
BReachD(P1, I1) :- BReachD(P2, I2), Exchange[Novel](P1, P2, I1, I2, _)
Dup :- BReachA(_, I), BReachD(_, I)
Query :- Dup
```

The first and the second rules collect the trace information regarding peer A.
The third and the fourth rules play the same role for peer D. The fifth rule
investigates whether peer A and peer D share a record id for identifying records.
If this rule is satisfied, it can be concluded that the sources are the same.

Query 4: Suppose we wish to determine whether a record (t1, a1) held by peer
A is the newest version. That is to say, we wish to determine whether some peer
which gave the record to peer A, either directly or indirectly, has subsequently
modified its record value. Suppose that peer A received the record (t1, a1)
from some peer and that peer A wishes to determine whether the record was
modified by any of the peers in the path from peer A to the originating peer.

The following query satisfies the requirement. BReach3 is identical to BReach,
shown above. The third rule detects if a record with id I1 has been modified.
The constraint that I2 is not empty ensures that I1 is not a deleted record.

```
BReach3(P,I1,T,A) :- Data[Novel]('t1', 'a1', 'A', I2),
                     Exchange[Novel](P, 'A', I1, I2, _),
                     Data[Novel](T, A, P, I1)
BReach3(P1,I1,T,A) :- BReach3(P2, I2,_,_),
                      Exchange[Novel](P1, P2, I1, I2, _),
                      Data[Novel](T, A, P1, I1)
Modified(P,T,A) :- BReach3(P, I1, T, A),
                   Change[Novel](P, I1, I2, _), I2 != NULL
Query(P,T,A) :- Modified(P,T,A)
```

4   The Physical Layer

4.1   Basic Idea

In the logical layer, queries are expressed using virtual views that unify all the
information in a P2P network. However, it is inappropriate to materialize the
views at a central server due to the following reasons.

- Each peer in a P2P network acts autonomously and functions cooperatively.
  A peer may not be interested in all the information in a network. Thus, it
  is not required to manage all the information in a central server.
- Although materialized views support efficient processing for tracing queries,
  such queries may be issued infrequently. Therefore, management of materialized views in this context may not be economical due to the high communication and processing cost.

In our framework, we assume that each peer in a network manages the information related to itself. Tracing queries in the logical layer are processed cooperatively by the peers in the physical layer.

4.2   Data Representation

The three relations in the logical layer are represented by four relations in the physical layer. The relations in Figs. 6 and 7 correspond to Data[Novel] and Change[Novel] in the logical layer, as shown in Figs. 3 and 4. These relations represent the information managed by peer A.

title	author	id
t1	a1	#A011
t5	a5	#A028

Fig. 6. Data[Novel] of Peer A

from_id	to_id	time
−	#A011	5/2/07
−	#A028	8/18/07

Fig. 7. Change[Novel] of Peer A

The content of Exchange[Novel] in the logical layer is partitioned and distributed among peers. The relation From[Novel] in the physical layer shown in Fig. 8 corresponds to Exchange[Novel] shown in Fig. 5 and contains records received by peer A. Figure 8 shows the record with id #A011 is copied from the record with id #B032 at peer B. In addition, each peer records information when it provides a record to another peer. Fig. 9 shows the To[Novel] relation of peer B, which corresponds to the From[Novel] relation of peer A shown in Fig. 8. It signifies the record with id #B032 at peer B was copied by peer A with id #A011.

id	from_peer	from_id	time
#A011	B	#B032	5/2/07

Fig. 8. From[Novel] of Peer A

id	to_peer	to_id	time
#B032	A	#A011	5/2/07

Fig. 9. To[Novel] of Peer B

From[Novel] and To[Novel] contain duplicated information but are stored by different peers. In the above example, when peer A copies the record from peer B, the following steps are performed:

1. Peer A sends a query to peer B. Peer A obtains a record set from peer B.
2. Peer A selects the record with id #B032 and registers it in its own local record management system.
3. Peer A assigns id #A011 to the record and inserts the information into its Data[Novel] relation. The information that peer A received the record from peer B is then recorded in From[Novel] relation.
4. Peer A transmits the update information "#B032 was registered with id #A011 and timestamp 5/02/07" to peer B.
5. Peer B records the information from peer A in its To[Novel] relation.

$$
\begin{array}{c|c}
\text{P} & \text{I} \\
\hline
\text{B} & \text{\#B032} \\
\text{C} & \text{\#C128} \\
\text{B} & \text{\#B093}
\end{array}
$$

Fig. 10. Relation BReach at Peer A

The reason for performing the process shown above is that peer B cannot know which records among those sent to peer A were actually registered at peer A, without the information provided in the message from peer A. The whole process is executed as a distributed transaction among peers A and B in order to ensure the consistency of the information stored.

4.3   Query Processing

In the logical layer, queries are described in datalog using virtual views that integrate all of the information in a P2P network. In order to process a tracing query, it is necessary to transform the given query to suit the organization of the physical layer. The approach for achieving this is described below.

Query Mapping to the Physical Layer. Consider again Query 1 and assume that the query is issued at peer A. In order to map logical relations into physical relations, the query is translated into the following physical layer query:

```
BReach(P, I1) :- Data[Novel]@'A'('t1', 'a1', I2),
                 From[Novel]@'A'(I2, P, I1, _)
BReach(P1, I1) :- BReach(P2, I2), From[Novel]@P2(I2, P1, I1, _)
Origin(P) :- BReach(P, I), NOT From[Novel]@P(I, _, _, _)
Query(P) :- Origin(P)
```

Notation of the form Data[Novel]@'A' indicates a physical relation at a specific peer, in this case the relation Data[Novel] at peer A. Similarly, From[Novel]@P represents the relation From[Novel] at peer P. Note that P is a variable. The interpretation of the transformed query is straightforward: it traverses the path from peer A to the originator of the record (t1, a1) using recursive processing.

There are several options for the mapping. For example, we may replace every occurrence of Exchange[Data] in the logical query with To[Novel]. Although this produces another correct datalog query, it is hard to execute because it requires a traversal from the originator of the record to peer A.

Local Execution. When seeking the local rule which can be processed at peer A, in this example, the first rule is a local rule and can be executed at peer A immediately. As a result, BReach will contain the information regarding which peers directly provided the record (t1, a1) to peer A. Assuming that the local evaluation result shown in Fig. 10 was obtained, we note that the contents of Fig. 10 do not show the correct result given the relations in Figs. 6 and 8, but are used below for ease of understanding.

Query Forwarding. Peer A forwards the following sub-query corresponding to the transformed query to the peers that can directly execute it using the current contents of BReach.

```
BReach(P1, I1) :- BReach(P2, I2), From[Novel]@P2(I2, P1, I1, _)
Origin(P) :- BReach(P, I), NOT From[Novel]@P(I, _, _, _)
```

In this case, peer B will receive the sub-query with a set $\{(B, \#B032), (B, \#B093)\}$, which is a subset of tuples in BReach at peer A corresponding to peer B. Peer C also receives the sub-query and a tuple set $\{(C, \#C128)\}$.

In the next step, peer B may, for example, attempt to execute the given sub-queries. If peer B is the originator of the record, the query reaches a *fixpoint* and the resulting relation Origin will have a tuple (B), and the result is then returned to peer A. Otherwise, peer B forwards the sub-query with its partial result to the descendant peers on the path to the originator. Peer C also performs similar processing.

Collection of Results. As shown above, the given query is processed in a recursive manner in the P2P network. Peer A ultimately receives the results of all the recursive processing via peers B and C. These results are merged and presented to the user as the final result.

The query processing strategy shown above is an extension of the *semi-naive evaluation* strategy in deductive databases [2,9]. The approach of extending the strategy to distributed networks was originally presented in a *declarative networking* project [8]. Although our approach is a variation of the method presented in [8], our query processing strategy has some differences in the framework which reflect the three-layer organization of record exchange and the structure of logical/physical relations for representing the information used for tracing.

5   Related Work

There are a variety of research topics regarding P2P databases, such as coping with heterogeneities, query processing, and indexing methods [1]. In this paper, we provided a different viewpoint to the research field. The proposed approach is based on the requirement for reliable information exchange in P2P networks. One of the features of our approach is to employ database technologies as the underlying foundation with which to support reliable P2P record exchange.

One related research field is data provenance. The term *data provenance*, or alternatively *lineage tracing*, previously referred to the process of tracing and recording the origins of data and its movement between databases [5,10]. The target field of data provenance is comparatively wide and covers data warehousing [4], uncertain data management [11], and other scientific fields such as bioinformatics [3]. However, the notion of data provenance has not previously been applied to P2P information exchange to the best of the authors' knowledge.

Some taxonomies have been produced regarding data provenance. [6] presents the notions of *where-provenance* and *why-provenance*. Since our framework treats problems such as where data came from and whether data was reproduced by other peers, it belongs under the heading where-provenance. Another taxonomy distinguishes between the *lazy* approach and the *eager* approach [10]. The former describes models in which queries tracing lineage are executed when necessary and the latter describes the case that metadata and/or annotations [3] representing lineage are maintained. Our approach to traceability is based on histories maintained at peers and thus belongs to the eager approach.

Another related field is *dataspace management* [7]. This is an emerging new research field in the area of databases and focuses on more flexible information integration over the network in an incremental, "pay-as-you-go" fashion. Our approach to P2P record exchange can also be seen as an extension of the traditional approach of data integration: peers in a P2P network can behave autonomously and exchange information when required. In this sense, our information integration framework is quite flexible, but the framework includes traceability functions which yield reliable record exchange.

Our query processing approach is a variation of *declarative networking* as described in [8]. In contrast to their approach, which focuses on efficient query processing in a network (e.g., a sensor network), our target is to represent tracing queries in a compact and clear manner. Since our framework shares the requirement for efficient query processing with their approach, it will be possible to extend our query processing method by considering this former work.

6   Discussions and Conclusions

In this paper, the concept of traceable P2P record exchange was presented and a framework implementing it was shown. The proposed framework consists of three layers, and tracing queries are written in datalog and executed in a recursive manner. We need to consider several issues for the practical implementation of our ideas, for example:

- *Joining/Leaving Facilities*: A dynamic P2P network should provide facilities for joining and leaving for peers, and the peers in the network need to preserve the consistency of the information contained in the network in a cooperative way. The problem is more serious for the leaving facility. One solution would be as follows: 1) the leaving peer A selects a voluntary peer B which can take over its data, 2) peer B copies all the data maintained by peer A, 3) peer B replies to all the subsequent queries on behalf of peer A. We assume that all peers will obey the joining/leaving protocols.
- *Search Facility*: For the search processing, we can apply the existing approaches for efficient P2P search. Some types of the P2P search methods focus on anonymity while exchanging information. In contrast, our method needs search function that explicitly preserves identities of peers which provide the records. That may require slight changes to existing search methods.

In addition, the following issues will be addressed in future work:

- Enhancement of the tracing facilities: It is intended to extend the tracing query language to represent more detailed information.
- Development of efficient query evaluation and optimization techniques: As described above, we will further develop strategies considering the existing methods for deductive databases and P2P databases.
- Prototype system implementation and experiments: We are planning to develop a prototype system for P2P record exchange over relational database management systems (RDBMSs). For this purpose, it is necessary to translate queries in datalog into SQL queries for execution using an RDBMS. It is also necessary to implement facilities for searching, registration, and modification of records.

Acknowledgements

This research was partly supported by a Grant-in-Aid for Scientific Research (19300027) from the Japan Society for the Promotion of Science (JSPS). In addition, this work was supported by grants from the Kayamori Foundation of Informational Science Advancement and the Hoso Bunka Foundation.

References

1. Aberer, K., Cudre-Mauroux, P.: Semantic overlay networks. In: VLDB, (tutorial notes) (2005)
2. Abiteboul, S., Hull, R., Vianu, V.: Foundations of Databases. Addison-Wesley, Reading (1995)
3. Bhagwat, D., Chiticariu, L., Tan, W.C., Vijayvargiya, G.: An annotation management system for relational databases. In: Proc. VLDB, pp. 900–911 (2004)
4. Cui, Y., Widom, J.: Lineage tracing for general data warehouse transformations. In: Proc. VLDB, pp. 471–480 (2001)
5. Buneman, P., Khanna, S., Tan, W.-C.: Data provenance: Some basic issues. In: Kapoor, S., Prasad, S. (eds.) FST TCS 2000. LNCS, vol. 1974, pp. 87–93. Springer, Heidelberg (2000)
6. Buneman, P., Khanna, S., Tan, W.-C.: Why and where: A characterization of data provenance. In: Van den Bussche, J., Vianu, V. (eds.) ICDT 2001. LNCS, vol. 1973, pp. 316–330. Springer, Heidelberg (2000)
7. Halevy, A., Franklin, M., Maier, D.: Principles of dataspace systems. In: Proc. ACM PODS, pp. 1–9 (2006)
8. Loo, B.T., Condie, T., Garofalakis, M., Gay, D.E., Hellerstein, J.M., Maniatis, P., Ramakrishman, R., Roscoe, T., Stoica, I.: Declarative networking: Language, execution and optimization. In: Proc. SIGMOD, pp. 97–108 (2006)
9. Ramakrishnan, R., Gehrke, J.: Database Management Systems, 3rd edn. McGraw-Hill, New York (2002)
10. Tan, W.-C.: Research problems in data provenance. IEEE Data Eng. Bull. 27(4), 45–52 (2004)
11. Widom, J.: Trio: A system for integrated management of data, accuracy, and lineage. In: Proc. CIDR, pp. 262–276 (2005)

ACORN : Towards Automating Domain Specific Ontology Construction Process

Eric Bae[1], Bintu G. Vasudevan[2], and Rajesh Balakrishnan[2]

[1] NICTA Victoria Laboratory, Department of Computer Science and Software Engineering, University of Melbourne, Australia
[2] SETLabs, Infosys Technologies Limited, Bangalore, India

Abstract. A number of ontologies have been recently developed in order to represent common knowledge in a structured manner. This allows users and agents involved in a particular domain to make inquiries and discover the underlying conceptual differences present in the data. However, currently the majority of ontology construction tools are heavily dependent on the human domain experts for selecting concepts and defining their relationships. In this paper, we would like to present a new tool called ACORN, which implements novel techniques for automatically extracting concepts and building concept-to-concept relationships. We first utilize the WordNet lexical database and term co-occurrence frequency for discovering domain specific concepts and introduce 'cluster mapping' and 'generality ordering' techniques for connecting these concepts. We apply our techniques to a widely available dataset and show that ACORN is able to produce high quality ontologies.

1 Introduction

The task of automatically acquiring knowledge from data is a critical research topic and a key component in many business. The importance of this endeavour has become more apparent recently with the advent of the Internet and the increasing amount of large and complex data being transacted online. However, without the provision of any external knowledge, it is difficult to automatically distinguish the underlying conceptual differences of natural text in data. Moreover, given a particular domain, there are various ways to represent common knowledge, which can create ambiguity between parties exchanging information.

To overcome the above difficulties (and for many other reasons), ontologies can be constructed, which arrange domain specific concepts and their relationships in a structured format. These ontologies then can serve as common repositories of knowledge for users and agents, which can be inquired and reasoned with to clarify the uncertain analysis of data and help valuable information to be extracted.

However, currently the construction of ontologies is predominantly performed by human domain experts. This manual process obviously requires much time and effort and for developing larger ontologies (i.e. SNOMED [4]), it is not uncommon to involve many contributors over a number of years. Although several research works have attempted at automating its construction, many are limited

Y. Zhang et al. (Eds.): APWeb 2008, LNCS 4976, pp. 487–498, 2008.

to just extracting concepts and defining relationships between them remains as a difficult task. In this paper, we would like to propose a new tool called ACORN[1], which includes the following novel features :

- Extracting domain specific concepts : Given an input corpus of documents, we first define a new keyword similarity function using JCn function [8] (based on the WordNet lexical database [5]) and term co-occurrence frequency values. Using this function allows us to discover concepts of higher quality than other techniques.
- Identifying concept-to-concept relationships : We explore two techniques to identify relationships between concepts. These are,
  - *Cluster mapping* method generates a cluster hierarchy from the input corpus and approximates the position of concepts in the final ontology.
  - *Generality ordering* technique determines how general each concept is and locates concepts in their ontology.

This paper is organized as follows. In the next section, we explore some of other relevant work which automates the ontology construction task. In section 3, we describe our tool and its components in details. We provide the results of our experiments in section 4 to demonstrate the benefits of using ACORN and finally we propose extensions to our tool for future work in section 5.

2   Related Work

We note here that the automatic construction of a complete and comprehensive ontology is an extremely difficult task. Ontologies often capture relationships, constraints and rules between concepts are specific to domain, which may not be directly attainable from input corpus. Commonly, many tools present a semi-automatic environment which provides a 'skeleton' ontology (i.e concepts and some relationships defined) upon which human experts can include further details regarding its domain. With this in mind, here we describe some of the related work.

In [7], authors introduce 'OntoGen' which uses singular value decomposition, k-means clustering algorithm and a SVM classifier to extract domain specific concepts from a document corpus. Once concepts are found, a human expert must identify relationships between these concepts through OntoGen's GUI. Although, the concepts extracted from this system are reasonable and its outlier detection feature can further enhance the quality of concepts, OntoGen still requires significant effort and time by human. In [2], COBWEB [6] algorithm was proposed as a tool to automatically generate ontologies. COBWEB algorithm is well known for extracting hidden concepts from data using its category utility function. However, results in [2] do not sufficiently support the direct adoption of COBWEB as an ontology construction tool and it was even found in [11] that other clustering techniques can outperform COBWEB.

---

[1] Automatic Construction For Ontologies.

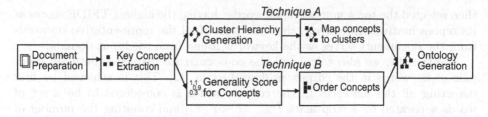

Fig. 1. ACORN Framework

In [14], a fuzzy formal concept analysis was used in order to extract key concepts from documents, which are subsequently clustered to form a hierarchical structure. However, from experiments, we have found that key concepts of higher quality can be extracted using simpler and common TFIDF weighting than using fuzzy formal concept analysis. Moreover, the technique is currently limited to work with only document citation information. On the other hand in [3], the author uses rough set theory to determine the relationships between concepts and implements a link-based clustering for further refining the concept groups. The technique, however, assumes that there is already a high level document clusters available and a number of initial thresholds, which need to be specified a priori, can vary the outcome. Finally in [13], *OntoEdit* is presented as a framework for managing ontology construction and maintenance. In relation to building an ontology, 'OntoEdit' uses its own lexical database to first extract concepts then uses association rule algorithm to build relationships between them. The problem here is that the lexical database itself may not contain the domain specific keywords, while the association rule algorithm works very similarly to simpler technique such as term co-occurrence measure [3].

From all the available methods, we observe that the ontology construction task involves two critical components - extracting concepts and identifying relationships between them. While many related works have included the former component, the latter is acknowledged as difficult and largely left to the experts. In ACORN we propose two different methods, which aim to identify meaningful relationships and we describe its process in detail in the following section.

3   ACORN Framework

We illustrate the overall process of ACORN in figure 1. Each of these components will now be described in detail.

Document Preparation. This stage processes the input corpus in such a way that appropriate data can be prepared for ontology construction. This requires first performing 'document cleansing' by removing all stop-words and applying stemming algorithm (i.e. Porter Stemmer) to prune unwanted words. Each term is then weighted with TFIDF score as described in [16]. From each document, we then selected the top k number of keywords[2] having the highest TFIDF scores as its representatives. Effectively, the collection of all the representative keywords from the documents serves as the keywords found from the input domain.

Additionally, we also calculated the co-occurrence frequency between every pair of keywords in the corpus as described in [12]. This is achieved by first detecting all the sentences in the corpus, which is considered to be a set of words separated by a stop mark (".", "?" or "!") and counting the number of times a pair of keywords occurs together in all sentences. The aim of finding the co-occurrence frequency is to determine the keywords, which are semantically similar (since they appear together frequently in sentences). We use this value in the later stages of ACORN to extract concepts and also to build relationships between them.

Concept Extraction. Once all the keywords are found, we would like to group together those which are conceptually similar. These groups of keywords will then form a particular concept present in the current domain. We define such similarity as below.

Definition 1 *KeywordSim(A,B). For a pair of words A and B, let the co-occurrence frequency value be given by $COOF(A,B)$ and $JCn(A,B)$ be the JCn [8] similarity measure according to the WordNet lexical database [5]. The overall conceptual similarity between a pair of words then can be given as below.*

$$KeywordSim(A,B) = \alpha(COOF(A,B) + JCn(A,B)) \qquad (1)$$

In the above definition, the α is the normalization factor which ensures that the sum of $COOF(A,B)$ and $JCn(A,B)$ gives a value between 0 and 1. The value of α was set to 0.5 in ACORN. The main emphasis with the equation 1 is that we are able to leverage the positive features of both measures. While JCn measure provides the general conceptual similarity between two words, the $COOF$ can identify the similarity, specific to the given domain. Moreover, the WordNet may not contain all the terms extracted, whereas $COOF$ will not be sufficient to detect conceptually similar, words which do not occur in the same sentences.

Given these keywords and their pairwise similarity values, we can present them in a graph $G = (V, E)$, called 'concept lattice' where $V = \{v_1, v_2, .., v_n\}$ is a set of vertices containing each keyword, and $E = \{e_1, e_2, .., e_m\}$ is a set of edges between vertices. Each edge connecting vertices v_i and v_j is weighted with the similarity and is expressed as $w_{ij} = KeywordSim(v_i, v_j)$. The illustration is given in figure 2.

Once the keywords and the pairwise similarity values are represented in the lattice , we can merge closely related keywords together to form concepts. We achieve this by setting a similarity threshold t where $0 \leq t \leq 1$ and combine the keywords (where the $w_{ij} > t$) and form a cluster of concepts (concept node). We

---

[2] We varied the value of k from 5 to 20 and found $k = 10$ to be the most reasonable.

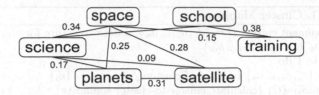

Fig. 2. Keywords and their similarity values represented in a concept lattice. The keywords are extracted from 'space' newsgroup in 20 newsgroup dataset.

| science space | spacewalk space | space mission age | school training | jet propulsion laboratory |

Fig. 3. Domain specific concepts extracted from the 'space' newsgroup dataset by merging semantically similar words together according to the threshold t

show some of the concepts generated from the 'space' newsgroup, when $t = 0.3$ in figure 3.

In section 4, we show the concepts extracted from ACORN and evaluate their quality. The next stage involves developing relationships between the key concepts, and as aforementioned, we provide two techniques - cluster mapping and generality ordering - to build a hierarchical domain ontology.

3.1 Ontology Construction Through Cluster Mapping

Although a number of research works [13,7,2] have applied some type of clustering in their ontology construction process, its use has been limited to keyword extraction and grouping them together for forming concepts. However, clusters generated from the input corpus, especially hierarchical clusters (or dendrograms) by agglomerative clustering algorithms, can provide valuable information for establishing relationships between concepts.

Although it could be argued that the resultant dendrogram can be directly used as an ontology, such work has had a limited success [10] and it does not appropriately capture the conceptual taxonomy information contained in the domain. Instead, in this section, we introduce a 'cluster mapping' technique, in which the document cluster hierarchy generated by the proposed algorithm aids concepts to be mapped to its approximate position in the ontology. The mapping is achieved by comparing the keywords in each concept group discovered in the previous step to the labels representing clusters. The assumption here is that the document clusters will place general topics towards the top of the hierarchy, while more specific topics are positioned near the bottom. The algorithm is shown in algorithm 1.

The clustering algorithm we implement is a *single-linkage* algorithm [9], which is a popular agglomerative clustering method and has been tested in [11] to be the best performing algorithm for producing a meaningful dendrogram. In this

---

Algorithm 1. Cluster Mapping

Require: document corpus $D = \{d_1, d_2, .., d_n\}$, a data structure for storing hierarchy T, concepts $K = \{k_1, k_2, .., k_m\}$

1: **for** $i = n$ to 1 **do**
2:     let $C = \{c_1, c_2, .., c_n\}$ be a set of clusters, where $c_i = \{d_i\}$
3:     $calculateSim(C)$ {calculate cluster-to-cluster similarity}
4:     $(c_1, c_2) = maxSim(c_i, c_j)\forall i, j$ where $1 \leq i, j \leq n$ {$maxSim$ finds a pair of clusters with maximum similarity (or smallest cosine angle)}
5:     $c_{merged} = \{c_1, c_2\}$
6:     $recalculateTFIDF(c_{merged})$
7:     let $labels_{c_{merged}} = \{w_1, w_2, .., w_k\}$ {select p highest TFIDF-weighted terms as labels}
8:     $calculateSim(C)$ {recalculate the cluster-to-cluster similarity}
9:     $insert(T, c_{merged})$
10: **end for**
    {For each concept k_i, we traverse through the cluster hierarchy tree T and find the cluster node with the highest similarity}
11: **for** $i = 0$ to m **do**
12:     $n_{closest} = max(similarNode(k_i, T))$ {traverse through the cluster hierarchy T and calculate the $KeywordSim$ between keywords of k_i and labels of every cluster in T}
13:     $map(n_j, k_i)$
14: **end for**

---

algorithm, each document is initially assigned to a single cluster (line 2) and a pair of most similar clusters are merged (lines 3 to 5). This process is continued until only one cluster remains. The similarity between clusters is calculated by the cosine angle between closest documents from each cluster (line 4) since each document is represented by k number of TFIDF-weighted words (as mentioned in section 3 and therefore can be projected onto a vector space. Additionally, after each cluster merge, we recalculate the TFIDF weights of terms in the new cluster and reselect the top k terms as cluster labels. We also insert each merged cluster into a tree data structure T for mapping concepts (line 9).

Once the hierarchy is built, we would like to map each concept k_i in K to a node n_j in T (line 11 to 13). We achieve this by traversing through the tree T and calculating the similarity between concepts and nodes. This similarity can be defined as below.

Definition 2 *ConceptNodeSim(k,n). Let $ConceptNodeSim(k, n)$ be a function, which calculates the similarity between a concept k and a node n. It is given by the average similarity between every pair of terms in the concept and terms in the tree node (labels of cluster node).*

$$ConceptNodeSim(k, n) = \frac{\sum_{i=1}^{|k|} \sum_{j=1}^{|n|} KeywordSim(k_i, n_j)}{|k| \cdot |n|} \quad (2)$$

where $|k|$ and $|n|$ indicate the number of terms contained in the concept and labels respectively

If more than one concept is mapped to a node n, we create an another child node to the parent node of n (effectively creating a sibling node) to assign extra concepts. On the other hand, all cluster nodes which do not have concepts assigned will be pruned away. The resultant tree forms an ontology, which will then only show cluster nodes with the concepts assigned.

3.2 Ontology Construction Through Generality Ordering

The second technique we introduce in ACORN for identifying the hierarchical relationships between concepts is through 'generality ordering'. In this technique, we quantitatively calculate the 'generality' of a concept, which effectively determines how general (or conversely, specific) the concept is within the given domain. Ordering the concepts by this generality score will provide information on their positions in the ontology. We define this generality as below.

Definition 3 *ConceptGenerality*. *Given a concept $k = \{w_1, w_2, .., w_p\}$, which is a set of semantically related keywords, ConceptGenerality(k) is a function, which averages the inverse document frequency (IDF) weights of each keyword w_i in k, given as follows.*

$$ConceptGenerality(k) = \frac{\sum_{i=1}^{p} log\frac{|D|}{n_{w_i}}}{p} \tag{3}$$

The IDF weighting scheme has been widely used for determining the 'generality/specificity' of a term within a corpus [15] and in equation 3, we calculate the average IDF value from all the keywords in the concept. The overall algorithm for generality ordering technique is described in algorithm 2.

After calculating and ordering concepts by the generality score (line 1 to 4), we also calculate concept-to-concept similarity (line 5-9), which is defined identically as the equation 2, except we are comparing keywords between two concepts. Since the linearly ordered concepts does not reflect the hierarchical structure of the final ontology, the concept-to-concept similarity can serve as additional information for specifying the exact location of a concept, which is achieved in the third step of the algorithm (line 10 to 16). Here, a concept k_i is inserted into the tree T under the node which contains the concept which is more general than k_i and also shares the highest similarity value (line 14). This ensures that each concept belongs to the most similar yet more general parent concept. The steps 2 and 3 in algorithm 2 are illustrated in figures 4 and 5 respectively.

3.3 Ontology Generation

The two techniques introduced above return a tree structure containing various domain specific concept nodes, which can be directly viewed as an ontology and

---

Algorithm 2. Ontology Construction through generality ordering

---

Require: concepts $K = \{k_1, k_2, .., k_m\}$, ordered concept array O, ontology tree T
{Step 1 : First calculate the generality of concepts}
1: **for** $i = 1$ to m **do**
2:     $g_{k_i} = Generality(k_i)$
3:     $insert(O, k_i, g_{k_i})$ {insert k_i into O in the descending order of generality score}
4: **end for**
{Step 2 : Calculate concept-to-concept similarity}
5: **for** $i = 1$ to m **do**
6:     **for** $j = 1$ to m **do**
7:         $ConceptSim(k_i, k_j) = \frac{\sum_q \sum_r KeywordSim(w_{q_i}, w_{r_j})}{|k_i|x|k_j|}$
8:     **end for**
9: **end for**
{Step 3 : Generate the ontology}
10: **for** $i = 1$ to m **do**
11:    **if** $i == 1$ **then**
12:        $root(T) = k_i$
13:    **else**
14:        traverse T and find a node n where the $generality(k_n) > k_i$ and $max(ConceptSim(k_n, k_i))$
15:        create a new node n_{new} and assign k_i
16:    **end if**
17: **end for**

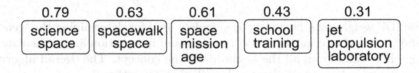

Fig. 4. Concepts ordered by their generality

we have described how ACORN is able to automate this process. As aforementioned, however, constructing a full featured ontology (i.e. including constraints and hidden rules between concepts) is a difficult task without any other external knowledge. Therefore, in ACORN, we represent its results as 'templates' and display them through a graphical user interface, from which the domain experts can make further enhancements.

4   Experiments

For our experiments, we have used 20 newsgroup dataset available from [17]. This dataset comprises of 20,000 messages from 20 newsgroups (2000 per each group). We have applied each newsgroup articles as input data to generate an ontology specific to the given newsgroup domain. The dataset was applied through the

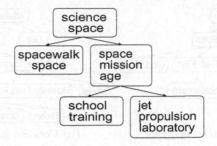

Fig. 5. Concepts are inserted into a hierarchy in the order of generality scores. Moreover, the task of selecting parent concept is determined by concept-to-concept similarity. A concept will become a child node to a concept which shares the highest similarity.

ACORN process described in figure 1. Due to the space constraints, we are not able to show all the experimental results.

Concept extraction. In table 4, we present the concepts extracted by ACORN and compare them to those generated by OntoGen [7]. From the table, it is quite clear that the concepts from ACORN is more meaningful than those from OntoGen. For example, OntoGen makes inappropriate groups of keywords such as (IBM, speak, kind) and (era, total, field), which do not represent a specific concept for the given domain. On the other hand, ACORN is able to find many of domain specific concepts such as (Islam, Muslims, Fatwa) where Fatwa is an Arabic term used in Islamic religion and (Lankford, pinch, hitter), which describes a player who was a 'pinch hitter' in the baseball league.

Table 1. Some of the concepts extracted from OntoGen and ACORN from the 'atheism' and 'baseball' newsgroups

dataset	OntoGen	ACORN
atheism	definition, killed, suppose	Islam, Muslims, Fatwa
	God, sex, Islamic	God, believe, evidence
	science, objective, values	Christian, morality
	IBM, speak, kind	satan, Lucifer
	morality, sgi, humanity	worth, importance, usefulness
baseball	hitter, pinch, hit	bat, hit
	game, Gant, football	clutch, play, hitting
	hit, won, lost	blacks, hispanic, players
	era, total, field	Lankford, pinch, hitter
	Dave, Smith, Eddie	pitching, box

Discovery of concept-to-concept relationships : Figures 6 and 7 display the parts of domain specific ontologies generated by both cluster mapping and generality ordering technique.

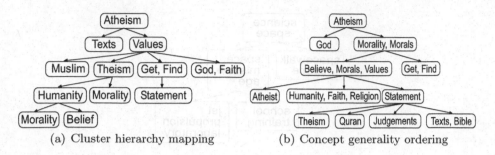

(a) Cluster hierarchy mapping   (b) Concept generality ordering

Fig. 6. Ontologies generated by two techniques on atheism newsgroup articles

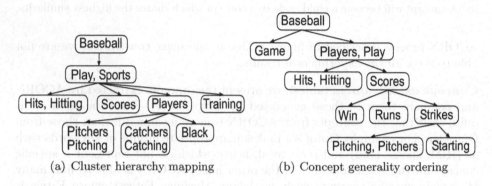

(a) Cluster hierarchy mapping   (b) Concept generality ordering

Fig. 7. Ontologies generated by two techniques on baseball newsgroup articles

As it can be observed, all four ontologies are able to establish meaningful relationships between majority of concepts. For instance, the generality ordering technique was able to join 'statement' as a parent and 'theism', 'quran' and 'bible' as its children. Such relationship particularly referred to the religious statements being made within the newsgroup documents. Furthermore, cluster mapping also correctly connected 'players' as a parent and 'pitchers', 'catchers' and 'black' as its children, indicating that the documents in the 'baseball' newsgroup categorizes players into their fielding positions (pitchers and catchers), but also by their races (black and hispanic). The key feature to note here is that we are creating ontologies, which are specifically relevant to the domain suggested by the input corpus.

We can also notice that the two techniques work slightly differently. First, we found that the cluster mapping technique tends to produce more shallow ontologies than the generality ordering technique. This is due to some cluster node labels not necessarily matching closely with some of the concepts while multiple concepts were assigned to one node. On the other hand, the generality score ordering ensures that all concepts have different scores and determining their position of level in the hierarchy is calculated by similarity between the nodes. This allows less number of concepts to be assigned to the same node, hence creating a taller tree.

Both techniques also showed some limitations when connecting concepts. Firstly, some concepts were assigned to the wrong places in the hierarchy. For cluster mapping technique, terms like 'texts' in figure 6(a) does not seem to belong near the top. The term 'atheist' in the figure 6(b) for the generality ordering technique can be also repositioned. Lastly, concepts such as 'get,find' (figure 6(a)) and 'starting' (figure 7(b)) are ambiguous and seem redundant to be included in the ontology.

Overall, the generality ordering technique constructed slightly better ontologies than the cluster mapping technique. We have observed that the cluster mapping technique is heavily dependent on the quality of document clusters for generating valid ontologies, which did not occur for all experiments.

Nevertheless, the experimental results above show that through two novel techniques, ACORN is able to extract domain specific ontologies, which can be further enhanced by the domain experts but at the same time provides a shortcut in constructing ontologies.

5 Future Work

The experimental results highlighted the need for high quality document clusters when using cluster mapping technique. We plan to explore a wider range of document clustering algorithms in the future. Moreover, applying alternate clustering algorithms like COALA [1] may provide a different technique for mapping concepts as it generates multiple clusterings from the same dataset. Finally, we aim to include filtering techniques in ACORN in order to prune ambiguous and redundant concepts from ontologies using techniques such as singular value decomposition.

6 Conclusion

In this paper, we have identified the importance of automating the ontology construction process and introduced a new tool called ACORN, which implements techniques for extracting concepts and building relationships between them. Through experimental results, we found that ACORN discovers domain specific concepts of high quality. Moreover, it was shown that both 'cluster mapping' and 'generality ordering' techniques are useful in defining relationships between concepts. Through the evaluation of our results, we have also identified the limitations of our methods and highlighted the areas to improve for future work.

References

1. Bae, E., Bailey, J.: COALA: A novel approach for the extraction of an alternate clustering of high quality and high dissimilarity International Conference on Data Mining, pp. 53–62 (2006)
2. Clerkin, P., Cunningham, P., Hayes, C.: Ontology discovery for the semantic web using hierarchical clustering Semantic Web Mining Workshop (2001)

3. Dey, L., Rastogi, A., Kumar, S.: Generating Concept Ontologies through Text Mining. In: Proceedings of Web Intelligence Conference, pp. 23–32 (2006)
4. Elkin, P., Brown, S.: Automated enhancement of description logic-defined terminologies to facilitate mapping to ICD9-CM. Journal of Biomedical Informatics 35, 281–288 (2002)
5. Fellbaum, C.: WordNet: An electronic lexical database. MIT Press, Cambridge (2000)
6. Fisher, D.: Knowledge acquisition via incremental conceptual clustering. Machine Learning 2, 139–172 (1987)
7. Fortuna, B., Grobelnik, M., Mladenic, D.: Semi-automatic Construction of Topic Ontology The Second International Workshop on Knowledge Discovery and Ontologies (2005)
8. Jiang, J., Conrath, D.: Semantic similarity based on corpus statistics and lexical taxonomy International Conference Research on Computational Linguistics (1997)
9. Johnson, S.: Hierarchical clustering schemes Psychometrika, vol. 2, pp. 241–254 (1967)
10. Khan, L., Wang, L.: Automatic ontology derivation using clustering for image classification International Workshop on Multimedia Information Systems, pp. 56–65 (2002)
11. Leouski, A., Croft, W.: An evaluation of techniques for clustering search results Technical Report IR-76, University of Massachusetts (1996)
12. Matsuo, Y., Ishizuka, M.: Keyword extraction from a single document using word co-occurrence statistical information International. Journal on Artificial Intelligence Tools (2003)
13. Maedche, A., Staab, S.: Mining Ontologies from Text Proceedings of the 12th European Workshop on Knowledge Acquisition. In: Modeling and Management, pp. 189–202 (2000)
14. Quan, T., Hui, S., Fong, A., Cao, T.: Automatic generation of ontology for scholarly semantic web. In: Proceedings of International Semantic Web Conference (2004)
15. Robertson, S.: Understanding Inverse Document Frequency: On theoretical arguments for IDF. Journal of Documentation 60, 503–520 (2004)
16. Salton, G., Buckley, C.: erm-weighting approaches in automatic text retrieval. Information Processing and Management 24, 513–523 (1988)
17. 20 Newsgroup dataset,
http://kdd.ics.uci.edu/databases/20newsgroups/20newsgroups.html
18. Zhao, Y., Karypis, G.: Comparison of agglomerative and partitional document clustering algorithms. In: SIAM Workshop on Clustering High-dimensional Data and its Applications (2002)

Ontological Knowledge Management Through Hybrid Unsupervised Clustering Techniques

Ching-Chieh Kiu[1] and Chien-Sing Lee[2]

Faculty of Information Technology,
Multimedia University,
Jalan Multimedia, 63100 Cyberjaya. Malaysia
{cckiu, cslee}@mmu.edu.my

Abstract. In the Semantic Web, ontology plays a prominent role to actualize knowledge sharing and reuse among distributed knowledge sources. Intelligently managing ontological knowledge (classes, properties and instances) enables efficacious ontological interoperability. In this paper, we present a hybrid unsupervised clustering model, which comprises of Formal Concept Analysis, Self-Organizing Map and K-Means for managing ontological knowledge, and lexical matching based on Levenshtein edit distance for retrieving knowledge. The ontological knowledge management framework supports the tasks of adding a new ontological concept, updating and editing an existing ontological concept and querying ontological concepts to facilitate knowledge retrieval through conceptual clustering, cluster-based identification and concept-based query. The framework can be used to facilitate ontology reuse and ontological concept visualization and navigation in concept lattice form through the formal context space.

Keywords: Ontology management, ontology editing, ontology update, clustering techniques, knowledge sharing and reuse.

1 Introduction

Ontology is added to knowledge management application design to enable knowledge sharing, reuse and fusion across knowledge communications [1]. To allow reliable knowledge exchange over the network, the ontology that holds the concepts needs to be defined explicitly and formally to the data in the knowledge management repositories, whereby each data is associated with ontological concepts and their corresponding attributes. Intelligently managing ontological knowledge (concepts, properties and instances) will contribute to efficient data retrieval from the knowledge repository, and facilitate knowledge reusability among knowledge seekers.

In this paper, we propose an algorithmic framework for managing ontological knowledge that adds semantics to the data in the knowledge repository. The framework comprises of three methods: conceptual clustering, cluster-based identification and concept-based query for managing ontological knowledge. The algorithmic framework employs unsupervised data mining methods incorporating Formal Concept Analysis (FCA) [2], Self-Organizing Map (SOM) and K-Means [3] to intelligently

Y. Zhang et al. (Eds.): APWeb 2008, LNCS 4976, pp. 499–510, 2008.

manage ontological knowledge without prior knowledge; and lexical matching based on Levenshtein edit distance [4] to resolve lexical similarity of the ontological concepts in order to effectively manage the ontological knowledge and also to facilitate knowledge retrieval in the knowledge repository [5].

The framework is able to capture ontological components, which form the metadata schemas in the form of attributes. Therefore, it may be easier to modify and update metadata schemas ontologically when the need arises. In the framework, the ontology is conceptualized into a concept lattice to enable effective ontological knowledge management (edit, update and query) by capitalizing the formal context space of the ontological concept through semantic clusters generated by unsupervised clustering techniques without the need for prior knowledge [5]. The ontological concept can be effectively created and the metadata of existing ontological concepts can be easily updated and altered within the specified related cluster. In addition, ontological representation in concept lattice form offers a scalable or composited environment for visualization of concepts and its properties and instances.

The rest of the paper is outlined as follow. Section 2 presents related work and Section 3 explains the definition of ontology and the clustering techniques used. The prototypical framework of ontological knowledge management is presented in Section 4. Meanwhile the ontological knowledge management tasks and algorithms of conceptual clustering method, cluster-based identification method and concept-based query method for managing ontological knowledge are presented in Section 5. Lastly, the paper is concluded with future work.

2 Related Work

According to [6], ontology management functions include facilities for browsing, searching, querying and managing ontologies and their instances. These functions are resembled in ontology editors. There are a lot of ontology editor tools, which can be found at [7]. However most of the tools provide environment for building ontologies but not functionalities for managing ontologies as stated by [6]. An existing tool that is closely related to our work is the Cmaptools Ontology Editor (COE) [8].

COE supports importing, editing and storing of OWL ontologies in the form of concept maps (Cmaps). COE incorporates agglomerative, hierarchical clustering algorithms to group similar OWL concepts into clusters based on similarity. It allows searching for concepts and properties in existing concept maps to locate potentially useful concepts, and locating clusters of contextually relevant concepts in the existing ontology. COE employs a two-stage search algorithm to query the neighborhood of concepts/ontologies in the repository. The first-stage finds concepts that match the query based on simple substring comparisons. The second-stage finds the focus clusters in which the matching concepts have stabilized whereas the smallest clusters that fully contain the matching concepts.

Our objective is to develop robust and scalable techniques to manage the knowledge repository's ontology and its instances, especially for managing large ontologies to enable knowledge sharing and reuse among the knowledge seekers. Our work is different from the review tool, whereby the ontological concept adding, editing, updating and querying are performed through formal context or concept lattice derived from the specific-defined clusters, instead of managing the ontology directly from the ontological context. Managing ontological concepts and their elements directly from its ontological context is difficult and time-consuming when the ontology is huge and large. In addition, with our hybrid model, prior knowledge is not required.

3 Methodologies for Ontological Knowledge Management

In this section, the definition of ontology is presented followed by the techniques used in ontological knowledge management.

3.1 Ontologies

An ontology is an explicit specification of a shared conceptualization [9]. In general, ontology consists of concepts, attributes and relations. The relationship between the concepts is defined by a set of relations and the hierarchical relationship of the concepts. Subconcept-superconcept is defined by the *is-a* relation. The subconcepts inherit the attributes of the superconcepts. Hence, ontology can be formalized as a tuple $O: = (C, S_C, R, S_R, is\_a, A, I)$, where C is Concepts of ontology and S_C corresponds to the hierarchy of Concepts. The relationship between the concepts is defined by Relations, R where S_R corresponds to the hierarchy of Relations. *is_a* is the hierarchical relationship between the concepts. I is instances of the concepts and A is axioms used to infer knowledge from existing knowledge.

3.2 Formal Concept Analysis

Formal Concept Analysis (FCA) is an unsupervised learning technique and also a conceptual clustering tool used for discovering conceptual structures of data. To allow significant data analysis, a formal context is first defined in FCA. Consequently, the concept lattice is depicted according to the context to represent the conceptual hierarchy of the data. The concept lattice is a structured graph depicted according to the context. Concept lattice can be used to discover the relationship between objects and its attributes through the conceptual hierarchy of the data. More extensive explanation on the formal context and concept lattice can be referred at [2].

3.3 Self-organizing Map and K-Means

The Self-Organizing Map (SOM) is an unsupervised neural network used to cluster the data set according to similarity. SOM compresses complex and high-dimensional data to lower-dimensional data, usually to a two-dimensional grid. The most similar data are grouped together in the same cluster.

K-Means clustering is applied on the learnt SOM to reduce the problem size of the SOM cluster and visibly divided semantic contexts. K-means iteratively divides a data set to a number of clusters and minimizes the error function. To compute the optimal number of clusters k for the data set, the Davies-Bouldin Validity Index (DBI) is used. Refer [3] for the detail of the SOM and K-Means clustering algorithms.

3.4 Levenshtein Edit Distance

Levenshtein edit distance is a well known edit distance method for measuring the similarity between two strings as demonstrated in [4, 10]. In edit distance (*ed*), the cost of operation to transform one string into the other is measured based on the minimum number of substitutions, deletions and insertions. Based on the Levenshtein edit distance, a Lexical Similarity Measure (LSM) for two entries of ontological elements, $O_{elementA}$ and $O_{elementB}$ is depicted in Eq. 1. The degree of similarity returned by LSM is in the range of 0 to 1. LSM = 0 means no similarity and LMS = 1 means perfect similarity.

$$LSM(O_{elementAi}, O_{elementBj}) = 1 - \frac{ed(O_{elementAi}, O_{elementBj})}{max(len(O_{elementAi}), len(O_{elementBj}))} \in [0,1] \quad (1)$$

4   Framework for Ontological Knowledge Management

Our prototypical framework for the ontological knowledge management, reuse and sharing is depicted in Fig. 1. The framework consists of two core functionalities, which are semantic clustering generation and semantic query matching to facilitate ontological knowledge management tasks. Both functions are independent from each other. Hybrid unsupervised clustering FCA, SOM and K-Means are applied in the semantic clustering generation function, while the semantic query matching function

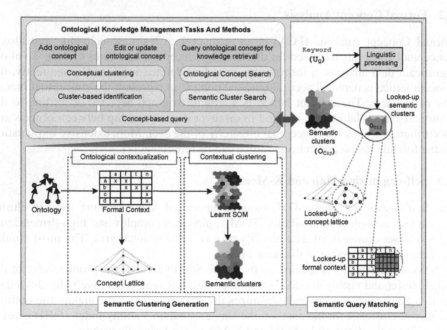

Fig. 1. Framework for ontological knowledge management

incorporates Levenshtein edit distance to resolve lexical similarity between a user-supplied query and ontological concepts in the knowledge repository.

The framework supports the adding, editing and updating of ontological concepts through conceptual clustering, cluster-based identification and concept-based query. Ontological concept editing includes changing the metadata of the concept, attribute and instance, and ontological concept updating includes adding new attribute, adding new instance and adding new metadata to the concept, attribute or instance. Concept-based query enables knowledge retrieval through ontology concept search and semantic cluster search. The algorithms for these tasks are explained in Table 1, Table 2 and Table 3.

Table 1. Adding new ontological concept (a) Conceptual clustering method, (b) Cluster-based identification method, (c) Concept-based query method

Graphical View:

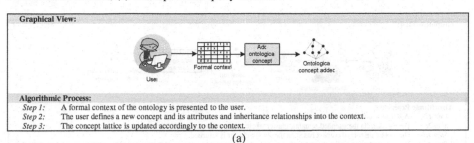

Algorithmic Process:
Step 1:   A formal context of the ontology is presented to the user.
Step 2:   The user defines a new concept and its attributes and inheritance relationships into the context.
Step 3:   The concept lattice is updated accordingly to the context.

(a)

Graphical View:

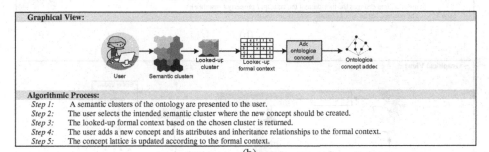

Algorithmic Process:
Step 1:   A semantic clusters of the ontology are presented to the user.
Step 2:   The user selects the intended semantic cluster where the new concept should be created.
Step 3:   The looked-up formal context based on the chosen cluster is returned.
Step 4:   The user adds a new concept and its attributes and inheritance relationships to the formal context.
Step 5:   The concept lattice is updated according to the formal context.

(b)

Graphical View :

Algorithmic Process:
Step 1:   A user inputs the keyword as a query.
Step 2:   The semantic query matching function is triggered.
　　　　　Step 2.1: If LSM < t; then proceed to Step 1 to refine the query or enter a new query.
　　　　　Step 2.2: If LSM ≥ t; then proceed to Step 3.
Step 3:   The matched semantic cluster (looked-up cluster) is defined.
Step 4:   The looked-up formal context corresponding to the defined looked-up cluster is presented to the user.
Step 5:   The user adds a new concept and its attributes and inheritance relationships to the formal context.
Step 6:   The concept lattice is updated according to the formal context.

(c)

Table 2. Editing or updating ontological concept (a) Conceptual clustering method, (b) Cluster-based identification method, (c) Concept-based query method

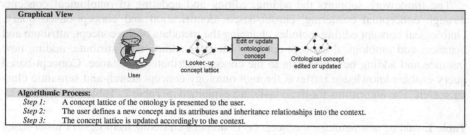

Algorithmic Process:
Step 1:   A concept lattice of the ontology is presented to the user.
Step 2:   The user defines a new concept and its attributes and inheritance relationships into the context.
Step 3:   The concept lattice is updated accordingly to the context.

(a)

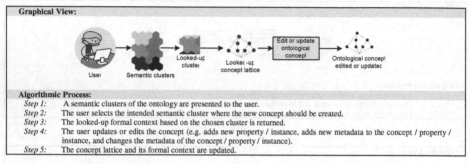

Algorithmic Process:
Step 1:   A semantic clusters of the ontology are presented to the user.
Step 2:   The user selects the intended semantic cluster where the new concept should be created.
Step 3:   The looked-up formal context based on the chosen cluster is returned.
Step 4:   The user updates or edits the concept (e.g. adds new property / instance, adds new metadata to the concept / property / instance, and changes the metadata of the concept / property / instance).
Step 5:   The concept lattice and its formal context are updated.

(b)

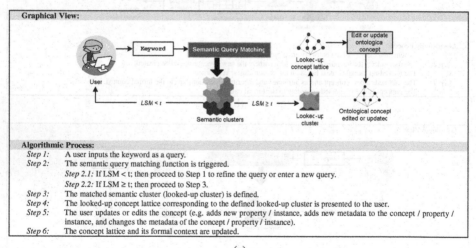

Algorithmic Process:
Step 1:   A user inputs the keyword as a query.
Step 2:   The semantic query matching function is triggered.
            Step 2.1: If LSM < t; then proceed to Step 1 to refine the query or enter a new query.
            Step 2.2: If LSM ≥ t; then proceed to Step 3.
Step 3:   The matched semantic cluster (looked-up cluster) is defined.
Step 4:   The looked-up concept lattice corresponding to the defined looked-up cluster is presented to the user.
Step 5:   The user updates or edits the concept (e.g. adds new property / instance, adds new metadata to the concept / property / instance, and changes the metadata of the concept / property / instance).
Step 6:   The concept lattice and its formal context are updated.

(c)

5 Ontological Knowledge Management Tasks

Ontological knowledge management tasks can be performed through conceptual clustering, cluster-based identification and concept-based query. The conceptual clustering

Table 3. Querying ontological concept (a) Knowledge retrieval through ontological concept (b) Knowledge retrieval through semanctic cluster

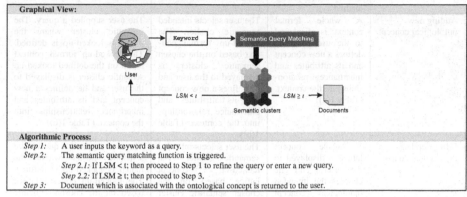

Graphical View:

Algorithmic Process:

Step 1: A user inputs the keyword as a query.

Step 2: The semantic query matching function is triggered.

Step 2.1: If LSM < t; then proceed to Step 1 to refine the query or enter a new query.

Step 2.2: If LSM ≥ t; then proceed to Step 3.

Step 3: Document which is associated with the ontological concept is returned to the user.

(a)

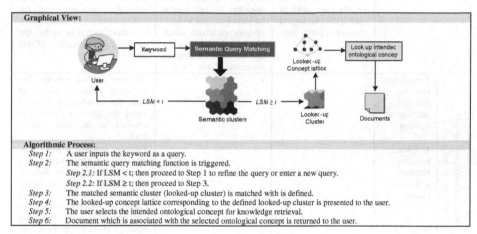

Graphical View:

Algorithmic Process:

Step 1: A user inputs the keyword as a query.

Step 2: The semantic query matching function is triggered.

Step 2.1: If LSM < t; then proceed to Step 1 to refine the query or enter a new query.

Step 2.2: If LSM ≥ t; then proceed to Step 3.

Step 3: The matched semantic cluster (looked-up cluster) is matched with is defined.

Step 4: The looked-up concept lattice corresponding to the defined looked-up cluster is presented to the user.

Step 5: The user selects the intended ontological concept for knowledge retrieval.

Step 6: Document which is associated with the selected ontological concept is returned to the user.

(b)

method displays the formal context to the user to create new ontological concepts or concept lattice to the user for editing or updating ontological concepts. This method is effective and efficient for managing small ontologies.

The cluster-based identification method displays semantic clusters to the user and if he or she chooses the desired cluster (looked-up cluster) from the semantic clusters, then the looked-up formal context associated with the looked-up cluster is returned to the user to create new ontological concepts. Meanwhile, the looked-up concept lattice associated with the looked-up cluster is returned to the user for editing or updating of ontological concepts and also to support knowledge retrieval.

The concept-based query method allows the user to supply a keyword as a query for finding the match or the nearest match based on semantic associations. The semantic matching query function is adopted to find the match. The summarization of ontological knowledge management tasks is shown in Table 4.

Table 4. Methodologies for the ontological knowledge management tasks

		Conceptual clustering method	Cluster-based identification method	Concept-based query method
Adding new ontological concept		A whole formal context is displayed to the user, and he defines a new concept and its attributes and inheritance relationships into the context. (Table 1(a))	The user selects intended semantic cluster. The looked-up formal context based on the chosen semantic cluster is displayed to the user and he defines a new concept and its attributes and inheritance relationships into the context. (Table 1(b))	The user supplied a query. The semantic cluster where the match is looked-up is defined. The looked-up formal context based on the defined looked-up semantic cluster is displayed to the user and he defines a new concept and its attributes and inheritance relationships into the context. (Table 1(c))
Edit or update ontological concept		A whole concept lattice is displayed to the user, and he chooses the intended ontological concept from the concept lattice which he intends to edit or update the ontological concepts. (Table 2(a))	The user selects intended semantic cluster. The looked-up concept lattice based on the chosen semantic cluster is displayed to the user and he chooses the intended ontological concept from the looked-up concept lattice which he intends to edit or update the ontological concepts. (Table 2(b))	The user supplied a query. The semantic cluster where the match is looked-up is defined. The looked-up concept lattice based on the chosen semantic cluster is displayed to the user and he chooses the intended ontological concept from the looked-up concept lattice which he intends to edit or update the ontological concepts. (Table 2(c))
Querying ontological concept (knowledge retrieval)	Knowledge retrieval through ontological concept	-	-	The user supplied a query. The document which is associated with the ontological concept where the query is matched is returned to the user. (Table 3(a))
	Knowledge retrieval through semantic cluster	-	-	The user supplied a query. The semantic cluster where the match is looked-up is defined. The looked-up concept lattice based on the chosen semantic cluster is displayed to the user and he chooses the intended ontological concept from the looked-up concept lattice which he intends to retrieve knowledge (document). (Table 3(b))

5.1 Semantic Clustering Generation

The ontology is conceptualized using FCA to a formal context, and then a concept lattice is depicted according to the context. SOM and K-Means are used to form semantic clusters. The concept lattice provides the visualization of ontological concepts to support ontological knowledge management and retrieval. Meanwhile, the semantic clusters provide a search context for the concept-based query matching task. The concept lattice and semantic clusters are stored in the database to facilitate ontological knowledge management and knowledge retrieval. Whenever the ontology is altered (e.g. add new ontological concepts, and update and edit the existing ontological concepts), the generation process will be triggered to regenerate a new concept lattice and semantic clusters.

Clustering the ontology (*merged ontology or local ontology*) into numbers of *k* clusters using SOM and K-Means effectively reduces the search space for creating ontological concepts, updating metadata schema of documents and querying documents by using the OntoVis visualization tool [8]. As the ontological concepts are clustered to several clusters, search with semantic clusters prior to navigating the concept space via the concept lattice can benefit and expedite ontological knowledge management and knowledge retrieval tasks. The graphical user interfaces (GUI) of the formal context, concept lattice and semantic clusters is illustrated Fig. 2.

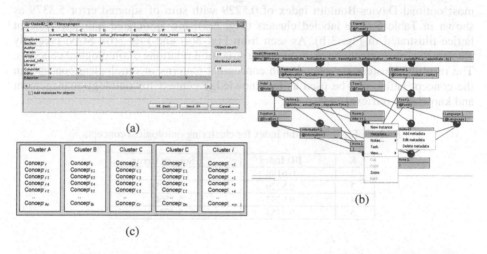

(a)

(b)

(c)

Fig. 2. (a) GUI of formal context interface in OntoVis [8], (b) GUI of concept lattice interface in OntoVis [8], (c) GUI of semantic clusters representation (example)

In general, the process of the concept lattice and semantic cluster generation consists of ontological contextualization and contextual clustering processes as shown in Fig. 1. The concept lattice is generated after applying step 1, and subsequently in step 2, the semantic clusters are generated as outlined in the semantic clustering generation algorithm below:

Input   : Ontology, O_S.
Step 1   : ***Ontological contextualization***
             Given an ontology $O_S : = (C, S_C, P, S_P, is\_a, A)$, O_S is contextualized into the formal context using FCA, K_S. The ontological concepts C are denoted as G (objects) and the rest of the ontology elements, S_C, P, S_P, is_a and A are denoted as M (attributes). The binary relation $I \subseteq G \times M$ of the formal context denotes the ontology elements, S_C, P, S_P, is_a and A corresponding to the ontological concepts C.
Output 1   : Concept lattice is formed based on the formal context.
Step 2   : ***Contextual clustering***
             SOM and k-means are applied to form semantic clusters of ontological concepts based on the conceptual pattern discovered in the

formal context, K_S in Step 1. Firstly, SOM is used to model the formal context K_S to discover the intrinsic relationship between ontological concepts of ontology O_S. Subsequently, K-Means is applied on the learnt SOM to reduce the problem size of the SOM cluster to the most optimal number of k clusters based on the Davies-Bouldin validity index.

Output 2 : Semantic clusters are formed.

Fig. 3 illustrates clustering of the ontology into three clusters as determined by the most optimal Davies-Bouldin index of 0.5229 with sum of squared error 5.3379 as shown in Table 5. The labeled clusters (Fig. 3(a)) are associated with the concept lattice illustrated in Fig. 3(b). As seen from Fig. 3(a) and Fig. 3(b), SOM precisely clusters the ontological concepts given the structural relations of ontological concepts. The resulting clusters associated with conceptual patterns of ontological concepts in the concept lattice can be used as the knowledge context for ontology management and knowledge retrieval.

Table 5. Davies-Bouldin index for clustering ontological concepts

K	BD Index	∑ Square Error
2	0.6188	12.0923
3	0.5229	5.3379
4	0.6293	5.1560
5	0.7754	5.4231

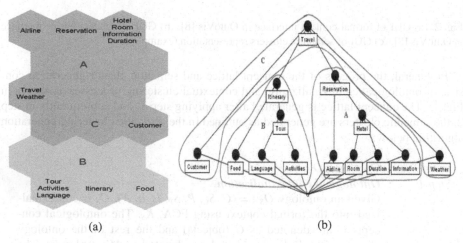

(a)   (b)

Fig. 3. Clustering clusters associated with conceptual patterns of ontological concepts (a) Cluster ontological concepts via SOM and K-means (b) Partition of concepts in concept lattice based on clusters in (a)

5.2 Semantic Query Matching

The semantic query matching function is to facilitate the concept-based query matching task to support ontological knowledge management. A keyword-based query is adopted in the framework.

Given the user-supplied query, U_Q and ontological concept in each semantic cluster, O_{Cij}, where semantic cluster is $i = 1, 2, 3, \ldots, n$, and a ontological concept in the semantic cluster is $j = 1, 2, 3, \ldots, n$. Linguistic processing is applied to discover Lexical Similarity Measure (LSM) between U_Q and O_{Cij} as depicted in Fig. 1.

Commonly, terminology of ontological concepts and their elements can be defined in the form *location*, *has_date*, *hasName*, *BookPart*, *3days_visa* and etc. In linguistic processing, string normalizations such as case normalization, digit normalization, link stripping, blank normalization, stopword filtering and namespace prefixes elimination are applied to normalize ontological concepts and user supplied query for Levenshtein edit distance to discover LSM between them. Threshold value (t) of 0.8, gives the most precise match between two strings as defined by the experimental result for mapping and merging ontologies in [4] using LSM.

If the LSM ≥ 0.8, the relevant semantic cluster where the ontological concept is defined is looked-up. The formal context or concept lattice corresponding to the looked-up cluster is returned to the user according to the ontological knowledge management task specified by the user.

6 Conclusion

In this paper, we have proposed an algorithmic framework for managing ontological knowledge and knowledge retrieval in the knowledge repository. Semantic clustering generation and semantic query matching are core functions in framework to facilitate ontological knowledge management and knowledge retrieval through unsupervised data mining techniques. The ontological knowledge can be managed through cluster-based identification, concept-based query and conceptual clustering methods. It allows the knowledge seeker to semantically retrieve knowledge from the ontological concept and looked-up concept lattice.

References

1. Sure, Y., Staab, S., Studer, R.: Methodology for Development and Employment of Ontology Based Knowledge Management Applications. SIGMOD Record 31(4), 18–23 (2002)
2. Ganter, B., Wille, R.: Applied Lattice Theory: Formal Concept Analysis (1997), http://www.math.tudresden.de/~ganter/psfiles/concept.ps
3. Vesanto, J., Alhoniemi, E.: Clustering of the Self-Organizing Map. IEEE Transactions on Neural Networks 11(3), 586–600 (2000)
4. Cohen, W., Ravikumar, P., Fienberg, S.: A Comparison of String Distance Metrics for Name-matching tasks. In: IIWeb Workshop held in conjunction with IJCAI (2003)
5. Kiu, C.C., Lee, C.S.: Ontology Mapping and Merging through OntoDNA for Learning Object Reusability. Educational Technology & Society 9(3), 27–42 (2006)
6. Ding, Y.: D17 v0.1 Ontology Management System, SW-Portal Working Draft August 31 (2004), http://sw-portal.deri.at/papers/deliverables/d17_v01.pdf
7. Denny, M.: Ontology Building: A Survey of Editing Tools (2002), http://www.xml.com/pub/a/2002/11/06/ontologies.html

8. Hayes, P., Eskridge, T.C., Mehrotra, M., Bobrovnikoff, D., Reichherzer, T., Saavedra, R.: COE: Tools for Collaborative Ontology Development and Reuse. Knowledge Capture Conference (K-CAP) 2005, Banff, Canada (2005)
9. Ehrig, M., Sure, Y.: Ontology Mapping - An Integrated Approach. In: Bussler, C.J., Davies, J., Fensel, D., Studer, R. (eds.) ESWS 2004. LNCS, vol. 3053, pp. 76–91. Springer, Heidelberg (2004)
10. Lim, W.C., Lee, C.S.: Knowledge discovery through composited visualization, navigation and retrieval. In: Hoffmann, A., Motoda, H., Scheffer, T. (eds.) DS 2005. LNCS (LNAI), vol. 3735, pp. 376–378. Springer, Heidelberg (2005)
11. Stoilos, G., Stamou, G., Kollias, S.: A String Metric For Ontology Alignment. In: Gil, Y., Motta, E., Benjamins, V.R., Musen, M.A. (eds.) ISWC 2005. LNCS, vol. 3729, Springer, Heidelberg (2005)

Semantic-Enabled Organization of Web Services

Khanh Nguyen[1], Jinli Cao[1], and Chengfei Liu[2]

[1] Department of Computer Science and Computer Engineering, La Trobe University
Bundoora, VIC 3086, Australia
tk11nguyen@students.cs.latrobe.edu.au, j.cao@latrobe.edu.au
[2] Faculty of ICT, Swinburne University of Technology, Australia
cliu@it.swin.edu.au

Abstract. Web services are the new paradigm for distributed computing. They offer interoperability of application and integration of large
scale distributed systems. Amongst the web service standards, service
registry is crucial to the overall utility of web service technology. However, the web services discovery mechanism pertaining with the UDDI
[1] registry is inefficient because it only supports keyword-based and
category-based matching, leading to irrelevant matches. Therefore, the
provision of the scalable, flexible and robust web services publication
and discovery system is crucial for the overall utility of web services.
In this paper, we develop a scalable, robust and efficient infrastructure
for web services publication and discovery amongst domain-based registries. More importantly, a novel algorithm is provided to classify web
services into domain-based registries based on their semantic description.
Experimental results show that our approach outperforms a centralized
approach in terms of service discovery time.

Keywords: Semantic Web services, specific domain ontology, semantic
publication, semantic discovery, domain-based registry

1 Introduction

Web services are a new dominant paradigm for distributed computing. A typical
web service architecture consists of three entities: service providers who produce
web services and publish the service advertisement to a registry; a registry which
is a public repository providing publication and discovery services; and service
requesters who query the registry for desired services. In the architecture, the
service registry is crucial to the ultimate utility of web services and must provide scalable, robust and effective discovery mechanisms. However, the current
registry specifications have following limitations:

- The current web services registry standard [1] is originally working with UBR
  [1] which is a master directory for all publicly available web services. To deal
  with the limitations of traditional centralized systems such as single point of
  failure and bottleneck performance, replication approaches were proposed.
  However, replication approaches add communication overhead to maintain

Y. Zhang et al. (Eds.): APWeb 2008, LNCS 4976, pp. 511–521, 2008.

the information consistency between registries and seem to be impractical when the number of web services and registries increase significantly. In addition these approaches do not improve the performance of discovery process because the matchmaker has to match the query to the whole service descriptions in the registry to find the desired services. Thus, finding a scalable approach for Web services organization and discovery has become crucial.

– The growth in the number of service registries brings new challenge for searching desired services to consumers. Searching through thousands of registries for desired services is time-consuming; it requires locating relevant registries first and then searching within located registries for desired services. Thus it is crucial to provide a searching environment in which users' queries are automatically routed to relevant registries without human interventions.

– The current searching mechanism pertaining with UDDI [1] only supports keyword-based and category-based search, which can lead to irrelevant matches. Furthermore, category-based discovery requires service requesters have to choose services domains they want to search. It is better if the system can automatically select relevant domains related to users' requirements based on extracted information from the queries.

In this paper, we will develop an infrastructure for web service publication and discovery which addresses the above limitations. In the infrastructure, registries are organized based on specific domains in which each registry is mapped to one or a group of domains. Web services publication or discovery will be automatically routed to domain-related registries based on their semantic information extracted from web services description or query. The main contributions of this work are summarized as follows:

– Proposing an approach for integrating multiple private/public registries, providing a unified, transparent view of underlying registries.
– Providing a mechanism for organizing web service registries based on specific domains in which each registry is mapped to one or a group of specific domains.
– Developing an algorithm for classifying web services into domain-based registries based on their semantic description.
– Devising an algorithm for selecting relevant registries containing required services based on the semantic features of service query.
– Tackling the identified problems of UDDI [1] standard such as single point of failure, performance bottleneck, semantic deficiency and lack of scalability by organizing web services in to multiple domain-based registries.

The rest of this paper is organized as follows: Section 2 discusses some related work on Web service organization. Section 3 presents our proposed framework for Web service organization. Section 4 discusses more detail about functionality of Web service publication and discovery. Section 5 evaluates the experimental results of our proposed framework. Finally in Section 6, we will outline our intention for future work.

2   Related Work

The current approaches to web services organization can be classified into centralized [1,2,3,4] and decentralized approaches [5,6,7,8,9,10,11,12]. In the centralized infrastructure, all published web services are stored in a common repository. As realized that replicating the UDDI is impractical due to the significant growth in the number of web services and registries in future. A number of studies in this field have put attention to decentralized approaches because they are believed as a promising solution to address the limitations of centralized approach. The approach [5,6] are most relevant to our work. In these approaches, web services are organized into distributed registries which each registry is mapped to a domain or a group of domains. However, none of them support automatic selection of registries related to domain of interest of requesters. To publish or discover a web service, these approaches require requester choosing a domain which can satisfy requester's requirements. This is very difficult for novice users due to the large number of domain used in the system. In addition, requiring human intervention is a barrier for automatic web service composition in which agents or programs are used to discover services rather than human. In our infrastructure, an algorithm for automatic selecting relevant registries, based on semantic relationship between users' query and specific domain ontology has been proposed.

Other approaches [7,8,9,10,11,12] based on peer-to-peer (P2P) platform that support Distributed Hash Table (DHT) interface such as CAN [13], Tapestry [14] and Chord [15] to organized service registries in distributed environment. In these approaches, each registry is seen as a peer and the web services are distributed to these peers . However, in these approaches web services are not classified on to peers based on service domains, so the query will be routed to a number of peers which eventually increases the time for service discovery.

3   Semantic-Enabled Organization of Web Services Framework

Our proposed framework for web service organization consists of four layers: Data layer, Semantic layer, Classification layer, and System interface layer. The layered architecture is shown in Fig 1.

Data Layer consists of multiple autonomy registries supporting semantic service publication and discovery. The registries are organized into specific domains

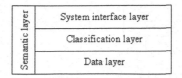

Fig. 1. The layered architecture

in which each registry is mapped to one or a group of domains. The purpose of this organization is that the system can automatically classify web services which have similar functionalities into one registry or a group of registries. Consequently, a query for desired services is only routed to registries mapped to the required service's domain. This approach addresses following outstanding problems of the traditionally centralized registry (e.g. UDDI [1]):

- By using multiple registries, this approach addresses the single point of failures and performance bottleneck issues pertaining with centralized approaches.
- Improving searching performance (in terms of time and reliability) because it reduces the number of services matchmaker has to match.

Semantic Layer is used to provide semantic service registration and discovery in our framework. The main purpose of this layer is to address the semantic deficiency of the current standard UDDI registry. The semantic layer consists of specific domain ontologies and the classification ontology.

Specific Domain Ontology. Each specific domain ontology comprises a set of concepts and relationship between the concepts in a specific application domain. In our framework, specific domain ontologies are used to support semantic service description and discovery. Specifically, each specific domain ontology is used for describing semantic features of services in a specific domain. At the classification layer, these domain ontologies are used as knowledge bases to classify web services into domain-based registries. At the data layer, these ontologies are used by matchmaker for semantic service discovery.

Classification Ontology. This ontology functions as a taxonomy for services classification in our framework. The classification ontology maintains the relationships between registries and specific domain ontologies. These relationships are captured as "mapTo" relationships in the classification ontology. Fig 2 shows the sample structure of the classification ontology.

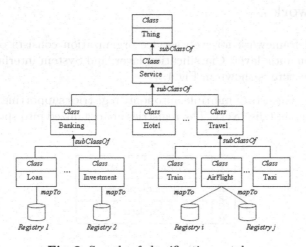

Fig. 2. Sample of classification ontology

Each node in the classification ontology also stores the details of the registry operator that manages the domain. Using the classification ontology, the web services will be automatically classified into relevant registries; consequently service discovery is only performed in the registries related to the requester' interested domain.

Classification Layer. The main functions of this layer are summarized as follows:

- Classify Web services description into relevant registries based on service domains.
- Route the service query to registries which contain requester' desired services.

To fulfill these functions, this layer comprises two kinds of peers: classifiers and registry operators.

Classifiers: The function of classifiers is to classify web services, at the publication time, to relevant registries based on semantic service descriptions. When receiving a request for service publication, the classifiers will refer to the classification ontology to get the addresses of registry operators. The query will be forwarded to the registry operators. Each registry operator will evaluate whether the query belongs to its domain and response the evaluation results to the classifiers. Based on these responses the classifiers will route the query to operator peers which maintain domain-related registries.

Registry Operators: Each registry operator is responsible for one specific domain and one registry or a group of registries mapped to that domain. The registry operator contains the access URLs of its specific domain as well as the URLs of all registries are mapped to its domain. More specifically, the registry operators have two main functions: (1) Receive the query in form of Q(Op, Is, Os), where Is and Os are sets of inputs and outputs of the operation Op, from classifier peers and check whether this query Q is related to their managed domain and then reply the responses to the classifier peers; (2) route the query to their maintained registries for services publication and discovery.

System interface layer provides interfaces for service providers and service requesters to interact with our service organization infrastructure. For web service registration, the service provider can connect to one of the known client in the system interface layer which is designed to provide an interface for the service provider to publish their semantic service description. Then the semantic service description will be routed to lower layer for registration or discovery. The service discovery is performed with the same manner. The details of our proposed infrastructure in shown in Fig 3.

4   Web Service Publication and Discovery

In this section, we focus on the process of mapping between registries and specific domains as well as service publication and discovery. Before discussing these functionality details, the following definitions should be formally defined.

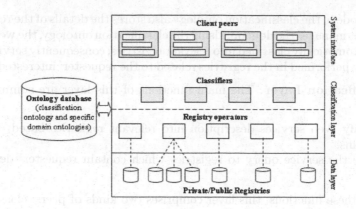

Fig. 3. Details of our proposed architecture

Definition 1. *A specific domain ontology SDO(C, R) comprises a set of concepts C and relationship R between the concepts in specific application domain. The concepts have two parts: Operation concepts C_O and data concepts C_D. The operation concepts C_O comprise concepts used for describing the types of operations which web services perform. The data concepts C_D are used for describing the data types of input and output of web services' operation.*

Definition 2. *A semantic web service is a triple $SWS = (Op, Is, Os)$, where Op is the operation; $I_s = \{I_1, I_2, \ldots, I_m\}$ is a set of inputs and $Os = \{O_1, O_2, \ldots, O_n\}$ is a set of outputs of the operation Op.*

Definition 3. *A semantic web service $SWS = (Op, Is, Os)$ is totally described by a specific domain ontology $SDO_k(C_k, R_k)$ if its operation Op, inputs Is and outputs Os are only described by using the set of concepts C_k of ontology SDO_k.*

Definition 4. *If a concept $C \in C_k$, where C_k is a set of concepts of the specific domain ontology $SDO_k(C_k, R_k)$, then the $match(C, SDO_k)$ function returns value 1, else $match(C, SDO_k)$ returns value 0.*

Definition 5. *A service template $ST(Op, Is, Os)$, where $Is = \{I_1, I_2, \ldots, I_m\}$ is a set of inputs and $Os = \{O_1, O_2, \ldots, O_n\}$ is a set of outputs of the operation Op respectively, is related to specific domain ontology SDO_k if the following formula:*

$$F(ST, SDO_k) = \frac{1}{1 + |Is| + |Os|} \left[match\,(Op, SDO_k) + \sum_{i=1}^{m} match\,(I_i, SDO_k) \right.$$

$$\left. + \sum_{j=1}^{n} match\,(O_j, SDO_k) \right] \tag{1}$$

returns a value x, with $x \in (0, 1]$, it means that the service template ST is semantically described by using a set of concepts in the domain ontology SDO_k.

4.1   Mapping Between Registries and Domains

The tasks of mappings between registries and domains in the classification ontology are required knowledge about services domains. In this paper, we assumed that people called registry providers are responsible to perform these mapping tasks. To map a registry to a specific domain in the classification ontology, the registry provider can connect to one of the client peers in the system interface. The registry provider's mapping request will be routed to one of the classifiers by the client peer. Then, the classifier retrieves the classification ontology from the semantic layer and sends back to the client peer. The client peer receives the classification ontology from the classifier and displays it in form of an ontology tree to the registry provider. The registry provider can map his registry to one or a group of domains in the classification ontology. In case of the registry provider's interested domains do not exist, they can add new nodes to the classification ontology and then map their registries to added nodes.

4.2   Web Service Publication

Service publication is performed in our framework through two steps: registry selection and service registration.

Registry selection: To publish a semantic web service, the services provider connects to one of the client peers in the system interface layer. The client peer will send the semantic web service description to the classifier peer. The classifier peer extracts the service's operation (Op) and its inputs (Is) and outputs (Os), then sends the publication query in form of $Q_P(Op, Is, Os)$ to operator peers. Each operator retrieves its managed domain ontology SDO_k, and checks whether this service belongs to it managed domain by calculating the $F(Q_P, SDO_k)$ function. The result of this calculation will be sent back to the classifier.

Service registration: Depends on the returned value of F, we group service registration into following categories:

– Automatic registration: If there is only one operator peer returns value 1 and all other operator peers return 0, it means that the web service is only described by concepts of one specific domain. In this case, the classier will route the web service description to the registry operator related to that domain for service registration. This process is depicted in Fig 4.
– Semi-automatic registration: If there are more than one operator peers return a value between 0 and 1. It means that the web service is described by using concepts of some domains or the set of concepts used to describe the service exist in more than one specific domain ontologies. In this case, the classifier peer will send a list of related domains to the client peer. The service provider will choose a domain which the service will be registered. The selected domain will send back to the classifier which then routes the web service description to the registry operator for service registration. Fig 5 shows the process of semi-automatic service registration.
– If all operator peers return value 0, it means that current domain which the web service belongs does not exist in the classification ontology. In this case,

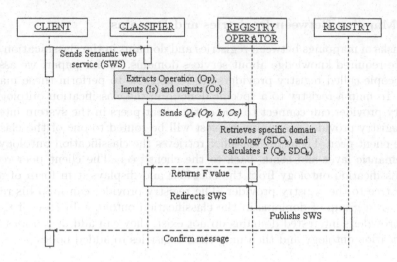

Fig. 4. Automatic service registration

the web services cannot be registered until the desired domain and registries for this domain have been added.

4.3 Web Service Discovery

Searching for a web service in our framework comprises two steps: locating relevant registries containing desired services and then searching within the located registries to retrieve requested services. The automatic service discovery is shown in Fig 6.

Locating relevant registries: The web service query in form of $Q_D(Op, Is, Os)$ with Is and Os are sets of inputs and outputs of the operation Op, from the client peer will be routed to one of the classifier peers. The selected classifier forwards the query to operator peers. Each registry operator determines whether this service query is related to its domain by calculating $F(Q_D, SDO_k)$ function, where SDO_k is its managed domain. If the value of F is greater than 0, the query will be routed to registries which are mapped to its domains for service discovery. The results returned from these registries will be received by these registry operators and then sent back to client peers that finally display the discovered services to the requester.

Searching within located registries: This step is done by individual registry. Actual search algorithms of registries are out of the scope of this paper.

5   Experimental Evaluation

We conducted a simulation to evaluate the performance of our proposed approach and a centralized approach such as UDDI [1]. The simulation was conducted with the number of web services increasing from 3,000 to 120,000. In

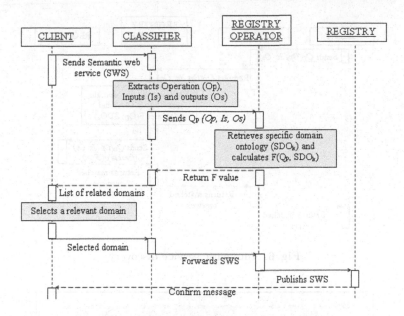

Fig. 5. Semi-automatic service registration

the simulation, six different specific domain ontologies are used. Each specific domain has a registry to store services belonging to the domain. The classification algorithm has been implemented to automatically route the service query to relevant registries which are mapped to the requesters' interested domain. The input parameters of the simulation can be summarized as follows:

Ontology database: contains six specific domain ontologies. Each specific domain ontology contains 10,000 concepts. These concepts are stored in *.owl file.

Web services: in our approach, web services are distributed in six different databases. For centralized approach, all services in six decentralized databases are combined into one centralized database.

The searching time of our approach and centralized approach in each case was measured. The results of experimental simulation is demonstrated in Fig.7.

As can be seen from Fig.7 that our approach is outperformed a centralized approach. The large growth of Web services can lead to the overwhelming of centralized registry and significantly increase the duration of services discovery, which eventually degrade the overall utility of Web services. Conversely, in our approach web services are classified and stored on domain-based registries. As a result, service discovery only performed in registries which are mapped to the requester' interested domains. Thus, it significantly reduces the number of web services the matchmaker has to match, which consequently minimizes the time for services discovery.

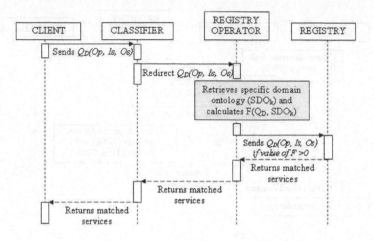

Fig. 6. Automatic service discovery

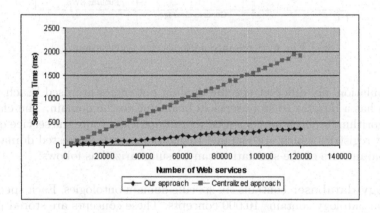

Fig. 7. Comparison of searching time

6   Conclusion

In this paper we have presented an infrastructure for semantic publication and discovery of web services addressing the outstanding problems of the UDDI registry. In the framework, web services which are semantically described are classified and stored in domain-based registries. This infrastructure provides a scalable and robust system for web services publication and discovery. In our infrastructure, the challenge of searching through multiple registries for a desired service is addressed by using classification ontology. More importantly, we have proposed a novel algorithm for automatically classifying web services into domain-related registries based on their semantic description. Consequently, the service requester's query will automatically routed to only registries which contain desired services. The experimental results show that our approach is outperformed compared with the centralized UDDI registry.

References

1. UDDI version 3.0.2. UDDI spec technical committee draft, dated 20041019, http://uddi.org/pubs/uddi-v3.0.2-20041019.htm
2. Dogac, A., et al.: Enriching ebxml registries with owl ontologies for efficient service discovery. In: Proceedings of the 14th International Workshop on Research Issues on DataEngineering: Web Services for E-Commerce and E-Government Applications(RIDE 2004) (2004)
3. Dong, X., et al.: Using categorization on further enhance the utilization of semantic webin uddi. In: Proceedings of the First International Multi-Symposiums on Computer andComputational Sciences (IMSCCS 2006) (2006)
4. Liu, J., et al.: Service registration and discovery in a domain-oriented uddi registry. In: Proceedings of the Fifth International Conference on Computer and InformationTechnology (CIT 2005) (2005)
5. Pilioura, T., et al.: PYRAMID-S: A scalable infrastructure for semantic web service publicationand discovery. In: Proceeding of the 14th International Workshop on Research Issue on Data Engineering: Web Services for E-Commerce and E-Government Applications(RIDE 2004) (2004)
6. Verma, K., et al.: METEOR-SWSDI: A scalable p2p infrastructure of registries for semantic publication and discovery of web services. Information Technology and Management 6, 17–39 (2005)
7. Gagnes, T., et al.: A conceptual service discovery architecture for semantic web services indynamic environments. In: Proceedings of the 22nd International Conference on Data Engineering Workshops(ICDEW 2006) (2006)
8. Guo, D., et al.: Enhance uddi and design peer-to-peer network for uddi to realize decentralizedweb service discovery. In: ISWS, pp. 59–65 (2005)
9. Lin, Q., et al.: DWSDM: A services discovery mechanism based on a distributed hash. In: Proceedings of the Fifth International Conference on Grid and CooperativeWorkshops (GCCW 2006), IEEE, Los Alamitos (2006)
10. Sivashanmugan, K., et al.: Discovery of web service in a federated registry environment. In: Proceedings of the IEEE International Conference on Web Services (ICWS 2004) (2004)
11. Zhong, J.: A semantic web based peer-to-peer service registry network. In: Proceedings of the First International Conference on Semantics, Knowledge,and Grid (SKG 2005) (2006)
12. Yin, L., et al.: edsr: A decentralized service registry for e-commerce. In: Proceedings of the 2005 IEEE International Conference on e-Business Engineering (ICEBE 2005) (2005)
13. Ratnasamy, S., et al.: A scalable content-addressable network. In: ACM SIGCOMM 2001 (August 2001)
14. Zhao, B., et al.: Tapestry: An infrastructure for fault-tolerantwide-area location and routing. UCB/CSD-01-1141 (April 2000)
15. Stoica, I., et al.: Chord: a scalable peer-to-peer lookup service for internet applications. In: Proceedings of ACM SIGCOMM 2001, San Diego (September 2001)

Rule Mining for Automatic Ontology Based Data Cleaning

Stefan Brüggemann

OFFIS - Institute for Information Technology
Escherweg 2, 26121 Oldenburg, Germany
brueggemann@offis.de

Abstract. Automatic detection and removal of inconsistencies in data are open challenges in the data quality management cycle. Specific knowledge is needed to clean invalid data, which often requires user interaction with domain experts. Domain specific classes and attributes can be described in ontologies. Attribute value combinations can be labeled as valid or invalid. Our approach on data cleaning allows for detection and removal of semantic errors in data. The analysis of replacements enables the creation of rules, which can minimize the required user interaction. We provide an algorithm which analyzes frequencies of replacement operations for invalid tuples in the ontology and generates rules, which are then applied in data cleaning environments automatically.

1 Introduction

(Semi-)automatic data cleaning is a difficult task in the data quality management cycle. Detection and removal of erroneous data like invalid attribute value combinations are important processes in that cycle [11]. Data cleaning depends on the involvement of domain experts [13], because domain specific knowledge is required for the detection and cleaning of semantic and domain specific anomalies. As described in [9], automatic cleaning is important due to the large amount of data which are normally processed during data validation. This amount requires a lot of time for an expert to clean the data manually.

We use ontologies [5] for modeling domains of discourse and to represent valid and invalid attribute value combinations[3].

Data validation consists of checking tuples for correctness [8]. When invalid tuples are being detected, they have to be modified using valid tuples stored in the ontology. Correction suggestions can be proposed by ontology-based data cleaning systems using the hierarchical ontology structure to infer possible corrections for this tuple. User can then select a valid tuple from these suggestions. User selections are being saved and used to identify replacement rules.

We propose an algorithm for rule mining in ontology-based data quality management environments. This algorithm depends on user-performed replacement operations. User identify invalid tuples and replace them with a valid one, which can be chosen by a set of correction suggestions made by the system. Frequencies in these user selections are being analyzed and, when defined tresholds are reached, used for rule creation.

Y. Zhang et al. (Eds.): APWeb 2008, LNCS 4976, pp. 522–527, 2008.

2   Rule Mining

This section describes how ontologies can be used for data cleaning purposes and how rules can be found in data cleaning ontologies. First, we describe our approach for ontology based data cleaning and rule mining informally. Then we provide an algorithm for automatic rule detection in data cleaning environments.

2.1   Basic Idea

The key idea of our approach is to model specific domains of discourses as ontologies. A discourse world can be described by a set of concepts, instances, and relations between both. Concepts of the discourse world often can be attributed manifold. We use concepts to model these attributes. In many real world scenarios like tumour classification many well defined attributes exist. We use instances to model the ranges of these attributes. In most cases more than one attribute is needed to describe a concept like a tumour. Then restrictions exists on combining attribute values. We use relations between instances of the classifying concepts. The International Classification of Diseases (ICD)[10] is a classification scheme which describes the topology of diseases. For instance, the range of gender consists of "male" and "female", the range of the ICD is much more complex and furthermore hierarchical. "Breast cancer (ICD-code 174.0)", for instance, can not be combined with "male".

Valid combinations are initially learned from a data source, e.g. a data warehouse, which is defined as containing only valid attribute value combinations. New valid and invalid combinations are learned by user interaction during data integration or data validation processes if a user classifies combinations as valid or invalid which hitherto were not modeled in the ontology.

A part of an ontology used for tumour documentation in epidemiological cancer registries is shown in figure 1. On the left it shows instances of ICD. "'C02'" describes a malignant neoplasm of the tongue, "'C02.1'" describes the borders of the tongue, "'C02.2'" describes the bottom of tongue, and "'C02.3'" describes the front part. The instances of dignity on the right desribe several benign and malign characteristica of tumours. The ICD values have a hierarchical structure, which means that the values C02.x are more specific than "C02". Valid (from "C02.3" to "1a"-"1d") and invalid (from "C02.3" to "1e") attribute value combinations are defined. Valid combinations can be used to describe a tumour correctly. Invalid combinations instead are impossible pairs of values. In a data validation process for a given tuple it has to be decided whether it is valid or not. Then these valid and invalid combinations are used .The valid combinations can be used for suggesting correcions when invalid tuple were found and when they have to be corrected.

The ontology can be used to suggest valid combinations when invalid combinations occur. The user can choose a valid combination which shall replace the invalid tuple. In most cases there is more than one possible valid combination, which can be presented to a user. The user then decides which combination has to be used to correct the invalid data. Data cleaning systems can store these

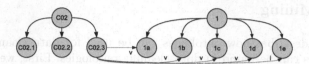

Fig. 1. Part of an ontology showing valid (v) and invalid (i) attribute value combinations

user selections and use them to mine rules. When a treshold is reached and it is obvious that in most cases the same valid combination is chosen, a rule can be created. This rule then will always be applied when the invalid combination is detected again and users must not be asked.

When a tuple has to be corrected, then the other data in the same row are used to correct the invalid combination. This is done by using a tuple as an input for the ontology, which means that each attribute value is searched in the ontology and then the ontology determines whether all combinations are valid or not. Then the invalid attributes are modified until the whole tuple is valid.

2.2 Rule Mining Algorithm

We now present a concrete rule mining algorithm. This algorithm identifies significant replacement operations for which rules can be created. We assume an ontology O with sets of concepts C and instances I. Relations between instances of different concepts define valid and invalid tuples. Let $INV(c)$ be the set of invalid tuples, where the tuples contain instances of concept c. Let $VAL(c)$ be defined analog with valid tuples.

A replacement operation is defined as an operation $r : i \rightarrow v$ with $i \in INV(c)$ and $v \in VAL(c)$. These operations represent how invalid tuples i were replaced with valid tuples v. User perform these operations during the data validation task in data integration environments.

We assume that a set REP of replacement operations r has been stored during data integration processes or data quality managment tasks. REP consists of operations r together with occurence values. These occurences store how often a replacement operation has been performed or selected.

We describe the number of occurences of a replacement operation as $supp(i \rightarrow v)$. A predefined treshold $minsupp$ defines a value used for rule creation as follows:

The significance sig of a replacement operation $r : i \rightarrow v$ is defined as

$$sig_v^{v'}(r) = \frac{supp(i \rightarrow v)}{supp(i \rightarrow v')}$$

and describes how often a replacement operation v has been performed in respect to another operation containing v'.

A rule $i \rightarrow v$ can be defined if

$$sig_v^{v'}(r) > minsupp$$

for all $v' \in VAL$ which occur in a replacement operation $i \rightarrow v'$.

Now we are able to construct an algorithm which identifies possible rules for invalid tuples from the concepts of the ontology:

```
for all concepts c in O:
    for all invalid tuples i in c:
        for all replacement operations r containing i:

            calculate sig (r)
            create rule if sig (r) > minsupp for
            all replacement operations
```

Now we provide an example which demonstrates how the algorithm identifies rules. The example is based on figure 1 and table 2. The figure describes a part of an ontology containing ICD (left side) and dignity values. Valid and invalid attribute value combinations are shown for the instance "C02.3".

The whole set of invalid tuples when focusing on ICD is defined as $INV(ICD)$ = $\{(C02.3, 1e),\ (C02.2, 1b),\ (C02.1, a)\}$. Valid tuples are $VAL(ICD)$ = $\{(C02.3, 1a),\ (C02.3, 1b),\ (C02.3, 1c),\ (C02.3, 1d)\}$. Not all tuples are shown in the figure.

The set of replacement operations REP is shown in table 2. The table lists invalid combinations and shows which valid tuple has been used to correct the tuple. The last column contains how often valid combinations have been chosen to replace the invalid tuple.

When the algorithm is being executed, it first iterates over the concepts of the ontology, starting at ICD. Then the set of invalid tuples is iterated, starting at (C02.3, 1e). Now, all replacement operations containing the tuple (C02.3, 1e) are identified: $\{\#1, \#2, \#3, \#4\}$. For each operation $sig(r)$ is calculated, shown in table 3. The value of $sig(1)$ is lower than $minsupp$ for all other replacement operations. This is true for $sig(2)$ and $sig(3)$. Only the calculated values for $sig(4)$ are greater than $minsupp$, so the rule

$$(C02.3,\ 1e) \rightarrow (C02.3,\ 1a)$$

can be created.

Furthermore the algorithm creates the following rules:

$$(C02.3,\ 1b) \rightarrow (C02.3,\ 1c)$$
$$(C02.3,\ 1a) \rightarrow (C02.3,\ 1d)$$

We identified $minsupp = 10$ as a practicable treshold.

3   Related Work

Our approach is related to the field of association rule mining (ARM) [1], the mining of rules existing in a data set. Rules are expressions $X \Rightarrow Y$, where X and Y are sets of items. $X \Rightarrow Y$ then means that when a transaction contains X,

Num-ber	Invalid Combination i	Replace-ment v	Occurence $supp(i \to v)$
1	(C02.3, 1e)	(C02.3, 1d)	12
2	(C02.3, 1e)	(C02.3, 1c)	7
3	(C02.3, 1e)	(C02.3, 1b)	34
4	(C02.3, 1e)	(C02.3, 1a)	520
5	(C02.2, 1b)	(C02.2, 1a)	9
6	(C02.2, 1b)	(C02.3, 1d)	3
7	(C02.2, 1b)	(C02.3, 1c)	477
8	(C02.2, 1b)	(C02.3, 1a)	11
9	(C02.1, 1a)	(C02.2, 1a)	19
10	(C02.1, 1a)	(C02.3, 1d)	642
11	(C02.1, 1a)	(C02.3, 1c)	12
12	(C02.1, 1a)	(C02.3, 1b)	6

Operation	sig	value
1	supp(1)/supp(2)	1.71
1	supp(1)/supp(3)	0.35
1	supp(1)/supp(4)	0.02
2	supp(2)/supp(1)	0.58
2	supp(2)/supp(3)	0.2
2	supp(2)/supp(4)	0.01
3	supp(3)/supp(1)	2.83
3	supp(3)/supp(2)	4.85
3	supp(3)/supp(4)	0.06
4	supp(4)/supp(1)	43.3
4	supp(4)/supp(2)	74.28
4	supp(4)/supp(3)	15.29

Fig. 2. Replacement Occurrences   **Fig. 3.** Calculated values for sig(r)

then it often contains Y. ARM is often performed on shopping cart data to find products that are bought together is many cases. As not to perform mining on all data, ARM approaches like [4,6,12] partition the search space into frequent and infrequent itemsets to minimize the search effort. The frequent itemsets then are these fulfilling *minsupp*. Contrary to ARM, we mine replacement rules based on sets of operations $i \to v$. When mining rules for an invalid tuple i, the search space can be partitioned into replacement operations containing i and into operations not containing i.

Ontologies are also being used for semantic information integration [7]. Existing approaches like [2] use ontologies to describe semantic similarities of concepts occuring in data sources. Therefore first a global ontology is being created. This acts as a global schema and can be equal to the target data schema. Second the existing data sources are integrated in the global ontology. Here the concepts of the sources are being described as derived concepts of concepts from the global ontology. The third step is used for query processing. Queries are being described as concepts, too. All concepts which are more specific than the querying concept are correct results.

4 Conclusion

Data validation and cleaning are important but expensive tasks. They are highly depending on domain specific knowledge. In many cases domain experts have to be involved for detecting and correcting erroneous data. Unfortunately, in some cases when large sets of data are being processed, manual correction of data is not possible. We use ontologies for semi-automatic data cleaning. Knowledge describing valid and invalid tuples can be modeled using ontologies. This knowledge can be used for detecting invalid tuples and for generation of correction suggestions. After a learning phase ontology-based data cleaning systems are able to mine rules. The presented algorithm uses a set of user-performed replacement operations. The rules can then be applied automatically when erroneous data is detected. The presented algorithm is a good tool for minimizing the effort for data cleaning, because large sets of manually performed replacement operations are being executed automatically.

References

1. Agrawal, R., Imieliński, T., Swami, A.: Mining association rules between sets of items in large databases. In: SIGMOD 1993: Proceedings of the 1993 ACM SIG-MOD international conference on Management of data, pp. 207–216. ACM Press, New York (1993)
2. Arens, Y., Hsu, C.-N., Knoblock, C.A.: Query processing in the sims information mediator, pp. 82–90 (1998)
3. Brüggemann, S., Aden, T.: Ontology based data validation and cleaning: Restructuring operations for ontology maintenance. In: Hitzler, P., Sure, Y. (eds.) GI Proceedings 109, Band 1, LNI, GI. vol. 94 (2007)
4. Brin, S., Motwani, R., Ullman, J.D., Tsur, S.: Dynamic itemset counting and implication rules for market basket data. In: Peckham, J. (ed.) SIGMOD 1997, Proceedings ACM SIGMOD International Conference on Management of Data, Tucson, Arizona, USA, May 13-15, 1997, pp. 255–264. ACM Press, New York (1997)
5. Gruber, T.R.: A translation approach to portable ontologies. Knowledge Acquisition 5(2), 199–220 (1993)
6. Han, J., Pei, J., Yin, Y.: Mining frequent patterns without candidate generation. In: Chen, W., Naughton, J., Bernstein, P.A. (eds.) 2000 ACM SIGMOD Intl. Conference on Management of Data, May 2000, pp. 1–12. ACM Press, New York (2000)
7. Leser, U., Naumann, F.: Informationsintegration. dpunkt.verlag (2007)
8. Milano, D., Scannapieco, M., Catarci, T.: Using ontologies for xml data cleaning. In: Meersman, R., Tari, Z., Herrero, P. (eds.) OTM-WS 2005. LNCS, vol. 3762, pp. 562–571. Springer, Heidelberg (2005)
9. Müller, H., Freytag, J.-C.: Problems, methods, and challenges in comprehensive data cleansing. Technical report, Humboldt University Berlin (2003)
10. W.H. Organization: ICD 10: International Statistical Classification of Diseases and Related Health Problems, 10th edn. American Psychiatric Association (1992)
11. Rahm, E., Do, H.H.: Data cleaning: Problems and current approaches. Bulletin of the IEEE Computer Society Technical Committee on Data Engineering 23(4), 3–13 (2000)
12. Savasere, A., Omiecinski, E., Navathe, S.B.: An efficient algorithm for mining association rules in large databases. In: VLDB 1995: Proceedings of the 21th International Conference on Very Large Data Bases, pp. 432–444. Morgan Kaufmann Publishers Inc., San Francisco (1995)
13. Wang, X., Hamilton, H.J., Bither, Y.: An ontology-based approach to data cleaning. Technical report, Department of Computer Science, University of Regina (June 2005)

Towards Automatic Verification of Web-Based SOA Applications

Xiangping Chen, Gang Huang*, and Hong Mei

Key Laboratory of High Confidence Software Technologies, Ministry of Education
School of Electronics Engineering and Computer Science, Peking University
Beijing, 100871, China
chenxp04@sei.pku.edu.cn, huanggang@sei.pku.edu.cn,
meih@pku.edu.cn

Abstract. Nowadays, developing web applications in a Service-Oriented Architecture (SOA) style is emerging as a promising approach for delivering services to end users. Such web-based SOA applications are likely to suffer correctness and reliability problems mainly because their runtime environments (including web browsers and service platforms) are heterogeneous and their service interactions and flows are complex without explicit specifications. In this paper, we propose a model-checking based approach for verifying web-based SOA applications. At first, the application behavior will be automatically specified by analyzing the web-side source codes. And it will be combined with the pre-defined environment behavior so that a precise and complete enough behavior model of the application can be generated automatically. With user-defined constraint and refinement specifications, the behavior model is automatically translated to the formal specification (Promela for Spin) as the input of the model checker. If the model is flawed, the application has correctness and reliability problems. The violation traces generated by the model checker will be visualized in the behavior model for helping developers to solve the detected problems in a user-friendly manner.

1 Introduction

Nowadays, developing web applications in a Service-Oriented Architecture (SOA) style is getting attention, since more and more services are delivered to end users [1]. These applications, called web-based SOA applications, take services as the basic constructions to provide new services in the web. Most of such applications are developed by mashup [4,5], which is a novel approach for combining data and function from multiple services.

When developing web-based SOA applications, the features of web and SOA bring some challenges to the assurance of correctness and reliability. First, a web-based SOA application usually has many diverse and complex interactions. Implementation of such interactions is error-prone. Second, the holistic view of service flows is implicit. Services interact with each other in a GUI element or through the interaction of

---

* Corresponding Author.

Y. Zhang et al. (Eds.): APWeb 2008, LNCS 4976, pp. 528–536, 2008.

GUIs. The functions and interactions of the GUI elements bring extra behavior and make the service flow more complex. Finally, web-based SOA applications are running in heterogeneous environments including the web browser and service platforms. It is difficult for developers to understand the execution semantics of the heterogeneous environments and analyze their impacts on the application behavior.

There are many work on verifying service composition [6, 8] and web applications [9,10,11]. But few of them can be directly applied to web-based SOA applications. In this paper, we propose a model-checking based approach for verifying web-based SOA applications. Our approach begins with generating the application behavior from the source code. The execution environment is abstracted as a reusable template and will be automatically combined with the previous application behavior to form a complete enough and formal behavior model. The behavior model is described using the notation of UML sequence diagram. With user-defined constraint and refinement specifications, the behavior model is translated to the formal specification (Promela for Spin) as the input of the model checker Spin [2]. If the model is flawed, the violation traces generated by Spin will be visualized in the behavior model. This approach is illustrated and demonstrated by mashup, one of the most popular methods for developing web-based SOA applications.

The remainder of the paper is organized as follows. Section 2 illustrates a sample of an incorrect mashup application. Section 3 presents the details of our approach. Section 4 discusses the scalability of the approach with experiment results. Section 5 and Section 6 discuss the related work and our future work, respectively.

2 Illustrative Sample

Fig. 1 shows a mashup application that provides travel information to users. This application has four components: (1) a travel planner, (2) a city selector, (3) a table of beauty spots and (4) a map. These four components act as the GUI of three services, respectively. In this section, we will show an inconsistency problem found in development of this application. This sample will be used in the following sections.

Suppose the user first chooses a travel plan provided by the USA Travel online after searching by the keyword "New York", and then he/she thinks it is better to look

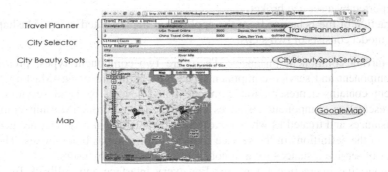

Fig. 1. An incorrect mashup application sample

through all the attractions of Cairo before making a decision. So he/she clicks the "Cities" and selects "Cairo" very quickly after the selection of the travel plan. The "City Beauty Spots" table shows all the beauty spots of Cairo, and the "Map" is expected to display the locations of these beauty spots. But in some cases, an inconsistency problem may occur! The "Map" will display the locations of Denver and New York, which is the result of the previous action, i.e. choosing the travel plan.

Due to the unpredictable network delay and server response time, the order of responses may not be the same as the order of requests. This inconsistency problem is very common in mashup applications.

3 Approach

3.1 Generating Behavior Model

Our approach begins with generating the behavior model of the applications. The behavior model includes application and environment behaviors. Aiming at supporting the verification in a user-friendly way, we provide a behavior model using notations of UML sequence diagram with formal semantics [12] as an intermediate layer. Thus, users can perceive the application behavior and specify the constraints using their familiar notations. Additionally, violations of the constraints can be traced back and animated in the behavior model to help users to solve the detected problems.

3.1.1 Generating Application Behavior Model from Source Code

Generation of the application behavior depends on the programming model. As web-based SOA application development has not been standardized yet, different component models exist. Here, we choose a mashup component model iMashup, which is used in developing the sample application, to illustrate this process.

An iMashup component has two interface types: User Interface and Programming Interface. User Interface is the interface that responds to users' actions and invokes the corresponding functions. Programming Interface is the interface by which iMashup components interact with each other. The interaction may be synchronous or asynchronous. The backend services are regarded as internal logic, and invisible to other components. Asynchronous invocations of the services and callback functions are supported by a service broker.

Extraction of the behavior from the application implemented using iMashup component model consists of two steps:

- Inferring component instances: There are two component types, presentation component and service component. The inner structure of an iMashup component contains a presentation component and several backend services. Thus, presentation component instances can be generated from iMashup component instances and treated as white-boxes. Service component instances are generated from the definitions in the service brokers and treated as black-boxes. The number of service instances for a remote service can be set by users.
- Generating interaction fragments: For every interface and callback function, a sequence of interaction related code is extracted to generate the interaction between components. The interaction between presentation components through

method call or event is modeled as synchronous. The interaction between a presentation component and a service component is modeled as asynchronous and alternatives between different service instances are added if multiple service instances exist.

Let us take the interface "searchTravelPlan" of the component "travelPlanner" as a sample. **Fig. 2** shows the source code extracted and the sequence diagram generated.

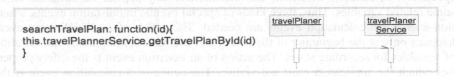

```
searchTravelPlan: function(id){
this.travelPlannerService.getTravelPlanById(id)
}
```

Fig. 2. SearchTravelPlan

3.1.2 Generating Environment Behavior Model

Environment modeling is a key problem in model checking open systems. In the web-based SOA applications, the environment behavior is the behavior of the user actions, service server and web browser. The user actions can be abstracted as random invocations of user interfaces. The return of message from the services is in parallel with the execution for user actions. Thus, the event stack of the browser is abstracted as random selection between the executions of user actions and callback functions.

The template and the complete behavior model of the example generated are shown in **Fig. 3**. For every presentation component, each sequence diagram corresponding to a user interface or a callback function is added as an "operand" of the first "alternative combined fragment". For every service component, a return message is added as an "operand" of the second "alternative combined fragment", which is parallel to the first one. The "guide" of the "operand" is added for simulating the event stack in the browser. We support the automatic generation of environment behavior information using a template.

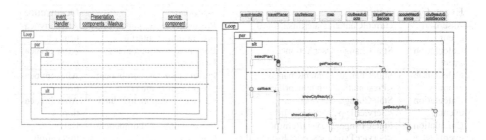

Fig. 3. The template for generation of runtime environment and fragment of example behavior Model after merged with the Template

3.2 Defining Constraint and Refinement Specification

The generated behavior model only includes interaction behaviors. Therefore, in the verification stage, extra constraints and refinements are needed. Developers can add them mutually. But for the properties that are required in most applications, we developed a template for experts to specify. When applied, the specification will support automatic generating the constraints and refinements in the model.

In the behavior model, constraints and refinements are occurrence specifications. Thus, specification should include content of the occurrence specifications and the position where they should be inserted.

Refinements and constraints are extra calculations and assertions of the model, respectively. In addition, assertions and calculations may depend on initialization of some related variables. Thus, three kinds of special events: initialization events, assertion events and calculation events are needed. The action of an initialization event happens only at the beginning of the model and includes definition and initialization of variables for recording states. The action of an assertion event is the safety properties which the application must not violate. The action of a calculation event is the change of the states of the variables in the model.

For adding the constraints and refinements to the model, we support adding occurrence specifications before/after invocation/return of the interfaces. The scope of interfaces for adding occurrences can be user interface/programming interface/callback of representation components and interface of service components.

With the requirement, a template for specification and automatic adding of the constraints and refinements is as follow.

> **Event type**<event type>
> **Action** <inline in Promela>
> **Scope** <scope>
> **Occurrence**<occurrence>

We use the consistency property in the illustrative sample to show how to use the template to define a reusable specification. The property is: the order of the contents displayed in the presentation components should be the same as the order of requests by the user actions. Its specification is defined in **Table 1**.

Table 1. Specification of the consistency property

Event Type	Action	Scope	Occurrence
Assertion event	currentContent.state[this.id]==0	Presentation component: Programming interface	After invocation
Calculation event	StateChange_UserAction()	Presentation component: User interface	After invocation
Calculation event	StateChange_ContentChange(<PresentationComponentID>)	Presentation component: Programming interface	After invocation
Calculation event	StateChange_Service(<ServiceType>,<ServiceInstanceID>)	Service component: Interface	After invocation
Calculation event	StateChange_ServiceReturn(<ServiceType>,<ServiceInstanceID>)	Presentation component: Callback	Before invocation
Calculation event	StateChange_RemoteCall(<ServiceType>)	Service component: Interface	Before invocation

1) Initialization event: The property requires variables for recording the version relationship between the contents of components.
2) Assertion event: The safety property is that the content for updating should be the result of the user action happening after the request for the current content. We

model it as an assertion that the variable of the version relationship should be valid.

3) Calculation event: It is easy to infer that calculation events happen after user action and interaction between components, because new requests or updates of contents in presentation components imply the change of the version relationship.

The template also contains tags that are place holders for the automatically generated functions, such as the tag <service Type> for recording service type.

3.3 Verifying Behavior Model

The behavior model uses the notation of UML sequence diagram with formal semantics, to describe the behavior of the applications. For existing model checker can fulfill our requirement in verification, we choose to integrate a widely-used model checker Spin to perform the verification.

The translation from behavior model with its specifications to verification model (Promela for Spin) is supported based on semantics of behavior model. After verification, the results can be traced back to the behavior model. This means that we use Spin to perform the verification, but the programmers do not need to know how to use Spin and Promela.

The translation from behavior model to verification model consists of two steps:

First, the specification is applied to the behavior model and occurrences of three types of events are added. For example, if the assertion event of the consistency property is defined with the scope of all programming interfaces of all presentation components, occurrences of the events will be added in all operations in the scope. **Fig. 3** shows the occurrences of the added events when the consistency property is applied to the behavior model of the illustrated sample. The red and yellow dots are occurrences of the assertion and calculation events, respectively.

Second, the behavior model is translated to Promela. After the first step, the model consists of two parts: the behavior model and its constraints and refinements. Occurrences in the behavior model are translated to "atomic" blocks in Promela. Occurrences of constraints and refinements are translated to be right before or after the occurrences in the behavior model, because the sentences of their action definitions should not be allowed to interleave with other ones. For the space limit, details of translation rules are omitted.

After the translation, the verification model can be checked by Spin. For tracing back the violation trace to the behavior model, a probe is implemented for keeping the traceability. The probe inserts "printf" sentences in all atomic blocks corresponding to the occurrences in the behavior model. The occurrence IDs generated by the violation trace will be matched to the occurrences in the behavior model so that the violation traces can be simulated in the behavior model. This work is under development.

With the specification of the consistency property, we use our tool to verify our example application. Violations of the consistency property in the component "Map" and "City Beauty Spots" are discovered. Expected animation of the trace will be the red dotted lines in **Fig. 3**. After verification, users can remove the constraints or modify the behavior model and its implementation to avoid the violations.

4 Discussion

Model checking has many advantages while its major limits, i.e. scalability and usability, do not exist when applying to web-based SOA applications. In the previous sections, we use the illustrative sample to show the usability of our approach. To evaluate the scalability, we carry out an experiment on the example behavior model.

We use the behavior model with different setting of the numbers of service instances for verification. And we chose the "do not stop at errors" option in Spin to explore all execution traces of the model. Therefore, the experiment result indicates the worst possible performance for this model, and is independent to the assertions.

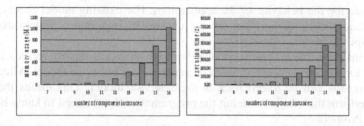

Fig. 4. Experiment results

Fig. 4 shows the memory usage and execution time in verifying the sample with different numbers of component instances on a Pentium 4 2.26GHz with 1.5G DDR memory. In fact, the scale of a web-based SOA application is small because its single page interface model and a browser cannot present too many components in one page. In practice, the number of instances in an application is usually between 10 and 15. Therefore, the result shows that the scalability is acceptable in our approach.

There are some limitations of our approach. The behavior generation relies on the component models. In our tool, it is based on iMashup component model. If there is a standard component model, our approach can be applied to more applications developed by different tools. In addition, the template for specifying constraints only provide basic set of options. It is still far from flexible and practical. And we only support the traceability from verification results to behavior model. The traceability from behavior model to source code may be more important in fault location.

5 Related Work

Many methods and tools have been developed for the rapidly growing community of web programmers who create applications in an SOA style. Some tools focus on facilitating end-user programming [4,7] and some emphasize component reuse[3,5]. However, they do not pay enough attention to the correctness and reliability problems.

There have been some efforts to apply model checking techniques for reliable service assembly [6,8]. Most of them are aiming at the verification of service composition implemented using web services flow specification language like BPEL4WS and WSCI. Our approach differs in that our work is for applications which perform service assembly in the presentation level. Therefore, the behavior model for verification includes user actions in the UI and invocations to backend services.

There are plentiful efforts on applying formal methods to verify the web applications. The work recognized the needs for modeling and verifying the user navigation [9] and instruction processing [10,11]. However, these approaches are built for classical web applications, which are based on a multi-page interface model with interactions in a page-sequence paradigm. Our approach is for the applications based on a single page interface model, and service navigation is also considered.

6   Conclusion and Future Work

Nowadays, development of web applications in an SOA style is receiving attention. However, some natures of web-based SOA make the resulting applications easy to suffer incorrectness and unreliability problems. In this paper, we present a model-checking based approach for verifying the applications in an automated manner.

In the future, we would study more cases to further demonstrate our tool. We expect that these will help us identify more properties and enhance our template for specification of the constraints. We would develop the visualization of animating violation sequences in the behavior model and provide more facilities such as automatic faults location in the source code.

Acknowledgements. This effort is sponsored by the National Key Basic Research and Development Program of China (973) under Grant No. 2005CB321805; the National Natural Science Foundation of China under Grant No. 90612011; and the IBM University Joint Study Program.

References

1. Whatcott,   J.:   SOA's   next   wave: Service-oriented   clients   (2006), http://www.cio.com/
2. Holzmann, G.J.: The SPIN Model Checker: Primer and Reference Manual. Pearson Educational (September 2003)
3. Yu, J., Benatallah, B., Saint-Paul, R., Casati, F., Daniel, F., Matera, M.: A Framework for Rapid Integration of Presentation Components. In: Proc. 16th International World Wide Web Conference (2007), Banff, Alberta, Canada, pp. 923–932 (2007)
4. Hong, J., Wong, J.: Marmite: end-user programming for the web. In: Proc. CHI 2006 extended abstracts on Human factors in computing systems, Canada, pp. 1541–1546 (2006)
5. Ankolekar, A., Krotzsh, M., Tran, T., Vandecic, D.: The Two Cultures: Mashing up Web 2.0 and the Semantic Web. In: Proc. 16th International World Wide Web Conference (2007), Banff, Alberta, Canada, pp. 825–834 (2007)
6. Nakajima, S.: Model-Checking Verification for Reliable Web Service. In: Proc. OOPSLA 2002 Workshop on Object-Oriented Web Services (2002), Seattle, Washington, USA (2002)
7. dataMashups, http://www.datamashup.com
8. Foster, H., Uchitel, S., Magee, J., Kramer, J.: Model-based Verification of Web Service Compositions. In: Proc. 18th IEEE International Conference on Automated Software Engineering (2003), Montreal, Canada, pp. 152–161 (2003)

9. Licata, D.R., Krishnamurthi, S.: Verifying interactive Web programs. In: Proc.19th International Conference on Automated Software Engineering (2004), Linz, Austria, pp. 164–173 (2004)
10. de Alfaro, L.: Model checking the World Wide Web. In: Proc. 13th International Conference Computer Aided Verification (2001), Paris, France, pp. 18–22 (2001)
11. Graunke, P., Findler, R.B., Krishnamurthi, S., Felleisen, M.: Modeling web interactions. In: Proc. 12th European Symposium on Programming (2003), Warsaw, Poland, pp. 238–252 (2003)
12. Knapp, A.: A Formal Semantics for UML Interactions. In: France, R.B., Rumpe, B. (eds.) UML 1999. LNCS, vol. 1723, pp. 116–130. Springer, Heidelberg (1999)

A Novel and Effective Method for Web System Tuning Based on Feature Selection*

Shi Feng, Yan Liu, Daling Wang, and Derong Shen

College of Information Science and Engineering, Northeastern University,
Shenyang 110004, P.R.China
wondertime@gmail.com, ufoly94@hotmail.com,
dlwang@mail.neu.edu.cn, shenderong@ise.neu.edu.cn

Abstract. Web has become the main platform for the interchange of information and the transaction of commerce. The performance of a Web system can be greatly improved by tuning its configuration parameters. However, there are dozens or even hundreds of tunable parameters in one Web system, and tuning can be the tough work even for the most experienced server administrators. Traditional Web tuning methods only focus on two or three specified parameters, and can not provide an effective solution to the tuning problem when the number of parameters is large. In this paper, we propose a feature selection algorithm based on Information Gain criterion to find the key parameters of a Web system. The algorithm can pick out the parameters that significantly affect Web system performance. Therefore, the tuning approach can be simplified dramatically. We have carried out extensive experiments with different Web systems. The results show that the algorithm is effective in searching the most important parameters under different conditions and reducing the time cost of next tuning steps.

1 Introduction

Our life and work are facilitated by the Web-based applications such as online stores and brokerage, human resource applications, and supply-chain management systems. The performance and availability of these Web-based systems play a very important role in the operation of an organization. However, the performance of a Web system depends heavily on appropriate configuration. Tuning a Web system means to find the correct parameter settings under the specified conditions. In general, there are dozens or even hundreds of tunable parameters in just one tier of a Web system. How to get the server's appropriate configuration is very difficult for a layman. With different hardware conditions and complex applications deployed, tuning can be the tough work even for the most experienced Web system administrators. For example, a BEA Weblogic server may have 30 tunable parameters, and in different hardware conditions, the response time bottleneck of the system may be caused by different bad tuned parameters. On the other hand, the application to be deployed may have

---

* This work is supported by National Natural Science Foundation of China (No. 60573090) and National Ministry of Education Project of China (GFA060448).

Y. Zhang et al. (Eds.): APWeb 2008, LNCS 4976, pp. 537–547, 2008.

dissimilar types, such as CPU-intensive or memory-intensive, which induce inequable parameter settings.

Most recently, with the fast development of Web 2.0[12], the concept of "blogging and the wisdom of crowds" is well known by common people. Nowadays, "dynamic websites" (i.e., database-backed sites with dynamically generated content) replaced static Web pages, which produces more pressure on Web system. Many personal homepages and small-middle scale web sites providing Web 2.0 services are emerging. Most of them are deployed on open-source Web servers, such as Tomcat, JBoss, Apache and etc. There is a lack of detailed documentations and technical support staffs for open-source Web servers. Faced with dozens of tunable parameters, a confused Web master doesn't know how to get started. Therefore, how to efficiently get the appropriate configuration for the Web system parameters has become a major concern in the age of Web 2.0.

There are many drawbacks in existing Web system tuning approaches. On one hand, a Web system may have a hundred parameters that can be modified. Previous approaches choose two or three important parameters for tuning [11; 14], but they do not tell how to find out these important parameters among the hundreds parameters. When the number of parameters is large, the tuning approaches become unfeasible. On the other hand, some of the previous approaches focus on only one type of Web system software such as Apache, and they are not the all-purpose ways for performance tuning [15].

Our work focuses on how to find the most important parameters of a Web system, which is seldom introduced in previous literatures. In this paper, we proposed a feature selection method based on Information Gain criterion to find the parameters that affect system performance dramatically. Comparing to previous approaches, we can get the following advantages: Firstly, we solve the tuning problem when the number of parameters is large. If the number of parameters is reduced to be acceptable, many mature approaches can be used for tuning. So, our method paves the way for the next tuning steps. Secondly, little domain knowledge is needed during the tuning, for the selection approach is based on sampling data and train set. Thirdly, our approach can quickly find the most important parameters without the limitation of the application deployed or running environment. Finally, the approach can be applied to all kinds of Web system, so it is an all-purpose tuning method.

The rest of the paper is organized as follows. Section 2 introduces the related work on Web system performance tuning. Section 3 presents the feature selection based method for finding the most important parameters. Section 4 presents the tuning methodology based on feature selection. Section 5 provides experimental results on different application deployed and on different Web servers. Finally we present concluding remarks and future work in Section 6.

2   Background and Related Work

There have has some previous work on the performance tuning of Web system. We can classify the existing approaches of parameter tuning for Web system into two categories: model-based approach and experiment-based approach.

2.1 Model-Based Approach

In this approach, an analytical model is built to predict and validate the performance of a Web system. There are some sophisticated mathematics models for system analyzing such as queuing system and control theory. The queue model is a widely used conceptual framework in which computing systems are viewed as networks of queues and servers. It has been proven quite effective at modeling the steady-state behavior of computing systems. In [10], a network queuing model is used to guide the search for the best combination of configuration parameters. Bhuvan Urgaonkar et al. [13] analyze the architecture and request processing method in multi-tier Internet services, and propose a queuing model to compute the response time of Internet services. In [5], a methodology is described to determine the optimal concurrency level of an application server. A simple benchmarking application is used to derive a queue model of the server, which is strictly related to the design pattern of the running application as well as the application server implementation. Besides, the control theory is also well studied and applied to many server tuning approaches. In [1], [9] and [15], the Web system is treated as a black box, thus the complexity of tuning is reduced. Gandhi et al. [6] use feedback control theory to model the Apache server, and the experiments shows that the model is efficient within the appropriate workload regions.

These model-based approaches give some good solutions to the tuning and prediction of server performance. However, these studies may simplify actual systems too much despite the fact that today's Web systems are quite complex. Moreover, building these models needs detailed knowledge of server's internal structure. Some online tuning methods have to change the source code of the server software. Still further, the models do not include the part of server's hardware and do not consider the type of the application deployed, which may make a great contribution to the overall performance of the system.

2.2 Experiment-Based Approach

In this approach, the Web system tuning problem is viewed as mathematical optimization problem and some heuristic search techniques are used to discover proper settings of parameters. Chung et al. [3] adjust the tunable parameters based on the observed performance results to improve the overall system performance. In [14], a Smart Hill-Climbing algorithm is proposed using the ideas of importance sampling and Latin Hypercube Sampling. The algorithm is tested using an online brokerage application running in a WebSphere environment, and the result shows that the algorithm is superior to traditional heuristic methods. In [11], the Quick Optimization via Guessing algorithm is proposed, which quickly selects one of nearly best configurations with high probability.

While the proposed experiment-based approaches do not need much knowledge about target parameters and internal server states, they have some limitations. Firstly, these approaches need many experiments to collect sample data, and cost quite a long time to find the optimal values. Secondly, most of these approaches focus on tuning one or two key parameters on a certain Web system. They do not explain why these two parameters are more important than others. Moreover, there may be dozens of tunable parameters in one server, and finding the influential parameters may be a domain knowledge concerned approach.

In the next section, we propose a feature selection based method to find the most important parameters. Our method can be the pre-step of many mature experiment-based tuning approaches and solve the tuning problem when the number of parameters is large.

3   Feature Selection Method for Searching Key Parameters

3.1   Key Parameters

There may be dozens of parameters that can be modified in one Web system, but in a certain running environment, only some of these parameters significantly affect server performance. Here we give the brief description of the key parameters.

Assuming that the server parameter set is P, and Web system performance is a function of P. The key parameter set is P_{kp}. P_{kp} is the subset of P, and contains the most influential parameters of a Web system. When P is large, it is a time-consuming task to get the appropriate configuration of all tunable parameters in P. However, we can tune parameters in P_{kp} instead of P to reduce the configuration searching space. To confirm the effectiveness of key parameters, we measured the performance of the Apache Web server using a Web benchmark (for experiment setting details, refer to Sect. 5). Figure 1 plots the server average response time for different *Threads-PerChild* and *Timeout* settings.

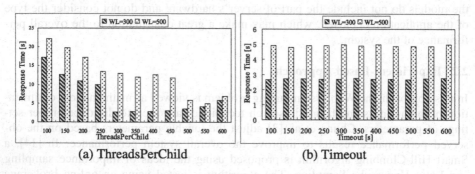

(a) ThreadsPerChild   (b) Timeout

Fig. 1. Performance Effects of ThreadsPerChild and Timeout

We can see that *ThreadsPerChild* dramatically affects server performance. In Figure 1(a), when *ThreadsPerChild* is set to 300, with workload 300, the response time de-grade down to 16.3% compared to the value of *ThreadsPerChild* 100. With the work-load 500, when *ThreadsPerChild* is set to be 550, the server response time is 18.9% compared to the value of *ThreadsPerChild* 100. However, Figure 1(b) shows that server performance does not change much according to *Timeout* settings. So, in this running environment, *ThreadsPerChild* is the key parameter. The system performance can be quickly improved by just tuning one parameter *ThreadsPerChild*. However, we can pick out *ThreadsPerChild* from others because it is a well-known influential parameter in performance tuning of Apache. In many other Web systems, the key parameters may change according to the type of application deployed and hardware conditions. In next section, we proposed an all-purpose key parameters searching method based on feature selection. With little domain knowledge, the method can quickly find out which parameters are important under certain circumstances.

3.2 Searching Key Parameter Based on Feature Selection

Feature selection reduces the data set size by removing irrelevant or redundant features. The goal of feature selection is to find a minimum set of features so that the resulting probability distribution of the data classes is as close as possible to the original distribution obtained using all features [8]. Given a feature set $F = \{F_1, F_2, F_3, ..., F_n\}$, n denotes the size of the feature set. A feature subset can be depicted by a binary vector: $S = \{s_1, s_2, ..., s_n\}$, $s_i \in \{0,1\}$, $i=1, 2, ..., n$, and $s_i = 1$ denotes the ith feature f_i is selected, otherwise, the ith feature f_i is not selected. We assign $E(S)$ to be the evaluation function for the feature set F, so the feature selection is to find the optimal value of the expression: max $E(S)$.

The performance of a Web server can be recognized as the function of tunable parameters, and different parameters have different weights to affect the system performance. By feature selection method, we can pick out the most important parameters from the others, therefore, the tuning approach can be simplified and the cost for finding the optimal parameter configuration can be reduced dramatically.

Basically, feature selection methods fall into three categories [7]: filter, wrapper, embedded. Among these categories, filter is the most comprehensively-studied methods, and it can be independent from learning. Therefore, we use filter feature selection algorithm for searching key parameters of the Web system.

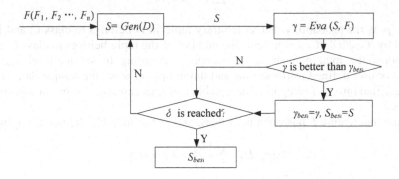

Fig. 2. Key Parameters Searching Algorithm Based On Feature Selection

We define the parameters settings of the Web system as the feature set F (F_1, F_2, ..., F_n), and size of the set n denotes n tunable parameters, in which there are key parameters. The general approach for searching these key parameters is shown in Figure 2. In Figure 2, the function Gen(F) gets the feature subset S according to generation algorithm. The evaluation function $Eva(S, F)$ checks the goodness of subset S, and returns γ. There is a loop to find the best S until the stopping criterion is reached.

The critical problems in this key parameters searching algorithm (KPSA) are discussed as follows:

Subset generation: The subset selection includes search direction and search strategy. For search direction, there are Forward Generation, Backward Generation and Random Generation [4]. Since there may be dozens of features (i.e. tunable parameters), forward direction is the appropriate way of generating subset. The procedure starts from an empty set. As the search starts, parameters are added one at a time. At each iteration, the best parameter among unselected ones is chosen based on the evaluation function. The subset grows until the stopping criterion is reached. For search strategy, a complete evaluation of the 2^N subsets is impractical. So, in our key parameters searching algorithm, we use a kind of depth-first search method guided by heuristics. The heuristic search strategy is very fast in producing results because the search space is only quadratic in terms of the number of parameters.

Evaluation function: Feature selection can be seen as an optimization problem, and the key of problem is to establish a standard to evaluate which feature is important and which feature is redundant. We use information theory to evaluate the features. In classical information theory, the higher the Information Gain (IG) in one feature is, the bigger its information amount is. During the key parameters searching, the IG of each feature is calculated, and the feature with highest IG in dataset is the most important parameter for performance tuning. The Information Gain [2] of the feature F in sample dataset D is defined as:

$$Gain(F) = Info(D) - Info_F(D) \qquad (1)$$

The $Info$ (D) is the entropy of the sample, which is defined as:

$$Info(D) = -\sum_{i=1}^{m} p_i \log_2(p_i) \qquad (2)$$

where p_i is the probability that an arbitrary tuple in D belongs to class C_i and is estimated by $|C_{i,D}|/|D|$. $|C_{i,D}|$ represents the number of the tuple belongs to class C_i and $|D|$ represents the total number of the sample. According to service-level agreement (SLA) or user defined response time and throughput criteria, the sample dataset D can be classified into two categories: acceptable and unacceptable. So in our algorithm, m equals 2.

Assume that feature F has v distinct values, and in Formula (3), $Info_F(D)$ is defined as:

$$Info_F(D) = \sum_{j=1}^{v} \frac{|D_j|}{|D|} \times Info(D_j) \qquad (3)$$

The term $|D_j|/|D|$ denotes the weight of the jth partition divided by v distinct values of feature F. $Info(D_j)$ can be defined as:

$$Info(D_j) = -\sum_{i=1}^{m} p_{ij} \log_2 p_{ij} \qquad (4)$$

where p_{ij} denotes the probability that a tuple in jth partition belongs to class C_i. With $Gain(F)$, we can evaluate the features in the sample dataset and pick up the key parameters.

Stopping criterion: Without a suitable stopping criterion the feature selection process may run exhaustively or forever through the space of subsets. The more parameters the algorithm produces, the more complex the tuning task is. Therefore, we can set a threshold to stop the iteration. When evaluation is less than the threshold τ, the search is stopped. Also, we can define the stopping criterion as a predefined number of features that are selected.

According to what we have discussed above, we proposed the key parameters searching algorithm (KPSA). The KPSA is effective in acquiring the most important parameters. In the next section, we discussed how to tune a Web system using KPSA.

4 Feature Selection Based Tuning Methodology for Web System

The KPSA can efficiently reduce the Web system configuration space, and pave the way for the next tuning steps. We propose a general Web system tuning methodology, which includes the entire process of parameter configuration (see Fig.3).
Fig.3. depicts several steps of tuning a Web system.

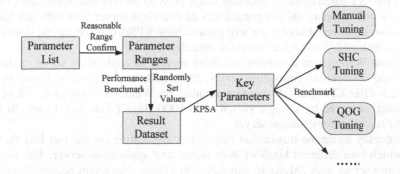

Fig. 3. Feature Selection Based Tuning Methodology

1) Collect the tunable parameter list of a Web system, and usually, this can be done by reading the software official guide documents.

2) For each configuration parameter, select a reasonable range. The range may be established on basic knowledge. For example, the JVM heap max value can not be set 1GB if the server only has 512MB physic memory. In most cases, it is useful to select a range from 50% to 300% of the parameter's default value.

3) Use benchmarks to test the system performance and record the response time or other criterion. For each test run, use random functions to generate parameter values in the reasonable ranges. The number of test runs varies according to the number of the parameters. Make a standard to check whether a response time is acceptable, and use this standard to add classifier tag to each tuple in result dataset. Therefore, the test run result dataset includes parameter values and a classifier tag.

4)   Based on result dataset, use KPSA to find the key parameters.
5)   KPSA reduces the parameter configuration space, and with key parameters, the next tuning steps become much easier. Rules-of-thumb may be used for manual tuning, and there are some effective statistics based method for Web system tuning, such as SHC tuning method [14] and QOG tuning method [11].

Our methodology has some outstanding advantages. Firstly, if there are a lot of tunable parameters, the experiment based tuning is a time-consuming task. Using key parameters, the further tuning method can get the result with less test runs. Secondly, different from some parameter specified method, our methodology is not related to the design pattern of running application as well as the server implementation. Thirdly, our methodology uses little domain knowledge, and it is especially useful for some open source web systems with few tuning guidelines.

5   Experiments

In this section, we test our method on different Web systems. We will first use KPSA to find the key parameters on different kinds of Web servers and application servers. Then, we will compare the key parameters on one same server with different kinds of application deployed. Finally, we will present how KPSA can reduce the time cost of experiment-based Web system tuning approaches.

Our test Web system consisted of a Web server, an application server, a database server and a workload generating server. Each node is equipped with two Pentium Xeon 2.8 GHz CPUs, 2 GB memory running on Windows 2000 SP4. All of these were interconnected by a single switch via a 1000 Base-T Ethernet. Oracle 9i (patch 9.2.0.6) is chosen for database server.

We deploy an online transaction processing application on our test bed Web system, which uses different kinds of Web server and application server. The workload generating server uses JMeter to simulate 300 clients concurrent access. Firstly, followed our feature selection based tuning methodology, we list the tunable parameters. Then, we set the range from 50% to 300% of the parameter's default value. Thirdly, we use the workload generating server to take the performance benchmark, and for one kind of server, the test runs 100 times. For each run, parameter configurations are generated by a random function within the range. In one result dataset, the tuple contains parameter values and response time. We classify the dataset into two categories according to whether the response time is acceptable. Finally, we employ the KPSA to find the key parameters, and the result is shown is Table 1.

Table 1. Key parameters of an online transaction processing application on different servers

Server	Parameters Number	Key Parameters
Apache	8	*ThreadsPerChild, KeepAliveTimeout*
Weblogic	15	*ThreadCount, AcceptBacklog*
Websphere	10	*ThreadPoolMaxSize*
JBoss	8	*MaxPoolSize*
Tomcat	8	*maxThreads, acceptCount*

In the test bed, Apache works as the Web server, and Weblogic, Websphere, JBoss and Tomcat work as application server. The KPSA pick out two from eight and fifteen with Apache and Weblogic respectively. With high workload, the OLTP Web system performance bottleneck is most likely to be caused by server's receiving request ability. Previous literatures confirm that *ThreadsPerChild* determines the maximum number of clients connecting to the server at the same time. The KPSA picks out the *ThreadsPerChild* validates that our algorithm can effectively find the parameters which affect the system performance dramatically.

With different type of applications deployed, the key parameters in one server may be different. The table 2 compares key parameters found by KPSA when different type of applications is deployed on Weblogic.

Table 2. The key parameters of different applications deployed on Weblogic

Application Type	Technique	Key Parameters
Transaction processing	JSP, Servlet	*ThreadCount, AcceptBacklog*
Online computing	EJB	*MaxBeansInFreePool,Mem_agrs-Xmx*

We test an EJB-based application for online graphics processing on Weblogic. Table 2 shows that our algorithm can pick out key parameters when different applications are deployed.

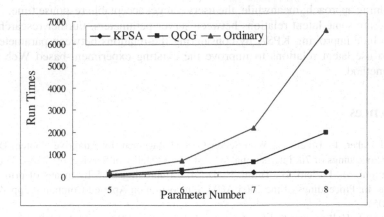

Fig. 4. The comparison of KPSA, QOG and Ordinary Tuning Method

The key parameters pave the way for the next step experiment-based Web system tuning. Three ways of further tuning are compared is Figure 4. The Y axis values denote the run times needed by a tuning experiment. Suppose there are N tunable parameters in one server, each parameter has three available values to be set. The ordinary tuning way needs 3^N test runs to complete the task. Quick Optimization via Guessing (QOG) can reduce the test time by guessing the system performance. However, in [11] experiment, only four parameters are picked out for configuration, and it does not explain how these four parameters are picked out from the others. At first, when there are five optional parameters to be tuned, KPSA produces as much run times as the others because there is fifty test runs to find key parameters. When parameter space increases, the run times of ordinary tuning way turns to be huge and becomes unacceptable. QOG also requires a lot of time to get the best configuration. However, KPSA can find the key parameters, so the test runs times can be reduced dramatically. During the experiment, the best response time of this three tuning methods with different tunable parameters has been recorded. It shows that when we only use key parameters to tune the system, the response time is only average 8% slower than the QOG and ordinary method. Compare to the benefit of much less test runs, we think this 8% slower response time is acceptable. Therefore, we can conclude that KPSA can dramatically reduce the test runs for experiment-based tuning approaches and still get acceptable system performance.

6 Conclusion and Future Work

Although there may be dozens of tunable parameters in one Web system, in previous literature, little attention is paid on how to find the most influential parameters of an online application. In this paper, we propose a feature selection method based on Information Gain criterion to find the key parameters of a Web system. Based on KPSA, the tuning methodology for Web system is introduced. Experiments show that our algorithm can effectively find the parameters that affect system performance most. Using key parameters, we can dramatically reduce the test run times of an experiment-based tuning approach; meanwhile, the users can get acceptable response time.

There are some latent relations between server parameters. Further research directions include improving KPSA by considering the relations between parameters. We can also use latent relations to improve the existing experiment-based Web system tuning method.

References

1. Abdelzaher, T., Bhatti, N.: Web Server QoS Management by Adaptive Content Delivery. In: Proceedings of 7th International Workshop on Quality of Service, pp. 216–225 (1999)
2. Baglioni, M., Furletti, B., Turini, F.: DrC4.5: Improving C4.5 by means of prior knowledge. In: Proceedings of the 2005 ACM Symposium on Applied Computing, pp. 474–481 (2005)
3. Chung, I., Hollingsworth, J.K.: Automated cluster-based Web service performance tuning. In: Proceedings of 13th IEEE International Symposium on High Performance Distributed Computing, pp. 36–44 (2004)
4. Dash, M., Liu, H.: Feature Selection for Classification. Intelligent Data Analysis 1, 131–156 (1997)
5. Dumitrascu, N., Murphy, S., Murphy, L.: A Methodology for Predicting the Performance of Component-Based Applications. In: Proceedings of 8th International Workshop on Component-Oriented Programming, pp. 61–68 (2003)
6. Gandhi, N., Tilbury, D.M., Diao, Y., Hellerstein, J., Parekh, S.: MIMO Control of an Apache Web Server: Modeling and Controller Design. In: Proceedings of American Control Conference, pp. 4922–4927 (2002)

7. Guyon, I., Elisseeff, A.: An Introduction to Variable and Feature Selection. Journal of Machine Learning Research, 1157–1182 (2003)
8. Han, J., Kamber, M.: Data Mining Concepts and Techniques, 2nd edn. Morgan Kaufmann, San Francisco (2005)
9. Li, B., Nahrstedt, K.: A Control-based Middleware Framework for Quality of Service Adaptations. IEEE Journal on Selected Areas in Communications 17(9), 1632–1650 (1999)
10. Menasce, D.A., Barbara, D., Dodge, R.: Preserving QoS of E-commerce Sites Through Self-Tuning: A Performance Model Approach. In: Proceedings of the 3rd ACM conference on Electronic Commerce, pp. 224–234 (2001)
11. Osogami, T., Kato, S.: Optimizing System Configurations Quickly by Guessing at the Performance. In: Proceedings of the 2007 ACM SIGMETRICS International Conference on Measurement and Modeling of Computer Systems, pp. 145–156 (2007)
12. O'Reilly, T.: What Is Web 2.0., http://www.oreilly.com/
13. Urgaonkar, B., Pacifici, G., Shenoy, P., Spreitzer, M., Tantawi, A.: An Analytical Model for Multi-tier Internet Services and Its Applications. In: Proceedings of the International Conference on Measurements and Modeling of Computer Systems, pp. 291–302 (2005)
14. Xi, B., Liu, Z., Raghavachari, M., Xia, C.H., Zhang, L.: A Smart Hill-climbing Algorithm for Application Server Configuration. In: Proceedings of the 13th international conference on World Wide Web, pp. 287–296 (2004)
15. Zhang, Y., Qu, W., Liu, A.: Automatic Performance Tuning for J2EE Application Server Systems. In: Proceedings of 6th International Conference on Web Information Systems Engineering, pp. 520–527 (2005)

Effective Data Distribution and Reallocation Strategies for Fast Query Response in Distributed Query-Intensive Data Environments*

Tengjiao Wang, Bishan Yang, Jun Gao, and Dongqing Yang

Key Laboratory of High Confidence Software Technologies (Peking University),
Ministry of Education, China
School of Electronics Engineering and Computer Science,
Peking University, Beijing, 100871, China
{tjwang, bishan_yang, gaojun, dqyang}@pku.edu.cn

Abstract. Modern large distributed applications, such as mobile communications and banking services, require fast responses to enormous and frequent query requests. This kind of application usually employs in a distributed query-intensive data environment, where the system response time significantly depends on ways of data distribution. Motivated by the efficiency need, we develop two novel strategies: a static data distribution strategy DDH and a dynamic data reallocation strategy DRC to speed up the query response time through load balancing. DDH uses a hash-based heuristic technique to distribute data off-line according to the query history. DRC can reallocate data dynamically at runtime to adapt the changing query patterns in the system. To validate the performance of these two strategies, experiments are conducted using a simulation environment and real customer data. Experimental results show that they both offer favorable performance with the increasing query load of the system.

Keywords: data distribution, data reallocation, query response.

1 Introduction

In recent years, more and more large distributed applications, such as mobile communications and banking services, require fast responses to enormous and frequent query requests. For example, when a telecom subscriber makes a phone call, a request is sent to a GSM switching system to query his or her location information. At the same time, there may be large amount of query requests waiting for responses when numbers of subscribers are asking for service. Each query must be responded with spectral efficiency, otherwise the phone call connection will be considered failed. This kind of application usually employs in a

---

* This work is supported by the NSFC Grants 60473051, 60642004, the National '863' High-Tech Program of China under grant No. 2007AA01Z191, 2006AA01Z230, and the Siemens - Peking University Collaborative Research Project.

Y. Zhang et al. (Eds.): APWeb 2008, LNCS 4976, pp. 548–559, 2008.

distributed query-intensive data environment, which is different from traditional distributed environment. How to provide fast response time with extremely high query load is an essential issue greatly concerned by the service providers.

Looking into the distributed query-intensive data environment, a query request is basically processed by a system node that contains the queried data. Besides, the query load varies from node to node for the query frequencies are different on the system nodes. As a result, load imbalance will occur when some nodes are facing high access traffic while others are rarely visited or even idle. In this case, the overall performance of the system will slow down due to the imbalanced use of the system resources. Therefore, query load balancing is a key concern for improving system's performance (query response time) in this environment.

For load balancing problem in traditional distributed systems, most previous works concern ways of job allocation to balance the workload. Static load balancing[1,2,3] strategies assign jobs to the system nodes when the jobs are generated. The assignment is computed based on the information about the average behavior of the system. Once the assignment is determined, it will not change throughout the lifetime of the system. Dynamic load balancing[4,5,6,7,8] strategies balance the workload at runtime to adapt the system changes, through transferring jobs or data from an overloaded node to an underloaded node. The transferring decisions can be made based on the state information of the current system, at the expense of communication overhead over the system nodes.

Although the above strategies help balance the workload, they can not be directly used in the distributed query-intensive data environment due to several reasons: (1) The system requires spectral response efficiency for querying the query-intensive data. The response time is highly sensitive to any additional communication and processing cost in the system. (2) Oftentimes, the data size is huge and the data may be distributed among the system nodes without sharing. Therefore the query tasks can not be simply transferred from one node to another. (3) The query load is unknown previously. However, it can be predicated based on the query history in the system, since the query requests have some inherent patterns in practice. If the data is frequently queried by a subscriber in the past, it will possibly be often queried in near future.

Note that in the distributed query-intensive data environment, the query load distribution significantly depends on how the data is distributed. Therefore, in this paper, we concern ways of data distribution to improve system's performance through load balancing. More specifically, we make the following contributions:

- We provide a static data distribution strategy DDH[1], that uses a hash-based heuristic technique to distribute data off-line according to the query history. It aims to distribute data reasonably and help balance the query load of the system.
- We design a dynamic data reallocation strategy DRC[2], that can reallocate data dynamically at runtime to adapt the changing query patterns in the

---

[1] DDH stands for Static Data Distribution using a Hash-based Heuristic Technique.
[2] DRC stands for Dynamic Reallocation with a Central Controller.

system. The query load can be balanced by using a central controller to make reallocation decisions and conduct data migration.

- A simulation environment is designed to test the effectiveness of the two strategies. Experimental results show that they both offer favorable performance improvements with increasing query load in the system.

The remainder of this paper is organized as follows. Section 2 describes the considered distributed query-intensive data environment. In Section 3, the static data distribution strategy DDH is introduced. Section 4 describes the dynamic data reallocation strategy DRC. Section 5 describes the experiment environment and presents the evaluation results of the two strategies. Finally, section 6 concludes the paper.

2   Model Description and Formulation

To better understand our proposed strategies, we consider a distributed query-intensive data environment, which is commonly used in mobile communications[9]. (Fig. 1 shows the system architecture) The considered system can be divided into three parts: the back-end (BE), the front-end (FE) and the application client (AC). The central controller (CC) in the gray rectangle is an additional part which is designed for employing the dynamic data reallocation strategy.

The back-end is comprised of several BE servers. The BE servers hold the permanent data about subscribers, including service profiles and location information. For BE servers in the same area, the stored data is the same (these BE servers are viewed as a whole and one of them is considered as a deputy). For BE servers over different locations, there is no overlap of data, and the data can be migrated among these BE servers. The database on each BE server has homogeneous structure, but the processing power of these BE servers may be different. When receiving a query request, a BE server will execute the query on the local database, and then return the result data.

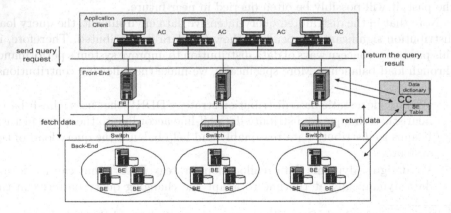

Fig. 1. A distributed query-intensive data environment

The front-end contains several FE servers. Each FE server can contact with all the BE servers in the system. The FE servers store selected administrative information about subscribers in order to service the visiting subscribers. The request submitted by a subscriber will first arrive at a FE server (the FE server is selected according to the subscriber's current location). If the needed data can be found on this FE server, the result can be directly returned to the subscriber, otherwise, the FE server routes the query to the BE server that contains the queried data, and waits for the result data and then return it to the subscriber.

The application client AC is carried by the subscriber. A query request is submitted from AC to one FE server and waits for the response. That is, subscribers only contact with FE servers in the system, and BE servers are transparent to them.

CC is a central controller which we design for employing the dynamic data reallocation strategy. In practice, CC can be implemented by a single server or run as a service in an existing server. More Details about the functionality of CC will be described in Sect. 4.

To better understand the proposed strategies in Sect. 3 and Sect. 4, we first express several definitions.

Definition 1. *Data unit T is defined as the smallest unit with a unique ID in the system. A data unit contains a record with a primary key in the relational database, or an entry with a DN in the LDAP database.*

Definition 2. *The query load associated with data unit T in time interval t is the total frequency of read execution on T during t, denote as $L(T)$.*

Definition 3. *Given the entire data-set $D = T_1, T_2, \ldots, T_n$, where $T_i (1 \leq i \leq n)$ is a data unit. A data group G is made up of k data units, $(1 \leq k \leq n)$. Then there are $\lceil n/k \rceil$ data groups in the system.*

Definition 4. *Given a data group $G = T_1, T_2, \ldots, T_k$, the query load associated with G is the sum of query loads associated with the data units contained in G, so $L(G) = \sum_{i=1}^{k} L(T_i)$.*

The basic idea of defining data groups is to make the management of the massive information more flexible. The control granularity of the system can be determined by the data group size. To better perform data distribution, we introduce a data structure called *data dictionary*, which stores the information of all data groups in the system including *data group ID, data group location and read load*. Each FE server keeps a data dictionary and CC also has one. For the dictionary in a FE server, the *read load* of a data group records the query frequency on the data group detected by the FE server (we only consider read load here since in this environment read is the main operation). For the dictionary in CC, the *read load* of a data group is the sum of the read load values of the data group recorded in all FE servers. *BE table* is another structure stored in CC which records the information of each BE server, including *BE ID, the process power(PC) and the query load(Load)*. The *BE Table* gives an overview of the global query load information which can help make better decisions to reallocate data.

3 DDH-A Static Data Distribution Strategy

DDH is designed to make reasonable data distribution off-line to help obtain query load balancing in the system. As the query behavior of subscribers often has some inherent pattern, the data frequently queried by a subscriber is very likely to be queried again. Therefore we believe that it is important to take the query load history into account when concerning a reasonable data distribution.

With the query history information, DDH distributes data using a hash-based heuristic technique. In the distribution, the entire data set is presents as numbers of data groups. The main idea is to distribute these data groups randomly to buckets through hashing. Each bucket contains similar query load instead of similar data size (The query load is measured by the query frequency on the data). Then through assigning buckets averagely to the BE servers, each server can be expected to have near-equal query load.

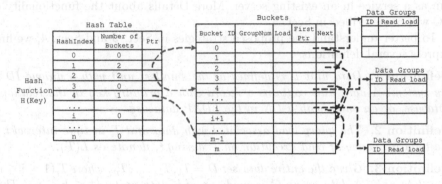

Fig. 2. Hashing process

Since the data size is usually very large in practice, a secondary-storage hash table is used. A hash function maps the search keys (the unique key feature of the data group) into range 0 to $n - 1$, where n is the length of the array which consists of pointers to the header of bucket lists (n is determined by the total data size, the data group size and the maximum capacity of a bucket). Data groups that are hashed to the same address will be distributed to the same bucket. The capacity of a bucket has a upper bound, and we denote the max number of data groups that can be held in a single bucket as GN_{max}. The mean load of all the data groups in the system is DL_{avg}. A bucket overflows when the total load of its containing data groups exceeds a threshold φ, which is calculated as $GN_{max} * DL_{avg}$. If a bucket overflows, a chain of overflow buckets can be added to the bucket. A bucket table is used to organize all the buckets by storing their links. The hashing process is given in Fig. 2. As the hash table and the bucket table records the pointers instead of specific data information, they can be expected to be maintained in memory. When the hashing process is finished, all the buckets are assigned to the BE servers averagely. In this way, each BE server contains data groups with approximately the same load. The algorithm is given in Algorithm 1.

Algorithm 1. Static Data Distribution DDH

1: Create hash table HT
2: **for all** data group G **do**
3:    Get the last bucket LB in the entry of $H(G.ID)$ /* H is the hash function */
4:    **if** $LB.Load \geq \varphi$ **then**
5:       Create a new bucket NB and $LB.NextBucket \leftarrow NB$
6:       Append G to NB
7:       $NB.Load \leftarrow G.ReadLoad$
8:       $NB.GroupNum \leftarrow 1$
9:    **else**
10:       Append G to LB
11:       $LB.Load \leftarrow LB.Load + G.ReadLoad$
12:       $LB.GroupNum \leftarrow LB.GroupNum + 1$
13: **for all** nonempty bucket B_j in HT **do**
14:    distribute the data groups in B_j to $BE[j\%n]$ /* n is the number of BE servers in the system */

In sum, DDH provides a static strategy to distribute data off-line to help balance the query load of the system. The data distribution is a hash-based heuristic process based on the information of the query load history. Furthermore, the distribution process is very efficient due to the effectiveness of hash, and the location index of the data groups can be built up simultaneously during the distribution process. Distributing data reasonably with DDH can greatly help obtain query load balancing in the system, and thus the query response time can be expected to be sped up.

4   DRC-A Dynamic Data Reallocation Strategy

Since the query behavior of subscribers is not invariable in practice, one can not expect the query load to be balanced throughout the lifetime of the system based on the initial data distribution. Therefore, we propose DRC, a dynamic data reallocation strategy, that can reallocate data dynamically at runtime to adapt the changing query patterns. The query load can be balanced by using a central controller to make reallocation decisions and conduct data migration. The reallocation decisions are made based on the global query load information. Similar to the dynamic load balancing strategy in [4], DRC is comprised of three parts: information rule, location rule and migration rule. Together with the migration rule, a data migration mechanism for ensuring data consistency is proposed.

Information rule. The information rule describes the collection and storing methods of the query load information of the runtime system. To avoid making decisions favoring only some system nodes, it needs to collect the global information of the system. Denote the read load associated with data group G_i on FE_j as $L_{FE}(i,j)$. Assume there are k FE servers in the system called $FE_1, FE_2, \ldots, FE_k$. The total load of data group G_i is the sum of read loads recorded in all the FE servers, expressed as $L(i) = \sum_{j=1}^{k} L_{FE}(i,j)$. Since the data groups have no overlap on BE servers, we can calculate the query load on BE_i as $L_{BE}(i) = (\sum_{Gj \text{ is in } BEi} L(j))/C(i)$, where m is the number of data groups stored on BE_i, and $C(i)$ is the query processing power of BE_i.

To avoid large expense of communication, CC initiatively notifies all the FE servers to collect the load information of each data group. Then the query load information of all the BE servers can be calculated. In order to avoid traffic, the collection can be done periodically or at a time when the system is relatively idle. When obtaining the query load of a BE server, CC stores it in the corresponding entry in the *BE Table*.

Location rule. The location rule aims to choose a sender node and a receiver node for data migration. The basic idea is to define a maximum deviation of the average query load to help determine whether the node is overloaded or under-loaded. We define $AvgL$ as the mean load of all BE servers in the system, $AvgL = (\sum_{i=1}^{n} L_{BE}(i))/n$, where n is the number of BE servers in the system. It can help to estimate whether the system is busy or idle currently. With deviation θ (can be set based on experience), $AvgL - \theta$ is defined as the lower bound and $AvgL + \theta$ the upper bound of query load for each BE server. For a BE server, if the load on it exceeds $AvgL + \theta$, then it will be chosen as a sender candidate (sender condition); if the load is under $AvgL - \theta$, then it will be chosen as a receiver candidate (receiver condition). Since $AvgL$ is the mean load of all BE servers, if there is a BE server satisfying the sender condition, there must be another BE server satisfying the receiver condition.

Migration rule. The migration rule determines when or whether or not to migrate data. First, the algorithm chooses the node with the max load among all the sender candidates as a sender node. Then it scans all the data groups on the sender node, in decreased order of their load values, to find whether there is one that do not make the load on the sender node decrease below $AvgL - \theta$ and the load on the receiver node increase upon $AvgL + \theta$ after data migration. The receiver node is the node with the minimum load among all the receiver candidates. Once the data group is found, the migration decision can be made. This procedure will be executed iteratively until there are no sender candidates in the system. The algorithm is given as algorithm 2. The data migration mechanism is described in the *Migration Mechanism* part.

Migration Mechanism. To ensure data consistency, a data migration mechanism is designed. Fig. 3 shows the mechanism. At the beginning of the migration,

---

Algorithm 2. Dynamic Data Reallocation DRC

/* T_{BE} is the BE table on CC, D is the dictionary stored on CC*/
1: **while** the set of sender candidates is not empty **do**
2:    Select sender node BE_s with the max load from the sender candidates
3:    **for all** data group G_i on BE_s (in decreased order on load) **do**
4:       Select receiver node BE_r with the minimum load from the receiver candidates

5:       **if** $T_{BE}[BE_r].Load + D[G_i].load/T_{BE}.[BE_r].PC \geq AvgL + \theta$ and
       $T_{BE}[BE_s].Load - D[G_i].load/T_{BE}[BE_s].PC \leq AvgL - \theta$ **then**
6:          Migrate data G_i from BE_s to BE_r
7:          Update the *Load* value for BE_r and BE_s in T_{BE}
8:          Update location information for G_i in data dictionaries on CC and all FE
          servers
9:       **if** $T_{BE}[BE_s].Load \leq AvgL + \theta$ **then**
10:          break

CC sends a command to the sender node BE_s for migrating data G to the receiver node BE_r (step 1). Then BE_s migrates the data G to BE_r (step 2). If BE_r successfully receives and stores G, it sends a success signal to CC (step 3). CC changes the location record of G from BE_s to BE_r in its own data dictionary (step 4), and then broadcasts messages to all FE servers to ask for location update in their data dictionaries (step 5). Each FE server performs location update (step 6), and then sends a success signal to CC (step 7). After receiving the success signals from all FE servers, which means the location information has been updated in all data dictionaries in the system, CC sends a command to BE_s for deleting G (step 8). Then BE_s performs the deletion (step 9). Only after these 9 steps are successfully done can the migration be considered accomplished. At each step, logs are created to record the state information of the current system. If a failure occurs at any step, the system can be recovered to its original state. For example, if the data dictionary on a FE server is unsuccessfully modified, then all the dictionaries on FE servers and CC will be restored to their original state and the migrated data will be deleted from the receiver node BE_r. During the migration process, CC plays an important part in controlling the message signals to and from the other system nodes and conducting the actions of these nodes.

The above strategy overlooks the distance between a FE server and a BE server in the system. A problem will arise when some FE servers are set at a remote site from BE servers, since the transform time between them should not be neglected. We define $D(FE_i, BE_j)$ to be the distance from FE_i to BE_j. Suppose the distance information can be obtained before taking the reallocation strategy. To tackle the distance problem, the reallocation algorithm can be modified by the following heuristic technique. For each sender candidate BE_s, the load values of its data groups are calculated as $\sum_{FE_j \in SFE} L_{FE}(i,j)\}$, where $L_{FE}(i,j)$ is the query load associated with data group G_i which is detected by

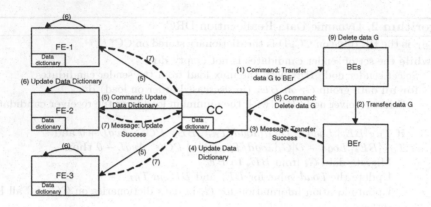

Fig. 3. Data Migration Process

FE_j, m is the number of data groups storing on BE_s, and SFE is a set storing k FE servers which are the nearest to BE_s.

In sum, DRC is developed to reallocate data dynamically at runtime for query load balancing in the changing system. The strategy is implemented by a central controller CC, which makes reallocation decisions based on the global query load information and controls data migration for protecting the system from migration failures. To avoid incurring large communication overhead, CC sends a collection request to the systems nodes at a suitable time rather than receives notifications from the system nodes all the time. Through balancing the query load at runtime, the query response time can be expected to be fast with the changing query load in the system.

5   Experiments

For the purpose of this study, we prepared real customer data from a mobile communication company. The subscriber database is stored in the LDAP database (LDAP database is commonly used in the query-intensive environment). It contains 200,000 subscribers with size 350MB, covering attributes like the basic DN, the current location and IMSI (International Mobile Subscriber Identification Number) of each subscriber. The simulation system model consists of a collection of six computers connected by a communication network with 100.0 Mbps bandwidth. The central controller CC ran on a AMD Athlon 64 3200+ processor with 1G RAM. One FE server and two BE servers used AMD Athlon 64 3200+ processor with 1G RAM, and one FE server and one BE server used 4 AMD Opteron 2.4G processors with 32G RAM.

In the experiments, a query was created by generating an id number, which was used as the key for searching the subscribers' location information in the database. Queries were generated 6000 times per second and randomly sent to one of the FE servers. Then the FE server routed the query to a corresponding BE server where the query can be processed.

First, we evaluated the effectiveness of the static data distribution strategy DDH. For comparison purpose, an algorithm NoLB was used. It distributes data based on the commonly used horizontal fragmentation method and does not consider any query load information. As for DDH, a hash function is chosen to take u modulo p, where p is a prime number that is approximately equal to n. The data group size is set to be 10 records. The system load level is measured by the ratio of the total query arrival rate in the system to the total processing power of all BE servers.

Fig. 4 compares the average response time of DDH to NoLB. The result shows that the average response time of DDH is much more faster than that of NoLB with the increasing system load level. This is because the data is distributed more fairly by DDH at the initial stage of the system, and the query load on the system nodes is more balanced. The query response time can be reduced due to the efficient use of the processing resources in the system. The improvement is more significantly when the system load is high.

To measure the fairness of data distribution, a fairness index I proposed in [10] is used. $I = \dfrac{[\sum\limits_{i=1}^{n} F_i]^2}{n \sum\limits_{i=1}^{n} F_i^2}$, where F_i is the expected response time of node i. If the expected response time is the same for all nodes, then $I = 1$, which means the system achieves fairness distribution. Otherwise, if the expected response time differs significantly from one node to another, then I decreases, which means the system suffers load imbalance to some extent. Fig. 5 shows that DDH decreases more slowly on fairness index with the increasing system load.

In the next, we evaluated the effectiveness of the dynamic data reallocation strategy DRC. In the experiment, CC collected query load information from all FE servers at a time when the system is idle (we control the time by detecting the performance of the runtime system). The deviation of the average query load θ is set at 10% of the average load. The data group size is the same as that used in the first experiment. The performance of NoLB, DRC and DRC* is compared. DRC* is the improved version of DRC that takes account of the distance factor. A FE server was set out of the inner-net to increase the communication distance. Form the result in Fig. 6, DRC and DRC* both show better performance than NoLB on the average response time. Because dynamically reallocating data can contribute to achieve load balancing and adapt the changing query patterns. When a system node is detected as overloaded, the queried data can be migrated from it to a relatively idle node. Therefore the overall performance of the system can be improved at runtime. As expected, DRC* spends less response time than DRC, since it makes better reallocation decisions by taking account of the distance factor.

We also compare the off-line probability of NoLB, DRC and DRC* versus the system load level. In mobile communication systems, the response time for a submitted query request must be within 0.4ms, otherwise, the system will consider the connection as off-line. In practice, subscribers require the off-line probability to be under 0.5%. Fig. 7 shows that DRC and DRC* both have lower

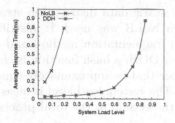

Fig. 4. Average response time(1)

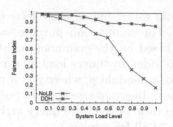

Fig. 5. Fairness Index

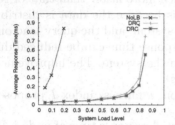

Fig. 6. Average response time(2)

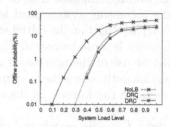

Fig. 7. Off-line probability

off-line probability than NoLB. This is because the query response time is sped up by using DRC and DRC*, and the off-line probability is significantly reduced.

6   Conclusion

To address the response efficiency challenge in the distributed query-intensive data environment, we have presented two effective strategies to speed up system's query response time through load balancing. One is DDH, a static data distribution strategy, that uses a hash-based heuristic technique to distribute data off-line according to the query history. The query load of the system can be more balanced due to DDH's reasonable data distribution. The other is DRC, a dynamic data reallocation strategy, that can reallocate data dynamically at runtime for query load balancing in the changing system. A central controller CC is used to make reallocation decisions based on the global load information for better decision quality. Much communication overhead can be avoided by using CC to collect query load information at a suitable time. Moreover, CC can control the data migration process and ensure the data consistency in the process. A simulation environment is designed to test the effectiveness of these two strategies, experimental results show that they both offer favorable performance with the increasing query load of the system.

Acknowledgments. We gratefully thank Ling Wu for her previous work on this paper.

References

1. Grosu, D., Chronopoulos, A.T., Leung, M.-Y.: Load Balancing in Distributed Systems: An Approach Using Cooperative Games. In: Parallel and Distributed Processing Symposium, pp. 52–61 (2002)
2. Li, J., Kameda, H.: Load balancing problems for multiclass jobs in distributed/parallel computer systems. IEEE Transactions on Computers 47(3), 322–332 (1998)
3. Kim, C., Kameda, H.: Optimal static load balancing of multi-class jobs in a distributed computer system. In: 10th International Conference on Distributed Computing Systems, pp. 562–569 (1990)
4. Lin, H.-C., Raghavendra, C.S.: A Dynamic Load-Balancing Policy with a Central Job Dispatcher (LBC). IEEE Transactions on Software Engineering 18(2), 148–158 (1992)
5. Sundaram, V., Wood, T., Shenoy, P.: Efficient Data Migration in Self-managing Storage Systems. In: IEEE International Conference on Autonomic Computing, pp. 297–300 (2006)
6. Zhang, Y., Kameda, H., Hung, S.-L.: Comparison of dynamic and static load-balancing strategies in heterogeneous distributed systems. Computers and Digital Techniques, IEE Proceedings 144(2), 100–106 (1997)
7. Qin, X., Jiang, H., Zhu, Y., Swanson, D.R.: A dynamic load balancing scheme for I/O-intensive applications in distributed systems. In: International Conference on Parallel Processing Workshops, pp. 79–86 (2003)
8. Kuo, C.-F., Yang, T.-W., Kuo, T.-W.: Dynamic Load Balancing for Multiple Processors. In: 12th IEEE International Conference on Embedded and Real-Time Computing Systems and Applications, pp. 395–401 (2006)
9. Feldmann, M., Rissen, J.P.: GSM Network Systems and Overall System Integration. Electrical Communication, 2nd Quarter (1993)
10. Jain, R.: The Art of Computer Systems Performance Analysis: Techniques for Experimental Design, Measurement, Simulation, and Modeling. Wiley-Interscience, New York (1991)

A Novel Chi2 Algorithm for Discretization of Continuous Attributes

Wenyu Qu[1,2], Deqian Yan[3], Yu Sang[3], Hongxia Liang[3], Masaru Kitsuregawa[2], and Keqiu Li[4]

[1] School of Computer Science and Technology
Dalian Maritime University
1, Linghai Road, Dalian, 116029, China
[2] Institute of Industrial Science
The University of Tokyo
4-6-1 Komaba, Meguro-ku, Tokyo, 153-8505, Japan
[3] Department of Computer Science and Engineering
Liaoning Normal University
850, Huanghe Road, Dalian, 116029, China
[4] Department of Computer Science and Engineering
Dalian University of Technology
2, Linggong Road, Dalian, 116023, China

Abstract. The Chi2 algorithm, together with the Modified Chi2 algorithm and the Extended Chi2 algorithm, is famous for discretization algorithms with the base of probability and statistics. After studying these algorithms and analyzing their drawbacks, we present a new Chi2 algorithm called Rectified Chi2 algorithm, which regards a new merging standard as the basis of interval merging and discretizes the real value attributes exactly and reasonably. We also present a new sequence method (DSM) to overcome of the drawbacks that the Modified Chi2 algorithm and the Extended Chi2, i.e., they only adopt the maximal difference as standard of interval merger. We evaluate the performance of the Rectified Chi2 algorithm and DSM over extensive experiments. The experiment results show the effectiveness of the proposed algorithms.

Keywords: Discretization algorithm, Chi2 algorithm, Rough sets, Difference sequence (DS), χ^2 statistic.

1 Introduction

Discretization is an effective technique to deal with continuous attributes for machine learning and data mining. In many algorithms such as rules selection and feature classification, continuous (real value) tributes must be discretized, which is especially useful in the research and application field of data mining with application of rough set theory [9]. The discretization of continuous attributes is to replace the real value attributes with the symbolic attributes.

Y. Zhang et al. (Eds.): APWeb 2008, LNCS 4976, pp. 560–571, 2008.
© Springer-Verlag Berlin Heidelberg 2008

The main essence of discretization is to keep the information as discrete as possible and reduce the information loss. Existing discretization methods can be generally classified into three tracks, i.e., global versus local, supervised versus unsupervised, and static versus dynamic [3]. Local methods are used for producing partitions that are useful for localizing regions of the instance space while the global discertization methods use the entire instance space to discretize. Unsupervised methods, such as equal width interval and equal frequency interval methods, do not utilize instance class labels in the discretization process. By contrast, supervised methods are referred to those discretization methods that utilize the class labels. Static methods, entropy-based partitioning for instance, perform one discretization pass of the data for each attribute and determine the value of m for each attribute independent of the other attributes, where m is the maximum number of intervals produced in discretizing an attribute. Dynamic methods conduct a search through the space of possible m values for all attributes simultaneously, thereby capturing interdependencies in attribute discretization. A number of entropy measure-based methods establish a strong group of works in the discretization domain. This concept uses class entropy as a criterion to evaluate a list of best cuts, which together with the attribute domain induce the desired intervals [11]. In [10], the authors considered a general genetic strategy-based algorithm of searching for an optimal set of separating hyper-planes by a genetic algorithm. Liu and Wang proposed a new measure of class heterogeneity of intervals from the view point of class probability itself in [7]. Other algorithms use class attributes interdependency information as the criterion [2],[4]. These methods try to maximize the interdependence between the discretized attributes and class labels based on information theory.

The ChiMerge algorithm introduced by Kerber in 1992 is a supervised global discretization method [4]. The method uses χ^2 test to determine whether the current point is merged or not. Liu et al. [6] proposed a Chi2 algorithm in 1997 based on the ChiMerge algorithm. In this algorithm, the authors increase the value of the χ^2_α threshold dynamically and decide the intervals' merging order according to the value of D, where $D = \chi^2_\alpha - \chi^2$ and χ^2_α is fractile decided by the significance level α. Francis et al. further improved the Chi2 algorithm and proposed the Modified Chi2 algorithm in [15]. The authors showed that it is unreasonable to decide the degree of freedom by the number of decision classes on the whole system in the Chi2 algorithm. Conversely, the degree of freedom should be determined by the number of decision classes of each two adjacent intervals. In [14], the authors pointed out that the method of calculating the freedom degrees in the modified Chi2 algorithm is not accurate and proposed the Extended Chi2 algorithm, which replaces D with $D / \sqrt{2v}$.

In this paper, the principle of that in Chi2 and related algorithms by using of χ^2 statistic and significance level to determine whether a node should be merged is analyzed. Based on the study the meaning of χ^2 statistic, a new modified algorithm called Rectified Chi2 algorithm is proposed. The new algorithm regards a new merging standard $D' = \chi^2_\alpha - (v / k)\chi^2$ as basis of interval merging and discretes the real value attributes exactly and reasonably. We ran respectively C4.5 and SVM on the discreted data, and the experiment results show that the proposed lgorithm is effective. Besides, a sequence method (DSM) is proposed to solve the problem that the algorithms of Modified Chi2 algorithm and Extended Chi2 have, i.e., the two algorithms only adopting maximal difference as standard of interval merger, and the new algorithm's advantage performance is manifested by experiments. Compared with Modified Chi2 algorithm based on DSM and Extended Chi2 algorithm based on DSM, the Rectified Chi2 algorithm based on DSM still have better performance.

Finally, the imprecise problem of E_{ij} value in the χ^2 statistic are analyzed and two modified schemes are proposed. The experiment results show the effectiveness of the above presented algorithm.

The rest of this paper is organized as follows. Section 2 introduces some basic concepts related to this paper. Sections 3 and 4 propose the Rectified Chi2 algorithm and DSM, respectively. Section 5 evaluates the proposed algorithm by experiments. Finally, Section 6 concludes this paper.

2 Correlative Concepts of Chi2 Algorithms

Several conceptions about discretization are introduced as follows:

1） Interval and Node

A single value of continuous attributes is a node, two nodes making up of an interval. Adjacent two intervals have a common node. Discretization algorithm of real value attributes actually is in the process of removing node and merging adjacent intervals based on definite rules.

2） χ^2 and χ^2_α [1]

χ^2 is a statistic in probability. χ^2 value of adjacent intervals needs to be computed in this discretization algorithm.

The formula for computing the χ^2 value is:

$$\chi^2 = \sum_{i=1}^{2}\sum_{j=1}^{k} \frac{(A_{ij} - E_{ij})^2}{E_{ij}} \quad (1)$$

where

k : number of system classes; A_{ij} : number of patterns in the ith interval, jth class;

$C_j = \sum_{i=1}^{2} A_{ij}$: number of patterns in jth class; $R_i = \sum_{j=1}^{k} A_{ij}$: number of patterns in ith

interval; $N = \sum_{i=1}^{2} R_i$: total number of patterns; $E_{ij} = R_i \times C_j / N$: expected frequency

of A_{ij}. χ^2_α is a threshold which is determined by degrees of freedom of adjacent two intervals and significance level α. In statistics, asymptotical distribution of χ^2 statistic with k degree of freedom is χ^2 distribution with $k-1$ degree of freedom, namely

$\chi^2_{(k-1)}$ distribution. χ^2_α is determined by selecting a desired significance level α, gven a α one can obtain the related fractile χ^2_α by consulting table or formula.

3 A Rectified Chi2-Based Algorithm

In formula （1）, C_j / N is the proportion of number of patterns in jth class accounting for total number of patterns, $E_{ij} = R_i \times C_j / N$ is number of patterns in the ith interval, jth class in such a proportion (probability). Therefore, statistic χ^2 indicates the equality degree of the jth class distribution of adjacent two intervals. The smaller χ^2 value is, the more similar is class distribution, the more unimportant node is. So, it should be merged.

For statistic χ^2, the smaller χ^2 value is, the more similar is class distribution, the more unimportant is node. Determining the importance of a nodes by $D = \chi^2_\alpha - \chi^2$ intend to farthestly enhance statistic of nodes in system under definite significance level. χ^2 value not only relates with the number of degrees of freedom of adjacent two intervals, but also relates with the number of system classes, namely: the higher the number of system classes are, the more number of classes of adjacent two intervals are, the greater degrees of freedom of adjacent two intervals are. Thus, χ^2 value calculated is different. So, significance of nodes should be determined by $D' = \chi^2_\alpha - (v / k)\chi^2$, because this important standard readjusts the order of interval merger after multiplying v/k in front of χ^2 and reduces χ^2 value pro rata, not decreasing prematurely significance level α and illuminating dequately rationality of node merging. Based on the study a new modified algorithm called Rectified Chi2 algorithm is proposed. The new algorithm regards a new merging criterion $D' = \chi^2_\alpha - (v / k)\chi^2$ as basis of interval merger.

The rectified Chi2 algorithm is described as follows:

Step1: Initialization. Set significance level α=0.5. Calculate inconsistency rate of information systems: $Incon\_rate$.

Step2: Sort data in ascending order for each attribute and calculate χ^2 value of each adjacent two intervals according to （1）, then using a table to obtain the corresponding χ^2 threshold. Calculate difference $D' = \chi^2_\alpha - (v / k)\chi^2$.

Step3: Merge.
    while(mergeable node)
    { Search node that have the maximal difference D', then merging it;
      If $Incon\_rate$ change [12]
        { withdraw merging;

```
        goto Step4; }
            else goto Step2;
    }
Step4•if α can not be decreased
                Exit procedure;
        else { α₀ = α ;
```

$\alpha_0 = \alpha$;

```
            decreasing the significance level by one level;
                goto Step2; }
Step5•do until no attribute can be merged
        { For each mergeable attribute i
            { Calculate difference D';
```

$\alpha = \alpha_0$;

```
            sign flag=0;
            while(flag= =0)
            { while( mergeable node )
                { Search node that have the maximal difference D', then merging it;
                    If Incon _ rate change
                        { withdraw merging;
                        flag=1;
                            break; }
                    else update difference D';
                }
                If α can not be decreased
                        break;
                    else { decreasing the significance level by one level;
                        update difference D';}
            }
        }
    }
```

4 Interval Merger Method Based on Different Sequence

The variational range of difference $D = \chi_\alpha^2 - \chi^2$, difference $D/\sqrt{2v}$ and difference $D' = \chi_\alpha^2 - (v/k)\chi^2$ are influenced by degrees of freedom. In fact, nodes shoule be merged with a measure of probability. Perhaps the node which has the maximal difference value can be merged with a greater probability. So, the original algorithm that only adopted maximal difference as standard of interval merger is unreasonable and unfair. Thus, a difference sequence method (DSM) is proposed to solve the problem that the algorithms of Modified Chi2 algorithm and Extended Chi2 have, i.e., the two algorithms only adopting maximal difference as standard of interval merger. When adjacent two intervals that have the maximal difference can not be merged, we should search in sequece until finding mergable node, resolving the unfair problem and given the chance of merger for all of the nodes.

For example: (see Table 1) :

Table 1. Decision table

U	a	b	c	d
1	0	0	0	1
2	0	1	0	1
3	0	1	1	2
4	1	1	1	1
5	1	1	2	2
6	1	2	2	3

As seen from Table 1, a, b, and c are condition attributes, d is decision attribute. The same value of condition attribute can be seen as a interval. There are five nodes and five groups of adjacent intervals in Table 1. Take Extended Chi2 algorithm of references [14] for example (merge criterion: $D/\sqrt{2v}$) : We may get significance level $\alpha = 0.9$ and difference value of adjacent two intervals of attribute a is maximal by calculating (degree of freedom $v_1 = 2$, table look-up: $\chi_\alpha^2 = 4.61$), namely: $D_1 = 1.6383$. For adjacent two intervals of attribute b (pattern 1 and pattern 2, 3, 4, 5), degree of freedom $v_2 = 1$, table look-up: $\chi_\alpha^2 = 2.71$, namely: $D_2 = 1.1858$. At this time, adjacent intervals of attribute a that have the maximal difference value can not be merged. Though $D_2 < D_1$, the adjacent intervals of attribute b can still be merged, which should be given the chance of merger, achieving better discretization effect. The experiment results show that Modified Chi2 algorithm based on DSM, Extended Chi2 algorithm based on DSM and Rectified Chi2 algorithm based on DSM have better performance.

In the following, we analyze the improvement of E_{ij} value of χ^2 statistic. In the χ^2 formula, if either R_i or C_j is 0, then E_{ij} is set to 0.1. In other words, when the number of classes of adjacent two intervals are less than the number of system classes, some of inner items of χ^2 equal to 0.1. There is unreasonable to do so. For instance, there are two groups of adjacent intervals. The number of classes of one is more than the number of classes of the other, and the two groups of adjacent intervals do not equal to the number of system classes. Thus, class distribution of adjacent intervals with the greater number of classes is not always fluctuation, while class distribution of adjacent intervals with the smaller number of classes is not always smoothness. There is unfair to plus 0.1 synchronously for missing class item of two adjacent intervals. So, we need to adjust E_{ij} value to achieve equitable competitive.

If the number of degrees of freedom of adjacent two intervals is greater, E_{ij} value should be decreased properly (decreasing properly χ^2 value). If the number of degrees of freedom of adjacent two intervals is smaller, E_{ij} value should be increased properly. The two groups of adjacent intervals which have different degrees of freedom can get to balance. So, we have the following two improved method:

1. Unified Standard χ^2: There is a unified E_{ij} standard to calculate χ^2.

$$E_{ij} = 0.3 * \left((k-v)/k \right) * \left((N-(s+t))/N \right)$$

where v is number of degrees of freedom of adjacent two intervals and $s+t$ is the sum of number of patterns of adjacent two intervals. Explanation for the above E_{ij} standard: Firstly, when significance level α is 0.5 and degrees of freedom v is 1, χ_α^2 is 0.4548. So, the upper limit of E_{ij} value is a proper value 0.3 we select which is less than 0.4548. Secondly, considering equitable factor, we can adjust E_{ij} value according to the number of degrees of freedom of adjacent two intervals. Namely, if the number of degrees of freedom of adjacent two intervals is greater, E_{ij} value should be decreased properly. If the number of degrees of freedom of adjacent two intervals is smaller, E_{ij} value should be increased properly. So, we select $\left((k-v)/k \right)$ as the member of E_{ij}. Finally, if the number of system classes is 3, here v in $\left((k-v)/k \right)$ is still a constant quantity which can not be distinguished. Considering the number of degrees of freedom of adjacent two intervals have connection with the size of adjacent two intervals, we also select $\left((N-(s+t))/N \right)$ as the member of E_{ij}.

2. Optimized χ^2: Finding E_{ij} value which can assist to achieve higher predictive accuracy and is usually between 0.0 and 0.4.

5 Experimental Results

To contrast experiment, we adopt the data sets of UCI machine learning database [8] (see Table 2) ,The UCI machine learning data sets are common used in data mining experiment.

Eight data sets were discreted by Rectified Chi2 algorithm which is proposed in this paper (shortened form Rec) and References [15] method (shortened form Mod) and References [14] method (shortened form Ext) and Rectified Chi2 algorithm based DSM (shortened form Rec DSM) and Modified Chi2 algorithm based DSM (shortened form Mod DSM) and Extended Chi2 algorithm based DSM (shortened form Ext DSM). Meanwhile, Eight data sets were discreted by the above six algorithms which adopt Optimized χ^2 and Unified standard χ^2, denoted by "shortened form (χ^2) ", for example : Mod (χ^2), Ext DSM (χ^2) etc.

Table 2. Data sets information

Data	Continuous attributes	Discrete attributes	Classes	Examples
Iris	4	0	3	150
Breast	9	0	2	683
Wine	13	0	3	178
Auto	5	2	3	392
Bupa	6	0	2	345
Machine	7	0	8	209
Pima	8	0	2	768
Glass	9	0	6	214

We ran C4.5 [13] on the discreted data. Choosing randomly 80 percent of Examples are training sets, the rest are testiing sets. The average predictive accuracy and the average numbers of nodes of decision tree and the average numbers of rules extracted are computed and compared by different algorithms （see Table 3 to Table 8). Meanwhile, discreted data is classified by multi-class classification method [17] of one to one of SVM. Choosing randomly 80 percent of Examples are training sets, the rest are testiing sets. Model type: C-SVC. Kernel function type: RBF function. Search range of penalty C: [1, 100]. Kernel function parameter γ: 0.5. Predictive accuracy (acc) and the number of support vector (svs) are computed and compared for the above twelve algorithms （see Table 9 to Table 12）. Considering computational complexity of kernel function depends on vector inner product of samples and attribute values that are bigger lead to computing complexity, attribute values need to be normalized:
$$\overline{x}_i = 2\frac{x_i - \min(x_i)}{\max(x_i) - \min(x_i)} - 1$$

Attribute value after normalization : $x_i \in [-1, +1]$. The same normalization method can be used in training sets and testing sets.

As seen from Tables 3 and 4, compared Rectified Chi2 algorithm with Modified Chi2 algorithm and Extended Chi2 algorithm, the average predictive accuracy are higher except for Glass of eight data sets, especially, for Wine, Auto and Pima the accuracy have raised greatly. The average numbers of nodes of decision tree of

Table 3. Using C4.5 with Different Discretization Algorithm

Data	Predictive accuracy			Number of nodes			Number of rules		
	Mod	Ext	Rec	Mod	Ext	Rec	Mod	Ext	Rec
Iris	0.9167	0.9167	0.9367	20.85	20.85	18.9	14.35	14.35	13.05
Breast	0.9255	0.9255	0.9327	89.1	89.1	76.8	57.5	57.5	48.25
Wine	0.8028	0.8028	0.9343	62.8	62.8	32.75	36	36	17.6
Auto	0.7627	0.7715	0.8157	138.15	139.65	111.9	96.75	99.35	80.05
Bupa	0.4529	0.4529	0.4567	236.4	236.4	204.65	183.55	183.5	160.85
Machine	0.7738	0.7738	0.7762	68	64.15	54.2	44.15	42.45	33.8
Pima	0.6182	0.6182	0.6497	448.4	448.4	458.2	342.1	342.1	359.6
Glass	0.4058	0.5116	0.3646	121.25	121.5	104.05	84.3	79.55	68.25

Table 4. Using C4.5 with Different Discretization Algorithm

Data	Predictive accuracy			Number of nodes			Number of rules		
	Mod DSM	Ext DSM	Rec DSM	Mod DSM	Ext DSM	Rec DSM	Mod DSM	Ext DSM	Rec DSM
Iris	0.9383	0.935	0.9383	18	19.2	18	12.15	10.25	12.35
Breast	0.9255	0.9255	0.9327	89.1	89.1	76.8	57.5	57.5	48.25
Wine	0.8028	0.8028	0.9343	62.8	62.8	32.75	36	29.8	17.6
Auto	0.793	0.793	0.8006	142.8	142.8	142.6	93.75	97.55	93.55
Bupa	0.5319	0.5319	0.5319	229.55	229.55	229.55	167.75	176.7	167.75
Machine	0.7833	0.7833	0.7929	53.55	53.55	55.2	35.2	31.8	33.15
Pima	0.6497	0.6497	0.6497	449.8	449.8	449.8	304.6	304.6	304.6
Glass	0.3872	0.2291	0.2558	124.7	126.3	121.3	74.05	77.6	70.15

Table 5. Using C4.5 with Different Discretization Algorithm (Optimized χ^2)

Data	Predictive accuracy			Number of nodes			Number of rules		
	Mod (χ^2)	Ext (χ^2)	Rec (χ^2)	Mod (χ^2)	Ext (χ^2)	Rec (χ^2)	Mod (χ^2)	Ext (χ^2)	Rec (χ^2)
Iris	0.9167	0.93	0.9367	20.85	21.5	18.9	14.35	14.65	13.05
Breast	0.9255	0.9255	0.9327	89.1	89.1	76.8	57.5	57.5	48.25
Wine	0.8194	0.8583	0.9343	61.35	48.7	32.75	35.5	29.5	17.6
Auto	0.788	0.7886	0.8157	120.8	127.6	111.9	88.5	91.35	80.05
Bupa	0.4529	0.4529	0.4567	236.4	236.4	204.65	183.55	183.55	160.85
Machine	0.7929	0.7929	0.8179	63.85	63.85	64.95	42.05.	42.05.	38.95
Pima	0.6182	0.6182	0.6497	448.4	448.4	458.2	342.1	342.1	359.6
Glass	0.4058	0.5267	0.4442	121.25	115.55	119.6	84.3	77.3	75.8

Table 6. Using C4.5 with Different Discretization Algorithm (Optimized χ^2)

Data	Predictive accuracy			Number of nodes			Number of rules		
	Mod DSM (χ^2)	Ext DSM (χ^2)	Rec DSM (χ^2)	Mod DSM (χ^2)	Ext DSM (χ^2)	Rec DSM (χ^2)	Mod DSM (χ^2)	Ext DSM (χ^2)	Rec DSM (χ^2)
Iris	0.9383	0.93	0.935	18	20.85	19.2	12.35	13.2	12.15
Breast	0.9255	0.9255	0.9327	89.1	89.1	76.8	57.5	29.5	48.25
Wine	0.8194	0.8583	0.9343	61.35	48.7	32.75	35.5	36	17.6
Auto	0.8006	0.8006	0.8157	148.55	148.55	111.9	97.55	97.55	80.05
Bupa	0.4529	0.4529	0.5319	236.4	236.4	229.55	183.55	183.55	167.75
Machine	0.7833	0.7833	0.8821	53.55	53.55	52.6	35.2	35.2	31.15
Pima	0.6182	0.6182	0.6497	448.4	448.4	449.8	342.1	342.1	304.6
Glass	0.4209	0.5058	0.2965	123.7	113.65	115.65	79.7	66.1	69.9

Rectified Chi2 algorithm and the average numbers of rules extracted of Rectified Chi2 algorithm have decreased apparently except for Pima data set, which the two aspects are exactly the results we expect, manifesting advantage of Rectified Chi2algorithm. Compared with the above three algorithms, the average predictive accuracy of the three algorithms based on DSM have essentially enhancement except for Glass data set and the average numbers of nodes of decision tree and the average numbers of rules extracted have decreased apparently for the majority of data sets, manifesting superiority of the three algorithms based on DSM. Besides, Rectified Chi2 algorithm based on DSM still has better performance.

Tables 5 and 6 have presented the results of the the above six discretization algorithms based on improved χ^2 (Optimized χ^2). Compared with unimproved χ^2, the average predictive accuracy of the six discretization algorithms based on Optimized χ^2 have enhanced greatly except the average predictive accuracy of Breast, Bupa and Pima data sets which have only two classes is the same as before. This is specially characterist of Optimized χ^2. The average numbers of nodes of decision tree and the average numbers of rules extracted of the six discretization algorithms presented in Tables 3 and 4 based on Optimized χ^2 have decreased apparently for the majority of data sets, manifesting advantage of the six algorithms based on Optimized χ^2. Besides, either Rectified Chi2 algorithm based on Optimized χ^2 or Rectified Chi2 algorithm based on DSM and Optimized χ^2 still has better performance.

Table 7. Using C4.5 with Different Discretization Algorithm (Unified Standard χ^2)

Data	Predictive accuracy			Number of nodes			Number of rules		
	Mod (χ^2)	Ext (χ^2)	Rec (χ^2)	Mod (χ^2)	Ext (χ^2)	Rec (χ^2)	Mod (χ^2)	Ext (χ^2)	Rec (χ^2)
Iris	0.93	0.93	0.9167	21.5	21.5	20.85	14.65	14.65	14.35
Breast	0.9255	0.9255	0.9327	89.1	89.1	76.8	57.5	57.5	48.25
Wine	0.8028	0.8028	0.9222	62.8	62.8	38.4	36	36	21.45
Auto	0.7797	0.7886	0.788	120.25	127.6	123.85	89.05	91.35	89.5
Bupa	0.4529	0.4529	0.4567	236.4	236.4	204.65	183.55	183.55	160.85
Machine	0.7905	0.7905	0.8036	67.05	63.55	58.3	43.6	42.15	35.2
Pima	0.6182	0.6182	0.6497	448.4	448.4	458.2	342.1	342.1	359.6
Glass	0.4942	0.3384	0.3267	109.95	118.2	109.35	73.25	78.25	72.15

Table 8. C4.5 with Different Discretization Algorithm (Unified Standard χ^2)

Data	Predictive accuracy			Number of nodes			Number of rules		
	Mod DSM (χ^2)	Ext DSM (χ^2)	Rec DSM (χ^2)	Mod DSM (χ^2)	Ext DSM (χ^2)	Rec DSM (χ^2)	Mod DSM (χ^2)	Ext DSM (χ^2)	Rec DSM (χ^2)
Iris	0.935	0.93	0.935	19.2	20.85	19.2	12.15	13.2	12.15
Breast	0.9255	0.9255	0.9327	89.1	89.1	76.8	57.5	57.5	48.25
Wine	0.8028	0.8028	0.9222	62.8	62.8	38.4	36	36	21.45
Auto	0.8006	0.8006	0.788	148.55	148.55	123.85	97.55	97.55	89.5
Bupa	0.4529	0.4529	0.5319	236.4	236.4	229.55	183.55	183.55	167.75
Machine	0.7833	0.8298	0.8821	53.55	49.75	52.6	35.2	29.5	31.15
Pima	0.6182	0.6182	0.6497	448.4	448.4	449.8	342.1	342.1	304.6
Glass	0.4477	0.486	0.3442	129.4	120.55	121.25	75.95	70.2	70.8

Tables 7 and 8 have presented the results of the six discretization algorithms presented in Table 3 and Table 4 based on improved χ^2 (Unified Standard χ^2). Compared with unimproved χ^2, the average predictive accuracy of Breast, Bupa and Pima data sets which have only two classes is the same as before. For the rest of data sets, the average predictive accuracy of Modified Chi2 algorithms and Extended Chi2 algorithms presented in Tables 3 and 4 based on Unified Standard χ^2 have greatly enhanced, especially Auto and Machine data sets. The average predictive accuracy of Machine data sets of Rectified Chi2 algorithms based on Unified Standard χ^2 have enhanced apparently. Meanwhile, the average numbers of nodes of decision tree and the average numbers of rules extracted of the six discretization algorithms presented in Tables 3 and 4 based on Unified Standard χ^2 have decreased apparently for the majority of data sets, manifesting superiority of algorithms based on Unified Standard χ^2.

6 Conclusions

Study of discretization algorithm of real value attributes operates an important effect for many aspects of computer application. Though many discretization algorithm were proposed, the methods and theory should be developed in practical. Series of algorithms of Chi2 algorithm based on probability statistics theory offer a new way of thinking to discretization algorithm of real value attributes. This paper gives a study on application sense of χ^2 statistic of these algorithms, improving Modified Chi2 algorithm and Extended Chi2 algorithm. And, a new modified algorithm called Rectified Chi2 algorithm is proposed. The new algorithm regards a new merging standard as basis of interval merging and proves through theory and experiment that it can discrete the real value attributes exactly and reasonably. Besides, a difference sequence method (DSM) is proposed to solve the problem that the algorithms of Modified Chi2 algorithm and Extended Chi2 have, i.e., the two algorithms only adopting maximal difference as standard of interval merger, and he advantage performance of DS over that the two algorithms is manifested by experiments. Finally, the imprecise problem of E_{ij} value in the χ^2 statistic are analyzed and two modified schemes are proposed. The experiment results show that algorithms based on DSM can greatly enhance the effect of the three discretization algorithm. Besides, Rectified Chi2 algorithm based on DSM still has better performance and discretization algorithms based on improved χ^2 are effective. There is practical function to promote the study of the aspects.

Acknowledgments. This work is supported by Science and Technology Plan of Dalian, China (2007A10GX117).

References

1. Bian, G.R., Wu, L.D., Li, X.P., Wang, J.G.: Probability Theory. In: Mathematical Statistics, vol. 2, People's Education Press, Beijing (1979)
2. Ching, J.Y., Wong, A.K.C., Chan, K.C.C.: Class-Dependent Discretization for Inductive Learning from Continuous and MixedMode Data. IEEE Trans. on Pattern Analysis and Machine Intelligence 17(7), 641–651 (1995)
3. Dougherty, J., Kohavi, R., Sahami, M.: Supervised and Unsupervised Discretization of Continuous Feature. In: Proc. of the 12th International Conference on Machine Learning, pp. 194–202 (1995)
4. Kerber, R.: ChiMerge: Discretization of Numeric Attributes. In: Proc. of the Ninth National Conference on Artificial Intelligence, pp. 123–128. AAAI Press, Menlo Park (1992)
5. Kurgan, L.A., Cios, K.J.: CAIM Discretization Algorithm. IEEE Trans. on Knowledge and Data Engineering 16(2), 145–153 (2004)
6. Liu, H., Setiono, R.: Feature Selection via Discretization. IEEE Trans. on Knowledge and Data Engineering 9(4), 642–645 (1997)
7. Liu, X.Y., Wang, H.Q.: A Discretization Algorithm based on Heterogeneity Criterion. IEEE Trans. on Knowledge and Data Engineering 17(9), 1166–1173 (2005)
8. Merz, C.J., Murphy, P.M.: UCI Repository of Machine Learning Database, http://www.ics.uci.edu/~mlearn/MLRRepository.html
9. Nguyen, H.S., Skowron, A.: Quantization of Real Value Attributes: Rough Set and Boolean Reasoning Approach. Bull. Int'l Rough Set Soc. 1(1), 5–16 (1997)
10. Nguyen, H.S.: Discretization of Real Value Attributes: A Boolean Reasoning Approach. PhD thesis, Warsaw Univ (1997)
11. Nguyen, H.S.: Discretization Problem for Rough Sets Methods. In: Proc. of the First International Conference on Rough Sets and Current Trend in Computing, pp. 545–552 (1998)
12. Pawlak, Z.: Rough Set. International Computer Information Scinence 11(5), 341–356 (1982)
13. Quinlan, J.M.: C4.5: Programs for Machine Learning. Moorgan Kaufmann, San Mateo, Calif (1993)
14. Su, C.T., Hsu, J.H.: An Extended Chi2 Algorithm for Discretization of Real Value Attributes. IEEE Trans. on Knowledge and Data Engineering 17(3), 437–441 (2005)
15. Tay, E.H., Shen, L.: A Modified Chi2 Algorithm for Discretization. IEEE Trans. on Knowledge and Data Engineering 14(3), 666–670 (2002)
16. Vapnik, V.: Statistical Learning Theory. Wiley Inter-science, Chichester (1998)
17. Weston, J., Watkins, C.: Muti-class Support Vector Machines. Technical Report, CSD-TR-98-04, Department of Computer Science, Royal Holloway University of London, England (May 1998)

Mining Multiple Time Series Co-movements

Di Wu, Gabriel Pui Cheong Fung, Jeffrey Xu Yu, and Zheng Liu

The Chinese University of Hong Kong
{dwu,pcfung,yu,zliu}@se.cuhk.edu.hk

Abstract. In this paper, we propose a new model, called co-movement model, for constructing financial portfolios by analyzing and mining the co-movement patterns among multiple time series. Unlike the existing approaches where the portfolios' expected risks are computed based on the co-variances among the assets in the portfolios, we model their risks by considering the co-movement patterns of the time series. For example, given two financial assets, A and B, where we know that whenever the price of A drops, the price of B will drop, and vice versa. Intuitively, it may not be appropriate to construct a portfolio by including both A and B concurrently, as the exposure of loss will be increased. Yet, such kind of relationship can not always be captured by co-variance(i.e traditional statistics). Apart from manipulating the risk, our proposed co-movement model also alters the computation of the portfolio's expected return out of the traditional perspective. Existing approaches for computing the portfolio's expected return are to combine the expected return of each individual asset and its contribution in the portfolio linearly. This formulation ignores the dependence relationship among assets. In contrast, our co-movement model would capture all dependence relationships. This can mimic the real life situation much better than the traditional approach. Extensive experiments are conducted to evaluate the effectiveness of our proposed model. The favorable experimental results indicate that the co-movement model is highly effective and feasible.

In the modern financial engineering discipline, even though there are numerous different portfolio management theories proposed where their contents and arguments may be criticizing each others, their underlying principles always converge to one single theme: all investment decisions should be made to achieve an optimal tradeoff between risk and return [11]. As a portfolio manager, her main task is therefore to identify an appropriate combination of assets from the existing opportunities, such that this particular combination of assets would result in the lowest level of expected risk according to a desired level of return. A combination of assets is known as a portfolio.

While the underlying principle of governing how the portfolios should be constructed is simple and intuitive – minimize the expected risk and maximize the expected return, the definition of expected risk and the definition of expected return are usually ambiguous. In fact, this is exactly what lead to the major differences among different portfolio management theories. Little consensus exists among different theories regarding to their meanings and their measurements [12,1,9,5,6].

Traditionally, the portfolio's expected return is computed by combining the multiplication of the expected return of each asset in the portfolio and its contribution in there in a linear manner. Under this formulation, the dependence relationship among the assets

Y. Zhang et al. (Eds.): APWeb 2008, LNCS 4976, pp. 572–583, 2008.

in the portfolio will be ignored. For instance, given two financial assets, A and B, where their expected returns are respectively $r_A = \$10$ and $r_B = \$10$. Furthermore, assume that we will formulate a portfolio, P, by buying equal shares of them (50% each). Consider for an ideal situation that A and B are complementary of each others – whenever the price of A rises, the price of B will drop, and vice versa. Obviously, the expected return of P would be $0, i.e. we should not expect to gain any profit by buying A and B simultaneously with equal shares. Unfortunately, if we compute the expected return of P by using the traditional approach, then it becomes $\$10 \times 0.5 + \$10 \times 0.5 = \$10$, which is nevertheless subject to debate.

In this paper, we proposed a model, called co-movement model, to compute the portfolio's expected return by considering the dependency relationships among assets in the portfolio. If we know the price of two assets will frequently rise together, then a portfolio constructed by including these two assets should be expected to result in a somewhat higher return than simply combining their individual expected return.

Apart from manipulating the portfolio's expected return, our proposed co-movement model also alters the computation of the portfolio's expected risk out of the traditional perspective. Traditionally, the portfolio's expected risk is defined as the *variance the portfolio's expected return*.[1] While we agreed that there is a strong relationship between the expected risk and the expected return of a portfolio, we claim that expected risk may not necessarily be tied with the variance of the expected return.

In fact, there are ample evidences that most investors characterize risk as failure to meet some target rate of return rather than variance[12]. Instead, the major essence of risk should be the magnitude of loss, the chance of loss and the exposure to loss [9,8]. Accordingly, in this paper, we try to quantify the expected risk of a portfolio as hazard of loss against a specified desired level of return rather than the distribution of all possible outcomes (i.e. variance) against a specified desired level of return.

Specifically, our co-movement model quantifies the expected risk by mining the co-movement patterns among the time series of the assets in the portfolio. Again, given two financial assets, A and B, where we know that whenever the price of A drops, the price of B will drop, and vice versa. One should agree that it is not that appropriate to construct a portfolio by including both A and B concurrently, as the exposure of loss will be increased (the price of one asset drops, the other must also drop). Yet, such kind of relationship cannot be captured by the co-variance between A and B. Details would be discussed in the later sections.

Up to our knowledge, none of the existing literatures proposed the direction of computing the portfolio's expected risk and its expected return by using the formulation discussed in this paper. We are the first mover in this area. In order to evaluate the appropriateness and the suitability of the proposed co-movement model, we have archived four years' financial data from the Morgan Stanley Capital International(MSCI) index. It consists of daily equity indices of seven countries. We compared our co-movement model with the standard mean-variance model. The favorable experimental results indicated that the co-movement model is highly effective and feasible.

---

[1] This is usually regarded as the mean-variance model (a.k.a. Markowitz model) [12], and is the pioneer of the modern portfolio management theories.

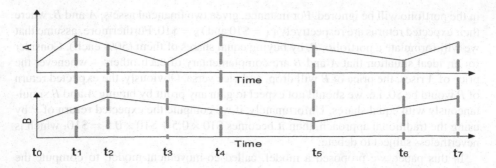

Fig. 1. The definition of co-movement. In the figure, there are two financial assets: A and B. With respect to A, there are four types of co-movements: (1) up-up ($t_0 - t_1$, $t_6 - t_7$); (2) up-down ($t_3 - t_4$); (3) down-up ($t_1 - t_2$, $t_5 - t_6$); (4) down-down ($t_2 - t_3$, $t_4 - t_5$).

The rest of this paper is organized as follows – Section 1 presents the proposed co-movement model in details. Section 2 evaluates the effectiveness of our proposed work and reports our findings. Section 3 addresses the major works that are directly related to the problem in this paper in brief. Section 4 summarizes this paper and discusses its possible extensions.

1   Proposed Work

We will present our proposed work in this section. Section 1.1 will define four different kinds of co-movements between two time series, whereas Section 1.2 and Section 1.3 will respectively present how the expected return and expected risk are computed based on the idea of co-movements. Finally, Section 1.4 will show how a portfolio will eventually be constructed by applying these novel measures and concepts.

1.1   Four Different Types of Co-movements

In order to account for different types of co-movements, let us first refer to figure 1. In Figure 1, it shows two time series: i and j. For time series i, there are totally five segments, whereas for time series j, there are totally three segments. With respect to time series i, four different types of co-movements can be identified: (1) up-up; (2) up-down; (3) down-up; and (4) down-down. Let us explain this in the next paragraph.

In figure 1, the co-movement between i and j forms a total number of seven co-movement patterns (1) Two up-up co-movements, from t_0 to t_1 and t_6 to t_7; (2) Two down-up co-movements, from t_1 to t_2 and t_5 to t_6; (3) Two down-down co-movements, which has the longest cumulative duration, from t_2 to t_3 and t_4 to t_5; and (4) one up-down co-movements, from t_3 to t_4.

1.2   Co-movement Based Expected Return

As discussed previously, the existing literatures compute the portfolio's expected return when they will ignore the dependence relationships among assets in the portfolio. Its computation is based on a linear combination of the multiplication between the expected return of each individual asset and its contribution in the portfolio. In this paper, we try to incorporate the dependence relationships among assets by using the concept of co-movement, which has been defined in the previous section. Given two financial assets, i and j, conceptually, our co-movement based expected return is as follows:

1. If whenever the price of i rises(drops), the price of j will rise(drop) (i.e. a lot of up-up (down-down) co-movements), then their co-movement based expected return should be higher(lower) than the sum of their independent expected return.
2. If whenever the price of i rises(drops), the price of j will drop(rise) (i.e. a lot of up-down(down-up) co-movements), then their co-movement based expected return should a mixture of their independent expect return considering their contribution to portfolio.

As a result, there are four different kinds of returns, which in-turn depends on the four different kinds of co-movements. Intuitively, the overall expected return of a portfolio should be the summation of these four kinds of returns. In the followings, we will provide the mathematical details for this formulation step by step.

In the most simplest case, let us consider there are only two financial assets, i and j, in the portfolio. The expected return of the portfolio is therefore:

$$r_{ij} = \sum_{s} r_{ij}^s \cdot P_{ij}^s, \qquad (1)$$

where s denotes different types of co-movements (up-up, up-down, down-up and down-down); r_{ij}^s denotes the expected return between asset i and asset j when their type of co-movement is s; P_{ij}^s is the probability that the type of co-movement between asset i and asset j is s. Hence, Eq. (1) computes the portfolio's expected return by summing the expected return of each type of co-movement with its probability of occurrence.

The probability, P_{ij}^s, in Eq.(1) is computed as follows:

$$P_{ij}^s = \frac{T_{ij}^s}{T}, \qquad (2)$$

where T_{ij}^s denotes the accumulative length of which the type of co-movement between asset i and asset j is s and T is the total length of the time series. Let us illustrate the idea of Eq. (2) with the help of Figure 1. In the figure, there are two periods for the up-up co-movement between asset A and asset B: from t_0 to t_1 and from t_6 to t_7. Assume further that the differences between t_0 and t_1 is 10, between t_6 and t_7 is 12 and between t_0 and t_7 is 50. Then, $P_{AB}^s = T_{AB}^s / T = ((t_1 - t_0) + (t_7 - t_6)) / (t_7 - t_0) = (12 + 10)/50$.

The expected return, r_{ij}^s, in Eq.(1) is computed as follows:

$$r_{ij}^s = \frac{1}{N_{ij}^s} \sum_{k \in s} \left(w_i r_i^k + w_j r_j^k \right), \qquad (3)$$

where w_i and w_j are respectively the weights of asset i and asset j in the portfolio; N_{ij}^s denotes the number of segments of type s in the time series; k is a segment belongs to the co-movement of type s; r_i^k and r_j^k are respectively the expected return of

576   D. Wu et al.

asset i and asset j in segment k. Again, let us use Figure 1 to illustrate the idea of this equation. Consider for the case where s denotes for the up-up co-movement. In this situation, there are two segments, from time t_0 to t_1 and from time t_6 to time t_7, belong to s. Assume further that the expected return of asset A in the two segments are respectively $r_A^1 = \$5$ and $r_A^2 = \$4$. Similarity, let $r_B^1 = \$6$ and $r_B^2 = \$3$. Hence, $r_{AB}^s = (1/2)((w_A \cdot \$5 + w_B \cdot \$6 + w_A \cdot \$4 + \$w_B \cdot \$3))$.

Now, we extend this simple portfolio with two assets only into a portfolio with n multiple assets. In the multiple assets situation, we can obtain the expected return of the portfolio, r_Θ, by a pairwise linear summation of every two assets:

$$r_\Theta = \sum_{i=1}^{n-1} \sum_{j \neq i}^{n} r_{ij}, \tag{4}$$

where r_{ij} is preciously defined in Eq. (1). In order for efficient computation by using linear programming[15], Eq. (4) can be formulized in this form:

$$r_\Theta = \sum_{i=1}^{n} w_i r_i, \tag{5}$$

$$r_i = \frac{1}{(n-1)} \sum_{j=1, j \neq i}^{n} \sum_s \frac{1}{N_{ij}^s} (\sum_{k \in s} r_i^k) \cdot P_{ij}^s, \tag{6}$$

where the component $1/(n-1)$ in Eq. (6) is added for the reason of normalization and r_i is the return generated from asset i in its co-movement with all the other assets.

1.3 Co-movement Based Expected Risk

In this section, we will present how we quantify and derive the expected risk of the portfolio, β_Θ. Unlike the approaches used in the modern portfolio theories where the portfolio's expected risk is usually defined as the variance of the portfolio's expected return (a statistical perspective), in this paper, we define the portfolio's expected risk as *the chance of exposure to loss*, by considering the co-movement among the time series of the assets in the portfolio (a data mining perspective).

In the beginning, let us use Figure 1 again to illustrate our idea. For simplicity, assume we only have two choices for formulating the portfolio: either buy asset i solely (i.e. $w_i = 1$ and $w_j = 0$) or buy equal shares of asset i and asset j (i.e. $w_i = 0.5$ and $w_j = 0.5$). Let β_1 and β_2 be the risk associated with buying asset i solely (the first option) and risk associated with buying asset i and asset j simultaneously (the second option), respectively. Suppose we have identified β_1. Now, if we switch to the second option (buy asset i and asset j simultaneously), would the chance of exposure to loss be increased or decreased? That is, would $\beta_2 > \beta_1$ or $\beta_2 \leq \beta_1$?

In Figure 1, from t_2 to t_3, the type of co-movement between asset i and asset j belongs to down-down. If this type of co-movement occurs frequently in the entire time series, we might have to expect the price drops of asset i would usually accompany with the price drops of asset j. In other words, *the chance of exposure to loss* is *increased*. So if two assets frequently involved in the down-down co-movement, they should generate a higher level of risk with respect to having either of themselves alone. To conclude, if the down-down co-movement frequently appears among the two time-series, then $\beta_2 > \beta_1$.

On the other hand, from t_3 to t_4 in Figure 1, the type of co-movement with respect to asset i is up-down. This means that when the price of asset i goes up, the price of asset j will goes down. From asset i's stand point of view, *the chance of exposure to loss* is also *increased* in this situation. Thus, the more frequently this situation happens, the higher the risk should be expected. Yet, it is worth noting that this type of risk is *not* symmetric. From asset j's angle, *the chance of exposure to loss* is *decreased*. This is because when the price of asset j decreases, the price of asset i will increase. This compensate effect will surely reduce the loss from asset j's perspective. As a result, the expected risk of the portfolio should take the considerations of both sides. Therefore, in our co-movement model, the underlying idea is to consider the frequency of this up-down co-movement as well as the relative magnitude of their movements. The mathematical details of how the expected risk should be computed will be discussed in the later paragraphs.

Finally, the co-movements with respect to asset i for the remaining time periods in Figure 1 are either up-up co-movement or down-up co-movement. Both of these co-movements imply *the chance of exposure to gain* increase (a kind of up-side risk). Hence, these two types of co-movements do not need to be included in computing the portfolio's expected risk, as our definition of risk is *the chance of exposure to loss* (a kind of down-side risk).

To summarize, following these discussions, the expected risk of a portfolio is computed by having a pairwise comparison of the assets in the portfolio against the frequencies and magnitudes of this down-down co-movement or up-down co-movement. Mathematically, the risk of the portfolio, β_Θ, is computed as follows:

$$\beta_\Theta = \sum_{i=1} w_i \sum_{j=1} \beta_{ij}, \qquad (7)$$

where β_{ij} is the risk of buying asset j with respect to asset i. Note that $\beta_{ij} \neq \beta_{ji}$ according to the previous discussion about the asymmetric relationship of risk. Note that the summation in Eq.(7) includes β_{ii}, which can be interpreted as risk of buying the asset i itself.

Now, the only remaining question here is how to quantify β_{ij}. Unfortunately, it is not a trivial task. To account for it, let us refer to Figure 2 which shows the down-down and up-down co-movements with different kinds of slopes. In the figure, the first row (A, B, C and D) compares four different cases of down-down co-movements, whereas the second row (E, F, G and H) compares four different cases of up-down co-movements.

In case A, both time series, $S1$ and $S2$, go down straightly, whereas in case B, only the time series $S2$ goes down straightly. Comparing case A and case B, one should agree that the risk (chance of exposure loss) associated with case B is *higher* than that of case A, because case B outlines a small drop of $S1$ will lead to big drop of $S2$. Similarly, in case C, a big drop of $S1$ will only lead to a small drop of $S2$, whereas in case D, a small drop in $S1$ will immediately accompany with the same amount of drop of $S1$. Accordingly, the risk (chance of exposure loss) associated with case C should be less because the relative magnitude of dropping in case C is less. Lastly, it is obvious that the risk associated with case D is less than that of case A, meanwhile case C is associated

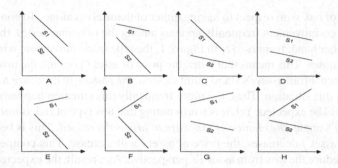

Fig. 2. Different level of risks generated by different kinds of co-movement

with less risk than case B. To summarize, the impact of β_{ij} in down-down correlation for these four cases should be: $B > A > D > C$.

By using an analysis similar to the above discussion, the impact of β_{ij} in the up-down co-movement for case E, case F, case G and case H should be: $E > F > G > H$. To conclude, in order to quantify the risk, we need to consider the relative magnitude of the slopes in each segment within the entire time series. Eventually, β_{ij} (the risk of buying asset j with respect to asset i) is formulated as follows:

$$\beta_{ij} = \sum_{k \in s} \delta_i^k \cdot \theta_j^k \cdot \left(\frac{\theta_j^k}{\theta_i^k} \right) \cdot (c_i^k c_j^k) \cdot P_i^k \tag{8}$$

where s belongs to either down-down or up-down co-movements. θ_i^k and θ_j^k are respectively the slope of asset i and asset j in segment k. Hence, the second component θ_j^k represents the magnitude of slope of asset j, the third component θ_j^k / θ_i^k accounts for the relative slope difference between asset i and asset j. c_i^k and c_j^k are respectively the coefficient of determination[13] for asset i and asset j in segment k, which denote how confident is the co-movement relationship of that segment for the two time series. P_i^k is the probability of segment k appears in asset i. $\delta_{i,k}$ is defined as:

$$\delta_{i,k} = \begin{cases} -1 & \theta_i^k < 0, \\ 1 & \text{otherwise.} \end{cases} \tag{9}$$

which identifies the type of co-movements (up-down or down-down) between i and j.

1.4 Co-movement Based Portfolio Construction

According to the discussions of the previous sections about how the portfolio's expected return and the portfolio's expected risk are derived, we now present how a portfolio could be constructed in our proposed co-movement model in this section.

Underlying for all of the portfolio management theories is that we should choose a portfolio that provides an expected return equals to a desired level of return, R, meanwhile it accumulates the lowest level of expected risk. Mathematically, we are trying to solve the following optimization problem by identifying all w_i based on a given R:

$$\min \ \beta_\Theta = \sum_{i=1} w_i \sum_{j=1} \beta_{ij}, \tag{10}$$

$$\text{subject to:} \ \sum_{i=1}^{n} w_i r_i = R. \tag{11}$$

where $\forall i$, $w_i \geq 0$ and their summation equals to 1. The r_i in Eq. (11) is defined in Eq. (6). This formulation is different from the traditional mean-variance model. In this new model, our target function (Eq. (10)) is a linear function but not a quadratic function. Finally, if short selling is allowed, then constraint $w_i \geq 0$ can be omitted.

2 Experimental Study

In this section, we evaluate the effectiveness of our proposed co-movement model by using a real life dataset, the Morgan Stanley Capital International G7 index (MSCI-G7).[2] This dataset, MSCI-G7, consists of equity indices of seven countries (Canada, France, Germany, Italy, Japan, United Kingdom and United States) that are recorded in every working day (5 working days per week). The period that we have archived is from 2003/1/3 to 2007/4/30 (around 4.3 years). Two experiments are conducted to compare and evaluate our proposed co-movement model:

1. **Different Portfolio Models Comparison.** We compare our proposed co-movement model with three different kinds of portfolio models. This experiment tries to identify the necessity of the assets' dependency relationships when computing the portfolio's expected return, and the possibility of having the non-traditional view of the portfolio's expected risk as the exposure of loss.
2. **Sensitivity of Segmentation.** As discussed in the previous section, since all financial time series contain high levels of noise, we need to smoothen the time series by using bottom-up segmentation [7] and regression analysis. In this experiment, we try to evaluate the sensitivity of our proposed co-movement model with different number of segments.

2.1 Different Portfolio Models Comparison

In this section, we compare the performance of different portfolio models with our proposed co-movement model. Here is a list of models that we have implemented:

1. **COM:** This is the proposed model that we have discussed in Section 1 in this paper.
2. **MVM:** This is the traditional mean-variance model [11] for portfolio construction. It serves as the benchmark approach in our experimental study. Mathematically details of this model is discussed in Section 3 – Related Work.
3. **COM-Mean:** This model is similar to our proposed co-movement model, except that the portfolio's expected return is computed by the traditional method, i.e. the dependency of the expected return of the assets in the portfolio is ignored. Mathematically, the r_i in Eq. (11) is simply the expected return of asset i in the portfolio, but *not* the one defined in Eq. (4).

---

[2] http://www.mscibarra.com

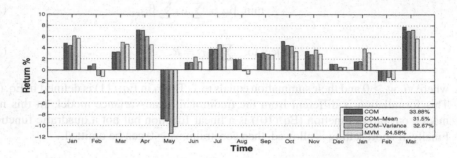

Fig. 3. The returns obtained by different portfolio models

4. **COM-Variance:** This model is similar to our proposed co-movement model, except that the portfolio's expected risk is computed by using the traditional method, i.e. the portfolio's expected risk, β_Θ, is the variance of the the portfolio's expected return. Mathematically, we change β_Θ in Eq. (10) with the following equation:

$$\beta_\Theta = \sum_{i=1}^{n-1} \sum_{j=i}^{n} Var(r_{ij}), \qquad (12)$$

where $Var(r_{ij})$ is the variance of the expected return.

In order to evaluate these four alternative allocation strategies in terms of their expected returns and risks, we evaluate their difference on realized portfolio performance. We use the first three years' data (784 days) of MSCI-G7 as training data, so as to compute the optimal weights, w_i, for each asset, i, in the portfolio. The rest of the data, 15 months, are served as the testing data for evaluation. The portfolio will be re-constructed each month in these 15 months, so as to capture the latest movements of the time series. The oldest month would be discarded whenever for the re-construction, so as to denote for ignoring the outdated information. The average segment number is set to 90 in this experiment. The sensitivity of number of segments will be discussed in the next section. The desired level of return, R, in Eq. (11) is set as the mean return of their corresponding efficient frontiers [14].

Figure 3 shows the returns that we get from each model in each month, as well as the cumulative return for the four approaches over the 15 months in the bottom right corner in the box. Although our proposed co-movement model, COM, is not a sole winner for all of the months, the cumulative return of COM does outperform the other approaches significantly. The highest cumulative return that we get is came from COM, with a profit of 33.88%, which is 11% higher than that of MVM (the traditional Mean-Variance model, the benchmark approach), which obtains only 24.5%.

Furthermore, it is worth noting that all the approaches that are related to co-movements (COM, COM-Mean and COM-Variance) outperform the MVM (the benchmark approach). It demonstrates clearly that the importance of identifying different types of co-movements is justifiable.

By comparing between COM and COM-Variance, one can justify the possibility of regarding risk as *the exposure of loss* rather than the traditional point of view – *vari-*

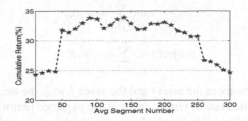

Fig. 4. The cumulative return of different number of segments

ance of return. Although COM-Variance performs better than COM in a few months, it performs inferior than COM most of the time. Similarly, COM-Mean performs inferior than COM, which demonstrates the importance of considering the dependency relationship when computing the portfolio's expected return.

2.2 Sensitivity of Segmentation

Since our proposed co-movement model requires data pre-processing by using bottom-up segmentation and regression analysis for identifying the trends of the time series, a question commonly asked would be: how sensitivity is the co-movement model with the number of segments? As a result, we try to evaluate how the average number of segments will affect the performance of our approach in this experiment.

The settings of this experiment is the same as the previous one (Section 2.1), except that we alter the number segments from 10 to 300, and then we record down the cumulative returns. Figure 4 shows the empirical result of this experiment. We address some interesting findings below.

1. The cumulative returns for all kinds of segments are always positive. This is a very encouraging sign, since it implies the number of segments will not change the cumulative return from positive to negative or vice versa.
2. The overall shape of the graph is concave. This suggests that the cumulative return and the number of segments have some kinds of dependency relationships, where the cumulative return is *not* distributed randomly with respect to the number of segments. Moreover, a concave shape also suggested that there should exist a "best" range against the number of segments, which is around 100 to 150, i.e. each segment lasts for around 5 to 7 days (a week).
3. When the number of segment is too few (less than 50) or too many (more than 300) the cumulative return would drops to around 25%. Nevertheless, it still performs better than MVM (the benchmark model, 24.5%).

3   Related Work

Most of the modern portfolio theory is motivated by the mean-variance model[12], which is also know as Markowitz model. It uses the minimum variance strategy to select the portfolio which can be summarized as below:

$$\min \sum_{i,j=1}^{n} w_i w_j \sigma_{ij} \qquad (13)$$

$$\text{subject to: } \sum_{i=1}^{n} w_i r_i = R. \qquad (14)$$

where σ_{ij} is the covariance of the asset i and the asset j, w_i is the weight of asset i in this portfolio, R is the desired rate of return, and r_i is the expected return of asset i. Hence, if r_i and σ_{ij} are identified and R is specified, we can identify the weight, w_i, of each asset in this portfolio. Intuitively, given a specified return, a portfolio with the minimum risk can always be formulated. It is obvious that the covariance of two asset i and j is taken as the measure of risk for the allocation of the two assets.

Yet, this mean-variance model perceives the expected risk is a combination of and is equal for both the *down-side* (the possibility of earning *less* than the desired level of return) and *up-side* (the possibility of earning *more* than the desired level of return). Obviously, this cannot mimic the real-life situations in the financial markets, where the investors always concern with the down-side risk only rather than the up-side risk. Accordingly, many different theories are proposed which are targeting for modeling the expected risk as down-side risk only [5,10,9,8]. Nevertheless, their concepts for modeling the down-side risk are the same – the variance of the portfolio's expected return below a specified desired level of return.

In respect to downside risk, even the pioneering work by Markowitz acknowledged the relevance of risk associated with failure to achieve a target return. Harlow and Rao [5] discussed asset allocation to minimize portfolio downside risk for any given level of expected return. The downside risk approach is more attractive than traditional mean-variance approach because of its consistency with the observation that investors are averse to downside results but not upside variability[3]. In this regards, Harvey et.al[2] indicated that the usual measure of correlation represents average co-movement in both up and down markets. Separate correlation estimates in different return environments would permit detection of whether correlation increases or decreases in down markets. Increased correlation in down markets reduces the benefit of portfolio diversification. However, different from the definition in our model, they define the down(up) market as the time with return below(upon) average return and compute the correlation in the same way as semi-variance measure[10].

In our preprocess for time series, we explore segmentation algorithm to discretize the time series[7,4]. Keogh [7] gives a good survey of segmentation algorithm on time series where the the bottom-up algorithm can achieve best performance on the selected financial dataset. Among the work of discovering correlation in time series, StatStream [16] also monitor thousands of time series data but focus on finding largest correlation of them.

4 Conclusion

In this paper, we describe a co-movement model for constructing financial portfolio by analyzing and mining the co-movement patterns among multiple time series. Different from traditional statistical approach in financial world, which takes the co-variance among the portfolio as risk and the summation of individual expect return as portfo-lio return, our approach models the risk from the co-movement patterns and computes portfolio returns by considering all dependency relationships among assets. As the first step of this area, the promising experiment results on real financial data show the new formulation of our return and risk can bring great benefits for effective asset allocation.

Acknowledgment. This work was supported by a grant of RGC, Hong Kong SAR, China (No. 418206).

References

1. Bawa, V.S., Lingenberg, E.B.: Capital market equilibrium in a mean-lower partial moment framework. Journal of Financial Economics 5(2), 189–200 (1977)
2. Erb, C.B., Harvey, C.R., Viskanta, T.E.: Forecasting international equity correlations. Finan-cial analysis Journal 50(5), 32–45 (1994)
3. Fama, E.F.: The behavior of stock-market prices. The Journal of Business 38(1), 34–105 (1965)
4. Ge, X., Smyth, P.: Deformable markov model templates for time-series pattern matching. In: KDD, pp. 81–90 (2000)
5. Harlow, W.: Asset allocation in a downside-risk framework. Financial analysis Journal 47(5), 28–40 (1991)
6. Miller, J.R.K.: Measuring organizational downside risk. Strategic Management Jour-nal 17(9), 671–691 (1996)
7. Keogh, E.J., Chu, S., Hart, D., Pazzani, M.J.: An online algorithm for segmenting time series. In: ICDM, pp. 289–296 (2001)
8. Henriksson, R.D., Leibowitz, M.L.: Portfolio optimization with shortfall constraints:a confidence-limit approach to managing downside risk. Financial analysis Journal 43(1), 34–41 (1989)
9. Kogelman, S., Leibowitz, M.L.: Asset allocation under shortfall constraints. Journal of Port-folio Management 17(2), 18–23 (1991)
10. Lewis, A.: Semivariance and the performance of portfolios with options. Financial analysis Journal 46
11. Luenberger, D.G.: Investment Science. Prentice-Hall, Englewood Cliffs (1997)
12. Markowitz, H.: Portfolio selection. Journal of Finance 7(1), 77–91 (1952)
13. Montogomery, D.C., Runger, G.C.: Applied Statistics and Probability for Engineers, 2nd edn. John Wiley & Sons, Inc., Chichester (1999)
14. Sharpe, A., Bailey,: Investments. Prentice-Hall, Englewood Cliffs (1999)
15. Rivest., R.L., Cormen, T.H., Leiserson, C.E., Stein, C.: Introduction to Algorithms. MIT Press and McGraw-Hill (2001)
16. Zhu, Y., Shasha, D.: Statstream: Statistical monitoring of thousands of data streams in real time. In: VLDB, pp. 358–369 (2002)

Effective Spatio-temporal Analysis of Remote Sensing Data*

Zhongnan Zhang, Weili Wu, and Yaochun Huang

Department of Computer Science, University of Texas at Dallas,
Richardson TX 75083, USA
{znzhang,weiliwu,yxh038100}@utdallas.edu

Abstract. Extracting knowledge and features from a large amount of
remote sensing images has become highly required recent years. Spatio-
temporal data mining techniques are studied to discover knowledge from
these images in order to provide more precise weather prediction. Two
learning granularities have been proposed for inductive learning from
spatial data: one is spatial object granularity and the other is pixel granularity. In this paper, we propose a pixel granularity based framework to
extract useful knowledge from the remote sensing image database by using SOM and association rules mining. A three-stage algorithm, named
as STARSI, is also proposed and used in this framework.

1 Introduction

Among the various forms of spatio-temporal data, remote sensing images play
an important role, due to the growing wide-spreading of outer space satellites.
Remote sensing is defined as the science and technology by which the characteristics of interesting objects can be identified, measured or analyzed without
direct contact [7]. The concept is illustrated in Figure 1.

Remote sensing satellites are currently the most significant source of new data
about our planet, and remote sensing image databases are the fastest growing
archives of spatial information. Data mining techniques are being studied to
discover knowledge from meteorological observation data in order to provide
more precise weather predictions [2,8,11,10]. Image data such as satellite images
and medical images often amount to several Terabytes, thus manual and detailed
analysis of these data becomes impractical [9]. Therefore an automated (or semi-
automated) process to exact knowledge from these data should be included in
the data mining from the image database [6].

Features of our studies applied to the images are summarized as follows: (1)
The application of data mining method applied to image classification; (2) Association rules generation from the classified data.

The organization of the rest paper is as follows. In Section 2, we present the
related works and our contribution. The problem definition is formally given in

---

* This work has been partially supported by the National Science Foundation under
grant IIS-0513669 and CCF-0514796.

Y. Zhang et al. (Eds.): APWeb 2008, LNCS 4976, pp. 584–589, 2008.

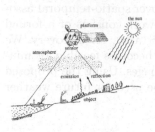

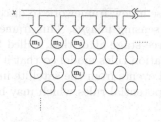

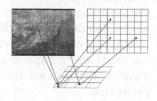

Fig. 1. Data collection by remote sensing

Fig. 2. Basic structure of Kohonen's self-organizing map

Fig. 3. Stage One of the STARSI algorithm

Section 3. In Section 4, we propose an algorithm for mining association rules on the pre-processed data. We present our experience with the algorithm on a real-life dataset in Section 5. Finally, we discuss the related future work and conclude with summary.

2 Related Works and Our Contribution

For similarity searching in multimedia data, there are two main families of multimedia indexing and retrieval systems [1]: *description-based retrieval systems* and *content-based retrieval systems*. In a content-based retrieval system, image feature specification queries specify or sketch image features like color, texture, or shape, which are translated into a feature vector to be matched with the feature vectors of the images in the database.

Kohonen's self-organizing map is a two layer network that organizes a feature map by discovering feature relations based on input patterns through iterative non-supervised learning [3]. Figure 2 presents its basic schematic structure.

The input variables $\{\xi_i\}$ of SOM is defined as a real vector $x = [\xi_1, \xi_2, \ldots, \xi_n]^T$ $\in \Re^n$. With each element in the SOM array we associate a parametric real vector $m_i = [\mu_{i_1}, \mu_{i_2}, \ldots, \mu_{i_n}]^T \in \Re^n$ that we call a model. The input signals x are classified into the activated (nearest) unit m_c of the input layer and projected onto the competition grids. The distance on the competition grids reflects the similarity between the patterns. After the training is complete, the obtained competition grids, i.e., the feature map, represents a natural relationship between the patterns of input signals entered into the network.

Spatial associations are rules that associate one or more spatial objects with other spatial objects. There have been two approaches to the problem. In the first one, Koperski and Han [4] defined spatial association rule of the form $X \Rightarrow Y(c\%)$, where X and Y are sets of spatial predicates or non-spatial predicates and $c\%$ is the confidence of the rule. More recently, Shekhar and Huang [5] tackled the co-location patterns problem, where the spatial information defines a new and flexible notion of transaction, able to cope with the non-transitivity of the closeness predicate.

In this paper, we propose a new framework to discover spatio-temporal association rules from remote sensing images. This framework consists of a formal model definition and a three-stage algorithm, called STARSI, for discovery. We modify the classical association rules model so that it becomes feasible for mining relations for sequential events from the satellite images. Using the proposed algorithm we can get the potential rules which may be used for future weather prediction.

3   Formal Model

In this section, we present the formal model definition for the remote sensing images analysis. In addition, some specifications on the parameters' selection are also discussed in this section. The problem is formally defined as follows:

Given:
  - A set of remote sensing images, $I = I_1, I_2, \ldots, I_n$.
  - The image partition parameters, M and N.
  - The SOM feature map parameters, X, Y, and K.
  - The minimum support factor and confidence factor, $minsup$ and $minconf$.

Find: Spatio-temporal association rules $A \Rightarrow B$ with the support factor $0 \leq s \leq 1$ and confidence factor $0 \leq c \leq 1$.

Objective: Completeness and Correctness.

Constraints:
  1. A and B are two image blocks from two consecutive images respectively.
  2. The position of A must be one of the positions of B's eight-way neighbors.
  3. $X \times Y \ll M \times N \times n$ and $X \times Y \ll n$.
  4. The support factor $s \geq minsup$ and the confidence factor $c \geq minconf$.

In the above formulation, the problem of rule mining has the following issues which are needed to be clarified:

(1). There're two simple choices for neighborhood topology of the SOM space: **hexagonal lattice** and **rectangular lattice**. In the hexagonal lattice, each node will have six neighbors. However in the rectangular lattice, each node can have four or eight neighbors. The selection of the lattice type and the neighbor numbers is based on the type of application.

(2). What's the feasible time intervals for these time series images? If the images are high definition images, we need to decrease the intervals because even after a small time slot, the images will show a big difference. On the other hand, for low definition images, the intervals can be increased since the differences are not that much.

4   Spatio-temporal Analysis on Remote Sensing Images (STARSI) Algorithm

In general, the STARSI algorithm consists of the following three stages:

Stage I: Clustering of image blocks

1. All images are divided into $M \times N$ blocks.
2. Store the physical position of each block.
3. Use SOM to generate the feature map for all image blocks, taking each block's raster-like scanned intensity vector as the input vector.

The physical position of each block contains two parts: (1)the image ID from which the block comes; and (2) the x-y position within the image after the image partition. After the third step, SOM generates a feature map. Using this map we can assign a feature ID to each block. At the same time, we can also get a feature matrix for each image. Each element of the matrix, represented by the feature ID, is just some type of representation of the block. The whole stage is illustrated by Figure 3.

Stage II: Clustering of images

1. Each image is represented by a matrix of the SOM address for blocks.
2. Generate the feature map of SOM for all images by taking the matrix of each image as the input.

The basic idea of the second stage is the same as the first stage, except that the input represents each image but not each block. After this stage, each image will also be assigned a feature ID. However this feature ID is totally different to the block's feature ID, even they may have the same value. Therefore, each block now has two feature IDs, one represents the type of the block and the other represents the type of the image it belongs to.

Stage III: Association rules generation

1. The neighborhood relationship for image I_i at block position $(m,\ n)$ can be represented as: $NB(I_i)_{(m,n)} = [I_{i(m,n)},\ I_{i(m-1,n-1)},\ I_{i(m-1,n)},$ $I_{i(m-1,n+1)},\ I_{i(m,n-1)},\ I_{i(m,n+1)},\ I_{i(m+1,n-1)},\ I_{i(m+1,n)},\ I_{i(m+1,n+1)}].$ Applying this relationship to two consecutive images I_k and I_{k+1}, we can generate the following type of candidate rules:

$$I_k \bowtie I_{k+1} = \{I_{k(u,v)} \Rightarrow I_{k+1(x,y)} | \forall u : 0 \le u \le M-2, \forall v : 0 \le v \le N-2,$$

$$|u - x| \le 1 \text{ and } |v - y| \le 1 \text{ and } |u - x| + |v - y| \ge 1\} \quad (1)$$

where $I_{i(a,b)} = [ID_1, ID_2]$, ID_1 represents the type of the block and ID_2 represents the type of the image it belongs to.

2. Use $minsup$ and $minconf$ to prune these candidate rules.

The last stage is to use the previous results to generate useful association rules. For the rules generation, we still use the Apriori-like algorithm that referenced before.

5   Experimental Evaluation

We assessed the correctness and effectiveness of our algorithm by experimenting with a real-life dataset. The satellite images were captured by GOES-E, which operated by the National Oceanic and Atmospheric Administration (NOAA) [1], USA. All images are in the same size with 640×480 pixels in gray-scale. The total number of images is 600. The time range of these data is from May 2007 to Aug. 2007.

5.1   Experiment Design

The goal of the experiment is to evaluate the effect of our model. As follows, the process can be divided into three parts: SOM map generation,Learning and Testing.

The first part is to generate two SOM maps: one for all blocks and the other for all images. All the images, including the images used only for testing later, need to be processed in this part. The reason is that we request the representation criteria for both learning data and testing data should be the same. M is set to 12 and N is set to 9. The sizes of the two SOM maps are the same: a 6×6 square. For the SOM learning iteration times, K, in stage one we set it as 50000 and in stage two we set it as 10000. After this part, we have two maps and two mappings: (1)represents each block with a SOM address; (2)represents each image with a SOM address. Two thirds of all the images, that is 400 images, were used as the input data for the learning data. The left images were used as testing data.

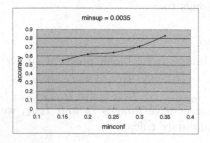

Fig. 4. Accuracy of the STARSI algorithm when *minsup* is fixed

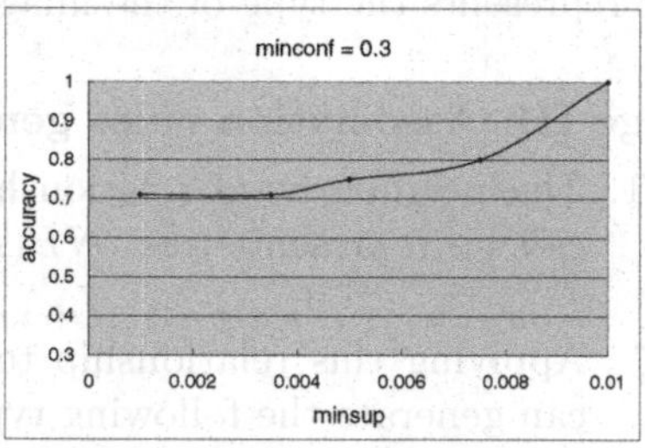

Fig. 5. Accuracy of the STARSI algorithm when *minconf* is fixed

5.2   Experiment Result

The experiment has two parts: (1)the value of *minsup* is fixed, the correctness is generated with five different *minconf* values; (2)the value of *minconf* is fixed, the correctness is generated with five different *minsup* values. The sample results are illustrated in the following two figures: Figure 4 and 5. From the Figure 4,

---

[1] We would like to thank NOAA for providing the satellite images for our experiments.

we can see that when the *minsup* is set to a fixed value, with the increase of the value of *minconf* the accuracy also has an increase. The increasing speed is higher when *minconf* has a higher value. From the Figure 5, when the *minconf* is set to a fixed value, with the increase of the value of *minsup* we can get almost the same situation.

6   Conclusion

In this paper, we presented STARSI, a new algorithm for mining spatio-temporal patterns from remote sensing images. First, we gave a formal model of this type of problems. We then discussed in detail our solution, the three-stage algorithm and explained it step-by-step. Finally, we used a real-life dataset to justify effectiveness of our algorithm. The experiments show that rules' correctness are high when the parameters are set properly.

One potential future work is to use other different classification or clustering methods to replace SOM in the first two stages or use two different methods in different stages. Another one is to use other different mining methods to generate knowledge, such as decision trees.

References

1. Han, J., Kamber, M.: Data Mining: Concepts and Techniques. Morgan Kaufmann, San Francisco (2000)
2. Honda, R., Takimoto, H., Konishi, O.: Semantic indexing and temporal rule discovery for time-series satellite images. In: The 1st Int. Workshop on Multimedia Data Mining (2000)
3. Kohonen, T.: Self-Organizing Maps. Springer, Berlin (2001)
4. Koperski, K., Han, J.: Discovery of spatial association rules in geographic information databases. In: Proc. of the 4th Int. Symp. Advances in Spatial Databases (1995)
5. Shekhar, S., Huang, Y.: Discovering spatial co-location patterns: A summary of results. In: Jensen, C.S., Schneider, M., Seeger, B., Tsotras, V.J. (eds.) SSTD 2001. LNCS, vol. 2121, pp. 236–256. Springer, Heidelberg (2001)
6. Shekhar, S., Schrater, P.R., Vatsavai, R.R., Wu, W., Chawla, S.: Spatial contextual classification and prediction models for mining geospatial data. IEEE Transactions on Multimedia 4(2), pp. 174–188 (2002)
7. Stein, A., Meer, F., Gorte, B. (eds.): Spatial Statistics for Remote Sensing. Kluwer Academic Publishers, Dordrecht (1999)
8. Tsoukatos, I., Gunopulos, D.: Efficient mining of spatiotemporal patterns. In: Proc. of the 7th Int. Symp. on Advances in Spatial and Temporal Databases, pp. 425–442 (2001)
9. Zaiane, O., Han, J., Li, Z., Chiang, J., Chee, S.: Multimedia-miner: A system prototype for multimedia data mining. In: Proc. of the 1998 ACM SIGMOD Int. Conference on Management of Data, pp. 581–583 (1998)
10. Zhang, Z., Wu, W., Deng, P.: Mining dynamic spatio-temporal association rules for local-scale weather prediction. In: The 5th Int. Workshop on Multimedia Data Mining (2004)
11. Zhang, Z., Wu, W., Huang, Y.: Mining dynamic interdimension association rules for local-scale weather prediction. Compsac 02, pp. 146–149 (2004)

Supporting Top-K Aggregate Queries over Unequal Synopsis on Internet Traffic Streams

Ling Wang, Yang Koo Lee, and Keun Ho Ryu*

Database/Bioinformatics Laboratory, School of Electrical & Computer Engineering, Chungbuk
National University, Chungbuk, Korea
{smile2867,leeyangkoo,khryu}@dblab.chungbuk.ac.kr

Abstract. Queries that return a list of frequently occurring items are important
in the analysis of real-time Internet packet streams. While several results exist
for computing Top-k queries using limited memory in the infinite stream model
(e.g., limited-memory sliding windows). To compute the statistics over a sliding
window, a synopsis data structure can be maintained for the stream to compute
the statistics rapidly. Usually, a Top-k query is always processed over an equal
synopsis, but it's very hard to implement over an unequal synopsis because of
the resulting inaccurate approximate answers. Therefore, in this paper, we focus
on periodically refreshed Top-k queries over sliding windows on Internet traffic
streams; we present a deterministic DSW (Dynamic Sub-Window) algorithm to
support the processing of Top-k aggregate queries over an unequal synopsis and
guarantee the accuracy of the approximation results.

Keywords: sliding window, Top-k query, synopsis data structure, DSW (Dynamic Sub-Window) algorithm, internet traffic streams.

1 Introduction

On-line data streams possess interesting computational characteristics, such as
unknown or unbounded length, a possibly very fast arrival rate, the inability to
backtrack over previously arrived items (only one sequential pass over the data is
permitted), and a lack of system control over the order in which the data arrive[1].
The real-time analysis of network traffic has been one of the primary applications
of data stream management systems, examples of which include Gigascope [2] and
STREAM [3]. A problem of particular interest, motivated by traffic engineering,
routing system analysis, customer billings, and the detection of anomalies such as
denial-of service attacks, concerns the statistical analysis of data streams with a
focus on newly arrived data and frequently appearing packet types[4, 5, 6]. For
instance, an ISP may be interested in monitoring streams of IP packets originating
from its clients and identifying those users who consume the most bandwidth
during a given time interval. The objective of these types of queries is to return a
list of the most frequent items (called Top-K Queries or hot list queries) or items
that occur above a given frequency (called threshold queries).

---

* Corresponding author.

Y. Zhang et al. (Eds.): APWeb 2008, LNCS 4976, pp. 590–600, 2008.
© Springer-Verlag Berlin Heidelberg 2008

A solution for removing stale data is to periodically reset all statistics. The sliding window model causes old items to expire as new items arrive. Initially proposed by Zhu and Shasha in [7], the objective of this method is to divide the sliding window into some sub-windows, and only store the synopsis of each sub-window in full. Thus, in the synopsis, if the sub-windows have the same or similar size, we call this kind of synopsis an Equal Synopsis; otherwise, it is an Unequal Synopsis. Usually, Top-k query is always processed over an equal synopsis, but it's very hard to implement over an unequal synopsis because of the resulting inaccurate approximate answers. Therefore, in this paper, we focus on periodically refreshed Top-k queries over sliding windows on Internet Traffic Streams. We present a deterministic DSW (Dynamic Sub-Window) algorithm to support the processing of Top-k aggregate queries over an unequal synopsis and guarantee the accuracy of the approximation results.

2 Related Work

The question of how to maintain an efficient synopsis data structure is very important for the stream in order to compute the statistics rapidly. There are many types of synopsis data structures that have been presented in recent years. The running synopsis as an unequal synopsis is good at subtractable aggregates [8] such as SUM and COUNT. An interval synopsis [9] is used to distributive aggregates that are not subtractable, such as MIN and MAX. The equal synopsis basic-window synopsis was proposed by Yunyue Zhu et al. in [7]. Paned-window synopsis, which was proposed by Jin Li et al. in [10] is an extended version of the basic-window synopsis in which the sub-windows are called "panes". Compared with the unequal synopsis, the paired-window synopsis, which was proposed by Sailesh Krishnamurthy et al. in [11], improves on the paned-window synopsis by using paired-windows which chop a stream into pairs of possibly unequal sub-windows. There have also been some special synopsis structures such as the holistic synopsis [12], wavelet synopsis [13] [14], etc. designed to minimize some appropriate error measure, given a space budget. For indexing and searching schema-less XML documents based on concise summaries of their structural and textual content, Weimin He et al. presented two data synopsis structures in [15] that correlate the textual with the positional information in an XML document and improve the query precision.

Queries that return a list of frequently occurring items are important in the context of traffic engineering, routing system analysis, customer billing, and the detection of anomalies such as denial-of-service attacks. There has been some recent work on answering Top-K queries over sliding windows [5]. Computing top-k queries from multiple sub-windows using the top-k list is similar to rank aggregation in conventional DBMSs [17]. The main idea is to combine multiple ranked lists, which are assumed to be available in their entirety, into a single list. The difference in the DSMS context is that a top-k query does not have access to complete sets of frequency counters in each sub-window and, therefore, must calculate the overall top-k list over the entire sliding window based upon limited information. In [18], the authors present a framework for distributed top-k monitoring, but their goal is to minimize the amount of data transferred from distributed sources to a central processing system. Furthermore, they do not consider the sliding window model.

Lukasz Golab et al. in [12] proposed a FREQUENT algorithm, which identifies the frequently occurring items in sliding windows and estimates their frequencies. They answer the frequent item queries using small-size Basic-Window synopses (sub-windows), because there is no obvious rule for merging the partial information in order to obtain the final answer. They store a top-k synopsis in each Basic-window and maintain a list of the k most frequent items in each window at the same time. Finally, they output the identity and value of each global counter over the threshold, δ. They proposed a classification of distribution models for sliding windows over data streams in [19]. They used the drifting distribution models to develop algorithms for answering frequent item queries over multinomial-distributed sliding windows.

3   Synopsis Data Structure

The Synopsis Data Structure can be stored in the allotted amount of memory and used to provide approximate answers to user queries along with some reasonable guarantees on the quality of the approximation. Such approximate, on-line query answers are particularly well suited to the exploratory nature of most data-stream processing applications, such as trend analysis and fraud or anomaly detection in telecom-network data, where the goal is to identify generic, interesting or "out-of-the-ordinary" patterns rather than provide results that are exact to the last decimal.

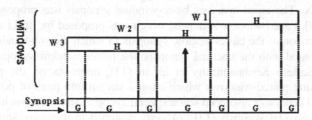

Fig. 1. A synopsis data structure containing two different size sub-windows where the whole synopsis process contains three steps: the partial aggregate step (support aggregate function G), final aggregate step (support aggregate function H) and the whole window aggregate step (where W1, W2, and W3 are the sliding windows)

In this section, we briefly discuss the basic scheme used for the Synopsis Data Structure. As shown in Fig.1, this synopsis contains two different size sub-windows. Generally, the processing of the Synopsis Data Structure contains three steps. In the first step, called the partial aggregate step, we apply the function, G, over all the tuples in each sub-window. Then, in the second step, called the final aggregate step, an overlapping window aggregate operator buffers these partial aggregates and successively applies the function, H, to these sets of sliced aggregates to compute the query results. Finally, in the third step, we let A be the aggregate function we are computing over the whole set of windows.

3.1   Top-K over Unequal Synopsis

Here, we classify the Synopsis Data Structure into two major types, Equal Synopsis and Unequal Synopsis. In the Equal Synopsis (e.g. Paned Window Synopsis), all the sub-windows have the same size. Otherwise, we call it the Unequal Synopsis (e.g. Paired Window Synopsis). The use of the synopsis can reduce both the space and computation cost of evaluating sliding window queries by sub-aggregating and sharing the computation. The Equal Synopsis is very easy to implement, but it's very hard to share the overlapping windows when we solve the multi-aggregate queries over the streams, because it always leads to more slices. The Unequal Synopsis can solve this problem well.

In this section, we use a paired-window synopsis as an example of an unequal synopsis. Fig.2 shows how paired-windows split a window into a "pair" of exactly two unequal slices. The characters *a*, *b* and *c* can be considered as three types of packet streams. *20, 10* and *5* can be considered as frequency counts. In this example, this synopsis contains 7 sub-windows (such as *S1, S2*) and we apply Top-3 as the partial aggregate. In the partial aggregate step, we output the Top-3 types in each sub-window. In the final aggregate step, if we want to output the most frequent item in the whole first window, type *c* as the most frequent item, rather than type *a*, will be the result, because its frequency count is *85*. Type *a* is dropped in the second sub-window *S2*, although its final frequency counts will be the highest in the whole first window.

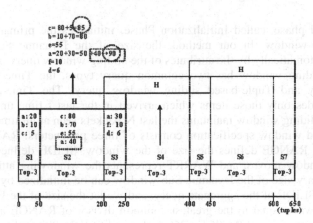

Fig. 2. A paired-window structure splits a window into a "pair" of exactly two unequal slices (such as S1 and S2)

So, in this paper, we present a new method named the DSW (Dynamic Sub-Window) method. The DSW method is not restricted by the form of the synopsis and resets the workload of each window, in order to ensure that false negatives occur very rarely.

4 DSW (Dynamic Sub-Window) Method

We propose the following simple Method, the DSW (Dynamic Sub-window) Method, which employs the Unequal Synopsis approach and stores a Top-k synopsis in each sub-window. In this section, we will explain the three phases of this method and how to implement this method in detail.

4.1 Three Phases of DSW Method

The whole Dynamic Sub-window method contains three phases. The first phase, called Redefinition Phase, is shown in Fig.3. We use an unequal synopsis which contains 7 sub-windows S1, S2, S3......S7. In our method, we design a new window called the Dynamic Sub-window that redefines the Long Sub-Window (such as S2, whose size is much larger than the others) into some new small sub-windows (the shaded area in Fig. 3.). Throughout this process, we can make all of the sub-windows have a similar size in order to reduce the size difference and to improve the accuracy of the Top-k results.

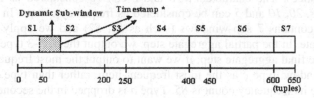

Fig. 3. A dynamic sub-window synopsis

The second phase, called Initialization Phase, initializes the primary size of the Dynamic Sub-window. In our method, the size of the Dynamic Sub-window is maintained automatically by the attributes of the sliding window query. As mentioned above, the sliding window has two common query types, the Time-based sliding window query and Tuple-based sliding window query. The Time-based sliding window includes only those items which arrived in the last t time units, while the Tuple-based sliding window maintains the last N packets seen at all times. Usually, a non-partitioned window specification consists of three parameters: RANGE, SLIDE and WARRT. RANGE defines the size of the window, SLIDE defines the steps at which the window moves, and WARRT represents the windowing attribute. In our case, the primary size of the Dynamic Sub-window can be initialized by the values of RANGE and SLIDE of the aggregate query, and we let the size of the Dynamic Sub-window be always equal to the greatest common divisor of RANGE and SLIDE of this query. For example, if RANGE= 250 tuples and SLIDE=200 tuples, the Dynamic Sub-window size is equal to 50 tuples.

After the size of the Dynamic Sub-window has been initialized, the next problem is how to implement this Dynamic Sub-window in the synopsis. So, we design the third phase: Maintain Phase. We use a function to sign the timestamp * into the synopsis, which can control the end and restart of the Dynamic Sub-window. This means that when the Dynamic Sub-window has been redefined in the synopsis, it will be controlled by the timestamp *. It will be ended and restarted when it meets the timestamp *(The idea of the punctuation * was first introduced by Jin Li, David Maier, and Kristin Tufte in [10].).

4.2 Theorem

In studying the implementation of the Dynamic Sub-window, we found that its size can be maintained by a useful function. That is to say, it is always controlled by three parameters which can be maintained by a useful function.

In our DSW method, we use the function $\theta= \{(N-1) / M\} + 1$ to sign the timestamp * into the synopsis. The Dynamic Sub-window will be controlled by the timestamp * in that it will end or restart when it meets the timestamp *. We propose this method of maintaining the size of the Dynamic Sub-window when the sub-windows are redefined. This function is defined by three parameters: θ which is the frequency count of Top-1 in each sub-window, N which is the number of arriving stream data and M which is the total number of types which have arrived in the synopsis. There three parameters are always changed in real-time, and when we did an analysis on the relationship among them using real data, we found that N is always bounded by two other parameters, θ and M, such that $\theta \leq N \leq (\theta -1) \times M + 1$. This means that if we set the values of θ and M, the limited-length of N can be easily calculated. In other words, when the tuples are arriving into the synopsis over the streams, we don't know when they will end, but we can use this function to calculate when the dynamic sub-window will have to be ended.

Theorem: Suppose θ is the frequency count of Top-K in each sub-window, N is the number of arriving stream data and M is the total number of frequency types which have arrived in the synopsis. Then these values are always changed by the balance function $\theta= \{(N-1) / M\} + 1$.

Proof: Suppose each Dynamic Sub-window contains M kinds of tuple types, and the frequency count of the most frequent item is θ', then the total number N' equals $\theta' \times M$. When M is fixed, $\theta= \theta'+1$, then $N= N'+1$ and $N'= \theta' \times M$. Therefore, since $\theta'= \theta-1$, and $N'= N-1$, then $N-1= (\theta-1) \times M$. Therefore, $\theta= \{(N-1) / M\} + 1$. This means that if θ increases by 1, according to the saturation state, it only needs add 1.

4.3 Dynamic Sub-window Algorithm

In the following DSW algorithm, we assume that a single DS-window fits in main memory, within which we may count the item frequencies exactly. We define the Dynamic Sub-window as a DS-window in our algorithm.

Input: the number of arriving stream data (N), the greatest common divisor of RANGE and SLIDE (n), the frequency count of Top-1 in each sub-window (θ), and the total number of types (M) which have arrived in the synopsis.

```
Repeat:
1. For each tuple e in the next DS-window N:
      If a local counter exists for the type of tuple
      e:
         Increment the local counter.
      Otherwise:
         Create a new local counter for this type and
         set it equal to 1.
2. Initialize let N equals n and let M' equals the M,
      then j = {(N-1) / M} + 1.
3. For each local counter in the next DS-window N, each
      kinds of tuple types M' in the next DS-window,
      DS-window restarts when it meets timestamp *:
      If any local counter equals j:
```

```
          Stop the DS-window and make a new timestamp*
          into the synopsis.
      Otherwise:
          Stop the DS-window when it meets any
          timestamp * in the synopsis.
      If the kinds of tuple types M' exceed M:
          Stop the DS-window.
      Otherwise:
          Stop the DS-window when it reach any
          timestamp* in the synopsis.
   4. Output the k most frequent items on each DS-window
          synopsis list.
```

5   Experiment and Evaluation

We first describe the experimental setup and then present and analyze the results.

5.1   Experimental Setup

We tested the DSW algorithm on count-based windows over TCP traffic data. The trace contains 199 and 1592 distinct source IP addresses for workloads (A) and (B), respectively, which can be treated as distinct item types. We set workload (A) to contain N = 10000 and workload (B) to contain N' = 80000. Experiments were conducted with two values of Θ: Θ = 3 (initialized sub-window size is 80) and Θ = 9 (initialized sub-window size is 400). The size of the top-k list, k, is varied from one to ten.

We built a prototype TCP/IP stream data in C++, and let the entire stream data run in real time. We imitate the attributes of the TCP/IP addresses to make two kinds of stream data set and, at the same time, we sign the item types into each TCP/IP address. Then, we design a TCP/IP address generator. According to this method, we can process theses queries on the synopsis in memory in real time.

5.2   Experimental Results

In this section, we use real data and some different performance environments to test the DSW algorithm. At the same time, we compare the accuracy rates for the 3 strategies based on the DS-window (Dynamic Sub-window) and Paired-window approaches discussed below.

Same Predicates, Diff. Windows in Workload (A)

Table 1. Performance environment (A) with initialized DS-window size equal to 80 tuples

Type	Name	Values (tuples)
Workload	A	[0,10000]
Distinct IP addresses	M	199
Predicate	Θ	Frequency: 3
Window	DS-window	Initialized size: 80
	Paired-window	RANGE: [0,400]
		SLIDE: [0,320]

In workload (A), we examine queries with identical selection predicates and different periodic windows over a real data set. We compare the execution time for 2 strategies (DS-window and Paired-window) based on the variation in the percentage of identified frequent item types with increasing Top-k. Table 1 shows the performance environment and Fig.4 (a) shows the percentage of IP addresses that were identified by the DSW algorithm. The general trend is that for k ≥ 2, at least 80% of the IP addresses are identified. But for the Paired window, at least k ≥ 4, 80% of the IP addresses can be identified. This is because when k increases, the space usage will be very high. Therefore, if we use an algorithm to improve the identification rate of the frequent items, we not only reduce the space usage, but also improve the accuracy rate. In the figure, we can see that the DS-window can identify IP addresses from approximately k = 2.

Table 2. Performance environment (A) with initialized DS-window size equal to 400 tuples

Type	Name	Values (tuples)
Workload	A	[0,10000]
Distinct IP addresses	M	199
Predicate	Θ	Frequency: 9
Window	DS-window	Initialized size: 400
	Paired-window	RANGE: [0,2400]
		SLIDE: [0,2000]

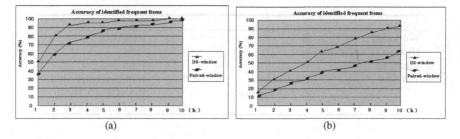

(a)   (b)

Fig. 4. Accuracy of identified frequent items with initialized DS-window size equals to 80 tuples in workload (A), the experimental results as shown in figure (a). Accuracy of identified frequent items with initialized DS-window size equals to 400 tuples in workload (A), the experimental results as shown in figure (b).

Fig.4 (b) also shows the percentage of IP addresses which were identified by the DSW algorithm, but for the performance environment defined in Table 2. In this Figure, we can see that although the size of RANGE is larger than before, from k = 7, the DS-window also keeps high accuracy, whereas the paired–window can only identify half of the frequent items.

Diff. Predicates, Same Windows in Workload (A)

In the DSW algorithm, the initialization of Θ is very important. If the value of Θ is too large, it can affect the size of the DS-window, as well as resulting in low

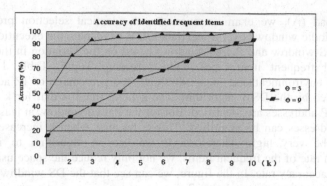

Fig. 5. Relationships between Accuracy and Value of Θ

accuracy. Otherwise, if the value of Θ is too small, it will result in high space usage. The relationship between the accuracy and the value of Θ is shown in Fig.5.

Workload (A) and (B)

Fig.6 shows the effect of the workload on the DSW algorithm. We used the performance environment listed in Table 3. In this test, we can see that the data speed

Table 3. Performance environments for different workloads (A) and (B)

Type	Name	Values (tuples)
Workload	B	[0,80000]
Distinct IP addresses	M	199
Predicate	Θ	Frequency: 9
Window	DS-window	Initialized size: 400
	Paired-window	RANGE: [0,2400]
		SLIDE: [0,2000]

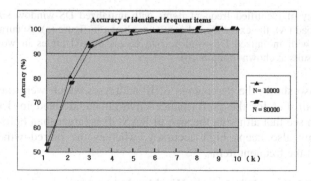

Fig. 6. Accuracy of approximation results for different workloads (A) and (B)

and the size of the data set are not the main factors affecting the DSW method. We designed a function $\theta = \{(N-1) / M\} + 1$ to maintain the DS-window size, and these parameters can be initialized by the query only once. In the following approaches, the DS-window can be maintained by itself. This is why the DSW algorithm suits both time-based and tuple-based sliding windows.

6 Conclusion

In this paper, we presented the deterministic DSW (Dynamic Sub-Window) algorithm to support the processing of Top-k aggregate queries over an unequal synopsis and guarantee the quality and accuracy of the approximation results. We classified the synopsis data structure into two major types, the Equal Synopsis and Unequal Synopsis. The Basic Synopsis (Equal Synopsis) is always applied by optimizing the sliding windows, but may incur large time and space computation costs in memory, especially for the multi-aggregates problem. Therefore, a method is needed in which the size of the synopsis is not decided based on the designers' subjective opinion but by the queries themselves, and which can be changed in real time in response to the changing attributes of the queries. Such a method would be more effective in solving the different kinds of aggregates. Based on this idea, we presented the DSW algorithm that not only guarantees the quality of the approximation results, but is also suitable for both the Equal Synopsis and Unequal Synopsis.

In a future work, we intend to solve more complex synopses which contain too many small sub-windows for multi-aggregate queries. We are also working on other aspects of processing streams, including the formalization of window semantics, the evaluation of window queries and the processing of disordered streams.

Acknowledgment. This research was supported by a grant (#07KLSGC02) from Cutting-edge Urban Development - Korean Land Spatialization Research Project funded by Ministry of Construction & Transportation of Korean government.

References

1. Golab, L., Ozsu, M.T.: Issues in data stream management. ACM SIGMOD Record 32(2), 5–14 (2003)
2. Cranor, C., Gao, Y., Johnson, T., Shkapenyunk, V., Spatscheck, O.: Gigascope: High performance network monitoring with an SQL interface. In: 2002 ACM SIGMOD international conference on Management of data, pp. 623–623. ACM Press, New York (2002)
3. Babcock, B., Babu, S., Datar, M., Motwani, R., Widom, J.: Models and issues in data streams. In: 21st ACM SIGMOD-SIGACT-SIGART symposium on Principles of database systems, pp. 1–16. ACM Press, New York (2002)
4. Demaine, E., Lopez-Ortiz, A., Munro, J.I.: Frequency estimation of internet packet streams with limited space. In: Möhring, R.H., Raman, R. (eds.) ESA 2002. LNCS, vol. 2461, pp. 348–360. Springer, Heidelberg (2002)

5. Mouratidis, K., Bakiras, S., Papadias, D.: Continuous monitoring for top-k queries over sliding windows. In: 2006 ACM SIGMOD international conference on Management of data, pp. 635–646. ACM Press, New York (2006)
6. Lee, Y.K., Jung, Y.J., Ryu, K.H.: Design and Implementation of a System for Environmental Monitoring Sensor Network. In: Chang, K.C.-C., Wang, W., Chen, L., Ellis, C.A., Hsu, C.-H., Tsoi, A.C., Wang, H. (eds.) APWeb/WAIM 2007. LNCS, vol. 4537, pp. 223–228. Springer, Heidelberg (2007)
7. Zhu, Y.Y., Sasha, D.: Statistical monitoring of thousands of data streams in real time. In: VLDB 2002, 28th International Conference on Very Large Data Bases, pp. 358–369. VLDB Press, Hong Kong (2002)
8. Cohen, S.: User-defined aggregate functions: bridging theory and practice. In: 2006 ACM SIGMOD international conference on Management of data, pp. 49–60. ACM Press, Chicago (2006)
9. Bulut, A., Singh, A.K.: A unified framework for monitoring data streams in real time. In: 21st International Conference on Data Engineering, ICDE 2005, pp. 44–55. IEEE Press, Tokyo (2005)
10. Li, J., Maier, D., Tufte, K., Papadimos, V., Tucker, P.A.: No pane, no gain: efficient evaluation of sliding-window aggregates over data streams. ACM SIGMOD Rocord 34(1), 39–44 (2005)
11. Krishnamurthy, S., Wu, C., Franklin, M.J.: On-the-fly sharing for streamed aggregation. In: 2006 ACM SIGMOD international conference on Management of data, pp. 623–634. ACM Press, Chicago (2006)
12. Toman, D.: On Construction of Holistic Synopses under the Duplicate Semantics of Streaming Queries. In: 14th International Symposium on Temporal Representation and Reasoning (TIME 2007), pp. 150–162. IEEE Press, Alicante (2007)
13. Garofalakis, M.N., Gibbons, P.B.: Wavelet synopses with error guarantees. In: 2002 ACM SIGMOD international conference on Management of data, pp. 476–487. ACM Press, Madison (2002)
14. Matias, Y., Uriel, D.: Optimal workload-based weighted wavelet synopses. Theoretical Computer Science 371, 227–246 (2007)
15. He, W.M., Fegaras, L., Levine, D.: Indexing and searching XML documents based on content and structure synopses. In: Cooper, R., Kennedy, J. (eds.) BNCOD 2007. LNCS, vol. 4587, pp. 58–69. Springer, Heidelberg (2007)
16. Golab, L., DeHaan, D., Demaine, E.D., Lopez-Ortiz, A., Munro, J.I.: Identifying frequent items in sliding windows over on-line packet streams. In: 3rd ACM SIGCOMM conference on Internet measurement, pp. 173–178. ACM Press, Miami Beach (2003)
17. Ilyas, I.F.: Rank-aware query processing and optimization. PhD thesis, Purdue University (2004)
18. Babcock, B., Olston, C.: Distributed top-k monitoring. In: 2003 ACM SIGMOD international conference, pp. 28–39. ACM Press, San Diego (2003)
19. Golab, L., DeHaan, D., Lopez-Ortiz, A., Demaine, E.D.: Finding Frequent Items in Sliding Windows with Multinomially-Distributed Item Frequencies. In: 16th International Conference on Scientific and Statistical Database Management (SSDBM 2004), pp. 425–425. IEEE Press, Santorini Island (2004)

ONOMATOPEDIA: Onomatopoeia Online Example Dictionary System Extracted from Data on the Web

Chisato Asaga[1], Yusuf Mukarramah[2], and Chiemi Watanabe[1]

[1] Graduate School of Humanities and Sciences
[2] Department of Information Sciences, Faculty of Science
Ochanomizu University
2-1-1 Otsuka, Bunkyo-ku, Tokyo 112-8610, Japan
{asaga,mukarramah}@db.is.ocha.ac.jp, chiemi@is.ocha.ac.jp

Abstract. Japanese is filled with onomatopoeia words, which describe sounds or actions like "click" or "bow-wow." In general, mastering onomatopoeia phrases is hard for foreign speakers, and example-based dictionaries are known to be useful for learning Japanese onomatopoeia. To construct such dictionaries, we need to collect as many examples as possible. This paper proposes an online onomatopoeia example-based dictionary named *ONOMATOPEDIA*, which comprises extensive example sentences collected from the Web. Inappropriate sentences are often included in web search results, for example, sentences that contain onomatopoeia words used as nick-names, or sentences that include uncommon usage patterns. We propose a model for extracting appropriate sentences as learning examples. Further, we propose a clustering algorithm for sentences having onomatopoeia that takes into account onomatopoeic words that could be used in different meanings depending on the context.

1   Introduction

Onomatopoeia is a word or a grouping of words that expresses sounds, action or status directly, such as "click" and "bowbow." Japanese is filled with onomatopoeic phrases and has more than other languages; and the phrases are widely used in news headlines, in conversation or in Manga (Japanese comic books), because they succinctly describe things perfectly. There are two categories: "giongo" and "gitaigo." Giongo are words that express voice or sounds. Gitaigo are words that express actions, states or human emotions.

Learners of Japanese language must master onomatopoeia to make their Japanese more descriptive and expressive. However, it is hard to master onomatopoeia use, even for advanced-level Japanese language learners. There are several reasons.

One is that Japanese has many "gitaigo," which express status or human emotions more than other languages. For instance, Japanese "barabara" is used to reflect an object's state of disarray or separation, and "shiiin" is the onomatopoeia form of absolute silence. Another reason is that most onomatopoeia are rich in meaning, depending on the context in which the phrase is used.

Y. Zhang et al. (Eds.): APWeb 2008, LNCS 4976, pp. 601–612, 2008.

An effective way to master onomatopoeia is to read many sentences that contain onomatopoeia.

We are, therefore, developing an online onomatopoeia example-based dictionary named *ONOMATOPEDIA*, which has extensive example sentences collected from the Web. This system targets advanced-level learners who can already communicate in Japanese, so the example sentences are written only in Japanese. Example sentences are collected by a search engine API using onomatopoeic phrases as the search keyword.

In this paper, we describe two important techniques for generating a good-quality onomatopoeia example-based dictionary: they are *collecting* appropriate sentences from the Web, and *organizing* them by onomatopoeic meaning.

Inappropriate sentences tend to be collected from search engines using onomatopoeic phrases. We introduce a model to collect and extract appropriate sentences as examples efficiently, using onomatopoeia's grammatical characteristics. That is, onomatopoeia's role is determined by an addition to the end of an onomatopoeic phrase.

Onomatopoeia have different meanings depending on the context in which they are used. We investigate a clustering method for collected sentences by meaning, based on a document vector model. Because sentences are collected on the Web, sentences exhibit varying qualities. To overcome this problem, we propose an incremental approach to clustering sentences.

2   Online Onomatopoeia Example-Based Dictionary *ONOMATOPEDIA*

We are developing an online-onomatopoeia example-based dictionary named *ONOMATOPEDIA*, which presents many examples of onomatopoeia for learners of Japanese. Because this system targets only advanced-level learners, the

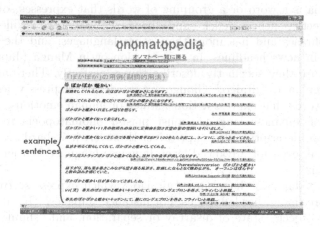

Fig. 1. Screen Image of ONOMATOPEDIA

sentences are written only in Japanese; they are not translated into other languages.

Sentences are collected from the Web using a web search engine, so we can expect them to contain practical and/or living Japanese sentences. Fig.1 shows example sentences for "pokapoka," which means "being nice and warm" or the "sound made by patting someone's head." In the current version, sentences are sorted by verb or noun, which is qualified by the onomatopoeia. A user can view many sentence examples, and if the user wants to see how the sentences are used on the web site, the user can also see the peripheral sentences by clicking on "show peripheral sentences," or the user can follow a link to the web page in which the sentence is included.

Fig.2 shows a workflow for collecting example sentences from the Web[2]. In advance, we prepare onomatopoeic phrases in the onomatopoeia database (Fig.2(1)). The process comprises the following three steps:

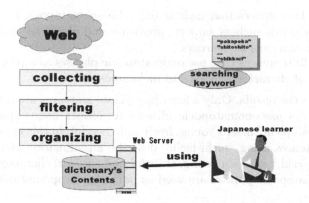

Fig. 2. Process Flow

- **Collection:** the system collects sentences by searching Yahoo API. It extracts sentences that contain the onomatopoeic phrase from each page. In this step, to collect appropriate sentences efficiently, web pages are retrieved by onomatopoeic phrases, and adjunct words such as "to," "na" or "da" are attached to the end of the phrase. For details, see the description of attaching adjunct word in section 3.
- **Filtering:** Sentences collected in the previous step contain inappropriate sentences as an example. In this step, the system selects appropriate sentences. The selection algorithm uses the analysis of sentences' dependency structures extracted by CaboCha[9], which is a dependency structure analyzer for Japanese sentences. See Section 3 for more details.
- **Organization:** As already stated, an onomatopoeic phrase has several meanings according to the context. To present sentences so that learners can understand how to use onomatopoeic phrases, we prepare several methods for grouping sentences. As shown in Fig.1, we first sort by verb or noun, which is qualified by the onomatopoeia. Beyond that, we generate document vectors

for sentences and their peripheral sentences, and we apply several clustering algorithms to these vectors to find the appropriate clustering method. Section 4 describes the investigation in more detail.

3   Sentence Collection from the Web

Sentences are collected from the search engine using onomatopoeic phrases as the keyword, but not all sentences are good examples for learners. An onomatopoeic phrase is generally used as an adverb, which qualifies a verb. Sometimes, however, onomatopoeic phrases are used as a product name or nickname of a person or pet, which means onomatopoeia can express impressions and intuition without a long description. As a preparatory experiment, we first collect sentences by search engine using 10 onomatopoeic phrases as keywords. We then analyze whether these sentences are appropriate, for example. We suppose that the following sentences are inappropriate.

- **pattern 1:** sentences that include only the onomatopoeic phrase and omit important words such as subject, predicate and object to understand what the sentence means for learners.
- **pattern 2:** sentences that use onomatopoeic phrases as a proper noun such as names of characters, goods and nicknames.

Fig.3 shows the results. Only a few appropriate sentences are collected. Many of the sentences use onomatopoeic phrases as proper nouns (pattern 2). For example, "pichi-pichi" means young, fresh and cheerful. In this experiment, most collected sentences using "pichi-pichi" describe an animation character named "pichi-pichi picchi," who is a young, cheerful and cute girl character. In this kind of case, onomatopoeic phrases are used as part of a compound noun.

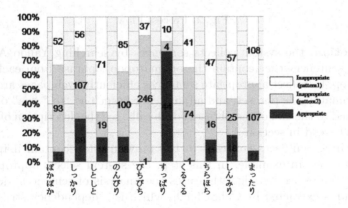

Fig. 3. Result of Preparatory Experiment

Many sentences grouped in pattern 1 are also found. These sentences use onomatopoeic phrases as headlines for blogs or columns. For example, a diary that describes a warm and peaceful holiday is entitled "poka-poka."

3.1   Extracting Appropriate Sentences for Examples

From the preparatory experiments, we found that inappropriate sentences have the following two features: (1) Nouns or verbs that are qualified by an onomatopoeic phrase are omitted in the sentence (2) onomatopoeic phrases are used as part of a compound noun.

To collect appropriate sentences, we introduced the following rules, according to grammatical characteristics for onomatopoeia.

- **rule I : a rule for search keyword**

  In general, onomatopoeia is mainly adverbial, which means it qualifies verbs. But it also takes on different classifications such as attribute, adjective verb and verb. The onomatopoeia part of speech depends on an adjunct word such as "to," "na," "no," "suru" and "da" (Table 1). Sentences that include an onomatopoeic phrase and the adjunct word become appropriate as examples because they do not tend to omit words qualified by the onomatopoeia.

  From the feature, we collect sentences using a pair of onomatopoeia and adjunct word as the keyword.

  However, it is also common for a sentence to omit an adjunct word. For example, "kare ha garigari ıto benkyou shita," which means he studied very hard, can omit the adjunct word "to" and this pattern is used as many times as the pattern using the adjunct word. We also collect sentences by the search keyword without adjunct words.

- **rule II: a rule for removing sentences**

  The rule is used in the filtering process. We regard sentences as inappropriate and remove them if they meet the two following conditions:

  - There are no additional particles after the onomatopoeic phrase and the onomatopoeia qualifies a noun. In this case the onomatopoeic phrase is used as part of a compound noun. For instance, "pichi-pichi picchi" described above. Sentences of this pattern are found in article titles.

  - A sentence omits a word the onomatopoeic phrase qualifies. Japanese sentences sometimes omits a word the onomatopeic phrase qualifies in the case that the sentences isn't lost mean in the flow of conversation even supposing that sentences omits the word. For instance, "ashita mo sikkari ne." which mean "Do (something) accurately tomorrow.". This sentence, although persons who talk with know means , the other person can't understand the mean. So the sentences with this pattarn are not available.

We experimented to verify effectiveness of the two rules introduced. Fig.4 shows the result. With the introduction of rule I, appropriate sentences for examples can be collected effectively. Beyond that, sentences that use an onomatopoeia phrase as part of a compound noun are rarely collected, because compound nouns don't include additional particles between the onomatopoeic phrase and the noun.

Beyond that, with the introduction of rule II, sentences in which important words are omitted are almost always removed. However, some inappropriate sentences remain. This happens because we automatically extract sentences from

Table 1. Relationship between the part of speech and the adjunct of onomatopoeia

adjunct	part of speech of onomatopoeia	example
to	adverb	"kamiga sarasarato yureru." (The hair is flowing in the wind)
na or no	attribute	"sarasarana kami." (Silky hair)
da	adjective verb	"kamiga sarasara da." (The hair is silky.)
suru	verb ot adjective	kamiga sarasara suru. (The hair is silky.)

web pages, followed by applying automatic dependency structure analysis. In general, elimination and dependency analysis of Japanese sentences is very difficult. Erroneous tasks take place and results are not always reliable.

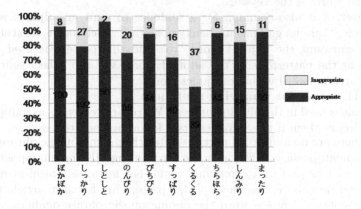

Fig. 4. The rate of Appropriate Sentences in Collected Sentences from the Web

3.2 CaboCha

CaboCha is a Japanese dependency analysis machine based on Support Vector Machines. If it receives a sentences , it shows the part of speech and qualifing of each word in the sentences. Fig.5 shows the result that CaboCha receives the sentences "ame ga shitoshito hutteiru," which means "It's been drizzling." For each word , there are imformation of part of speech and qualifing. For instance, 'shitoshito' whicih is onomatopoeia word are used as part of a adverb and it's qualifing is showed by '2D', '2D' means that 'shitoshito' qualifies second word 'hutteiru'.

4 Clustering by Onomatopoeia's Meaning

As described above, an onomatopoeia's part of speech is decided by that of the following word. In addition, the meaning of onomatopoeia may vary depending

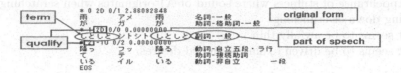

Fig. 5. Image of CaboCha

on the context. The order of example sentences is important for understanding how to use onomatopoeia.

In the current version, we classified sentences into three groups according to onomatopoeias' part of speech: adverb, adjective, and adjective verb. It is important to notice that there is an order among the groups because examples in the adverb group are most important in the sense that they are frequently used. In each group, sentences are also classified according to the word the onomatopoeia qualifies. These groups are sorted in the order of the number of sentences. Consequently, sentences are ordered by popularity.

Fig.6 shows part of a sentence list of onomatopoeia "garigari." In Fig.6(a), the first group shows usage examples as an adverb, and the second group shows usage as an adjective.

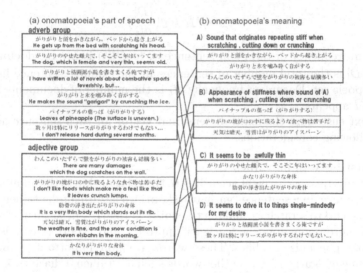

Fig. 6. Two Patterns for Assorting Sentences

This section proposes an additional classification method, taking into account that onomatopoeia usually has several meanings, depending on the context.

For example, onomatopoeia "garigari" has the following four meanings according to an onomatopoeia-dictionary[7] .

A) Sound that originates the repetition of something stiff scratching, cutting down or crunching.

B) Appearance of stiffness where sound of A) originates when scratching, cutting down or crunching.

C) It seems to be awfully thin.

D) It seems to be driven to single-minded desires.

Fig.6(b) shows the group classified manually according to the above-mentioned meanings. These classification methods make the onomatopoeia's meaning understandable.

In the following section, we investigate a clustering method based on the meaning of the onomatopoeia.

4.1   Weight Specification for Making a Sentence Vector

To classify sentences, we use the vector space model, which is a well known concept in the area of information retrieval and each weight is specified by the TF/IDF weight model.

The weight vector V_s for a sentence s is defined

$$V_s = w_{1,s}, ..., w_{N,s}$$

where

$$w_{t,d} = tf_t, log(\frac{|D|}{df_t})$$

tf_t is term frequency of term t in sentence s, $log(\frac{|D|}{df_t})$ is inverse document frequency. $|D|$ is the total number of sentences in the example sentence list for an onomatopoeia set. Document frequency df_t is the number of documents in which the term t appears.

We next customize the multi-dimensional vector for each sentence according to the following considerations.

- **Adjusting weight values**
  We then adjust each weight according to the relationship to the onomatopoeia. For instance, in a sentence "garigarito koori wo kezurukotoha taihen da." ("It is hard to scratch ice making the noise 'garigari.'"), "kezuru"("scratch") is the word onomatopoeia "garigari" qualifies, and "koori"("ice") is one the onomatopoeia also qualifies. These words are more important to specify the meaning of "garigari" than other words such as "taihen"("hard task"). So we give greater weight to the words "kezuru" and "koori" than to other words. How to know the word qualifing the specific word is to use CaboCha described above.

- **Adding peripheral sentences**
  Sometimes a sentence is not long enough to explain the context within the sentence itself. For example, the sentence "Kyou mo garigari shiteita" ("It scratches noisily today.") lacks important information: who scratches, what is scratched, and so on. In general, the information appears in peripheral sentences. Peripheral sentences of the above sentence are as follows: "Nekoga

ieno kabewo hikkaiteite komaru. Kyou mo garigari shiteita. Kabeni hikkakik-
izuga takusan dekita." ("I am embarrassed that my cat is always scratching
the wall in the house, making the sound 'garigari.' Today again, the sound
'garigari' was heard. The result is a wall with many scratches.") Several peripheral sentences are also used for the sentence vector. The word weights in
the peripheral sentences lessen as distance from the main sentence increases.
In the current version, words of sentences that are before and behind example
sentences is added a half the weight of words in the example sentence.
- **Compression of sentence vector**
  Sentence vectors that generate the above method are very sparse because
  a sentence has about 10 to 15 words while the dimension of the term can
  be several hundred. We compress sentence vectors using Latent Semantic
  Indexing (LSI),

4.2 Clustering Methods for Sentence Vector

To cluster sentences according to an onomatopoeia-dictionary as described in
section 4, we should consider sentence quality. Because we collect sentences from
the web, every sentence is not necessarily grammatical and semantically correct.
The quality of some sentences is not particularly high, especially if sentences are
collected from blogs or from pages containing many pictures.

We propose an incremental approach to clustering sentences. The clustering
method consists of four steps: sentence selection, clustering, weight re-calculation
and sentence re-selection. The re-selection process continues the clustering step
again and the process flow is repeated until all sentences are clustered.

1. **sentence selection:** We select sentences that meet the following conditions
   as good-quality sentences for the first clustering process.
   - In the sentence, onomatopoeia is used as an adverb (that is, it qualifies
     a verb term)
   - The verb terms the onomatopoeia qualifies are frequently used in col-
     lected sentences. We sort verb terms by order of frequency and regard
     the top 10 verb terms as important.
2. **clustering:** Sentences selected in the previous step are clustered. We apply
   k-means clustering in the current version.
3. **weight re-calculation:** It calculates the degree of importance of the term
   imp_t using the following formula.

$$imp_t = \alpha \sigma_{CI_t} = \alpha \sqrt{\frac{1}{n}(\Sigma_{i=1}^n CI_{i,t}^2 - \frac{1}{n}(\Sigma_{i=1}^n CI_{i,t})^2)}$$

where $CI_t = (CI_{0,t}, ..., C_{n,t})$ and α is a parameter. $CI_{i,t}$ means the degree
of importance of the term t in cluster C_i and it is calculated by

$$CI_{i,t} = \frac{cf_{C_i,t}}{|C_i|}$$

where C_i is i-th cluster, $cf_{C_i,t}$ is cluster frequency of term t in cluster C_i. imp_t shows the variance of $CI_{i,t}$. If sentences which includes term t are clustered a particular cluster, the value of imp_t becomes high and term t is a characteristic word for clustering. From the degree of importance imp_t, it re-calculate weight of each term w_t by the formula.

$$w_t = imp_f t f_t idf_t$$

4. **sentence re-selection:** Based on the weight w_t and importance imp_t, it re-selects sentences to cluster again. It first selects ten terms whose importance value imp_t is high except for the previously selected terms. It next selects sentences that include one of the selected terms with high tf_t values and the ones used in the previous clustering. Re-selected sentences are clustered and the process is repeated. Clustered sentences are incrementally added. By clustering these processes, each sentence is classified into appropriate groups.

5   Experiment

We investigate our proposed clustering method. We cluster 1342 sentences for an onomatopoeic phrases "gari-gari". In these sentences, 814 sentences are examples as adverb, the others are the one as adjective. Vectors of sentences are 2056 dimensions.

We use k-means algorithm for clustering, and a number of cluster K is 10. Although in the session 4 we argue that there are four meanings according to a dictionary, but a meaning group can be subtilized. Then we subtilize first, and then merge them.

We divide meaning groups of "gari-gari" into 7 types as follows manually.

(A) Sound of originate when scratching, cutting down or crunching.
  (A-1) The sound made by rasping and scratching with blemish.
  (A-2) The sound made by scratching softly their body by themselves.
  (A-3) The sound made by eating hard food.
  (A-4) The sound made by doing something but it isn't categorized to neither A-1, A-2 nor A-3.
(B) Appearance of stiffness where sound of originate when scratching, cutting down or crunching.
(C) It seems to awfully thin.
(D) It seems to driven to single-minded desires.

First we cluster all sentences by k-means at once. The results that only 4 small clusters (80 sentences) can collect similar sentences. Remains of cluster has different types of sentences, that is, this clustering method doesn't success.

Next, we choose sentences as described in the previous section. Fig.7(a) shows the result of clustering using only good-quality sentences by *sentence selection* in the clustering flow. Almost all cluster except the cluster No.13 successfully collects similar sentences.

Based on the result, we calculate the degree of importance imp_t for term t. Fig.7(b) shows the top 10 terms of total number, and Fig.7(c) show the 10 terms of imp_t. From these tables, we can understand that "suru"(do something) is used in most number of sentences, but this term is general, and it is used in most cluster. And, the terms listed in Fig.7(c) such as "hito"(person), "yuki"(snow) are characteristic for a particular cluster. Fig.7(d) show the re-clustering result. While 4 clusters can collect similar sentences by applying k-means clustering at once, almost cluster can gather similar sentences by clustering after weight re-calculation.

(a)

クラスタ	A1	A2	A3	A4	B	C	D
1	0	0	0	1	0	0	17
2	0	0	0	0	0	43	0
3	0	0	9	0	0	2	0
4	17	4	5	0	2	4	3
5	0	0	0	0	0	0	30
6	0	28	0	0	0	0	1
7	0	0	0	0	21	0	0
8	0	0	0	0	0	62	0
9	0	0	0	0	0	0	9
10	0	0	11	0	0	0	0
11	0	0	0	0	14	0	0
12	0	0	0	6	0	2	1
13	48	9	37	6	20	42	22
14	0	0	0	0	0	10	0
15	0	17	0	0	0	0	0
16	22	0	0	0	0	0	0
17	2	0	2	0	0	17	0
18	0	8	0	0	0	0	0
19	0	0	0	0	0	4	0
20	35	0	0	0	2	0	0

(b)

word	sum (e.count)
する (suru)	142
痩せる (yaseru)	83
削る (kezuru)	68
頭 (atama)	61
なる (naru)	50
やせる (yaseru)	48
描く (kaku)	46
描く (egaku)	46
凍る (koru)	35
書く (kaku)	33

(c)

word	cimp
人 (hito)	1.2039298
削る (kezuru)	1.1843337
書く (kaku)	1.1778746
痩せる (yaseru)	1.1756898
凍る (koru)	1.1719463
描く (egaku)	1.1704186
描く (kaku)	1.1662832
食べる (taberu)	1.1610492
頭 (atama)	1.0961364
する (suru)	1.0901629

(d)

クラスタ	A1	A2	A3	A4	B	C	D
1	0	40	0	0	0	2	0
2	0	0	0	0	0	37	0
3	1	1	0	0	1	26	1
4	0	0	0	0	0	0	10
5	3	0	20	1	0	5	2
6	3	0	0	0	2	1	2
7	0	0	0	0	0	0	19
8	10	0	1	1	2	9	7
9	22	0	0	0	0	0	0
10	0	0	0	0	0	60	0
11	0	18	0	0	0	0	0
12	0	0	0	0	0	9	0
13	107	4	43	11	65	106	68
14	0	0	0	0	14	0	0
15	43	5	5	4	12	9	25
16	0	0	0	0	0	9	55
17	0	0	0	6	1	1	1
18	38	0	0	0	1	0	0
19	32	1	5	3	5	7	14
20	3	0	1	3	1	5	3

Fig. 7. Experiment Results, (a) clustering result

6   Related Work

The information extraction from the huge Web space focuses attention in recent years. Fujii, et al. are developing a cyclopedia named "cyclone"[1] which is providing a huge content of encyclopedia by extracting sentences from the Web that are considered appropriate to use as explanations of a term. And, there are also several researches about extraction of famous people's information[4] and person's nickname[6]. These researches use heuristics to extract appropriate sentences. Like them, we also apply heuristics to the example sentences using onomatopoeia. However, in the case of using heuristics, we may hope getting data with some level of accuracy, but there is no guarantee for the credibility. In our case, we may get some inappropriate sentences, so we also consider combining user participatory model like "wikipedia" to edit the sentences.

Talking about e-learning for Japanese education, there is "asunaro"[3] which is provided by International Student Center of Tokyo Institute of Technology.

Asunaro displays the polyglot translation of sentences with scientific and technical terms, grammar of Japanese sentences and explanation of a term.

And about onomatopoeia online dictionary or cyclopedia, Koubayashi, et al. made an onomatopoeia dictionary[5] whose onomatopoeia's example sentences are extracted by hands from a book and shown in three languages. Also there are some Web pages, for instance, "alc"[8] developed by space arc or page[10] developed by The National Institute for Japanese Language. But these pages' example sentences are collected by hands by those service providers, so the number of sentences is limited. Therefore, if we are able to extract appropriate sentences from Web in huge number, that will help those system's construction or management.

7   Summary and Future Work

We are developing an onomatopoeia dictionary, Onomatopedia, which automatically extracts examples of onomatopoeia from the Web, and presents them to learners online. This paper describes efficient techniques of extracting appropriate examples and clustering examples based on the onomatopoeia's meaning. Because it is very difficult to collect and organize good-quality sentences for learning automatically, after we apply the collection and organization program, we currently must check and modify sentences manually.

As future work, we need to improve the precision of filtering and organization of this system. We are planning to make a portal site for Japanese onomatopoeia based on "Onomatopedia," and will introduce a mechanism for users of the portal site to improve sentences using folksonomy techniques.

References

1. Fujii, A., Ishikawa, T.: Extraction and Organization of Encyclopedic Knowledge Information Using the World Wide Web. The Japan Society of Information and Communication Research J85-D-U(2), 300–307 (2002)
2. Chang, G., Healey, M.J., McHugh, J.A.M., Wang, J.T.L.: Mining the World Wide Web -An Information Search Approach
3. Nishina., K., Okumura, M.: Japanese Reading System of Multi-Lingual Environment "ASUNARO". The Society of Technical Japanses Education 7(1), 16–17
4. Kimura., R., Oyama., S., Tanaka, K.: Generating dictionary about people based on automatic collection of timeline information. In: DBS No.2006-DBS-140(ll), pp. 51–58 (2006)
5. Koubayashi, T.: onomatope no onrain tagengo zisho no kouchiku. DEWS2002 A4-4 (March 2002)
6. Hokama., T., Hiroyuki, K.: Mnemonic Name Extraction from the Web. DBSJLetters 5(2), 49–52 (2006)
7. Yamaguchi, N.: gionEgitaigo ziten. ISBN: 4-06-265330-3
8. SPACE ALC, http://home.alc.co.jp/
9. CaboCha, http://chasen.org/~taku/software/cabocha/
10. giongoEgitaigo - nihongo wo tanosimou! - (in Japanses), http://jweb.kokken.go.jp/gitaigo/index.html

Connectivity of the Thai Web Graph

Kulwadee Somboonviwat[1], Shinji Suzuki[2], and Masaru Kitsuregawa[2]

[1] Graduate School of Information Science and Technology,
The University of Tokyo
7-3-1 Hongo, Bunkyo-ku, Tokyo, 113-0033, Japan
[2] Institute of Industrial Science, The University of Tokyo,
4-6-1 Komaba, Meguro-ku, Tokyo 153-8505, Japan
{kulwadee,suzuki,kitsure}@tkl.iis.u-tokyo.ac.jp

Abstract. The study of a national Web graph is challenging and can provide insight into social phenomena specific to a country. However, because there is no country border in the Web, deciding whether a web page belongs to that country or not is difficult. In this paper we aim at studying the characteristics of the Thai Web graph. We first address the challenge of gathering Thailand-related web pages from the borderless Web by proposing a set of criteria for defining Thailand-related web pages. Three Thai web snapshots have been collected during July 2004 (18M web pages), January 2007 (550K web pages), and May 2007 (1.4M web pages) respectively. We then analyze and report various statistical properties related to connectivity of the associated Thai Web graphs.

1 Introduction

The World Wide Web consists of huge amount of interconnected web pages. Mathematically, the Web can be represented as a graph whose nodes correspond to web pages and whose edges correspond to hyperlinks. Study of graphical properties and characteristics of the Web graph is not only theoretically challenging but also practically useful in the development of efficient algorithms for Web applications such as web crawling [11,12], web searching [9,17], and web community discovery [18].

This paper studies graphical properties of the Thai Web. We define the Thai Web as a set of web pages related to a country, Thailand. As a subgraph of the global Web, the Thai Web graph is a graph induced from these Thailand related web pages. Study of the Web of a country (or the national Web) has been conducted at many diffrent scales by various countries e.g. African [8], China [19], Korea [16], Spain [3], and Thailand [21]. While confirming many phenomena already observed in the global Web graph (e.g. 'small world, power-law degree distribution, and bow-tie structure phenomena as reported in [1,10]), statistics of the national Web graphs also reveal many characteristics that are specific to each national Web graph. The characteristics of these local national Web graphs have emerged as a result of many individual factors peculiar to each country such as penetration of the internet, internet usage behavior, social values, education levels, language, and culture etc. Thus, the study of a national Web

Y. Zhang et al. (Eds.): APWeb 2008, LNCS 4976, pp. 613–624, 2008.

graph of a country poses not only intriguing mathematical problems but can also potentially provides insight into society, culture, and economic of a country. In addition, many link-based algorithms and web applications can also be improved by exploiting link information specific to a country. Examples of such applications include language-specific resource discovery [22,23], topic-specific resource discovery [11,12,20], and localized web search.

The first challenge in the study of the Web of a country is how to decide whether a web page belongs to that country or not. The most commonly used criteria for deciding the nationality of a web page are country-code top-level domains (ccTLDs) and physical location of the web server containing that web page as determined by an IP address of the server [3,8,21]. Based on these two criteria, a web page will be assigned as belonging to a country if (1) the domain part of its URL matches the country-code of that country e.g. '.th' for Thailand and '.jp' for Japan, or (2) the IP address of the web server containing the web page is physically assigned to a geographical location in that country.

However, according to the statistics of the Thai Web observed in [22,23], there are many Thai web servers registered under the international domain names especailly '.com' and '.net' domains, and/or physically located outside Thailand. As a result, deciding if a web page belongs to Thailand based only on the two aforementioned criteria will result in low coverage of the Thai Web and thus is inappropriate for the purpose of studying the Web of Thailand. In order to obtain a higher coverage of the Thai Web, we need to address the problem of how to effectively collect Thai web pages outside the '.th' domain name. One way to do this is to make use of a characteristics unique to Thai web pages i.e. the Thai language, an official language of Thailand (Thai language is used in the schools, the media, and government affairs in Thailand). A web page will be assigned as belonging to the Thai Web if it contains some information written in the Thai language. We propose that in order to obtain a more complete snapshot of the Thai Web, it is necessary to gather not only web pages inside '.th' domain but also those web pages written in Thai language outside '.th' domain.

In this paper, we focus our study on the connectivity of the Thai Web. Statistics about degree distribution and connected components were extracted from three Thai web snapshots. The three Thai web snapshots were crawled using the language specific web crawling method proposed in [22,23], coupled with the checking of ccTLDs and geographical locations of the IP addresses. Thai web crawls were conducted during July 2004 (18M web pages crawled), January 2007 (550K web pages crawled), and May 2007 (1.4M web pages crawled) respectively. For each dataset, we conducted experiments to measure the following properties of the Thai Web graph: (1) in-degree and out-degree distributions, (2) weakly and strongly connected components (SCC and WCC), (3) the macroscopic structure, and (4) connectivity between domains under the Thailand national domain. From the experimental results, we observed the ubiquitous of power-law in the Thai Web graph. Furthermore, anomalies (or outliers) in the log-log plots of the degree distributions were also observed and a major root of these anomalies has also been uncovered. The analysis of the connected components in the Thai Web graph reveals an asymmetric bow-tie large-scale structure of the Thai Web. And, the statistics of connectivity between domain names indicate that linkage in the Web is, to a certain extent, reflecting the relationship between organizations in the real world.

The rest of this paper is organized as follows. Section 2 reviews the related work. Section 3 gives the definition of the Thai Web and describes characteristics of the Thai Web datasets. Section 4 presents our experimental results. Section 5 concludes the paper with directions for further work.

2   Related Work

The information available on the Internet is associated together by hyperlinks between pairs of web pages, forming a large graph of hyperlinks or the Web graph. The Web graph shares various graphical properties with other kinds of complex networks e.g. a citation network ,a power grid network, etc. There are several previous works on the empirical studies of the Web graph. These studies consistently reported emerging properties of the Web graph at diffrent scales. One of the most notable emerging property is a power-law connectivity in which the number web pages having k number of connections decays polynomially as $k^{-\gamma}$, with $\gamma > 1$. These results have remarkable impact in graph theory, and the design of efficient algorithms for applications such as web crawling, web search, and web mining. A power-law connectivity were observed in various scales of the Web graphs [1,5,10]. The study results in [1] show that the distribution of links on the World Wide Web follows the power-law, with power-law exponent of 2.1 and 2.45 for the in-degree and the out-degree distribution respectively. [10] reported on the power-law connectivity of the Web graph having exponent of 2.09 and 2.72 for the in-degree and the out-degree distribution respectively.

Another emerging property of the Web graph is its bow-tie structure. [10] conducted a series of experiments including analyses on WCC, SCC, and random-start BFS using large-scale Web graphs induced from two AltaVista crawls. Based on the results of these experiments, [10] inferred and depicted macroscopic structure of the Web graph as a "bow-tie" in which more than 90% of nodes reside in the largest weakly connected component. Because there is a disconnected component in the bow-tie structure, it is clear that there are a sizable portions of the Web that cannot be reached from other portions. Interestingly, the sizes of the WCC and SCC in this study also follow a power-law distribtuion. [14] studied the bow-tie structure in subgraphs of the Web graph associated with a particular topic.

The study of a Web graph of a country has been done by several countries such as [3,8,16,19,21].

African. [8] crawled the African Web and analyzed its content, link, and interconnection between domains of countries in the African Web. The reported power-law exponent of in-degree distribution is 1.92. The macroscopic structure of the African Web consists of a single giant SCC pointing to many small SCCs.

China. [19] measured many properties and evolution of the China Web graph. The reported power-law exponent for the in-degree and out-degree distributions

of the China Web graph are 2.05 and 2.62 respectively. The bow-tie structure
of the China Web has a very large MAIN SCC component which consists of
approximately 4/5 of the total number of web pages in the China Web graph.

Korea. [16] reported the power-law distribution of the number of connectiv-
ities per node for the Korea Web, the power-law exponents for in-degree and
out-degree distributions are 2.2 and 2.8 respectively. Like the China Web graph,
The bow-tie structure of the Korea Web has a very large MAIN SCC component.

Spain. [3] comprehensively analyzed the characteristics of the Web of Spain
in terms of content, link, and technology usage at three levels i.e. web pages,
sites, and domains. The reported power-law exponent for the in-degree and out-
degree distributions are 2.11 and 2.84 respectively. [3] depicted link structure
among web sites in the Web of Spain using the extended notion of the bow-tie
structure, proposed in [2].

Thailand. [21] conducted a study of the Thai Web (i.e. the Web of Thailand).
[21] presented quantitative measurements and analyses of various properties of
web servers and web pages of Thailand. Their dataset consists of 700K web pages
downloaded from over 8,000 web servers registered under '.th' domain on March
2000. The study in [21] presents several statistics regarding to the content of the
Thai Web. There is no statistics about the structure and characteristics of the
Thai Web graph given in [21].

3   Definition of the Thai Web and the Thai Web Datasets

As stated earlier in the first section, most studies on the properties of the Web of
a country usually define the Web of a country as a set of web pages of all Web sites
that are registered under the country top-level domain or that are hosted at an
IP associated with that country. We argue that this definition is not appropriate
for defining the Web of Thailand. Based on the language identification result
of web pages in our Jul2004 dataset which was obtained by breadth-first-search
crawling in July 2004, we found that more than half of the web pages written in
the Thai language are web pages of Web sites registered outside '.th' top-level
domain of Thailand (see Table 1).

Consequently, if we crawl only web pages with the corresponding Thai top-
level domain and/or the physically assigned location of the IP address of the
Web sites then we will fail to collect a large portion of Thai-language web pages.
Therefore, to increase the completeness of the Thai Web dataset, it is necessary
to add to the definition of the Thai Web a criterion which is based on the
language of a web page. Formally, we propose to use the following criteria to
decide whether a web page is Thai.

(1) Top-level domain of the web page is '.th'.
(2) IP address of its web server is physically assigned in 'Thailand'.
(3) Language of the web page is 'Thai'.

The first criterion can be easily implemented by adding a predicate function
to check the value of the top-level domain of each URL before adding it into the

Table 1. Language identification result of the Jul2004 dataset categorized by domain (in number of pages). More than half of the Thai-language web pages belong to Web sites registered outside Thailand's country top-level domain (i.e. '.th' domain).

Languages	'.th' domain	other domains	Total
Thai	591,683	1,131,088	1,722,771
non-Thai	263,777	16,357,579	16,621,356
Total	855,460	17,488,667	18,344,127

Table 2. Number of vertices and directed edges of the Thai Web Graphs

	Jul2004	Jan2007	May2007
number of vertices	39,078,795	5,785,349	18,864,382
number of directed edges	123M	12M	70M

URL queue of a crawler. For the second criterion, we need to check a geographical location of an IP address of each web server. The third criterion states that a web page should be included into the dataset if it is written in Thai regardless of its top-level domain. We achieved this by using a language-specific web crawling method proposed in [22,23].

In this study, we conducted experiments on three snapshots of Thai web crawls. The first snapshot (Jul2004 dataset) was crawled in July 2004. For the first snapshot, we used a naive BFS web crawling strategy to get a sample of the Thai Web. The second and the third snapshots (Jan2007 and May2007) were crawled in January, and May 2007. For these two datasets, we have implemented the three criteria as described earlier by applying a language-specific web crawling method [22,23] in our Thai web crawling. The start seed sets for all datasets consists of a number of popular websites and web portals in Thailand. The number of crawled web pages for Jul2004, Jan2007 and May2007 datasets are 18,344,127 pages, 551,233 pages and 1,402,206 pages respectively.

4   Connectivity of the Thai Web graph

For each Thai web dataset, we have constructed a link database which provides access to inlink and outlink information of a web page corresponding to an input URL address. The number of vertices and directed edges of the Thai Web graphs induced from Thai web datasets are as shown in Table 2. In the following subsections, we will present various statistical results and discuss about the characteristics of the connectivity of the Thai Web graph.

4.1   Indegree and Outdegree Distributions

The distribution of the number of connectivity per node or degree distribution of many subgraphs of the Web has been consistently reported to follow a power-law distribution. The power-law distribution has been described by the "rich

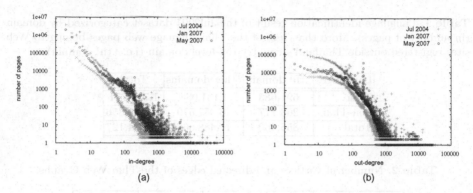

Fig. 1. Indegree and Outdegree distribution of the Thai Webgraphs

gets richer" phenomenon (or preferential attachment) where new links are more likely to point to web pages that already have many links pointing to them. We have plotted the degree distribution of the Thai Webgraphs. The in-degree distribution is shown in Fig. 1(a), and the out-degree distribution is shown in Fig. 2(b).

As can be seen from Fig. 1(a), in-degree distributions of all Thai Webgraphs in the log-log plots can be approximated by a straight line, a signature of the power-law distribution. After examining the web pages with large number of inlinks, we found that those web pages are homepages of popular Thai Web sites providing services such as free online-diary, blogs, and online communities. However, we also observed spam pages with very high number of inlinks. In the case of out-degree distributions in Fig. 1(b), all log-log plots show approximately straight lines with concave in the first portion. By examining our crawled data, we found that most web pages with tremendously large number of outlinks are web pages from pornographic and spam sites. Note that, we observe anomalous bumps in both in-degree and out-degree distributions of the log-log plots in Fig. 1. Manual inspection reveals that most of the web pages corresponding to these anomalies are spam pages.

In Fig. 2, we have separately plotted the degree distributions of total, internal-only, and remote-only inlinks and outlinks. An internal link is a hyperlink between web pages within the same website. Conversely, a remote link is a hyperlink between web pages residing in different websites. According to Fig. 2, degree distributions of remote links are better fit with the power-law and contain a little number of anomalous bumps. This demonstrated that the anomalous bumps found in the degree distributions are largely caused by the internal links. Another point that is worth mentioning is the absence of concavity part of the out-degree distribution in the plot of remote-only links. Obviously, the concavity in the out-degree distributions is caused by the characteristics of internal linkage. Consequently, while the process of hyperlinking between web sites can be described by the "rich get richer" model, another different model or a modified version of "rich get richer" model is needed for explaining the phenomenon found in linkage between web pages within the same web sites.

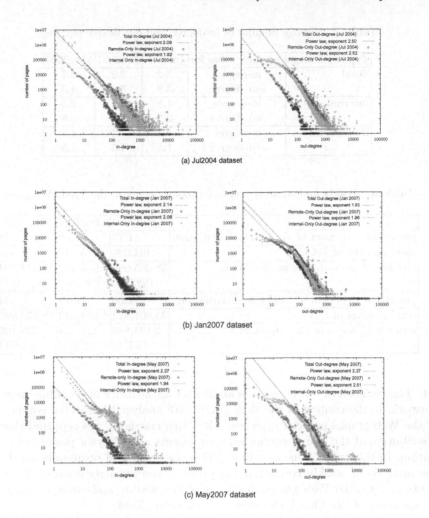

(a) Jul2004 dataset

(b) Jan2007 dataset

(c) May2007 dataset

Fig. 2. Internal-only vs. remote-only degree distribution

Table 3 shows average number of inlinks and outlinks for all datasets. According to Table 3, the average number of connectivity per node is 2–4 connections per node. So, the Thai Web graph is a sparse graph with some densely connected regions which may corresponding to homepages of some popular web sites, spam pages, or web pages with pornographic content.

4.2   Weakly and Strongly Connected Components

In graph theory, a weakly connected component (WCC) is a set of nodes such that all nodes can be reached by all other nodes in the set by traversing a set of undirected links. A strongly connected component (SCC) is a set of nodes in a directed graph such that all nodes can be reached by all other nodes in the

Table 3. Average number of in-degree and out-degree per page

Type of linkage	Average value for	Jul2004	Jan2007	May2007
Total	in-degree	3.2	2.2	3.7
	out-degree	3.8	3.8	4.3
Internal-only	in-degree	1.8	1.1	2.5
	out-degree	2.3	2.3	2.8
Remote-only	in-degree	1.4	1.1	1.2
	out-degree	1.5	1.5	1.5

Table 4. Weakly and strongly connected components in the Thai Web graph

	Jul2004	Jan2007	May2007
power-law exponent of WCC size distribution	1.73	1.60	1.48
total number of WCC	404,086	75,215	251,638
number of nodes in the largest WCC	38,055,065 (98.5%)	5,665,462 (98.0%)	18,314,704 (97.0%)
power-law exponent of SCC size distribution	2.00	2.12	1.34
total number of SCC	35,801,812	5,624,313	18,134,695
number of nodes in the largest SCC	2,193,836 (5.7%)	121,422 (2.1%)	539,195 (3.0%)

set. The connectivity of a graph is an important measure of its robustness as a network. In the context of the Web graph, [10] analyzed connected components of the Web graphs induced from two AltaVista crawls. It was reported that the distribution of the size of connected components also obeys a power-law distribution. In this subsection, we will describe the statistical results obtained from the analysis of weakly connected components and strongly connected components in the Thai Web graph. Statistics of the weakly and strongly connected components of the Thai Web graphs are shown in Table 4.

According to Table 4, the largest WCC components for Jul2004, Jan2007, and May2007 datasets comprise of 98.5%, 98%, and 97% of the total number of nodes respectively. Meanwhile, the total number of WCC components in each graph are 404,086 (Jul2004), 75,215 (Jan2007), and 251,638 (May2007). The topology of the Thai Web graph can be roughly seen as consisting of a single large giant connected component and several much smaller disconnected components. As a result, it can be implied that there are substantial portions of the Web that are not reachable from the other portions of the Web. The largest SCC components for Jul2004, Jan2007, and May2007 datasets comprise of 5.7%, 2.1%, and 3.0% of the total number of nodes respectively. Our largest SCCs are very small compared to the largest SCC observed in the global Web [10] ([10] reported the value of 28%). This is due to the effects of uncrawled nodes included in our link databases. We have tried running the WCC and SCC algorithms again on Jan2007 dataset, this time we consider only those nodes which are corresponding to crawled web pages. There are 551,233 nodes in Jan2007 dataset that are

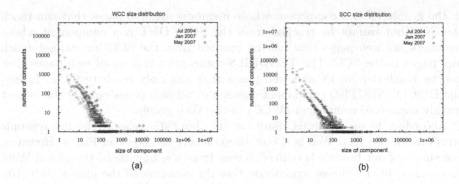

Fig. 3. Distribution of the size of connected components

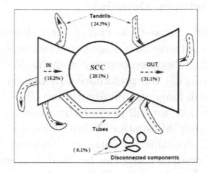

Fig. 4. Asymmetric bow-tie structure of the Thai Web graph (Jan2007 dataset)

corresponding to crawled web pages. The Web graph induced from crawled web pages in Jan2007 dataset has the largest WCC containing (517,934 nodes [94.0% of total number of nodes]) and the largest SCC containing (110,787 nodes [20.1% of total number of nodes]). This result is more consistent with what has been observed in the global Web graph. The distributions of the size of WCC and SCC of the Thai Web graphs also adhere to the power-law. The values of the power-law exponents are given in Table 4.

4.3   Macroscopic Structure of the Thai Web

In the following experiment, we would like to discover the macroscopic structure of the Thai Web by using the Web graph associated with crawled web pages of Jan2007 dataset. [10] conducted a series of SCC, WCC, and random-start BFS experiments on a large Web graph derived from AltaVista web crawls. Based on the interpretation of the experimental results, they depicted the structure of the Web as a bow-tie like structure consisting of five components: SCC, IN, OUT, TENDRILS, and DISCONNECTED. SCC is a large core component of the bow-tie. It consists of web pages in the largest strongly connected component in the graph. IN is a component whose members are web pages that can reach the SCC but cannot be reached from the SCC. OUT is a component whose members are web pages that can be reached from the SCC but cannot reach any pages in the SCC. The TENDRILS component is a set of web pages that can be reached from IN and web pages that can only reach to OUT. Lastly, the DISCONNECTED component consists of all web pages outside the largest weakly connected component (WCC) in the Web graph.

According to our analysis result on the Jan2007 dataset, the macroscopic structure of the Thai Web is a bow-tie structure as shown in Fig. 4. However, the shape of our bow-tie is quite different from the bow-tie of the global Web. In contrast to the almost symmetric bow-tie structure of the global Web [10], the bow-tie of the Thai Web is asymmetric with large OUT component.

4.4   Connectivity Between Domains

[7] studied linkage affinity between several country domains. The result of this study shows interesting patterns of linkage between country domains. For example, (1) asymmetrc linkage between country domains (e.g. links between China, Hong Kong and Taiwan), (2) the preference of language over geographic location (e.g. a strong English language affinity among US, UK, Australia, and New Zealand). In this subsection, we will present the domain linkage statistics of the Thai Web by using the link information of Jan2007 dataset. First let us consider the linkage within '.th' domains (or intra '.th' domain links). A second-level domain of a registered domain name can be used to identify types of the organizations who own the domain name. The second level domains under the '.th' domain include 'ac.th' (academic), 'co.th' (commercial), 'go.th' (governmental), 'in.th' (individuals), 'mi.th' (military), 'net.th' (internet provider), and 'or.th' (non-profit organizations).

Table 5. Links between second-level domains under '.th' domain (Jan2007 dataset)

	ac.th	co.th	go.th	in.th	mi.th	net.th	or.th
ac.th	**12366**	2513	3604	234	49	243	1977
	58.9%	12.0%	17.2%	1.1%	0.2%	1.2%	9.4%
co.th	208	**3887**	698	151	3	31	496
	3.8%	**71.0%**	12.8%	2.8%	0.1%	0.6%	9.1%
go.th	1689	2098	**10322**	725	97	122	1757
	10.0%	12.5%	**61.4%**	4.3%	0.6%	0.7%	10.5%
in.th	557	**15990**	1367	5642	18	39	514
	2.3%	**66.3%**	5.7%	23.4%	0.1%	0.2%	2.1%
mi.th	112	131	**547**	5	475	20	181
	7.6%	8.9%	**37.2%**	0.3%	32.3%	1.4%	12.3%
net.th	25	27	**92**	4	3	60	41
	9.9%	10.7%	**36.5%**	1.6%	1.2%	23.8%	16.3%
or.th	353	425	818	44	7	35	**1239**
	12.1%	14.5%	28.0%	1.5%	0.2%	1.2%	**42.4%**

In Table 5, we list the number of links between all pair of source and destination second-level domains. A domain in each row represents the source domain, and a domain in each column in Table 5 represents the destination domain. For each row, the column with the largest number of links is written in boldface. According to the results in Table 5, It can be seen that in almost every case the number of links within the same domain is larger than links to different domains. And there is a strong relationship between governmental, military, and non-profit organizations. Note that, according to analysis results on Jan2007 dataset, most links going out of '.th' domain is pointing to '.com' and '.net' domains. Symmetrically, most links going into the '.th' domain is coming from '.com' and '.net' domains.

5 Conclusion

In this paper we have addressed the challenge of collecting Thailand-related web pages by defining a set of criteria for deciding whether a web page belongs to the Thai Web or not. Based on three Thai web snapshots, we have observed many interesting characteristics of the Thai Web graph. First, although three datasets used in our experiments are all different in scale and acquisition time, statistics derived from each dataset show similar trends and phenomena. Second, we have identified the internal links as one of the possible causes of anomalies bumps frequently found in the log-log plot of in-degree and out-degree distributions. Third, our analysis result reveal the bow-tie structure of the Thai Web as an asymmetric bow-tie. Lastly, linkage between domains reflect relationships between organizations in the real-world. For the future work, we would like to (1) study the structure of the Thai Web graph in more detail, and (2) study the evolution of the Thai Web graph.

References

1. Albert, R., Jeong, H., Barabasi, A.: The diameter of the world wide web. Nature 401, 130 (1999)
2. Baeza-Yates, R., Castillo, C.: Relating web characteristics with link based web page ranking. In: Proc. of the 8th Int'l Symposium on String Processing and Information Retrieval (SPIRE 2001), pp. 21–32 (2001)
3. Baeza-Yates, R., Castillo, C., Lopez, V.: Characteristics of the web of spain. International Journal of Scientometrics, Informetrics and Bibliometrics 9(1) (2005)
4. Baldi, P., Frasconi, P., Smyth, P.: Modeling the Internet and the Web: Probabilistic Methods and Algorithms. John Wiley & Sons, Ltd., Chichester (2003)
5. Barabsi, A., Albert, R.: Emergence of scaling in random networks. Science 286(5439), 509–512 (1999)
6. Barabsi, A., Albert, R., Jeong, H., Bianconi, G.: Power-law distribution of the world wide web. Science 287(5461), 2115 (2000)
7. Bharat, K., Chang, B.-W., Henzinger, M.R., Ruhl, M.: Who links to whom: Mining linkage between web sites. In: Proc. of the 2001 IEEE Int'l Conf. on Data Mining (ICDM 2001), pp. 51–58 (2001)

8. Boldi, P., Codenotti, B., Santini, M., Vigna, S.: Structural properties of the african web. In: Poster Proc. of the 11th Int'l Conf. on World Wide Web (WWW 2002) (2002)
9. Brin, S., Page, L.: The anatomy of a large-scale hypertextual web search engine. In: Proc. of the 7th Int'l Conf. on World Wide Web (WWW 1998), pp. 107–117 (1998)
10. Broder, A.Z., Kumar, R., Maghoul, F., Raghavan, P., Rajagopalan, S., Stata, R., Tomkins, A., Wiener, J.L.: Graph structure in the web. Computer Networks 33(1–6), 309–320 (2000)
11. Chakrabarti, S., van den Berg, M., Dom, B.: Focused crawling: a new approach to topic-specific web resource discovery. In: Proc. of the 8th Int'l Conf. on World Wide Web (WWW 1999), pp. 1623–1640 (1999)
12. Cho, J., Garcia-Molina, H., Page, L.: Efficient crawling through url ordering. In: Proc. of the 7th Int'l Conf. on World Wide Web (WWW 1998), pp. 161–172 (1998)
13. Davison, B.D.: Topical locality in the web. In: Proc. 23rd Intl. ACM SIGIR Conf. on Research and Development in Information Retrieval (SIGIR 2000), pp. 272–279 (2000)
14. Dill, S., Kumar, S.R., McCurley, K.S., Rajagopalan, S., Sivakumar, D., Tomkins, A.: Self-similarity in the web. In: Proc. of 27th Int'l Conf. on Very Large Data Bases (VLDB 2001), pp. 69–78 (2001)
15. Fetterly, D., Manasse, M., Najork, M.: Spam, damn spam, and statistics: Using statistical analysis to locate spam web pages. In: Proc. of the 7th Int'l Workshop on the Web and Databases (WebDB 2004), pp. 1–6 (2004)
16. Han, I.K., Lee, S.H., Lee, S.: Graph structure of the korea web. In: Kotagiri, R., Radha Krishna, P., Mohania, M., Nantajeewarawat, E. (eds.) DASFAA 2007. LNCS, vol. 4443, pp. 930–935. Springer, Heidelberg (2007)
17. Kleinberg, J.M.: Authoritative sources in a hyperlinked environment. J. ACM 46(5), 604–632 (1999)
18. Kumar, R., Raghavan, P., Rajagopalan, S., Tomkins, A.: Trawling the web for emerging cyber-communities. In: Proc. of the 8th Int'l Conf. on World Wide Web (WWW 1999), pp. 1481–1493 (1999)
19. Liu, G., Yu, Y., Han, J., Xue, G.-R.: China web graph measurements and evolution. In: Zhang, Y., Tanaka, K., Yu, J.X., Wang, S., Li, M. (eds.) APWeb 2005. LNCS, vol. 3399, pp. 668–679. Springer, Heidelberg (2005)
20. Menczer, F., Pant, G., Srinivasan, P.: Topical web crawlers: Evaluating adaptive algorithms. ACM Trans. Inter. Tech. 4(4), 378–419 (2004)
21. Sanguanpong, S., Piamsa-nga, P., Poovarawan, Y., Warangrit, S.: Measuring and analysis of the thai world wide web. In: Proc. of the Asia Pacific Advance Network conference, pp. 225–330 (2000)
22. Somboonviwat, K., Tamura, T., Kitsuregawa, M.: Finding thai web pages in foreign web spaces. In: ICDE Workshops, p. 135 (2006)
23. Tamura, T., Somboonviwat, K., Kitsuregawa, M.: A method for language-specific web crawling and its evaluation. Systems and Computers in Japan 38(2), 10–20 (2007)
24. Thelwall, M., Wilkinson, D.: Graph structure in three national academic webs: power laws with anomalies. J. Am. Soc. Inf. Sci. Technol. 54(8), 706–712 (2003)

On the Trustworthiness and Transparency of a Web Search Site Examined Using "Gender-equal" as a Search Keyword

Naoko Oyama[1] and Yoshifumi Masunaga[2]

[1] Institute for Gender Studies, Ochanomizu University, 2-1-1 Otsuka, Bunkyo-ku, Tokyo, 112-8610 Japan
oyama.naoko@ocha.ac.jp
[2] Research Center for Information Science, Aoyama Gakuin University, 5-10-1 Fuchinobe, Sagamihara-shi, Kanagawa, 229-8558 Japan
masunaga@irc.aoyama.ac.jp

Abstract. It has been believed that Web search sites are organic, i.e., algorithmic, fair, and unbiased. However, in recent years, many people are of the opinion that there might be a problem with the trustworthiness of the search site Google; but, no studies to date have provided any evidence on this issue. This paper investigates the trustworthiness of the SERP (search engine result page) of Google in response to a search keyword "gender-equal" in Japanese. Although there were Web pages on Google's SERP in response to this search keyword, it was difficult to explain why the web pages pertaining to the gender-bashing sects were ranked consistently higher than those pertaining to the gender advocates. This observation was made when a SERP of Google was compared with the SERPs returned by its six partner search sites such as Nifty, Livedoor, Excite, Biglobe, Infoseek楽天, and goo as well as Yahoo!Japan. Further, it was also revealed that explaining the ranks of such Web pages in terms of both PageRank™ and the number of backlinks was difficult. In order to restore the trustworthiness of Web search sites, it is strongly recommended for Google to be aware of the "social responsibility" to achieve a comprehensive level of transparency in the SERP strategy.

1 Introduction

Web search sites comprise search engines and strategies of representing SERPs (search engine result pages), wherein SERP implies the listing of Web pages returned by a search engine in response to a keyword query. Usually, the earlier a Web site or a Web page is presented in the SERP, or the higher it ranks, the more searchers visit that site. According to a study [1], the searcher's eyes focus on a part at the top left corner of a screen when a Google's SERP is displayed; this part of the Web page is called as the "Golden Triangle." It shows that a Web site cannot maintain the visual check level of at least 50% if it is not within the seventh result. Therefore, if a business organization intends to advertise commercial products effectively or if a political organization launches a strong campaign on the Web, it becomes very important for their Web sites to be ranked within the top seven of the SERP in response to the

Y. Zhang et al. (Eds.): APWeb 2008, LNCS 4976, pp. 625–630, 2008.
© Springer-Verlag Berlin Heidelberg 2008

intended search keywords. Under such a situation, the trustworthiness of Web search sites emerges as one of the most contentious but important issues to be investigated. In this paper, we pay special attention to a search keyword "gender-equal" to check the trustworthiness of SERPs when the keyword is typed. This keyword was adopted because it acquired a strong political coloration under the environment in which the Japanese Cabinet Office began to revise the Basic Plan for Gender Equality in 2005 to formulate the Second Basic Plan for Gender Equality (next five-year plan). A conflict between the gender-bashing sects and gender advocates appeared to be a war on the Web, where the gender-bashing sects were keen to advertise the danger of gender equal education in elementary and secondary schools. Note that this advertisement has some degree of sympathy because the translation of "gender-equal" into a Japanese-English word "gender-free" often evokes promiscuity among adolescents.

2 Analytical Method of Trustworthiness and Transparency of a Web Search Site

By the trustworthiness of a SERP or Web search site, we imply that the display order of Web pages presented in a SERP returned by a Web search engine in response to a search keyword is true, correct, or reliable. To analyze the trustworthiness of a SERP returned by Google, we performed the following observation steps:

(a) Check a SERP returned by Google in response to "gender-equal" as a search keyword.
(b) Check the SERPs returned by Google's six partner search sites (at that point in time [2])—namely Nifty, Livedoor, Excite, Biglobe, Infoseek楽天, and goo—in response to the same search keyword.
(c) Check a SERP returned by Yahoo!Japan in response to the same.
(d) Compare the above SERPs from various viewpoints, including ranking.

Yahoo!Japan was taken into consideration because it used its own search engine called Yahoo! Search Technology (YST). The top fifteen pages of each SERP were compared and taken into consideration.

Needless to say, the transparency of a Web search site is necessary for its trustworthiness. To analyze the transparency of a SERP returned by Google, (hereafter, the transparency of Google), we examined the PageRank^{TM} and the number of backlinks for each Web page in a SERP returned by Google and Yahoo!Japan. Moreover, we examined the number of times the keywords appeared on a landing page. These statistics were obtained using reproducible tools such as Google PageRank Checker to obtain PageRank^{TM}, Yahoo! Site Explorer (http://siteexplorer.search.yahoo.com/) to specifically explore the number of backlinks of the Web pages indexed by Yahoo! Search, and a keyword appearance frequency analysis tool (http://www.searchengineoptimization.jp/tools/keyword_density_analyzer.html) to determine the number of times a keyword appeared on a page. In order to obtain further data on the number of backlinks, we used a Web tool named Link Popularity Check (http://www.uptimebot.com/).

Table 1. Analysis Results on Google's SERP in Response to Search Keyword "Gender-equal"

(Final Check: Jan 8, 2006)

Column 1 (YahooJapan Category Name and URL, if Registered)	Col 2 No. of Backlinks	Col 3 Ranking at Yahoo!J. Japan	Col 4 Shows inlinks from all pages	Col 5 Shows inlinks except from this subdomain	Col 6 Yahoo Backlinks	Col 7 MSN Backlinks	Col 8 Yahoo Linkdomain	Col 9 No. of Backlinks	Col 10 Google SERP Jan/7/2006	Col 11 PageRank	Col 12 Page Title (in Japanese)	Col 13 URL	Col 14	Col 15	Col 16	Col 17	Col 18 Google's SERP 2005/12/7	Col 19 Google SERP 12/27	Col 20	Col 21	Col 22	Col 23	Col 24	Col 25
	92	9	72	71	48	27	2610	2	1	3/10	ジェンダーフリー	www.jayu-shikan.org/teachers/hattori/001_1.html	31	3588	0.86%	15	1	1	1	1	1	1	1	1
	636	4	348	31	258	14	704000	27	2	5/10	ジェンダーフリー	www.worldtimes.co.jp/w-top/education/main-b.html	35	1373	2.55%	10	2	3	3	3	3	3	3	3
X	118	203	94	33	35	17	8190	3	3	2/10	ジェンダーフリーにについて	www.tohma.jp/enoku/	10	373	2.68%	0	3	2	5	5	5	5	5	5
	6790	3	6744	6466	3490	988	3560	142	4	3/10	ジェンダーフリーとは～Q&A	seikotop.hp.infoseek.co.jp/gender4 eaQandAhtml	84	9691	0.87%	29	6	4	2	2	2	2	2	2
	1260	8	文字化け 要対応	文字化け 要対応	761	25	1130000	20	5	3/10	ジェンダーフリーと比	d.hatena.ne.jp/keyword/	20	517	3.87%	6	9	5	4	4	4	4	4	4
X	63	10	42	41	46	20	402000	1	6	2/10	ヘルジェンダーフリー教育	homepage1.nifty.com/101/ende nh.html	27	4507	0.60%	25	4	6	8	8	8	8	8	8
X	18	2005DEC末現	16	16	12	9	378000	0	7	2/10	ぶっ飛びセ ジェンダーフリー	homepage1.nifty.com/newp sisers3.htm	139	20202	0.60%	73	5	7	9	9	9	9	9	9
	710	1	425	379	246	32	941000	65	8	3/10	ジェンダーフリーとは男の論理	plaza.rakuten.co.jp/hisahito/	31	4716	0.66%	17	7	8	6	6	6	6	6	6
X	37	18	28	13	14	4	299000	6	9	2/10	ジェンダーフリー教育	www.hi-ho.ne.jp/taku_anzai/gender/	4	112	3.57%	0	10	10	7	7	7	7	7	7
	32	23	28	27	21	8	276	4	10	2/10	ジェンダーフリー教育の基礎知識	www.seikyokyo.net/ronbun/specia l16.html	7	2087	0.34%	29	8	9	12	12	12	12	12	12
	173	5	77	文字化け	108	20	3580000	9	11	4/10	ジェンダーフリー - Wikipedia	ja.wikipedia.org/wiki/ジェンダーフリー	56	2773	2.02%	26	13	13	10	10	10	10	10	10
	10	7	10	4	4	5	1220000	1	12	3/10	***** ジェンダーフリーな街 *****	www.geocities.co.jp/HeartLand-Oak/4177/	6	88	6.82%	0	11	11	14	14	14	14	14	14
	24	2	21	20	10	14	2340	2	13	3/10	ジェンダーフリーのワード	www7.ocn.ne.jp/~kunitachi/kyouiku/gender.htm	0	22	0.00%	4	14	14	11	11	11	11	11	11
X	51	16	33	6	10	4	56300	7	14	3/10	「勝海舟とジェンダーフリー」	www7.ocn.ne.jp/~gender/	22	3130	0.70%	28	15	15	13	13	13	13	13	13
X	19	2005DEC末現	4	1	6	6	1160	0	15	3/10	ジェンダーフリーって	www.kyoikuen.net/_school/sara4	7	1187	0.59%	1	16	16	23	23	23	21	23	23

Yahoo!Japan (2006/Jan/7)

Tool 2: Yahoo! SiteExplorer (http://siteexplorer.search.yahoo.com/) (Jan/7/2006)

Tool 1: Link popularity check (http://www.uptimebot.com/) (Jan/7/2006)

Google's SERP (SearchEngineResultPage) in Response to Search Keyword "Gender-equal" (Jan/7/2006)

Tool 3: Landing Page Statistics (Jan/7/2006)

Google's SERP's (Dec 2005)

SERP's by Six Google's Partner Search Sites (Jan/7/2006)

Backlink Analysis Tool "Link popularity check" (http://www.uptimebot.com/)

Google Backlinks - This value shows how many links, which are pointing your page does Google considers influencing your ranking. Google doesn't shows backlinks from pages having PR (see Google Page Rank) value lesser than 3. Note: Google has recently stopped displaying all backlinks it knows. Webmasters can have general opinions upon those links, which are now displayed as backlinks: 1. Google randomly shows backlinks. 2. Google shows that backlinks, which are considered relevant for this page (relevant means that a page, which contains a link to your site is somehow related to yours in subject (content)). Click the value to see you backlinks. Sample Input: link:http://www.igs.ocha.ac.jp/

Yahoo Backlinks - quantity of links pointing your page, which are indexed by yahoo.

Yahoo Linkdomain - quantity of links in whole internet indexed by Yahoo, which are pointing all instances of a domain name you search for. For instance if you check www.yoursite.com, www.yoursite.com/page.html and so on. Sample Input: linkdomain:ocha.igs.ac.jp

Linkdomain will also show links pointing subdomain, yoursite.com, www.yoursite.com/page.html and so on. Sample Input: linkdomain:ocha.igs.ac.jp

MSN Backlinks - quantity of links pointing your page, which are indexed by MSN. Sample Input: link:www.igs.ocha.ac.jp/

Verification tool-1 of search engine

Verification tool-2 of search engine   "Yahoo! Site Explorer" to check inlinks   http://siteexplorer.search.yahoo.com/

Verification tool-3 of search engine

Key Word Appearance Frequency Analysis:
http://www.searchengineoptimization.jp/tools/keyword_density.html
http://www.searching.neoptimization.jp/tools/keyword_density_analyzer.html

Integrated Search Engine N-Search http://www.n-search.net/page_check.php

3 Analysis Result of Trustworthiness and Transparency of a Google's SERP

3.1 An Overview of Analysis Result

The SERPs returned by Google, Google's six partner sites and Yahoo!Japan, in response to the same search keyword gender-equal in Japanese were recorded on January 7, 2006, immediately after the war of words between the gender-bashing sects and gender advocates had concluded, i.e., the Cabinet Office's final decision to revise the Basic Plan for Gender Equality to formulate the Second Basic Plan for Gender Equality (next five-year plan) was completed on December 27, 2005. Table 1 presents an overall analysis result. It comprises twenty-five columns. The center part of Table 1 illustrates the SERP of Google in response to the search keyword gender-equal in Japanese, wherein the top fifteen Web pages are shown along with the relevant statistics. Table 2 shows the meanings of the twenty-five columns.

Table 2. Explanation of twenty-five columns of Table 1

Column Number	Meaning	Column Number	Meaning
1	Yahoo!Japan's URLs (if available)	14	Number of times that "gender-equal" appears on the landing page.
2	Number of Yahoo!Japan's backlinks	15	Total number of times that "gender-equal" appears on the landing page.
3	Yahoo!Japan's ranking	16	Ratio of 14 and 15
4	Number of backlinks of the Web page in Google's SERP including backlinks Google does not show, recorded using Yahoo! Site Explorer.	17	Number of times that the keyword "gender" appears.
5	Number of backlinks shown in column 4 except those from this subdomains.	18	Google's ranking on December 7, 2005
6	Number of Yahoo! backlins using Link Popularity Check.	19	Google's ranking on December 27, 2005
7	Number of MSN backlins using Link Popularity Check.	20	Nifty's ranking
8	Number of Yahoo! linkdomains using Link Popularity Check.	21	Livedoor's ranking
9	Number of backlinks shown by Google.	22	Excite's ranking
10	Google's ranking	23	Biglobe's ranking
11	Google's PageRank^{TM}	24	Infoseek 楽天's ranking
12	Page titles	25	goo's ranking
13	URLs (i.e. landing pages)		

3.2 Analysis Result of Trustworthiness and Transparency

First, we paid special attention to columns 10, 12, and 13. According to Golden Triangle, we carefully examined the characteristics of the first seven pages. The gender specialists in our research team concluded that the pages ranked 1st, 2nd, 3rd, 6th, and 7th were those pertaining to the sites that belonged to the gender-bashing sects, while the pages ranked 4th and 5th were those pertaining to the sites that belonged to the gender advocates. Second, we paid special attention to columns 20 to 25. It is highly interesting to state that the rankings in these six Web sites were completely identical up to the 15th placement with a minor mismatch. Based on this identification, we presumed that the ranking appearing in these columns was the "original" ranking according to Google's ranking strategy. Third, we compared the ranking of Google's SERP shown in column 10 to those of the SERPs of Google's six partners. Obviously, they were not identical except for the top rank page, although they were recorded on the same date. Thus, we presumed that there might be a "time delay" in the sense that Google provided "old" data to its partners for calculating SERPs. In order to confirm the credibility of this presumption, we prepared the SERPs of Google recorded on December 7 and 27, 2005, one month and two weeks prior to recording the objective data for comparison, i.e., January 7, 2006. However, the conclusion was negative. In other words, it was observed that neither the SERP recorded on December 7, 2005, nor the one recorded on December 27, 2005, was identical or even similar to the SERPs of Google's six partners recorded on January 7, 2006. These examinations clarified that the SERPs of Google's six partner sites were neither identical nor similar to Google's SERP recorded on the same date, the one recorded two weeks ago, or the one recorded one month ago. Thus, we concluded that the SERPs of Google's six partner sites could be the "original" SERP calculated by Google on that date. Fourth, we attempted to provide some explanations to resolve this disagreement. Through trial and error, we noticed a highly unexpected agreement. This agreement was noted when we carefully examined the rank of each page in the SERPs: The page that ranked first on Google's SERP was also placed first on the SERPs of its six partners. This identification is not disputed as much. However, the page that ranked second on Google's SERP was placed third on the SERPs of Google's six partners, i.e., this page was placed a rank above the one that might be originally calculated in Google and was delivered to Google's partner sites. The difference was merely one rank. However, we paid special attention to identify the page on Google's SERP that was ranked higher in comparison with the rank of that page on the SERPs of Google's partners. The other results are as follows: The page that ranked third on Google's SERP was two ranks higher, i.e., it might be ranked fifth in the original SERP. The page that ranked sixth on Google's SERP was two ranks higher as compared with that on the partners' SERPs, and the page that ranked seventh on Google's SERP was two ranks higher as compared with that on the partners' SERPs. As mentioned previously, the pages that ranked 1st, 2nd, 3rd, 6th, and 7th pertained to those sites that belonged to the gender-bashing sects. In other words, on studying the first seven pages of Google, we observed that all the pages of Google's SERP whose ranks were higher in comparison with their ranks on its partner sites belonged to the gender-bashing sites. In other words, the ranks of the pages pertaining to the gender-bashing sites were equally ranked higher in Google's SERP if they were compared with the SERPs of its partner sites. In contrast, we compared the rank of the pages on Google's SERP that were operated by the gender advocates sites, with their ranks on its partners' SERPs. We discovered that the page that ranked fourth on Google's SERP was two ranks lower as compared with that on the SERPs of Google's partner sites, and the page that ranked fifth on Google's SERP was one rank lower as compared with its rank on the SERPs of Google's partner sites. In other words, the ranks of pages pertaining to the gender advocates were equally ranked lower in Google's SERP if they were compared with the SERPs of its partner sites. If we combine the abovementioned discussions, the following unexpected agreement can be found: The ranks of the pages pertaining to the gender-bashing sects and gender advocates were consistently ranked higher and lower, respectively, on Google's SERP if they were compared with the SERPs of its partner sites.

By the transparency of a Web search site, we imply that the page ranking strategy for the SERP of that site is clear to searchers. Google uses Google's PageRank^{TM} algorithm which uses backlinks to determine a page rank. However, as is evident from Table 1, the correlation between the ranking of the Google SERP (column 10) and its PageRank (column 11) was not found. Since it is clarified that Google does not show backlinks from pages having PageRank value lesser than three and that it has stopped displaying all backlinks it knows, we paid special attention to Yahoo! search site and open source Web tools to obtain further accurate information pertaining to the backlinks. As an observation result, although the number of backlinks presented in columns 2, 4, 6, and 7 are different, the tendency can certainly be interpreted. Again, we observe a very weak relation between the ranking of a SERP and the number of backlinks.

4   Concluding Remarks and Acknowledgements

At the present when the gender-bashing war terminated, we cannot find any inconsistency between Google's SERP and the SERPs of Google's partner sites when we input the search keyword "gender-equal." The authors are thankful to Professor Kaoru Tachi of the Institute for Gender Studies, Ochanomizu University who gave us valuable comments in this research. This research was partly supported by a Grants-in-Aid for Scientific Research of MEXT of Japan in the Category of Scientific Research (B) (Grant number 19300025) on "Construction of an Integrated Web Mining Environment as a New Methodology for Social Science Studies" (2007-2009).

References

1. Eyetools Research and Reports: Eyetools, Enquiro, and Did-it uncover Search's Golden Triangle,
   http://www.eyetools.com/inpage/research_google_eyetracking_heatmap.htm
2. SEM-ch Search engine marketing information channel, AUN Consulting, Inc.,
   http://www.sem-ch.jp/

SemSearch: A Scalable Semantic Searching Algorithm for Unstructured P2P Network

Wei Song[1], Ruixuan Li[1,*], Zhengding Lu[1], and Mudar Sarem[2]

[1] College of Computer Science and Technology, Huazhong University of Science and Technology, Wuhan 430074, Hubei, P.R. China
[2] School of Software Engineering, Huazhong University of Science and Technology, Wuhan 430074, Hubei, P.R. China
weisong@smail.hust.edu.cn, {rxli,zdlu}@hust.edu.cn,
mudar66@hotmail.com

Abstract. Resource searching in the current peer-to-peer (P2P) applications is mainly based on the keyword match. However, more and more P2P applications require an efficient semantic searching based on the contents. In this paper, we propose a novel scalable semantic searching algorithm named SemSearch for the unstructured P2P networks. For the consistency and flexibility of semantic analysis, we integrate global and local semantic information to do the semantic analysis. Moreover, SemSearch transfers the searching requests to the peers whose shared resources are more semantic similar to implement semantic searching. We further evaluate the performance of SemSearch through the simulation experiments.

Keywords: SemSearch, Peer-to-peer, Semantic analyzing, Semantic searching.

1 Introduction

The resource searching in most current P2P systems is based on the keyword match. However, some P2P applications need an efficient resource searching over the contents of resources. There is not a scalable semantic searching algorithm in the real P2P applications. So, improving the semantic analysis and implementing a scalable P2P semantic searching is the main motivations of this paper. In this paper, we integrate global and peer's local semantic information to analyze the resource. Moreover, we propose SemSearch, a novel scalable semantic searching algorithm for the unstructured P2P networks.

The main contributions of this paper include: 1) Proposing a scalable semantic analysis method to achieve the consistency and scalability of the semantic analysis; 2)

---

* This work is supported by National Natural Science Foundation of China under Grant 60403027, 60773191, 70771043, National High Technology Research and Development Program of China under Grant 2007AA01Z403, China Postdoctoral Science Foundation under Grant 20060400846, Natural Science Foundation of Hubei Province under Grant 2005ABA258.

Y. Zhang et al. (Eds.): APWeb 2008, LNCS 4976, pp. 631–636, 2008.

Proposing an efficient P2P semantic searching algorithm, called SemSearch; 3) Carrying out the simulation experiments to evaluate the performance of SemSearch.

The rest of the paper is organized as follows. In the next Section, we present some related work in this background. Section 3 presents the details of the SemSearch protocols. Section 4 evaluates the SemSearch performance through the simulation experiments. Finally, the paper is concluded in Section 5.

2 Related Work

The current P2P resource searching can be classified into two categories. One is based on the resource description (keyword, file name, etc.) match such as Gnutella, Bitcomet, and other popular file sharing P2P applications. Some searching algorithms in the structured P2P networks include: Chord [1], Tapestry [2], and CAN [3] also belong to this category. In this category, the keyword match is used to judge whether resources satisfy a searching request or not. Many P2P applications have employed this method. However, the keywords are difficult to represent the resource contents and users' interests. So, the searching results can not satisfy the users.

The second category is based on the resource contents. Semantic Overlay Network (SON) [4] first introduces the concept of the semantic searching. We can classify the P2P semantic searching into two sub-categories: 1) Semantic searching based on the static semantic analysis, in which peer extracts the resource semantic information from a static semantic model. pSearch [5] used a semantic vector to describe the resources. SemreX [6] was a semantic P2P overlay based on a concept tree. The static semantic analyses are easy implemented. However, this method is difficult to be extended. Hence, such method only adapts to the special applications. 2) Semantic searching based on the scalable semantic analysis. In this sub-category, the semantic analysis is extendable by the self-study. H.T. Shen et. al. proposed a semantic P2P framework [7] based on Hierarchial Summary Structure. RS2D [8] was a risk driven semantic P2P research search service. FCAN [9] implemented the content query over structured P2P overlay. And Semantic Small World [10] researched the semantic searching from the views of small world theory and semantic clustering. These semantic searches are scalable and flexible. Nevertheless, with the network size growing, it is difficult to keep the consistency of semantic analyzing and search.

3 SemSearch Protocols

3.1 Semantic Description in SemSearch

Resnik has proposed that the key to the similarity of two concepts is the extent to which they share information[11]. Therefore, in SemSearch, we measure the word frequency in the resources to describe the resources. SemSearch has a global semantic knowledge base which is a keyword matrix, named Global Semantic Keyword Matrix (GSKM). A GSKM row describes a semantic cluster, for example, GSKM shown in (1) describes four semantic clusters.

$$\begin{bmatrix} \text{operation system} & \text{process} & \text{priority} & \text{scheduling} & ... \\ \text{database} & \text{view} & \text{oracle} & \text{pattern} & ... \\ \text{semantic web} & \text{ontology} & \text{OWL} & \text{meta-model} & ... \\ \text{distributed system} & \text{replication} & \text{P2P} & \text{Grid} & ... \end{bmatrix} \quad (1)$$

While a peer P joins SemSearch, it first loads the GSKM $A^{m\times n}$. Afterwards, P measures the frequency of A's elements in each resource r_i to build r_i's keyword frequency matrix $B_i^{m\times n}$. B_i describes the keyword's distribution in r_i's contents. We further normalize B_i to build a relative keyword frequency matrix $P_i^{m\times n}$, which describes the keyword frequency proportion in the resource r_i.

The sum of P_i's elements in a row describes r_i's keyword frequency proportion in a semantic cluster. So, we introduce the resource semantic vector $C=(1,1,...,1)^{1\times n}P^T$ to represent the resource contents' matching degree with each semantic cluster. Usually, the shared resources in a peer are similar. Consequently, we describe the semantic information of a peer by its local shared resources. Hence, we introduce the peer's semantic vector $R=\dfrac{1}{t}\sum_{i=1}^{t}C_i$ to represent the peer's semantic information.

3.2 Extended Semantic Analysis in SemSearch

The global GSKM can not reflect the comprehensive resource contents. Moreover, an exhaustive GSKM is quite difficult for a P2P system. Therefore, we build peer's Local Semantic Keyword Matrix (LSKM) to make semantic analysis scalable.

A peer P with t shared resources constructs a word frequency vector $B(b_1,b_2,...,b_t)$ for GSKM $A^{m\times n}$'s each keyword. Each component in B is a keyword's frequency locally. The similar meaning words usually have the similar distributions, so P looks for the new similar meaning words to extend LSKM. While P extends its LSKM $L^{m\times N}(N>n)$, it first copies A's elements into L and set the elements $l_{ij}(n<j\le N)$ as null. Moreover, we use the average word frequency vector to judge whether a new word is a semantic cluster's synonyms. This paper uses the cosine distance of two vectors to measure their similar degree. The cosine distance of two vectors x and y is defined as follows: $\theta(x, y)=\cos^{-1}\dfrac{x\cdot y}{|x||y|}$.

P full-text retrieves its local resources and collects a keyword candidates list. Afterwards, it computes the similar degree of every candidate and semantic cluster. Each semantic cluster in L has a cosine distance threshold $\hat{\theta}$. If the cosine distance of a candidate keyword and a semantic cluster's average word frequency vector is less than the threshold, then the keyword joins this semantic cluster.

When a new keyword S attempts to join the ith semantic cluster, if the ith row of L has null values, then S replaces a null value. Otherwise, S replaces the keyword which has the maximal cosine distance to the semantic cluster. And we set $\hat{\theta}$ as the elements' maximal cosine distance to the average word frequency vector. SemSearch peer periodically calls the extension algorithm to extend LSKM.

3.3 Semantic Search of SemSearch

A SemSearch searching request $q(g,\beta)$ is made up of a query vector g and a cosine distance threshold β. The searching request $q(g,\beta)$ searches the semantically similar resources whose cosine distance to g is less than β. The searching source peer P selects a set of keywords from its interests to construct the query vector. P constructs a matrix $V^{m \times t}$ in which the element v_{ij} is the jth resources' frequency in the ith semantic cluster. Moreover, for a set of inquired keywords U, P constructs an inquired keyword frequency vector $W(w_1, w_2, ..., w_t)$ in which w_i represents U's total frequency at the ith resources. The vector W represents the distributions of the inquired keywords at local resources. So, we use the matrix V and vector W to build query vector g.

$$g = \frac{WV^T}{\left| WV^T((1,1,...,1)^{1 \times m})^T \right|} \tag{2}$$

The peers return local semantically similar resources whose cosine distance to g is less than β. Furthermore, peers select no more than r neighbors whose peer semantic vectors are near to g to transfer the searching request.

The searching results of SemSearch depend on the threshold β. Therefore, we use an adaptive method to determine β value. When a peer first launches a searching request, it uses β_{init} as the initial threshold. Afterward, the peer adjusts the β value from β_{min} to β_{max} based on the previous return responses.

4 Simulation Experiments

In this Section, we evaluate the performance of SemSearch by simulations. The network size is 3000-5000, and the peer's connection degree follows the power law distribution of exponent $A=3.0$ and $Degree_{average}=4.0$. The shared resources are the documents in ACM database and classified based on ACM Computing Classification System 98 [12]. We select 3000 documents from each top catalog except General Literature and Computing Milieux. Moreover, the GSKM is a 9×5 matrix and LSKM is a 9×10 matrix. In the simulations, a peer launches two searching requests for each minute. More experimental parameters are show in Table 1.

Table 1. Parameter and settings in the simulation experiments

Parameter	Parameter Meaning	Default Value
β	cosine distance threshold	0.7
TTL	search radius	2-4
r	Max peers that searching request is transferred to	3

4.1 SemSearch Recall Rate

We do the simulation experiments to compare the average recall rate of SemSearch and Gnutella in various searching TTL and network size. The experimental results are shown in Figure 1.

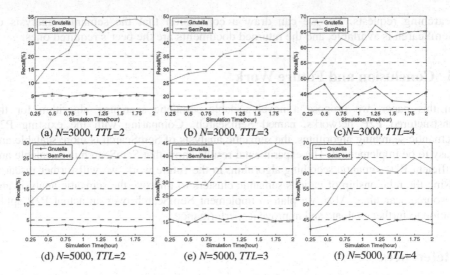

Fig. 1. Recall Comparison between Gnutella and SemSearch

As shown in the experimental results, SemSearch gets a higher recall rate. Furthermore, we can easily discover that the SemSearch recall rate rise over time. Analyze this phenomenon, we have found that the peer's semantic analysis is more accurate with the LSKM extending. So, the recall rate of semantic searching rises.

4.2 SemSearch Precision

We have measured the precision to evaluate the consistency of semantic searching. Based on the ACM CSS classification, a peer launches a searching request for the resources in one catalog. If the return resources are in the same catalog, we consider the returned resources are semantically correct. The experimental results are shown in Figure 2.

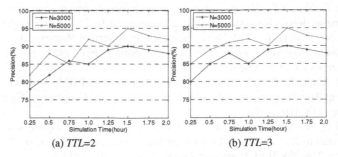

Fig. 2. Precision ratio of SemSearch over time

The experimental results show that SemSearch keeps a high precision (about 90%) in the experiments. Further, analyzed the experimental results, we have found that the 40%-50% semantically inapprehensive returned results still are similar to the initial searching requests. So, we can draw a conclusion that the semantic analysis of SemSearch is available, and the returned documents are the peer's required ones.

5 Conclusion and Future Work

In this paper, we propose a novel scalable semantic searching algorithm for the unstructured P2P networks, named SemSearch. Comparing with the existing P2P semantic searching algorithms, the semantic analysis of SemSearch is more scalable and easy to be implemented. The experimental results show that SemSearch is a scalable and efficient P2P semantic searching algorithm. In the future work, we will further research using the relationship of various semantic clusters to improve the semantic analysis and resource searching. Also, we plan to implement SemSearch over an actual P2P application to further evaluate and improve its performance.

References

1. Stoica, I., Morris, R., Karger, D., Kaashoek, M.F., Balakrishnan, H.: Chord: a scalable peer-to-peer lookup service for internet applications. In: ACM SIGCOMM Computer Communication Review, vol. 31, pp. 149–160. ACM Press, New York (2001)
2. Zhao, B.Y., Huang, L., Stribling, J., Rhea, S.C., Joseph, A.D., Kubiatowicz, J.: Tapestry: A Resilient Global-Scale Overlay for Service Deployment. J. IEEE Journal on Selected Areas in Communications 22, 41–53 (2004)
3. Ratnasamy, S., Francis, P., Handley, M., Karp, R., Shenker, S.: A Scalable Content-Addressable Network. In: The ACM SIGCOMM 2001, pp. 161–172. ACM Press, New York (2001)
4. Crespo, A., Garcia-Molina, H.: Semantic overlay networks for P2P systems. Technical report, Stanford University (2002)
5. Tang, C., Xu, Z., Mahalingam, M.: pSearch: Information retrieval in structured overlays. ACM SIGCOMM Computer Communications Review 33(1), 89–94 (2003)
6. Chen, H.H., Jin, H., Ning, X.M., Yuan, P.P., Wu, H., Guo, Z.X.: SemreX: A semantic similarity based P2P overlay network. J. of Software 17(5), 1170–1181 (2006)
7. Shen, H.T., Shu, Y.F., Yu, B.: Efficient semantic-based content search in P2P network. IEEE Transaction on Knowledge and Data Engineering 16(7), 813–826 (2004)
8. Klusch, M., Basters, U.: Risk Driven Semantic P2P Service Retrieval. In: The 6th IEEE International Conference on Peer-to-Peer Computing (P2P 2006), pp. 161–170. IEEE Computer Society, Los Alamitos (2004)
9. Wang, J., Yang, S., Gao, Y., Guo, L.: FCAN: A Structured P2P System Based on Content Query. In: The 5th International Conference on Grid and Cooperative Computing (GCC 2006), pp. 113–120 (2006)
10. Li, M., Lee, W., Sivasubramanian, A.: Semantic small world: an overlay network for Peer-to-Peer search. In: The 12th IEEE International conference on Network Protocols (ICNP 2004), pp. 228–238. IEEE Computer Society, Los Alamitos (2004)
11. Resnik, P.: Semantic similarity in a taxonomy: An information-based measure and its application to problems of ambiguity in natural language. J. of Artificial Intelligence Research 11, 95–130 (1999)
12. ACM CCS, http://www.acm.org/class/1998/

Web Image Annotation Based on Automatically Obtained Noisy Training Set

Mei Wang, Xiangdong Zhou*, and Hongtao Xu

Department of Computing and Information Technology,
Fudan University, China, 200433
{051021052,xdzhou}@fudan.edu.cn

Abstract. Training data acquisition is a problem in large scale statistical learning based web image annotation. A common idea is to build a large training set by analyzing the web content automatically. However, the noisy data is unavoidable involved in this kind of approach. In this paper, we present a novel web image annotation method based on noisy training set using Mixture Component based Local Fisher Discriminant Analysis (MLFDA). In our method, image annotation is viewed as a multiple class classification problem. To alleviate the influence of the noisy data, the separating hyper planes between different classes are learned by kernel-based local fisher discriminant analysis. Then the mixture components for each class are estimated in the subspace, where the noisy modals will gain small weights and play less important role in classification. The experimental results on a real-world web data set of 4000 images show that our method outperforms MBRM [3] and SVM-based method with F_1 measure improving 83% and 18% respectively.

1 Introduction

Nowadays, the digital images have become widely available on World Wide Web (WWW), which has brought great challenges for organizing and searching a large volume of available images. Well-known commercial systems including Google, MSN, and Yahoo! rely on surrounding descriptions of images embedded in the web pages for the image retrieval. Images without clear context descriptions will either be returned as false positives or be totally discarded during the retrieval. Image auto-annotation techniques [1,2,3,6,7,8,12] provide an attainable way to associate the "visuality" of the images with their semantics, which can be used to search unlabeled image collections, and return more relevant images to the users.

In the previous work, both the generative model [1] and the supervised classification methods [3,12] have been applied to improve the performance of image annotation. Such methods rely heavily on the quality of the present training set.

---

* This work was partially supported by the Natural Science Foundation of China Grant No.60403018 and No.60773077, Fundamental Research 973 Program of China Grant No.2005CB321905. Corresponding author.

Y. Zhang et al. (Eds.): APWeb 2008, LNCS 4976, pp. 637–648, 2008.

Fig. 1. Image classes "Coral"(first row) and "Solar Eclipse"(second row). Images in the left four columns are the real samples about "Coral" and "Solar Eclipse" respectively. Images in the right two columns are noisy images. Due to the visual diversity of the class, images in the same class have different visual contents. It is difficult to distinguish the noisy data from the real ones according to the visual difference. The noise data can be treated as "special" modals of the class.

In the web image annotation, one feasible way to obtain sufficient training data is to parse the web content automatically. However, due to the well-known problem of complexity and variety of the web pages, it is difficult to keep the quality of the training data high. Although some efforts have been made to parse the web content intelligently, such as [13,14]. However, the light weight methods, such as DOM, SAX are more widely used in real applications. Due to their unsupervised acquiring processes, the training set obtained automatically is usually impurer than what is required for traditional annotation problems. For example, Fig. 1 lists some training images with labels "coral" and "solar eclipse" respectively, which are automatically parsed out from the images' surrounding text in web pages. It is obvious that images in the right two columns should not be associated with "coral" and "solar eclipse" respectively, so called noise images in this paper. In generative model based annotation, if the unlabeled images have high visual similarities with noisy training images, they tend to be wrongly labeled. The previous work [11] demonstrates that the label noise will also impair the classification performance. Some efforts have been made by adopting some data cleaning procedure [11] or predicting the noisy data probabilistically [10]. However, due to the visual diversity of the web image, images in the same class often show different visual content. It is difficult to distinguish the noisy data from the real ones according to the visual difference.

To address this problem, we present a novel web image annotation approach based on multi-modal classification. In our method, image annotation is posed as a multiple class classification problem. We analysis the influence of the noisy data conducted on the multiple hyperplanes, then treat the noisy data as special kind of modals of the class. To implement the multi-modal classification, mixture component based local fisher discriminant analysis (MLFDA) is proposed in our approach. Our method works as follows: Firstly, an impure training set is automatically obtained by heuristically judging the relevance between the context of images and their semantics. To reduce the influence of the noisy data and achieve better separability between different classes, kernel-based LFDA is exploited to find an "optimal" separable subspace by preserving the local structure between noisy data and the true class samples. Then we learn mixture components for each class in the subspace, where the noisy modals will gain small weights and play less important role in classification. In summary, we make the following contributions in this paper:

- Based on the noisy training set automatically obtained from web, we propose a new web image annotation approach to label new images by exploiting the visual contents of the web images.
- By using MLFDA-based technique, our method can effectively reduce the bad influence of noisy data on annotation.
- Experimental results on the real data set of 4000 web images demonstrate that our annotation method outperforms state-of-the-art MBRM and SVM-based method with F_1 measure improving 83% and 18% respectively.

The rest of the paper is organized as follows. Section 2 introduces the related work. Section 3 presents our MLFDA-based annotation method. We discuss the experimental results in Section 4. Section 5 concludes this paper.

2   Related Work

Statistical Annotation Models: Generative model based image annotation approaches such as Relevance Model[1] has shown significant performance improvements. However, such unsupervised labeling process is strongly influenced by the image with similar visual feature but different semantics. Taking each semantic label as a class, image annotation could be implemented by supervised learning based classification techniques. Model-based [2] and SVM-based approaches [3] have been applied in image annotation. However, the model-based methods usually suffer from the problem of insufficient training data [3]. The SVM based methods need to learn the separate hyper plane for each class, thus may lead to a high computation cost. Furthermore, they often suffer from the imbalanced training set [12]. Namely, the number of negative samples is much larger than that of positive ones, which is very severe in web image annotation. Fisher Discriminant Analysis (FDA) [4] is a traditional statistical method that has proved successful on classification problems. Compared with previous generative model and classification method based image annotation, it has the following advantages: 1). FDA incorporates the discriminant information between different classes without learning different separating hyper-planes for each class. 2). There are no negative samples, and it is not affected by the imbalanced training samples. In this paper, we exploit the FDA as our basic multi-class classifier.

Web Image Annotation: Wang et al. [6] proposed a web image annotation method AnnoSearch using search and data mining techniques. However, in their framework, at least one accurate keyword is required in advance. Hua et al. [7]

proposed a system which can automatically acquire semantic knowledge such as description, people, temporal and geographic information for web images. Nevertheless, they did not explicitly exploit the visual similarity to label new images. Li et al. [8] proposed a real time computerized annotation system named ALIPR, which uses the available data set: Corel data set as their training set.

3   The Proposed Method

Fig. 2 presents the overview of our proposed scheme for learning semantics of web images. Two major steps of the proposed method are: web data processing, web image semantic annotation. Web data processing analyzes the relevance between the context of images and their semantics, and then gets a high confidence training sets. Based on the training images, we perform the annotation for the rest web images using MLFDA-based annotation method.

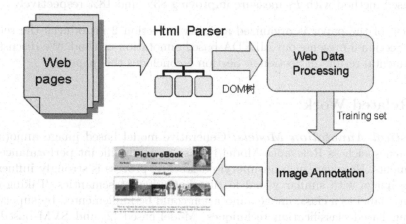

Fig. 2. The framework of the proposed method

3.1   Web Data Processing

In our system, we use HTML Parser [8] to transform html documents into DOM tree, and then extract the text and images by traversing through the DOM tree. During this process, we also obtain the correlation information between text and images if the correlation is clearly shown in the web page.

First, we generate the DOM tree for each web page containing the web images. From the bottom, visual objects like image, text paragraph are identified as basic elements. The tags such as <TABLE>, <TD>, <TR> and <HR> are used to separate the different content passages. Fig. 3 shows an example. The left part of Fig. 3 shows two paragraphs and one image of a web page about "Pyramid". The right part illustrates part of the DOM tree parsed from this web page.

Second, we extract the semantic text information for the images. The extracted information includes:

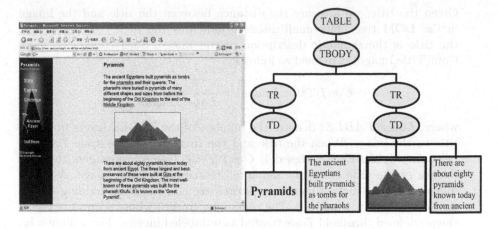

Fig. 3. An example of HTML DOM Tree

- ALT (Alternate) Text
- The closest title with the images in the DOM tree.

The ALT text in a web page is used for displaying to replace the associated image in a text-based browser. Hence, it usually represents the semantics of the image concisely. We can obtain the ALT text from ALT tag directly. Unfortunately, this information may sometimes be unavailable since many editors are too lazy to fill this field while they were designing the web pages[]. A feasible way is to analysis the context of the web image to obtain the semantic text information of the images. Fig.3 shows that the image and its "nearest" title "Pyramid" have clear semantic correlation. Notice that the title in this paper is used in a general sense, including both paragraph title and the web page title. It indicates that we could measure the "distance" between the title and the image to judge their semantic correspondence. The DOM tree shows the relevant title and the image are in the same <TABLE> tag in the web page. So we evaluate the "distance"by measuring the level difference of the TABLE or DIV tag between the image and the title in the DOM tree. The detailed web data processing steps are as follows:

- From the TABLE field containing the images, we traverse the DOM tree upward to obtain the title information. The visual features such as font, bold and position available in web pages are important cues for a phrase's importance. Such visual features can be combined into one single score according to the following formula:

$$Score(Text) = \frac{Is(Bold) + Is(Larger)}{Length} \qquad (1)$$

Here, if the text is bold, $Is(Bold) = 1$. $Is(Larger)$ indicates whether the text has larger font size than the surrounding text. $Length$ denotes the length of the text. If the score is higher than a given threshold C, the text is decided as a title.

– Given the title, we measure the distance between the title and the image
  in the DOM tree. The small distance indicates the high confidence that
  the title is the semantic description of the image. Such confidence score
  Conf(Title,Image) is defined as follows:

$$Conf(Title, Image) = \frac{1}{Layer(Table)}, \tag{2}$$

where, $Layer(TABLE)$ denotes the number of the TABLE layers from the
title to the image. When the title and the image lie in the same $TABLE$
field, $Lay(TABLE)$ is assigned 1. $Conf(Title, Image)$ represents the confidence of the title being the semantic description of the image. A set of
images with the highest confidence scores are chosen as the training set for
the image semantic analysis. The remaining images whose scores are below
the predefined threshold T are treated as unlabeled images. The semantic labels of these unlabeled images will be assigned by our automatic annotation
method, which will be discussed in the next subsection.

3.2   MLFDA-Based Web Images Semantic Annotation

After web data processing, we obtain two set of images: training set L, where
the correlation between the semantic keywords and the images is already known;
the remaining set of images U, where the correlation is unknown. We propose
the MLFDA-based image semantic annotation technique to assign the semantic
textual labels to the images in U. As we use FDA as our basic multiple classification classifier, we first give a brief introduction to FDA and analyze how
the noisy data damage the separation between different classes in FDA, then
introduce our MLFDA-based annotation algorithm.

Fisher Discriminant Analysis. Generally, each image has more than one
annotation words, the image annotation can be posed as multi-class learning
problem. Let x denote a training sample, y denote the class label, n_i be the
number of image samples in class i, c be the number of the classes. We can
define c discriminant function: $g_k(x), k \in \{1, \ldots, c\}$. If $i = max_k g_k(x)$, sample x
will be assigned to the ith class. Fisher discriminant analysis finds an "optimal"
discriminant subspace for separating the different groups, which is defined as:

$$\max_{\alpha} \frac{\alpha^T S_B \alpha}{\alpha^T S_W \alpha} \tag{3}$$

Where $\alpha^T S_W \alpha$ is the within class scatter matrix, $\alpha^T S_B \alpha$ is the between class
scatter matrix, which are all defined in the discriminant subspace. S_B and S_W
correspond to the within and between class scatter matrix in the origin space.
$g(x)$ can be defined as the monotonic decreasing function of the Mahalanobis
distance, which assigns the new samples to the class with centroid closest to
it in the subspace. Solving Eqn. 3 is equal to solving a generalized eigenvalue
problem.

Eqn. 3 implys that FDA is to maximize the between-class scatter as well as minimize the intra-classes scatter. However, when there are noisy data which have different visual contents with the true class samples as shown in Fig. 1, the above criteria indicates that the noisy training samples and the true samples in the same class will be merged together. Thus, the "optimal" separation of the subspace will be weaken. Due to the visual diversity of the image class, images in the same class often have different visual contents. It is hard to distinguish and eliminate the noisy data in advance. Instead of removing the possible noisy data directly, we just regard them as some special modals of the class. It is obvious that keeping the original structure of these modals of the class will increase the separability between different classes when mapping the origin data to the discriminant subspace. We exploit LFDA to achieve this goal, which is originally proposed by Masashi [9] to perform the multi modal dimension reduction.

Noise Processing. To preserve the local structure in the class, construct the local structure preservation matrix A. We take a simple way to define A. Specifically, let $A_{i,j} = 1$, if x_i and x_j are in the same class and x_j is the k-nearest neighbor of x_i. The noise sample x_k has different visual features with the other samples of the class they related to, so $A_{.,k}$ tends to be zeros. In the next part, we will show that such images have less influence on defining the separable hyper plane between different classes.

Local Fisher Discriminant Analysis. Review that in LFDA[9], the local with-class scatter matrix $\bar{S}_B$ and $\bar{S}_W$ are redefined as follows:

$$\bar{S}_W = \frac{1}{2} \sum_{i,j=1}^{n} \bar{A}_{i,j}^W (x_i - x_j)(x_i - x_j)^T, \tag{4}$$

$$\bar{S}_B = \frac{1}{2} \sum_{i,j=1}^{n} \bar{A}_{i,j}^B (x_i - x_j)(x_i - x_j)^T \tag{5}$$

Applying the local structure preservation matrix A:

$$\bar{A}_{i,j}^{(W)} = \begin{cases} A_{i,j}/n_k, & if \quad y_i = y_j = k \\ 0, & if \quad y_i \neq y_j \end{cases}, \tag{6}$$

$$\bar{A}_{i,j}^{(B)} = \begin{cases} A_{i,j}/(1/n - 1/n_k), & if \quad y_i = y_j = k \\ 1/n, & if \quad y_i \neq y_j \end{cases}, \tag{7}$$

Because the noisy data is far away with true samples in the class, they have less influence in $\bar{S}_B$ and $\bar{S}_W$. Using $\bar{S}_B$ and $\bar{S}_W$, the modified optimal discriminant subspace of LFDA is given by:

$$\max_{\alpha} \frac{\alpha^T \bar{S}_B \alpha}{\alpha^T \bar{S}_W \alpha} \tag{8}$$

Similarly, the subspace can also be obtained by solving a generalized eigenvalue problem.

Classification Based on Mixture Components. Fisher discriminat analysis can be viewed as a prototype classifier. Each class is represented by its centroid, and the unlabeled sample is classify to the closest one according to Mahalanobis distance measure. It is not difficult to imagine that the noisy data usually forms isolated and small clusters in the discriminant subspace obtained above, thus a single prototype is not sufficient to represent the complex classes, and mixture models are more appropriate. Assume kth class has a mixture Gaussian density, that is:

$$P(X|G = k) = \sum_{r=1}^{R_k} \pi_{kr}\phi(X; \mu_{kr}, \sigma), \qquad (9)$$

where $\{\pi_{kr}, \mu_{kr}, \sigma\}$ is parameter set, including weight set $\{\pi_{kr}\}$ and mixture parameter $\{\mu_{kr}, \sigma\}$. R_k is the number of the prototypes of the kth class. Given such a model for each class, the class posterior probabilities are given by:

$$P(G = k|X = x) = \frac{\sum_{r=1}^{R_k} \pi_{kr}\phi(X; \mu_{kr}, \sigma)\Pi_k}{\sum_{l=1}^{c} \sum_{r=1}^{R_l} \pi_{lr}\phi(X; \mu_{lr}, \sigma)\Pi_l}, \qquad (10)$$

where Π_k represent the class prior probabilities. We estimate the parameters with a maximum likelihood estimates (MLEs) as follows:

$$\sum_{k=1}^{c} \sum_{g_i=k} log\, P(x_i|G = k)\Pi_k. \qquad (11)$$

We followed the EM algorithm to estimate the model parameters with ML criterion. EM alternates between the two steps:

E-step: Given the current parameters, compute the weights of the subclasses in each class observations.

M-step: Compute the weighted MLEs for the parameters of each of the component Gaussians within each of the classes, using the weights from the E-step.

A $k-$means clustering method is applied to fit the data in each class to create the initial weight matrix.

The Summarization of Our Annotation Approach. Kernel function can be used to overcome the limitation of linear decision function of FDA. Then, our MLFDA-based annotation approach is summarized in the following steps:

1. Extract the global feature of the images to get the original feature space. The images which contain the same keyword are grouped as the same class.
2. Use a nonlinear mapping function $\phi : X \rightarrow F, x \rightarrow \phi(x)$ to project the original space into high-dimensional space F. The popular Gaussian kernel is chosen, which is $k(x, y) = \exp(-||x - y||^2/\sigma)$.
3. Find the optimal discriminant subspace according to Eqn. 8.
4. In the discriminant subspace, learn the mixture components for each class.
5. For unlabeled images, calculate and rank the posterior probability according to Eqn. 10 to obtain the semantic labels.

4   Experiments

All the data used in our experiments are crawled from Internet. The image data set is obtained by HTML parsing and meaningless small icons are filtered out. The size of the image data set is 4000. After web data processing, we obtain 1600 training images, and the remaining images with low confidence scores are used to be annotated. There are 70 semantic keywords in the data set. Instead of taking each keyword as the class, we group the most co-occurred keywords as our class. For example, the keyword group "mobile phone" is regarded as one class. The keywords in the class with largest posterior probability are assigned as the annotation of the new images. The global features are 528 dimensions with color and texture features extracted according to MPEG7. We compute the average recall, precision and F_1 to measure the quality of the algorithm. Given a query word w, let $|W_G|$ denote the number of human annotated images with label w in the test set, $|W_M|$ denote the number of annotated images with the same label by our algorithm, and $|W_C|$ denote the correct annotations by our algorithm. The recall, precision and F_1 are defined as:

$$Recall = \frac{|W_C|}{|W_G|}, \quad Precision = \frac{|W_C|}{|W_M|}$$

$$F_1 = \frac{2*Recall*Precision}{Recall+Precision}$$

4.1   The Effectiveness of Web Data Processing

By using our web data processing method to get the training set, there are two adjustable parameters to be decided: C, the threshold to decide whether the text is a title; T, to judge whether the title and the image in the same web page have high semantic correspondence. In our experiments, we choose $C = 0.4$. which is because when the text is bold, with larger font size and the length no more than 5, $Score(text) = 0.4$. When the title and the image lie in the same $TABLE$ field, which means $Lay(TABLE)$ is 1, as well as $Count(TR) \leq 10$, we assign this title as the semantic label of the given image. Accordingly, we choose $T = 1$. In this manner, we obtain the training set with 1600 images. Because of the complicated structures of web pages, some incorrect label information is usually introduced when correlation is set up for an image where the title of it is not related to its semantic meanings. Our training set contains 340 noise images with false class labels, makes up to about 20% percent. It is obvious that the quality of our training set automatically obtained by our method is below than the manually annotating ones, which bring great challenges for our annotation algorithm.

Because LFDA is effective on multi-modal problem, in order to differentiate that the performance of image annotation based on MLFDA is improved by alleviating the bias of noise data or by addressing the multi-modal problem, a sub data set with more noise is selected from the global training set. The sub training set consists 6 semantic classes with relatively less modal, and 296 training samples. The noise data makes up to 30% percent, and the noise images all have very different visual features with the true samples in the class. In the following, we denote the global training set as DATA1, the sub training set as DATA2.

4.2   The Effectiveness of MLFDA-Based Image Annotation

We compare the annotation performance of our MLFDA-based annotation method with generative modal based MBRM[1], SVM, and GDA(Generalized Discriminant Analysis)[4]. In the proposed MLFDA-based annotation method, we set $k = 15$ for the best performance. The experimental results on two training sets are shown in Table 1 and Table 2 respectively.

Table 1. The effectiveness of our method compared with MBRM, SVM and GDA on DATA1

	DATA1			
	MBRM	SVM	GDA	MLFDA
Recall	15.9%	25.4%	19.9%	30.1%
Precision	20.8%	30.8 %	35.1%	35.5%
F_1 measure	0.18	0.28	0.25	0.33

Table 2. The effectiveness of our method compared with MBRM, SVM and GDA on DATA2

	DATA2			
	MBRM	SVM	GDA	MLFDA
Recall	16.5%	31.3%	33.8%	38.3%
Precision	24.2%	43.9 %	37.3%	43.2%
F_1 measure	0.20	0.37	0.35	0.41

It is obvious that the F_1 measure of MBRM is much lower than that of other three methods on both training sets, which because MBRM does not exploit the discriminant information between classes and is strongly influenced by the noisy data. When the test images have high visual similarity with noisy data, the class label is easily to be mistaken. Our proposed MLFDA-based method generally performs best on both training set. From Table 1 and Table 2, we can observe that compared with other three methods, the improvement of the precision measure by using MLFDA in DATA1 is not as significant as that in DATA2, while the F_1 measure is much more improved in DATA1. This is because in DATA1, there are noisy data as well as the multi-model data, our MLFDA can deal with both problems effectively. The excellent performance on DATA2 with more noise demonstrates that our proposed MLFDA-based method can effectively alleviate the influence of noisy data on defining the separable subspace by regarding the them as the multi modal samples of the class. The results in

Table 1 and Table 2 also demonstrate that the SVM and the GDA have similar performance, and SVM is slightly better than GDA in our data sets.

Fig. 4 shows some web image examples whose semantic keywords can not be obtained from their context directly, the correct semantic labels provided by our method are also listed in the figure. From Fig. 4, we can see that by using our proposed method, more web images can get their semantic descriptions, which will provide more relevant results for web users.

hamburger   F1 car   monkey mountain snow mountain sky pyramid sphinx Egypt

hurricane   flower plant   mobile telephone   Mickey Mouse   mountain Great Wall

Fig. 4. Some image examples

5   Conclusions

In this paper, we proposed a novel web image auto-annotation framework, in which the method of automatically extracting the training samples and a MLFDA-based image annotation algorithm are proposed. The experimental results indicated that our method is effective and outperforms other related methods. In the future, we will design more effective methods to address the noisy data with similar visual content but false labels. We also plan to solve the computational problems associated with large data sets.

References

1. Feng, S.L., Manmatha, R., Lavrenko, V.: Multiple Bernoulli Relevance Models for Image and Video Annotation. In: Proc. CVPR 2004, pp. 1002–1009 (2004)
2. Barnard, K., Forsyth, D.: Learning the semantics of words and pictures. In: Proc. ICCV, pp. 408–415 (2001)
3. Gao, Y.L., Fan, J.P., Xue, X.Y., Jain, R.: Automatic Image Annotation by Incorporating Feature Hierarchy and Boosting to Scale up SVM Classifiers. In: Proc. ACM Multimedia, pp. 901–910 (2006)
4. Baudat, G., Anouar, F.: Generalized Discriminant Analysis Using a Kernel Approach. Neural Computation 12(10), 2385–2404 (2000)
5. HTML Parser, http://htmlparser.sourceforge.net

6. Wang, X.J., Zhang, L., Jing, F., Ma, W.Y.: AnnoSearch: Image Auto-Annotation by Search. In: Proc. CVPR, pp. 1483–1490 (2006)
7. Hua, Z.G., Wang, X.J., Liu, Q.S., Lu, H.Q.: Semantic Knowledge Extraction and annotation for Web Images. In: Proc. ACM Multimedia, pp. 467–470 (2005)
8. Li, J., Wang, J.Z.: Real-Time Computerized Annotation of Picture. In: Proc. ACM Multimedia, pp. 911–920 (2006)
9. Sugiyama, M.: Local Fisher Discriminant Analysis for Supervised Dimensionality Reduction. In: Proc. ICML, pp. 905–912 (2006)
10. Neil, D., Lawrence, B.: Scholkopf, Estimating a Kernel Fisher Discriminant in the Presence of Label Noise. In: Proc. ICML, pp. 306–313 (2001)
11. Zhu, X.Q., Wu, X.D.: Class Noise vs. Attribute Noise: A Quantitative Study of Their Impacts. Proc. Artificial Intelligence Review 22, 177–210 (2004)
12. Carneiro, G., Vasconcelos, N.: Formulating Semantic Image Annotation as a Supervised Learning Problem. In: Proc. CVPR, pp. 163–168 (2005)
13. Deng, C., Yu, S., Wen, J., et al.: VIPS:A Vision-Based Page Segmentation Algorithm. Microsoft Technical Report, MSR-TR-2003-79 (2003)
14. Cai, D., He, X.F., Li, Z.W., Ma, W.Y., Wen, J.R.: Hierarchical Clustering of WWW Image Search Results Using Visual, Textual and Link Information. In: Proc. ACM Multimedia, pp. 952–959 (2004)

An Effective Query Relaxation Solution for the Deep Web[*]

Ye Ma, Derong Shen, Yue Kou, and Wei Liu

Department. of Computer, Northeastern University, Shenyang, 110004, China
maye0429@163.com, Shendr@mail.neu.edu.cn

Abstract. The information on the deep web is much more abundant than the surface web, so it is important to make the best use of it. However, in the process of query, it is difficult to avoid the so-called failed queries that make no result. Instead of notifying the user that there is no result, it is more cooperative to modify the raw query to return non-empty result set. Inspired by the observations on the deep web, this paper presents a query relaxation solution. Firstly, it applies the technique of query probing to obtain data samples from the underlying deep web databases. Based on these data samples, the important degree of attributes are obtained by employing approximate functional dependence. Secondly, the databases matching the query better are chosen and divided into some groups in terms of their schemas. Then the groups are organized into a directed acyclic graph called database relationship graph (DRG) to implement query relaxation. Finally, it returns some results satisfying the query better. We have conducted experiments to demonstrate the feasibility and the efficiency of the solution.

1 Introduction

Along with the explosion of information resource, searching on the web is becoming the best way to obtain valuable knowledge. While not all the users are professional, they may not know how to exactly describe what they want and sometimes they type a wrong word into the query interface. Therefore, query systems may possibly return some results the users don't want or even no result at all. Such a query is called failed query by us. Sometimes the case emerges, especially for web databases. Since web databases only provide Web access interfaces to users, and their data and schema information are transparent to them.

Query relaxation aims to modify the failed query to obtain more satisfying results. Although the results obtained via query relaxation may not satisfy the user's query completely, some methods can be adopted to evaluate and filter them for obtaining a fixed quantity of results satisfying the query better to the user. In this way, query relaxation can avoid the occurrence of failed query. In other words, no matter what the user inputs, the system will always return a nonempty result set.

---

[*] This research is supported by the National Natural Science Foundation of China under Grant No. 60673139, 60573090.

Y. Zhang et al. (Eds.): APWeb 2008, LNCS 4976, pp. 649–659, 2008.

This paper focuses on query relaxation on the deep web and an effective query relaxation solution is proposed. When a failed query comes up, the query is relaxed transparently to return better results to users without manual intervention.

The main contributions of the paper are:

(1) An effective query relaxation solution is proposed, in which, DRG is provided to organize the candidate databases. Based on DRG, it is easy to specify the order of relaxing a failed query.

(2) An optimization strategy on DRG is presented to enhance the efficiency of query relaxation further.

(3) Experiments manifest the efficiency and feasibility of our solution.

2   Related Work

When a query fails, it is more cooperative to identify the causes, rather than just to report the empty result set. Information system with such capability is known as Cooperative Information System. Query relaxation and its related techniques have been investigated in both IR and DB areas, but most of the researches [1-7] focus on query relaxation in the local database which can provide enough information to assist the query relaxation. And it is not necessary to consider about the time cost of network transmission.

In deep web environment, all the data sources we faced are autonomous web databases, and there is no way to access data from them except to query through their query interfaces. Therefore the query relaxation methods for the local database are not appropriate here. The methods presented by Muslea and Nambiar are the familiar query relaxation methods for web databases.

Ion Muslea [8] [9] respectively used Decision Tree and Bayesian Network on the data samples of target databases to get candidate queries. The candidate queries are similar to the original query, and they are sure to get nonempty results because the candidate queries are actually some tuples in the data samples. The shortcoming of this method is that it will execute the complex data mining algorithm on every data sample for each arriving query, so its time cost is too high.

Ullas Nambiar [10] [11] employed approximate functional dependency to get the important degree of the schema attributes in a database, according to which the order of the relaxed attributes is specified, and the data samples used to compute the important degree are chosen randomly. Nambiar just thought there was only one autonomous web database and generated a query relaxation plan for this single data source.

The problem we faced is to make a query relaxation plan for multiple target databases, so it is necessary to improve the primary method.

3   Observation and Motivation

Generally, there are two kinds of methods to relax a query: value relaxation and attribute relaxation. Value relaxation is to lessen the constraints on an attribute, that is, to widen the range of acceptable data. However, the value range is difficult to acquire from an autonomous database on the web, in addition, not only the cost of query probing usually adopted for getting the value range metadata is higher, but also but also with lower accuracy. Moreover, the data in web databases may be changed frequently. Therefore the method of value relaxation is not a good choice for the deep web.

Attribute relaxation discards the constraints on an attribute which is easy to implement. But it may bring more inaccurate results. However, the deficiency can be compensated by filtering the results, and its cost is smaller than that of query probing. Hereby attribute relaxation method is chosen to relax query on the deep web by us.

On the deep web, when a user inputs a query, usually some data sources satisfying the query better are executed to return the results to the user, in the same way, more data sources are involved for relaxing a failed query. According to some intuitive methods, failed queries are relaxed in the following two ways.

Method 1: The executing space for the query is the whole of the chosen databases, that is to say, the ordinary query may be relaxed to many forms of queries, and each relaxed query will be sent to all the chosen databases. If the query interface of a data source does not cover all the attributes of the query, the query is actually relaxed automatically, and the more than one relaxed queries containing more attributes than an interface will be executed in the database in the same way. The repeated executions cost much time and waste system resources.

Method 2: The executing space for a query is limited in a subset of the chosen databases whose attributes satisfy the query completely. In this way, we can get the relaxed results, The executing space for a query is limited in a subset of the chosen databases whose attributes satisfy the query completely. In this way, we can get the relaxed results, but lots of databases out the subset satisfying the query may be excluded.

Considering the above solutions, the idea of our query relaxation solution is as follows: (1) the relaxed query will be sent to some proper data sources in terms of its requirement. (2) According to most of the popular relaxation methods, the least important attribute will be relaxed first, and (3) the databases whose schemas best satisfying the relaxed query will be accessed. (4) Since a query sometimes need to be relaxed in many times and the cost of finding the eligible databases is higher for each time, it is better to organize the chosen databases together first according to the matching degree between the database schema and the query. And then the order of databases to be relaxed is easy to specify, so that the time cost is decreased. Finally the query will be relaxed based on the specific databases.

4 Problem Statement

A Deep web data source is a website that is composed of a web query interface and an underlying database. Each data source has its particular query interface and result page which are called input schema and output schema respectively. In the process of query relaxation we focus on the input schema, therefore a data source can be described as: $ds_i = \{a_1, a_2, ... a_n\}$, where $a_j (1 \leq j \leq n)$ is an attribute that appears in the input schema of ds_i.

Most underlying databases of the deep web sites are relational databases, so in the paper we consider a result tuple as $t = \{a_1 = x_1, a_2 = x_2, ..., a_n = x_n\}$. Similarly the query model is denoted as $Q = \{a_1 = x_1, a_2 = x_2, ..., a_m = x_m\}$.

Given a deep web grid system G and a user query Q_0, many data sources ds_i ($1 \leq i \leq m$) providing query services are registered in G, including their accessing addresses, interface schemas together with the mapping information from the interface schemas to the global schema S_g. The system G provides a user interface in terms of the global schema, and the user query Q_0 is composed of attribute-value pairs.

Our goal is to relax a failed query to return as many as possible satisfying results to the user with lower cost. How much a result tuple t satisfying a query Q depends on the similarity degree $Sim(Q,t)$, and the time cost T is estimated by the times of sending the query to the actual data sources. The similarity degree and the time cost will be discussed in details in Section 5.

5 Solution

5.1 Preprocessing

Before the query relaxation, the system need to know the important degree of the attributes involved in the query. We call the phase of getting the attribute important degree as preprocessing phase. The attribute important degree is calculated based on the deep web databases, so it is necessary to get some data samples from the underlying databases by query probing.

5.1.1 Data Sampling

Via query-based sampling [12] [13], we can get the data samples from autonomous databases on the web. To manifest the data characteristics, the probing process is adopted by us as follows: Firstly, an initial query with a single-attribute query is chosen and executed in some databases, and the returned results are as part of data samples. Secondly, a new single-attribute query with the sample values got before is created and repeats the sampling process on other web databases until the termination condition is satisfied. The data samples obtained are better for summarizing the data in a specific domain.

5.1.2 Attribute Important Degree Analysis

The approximate functional dependency is adopted to calculate the important degree of the database attributes, and the concepts about approximate functional dependency are as follows.

Approximate Functional Dependency (AFD): The functional dependency $X \rightarrow A$ is an approximate functional dependency if it does not hold over a small fraction of the tuples. The error rate $error(X \rightarrow A)$ is the ratio of the number of the tuples that doesn't satisfy the relationship to the number of all the tuples.

Approximate Key (AK): An attribute set X is a key if no two distinct tuples agree on X. Let error(X) be the minimal fraction of tuples that need to be removed from relation r for X to be a key. The error rate error(X) is the ratio of the number of the tuples that need to be removed to the number of all the tuples.

One subtract the error rate gives the support rate of AFD or AK. It is convenient to get all the AFDs and AKs whose error rate is lower than a threshold T_{err} ($0 < T_{err} < 1$) by

Huhtala's method TANE [14], then the AK with the maximum support rate is obtained, and divides the attributes into two groups in terms of the AK, namely, the decide set consists of the attributes in the AK and the depend set consists of the other attributes except AK.

The weight of attributes in the decide set is defined as follow:

$$Weight_{decides}(k) = \sum \frac{1 - error(A \rightarrow k^{'})}{size(A)} \tag{1}$$

Where A is the set of all the AFDs' decide attributes including the attribute k, $k^{'}$ is an attribute that doesn't appear in A.

The weight of attributes in the depend set is defined as follow:

$$Weight_{depends}(j) = \sum \frac{1 - error(A \rightarrow j)}{size(A)} \tag{2}$$

Where A is the set of all the AFDs' decide attributes including the attribute j.

5.2   Query Execution

It is obviously unreasonable to send the query to all the databases registered on the system, especially for the databases whose schemas do not match with the query. Therefore, the first step of the query execution phase is to choose all the databases satisfying the query as the candidate databases, and then some proper databases are selected from the candidate databases each time for executing the relaxed query dynamically.

5.2.1   Databases Selection
Firstly, the candidate databases whose schemas satisfy the query are chosen from all of the deep web databases in the registry of system G, and the formula to calculate the matching score of a database is as follows:

$$Score_j = \frac{\sum Weight(a_i) \times isContain(a_i)}{\sum Weight(a_i)} \tag{3}$$

Where a_i is an attribute included by the query Q_0, if the query interface of the data source ds_j includes a_i, $isContain(a_i)$ is 1, otherwise $isContain(a_i)$ is 0.

5.2.2   Databases Relationship Graph (DRG) Construction
To accomplish the process of query relaxation efficiently, we introduce a directed acyclic graph as databases relationship graph.

First, the candidate databases are divided into some groups in each of which the databases have the same interface schema, and we call them databases candidate groups. Then the databases relationship graph is constructed, which is denoted as a directed acyclic graph. In the graph, each node n_i represents a candidate group, and it has enough information about which nodes it contains directly and which nodes are directly contained by it. For example, if the attribute set of node n_a is the superset of the attribute set of n_b and the attribute set of n_b is the superset of the attribute set of n_c, then n_a directly contains n_b, n_b directly contains n_c, and there is no directly containing relationship between n_a and n_c. The child nodes of node n_i are the nodes contained by n_i directly and they are arranged in the reverse order in terms of their matching scores, and the parent nodes of n_i are the ones directly containing n_i.

During query relaxation, Q_0 is first executed in a candidate group that matches the query completely. If there is no result or there is not such a group at all, the relaxation will occur on the least important attribute. Then Q_0 becomes a new query Q_1. Q_1 is executed in the candidate groups which totally match Q_1 based on DRG. The above procedure will be processed repeatedly until some results are obtained.

The DRG is constructed as follow:

(1) The candidate groups are ordered reversely in terms of their matching scores.

(2) Create an initial node n_s that includes all the attributes involved in the query Q_0 if there is no candidate group corresponding to this node.

(3) Insert a candidate group cg_i into the DRG as follow: Let node n_i include all the attributes of the group cg_i and create n_i's temporary parent nodes array $n_t[]$.

(4) Traverse the DRG from initial node n_s, if n_i's attributes are the same as n_s, replace n_s with n_i, if node n_j contains n_i, add n_j to $n_t[]$ and the nodes containing n_j are removed from $n_t[]$. When all the nodes in DRG have been traversed, add all the nodes in $n_t[]$ to the parent nodes of n_i, and add n_i to those nodes' child nodes.

Example 1: Let a query include 5 attributes denoted as a, b, c, d, e, and there are some candidate databases satisfying some of the attributes are selected from the G, based on the intersections between each candidate database attribute set and the attribute set of the query, suppose these databases are divided into 10 candidate groups and organized as figure 1 according to the above construction rules, and s0 is the initial node, s11, s12, s13 are the child nodes of s0 in descending order of their matching scores.

5.2.3 Query Relaxation

The whole query relaxation can be summarized as two aspects: moving the pointer to traverse the DRG and sending the query to the proper nodes.

The pointer is moved according to the following rules: When the pointer points to node n_i, and n_i must be a child node except that n_i is the initial node. If n_i is the last one of the child nodes, the pointer's next position is the first child node of the first sibling node of n_i, for example, in Fig. 1, s13(a, b, c) is the last child node of s0(a,b, c, d, e), and s11 (a, b, d, e) is the first sibling node of s13 (a, b, c), so the pointer next points the node s112 (a, d, e). If n_i is not the last child node, the pointer's next position is its next sibling node. If n_i is the initial node, the pointer's next position is its first child node.

The query relaxation rules are as follows: When the pointer points to node n_i, relax the raw query Q_0 to Q_i which has the same attributes as n_i, and send Q_i to n_i

and all the nodes containing n_i, that is to say, find all the paths from n_i to initial node n_s, and send Q_i to all the nodes in the paths, including n_i and n_s. Those nodes include all the attributes in Q_i, and it is possible to get some results from those nodes even though they have been traversed before by the query with more constraints.

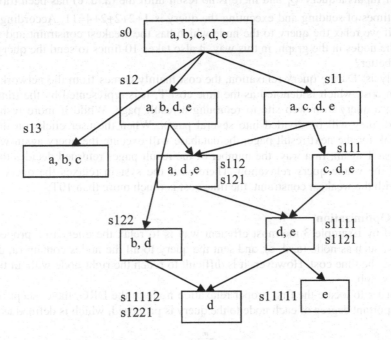

Fig. 1. The DRG of candidate databases in Example 1

Example 2: let the query user inputs be (a=10, b=20, c=15, d=30, e=20) and the chosen candidate databases be divided into 10 candidate groups, then the candidate groups construct a DRG as Fig. 1, so the process of query relaxation is as follows:

Step 1: Send (a=10, b=20, c=15, d=30, e=20) to node (a, b, c, d, e), and there is no result.

Step 2: Since the child nodes of (a, b, c, d, e) are (a, b, d, e), (a, c, d, e), (a, b, c), then the query is relaxed as (a=10, b=20, d=30, e=20), and send it to (a, b, d, e) and (a, b, c, d, e), and there is no result.

Step 3: The query is relaxed as (a=10, c=15, d=30, e=20), and send it to (a, b, d, e) and (a, b, c, d, e). There is no result.

Step 4: The query is relaxed as (a=10, b=20, c=15), and send it to (a, b, c) and (a, b, c, d, e), and there is no result.

Step 5: The query is relaxed as (a=10, d=30, e=20), and send it to (a, d, e), (a, b, d, e), (a, c, d, e) and (a, b, c, d, e), now there are some results, and the process of query relaxation is terminated.

If the whole DRG has been traversed and there is no result, go back to the phase of data source selection to choose some new databases to insert into the DRG, and the relaxation continues until there are enough results or time out.

Example 3: Let the candidate groups are constructed as a DRG shown in Fig. 1, and the user inputs a query Q_1 and there is no result until the (a, d, e) has been traversed, so the times of sending and executing the query is 1+2+2+2+4=11. According to the DRG, if we relax the query to the node which has the weakest constraint and send it to all the nodes in the graph, in this way, it also takes 10 times to send the query only, Is that better?

Analysis: During query relaxation, the cost mainly comes from the network transmission cost, which is denoted as the time cost T, it is represented by the time from sending a query to a web site to returning a result page. While if more results are selected, they will be divided into several pages. When the user clicks on the next page link for the next result page, the database will execute the query again with the time cost T again. In a way, the number of the result pages returned decides the time cost of the whole query relaxation. Thereby, if the system relaxes the query as the node with the weakest constraint, the time cost is much more than 10T.

5.2.4 Optimization

Inspired by Example 3 the most efficient way is to relax the query to a proper node directly, such as node (a, d, e), and sent the query to all the nodes contain (a, d, e) to decrease the time cost. However, it is difficult to reach the right node without traversing the graph.

In order to locate the first proper relaxation node in the DRG, the concept of relative important degree of each node to the query is proposed, which is defined as:

$$RW = \frac{\sum_{i=1}^{m} w_i}{\sum_{j=1}^{n} w_j} \qquad (4)$$

Where n is the number of attributes in the query and m is the number of attributes in a node, and w_i is the important degree of the i_{th} attribute.

Given a relative important degree threshold RW_0, then traverse the DRG to find the first node whose relative important degree is less than RW_0 as the start node.

In addition, given a result threshold M, it can be decided whether the returned results are sufficient or not, and two methods are adopted to deal with the large quantity of results returned. One is to execute the strict local query further, and the other is to tighten the relaxed query again.

5.3 Result Selection

The goal of this section is to choose the most similar results from the result set to the user. The similarity degree between a tuple t and a query Q is denoted as follow:

$$Sim(Q,t) = \sum_{i=1}^{n} Weight(a_i) \times Sim_{a_i}(Q,t) \qquad (5)$$

(1) If the attribute a_i is numerical, then $Sim_{a_i}(Q,t)$ is defined as:

$$Sim_{a_i}(Q,t) = 1 - \frac{|Q.a_i - t.a_i|}{\max diff_{a_i}} \qquad (6)$$

Where the $\max diff_{a_i}$ is the maximum difference on the attribute a_i in the data sample. If $|Q.a_i - t.a_i|$ is more than $\max diff_{a_i}$, the $Sim_{a_i}(Q,t)$ is 0, and then save the $|Q.a_i - t.a_i|$ to update $\max diff_{a_i}$.

(2) If the attribute a_i belongs to character type, then $Sim_{a_i}(Q,t)$ is defined as:

$$Sim_{a_i}(Q,t) = 1 - \frac{ed(Q.a_i, t.a_i)}{\max ed_{a_i}} \qquad (7)$$

Where the $\max ed_{a_i}$ is the maximum edit distance on the attribute a_i in the data sample, and $ed(Q.a_i, t.a_i)$ is the edit distance between $Q.a_i$ and $t.a_i$. If the $ed(Q.a_i, t.a_i)$ is more than $\max ed_{a_i}$, the $Sim_{a_i}(Q,t)$ is 0, and then save the $ed(Q.a_i, t.a_i)$ to update $\max ed_{a_i}$.

(3) If the attribute a_i is Boolean, then the similarity degree between two values is 1 if they are the same, otherwise it is 0.

6 Experiments

We conducted a simulation experiment to evaluate the feasibility and efficiency of our solution. In the experiment, 100 data sources are used, each of which is described by at most 7 attributes, and 100 queries are defined. All the tests run on a computer with Intel(R) Core(TM) 2 CPU with 1G RAM.

Nambiar's AIMQ [10] [11] is an existing query relaxation method for autonomous web database, but it is designed for only one data source. If we want use it in the multi-databases environment, the intuitive opinion is to apply it on each of databases and to integrate the results from them. The query relaxation on each database is independent. We call this method as intuitive method. We compare the relaxing performance by using both DRG and intuitive method to data sources.

Fig. 2 shows the average query times of DRG and intuitive method with different numbers of candidate databases. The times of queries are the times of T which is defined in 5.2.3. It demonstrates that the DRG method is more efficient than the intuitive method, since the average query times of DRG is less than that of intuitive method no matter what the number of candidate databases is.

Fig. 3 shows the average similarity degree between the query request and the results returned by DRG and the intuitive method with different numbers of candidate databases. It demonstrates that DRG method is more available than the intuitive method, since the similarity degree of the results returned by DRG is bigger than the intuitive method.

Fig. 4 shows the average query times of optimized DRG and plain DRG with 30 candidate databases and different relative important degree. If 10 is the average query time of the plain DRG with 30 candidate databases, while the average query times of the optimized DRG are less than 10, so it demonstrates that the optimized method is more effective.

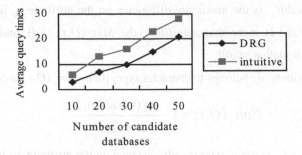

Fig. 2. Average query times of DRG and intuitive method

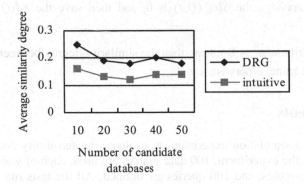

Fig. 3. The average similarity degree between the query and the results

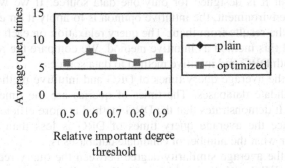

Fig. 4. The average query times of optimized DRG and plain DRG

7 Conclusion

In this paper, we present a query relaxation solution to enlarge quantity of heterogeneous databases on the deep web, Firstly, based on the data samples, the important degree of the attributes are obtained by employing the approximate functional dependence. Secondly, some of the candidate databases which match the user's query are chosen and organized into a database relationship group (DRG) to implement query relaxation. Thirdly, the results are ranked from the results returned according to their matching degrees. Finally, the experiments demonstrate the feasibility and the efficiency of the solution.

Next, we will focus on the optimization measures of query relaxation further to decrease the cost further.

References

1. Kaplan, S.J.: Cooperative Aspects of Database Interactions. Artificial Intelligence, 165–187 (1982)
2. Motro, A.: FLEX: A Tolerant and Cooperative User Interface to Databases. IEEE Trans. On Knowledge and Data Engineering (TKDE), 231–246 (1990)
3. Chu, W.W., Yang, H., Chiang, K., Minock, M., Chow, G., Larson, C.: CoBase: A Scalable and Extensible Cooperative Information System. J. Intelligent Information Systems (JIIS), 223–259 (1996)
4. Gaasterland, T.: Cooperative answering through controlled query relaxation. IEEE Expert: Intelligent Systems and Their Applications, 48–59 (1997)
5. Godfrey, P.: Minimization in Cooperative Response to Failing Database Queries. In: Int'l J. Cooperative Information Systems (IJCIS), pp. 95–149 (1997)
6. Chaudhuri, S.: Generalization and a Framework for Query Modification. In: IEEE ICDE, Los Angeles, CA, pp. 138–145 (1990)
7. Koudas, N., Li, C., Tung, A., Vernica, R.: Relaxing Join and Selection Queries. In: VLDB, pp. 199–210 (2006)
8. Muslea, I.: Machine learning for online query relaxation. In: KDD, pp. 246–255 (2004)
9. Muslea, I., Lee, T.J.: Online query relaxation via Bayesian causal structures discovery. In: AAAI, pp. 831–836 (2005)
10. Nambiar, U., Kambhampati, S.: Answering Imprecise Queries over Autonomous Web Databases. In: ICDE, p. 45 (2006)
11. Nambiar, U., Kambhampati, S.: Mining Approximate Functional Dependencies and Concept Similarities to Answer Imprecise Queries. In: WebDB, pp. 73–78 (2004)
12. Callan, J.: Query-Based Sampling of Text Databases. ACM Transactions on Information Systems, 97–130 (2001)
13. Callan, J.: Automatic Discovery of Language Models for Text Databases. SIGMOD, 479–490 (1999)
14. Huhtala, Y., Krkkinen, J., Porkka, P., Toivonen, H.: Efficient discovery of functional and approximate dependencies using partitions. In: Proceedings of ICDE, pp. 392–401 (1998)

A Framework for Query Capabilities and Interface Design of Mediators on the Gulf of Mexico Data Sources[*]

Longzhuang Li, John Fernandez, and Hongyu Guo

Department of Computing Sciences, Texas A&M University-Corpus Christi,
Corpus Christi, TX 78412, USA
{Longzhuang.Li,John.Fernandez,Hongyu.Guo}tamucc.edu

Abstract. The integration of various data sources collected from the Gulf of Mexico (GOM) will become a valuable resource for the public, local government officials, scientists, natural resource managers, and educators. Due to the exclusive and distributive nature of these data, a new framework is developed to retrieve partial results from the underlying data sources to answer more user queries for the union operator. In addition, the user interface of mediators is considered when computing the query capabilities.

Keywords: Data integration, mediator query capability, base view.

1 Introduction

Historically, environmental, hydrographic, meteorological, and oceanographic data have been collected and made available on the Internet by numerous local, state, and federal agencies as well as by universities around the Gulf of Mexico (GOM). Currently, users have to manually interact with these large collections of Internet data sources, determine which ones to access and how to access, and manually merge results from different sources. Without an adequate system and personnel for managing data, the magnitude of the effort needed to deal with such large and complex data sets can be a substantial barrier to the GOM research. We intend to answer complex queries by developing mediator-based data integration systems which using a uniform interface to access the underlying distributed and heterogeneous sources.

In mediator systems, the data sources to be integrated generally have different limitations and ways that users have to follow to query on their Web interface. For example, to retrieve data from TCOON[1] and TABS[2], TCOON requires at least one station from a station list and at least one field from a list of series identifiers, while TABS mandates only one station ID.

All the current methods [3,6,7] demand that the mediator queries satisfy the query conditions of all the underlying data sources for the union operator. This requirement is impractical in our data integration on the GOM. Unlike the simple bookstore database

---

[*] Research is supported in part by the National Science Foundation under grant CNS-0708596.
[1] http://lighthouse.tamucc.edu/pq/
[2] http://tabs.gerg.tamu.edu/Tglo/

Y. Zhang et al. (Eds.): APWeb 2008, LNCS 4976, pp. 660–671, 2008.

examples in [3,6,7], each GOM database measures and collects various environmental, biological, or ecological information at different locations and different sample rates. Although some attributes are measured by multiple GOM databases, no two GOM databases contain the same values for these attributes because they are measured at different locations and different times. In addition, the data collected from the GOM is usually limited and incomplete. As a consequence, even partial results are highly desirable for the GOM research, if not all the databases satisfy the query conditions for the union operator. For example, to retrieve data from TCOON and TABS, we must select at least one TCOON station and only one TABS station, but each station is exclusive, belonging to TCOON or TABS but not both.

The framework in this paper is developed from that proposed by Yerneni, et al. in [7]. Our framework extends the method in [7] in several aspects. First, it allows users to select single or multiple values for an attribute. For example, the method in [7] only permits the single value selection for an attribute, which can not emulate multiple value selections for TCOON's station ID or series identifier attribute. Second, it allows the partial answer from a subset of data sources, if not all data sources satisfy the query conditions for the union operator. Correspondingly, we develop the preprocessing technique to handle the partial answer. And finally, we employ the new attribute adornments (see Section 3) for the design of a mediator's user interface, which is not discussed in any of the existing methods.

2 Related Work

Most research projects in the area of capability description in the mediator systems are focusing on the capability description of the underlying data sources. For example, TSIMMIS [1] and Information Manifold [2] only allow the capability description of data sources in terms of conjunctive queries, while DISCO [5] and Garlic [4] allow queries in both the disjunctive and conjunctive expressions. But these systems do not compute the query capabilities of the mediators. So users have to follow the trial and error process to figure out the queries that can actually be answered by the mediator. In addition, it is impossible to take advantage of treating the existing mediators as sources for new ones [3].

There are relatively fewer research efforts on the capability description of the mediators, although it is an important research issue. Tang, et al. [6] developed capability object integration algebra to compute the mediator capabilities from the capabilities of underlying data sources. Since the proposed algebra can not distinguish between mandatory and optional attribute values/list of values, their method lacks of expressive power in real-world sources. Pan, et al [3] improved [7] by permitting multiple values and several comparison operators instead of only "=" operator for attributes. The method in [3] still does not allow the partial results as we do for the union operator and do not consider the interface design issue for the mediators.

Our framework is most similar to the one proposed in Yerneni, et al. [7]. But the framework in [7] requires such stringent mediator query conditions that some queries not satisfying any mapping function are still answerable. In addition, the framework in [7] offers a very limited mechanism to represent the query capabilities of sources and mediators.

3 The Framework for Computing Query Capabilities

In this section, we present our framework to describe the query capabilities of data sources and mediators. We extend the method proposed in [7] to tackle more complicated real-world situations. We allow users to select multiple values for an attribute, and allow the answers to come from a subset of the data sources for the union operator. The supported queries of a mediator are based on the query processing capabilities of data sources to integrate. When a query is submitted, the mediator translates the query into a set of relevant sub-queries associated with underlying data sources, and then the mediator merges the sub-query results via the pre-defined operations, union, join, selection, or projection. In addition, three techniques affecting the query capability of a mediator are preprocessing, post-filtering, and pass bindings. The details of the above three techniques can be found in Section 4, 5, and 6, respectively.

On the Web, data sources typically publish their query processing capabilities through query forms or base-view templates, which are filled out by users and submitted to the data sources as queries. Each data source may have more than one base view. In this paper we use *data source* and *base view* interchangeably. We use attribute adornments to specify how the attributes participate in supported queries. We adapt and extend the five attribute adornments (f, u, b, c, o) developed in [7] to the following eleven adornments:

- f: the attribute value is optional in the query;
- u: the attribute value can not be specified in the query;
- b: one attribute value must be specified in the query;
- $sc[l]$: one attribute value must be chosen from the list l;
- $mc[l]$: one or more attribute values must be chosen from the list l;
- $so[l]$: the attribute is optional in the query, and if specified, one attribute value must be chosen from the list l;
- $mo[l]$: the attribute is optional in the query, and if specified, one or more attribute values must be chosen from the list l.
- $bsc[l]$: one attribute value must be specified or chosen from the list l, which is equivalent to b *or* $sc[l]$;
- $bmc[l]$: one attribute value must be specified in the query, or one or more attribute values must be chosen from the list l, which is same as b *or* $mc[l]$,
- $fso[l]$: the attribute is optional in the query, and if specified, it may be a user specified value or be chosen from the list l, which is equivalent to f *or* $so[l]$;
- $fmo[l]$: the attribute is optional in the query. it may be a user specified value or multiple values are chosen from the list l, which is same as f *or* $mo[l]$.

The first seven adornments are used to model both the data source and mediator's query capabilities, while the last four adornments are employed mostly for the capability interface design of the mediators. In particular, $bsc[l]$ and $bmc[l]$ can be simplified to b, and $fso[l]$ and $fmo[l]$ to f, respectively, if the list $l \subset b$ or $l \subset f$. But we choose to keep the last four adornments for two reasons: (1) we intend to emulate the underlying data sources' user interface as much as possible; (2) the users do not have to remember the mandatory/optional list of values in the underlying data sources. On the other hand, if the list l contains all the eligible values for the attribute, such that $l \equiv b$ or $l \equiv f$, $bsc[l]$ and $bmc[l]$ can be simplified to $sc[l]$ and $mc[l]$, and $fso[l]$ and $fmo[l]$ to $so[l]$ and $mo[l]$, respectively. But in the following sections, we assume that $l \subset b$ or $l \subset f$.

Next, we focus on computing mediator view templates for two base views. Both mediator views and base views as well as the associated adornments are represented using the notation $R(ATTR_1$ $(adorn_1)$, $ATTR_2$ $(adorn_2)$, ..., $ATTR_n$ $(adorn_n))$. For example, a data source view $D(X$ (b), Y (f), Z $(u))$ means the data source D has three attributes X, Y, and Z, and the associated adornments are b, f, and u, respectively. An instance of the source is $D(x_i, y_i, z_i)$.

4 Mediators Using Preprocessing Technique

Because a mediator query does not have to satisfy the query requirements of all the underlying data sources for the union operator, preprocessing means we break down the query into sub-queries and submit them to the appropriate data sources, but do not carry out any post-processing such as post-filterings and pass bindings. The same sub-query may be submitted more than once because we allow the users to choose multiple values for an attribute.

4.1 Union Views

The symmetric Table 1 presents the union-view attribute adornment template computed from the two base-view attribute adornments. For example, the combination of f and b is f because even if the data source with the b adornment can not execute without specifying a value, the second data source with f adornment can still return some partial results. Another example is that the combination of u and b is f because we can still obtain some partial answer from one or both two data sources whether or not we specify a value for the attribute. If we specify a value, the partial answer comes from the data source with the adornment b only, otherwise the partial answer is returned by another data source with the adornment u. The combination of u and f is also f because the sub-query can be submitted to both data sources without specifying an attribute value; otherwise the sub-query is submitted to the data source with the adornment f only.

When only one attribute adornment contains a list of single or multiple choices, users are able to specify a value or choose value(s) from the list. The combination of f and $sc[l5]$ is $fso[l5]$, which means for the attribute users have three options: either specify a value not in the list $l5$, leave it blank, or choose a value from the list $l5$. For the first two options, the query is submitted only to the data source with the f adornment, while for the third option, the query is submitted to the both data sources. Union of u and $sc[l5]$ is $so[l5]$ because if we select a value from the list $l5$, the sub-query is submitted only to the data source with the adornment $sc[l5]$, otherwise the sub-query is submitted only to the data source with the adornment u.

Table 1. Union-view template using preprocessing technique

	f	u	b	$sc[l_5]$	$mc[l_6]$	$so[l_7]$	$mo[l_8]$
f	f	f	f	$fso[l_5]$	$fmo[l_6]$	$fso[l_7]$	$fmo[l_8]$
u	f	u	f	$so[l_5]$	$mo[l6]$	$so[l_7]$	$mo[l_8]$
b	f	f	b	$bsc[l_5]$	$bmc[l6]$	$fso[l_7]$	$fmo[l_8]$
$sc[l_1]$	$fso[l_1]$	$so[l_1]$	$bsc[l_1]$	$sc[l_1 \cup l_5]$	$mc[l_1 \cup l_6]$	$so[l_1 \cup l_7]$	$mo[l_1 \cup l_8]$
$mc[l_2]$	$fmo[l_2]$	$mo[l_2]$	$bmc[l_2]$	$mc[l_2 \cup l_5]$	$mc[l_2 \cup l_6]$	$mo[l_2 \cup l_7]$	$mo[l_2 \cup l_8]$
$so[l_3]$	$fso[l_3]$	$so[l_3]$	$fso[l_3]$	$so[l_3 \cup l_5]$	$mo[l_3 \cup l_6]$	$so[l_3 \cup l_7]$	$mo[l_3 \cup l_8]$
$mo[l_4]$	$fmo[l_4]$	$mo[l_4]$	$fmo[l_4]$	$mo[l_4 \cup l_5]$	$mo[l_4 \cup l_6]$	$mo[l_4 \cup l_7]$	$mo[l_4 \cup l_8]$

Example 1. Given two sources, $D_1(X\ (b),\ Y\ (f),\ Z\ (u))$ and $D_2(X\ (sc[l_5]),\ Y\ (f),\ Z$ $(so[l_7]))$, the union of the two data sources is $M(X\ (bsc[l_5]),\ Y\ (f),\ Z\ (so[l_7]))$. Suppose the user selects a value $x_i \in l_5$ for X and leaves both Y and Z unspecified in the union view, the sub-query is submitted to both data sources. For all the other values of X, Y, and Z, at most one of D_1 or D_2 can answer the sub-queries.

When both base-view attribute adornments have the lists, we union the two lists. The two rules are (1) the combination of s (single) and m (multiple) becomes m, and (2) the combination of c (mandate) and o (option) becomes o. For example, the combination of $sc[l_1]$ and $mc[l_6]$ is $mc[l_1 \cup l_6]$. The union-view mediator queries are broken down to the sub-queries, which are passed over to the data source with the adornment $mc[l_6]$ at most once, and the data source with the adornment $sc[l_1]$ zero or multiple times.

Example 2. Given two sources, $D_1(X\ (sc[l_1]),\ Y\ (f),\ Z\ (u))$ and $D_2(X\ (mc[l_6]),\ Y\ (u),\ Z$ $(f))$, and l_1 with constants $\{n_1, n_2, n_3\}$ and l_6 with constants $\{n_3, n_4, n_5\}$, the union of the above two data sources is $M(X\ (mc[l_1 \cup l_6]),\ Y\ (f),\ Z\ (f))$, where $l_1 \cup l_6$ contains $\{n_1, n_2, n_3, n_4, n_5\}$. Suppose the user selects the following three constants, n_2, n_3, and n_4, then the sub-query with the values n_3 and n_4 is submitted to D_2 once, and the sub-query is submitted to D_1 twice with the value n_2 and n_3, respectively. If the user selects n_4 and n_5, then the sub-query with the values is only submitted to D_2.

4.2 Join Views

A mediator with the preprocessing technique treats a query on a join view as follows. First, the mediator query is broken down to the sub-queries and passed to the each corresponding joining base view, then the results from each base view are joined. For the mediator on a join view, the join attributes need to appear in every joining base views and the non-join attributes appears in only one of the base views. In addition, answers are required from both joining base views, which is different from the union-view templates in Table 1 because a union-view template allows the partial answer from only one base view. As a result, the computation of attribute adornments in a join-view template is different from that in a union-view template.

To join two base views, the adornments of all the non-join attributes are simply copied over from the base view templates. The adornment computation of the join attributes in two base view templates is shown in Table 2.

Table 2. Join-view template using preprocessing technique

	f	u	b	$sc[l_5]$	$mc[l_6]$	$so[l_7]$	$mo[l_8]$
f	f	f	b	$sc[l_5]$	$mc[l_6]$	$fso[l_7]$	$fmo[l_8]$
u	f	u	b	$sc[l_5]$	$mc[l_6]$	$so[l_7]$	$mo[l_8]$
b	b	b	b	$sc[l_5]$	$mc[l_6]$	$bsc[l_7]$	$bmc[l_8]$
$sc[l_1]$	$sc[l_1]$	$sc[l_1]$	$sc[l_1]$	$sc[l_1 \cap l_5]$	$mc[l_1 \cap l_6]$	$sc[l_1]$	$sc[l_1]$
$mc[l_2]$	$mc[l_2]$	$mc[l_2]$	$mc[l_2]$	$mc[l_2 \cap l_5]$	$mc[l_2 \cap l_6]$	$mc[l_2]$	$mc[l_2]$
$so[l_3]$	$fso[l_3]$	$so[l_3]$	$bsc[l_3]$	$sc[l_5]$	$mc[l_6]$	$so[l_3 \cup l_7]$	$mo[l_3 \cup l_8]$
$mo[l_4]$	$fmo[l_4]$	$mo[l_4]$	$bmc[l_4]$	$sc[l_5]$	$mc[l_6]$	$mo[l_4 \cup l_7]$	$mo[l_4 \cup l_8]$

In Table 2, when both base-view attribute adornments must choose from their given lists, the two lists are intersected. On the other hand, when both base-view adornments offer the optional choices from the given lists, the two lists are unioned together. For example, the combination of $sc[l_1]$ and $mc[l_6]$ is $mc[l_1 \cap l_6]$ because the multiple chosen values from $l_1 \cap l_6$ can be sent one by one to the base-view with the adornment $sc[l_1]$, then results are joined with those from the base-view with the adornment $mc[l_6]$. Another example is that the base-view adornments $so[l_3]$ and $so[l_7]$ generates the join-view adornment as $so[l_3 \cup l_7]$ because if the users choose a value that only appears in one of the two base-view lists, another base view templates can still be queried by specifying no value for the attribute.

When one base-view adornment has the mandatory list and the second base-view adornment has the optional list (either single choice s or multiple choice m), the combined adornment is the same as the base-view adornment with the mandatory list. For example, the merge of $sc[l_1]$ and $mo[l_8]$ is $sc[l_1]$ because no matter what value is chosen from l_1, we can always conduct the sub-query on another base view with the attribute adornment $mo[l_8]$ by disregarding any value from l_8.

When combining f with $so[l_7]$ and $mo[l_8]$, the resulting join-view adornments are $fso[l_7]$ and $fmo[l_8]$, respectively. Because if a specified value is not in l_7 or l_8, then the base view with the adornment $so[l_7]$ or $mo[l_8]$ is treated with no value specified for the attribute. Similarly, the computation for the combination of b with $so[l_7]$ and $mo[l_8]$ are $bsc[l_7]$ and $bmc[l_8]$, respectively, and the combination for the combination of u with $so[l_7]$ and $mo[l_8]$ are $so[l_7]$ and $mo[l_8]$, respectively.

4.3 Projection and Selection Views

Unlike the union or join attributes which occur in both the underlying data sources, the projected attributes only appear in one of the two underlying data sources. When receiving a query on a projection view, the mediator passes the derived sub-queries down to the corresponding base views without specifying values for the hidden base attributes, which only appear in the base views. So during the projection, we do not produce a projection-view template if any of the hidden attributes has a b, sc, and mc adornment in a base-view template. We can create projection-view templates as long as the hidden attributes have the f, so, mo, or u adornments in the base-view templates. The created projection-view template simply copies the projected attribute adornments from the base-view template.

Similar to the projection view, the selection view attributes only appear in one of the two base views. A selection-view query is processed by passing it down to the underlying base view and applied the selection predicate on the results of the base view. Therefore, a selection-view is generated by copying the corresponding base-view template.

5 Mediators Using Preprocessing and Post-filtering Techniques

In this section, we present a post-processing technique, post-filtering, which enables the mediators to support more queries than the ones that only employ a preprocessing technique. Post-filtering means that the returned results from a base view can be filtered according to the pre-specified attribute values.

Table 3. Union-view template using both preprocessing and post-filtering techniques

	f	u	b	$sc[l_5]$	$mc[l_6]$	$so[l_7]$	$mo[l_8]$
f	f	f	f	$fso[l_5]$	$fmo[l_6]$	$fso[l_7]$	$fmo[l_8]$
u	f	f	f	$fso[l_5]$	$fmo[l_6]$	$fso[l_7]$	$fmo[l_8]$
b	f	f	b	$bsc[l_5]$	$bmc[l_6]$	$fso[l_7]$	$fmo[l_8]$
$sc[l_1]$	$fso[l_1]$	$fso[l_1]$	$bsc[l_1]$	$sc[l_1 \cup l_5]$	$mc[l_1 \cup l_6]$	$fso[l_1 \cup l_7]$	$fmo[l_1 \cup l_8]$
$mc[l_2]$	$fmo[l_2]$	$fmo[l_2]$	$bmc[l_2]$	$mc[l_2 \cup l_5]$	$mc[l_2 \cup l_6]$	$fmo[l_2 \cup l_7]$	$fmo[l_2 \cup l_8]$
$so[l_3]$	$fso[l_3]$	$fso[l_3]$	$fso[l_3]$	$fso[l_3 \cup l_5]$	$fmo[l_3 \cup l_6]$	$fso[l_3 \cup l_7]$	$fmo[l_3 \cup l_8]$
$mo[l_4]$	$fmo[l_4]$	$fmo[l_4]$	$fmo[l_4]$	$fmo[l_4 \cup l_5]$	$fmo[l_4 \cup l_6]$	$fmo[l_4 \cup l_7]$	$fmo[l_4 \cup l_8]$

5.1 Union Views

Table 3 shows the new mapping function for the union of the attribute adornments in two base-view templates. In Table 3, we follow the same notation defined in Section 3. The essential differences between the mapping function of Table 3 and the mapping function used in Table 1 are the treatment of the u adornment as well as so and mo adornments. When the attribute adornment of a data source is u, the mediator can start a query on the data source without specifying a value for this attribute and then filter out the results that are not supported by a value optionally specified by the mediator query for this attribute in the post-processing step. As a result, the u adornment is treated the same way as the f adornment. For example, the combination of u and $sc[l_7]$ is $fso[l_7]$. For this case, no matter what value is specified or chosen and whether or not the value is in l_7, the results from the data source with the u adornment can be filtered in the post-step using the given value. Similarly, the union-view combination of so or mo adornments with other adornments can be computed.

5.2 Join Views

The computation of join-view templates (see Table 4) for two base views using preprocessing and post-filtering techniques is similar to that (see Table 2) for two base views using only preprocessing technique. The differences between Table 4 and Table 2 lie in the treatment of u, so, and mo. When both base view adornments are so

or *mo*, the mediator can start a query on base views without specifying a value for this attribute and then filter out the results using the optional specified attribute value in the post-filtering step. The treatment of *u* with *so* or *mo* is similarly defined.

Table 4. Join-view template using both preprocessing and post-filtering techniques

	f	u	b	$sc[l_5]$	$mc[l_6]$	$so[l_7]$	$mo[l_8]$
f	f	f	b	$sc[l_5]$	$mc[l_6]$	$fso[l_7]$	$fmo[l_8]$
u	f	f	b	$sc[l_5]$	$mc[l_6]$	$fso[l_7]$	$fmo[l_8]$
b	b	b	b	$sc[l_5]$	$mc[l_6]$	$bsc[l_7]$	$bmc[l_8]$
$sc[l_1]$	$sc[l_1]$	$sc[l_1]$	$sc[l_1]$	$sc[l_1 \cap l_5]$	$mc[l_1 \cap l_6]$	$sc[l_1]$	$sc[l_1]$
$mc[l_2]$	$mc[l_2]$	$mc[l_2]$	$mc[l_2]$	$mc[l_2 \cap l_5]$	$mc[l_2 \cap l_6]$	$mc[l_2]$	$mc[l_2]$
$so[l_3]$	$fso[l_3]$	$fso[l_3]$	$bsc[l_3]$	$sc[l_5]$	$mc[l_6]$	$fso[l_3 \cup l_7]$	$fmo[l_3 \cup l_8]$
$mo[l_4]$	$fmo[l_4]$	$fmo[l_4]$	$bmc[l_4]$	$sc[l_5]$	$mc[l_6]$	$fmo[l_4 \cup l_7]$	$fmo[l_4 \cup l_8]$

5.3 Projection and Selection Views

The computation of projection-view templates is similar to what we have discussed in Section 4.3. The differences are when we use the post-filtering technique to handle the projected attribute adornments *u*, *so*, and *mo*, which are changed to *f*, *fso*, and *fmo* adornments respectively in the projected-view template.

In the computation of selection view templates, the selected attribute adornments *u*, *so*, and *mo* are changed to *f*, *fso*, and *fmo* adornments, respectively. Other selected attribute adornments, such as *f*, *b*, *sc*, and *mc*, are copied from the base-view template.

6 Mediators Using Preprocessing, Post-filtering, and Pass Binding

Pass binding is a technique that passes values from one base view to another base view for the common attributes in both base views. By combining the preprocessing, post-filtering, and pass binding techniques, we can answer more user queries as well as obtain more comprehensive query results. Since the attribute values returned from one base view query can be passed to the next base view query to satisfy the join requirements, the order to execute the sub-queries is important.

6.1 Union Views

In the union view, union by pass binding uses the same mapping functions as presented in Table 3.

Example 3. Given two data sources $D_1(X\ (b), Y\ (f), Z\ (sc[l_1]))$ and $D_2(X\ (b), Y\ (u), Z\ (u))$, the union-view mediator M of D_1 and D_2 is defined as $M(X\ (b), Y\ (f), Z\ (fso[l_1]))$. With the preprocessing and post-filtering techniques, the query $M(x_1, y_1, Z)$ can only retrieve partial result from D_2 because the query condition for D_1 is not satisfied. With the pass binding, results can be obtained from both D_1 and D_2. To retrieve the answer for query $M(x_1, y_1, Z)$, the mediator first invokes the feasible subquery $D_2(x_1, Y, Z)$, then applies the condition $(Y = y_1)$ on the result of $D_2(x_1, Y, Z)$ to get $D_2(x_1, y_1, Z)$.

668   L. Li, J. Fernandez, and H. Guo

Furthermore, the mediator can pass the Z attribute values from D_2 to D_1. In particular, for each value z_i of Z from the results of $D_2(x_1, y_1, Z)$, the mediator invokes the sub-query $D_1(x_1, y_1, z_i)$ if $z_i \in l5$. The union of the results from D_1 subqueries and D_2 sub-query provides the answer to the query $M(x_1, y_1, Z)$.

The query in Example 3 is not answerable by the existing methods [3,6,7] because the existing methods require the mediator queries to satisfy the query conditions of all the underlying data sources, but can be answered by using the combination of preprocessing technique and partial results from the underlying data sources.

6.2 Join Views

The mapping function for computing the join attribute adornments is presented in Table 5. In particular, the changes come from joining a mandatory adornment b, sc, or mc with an optional adornment f, so, or mo as well as u.

Table 5. Join-view template using preprocessing, post-filtering, and pass binding techniques

	f	u	b	$sc[l_5]$	$mc[l_6]$	$so[l_7]$	$mo[l_8]$
f	f	f	f	$so[l_5]$	$mo[l_6]$	$fso[l_7]$	$fmo[l_8]$
u	f	f	f	$so[l_5]$	$mo[l_6]$	$fso[l_7]$	$fmo[l_8]$
b	f	f	b	$sc[l_5]$	$mc[l_6]$	$fso[l_7]$	$fmo[l_8]$
$sc[l_1]$	$so[l_1]$	$so[l_1]$	$sc[l_1]$	$sc[l_1 \cap l_5]$	$mc[l_1 \cap l_6]$	$so[l_1]$	$so[l_1]$
$mc[l_2]$	$mo[l_2]$	$mo[l_2]$	$mc[l_2]$	$mc[l_2 \cap l_5]$	$mc[l_2 \cap l_6]$	$mo[l_2]$	$mo[l_2]$
$so[l_3]$	$fso[l_3]$	$fso[l_3]$	$fso[l_3]$	$so[l_5]$	$mo[l_6]$	$fso[l_3 \cup l_7]$	$fmo[l_3 \cup l_8]$
$mo[l_4]$	$fmo[l_4]$	$fmo[l_4]$	$fmo[l_4]$	$so[l_5]$	$mo[l_6]$	$fmo[l_4 \cup l_7]$	$fmo[l_4 \cup l_8]$

Example 4. Given two data sources $D_1(X\ (b), Y\ (f), Z\ (b))$ and $D_2(Z\ (so[l_7]), O\ (u), P\ (b))$, the join-view mediator M of D_1 and D_2 is defined as $M(X\ (b), Y\ (f), Z\ (fso[l_7]), O\ (u), P\ (b))$. For the attribute Z in the above join-view mediator M, the user may choose a value $z_i \in l_7$, specify a value $z_i \notin l_7$, or specify no value. For the first option, the answer for the mediator query $M(x_1, Y, z_i, O, p_1)$, $z_i \in l_7$ can be obtained by joining results from two sub-queries $D_1(x_1, Y, z_i)$ and $D_2(z_i, O, p_1)$. In the second option, to answer the query $M(x_1, Y, z_i, O, p_1)$, $z_i \notin l_7$, the query result from $D_2(Z, O, p_1)$ should be filtered by removing tuples with different Z values before D_2 results joins D_1 results. For the third option, the query $M(x_1, Y, Z, O, p_1)$ is answered in the following three steps: (1) The mediator first runs the sub-query $D_2(Z, O, p_1)$; (2) then another sub-query $D_1(x_1, Y, Z)$ can be executed multiple times by bind passing each value z_i of Z from the result of $D_2(Z, O, p_1)$; (3) join and combine the results from $D_2(Z, O, p_1)$ in step 1 and each running of $D_1(x_1, Y, z_i)$ in step 2.

7 A Case Study

In our study, we combine two existing coastal wide data collection platform (DCP) networks: The Texas Coastal Ocean Observation Network (TCOON) and The Texas Automated Buoy System (TABS).

7.1 Mediator Schema for TCOON and TABS

There is only one table in the TCOON database, *Tcoon*(*sid, date, time, ser, smv*), where *sid* is the station identification, *date* and *time* are the month/date/year and hour: minute of the measurement, *ser* is the series identifier, and *smv* is the value of a series identifier. The *ser* attribute takes one of the 12 values, *pwl* (primary water level), *sig* (water level standard deviation), *out* (water level outlier), *atp* (air temperature), *wtp* (water temperature), *wsd* (wind speed), *wgt* (wind gust), *wdr* (wind direction), *bpr* (barometric pressure), *bat* (battery voltage), *cla* (calibration temperature A), *clb* (calibration temperature B).

The TABS database contains three tables, velocity, meteorology, and buoy system. The three tables are defined as: *Velocity*(*sid, date, time, wtd, wdr, wtp*), *Meteorology*(*sid, date, time, wsd, atp, bpr, wgt, compass, tx, ty, par, relhum*), *Buoy*(*sid, date, time, bat, sigstr, compass, nping, tx, ty, adcpv, adcpcur, vbatt2*). Due to the page limit, we do not explain every attribute in the TABS database. Interested readers please refer to TABS (*http://tabs.gerg.tamu.edu/Tglo/*).

From the above four tables from two databases, it is easy to identify that both databases share the following ten attributes/values: *sid, date, time, atp, bat, bpr, wdr, wgt, wsd, wtp*. Especially, *atp, bat, bpr, wdr, wgt, wsd,* and *wtp* are attributes in TABS database but are a subset of a list of values that *ser* may take in TCOON database.

The mediator schema of TCOON and TABS consists of two virtual tables *Observation* and *Facility* based on the Global-As-View (GAV) approach [1]. The first table contains the oceanic observational data and the second table the hardware device information. The *Observation* is obtained by union of *Velocity* and *Meteorology* tables with *Tcoon* table, and the *Facility* by union of *System* table with *Tcoon* table. *Velocity* and *Meteorology* tables are inner joined together.

> *Observation*(*sid, date, time, ser, smv, wtd, par, relhum*) :-
>   (*Velocity*(*sid, date, time, wtd, wdr, wtp*) JOIN
>   *Meteorology*(*sid, date, time, wsd, atp, bpr, wgt, compass, tx, ty, par, relhum*))
> UNION
>   (*Tcoon*(*sid, date, time, ser, smv*) AND (*ser* in ('pwl', 'sig', 'out', 'atp', 'wtp',
>   'wsd', 'wgt', 'wdr', 'bpr')))

> *Facility*(*sid, date, time, ser, smv, sigstr, compass, nping, tx, ty, adcpv, adcpcur,*
>   *vbatt2*) :-
>   *System*(*sid, date, time, bat, sigstr, compass, nping, tx, ty, adcpv, adcpcur, vbatt2*)
> UNION
>   (*Tcoon*(*sid, date, time, ser, smv*) AND (*ser* in ('bat', 'cla', 'clb'))

7.2 Query Capabilities of the Mediator for TCOON and TABS

To retrieve data from TCOON, users must choose at least one station from a station list l_1 and at least one field from a list of series identifiers $l_2 \cup l_3$, respectively. l_2 is {'pwl', 'sig', 'out', 'atp', 'wtp', 'wsd', 'wgt', 'wdr', 'bpr'} and l_3 is {'bat', 'cla', 'clb'}. In addition, the *date* attribute is optional, and the attribute *time* and *smv* can not be specified. As a result, the query capability of the base-view TCOON is expressed as *Tcoon (sid (mc[l₁]), date (f), time (u), ser (mc[l₂ ∪ l₃]), smv (u))*.

The TABS website provides a query form that allows users to extract data by specifying one station ID from the station list l_4. The attribute *date* is optional and all the other attributes can not be specified. So the query capabilities of base-view TABS are: *Velocity(sid (sc[l₄]), date (f), time (u), wtd (u), wdr (u), wtp (u))*, *Meteorology(sid (sc[l₄]), date (f), time (u), wsd (u), atp (u), bpr (u), wgt (u), compass (u), tx (u), ty (u), par (u), relhum (u))*, and *System(sid (sc[l₄]), date (f), time (u), bat (u), sigstr (u), compass (u), nping (u), tx (u), ty (u), adcpv (u), adcpcur (u), vbatt2 (u))*.

To compute the query capabilities of the mediator, we apply the union-view template (Table 3 in Section 5) to the attributes occurring in both databases (such as *sid*) and the projection-view template (see Section 5.3) to the attributes appearing in only one of the two databases (such as *par*). The computed results are as follows: *Observation(sid (mc[l₁ ∪ l₄]), date (f), time (f), ser (fmo[l₂]), smv (f), wtd (f), par (f), relhum (f))*, *Facility(sid (mc[l₁ ∪ l₄]), date (f), time (f), ser (fmo[l₃]), smv (f), sigstr (f), compass (f), nping (f), tx (f), ty (f), adcpv (f), adcpcur (f), vbatt2 (f))*.

8 Answerable Mediator Queries

In the previous sections, we have presented the computation of a mediator's template in order to determine answerability of the user-issued mediator queries. However, as illustrated in Example 5, a mediator query may still not be answerable even if the query satisfies the mediator template for the union view.

Example 5. Given two data sources $D_1(X\ (b),\ Y\ (f),\ Z\ (f))$ and $D_2(X\ (f),\ Y\ (f),\ Z\ (sc[l_5]))$, the mediator view M according to Table 3 is $M(X\ (f),\ Y\ (f),\ Z\ (fso[l_5]))$. Query $M(X, Y, Z)$ is not answerable even if it satisfies the template of M. Because D_1 requires at least one value for X to support a query and D_2 needs at least one value from the list l_5 to provide some answer.

From the above example, we can see that the minimum requirement to answer a union-view mediator query is to satisfy the query conditions of at least one data source. For example, to answer Example 5, we should at least specify a value for X in D_1 or choose a value from the list l_5 in D_2.

The mapping functions computed in [7] are the sufficient conditions, which guarantee a query to be answerable if it satisfies the union-view template. But those conditions are too stringent such that some queries not satisfying a template are still answerable. The mapping functions presented in the tables of this paper can not guarantee that a mediator query is answerable because the presented mapping functions only consider the union of two attribute adornments but not the combination of multiple attribute adornments. But those functions give the necessary conditions to answer a mediator query.

On the other hand, the mapping functions for the join-view templates in this paper present the minimum conditions to guarantee the answerability of a mediator query, while the functions defined in [7] are either too stringent (such as *conservative templates*) or too loose (such as *liberal templates*).

9 Conclusion

In this paper, we develop a framework to handle the partial results from the underlying data sources using the preprocessing technique. In addition, we design new attribute adornments to handle interface design when computing query capabilities of mediators. The framework is proposed to integrate a large collection of data sources from the GOM. Although we do not present the computation of mediator view templates for three or more base views, it can be realized by considering two base views at a time, which is similar to [3,7].

References

1. Garcia-Molina, H., Papakonstantinou, Y., Quass, D., Rajaraman, A., Sagiv, Y., Ullman, J., Vassalos, V., Widom, J.: The TSIMMIS Approach to Mediation: Data models and Languages. Journal of Intelligent Information Systems 8(2), 117–132 (1997)
2. Levy, A.Y., Rajaraman, A., Ordille, J.J.: Querying Heterogeneous Information Sources Using Source Descriptions. In: International Conf. on Very Large Data Bases (1996)
3. Pan, A., Montoto, P., Molano, A.: A Model for Advanced Query Capability Description in Mediator Systems. In: Int'l Conf. on Enterprise Information Systems (2002)
4. Roth, M., Schwarz, P.: Don't Scrap it, Wrap it! A Wrapper Architecture for Legacy Data Sources. In: International Conference on Very Large Data Bases (1997)
5. Tomasic, A., Raschid, L., Valduriez, P.: Scaling Heterogeneous Databases and the Design of Disco. In: International Conference on Distributed Computing Systems (1996)
6. Tang, J., Zhang, W., Xiao, W.: An Algebra for Capability Object Interoperability of Heterogeneous Data Integration Systems. In: Asia-Pacific Web Conference (2005)
7. Yerneni, R., Li, C., Garcia-Molina, H., Ullman, J.D.: Computing Capabilities of Mediators. In: SIGMOD Conference (1999)

Process Mediation Based on Triple Space Computing

Zhangbing Zhou[1], Brahmananda Sapkota[1], Emilia Cimpian[2], Doug Foxvog[1],
Laurentiu Vasiliu[1], Manfred Hauswirth[1], and Peng Yu[3]

[1] DERI, National University of Ireland at Galway, Ireland
{zhangbing.zhou,brahmananda.sapkota,doug.foxvog,
laurentiu.vasiliu,manfred.hauswirth}@deri.org
[2] STI International, University of Innsbruck, Austria
emilia.cimpian@deri.org
[3] College of Computer Science and Technology, Jilin University, China
yupeng79@gmail.com

Abstract. Web services are inherently heterogeneous at both data and behavioral levels because of the nature of the Web, which is the main obstacle to the usability of Web services. The heterogeneity at a behavioral level is generally addressed by process mediation, in which the message flow is adjusted to suit the behavior of Web services involved in a given interaction. In this paper, we present a novel approach for process mediation, and propose an architectural for process mediation based on Triple Space Computing to solve resolvable message sequence mismatches. These resolvable mismatches can be classified into five classes for unveiling their essence. This work provides a basis for the generalization of mismatches themselves, as well as a potentially uniform solution to address these mismatches.

1 Introduction

Web services act as computational entities and the main pillar for Service-Oriented Architecture, and aim at supporting interoperable machine-to-machine interactions over the Web. Because of inherent *autonomy* and *heterogeneity*, Web services are heterogeneous at both data and behavioral levels. In general, the messages are often different in *format* and *granularity*, and public processes [9] are often diverse in activities and messages in terms of *form* and *sequence*. Data heterogeneities can be mitigated with the help of data mediation [1], while process mediation [12] aims at aligning different interaction patterns by adjusting bi-directional flows of messages. Process mediation is very valuable for complex service interactions in which several Web services may involve. However, process mediation is optional for RPC-style service interactions, where there is only a single request-response message exchange.

Based on the message exchange patterns specified for a Web service, called as the public process, process mediation aims at resolving message sequence mismatches for: (1) *service discovery and selection*: to ensure that discovered and/or selected Web services are behaviorally compatible [15] with a given goal. A functional aspect is currently the focus for service discovery and selection. Process mediation would be used for identifying whether a goal and the services are compatible from a behavioral

Y. Zhang et al. (Eds.): APWeb 2008, LNCS 4976, pp. 672–683, 2008.
© Springer-Verlag Berlin Heidelberg 2008

aspect, (2) *service composition*: process mediation would guarantee that there are only resolvable behavioral mismatches [12] among a goal and Web services, and (3) *service execution*: process mediation would instruct the exchange of messages among Web services, and thus smooth the interaction.

In this paper, we propose a novel approach of process mediation for dealing with message sequence mismatches. We apply our method to WSMO[1] based Web services. However, our method is general and can be applied to other semantic Web services (SWSs) models like OWL-S[2] at ease. A public process can be described by these SWSs conceptual models. In addition, we propose an architecture for process mediation based on Triple Space Computing (TSC) [8]. Potential solutions are presented for resolvable mismatches. Furthermore, we classify these mismatches into five classes. The main contributions of this paper are four-fold: (1) a novel approach of process mediation for dealing with behavioral mismatches, (2) an architecture for process mediation based on TSC, (3) potential solutions for resolvable message sequence mismatches, and (4) five classes for these resolvable mismatches.

The rest of the paper is organized as follows: In Section 2, we give an introduction to WSMO and TSC. In Section 3, we propose our TSC-based architecture for process mediation. In Section 4, we present potential solutions for resolvable message sequence mismatches and categorize them into five classes. In Section 5, we discuss related works. In Section 6, we conclude this paper and indicate our future work.

2 Background

Due to the space limitation, we give a brief introduction to WSMO in Section 2.1 and Triple Space Computing in Section 2.2.

2.1 WSMO

WSMO is one of the major SWSs conceptual models initiated by the Web Service Modeling Ontology working group[3] of the ESSI cluster[4]. WSMO defines four major components: *ontology*, *Web services*, *goal*, and *mediator*, following the framework proposed in WSMF [9]. Web services and goals have a common component: *an interface*, which specifies how their functionality can be achieved though a two-fold view of operational competence: *choreography* [10] and *orchestration* [14].

Choreography describes the behavioral interface of Web services by which a client can consume its functionality [3]. This means that it presents an interface from a user's point of view. A user can be a person, an application, or another Web service.

Orchestration defines the behavioral interface of a Web service for achieving its functionality by aggregating other Web services [18]. An orchestration could be regarded as a "*composition*" of several "*sub-goals*". Each "*sub-goal*", acting as a user, consumes another Web service through its choreography. However, an orchestration

---

[1] http://www.wsmo.org/TR/d2/v1.3/.
[2] http://www.ai.sri.com/daml/services/owl-s/1.2/overview/.
[3] http://www.wsmo.org/.
[4] http://www.essi-cluster.org/.

is optional if the functionality of a Web service could be achieved completely by its choreography.

2.2 Triple Space Computing

TSC is a persistent communication and coordination paradigm for application and service integration on the Web [21]. It is based on the convergence of Semantic Web [22] and tuple-space computing [7] technologies. TSC acts as a *globally accessible, Web-scaled* and *space-like* middleware to enable so-called Web paradigm: *information is persistently written to a globally shared space where other processes can smoothly access it without starting a cascade of message exchanges* [8]. Triple Spaces introduce an infrastructure that enables machines to use an equally powerful communication medium in the same way as the humans use the Web [21]. The main advantages that TSC brings are four-fold: (1) *time autonomy*: the only time dependency is that RDF triples must be written before they can be read, (2) *location autonomy*: storage location provided by Triple Space is independent to that of the provider or the requester, (3) *reference autonomy*: the provider and the requester do not need to know each other and there is no explicit communication channel between them. They exchange information by writing and reading RDF triples to and from a Triple Space, and (4) *data schema autonomy*: data written to and read from a Triple Space would follow TSC data schema, which follows RDF specification. This makes the provider and the requester independent of their internal data schemas.

Besides the functionality of space-based computing, TSC offers more features such as transaction support, distribution, and query using RDF format etc. For more information about TSC, the reader can visit the web page of TripCOM Project[5].

3   Process Mediation Integration with TSC

In this Section, we propose an approach of process mediation addressing behavioral mismatches for WSMO-based Web services, and present a TSC-based architecture for process mediation.

3.1 Process Mediation Framework

Process mediation bridges potential behavioral mismatches between a user and a Web service chorography, or between a *"sub-goal"* of a Web service orchestration and the choreography of another Web service. Data mediation is necessary to support process mediation if the goal and the services are represented using different ontologies. In order to support this requirement, we propose a framework for process mediation as shown in Figure 1. Process mediation would concentrate on behavioral compatibility [5] between two partners since the majority of interactions are often related to two partners, and an interaction involving multiple partners can often be decomposed into several pair-wise interactions.

---

[5] http://tripcom.org/deliverables.php.

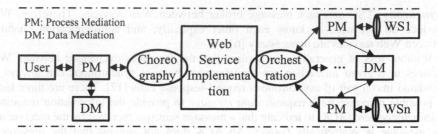

Fig. 1. Process Mediation Framework Based on WSMO

3.2 Architecture

TSC-based process mediation architecture is presented in Figure 2. Data mediation addresses potential data heterogeneity problems if a user and a Web service are described by different ontologies. TSC brings machine-to-machine Web service interaction to Web scale, and acts as a communication middleware between a user and a Web service (actually the choreography of Web services). There is a virtual data space within TSC for a given interaction, which includes two sub-spaces: one for the user and another for the Web service. Data would follow TSC data schema. The user and the Web service send requests to and retrieve responses from TSC. They do not communicate directly. Based on data stored in TSC, process mediation could handle possible behavioral mismatches between the user and the Web service.

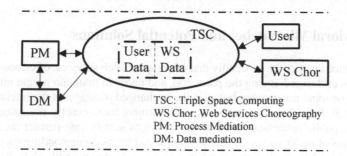

Fig. 2. Integrating Process Mediation with Triple Space Computing

Typically, Web service interactions are based on the message exchange paradigm, and often need to establish *synchronous* and *stateful* conversations. Therefore, they require a strong coupling in terms of reference and time [16]. However, these strong couplings do not exist in our architecture. The major advantages that TSC brings for service interactions are:

Backend storage: TSC provides a global, Web-scale space middleware for service interaction as well as process mediation. All data to be exchanged in an interaction is available in a shared virtual data space.

Asynchrony: TSC acts as a message broker between Web services. Therefore, Web services do not need to know each other explicitly, and communication channels between Web services are unnecessary [6].

Without loss of generality, we assume that the messages exchanged between Web services in a given interaction are asynchronous because synchronous can always be translated into a pair of asynchronous request-response calls [17]. There are three kinds of possible response for a request: *data message* to provide the information requested, *acknowledgement (ACK)* to indicate that a message sent was received by the receiver and the message is syntactically correct. An ACK does not suggest that the message is semantically valid. Or *negative acknowledgement (NAK)* to indicate that the data received is invalid in syntax.

State archiving for Web service interactions: TSC allows the storing of the *history* of the interactions, which is a flow of messages to and from TSC. This provides three main advantages: (1) this enables the monitoring of communicating applications, therefore helping in reuse of available Web services [6], (2) the archived messages represent the observed behavior of Web services. Therefore, we can check whether this observed behavior is consistent with the modeled behavior for a given Web service, and (3) the archived messages represent the locale of a given interaction. If the interaction fails for some reason, the interaction could resume from the point of failure with a possible manual adjustment, but does not need to start again from scratch. This is critical for long-running or non-repeatable interactions.

Semantic autonomy: TSC data model acts as intermediate model for the partners involving in a given interaction. They do not need to agree on a common data representation. Therefore, data mediation is not mandatory.

4   Behavioral Mismatches and Potential Solutions

A Web service interaction, especially that of complex Web services, often requires a flow of messages exchanged among the partner(s), and needs to maintain a state internally [2]. There may be some mismatches between the exchanged messages if the interaction is not perfect-match. Perfect-match means that the partners have exactly the same pattern to realize their public processes, and thus the messages sent by one partner are exactly the same in terms of *order* and *granularity* as requested by its corresponding partner [12]. In this case, only data mediation would be used to address possible data heterogeneity problems.

In this context, the mismatches could be classified into two categories: irresolvable message sequence mismatches (Section 4.1) and resolvable message sequence mismatches [12] (Section 4.2 and 4.3).

4.1   Irresolvable Message Sequence Mismatches

Irresolvable message sequence mismatches means that these mismatches cannot be handled automatically. As stated in [12], the following two scenarios are presented as irresolvable message sequence mismatches: (1) one partner wants to receive a message that the other does not want to send, thus the interaction fails because of lacking a required message, and (2) one partner expects an ACK for a certain message, but the other does not want to receive this message. Process mediation cannot generate such an ACK. Otherwise the entire communication is changed.

These types of mismatches, as stated in [9], could only be resolved by either of the following two solutions: (1) to change the interface definition for the goal and/or the Web service to avoid mismatches, or (2) to operate manually (such as skipping the activity that causes a fault) to bypass the failure.

However, Web service interface is often not allowed to be changed arbitrarily, and it is inappropriate to allow unexpected human interventions during service executions. This indicates that it is inappropriate that irresolvable mismatches exist in the public processes, and irresolvable mismatches are out of the scope process mediation.

4.2 Resolvable Mismatches and Potential Solutions

The left side of Figure 3 presents five scenarios for resolvable message sequence mismatches taken from [12]. These five scenarios are atomic mismatches, and complex mismatches can be built recursively by applying these five atomic ones. The right side presents corresponding data transformation in TSC data space. Process mediation has a *priori* knowledge of ontologies used by the business partners. Below we introduce these mismatch scenarios and their potential solutions:

Scenario A: suppose that BP1 sends messages "a" and "b" to BP2, but only "b" is expected by BP2. Process mediation should retain and store "a" for possible later use.
Potential Solution: "a" and "b" are sent to TSC and stored in the sub-space for BP1. Based on BP2's ontology, process mediation knows that only "b" is expected, process mediation adds or updates data instance for "b" in the sub-space of BP2.

Scenario B: suppose that BP1 sends messages "a" and then "b" to BP2, while BP2 expects to receive "b" and then "a". Process mediation reverses the ordering of these two messages.
Potential Solution: "a" and "b" are sent to TSC and stored in the sub-space of BP1. Based on the ontology of BP2, process mediation knows that both "a" and "b" are expected by BP2, and process mediation adds or updates data instances for both "a" and "b" in the sub-space of BP2.

Scenario C: suppose that BP1 sends both "a" and "b" in a single message to BP2, while BP2 expects to receive "a" and "b" separately. According to BP2's request, process mediation should split this single message from BP1.
Potential Solution: There are two kinds of possible reasons for this mismatch:
- BP1 and BP2 use different ontologies: ("a" + "b") is modeled by one concept in the ontology of BP1, while "a" and "b" are modeled by different concepts in the ontology of BP2.
- The ontologies of BP1 and BP2 are the same for "a" and "b": one concept for "a" and another concept for "b". But the messages are coded in different granularities. The granularity of a message indicates the number of the concepts implied by the data instances in this message. The *less* the number of concepts in a message, the *finer* the message is. In Scenario C, the messages of BP2 are finer than that of BP1, because the message of BP1: ("a" + "b") implies two concepts, while the messages of BP2: "a" and "b" imply only one concept.

Data mediation knows the mapping for BP1 ("a" + "b") to BP2 ("a") and BP2 ("b"). Process mediation adds or updates data instances for "a" and "b" in the sub-space of BP2.

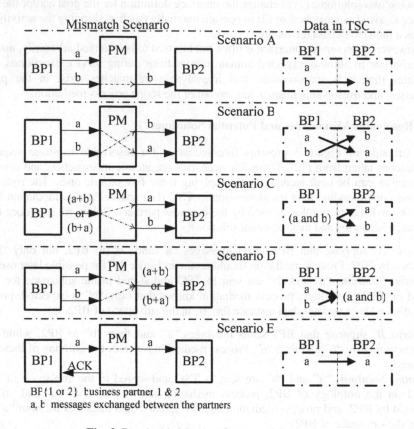

Fig. 3. Resolvable Message Sequence Mismatches

Scenario D: suppose that BP1 sends two messages "a" and "b" to BP2, while BP2 expects to receive them in one single message. Process mediation should combine two messages from BP1 into one message for BP2.
Potential Solution: This is the reverse case of Scenario C, and could be solved in the similar way as presented in Scenario C.

Scenario E: suppose that BP1 sends a message "a" to BP2 and expects an ACK for "a". If "a" is expected by BP2, and BP2 is not willing to send an ACK back, process mediation should generate this ACK and send it back to BP1.
Potential Solution: "a" is sent to TSC and stored in the sub-space for BP1. Process mediation knows that "a" is expected by BP2, so it updates the sub-space of BP2 for "a" and generate an ACK for BP1. If "a" is either wrong in format or invalid in syntax, a NAK is returned. Process mediation has the knowledge of message formats for both BP1 and BP2.

4.3 Discussion

In this section, we analyze these resolvable mismatches shown in Figure 3. They can be categorized into five classes:

Class One: Extraneous data. Scenario A in Figure 3 is a good example of this class: one partner provides more data than what its partner wants to receive. However, this is only applicable to asynchronous communication. If the interaction is synchronous, this falls into irresolvable mismatches. An example is illustrated in Figure 4, which is synchronous counterpart of Scenario A:

BP1 expects a response for "a". "a" is not expected by BP2, which means that BP2 has not a *priori* knowledge about "a", and is uncertain whether "a" is necessary or not. If process mediation generates an ACK (or NAK) to BP1 for "a", ACK (or NAK) indicates that "a" is expected by BP2 (while NAK indicates that "a" is expected by BP2, but "a" is not correct in terms of *format* or *syntax*). Obviously, ACK (or NAK) conveys a misleading indication to BP1 and changes the interaction. Therefore, the process mediation could not generate and send an ACK (or NAK) back to BP1, and this is a scenario of irresolvable mismatches.

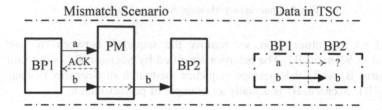

Fig. 4. Synchronous Communication for Extraneous Data

Class Two: Interaction with synchronous and asynchronous communication. Scenario E in Figure 3 falls into this class, which means that one partner is synchronous while the other is asynchronous. Based on the discussion in Class One, it is clear that Class Two is resolvable only if the data sent is expected by the receiver.

Class Three: Data heterogeneity. Scenarios C and D in Figure 3 fall into this class, which means that the partners use different ontologies, and some mismatches exist between these ontologies. Data mediation aims at supporting process mediation for solving these data heterogeneity problems.

Class Four: Message granularity heterogeneity. Scenarios C and D in Figure 3 belong to this class. This means that the ontologies used by the partners are similar, but the messages of different partners are of different granularity. Process mediation aims at solving this message granularity heterogeneity by combining or splitting the messages.

Class Five: Unnecessary message sequence dependencies. Scenarios B and D in Figure 3 fall into this class. Message sequence dependency means the ordering of the messages. The necessary message sequence dependency means that message ordering should be held during execution phases. We further explain this by Scenario B in

Figure 3. Firstly, we rename message "a" sending as MsgA, while message "b" sending as MsgB. The purpose of messages renaming is to indicate that this concept applies to both incoming and outgoing messages. Necessary message sequence dependency of MsgA and MsgB indicates that: MsgA should be sent or received before MsgB, while MsgB should not be sent or received before MsgA.

Based on Scenario B, Figure 5 gives an example for unnecessary message sequence dependency, and shows that an unnecessary dependency can be removed. In scenario B, BP1 is willing to send "a" and then "b". However, BP1 could reverse the sequence by sending "b" and then "a". The changed sequence has no impact to the behavior of BP1 because "b" does not depend on "a". It is the same for messages "b" and "a" in BP2.

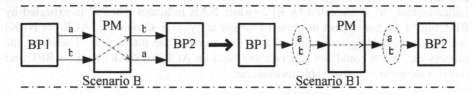

Fig. 5. Unnecessary Message Sequence Dependency

Based on this observation, we remove the sequence between "a" and "b" as presented in Scenario B1. The behavior specified by Scenario B1 is the same as that of Scenario B, while the message sequence mismatch in Scenario B disappears in Scenario B1. Scenario B1 is actually an example of perfect-match.

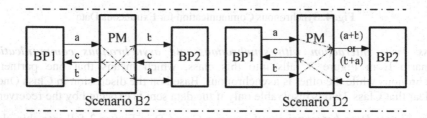

Fig. 6. Necessary Message Sequence Dependency

Figure 6 presents examples for necessary message sequence dependency in Scenario B2 and D2, which are based on Scenario B and D in Figure 3 with some change. These mismatches are irresolvable.

Scenario B2: BP1 is willing to send "a" then wait for "c", and after that send "b". "c" may be the response of "a", therefore the content of "c" may depend on the content of "a". The same is for "c" and "b". Consequently, the sequence of "a", "c" and "b" is necessary and cannot be changed. Similar for BP2 that the sequences among "b", "c" and "a" are necessary and need to preserve. A failure occurs during this interaction because BP1 waits for "c" while BP2 waits for "b". Process mediation cannot generate "c" for BP1, as well as "b" for BP2.

Scenario D2: BP1 wants to send "a" then wait for "c", after that send "b". Based on the analysis of Scenario B2, the sequences among "a", "c" and "b" are necessary and cannot be changed. The same for BP2 that the sequence between ("a+b") or ("b+a") and "c" is necessary and should be preserved. The interaction between BP1 and BP2 fails because BP1 waits for "c" while BP2 waits for ("a+b") or ("b+a").

5 Related Works

The requirement of process mediation for supporting complex Web service interactions has been widely accepted as an important research topic. In [11], the authors present the scope of process mediation, list resolvable message sequence mismatches, propose an approach to integrate process mediation as a component into WSMX [19], and specify the interaction mode of process mediation with other components. In addition, they argue that process mediation would be used to support service invocation [13]. This work has been used in DIP project[6]. This work is a starting point of process mediation in WSMX. This paper benefits much from this work, especially message sequence mismatches in Section 4. However, the proposed approach is simple and suits the simplest workflow pattern: *sequence*, but would fail for complex ones such as *choice* or *loop* [20]. The work does not mention the importance of process mediation for complex service interactions, in which process mediation needs to support service discovery and selection to guarantee that the goal and the services are behaviorally compatible.

An adapter-based approach is proposed in [4] intending to semi-automatically resolve Web service differences at interface and business protocol levels. Possible differences between Web services are identified and captured by *mismatch patterns*. A pattern includes a business logic template, and can be used as a type of mismatch addressed by an adapter. However, *mismatch patterns* at the interface level are actually data heterogeneity problems covered by data mediation. *Mismatch patterns* at the business protocol level are the same as our resolvable mismatches listed in Figure 3. This work aims to formalize Web service protocols and interface/protocol *mismatch patterns*, and thus to provide a high-level framework as well as a uniform mechanism to address these mismatches.

In [23], the authors present the purpose of a process mediator within WSMX, which is a message broker among the partners. Process mediator needs to decide which data belongs to which partner(s) based on choreography and ontology of the partner(s). This work extends process mediation to multi-lateral interactions, and focuses on message forwarding among the partners. However, this data distribution among the partners is actually, only a part of task that should be addressed by process mediation.

6 Conclusion and Future Works

Process mediation is a complex task and important for complex service interactions in which behavioral heterogeneity problems may exist in public processes. We argue that process mediation would aim at pair-wise interactions only. We propose a

---

[6] http://dip.semanticweb.org/deliverables.html.

process mediation architecture based on TSC, and present potential solutions for resolvable message sequence mismatches. In addition, we categorize these resolvable mismatch scenarios into five classes. This analysis generalizes the resolvable message sequence mismatches, provides the basis for checking Web service compatibility from the behavioral aspect, and offers an opportunity to have a uniform solution to address these mismatches.

Process mediation in the context of SWSs is still in its infancy [11]. The related work is based on the exchanged messages, which represent a part of service behavior. It is a common sense that two public processes, which are locally compatible, do not necessarily mean that they are globally compatible. In the future, process mediation needs to consider compatible [5] on the public process level, and to check whether the mismatches are resolvable. Transitional support is another direction for ensuring the integrity of the interaction and the recovery in case of failure. Also, we aim to implement this proposal for evaluating it against the real data sets taken from the real-life use cases.

Acknowledgments. The work presented in this paper was supported (in part) by the EU funded TripCom Specific Targeted Research Project under Grant No. FP6-027324, and (in part) by the Lion project supported by Science Foundation Ireland under Grant No. SFI/02/CE1/I131.

References

1. Mocan, A., Cimpian, E., de Bruijn, J.: D13.3v0.3 WSMX Data Mediation WSMX Working Draft 11 (2005), http://www.wsmo.org/TR/d13/d13.3/v0.3/
2. Wombacher, A.: Decentralized Consistency Checking in Cross-organizational Workflows E-Commerce Technology. In: Proc. of the 8th IEEE International Conference on and Enterprise Computing, E-Commerce, and E-Services (2006)
3. Norton, B., Pedrinaci, C., Lemcke, J., Kleiner, M., Henocque, L., Vulcu, G.: DIP Delivery: D3.9 An ontology for web services choreography and orchestration V3 (2006), http://dip.semanticweb.org/deliverables.html
4. Benatallah, B., Casati, F., Grigori, D., Nezhad, H.R.M., Toumani, F.: Developing Adapters for Web Services Integration. In: Pastor, Ó., Falcão e Cunha, J. (eds.) CAiSE 2005. LNCS, vol. 3520, Springer, Heidelberg (2005)
5. Benatallah, B., Casati, F., Toumani, F.: Representing, analysing and managing web service protocols. Data & Knowledge Engineering 58, 327–357 (2006)
6. Sapkota, B., Kilgarriff, E., Bussler, C.: Role of Triple Space Computing in Semantic Web Services. In: Zhou, X., Li, J., Shen, H.T., Kitsuregawa, M., Zhang, Y. (eds.) APWeb 2006. LNCS, vol. 3841, Springer, Heidelberg (2006)
7. Gelernter, D.: Generative Communication in Linda. ACM Transactions on Programming Languages and Systems 7, 80–112 (1985)
8. Fensel, D.: Triple-space computing: Semantic Web Services based on persistent publication of information. In: Proc. of IFIP Int'l Conf. on Intelligence in Communication Systems (2004)
9. Fensel, D., Bussler, C.: The Web Service Modeling Framework WSMF Electronic Commerce Research and Applications, pp. 113–137 (2002)

10. Roman, D., Scicluna, J., Nitzsche, J., Fensel, D., Polleres, A., de Bruijn, J., Heymans, S.: D14v0.4. Ontology-based Choreography. WSMO Working Draft 15 (2007), http://www.wsmo.org/TR/d14/v0.4/

11. Cimpian, E., Mocan, A.: WSMX Process Mediation Based on Choreographies. In: Bussler, C.J., Haller, A. (eds.) BPM 2005. LNCS, vol. 3812, pp. 130–143. Springer, Heidelberg (2006)

12. Cimpian, E., Mocan, A., Scicluna, J.: D13.7 v0.2 Process Mediation in WSMX WSMX Working Draft (2005), http://www.wsmo.org/TR/d13/d13.7/v0.2/

13. Cimpian, E., Mocan, A., Stollberg, M.: Mediation Enabled Semantic Web Services Usage. In: Mizoguchi, R., Shi, Z.-Z., Giunchiglia, F. (eds.) ASWC 2006. LNCS, vol. 4185, Springer, Heidelberg (2006)

14. Haas, H., Brown, A.: Web Services Glossary W3C Working Group (2004), http://www.w3.org/TR/ws-gloss/

15. Bordeaux, L., Salaün, G., Berardi, D., Mecella, M.: When are Two Web Services Compatible? In: Shan, M.-C., Dayal, U., Hsu, M. (eds.) TES 2004. LNCS, vol. 3324, Springer, Heidelberg (2005)

16. Nixon, L.J.B., Bontas, E.P., Scicluna, J.: D2.4.8.1: Technical and ontological infrastructure for Triple Space Computing. Delivery of Knowledge Web project (2006), Available at: http://knowledgeweb.semanticweb.org/semanticportal/sewView/frames.jsp

17. Pistore, M., Traverso, P., Bertoli, P.: Automated Composition of Web Services by Planning in Asynchronous Domains. In: Proc. of the 15th Interational Conferece on Automated Planning and Scheduling (ICAPS 2005), Monterey, California (2005)

18. Stollberg, M.: Reasoning Tasks and Mediation on Choreography and Orchestration in WSMO. In: Proc. of the 2nd International WSMO Implementation Workshop (WIW 2005), Innsbruck, Austria (2005)

19. Zaremba, M., Moran, M., Haselwanter, T., Lee, H.K., Han, S.K.: D13.4v0.3. WSMX Architecture WSMX (2005), http://www.wsmo.org/TR/d13/d13.4/v0.3/

20. Russell, N., ter Hofstede, A.H.M., van der Aalst, W.M.P., Mulyar, N.: Workflow Control-Flow Patterns: A Revised View. BPM Center Report BPM-06-22, BPMcenter.org (2006)

21. Krummenacher, R., Hepp, M., Polleres, A., Bussler, C., Fensel, D.: WWW or What is Wrong with Web Services. In: Proc. of 3rd European Conf. on Web Services (ECOWS 2005) (2005)

22. Berbers-Lee, T., Hendler, J., Lassila, O.: The semantic web. Scientific America 284, 34–43 (2001)

23. Kotinurmi, P., Vitvar, T., Haselwanter, T., Moran, M., Zaremba, M.: WSMX: A Semantic Service Oriented Middleware for B2B Integration. In: Dan, A., Lamersdorf, W. (eds.) ICSOC 2006. LNCS, vol. 4294, pp. 477–483. Springer, Heidelberg (2006)

An Efficient Approach for Supporting Dynamic Evolutionary Change of Adaptive Workflow

Daoye Zhang, Dahai Cao, Lijie Wen, and Jianmin Wang

School of Software, Tsinghua University, Beijing, 100084, P.R.China
Key Laboratory for Information System Security, Ministry of Education
Tsinghua National Laboratory for Information Science and Technology (TNList)
{zhangdy05,wenlj00}@mails.tsinghua.edu.cn,
caodahai@tsinghua.org.cn, jimwang@tsinghua.edu.cn

Abstract. Transferring multiple active instances from an original workflow model to a new one remains a big challenge for dynamic evolutionary change. Few approaches have been proposed to propagate model change to running instances. Existing ones adopt so strict criteria that too many transferable circumstances are left unmigrated. This paper proposes an efficient approach to smoothly migrate running instances to modified model. The approach is based on Dynamic Change Region (DCR) and Reduced Trace Set (RTS), which are introduced to identify transferable instances according to a newly proposed correctness criterion. A migration algorithm is provided to perform marking adaptation after transfer, and several examples are given for illustration.

Keywords: Adaptive Workflow, Dynamic migration, Dynamic Change Region, Reduced Trace Set.

1 Introduction

Due to changing business requirements and continuous process optimization, workflow model are subjected to increasingly fast changes. In order to achieve flexible business process and make quick response to varying requirements, it is extremely important to propagate the change to the in-progress instances. This is the business background of dynamic evolutionary change. As to the technical aspect, dynamic evolutionary change refers to the change that involves modification of workflow model, which results in migration of multiple running instances to the modified model.

As addressed by Aalst [1], the success of the next generation of Workflow Management System (WfMS) largely depends on the adaptability of workflow supporting dynamic change. The biggest advantage of adaptive WfMS over traditional ones lies in the ability to support dynamic change, including both ad-hoc and evolutionary ones. There are already some articles and systems supporting ad-hoc change, but few concrete techniques have been proposed to deal with evolutionary change, which has more pressing need for automatic processing in reality. Feasible solutions to the dynamic evolutionary change (dynamic migration in particular) are required in order to achieve real adaptive WfMS.

Y. Zhang et al. (Eds.): APWeb 2008, LNCS 4976, pp. 684–695, 2008.

1.1 Problem Description

Ever since Ellis first proposed the concept of dynamic change and pointed the dynamic change bug, great attentions have been paid to the hot topic. In [2], Ellis introduced the notion of Change Region and adopted a strategy of delayed change called Synthetic Cut-Over Change. Casati defined a complete and minimal set of modification primitives to ensure the correctness of migration [3].Sadiq provided a new method involving construction of Compliance Graph between the old and new workflow versions [4]. Aalst proposed an advanced inheritance concepts to support flexibility, but it is only suitable for circumstances when the new version is a subclass of the original one [5]. These ideas have been implemented by some WfMSs, such as Flow Nets, WIDE, TRAMs, ADEPT[6], JBees, etc.

All of these researches and systems have provided feasible strategies and even implementation solutions to the dynamic change issue, specifically the ad-hoc type. However, few of them have provided satisfying solutions to evolutionary change. How to correctly and efficiently migrate a large number of running instances from old model to a new one remains an unsolved problem, which is actually quite a common requirement in reality.

According to [7], ADEPT has the highest overall evaluation among different systems in aspect of dynamic change. It can perfectly solve four of the five typical dynamic change problems. However, it mainly focuses on the demonstration of the ad-hoc flexibility and its changing policy and criteria can hardly be extended to evolutionary change. It adopted the approach of Modification Operations Set (MOS), which imposes limitations on both the method and result of modification. In reality, BPR experts or business analysts usually provide a ready-made new workflow model rather than a series of modification steps. This makes the MOS approach unrealistic, for it tries to build the new model step by step with given modification operations. What we actually need is migration strategy, rather than steps of building a new workflow model based on an old one.

We agree with Sadiq's point that modifying a WF should be as flexible as building a new one [4], and dynamic modification can be considered as a substitution of sub-graph rather than the implement of a series of predefined operations. We then turned to Aalst's Change Region approach [8]. Based on change region, Aalst proposed an algorithm to enable dynamic evolutionary change. However, too many transferable instances are excluded from migration [9]. Maybe the region is too large or the criterion is too strict, in other words, either the region needs to be downsized or a reasonable and applicable criterion should be proposed.

1.2 Contribution

As mentioned above, the Change Region approach in [8] excluded too many transferable instances from automatic migration. We try to find a solution to this problem so as to achieve a correct and efficient way to migrate running instances. Instead of "reinventing the wheel", our research is based on many previous works, which are combined and improved to be made the best use of each. Specific features of different approaches can be found in [7]. Change region [2,7,8] and trace equivalence [6,9] are the two major strategies adopted by us for checking correctness and compliance. A new correctness criterion based on the consolidated view of reduced history [11] is put forwarded to perform compliance checking within change region, so that as many instances as possible can be compliant and migrated. A practical case of graduation defense application is also introduced to exemplify our approach.

Our approach can smoothly solve the five major typical dynamic change problems in [7] except the first one which allows the modification of history, accomplished tasks. The typical dynamic change bug mentioned in [2, 7] can also be settled without sacrificing efficiency. According to our experiment results, on average 66.7% instances within DCR are identified as transferable ones, which were totally neglected by the original approach of Aalst [8].

The rest of the paper is arranged as follows. Section 2 presents our approach combing DCR and RTS. We begin with the introduction of a case study. Then DCR, RTS are introduced and correctness criterion is redefined. A migration algorithm to compute the marking of the instance according to the new WF model after migration is also provided to perform the transfer. In section 3, we apply our approach to several examples and compare the results with Aalst's. Finally, we end up with a summarization of our approach and a plan for the future work.

2   An Integrated Approach Combing DCR and RTS

As to the dynamic changes of workflow, modification of many aspects should be considered. However, supporting flexible and correct structural changes is the primary issue. Our discussion within this paper is only confined to the structural changes, other modifications such as changes of data, resources and external applications are beyond the scope of this paper. A case study of dynamic evolutionary change is introduced first for illustration.

2.1   A Case Study

Typical applications of WfMS such as loan and insurance processes might be subject to unpredictable changes due to variation of policies. Such scenarios might best utilize and display the feature of adaptive workflow. However, they are usually too complicated for research and illustration, and we choose a simple but practical process, Graduation Thesis Dissertation Application, for case study.

Fig.1 shows the process of Graduation Thesis Dissertation Application. It is a simplified version and some tasks have been removed so as to focus on the parts related to modification. The process begins with application for graduation initiated by a graduate student. Once the application is submitted, both his/her advisor and officer of the Education Administration will receive a copy and decide whether to approve or reject (the rejection branch is removed here for simplification). If the application is approved by both actors, then the student is asked to submit his/her thesis. The thesis will go through two rounds of review, a preliminary review by advisor and a second round of review by an invited professor. Then, comments of the two reviewers will be sent to both the graduate student and the dissertation secretary.

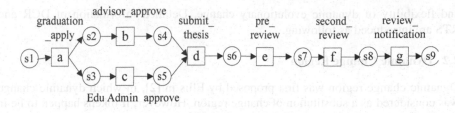

Fig. 1. Graduation Thesis Dissertation Application (original)

A new regulation is issued that a certain percent of graduation theses should be reviewed by an Advising Committee appointed by the University. Therefore, a parallel branch is inserted after the preliminary review. Meanwhile, another modification is made to the graduation application approval process, which is changed from parallel to sequential. The reason for change is as follows. According to the history records, most of the rejected applications are rejected by advisors rather than the Education Administration. In order to lighten the burden of the Education Administration and enhance the effectiveness of their work, they will only consider those applications already approved by advisors. Fig.2 shows the modified process.

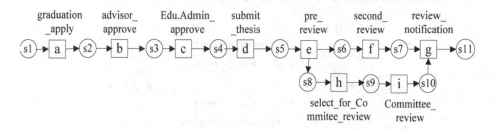

Fig. 2. Graduation Thesis Dissertation Application (modified)

How to deal with the already running instances? Obviously, we should adopt the progressive strategy rather than abort and flush ones proposed in [3]. All the running instances have to proceed together with new coming ones. A popular solution to this co-existing problem is to let the old ones follow their original model and new instances follow the new model. However, this solution can not satisfy circumstances requiring immediate response to the policy change. We need to propagate the changes to running instances without delay. The modification of the model involves two types of changes, a parallel to sequential change (Fig.3) and a parallel insertion (Fig.5), which happened to be two of the typical problems regarding dynamic change [7].

The following part of this section will concentrate on our solution to the migration issue. We utilize the algorithm of DCR proposed by Aalst to circle the affected change region, and introduce the concept of RTS for compliance checking to identify those transferable instances within the region. In this way, we can ensure both validity and flexibility of dynamic evolutionary change. Detailed information of DCR and RTS are introduced as following.

2.2 Dynamic Change Region

Dynamic change region was first proposed by Ellis in [2], in which dynamic change was considered as a substitution of change region. However, if tokens happen to be in the region, the substitution will be complex and dynamic change bug might arise. Aalst made a further contribution toward this approach by providing a revised definition of change region and an algorithm to calculate DCR [8], which is circled with dashed line in Fig.3.

Aalst adopted Petri nets (PN) to illustrate the problem and conduct the calculation [8]. There are three major components of Petri net: places to store tokens (represented by circles), transitions to execute tasks (represented by rectangles) and arcs to connect places and transitions. We also adopt Petri net, which has more advantages over other notations [10]. More introductions about Petri nets can be found in [8].

In [8], the calculation of DCR is symmetric, i.e. if the roles of the new and old are reversed, DCR remains the same. We only consider the one-way change as marked in the figure (from PN^O to PN^N). For all the following figures, the nodes mentioned in the DCR captions only refer to those that belong to PN^O.

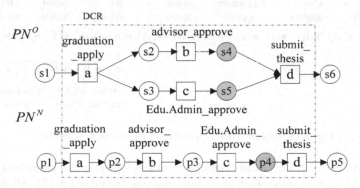

Fig. 3. Parallel to sequential change
DCR = {s2, s3, s4, s5 ,graduation_apply,
advisor_approve, Edu.Admin_approve, submit_thesis }

According to [8], valid transfer is permitted only when token is outside DCR. But the truth is 3 out of 4 cases within DCR could be migrated correctly, as shown in table 1. In Table 1, column 1 lists all the 4 possible states within DCR; column 3 shows that all of the states are prohibited from migration according to the pure DCR approach in [8]; column "Expect" specifies whether they should be transferred theoretically. Take the running instance with tokens in s4 and s5 for example. Obviously it can be migrated to an instance with token in p4 in the new model without causing error.

Is the change region too large? Sufficient proof has been provided to testify its completeness and correctness in [8], and we can hardly downsize it while still keeping the two features. Maybe the problem lies in the correctness criteria, which exclude all the instances with token within the region from migration. The thing left to be done is to propose a proper correctness criterion for migration and a feasible algorithm for identifying and migrating transferable instances within DCR as expected.

2.3 Correctness Criterion

Different approaches vary primarily in the correctness criterion on which they are based and the way to check the criterion. A survey on seven different criteria and comparison of respective approaches is given in [7]. Some approaches adopt a straightforward but restrictive compliance criterion, either requiring identical marking state or same execution history between the original and new one. But actually these criteria may not work well in circumstances involving deletion of an And-join/And-split or exchange of already finished tasks. For example, the case shown in Fig.3 cannot comply with the Aalst's criteria in [8] which require places with token cannot be removed. However, even if s4 and s5 become one place, a valid transfer can be made. The well known workflow model ADEPT adopted a comprehensive compliance criterion, which introduced the concept of Reduced Execution History (REH) [11]. This REH only discards the entries related to previous loops. In the absence of loops, it is exactly the same as complete history. For real-world applications with hundreds up to thousands of instances, it is too expensive and unnecessary to use such extensive execution data only to check compliance.

We aim to identify the useful part of execution history to decrease the performance burden, and what's more important, to enable a number of misclassified instances to migrate automatically. Having studied many approaches, we decided to adopt the DCR approach in [8] to circle the affected region and introduce Reduced Trace Set (RTS) to check compliance within DCR. As far as we know, no article has ever proposed any criterion or algorithm to perform compliance checking within DCR, and it is the first time to apply RTS to identify transferable workflow instances. In the following part, we will introduce RTS and our correctness criterion respectively.

Before introducing Reduced Trace Set. Trace Set, we first give a brief introduction of Trace Set (TS). TS is based on trace equivalence, but it is different from either event log or trace sequence (firing sequence). Unlike a complete event log, TS only record active parts of the net (i.e., the transitions) rather than passive elements such as places [12]. Unlike common trace sequence, TS is a collection of finished transitions, whose execution sequences are ignored. Take Fig.3 for example, suppose tokens are in s4 and s5 of PN^O, then TS = {a,b,c}, no matter whether b or c is executed first.

Definition 1 (Trace Set). Let $PN = (P, T, F)$ and i be an instance on PN and t_s be the state of transition, $t_s \in \{$enabled, completed$\}$. The trace set of i can be marked as TS , $TS = \{t \mid t \in T \ \& \ t_s = completed\}$.

Definition 2 (Reduced Trace Set). Let TS be the trace set of instance i and DCR be the dynamic change region on $PN = (P, T, F)$. RTS of i is obtained by discarding the transitions beyond the scope of DCR and those included in the previous loop iterations other than the completed or the current one.

RTS is different from TS in two respects. Firstly, it only includes those transitions within DCR; secondly, it excludes the records of past loops except the current one. For example, Fig.4 depicts a change involving insertion of task into loop. The instance is on its 2nd iteration with a token in s3. RTS excludes two transitions from TS: 'a' which is outside DCR and 'c' which belongs to the 1st iteration.

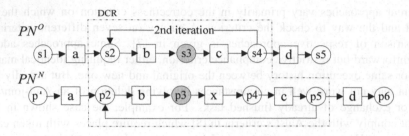

Fig. 4. Loop insertion
DCR={s2,s3,s4,b,c}; TS={a,b,c}; RTS = {b}

Definition 3 (Correctness Criterion). Let PN^O be the original process model and i^O be an instance on PN^O with reduced trace set RTS^O. Let further PN^O be transformed into new model PN^N. Then i^O can be correctly migrated to PN^N if a reduced trace set RTS^N which equals to RTS^O can be found on PN^N.

It is important to note that our compliance checking is only confined to DCR and the execution sequence of tasks inside DCR can be neglected. We only care about the current state (accomplished transitions) rather than the history (execution sequence). Take Fig.3 for example, since a RTS^N ($RTS^N = RTS^O$ ={a,b,c}) can be found on PN^N, the instance with token in s4 and s5 can be migrated to sequential, no matter whether b or c is executed first.

The criterion is based the notion that if the new instance can do whatever the old one has done, they are compatible and the migration is valid. Since only the control flow but not the data flow is discussed here, each transition is considered as non-related task. Then trace set is the proper way to record execution information and compare the old and new instance.

Running instances are classified into three types: i) Smoothly transferable; ii) Compliance achievable; iii) External intervention required. The smoothly transferable ones refer to those outside DCR and can be transferred directly. For those inside DCR, compliance checking is performed using RTS to identify the compliant ones, which belong to the second type and can be transferred automatically. The incompliant ones are left to be handled by external intervention. We are making efforts on performing more accurate compliance checking so that more instances originally classified to the third type can be reclassified to the second type.

2.4 Dynamic Migration Algorithm

Having introduced the concepts of DCR, RTS and Correctness Criterion, we then put forward an algorithm based on these definitions to perform dynamic migration. The algorithm can tell whether the instances can be transferred and further calculate the state of marking after migration.

The initial marking of PN^N inside DCR is denoted as Mi, and Mo denotes the marking after migration. If the instance is transferable, a firing sequence σ =(t1,t2,…tn) could be found to enable $Mi \xrightarrow{\sigma} Mo$. For a given state M, the set of enabled transitions is denoted as $Te(M)$. Te is a temp set to store the enabled transitions. RTS^N stores accomplished transitions.

Algorithm 1 (Dynamic Migration Algorithm)

```
Input Parameters: Mi , RTS^O
Output Parameter: Mo
Begin
01  Te  := Te(Mi)
02  RTS^N :=
03  M := Mi
04  begin
05  while Te ≠   do
06                  begin
07                  for each transition t in Te
08                  if t ∈ RTS^O
09                      RTS^N =  RTS^N ∪ {t}
10                      M = M − •t + t•
11                          Te  := Te(M)
12                  else
13                          Te = Te − {t}
14          end
15  end
16  if RTS^N = RTS^O
17                  Mo = M
18                  output Mo
19  else
20                  output "Transfer is not valid"
end
```

The loop stops till there is no element in Te (*Step5*). For each enabled transition t in PN^N, it can be fired if it belongs to RTS^O (*Step 8*) as well. Then the marking will shift to a new state and Te is updated (*Step 11*). If transition t does not appear in RTS^O, t needs to be removed from Te and no further consideration is needed for this instance (*Step 13*). When the loop is done and $RTS^N = RTS^O$, a valid transfer is permitted (*Step 16*) and the marking after migration is found and output (*Step 18*).

A pre-process and a post-process are needed to ensure the correct firing of transitions with DCR. If the first/last element of DCR is a transition rather than a place, a virtual starting/ending place is introduced to be the container of initial/end token. Pre-process aims to add virtual places, and post-process aims to process the output marking and remove the added virtual places. Suppose an instance with tokens in s4 and s5 (places in shade) will migrate to PN^N as shown in Fig.3. A virtual starting place should be inserted before the transition graduation_apply, then $Mi = (1,0,0,0,0)$. In addition, a virtual ending place is inserted after submit_thesis. The following steps demonstrate the results after each loop.

Initial values: $Te = \{a\}, RTS^O = \{a,b,c\}, RTS^N = \Phi, M = Mi = (1,0,0,0,0)$

i. $Te = \{b\}, RTS^N = \{a\}, M = (0,1,0,0,0)$

ii. $Te = \{c\}, RTS^N = \{a,b\}, M = (0,0,1,0,0)$

iii. $Te = \{d\}, RTS^N = \{a,b,c\}, M = (0,0,0,1,0)$

iv. $Te = \Phi, RTS^N = \{a,b,c\}, M = (0,0,0,1,0)$

v. $RTS^N = RTS^O = \{a,b,c\}. Mo = (0,0,0,1,0)$

Since the first and the last place are inserted virtual nodes, the output $Mo = (0,0,0,1,0)$ should be post-processed. The two virtual places should be removed from the output and the marking on PN^N should be (0,0,1), with token in s4.

3 Examples and Comparison

In this section, we will apply our method to more examples discussed by Aalst in [8]. They are chosen for two reasons. Firstly, they are simple but typical, and can be found in many complex scenarios. Secondly, they are the most suitable examples for comparison with Aalst's approach, which does not allow migration of any instance with token inside DCR. The comparison results can best illustrate that our approach can downsize the un-transferable instances pool by a large extent.

Example 1 (Parallel to sequential). Take Fig 3 for example. It can be used to compare with the parallel to sequential example used by Aalst. According to our approach, all the cases within DCR can be transferred except the one with tokens in s2 and s5, whose RTS = {a,c} cannot be found in PN^N. In fact, this exceptional case does lead to dynamic change bug [8]. The table below demonstrates that the result of our approach is the same as expected.

Example 2(Parallel insertion). As is shown in Fig.5, an instance i^O with token in s8 transfers from PN^O to PN^N. Since $RTS^N = RTS^O = \{e,f\}$ can be met, i^O can be transferred correctly. Tokens of the new instance should be in p7 and p8.

Table 1. Transferable States Comparison (Fig.3)

States	RTS	DCR	Expect	DCR+TS
s2, s3	{a}	×	√	√
s2, s5	{a,c}	×	×	×
s3, s4	{a,b}	×	√	√
s4, s5	{a,b,c}	×	√	√

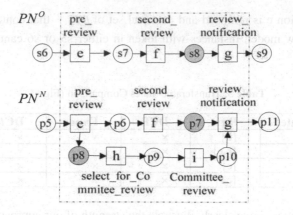

Fig. 5. Parallel insertion
DCR = {s7,s8,e,f,g }

According to the pure DCR, this instance cannot be migrated at all. In this case, our approach can identify 100% of transferable instances within DCR. (Table 2)

Table 2. Transferable States Comparison (Fig.5)

States	RTS	DCR	Expect	DCR+TS
s7	{e}	×	√	√
s8	{e,f}	×	√	√

Example 3(Change involving XOR and Parallel). Consider Fig.6, instance with token in s4 in PN^O can be migrated to PN^N, for $RTS^N = RTS^O =$ {b,c} can be found in PN^N. After migration to PN^N, tokens should be in p4 and p5.

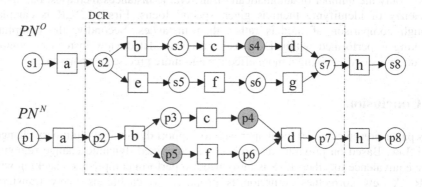

Fig. 6. Change involving XOR and Parallel
DC = {s2,s3,s5,s6,s7,b,c,d,e,f,g }

While transition e is removed and no equal set of RTS^o that contains {e} can be found in the new model, instances with token in either s5 or s6 cannot be migrated (Table 3).

Table 3. Transferable States Comparison (Fig.6)

States	RTS	DCR	Expect	DCR+TS
s3	{b}	×	√	√
s4	{b,c}	×	√	√
s5	{e}	×	×	×
s6	{e,f}	×	×	×

All the examples above clearly illustrate the strength of our approach, which lies in its ability to identify transferable instances within DCR. As is shown in Fig.7, the identified transferable instances within DCR amount to 66.7% on average. They were completely forbidden to migrate by the pure DCR approach, but now can be migrated exactly the same as expected.

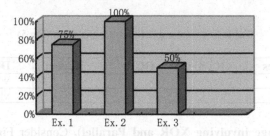

Fig. 7. Percentage of compliant instances within DCR

Not only the number of automatically transferable instances increases, but also the efficiency of identifying them is given special focus. Firstly, DCR is calculated through comparison of models rather than instances. Secondly, the compliance checking is performed within DCR using a consolidated view instead of complete execution history, enabling a more effective and more precise comparison.

4 Conclusion

This paper presented an efficient approach to support dynamic evolutionary change of workflow. Based on previous researches, we combined dynamic change region and trace equivalence together. RTS is introduced to perform compliance checking within DCR. A new correctness criterion is proposed to enable as many transferable instances as possible to migrate. The algorithm to perform migration and calculate the marking after transfer is also provided. The main contribution of this paper is a feasible solution to structural dynamic change, and enables the large number of transferable instances within DCR to migrate.

In the future, we plan to apply our approach to more complicated scenarios for further verification. We also plan to implement the approach in the WfMS developed by our team, and then extend the open source WfMS jBPM with a feature to calculate DCR and support migration within DCR using our algorithm. In addition, dynamic modifications of other aspects such as data, resource and external applications need further research.

Acknowledgement

This work is supported by National Basic Research Program（973 Plan, No. 2002CB 312006），NSFC（Adaptive Workflow, No.60473077）and Program for New Century Excellent Talents in University.

References

1. van der Aalst, W., Basten, T., et al.: Adaptive Workflow: On the Interplay between Flexibility and Support. Enterprise Information Systems, 63–70 (2000)
2. Ellis, C., Keddara, K., et al.: Dynamic Change Within Workflow Systems. In: Proceedings of the Conference on Organizational Computing Systems, Milpitas, California, August 1995, pp. 10–21 (1995)
3. Casati, S., Ceri, et al.: Workflow Evolution. In: Thalheim, B. (ed.) ER 1996. LNCS, vol. 1157, pp. 438–455. Springer, Heidelberg (1996)
4. Sadiq, S., Orlowska, M.: Architectural considerations in systems supporting dynamic workflow modification [A]. In: Proceedings of the Workshop on Software Architectures for Business Process Management, The 11th Conference on Advanced Information Systems Engineering [C], Heidelberg, Germany, pp. 14–18. Springer, Heidelberg (1999)
5. van der Aalst, W., Jablonski, S.: Dealing with workflow change: identification of issues and solutions[J]. International Journal of Computer Systems, Science, and Engineering 15(5), 267–276 (2000)
6. Reichert, M., Dadam, P.: ADEPTflex-supporting dynamic changes of workflows without losing control. JIIS 10(2), 93–129 (1998)
7. Rinderle, S., Reichert, M., Dadam, P.: Correctness criteria for dynamic changes in workflow systems:a survey. Data & Knowledge Engineering 50(1), 9–34 (2004)
8. van der Aalst, W.: Exterminating the Dynamic Change Bug: A Concrete Approach to Support Workflow Change. Information Systems Frontiers 3(3), 297–317 (2001)
9. Ehrler, L., Fleurke, M., Purvis, M.A., Savarimuthu, B.T.R.: Agent-Based Workflow Management Systems (WfMSs): JBees- A Distributed and Adaptive WFMS with Monitoring and Controlling Capabilities. Information Systems and E-Business Management 4(1), 5–23 (2006)
10. van der Aalst, W.: The Application of Petri Nets to Workflow Management. The Journal of Circuits, Systems and Computers 8(1), 21–66 (1998)
11. Rinderle, S., Reichert, M., Dadam, P.: Flexible support of team processes by adaptive workflow systems. Distributed and Parallel Databases 16(1), 91–116 (2004)
12. van der Aalst, W.M.P., de Medeiros, A.K.A., Weijters, A.J.M.M.: Process Equivalence: Comparing Two Process Models Based on Observed Behavior. Business Process Management, 129–144 (2006)

Author Index

Lecture Notes in Computer Science

Sublibrary 3: Information Systems and Application, incl. Internet/Web and HCI

For information about Vols. 1– 4566
please contact your bookseller or Springer

Vol. 4976: Y. Zhang, G. Yu, E. Bertino, G. Xu (Eds.), Progress in WWW Research and Development. XVIII, 699 pages. 2008.

Vol. 4956: C. Macdonald, I. Ounis, V. Plachouras, I. Ruthven, R.W. White (Eds.), Advances in Information Retrieval. XXI, 719 pages. 2008.

Vol. 4952: C. Floerkemeier, M. Langheinrich, E. Fleisch, F. Mattern, S.E. Sarma (Eds.), The Internet of Things. XIII, 378 pages. 2008.

Vol. 4947: J.R. Haritsa, R. Kotagiri, V. Pudi (Eds.), Database Systems for Advanced Applications. XXII, 713 pages. 2008.

Vol. 4936: W. Aiello, A. Broder, J. Janssen, E.. Milios (Eds.), Algorithms and Models for the Web-Graph. X, 167 pages. 2008.

Vol. 4932: S. Hartmann, G. Kern-Isberner (Eds.), Foundations of Information and Knowledge Systems. XII, 397 pages. 2008.

Vol. 4928: A. ter Hofstede, B. Benatallah, H.-Y. Paik (Eds.), Business Process Management Workshops. XIII, 518 pages. 2008.

Vol. 4903: S. Satoh, F. Nack, M. Etoh (Eds.), Advances in Multimedia Modeling. XIX, 510 pages. 2008.

Vol. 4900: S. Spaccapietra (Ed.), Journal on Data Semantics X. XIII, 265 pages. 2008.

Vol. 4892: A. Popescu-Belis, S. Renals, H. Bourlard (Eds.), Machine Learning for Multimodal Interaction. XI, 308 pages. 2008.

Vol. 4882: T. Janowski, H. Mohanty (Eds.), Distributed Computing and Internet Technology. XIII, 346 pages. 2007.

Vol. 4881: H. Yin, P. Tino, E. Corchado, W. Byrne, X. Yao (Eds.), Intelligent Data Engineering and Automated Learning - IDEAL 2007. XX, 1174 pages. 2007.

Vol. 4877: C. Thanos, F. Borri, L. Candela (Eds.), Digital Libraries: Research and Development. XII, 350 pages. 2007.

Vol. 4872: D. Mery, L. Rueda (Eds.), Advances in Image and Video Technology. XXI, 961 pages. 2007.

Vol. 4871: M. Cavazza, S. Donikian (Eds.), Virtual Storytelling. XIII, 219 pages. 2007.

Vol. 4858: X. Deng, F.C. Graham (Eds.), Internet and Network Economics. XVI, 598 pages. 2007.

Vol. 4857: J.M. Ware, G.E. Taylor (Eds.), Web and Wireless Geographical Information Systems. XI, 293 pages. 2007.

Vol. 4853: F. Fonseca, M.A. Rodríguez, S. Levashkin (Eds.), GeoSpatial Semantics. X, 289 pages. 2007.

Vol. 4836: H. Ichikawa, W.-D. Cho, I. Satoh, H.Y. Youn (Eds.), Ubiquitous Computing Systems. XIII, 307 pages. 2007.

Vol. 4832: M. Weske, M.-S. Hacid, C. Godart (Eds.), Web Information Systems Engineering - WISE 2007 Workshops. XV, 518 pages. 2007.

Vol. 4831: B. Benatallah, F. Casati, D. Georgakopoulos, C. Bartolini, W. Sadiq, C. Godart (Eds.), Web Information Systems Engineering - WISE 2007. XVI, 675 pages. 2007.

Vol. 4825: K. Aberer, K.-S. Choi, N. Noy, D. Allemang, K.-I. Lee, L. Nixon, J. Golbeck, P. Mika, D. Maynard, R. Mizoguchi, G. Schreiber, P. Cudré-Mauroux (Eds.), The Semantic Web. XXVII, 973 pages. 2007.

Vol. 4823: H. Leung, F. Li, R. Lau, Q. Li (Eds.), Advances in Web Based Learning - ICWL 2007. XIV, 654 pages. 2008.

Vol. 4822: D.H.-L. Goh, T.H. Cao, I.T. Sølvberg, E. Rasmussen (Eds.), Asian Digital Libraries. XVII, 519 pages. 2007.

Vol. 4820: T.G. Wyeld, S. Kenderdine, M. Docherty (Eds.), Virtual Systems and Multimedia. XII, 215 pages. 2008.

Vol. 4816: B. Falcidieno, M. Spagnuolo, Y. Avrithis, I. Kompatsiaris, P. Buitelaar (Eds.), Semantic Multimedia. XII, 306 pages. 2007.

Vol. 4813: I. Oakley, S.A. Brewster (Eds.), Haptic and Audio Interaction Design. XIV, 145 pages. 2007.

Vol. 4810: H.H.-S. Ip, O.C. Au, H. Leung, M.-T. Sun, W.-Y. Ma, S.-M. Hu (Eds.), Advances in Multimedia Information Processing - PCM 2007. XXI, 834 pages. 2007.

Vol. 4809: M.K. Denko, C.-s. Shih, K.-C. Li, S.-L. Tsao, Q.-A. Zeng, S.H. Park, Y.-B. Ko, S.-H. Hung, J.-H. Park (Eds.), Emerging Directions in Embedded and Ubiquitous Computing. XXXV, 823 pages. 2007.

Vol. 4808: T.-W. Kuo, E. Sha, M. Guo, L.T. Yang, Z. Shao (Eds.), Embedded and Ubiquitous Computing. XXI, 769 pages. 2007.

Vol. 4806: R. Meersman, Z. Tari, P. Herrero (Eds.), On the Move to Meaningful Internet Systems 2007: OTM 2007 Workshops, Part II. XXXIV, 611 pages. 2007.

Vol. 4805: R. Meersman, Z. Tari, P. Herrero (Eds.), On the Move to Meaningful Internet Systems 2007: OTM 2007 Workshops, Part I. XXXIV, 757 pages. 2007.

Vol. 4804: R. Meersman, Z. Tari (Eds.), On the Move to Meaningful Internet Systems 2007: CoopIS, DOA, ODBASE, GADA, and IS, Part II. XXIX, 683 pages. 2007.